Large fluvial lowlands are among Earth's most unique and productive environments but are being rapidly degraded by human activities. Pressure on large rivers and deltas has increased tremendously over past decades because of flood control, urbanization, and increased dependence for agriculture and food production. This book examines human impacts on lowland rivers and discusses how these changes affect different types of riverine environments and flood processes, and how these "lessons" can be used to more sustainably manage large rivers and deltas. Surveying a global range of large rivers, including the Mekong, Nile, Sacramento, Danube, Huanghe, among others, the book especially examines management of the Rhine River in the Netherlands and the lower Mississippi in Louisiana. A particular focus of the book is on sedimentology and hydraulic engineering, which is described in a straightforward writing style accessible to a broad audience of researchers, practitioners, and advanced students in physical geography, fluvial geomorphology and sedimentology, and flood and river management.

PAUL F. HUDSON is Associate Professor of Physical Geography and Sustainability at Leiden University in the Netherlands. Hudson relocated to the Netherlands after serving twelve years on the faculty at the University of Texas at Austin, where he is appointed as a Faculty Affiliate in Geography and the Environment. His main scholarly interests involve the study of environmental change of large coastal plain rivers through the lens of physical geography, and in particular, geomorphology and hydrology. Hudson's research investigates flooding, river adjustment, sediment transport, and management of floodplain environments. He has provided expert advice concerning environmental water resources across a range of governmental scales: community, state, and national, including the Dutch parliament.

Flooding and Management of Large Fluvial Lowlands

A Global Environmental Perspective

Paul F. Hudson
Leiden University

CAMBRIDGE UNIVERSITY PRESS

CAMBRIDGE
UNIVERSITY PRESS

University Printing House, Cambridge CB2 8BS, United Kingdom

One Liberty Plaza, 20th Floor, New York, NY 10006, USA

477 Williamstown Road, Port Melbourne, VIC 3207, Australia

314–321, 3rd Floor, Plot 3, Splendor Forum, Jasola District Centre, New Delhi – 110025, India

103 Penang Road, #05–06/07, Visioncrest Commercial, Singapore 238467

Cambridge University Press is part of the University of Cambridge.

It furthers the University's mission by disseminating knowledge in the pursuit of education, learning, and research at the highest international levels of excellence.

www.cambridge.org
Information on this title: www.cambridge.org/9780521768603
DOI: 10.1017/9781139015738

© Paul F. Hudson 2021

This publication is in copyright. Subject to statutory exception and to the provisions of relevant collective licensing agreements, no reproduction of any part may take place without the written permission of Cambridge University Press.

First published 2021

Printed in the United Kingdom by TJ Books Limited, Padstow Cornwall

A catalogue record for this publication is available from the British Library.

ISBN 978-0-521-76860-3 Hardback

Cambridge University Press has no responsibility for the persistence or accuracy of URLs for external or third-party internet websites referred to in this publication and does not guarantee that any content on such websites is, or will remain, accurate or appropriate.

To Mom and Dad, to my family,
voor Marie-Louise.

Contents

List of Figures   page ix
List of Tables   xv
Preface and Acknowledgments   xvii

1 Fluvial Lowlands and the Environment: Cause for Concern   1
     1.1 Introduction and Scope of Work   1
     1.2 Extent and Scale of Fluvial Lowlands   2
     1.3 Human Impacts to Fluvial Lowlands   3
     1.4 A Tale of Two Rivers, and Beyond   5
     1.5 Outline and Narratives   6

2 Rivers and Landscapes: A Drainage Basin Framework   9
     2.1 The Fluvial System: A Large River Perspective   9
     2.2 Downstream Implications of Land Cover Change   13
     2.3 River Channel Patterns   19
     2.4 Fluvial Adjustment to Environmental Disturbance   27

3 Hydrologic and Geomorphic Processes in Fluvial Lowlands and Deltas   31
     3.1 Context of Large Rivers   31
     3.2 Hydroclimatology of Large Rivers   32
     3.3 Channel Bank Erosion along Lowland Rivers   43
     3.4 Flooding across Fluvial Lowlands   48
     3.5 Lowland Floodplain Geomorphology   49
     3.6 Neotectonic Influences on Lowland Rivers   62
     3.7 Deltaic Geomorphology   64
     3.8 Closure: Looking Forward with an Eye on the Past   78

4 Dams, Rivers, and the Environment   80
     4.1 Hydraulic Infrastructure and Society   80
     4.2 Global Extent of Dams   82
     4.3 Environmental Impact of Dams and Reservoirs   89
     4.4 Dam Removal   115
     4.5 Managing Reservoir Deposits   130

5 Hydraulic Engineering for Channel Stability and Flood Control: The Sequence of Geomorphic Adjustment   138
     5.1 Overview of Approaches to Channel Engineering   138
     5.2 Influence of Sedimentology on Channel Engineering   143
     5.3 Engineering for Flood Control and Navigation   144
     5.4 Impacts of a Massive Flood Control System: Lower Mississippi Case Study   156
     5.5 Closure: Linkages between Channel Engineering and Floodplain Embankment   164

6 Dikes and Floodplains: Impacts of Flood Control on Lowland Rivers   166
     6.1 Dikes and Floodplains   166
     6.2 Flood Control Systems and the Cycle of Dike Management   167
     6.3 Scale and Pattern of Floodplain Embankment   177
     6.4 Hydrography of Embanked Floodplains   181
     6.5 Between the Dikes: Sedimentology of Embanked Floodplains   188
     6.6 Beyond the Dikes: Sand Boils, Dike Breach, and Sedimentation   193
     6.7 Embanked Floodplain Evolution and End-Member Regulated Rivers   201
     6.8 Closure: Toward Flood Basin Management   204

7 Flood Basins and Deltas: Subsidence, Sediment, and Storm Surge   205
     7.1 A Challenging Environment   205
     7.2 Flood Diversion Structures: Environmental Impacts and Stakeholder Controversy   206
     7.3 Lower Lowlands: Human Activities, Sedimentology, and Subsidence   218
     7.4 Avulsion and River Management   225
     7.5 Urban Flooding and Management Challenges   234

7.6 Hurricane Katrina Flooding of New Orleans: Cumulative Effects of River Basin Mismanagement	239
7.7 Closure: Reflection on Urban Flood Disasters	248
8 Toward Integrated Management of Lowland Rivers	**249**
8.1 Need for a New Approach	249
8.2 *Real* Channel Improvement	251
8.3 Embanked Floodplain Measures	258
8.4 Flood Basins and Deltas: Management Challenges	264
8.5 The Louisiana Experience: Subdeltas and Sediment Diversion	265
8.6 Integrated Management: EU Water Framework Directive and Room for the River	275
8.7 Managing Expectations: Integrated River Basin Management in Context	279
9 Lessons Learned: Future Management of Fluvial Lowlands	**284**
9.1 Lowland River Management in Review	284
9.2 Key Lessons Learned	284
9.3 Looking Forward: Concern and Optimism in Integrated Floodplain Management	291
9.4 It's Not Too Late	297
Bibliography	298
Index	326

Figures

1.1 Flood protection in a densely populated region. Waal River (Dutch Rhine) high-water in 1993 *page* 2
1.2 Failure of 17th Street Canal and catastrophic flooding of New Orleans, Louisiana associated with Hurricanes Katrina and Rita in 2005   2
1.3 Exogenic and endogenic influences on fluvial–deltaic vulnerability, emphasizing anthropogenic drivers   4
1.4 Delta size in relation to sediment needed to protect against sea level rise   4
1.5 Swamp forest in the Atchafalaya basin, vital to lower Mississippi flood management   5
1.6 Rice farming and aquaculture (shrimp) in the Mekong River delta, Vietnam   5
1.7 Global map of rivers, including many herein examined or referenced   7
2.1 Model of downstream (spatial) changes in fluvial characteristics of a large drainage basin displaying fluvial hinge-line   10
2.2 Land degradation by soil erosion and gullying in the Chinese Loess Plateau   11
2.3 Downstream translation and attenuation of a flood wave   12
2.4 Subsiding Sacramento–San Joaquin River Delta in California   13
2.5 Global area of cropland and grazing since 100 AD   13
2.6 Influence of varying land surface conditions on surface hydrologic processes   14
2.7 Post-settlement alluvium in Driftless region of Wisconsin   15
2.8 Reduction in annual sediment inputs (billions of tons) to middle Huanghe River due to soil conservation   16
2.9 Landscape and channel adjustment in response to land cover change   16
2.10 Fluvial response for three different settings caused by intensive agriculture   17
2.11 Hydraulic gold mining in Sierra Nevada foothills, California, in the 1870s   18
2.12 Changes in low flow levels (height) for three stations in the Sacramento basin   18
2.13 Projections to global agricultural land cover until 2050   19
2.14 Channel pattern distinguished by slope–discharge relation   20
2.15 Characterization of single-channel and multichannel (anabranching) river patterns   21
2.16 Meandering river channel morphologic indices   22
2.17 Relation between pool-to-pool spacing and channel width for a range of river sizes   22
2.18 Empirical relations between indices of planform and cross-sectional channel profiles   23
2.19 Pointbar stratigraphy and overbank flood deposits for a meandering river   24
2.20 Geomorphic and hydraulic features of mid-channel bars   25
2.21 Satellite image of dynamic anabranching middle Rio Paraná, northern Argentina   26
2.22 Relation between stream power and sediment relative to erosion and deposition   27
2.23 Adjustment of channel sinuosity and slope, from natural to engineered river   28
2.24 Knickpoint incision caused by channelization for flood control   28
2.25 Range of channel morphologic response to changes in streamflow and sediment   29
3.1 Ganges–Brahmaputra delta, largest tidal-dominated subaerial delta   32
3.2 Rural settlement along channel bank and natural levee of lower Amazon   32
3.3 Hydroclimatic flood mechanisms across a range of temporal and spatial scales   33
3.4 Earth's general atmospheric circulation showing major winds and pressure systems   33
3.5 Zonal and meridional flow over the northern hemisphere with changes in jet stream   33
3.6 Annual flood pulses for a range of large rivers that vary in size and climatic controls   34
3.7 Mean annual runoff for large rivers in relation to annual peak discharge variability   35

3.8 Discharge regime for the Mississippi basin and major tributaries 35
3.9 Discharge regime for the Rhine River and major tributaries 36
3.10 Hunger stone exposed along the Elbe River bed during the severe 2018 European drought 36
3.11 Historic hydroclimatology of the upper Mississippi River and changes to zonal flow 37
3.12 Average monthly discharge for the La Plata basin with different ENSO phases 41
3.13 Nilometer on the Isle of Rhoda, Cairo, Egypt, which dates to 861 AD 41
3.14 Influence of ENSO and PDO phases on the Columbia River streamflow 42
3.15 Murray–Darling streamflow in relation to ENSO and the millennium drought 42
3.16 River bank erosion on heavily mined section of the Mekong River in Cambodia 43
3.17 Channel bank erosion mechanisms along large alluvial rivers 45
3.18 Bank caving by arcuate slumping along the lower Amazon River, Brazil 46
3.19 Channel planform adjustment to cohesive floodplain deposits 46
3.20 Influence of cohesive sedimentology on Mississippi River meander bend migration 47
3.21 Hydrologic pathways for a lowland floodplain with multiple meander belts 49
3.22 Channel bank elevation profile in relation to varying flow levels, lower Mississippi 50
3.23 Local scale floodplain water bodies along the lower Mississippi 50
3.24 Valley profile across the lower Brazos River floodplain, near apex of delta 51
3.25 Spillage sedimentation model for large river floodplains 53
3.26 Sand sheet deposition from the record 2018–2019 flood along the lower Mississippi 54
3.27 Relation between lateral migration rates and natural levee development 54
3.28 Downstream pattern in natural levee width, lower Rio Pánuco, Mexico 54
3.29 Downstream profile in natural levee dimensions, Mississippi delta 55
3.30 Evolution of oxbow lake environments and sedimentary infilling 57
3.31 Model of clay plug development and natural levee burial in meander neck 58
3.32 Perirheic zone floodplain hydrology and inundation mechanisms 59

3.33 Bioturbation of sedimentary structure for natural levee and crevasse deposits 61
3.34 Soil development atop two meander belts, lower Mississippi 62
3.35 Adjustment to channel sinuosity to changes in valley slope by neotectonic faulting 63
3.36 Fluvial adjustment to uplift domes and faulting 63
3.37 Changes to channel and floodplain water bodies in response to neotectonic warping 64
3.38 Profile of active Mississippi channel belt displaying neotectonic influences 64
3.39 Lower Mississippi alluvial valley and delta plain 65
3.40 Rhine–Meuse drainage basin, including geomorphic sections and locations 66
3.41 Quaternary geomorphic units and tectonic structure of the Rhine delta region 67
3.42 Influence of sea level rise on Holocene backfilling of the Rhine delta 67
3.43 Delta area in relation to drainage area 69
3.44 Delta classification according to fluvial, wave, or tidal influence 71
3.45 Continuum of wave to fluvial deltas in relation to sediment and wave energy 73
3.46 Model of fluvial–deltaic sedimentation in marine receiving basin 73
3.47 Model of fluvial-dominated prograding deltaic deposits 74
3.48 The delta cycle, initiation to abandonment, fluvial–deltaic environments 75
3.49 Mississippi delta plain and chronology of individual delta lobes 75
3.50 Fluvial–deltaic environments associated with a subsiding delta lobe 76
3.51 Annual composite Mississippi delta fluid production and wetland loss 78
4.1 Red Bluff Diversion Dam on the Sacramento River in California 81
4.2 Global water withdrawals by sector since 1900, steep increase after the 1950s 81
4.3 Global tally of large dams 84
4.4 Global growth of reservoir capacity between 1900 and about 2010 85
4.5 Roman era Cornalvo dam in Extremadura, Spain; adjacent aqueduct 85
4.6 Height of large dams in Spain 86
4.7 Large dams in Australia organized by height 86
4.8 Characteristics of US dams in National Inventory of Dams 87
4.9 Global acceleration of hydropower dam construction since 1900 88

LIST OF FIGURES

4.10 Global trend and trajectory (to 2030) of river fragmentation and regulation   88
4.11 Conceptual model of annual flow regime for natural and impounded rivers   90
4.12 Flow regime for two large European rivers, before and after impoundment   91
4.13 Satellite image of the Nile River, extending ~1,000 km from Aswan High Dam to delta   92
4.14 Satellite image of Aswan High Dam, Lake Nasser, and Nile River in arid Egypt   92
4.15 Changes to Nile River streamflow below Aswan High Dam, between 1950 and 1985   92
4.16 Changes to Nile River suspended sediment loads after dam closure in 1964   96
4.17 Changes to Yangtze River streamflow and sediment after closure of Three Gorges Dam   96
4.18 Stair-step profile of a ~1,000 km segment of upper Mekong (Lancang) River   97
4.19 Large dams in the Mekong River basin   98
4.20 Longitudinal stair-step profile of the Rhine River, including geomorphic provinces   99
4.21 Gavins Point Dam and Lewis and Clarke Lake reservoir, Missouri River   100
4.22 Changes to suspended sediment along the Missouri and Mississippi Rivers   101
4.23 Ratio of pre-dam and post-dam suspended sediment loads along the Missouri River   102
4.24 Change to suspended sediment of lower Mississippi River at Tarbert Landing   102
4.25 Downstream channel bed incision for several US rivers after dam closure   103
4.26 Channel bed incision for Europe's two largest rivers   104
4.27 Changes to Rio Grande/Bravo at Presidio, Texas/Ojinaga, Chihuahua   105
4.28 Changes to channel width of the Platte River, NE, between 1880 and 1995   106
4.29 Changes to channel width and streamflow for the Platte River after dam closure   107
4.30 Photos documenting historic changes to riparian environments of the Platte River   107
4.31 Three different downstream responses of riverine fisheries to dam: Yangtze, Volga, and Nile Rivers   109
4.32 Chemical and physical characteristics of reservoir water quality relevant to aquatic ecology   110
4.33 Reservoir stratification (temperature, dissolved oxygen) in the Mekong basin   111
4.34 Comparison of pre-dam and post-dam streamflow temperature, lower Mekong basin   113
4.35 Tennessee River basin, showing TVA dams and drainage network   114

4.36 Photo of Glines Canyon Dam removal on Elwha River   116
4.37 Dam removal projects in the United States between 1900 and 2019   118
4.38 Sequence of steps for removal of reservoir sedimentary deposits   121
4.39 Arsenic, copper, and suspended sediment, Clark Fork River, Montana, 1996–2015   123
4.40 Reservoir sediment compaction for different sedimentary deposits   123
4.41 Photo of the Elwha River channel and bed material before and after dam removal   127
4.42 Channel changes to Elwha River after dam removal   127
4.43 Conceptual model of fluvial response to dam emplacement and removal   128
4.44 Conceptual model of fluvial response to dam removal, abiotic and biotic trajectories   129
4.45 Global distribution of reservoir storage lost to sedimentation   130
4.46 Model of reservoir storage loss due to sedimentation   130
4.47 Classification of sediment management strategies for reservoirs   133
4.48 Profile of dam and reservoir operation during sediment sluicing event   133
4.49 Seasonal pool operation at Three Gorges Dam, Yangtze River, China   133
4.50 Profile of dam and reservoir operation during a sediment flushing event   135
4.51 Reservoir sediment infilling and erosion of Sanmenxia Reservoir, Huanghe River   136
4.52 Photo of sediment flushing event, Sanmenxia Dam, Huanghe River   136
4.53 Models of dam and reservoir design related to sediment management   137
5.1 Intensively engineered Danube River in Romania, groynes and straightening   140
5.2 Cutterhead teeth of dredge, Ohio River   140
5.3 Lidar DEM of dredge spoil landforms, Apalachicola River floodplain, Florida   140
5.4 Dredge spoil and new Mississippi delta wetlands near Head of Passes, Louisiana   141
5.5 Sand mining on the Mekong River, Vietnam   141
5.6 Degradation of Arno River, Italy, over ~150 years due to engineering and land use   142
5.7 Photo of buried Holocene backswamp along the channel bank of the lower Mississippi   144
5.8 Channel cut-off types: chute, meander neck, mobile bar, river course shift   144
5.9 Planview perspective of meander bends, natural baseline, and artificial cut-offs   145

5.10 Photo of meander neck cut-off along the lower Mississippi in the 1930s 145
5.11 Extensive meander neck cut-off along Ucayali River, Peru, by local people 147
5.12 Channel straightening by Tulla cut-off plan (1833), Rhine River, Germany 148
5.13 Historic willow revetment construction along the lower Mississippi from the 1920s 149
5.14 Concrete revetment casting field, lower Mississippi at St. Francisville, Louisiana 149
5.15 Installation of modern articulated concrete revetment, lower Mississippi 150
5.16 Articulated concrete revetment emplacement along channel bank, lower Mississippi 150
5.17 Channel cross section changes after stabilization, Brahmaputra–Jamuna Rivers 150
5.18 Degradation of modern articulated concrete revetment, lower Mississippi 151
5.19 Channel change caused by concrete revetment, channel incision, and bank failure 151
5.20 Groynes along the Dutch Rhine (Lek-Nederrijn) 151
5.21 Patterns of groyne field flow and sediment deposition 152
5.22 Groyne field erosion from 1995 to 1998, subsequent sand deposition, Waal River 153
5.23 Annual number of tree snags removed from US rivers, late 1800s to early 1900s 154
5.24 View of historic Red River Raft from the 1870s 155
5.25 Historic map (1764) displaying altered channel of Red River, Louisiana 155
5.26 Velocity change at Vicksburg, MS, after meander cut-offs, lower Mississippi 158
5.27 Historic changes to river stage for specific discharge magnitudes, lower Mississippi 159
5.28 Average stage levels of lower Mississippi, before and after cut-offs 159
5.29 Adjustment to lower Mississippi stage levels, upstream and downstream of cut-offs 160
5.30 Annual high and low stage levels 1901–2016, lower Mississippi at Natchez, Mississippi 161
5.31 Incision of channel thalweg, lower Mississippi, between 1948 and 1999 162
5.32 Historic aerial photo (1967) of channel bar and island, lower Mississippi 162
5.33 Recent vegetated islands within the lower Mississippi 163
5.34 Vegetated island atop channel bar in the lower Mississippi 163
5.35 Historic increase in vegetated islands, lower Mississippi, 1965–2015 163
5.36 A lake within an island in the lower Mississippi, Bayou Goula towhead 164

6.1 Historic painting depicting dike breach flood by ice jam mechanism, Waal River (Rhine delta), 1799 167
6.2 Dike breach flooding in 1931, lower Yangtze near Nanjing, China 167
6.3 Hydrologic pathways related to flood control and sedimentology 168
6.4 Embanked floodplain evolution and subsidence after diking and land drainage 168
6.5 Flood basin subsidence in Rhine delta (Rijnland), the Netherlands 170
6.6 Upgrading Rhine River dikes as part of Room for the River program 170
6.7 Levee (dike) construction in the United States, from 1920 to 2018 171
6.8 The five stages in the cycle of dike management 172
6.9 The Tower Excavator method of dike (levee) construction, lower Mississippi 174
6.10 (A) Failure of peat dike during drought (B) Wetting peat dikes to reduce cracking 175
6.11 Organization of flood fighters and sand bag operations, 1995 Waal River flood (Rhine delta) 175
6.12 Flood wall failure along the London Avenue Canal, New Orleans, 2005 176
6.13 Dike failure after large flood event along the lower Mississippi River, 1983 176
6.14 Evolution of main-stem Mississippi River levees (dikes) for New Orleans 177
6.15 Changes in New Orleans levee (dike) height by settlement and sea level rise, 2023 and 2073 178
6.16 Conceptual comparison of inundated floodplain area with dikes and without dikes in relation to flood recurrence intervals, flood risk and human activities 178
6.17 Pattern of dike embankment due to physical, environmental, and human factors 179
6.18 Variation in floodplain embankment along several larger US rivers 180
6.19 Variation in floodplain embankment, lower Mississippi valley and delta 180
6.20 Changes in flood depth for Po River, Italy, due to dikes and channel change 181
6.21 Impact of embankments on discharge-stage along Wabash River, Illinois 182
6.22 Comparison of natural and artificial oxbow lake infilling, lower Mississippi 184
6.23 Borrow pit pond adjacent to main-stem dike (levee), lower Mississippi 185
6.24 Tally of borrow pits larger than 1 acre (0.4 ha), lower Mississippi 185
6.25 Embanked floodplain evolution along the lower Mississippi River 186
6.26 Historic map showing dike breach (wiel), Merwede River (Rhine delta), in 1729 187

LIST OF FIGURES

6.27 Dike breach pond (wiel) along Lek River from St. Elizabeth flood of 1421   188
6.28 "Patchy" floodplain landscape along the Rhine delta, particle size trend   190
6.29 Embanked floodplain aggradation, Huanghe River, China   191
6.30 Sediment flux into Raccourci Lake by batture channel (tie-channel), lower Mississippi   191
6.31 Influence of embankment on oxbow lake infilling, lower Mississippi   193
6.32 Adjustment to two oxbow lakes along the upper Rhine, Germany   193
6.33 Influences on floodplain sedimentology and hydrography due to flood control   194
6.34 Conceptual diagram showing progression of dike breach event   194
6.35 Photo of sand boil along the Rhine dike during the 1995 flood   195
6.36 Sand boil sedimentology along the Mississippi River after the 1993 flood   195
6.37 Dike breach events within the Dutch Rhine delta between 1700 and 1950   197
6.38 Beginning of dike breach at Mounds Landing, Mississippi, 1927 flood   198
6.39 Topography and hydrography of Mound Crevasse (1927), lower Mississippi   198
6.40 Response of river stage to ice jam events in relation to exceedance probability   198
6.41 Satellite image of the Great Flood of 1993, Missouri and Mississippi Rivers   199
6.42 Composite floodplain deposits associated with the Great Flood of 1993   201
6.43 Continuum of lowland floodplain adjustment to sequence of hydraulic engineering   202
7.1 Floodwater diversion by Bonnet Carré Spillway, 2016 flood, lower Mississippi   206
7.2 Historic map of Alblasserwaard and Vijfheerenlanden polders, the Netherlands   208
7.3 Photo of historic windmills and drainage canal at Kinderdijk, Rhine delta   209
7.4 Modern Archimedes screw to drain Alblasserwaard polder into Lek River   209
7.5 Intensity of engineering and management within Yazoo basin, lower Mississippi   210
7.6 Backwater flooding of Yazoo basin from record 2011 flood, lower Mississippi   212
7.7 Repair of channel erosion, Steele Bayou drainage, Yazoo basin, Mississippi   212
7.8 (A) Operation of Bonnet Carré Spillway for floodwater diversion, 2011 flood. (B) Sand splay in Bonnet Carré flood diversion spillway from historic 2011 flood   214
7.9 History of Bonnet Carré floodwater operation since construction in 1937   215
7.10 Discharge and suspended sediment, lower Mississippi, at Baton Rouge, LA, 1973 flood   216
7.11 Flooded Yolo Bypass along the lower Sacramento River (February 23, 2017)   216
7.12 Schematic of spillway structure sedimentation, Fremont Weir, Sacramento River   217
7.13 Peat compaction (%) for two 6 m-thick sequences over several thousand years   221
7.14 Flood basin evolution by land subsidence due to agriculture and drainage   222
7.15 Photo of petroleum infrastructure in the lower Mississippi valley   223
7.16 Land subsidence model for fluid withdrawal associated with oil and gas extraction   223
7.17 Wetland degradation and hydrologic alteration for fossil fuel extraction   224
7.18 Topographic profile across Lek River (Rhine delta) showing subsidence   224
7.19 Subsidence of Shanghai, Yangtze River delta, associated with groundwater withdrawal and urbanization   225
7.20 Change in pore space with sediment loading in Shanghai, Yangtze River delta   225
7.21 Different avulsion styles, including (A) partial, (B) full, and (C) nodal   226
7.22 Recent avulsion history of Huanghe River, China   227
7.23 Satellite image of the Huanghe River delta, showing rapid fluvial deltaic evolution   227
7.24 Channel thalweg profiles of Huanghe River avulsion, from 1976 to 2015   227
7.25 Dike breach caused by St. Elizabeth Day Flood of 1421 in Brabant, the Netherlands   228
7.26 Historic map (~1755) showing avulsion development from the 1421 dike breach event   228
7.27 Air photo of Biesbosch, NL, showing typical channel bifurcation and polder lands   229
7.28 Vicinity of upper Atchafalaya and Old River, including failed channel avulsions   230
7.29 Sedimentology of Atchafalaya avulsion, downstream of Red River Landing   231
7.30 Atchafalaya River cross section showing natural levee development   231
7.31 Head of Passes, Louisiana, bifurcation at the mouth of the lower Mississippi River   233
7.32 Avulsion node bifurcation (splitsingspunt) of Rhine River, 5 km downstream of delta apex   233
7.33 Historic map (1749) of Pannerdensch Kanaal, the Netherlands   233
7.34 Bangkok flooding in 2011 by Chao Phraya River   236

7.35 (A) Canal and drainage system to manage flooding for Bangkok, Chao Phraya River and (B) Topographic cross section of Bangkok in relation to 2011 flood stage 237
7.36 Flood control system of New Orleans metropolitan area 239
7.37 Geomorphic profile of New Orleans, from the lower Mississippi to Lake Pontchartrain 240
7.38 Subsidence in suburban New Orleans due to drainage and urban development 241
7.39 Historic map showing flood area from Pierre Sauvé Crevasse, 1849, New Orleans 242
7.40 Influence of 2005 Hurricane Katrina storm surge on Mississippi River stage levels 244
7.41 Subsidence in the Mississippi delta showing drowned geodetic benchmark and tree 245
7.42 View of 2005 failed floodwalls along the 17th St. canal, New Orleans 245
7.43 Strength of sediment relative to safety margins for samples underlying floodwall of the 17th Street Canal in New Orleans 246
7.44 Aerial photo of newly constructed pumps and floodgate at London St. Canal, as it flows into Lake Pontchartrain 248
7.45 Aerial view of new Inter-Harbor Lake Borgne Surge Barrier 248
8.1 Construction phase of Room for the River program, Waal River at Nijmegen, NL 250
8.2 Structural measures of Room for the River program 251
8.3 Groyne lowering along Waal River as part of the Room for the River program 252
8.4 Sediment replenishment along the Rhine River near Freiburg, DE 253
8.5 Role of sediment dumping and groynes in sediment budget, upper Rhine 254
8.6 Environmental river management: geotextile and vegetation, large woody debris 254
8.7 Bank swallow (*Riparia riparia*) habitat within alluvial channel banks of the Sacramento River 256
8.8 Sacramento River revetment installation and Bank Swallow (*Riparia riparia*) habitat 256
8.9 Channel reconnection of Kissimmee River, Florida, for environmental restoration 256
8.10 Comparison of hydrographs before and after flood attenuation measures 258
8.11 Side channel construction along the upper Rhine, downstream of Heidelberg, DE 259
8.12 Floodwater retention along Danube River, Austria 261
8.13 Changes in water surface elevations for ten-year RI events along the lower Missouri due to dike adjustment and channel engineering 263

8.14 Change in hydraulic roughness and flood depth for different floodplain surfaces 263
8.15 Modeled changes to flood stage in relation to floodplain vegetation, Maas River, the Netherlands 264
8.16 Floodplain and riverine landscape along the IJssel River, the Netherlands 264
8.17 Aerial view of dike breach, Feather River, California (Sacramento basin) 265
8.18 Change in land area of coastal Louisiana between 1932 and 2010 266
8.19 Satellite imagery depicting land growth of the Atchafalaya delta complex 267
8.20 Subdelta formation along the active Mississippi delta, year of subdelta initiation 268
8.21 Evolution of Cubits Gap subdelta, Louisiana 268
8.22 Caernarvon Freshwater Diversion structure, Mississippi delta, Breton Sound estuary 269
8.23 Modeled land gain 2030–2100 by sediment diversion, varying sea level rise scenarios 271
8.24 Location of proposed sediment diversion structures, mid-Baratarria and Breton Sound 271
8.25 Hydrograph typology in the context of environmental flows and ecological maintenance to support wetlands and oyster habitat 271
8.26 Delta lands retained and gained over fifty years with Louisiana's Coastal Master Plan 272
8.27 Sub-cost to implement different approaches of Louisiana's Coastal Master Plan 273
8.28 Range of sediment diversion approaches to restore different types of deltas 274
8.29 Floodwater retention along the Rhine River, illustrating improvement since the late 1970s 275
8.30 River and floodplain restoration measures along the Rhine River, 2005–2012 276
8.31 Atlantic salmon (*Salmo salar*) counts, 1990–2015, Rhine basin 277
8.32 Fish ladder along Nederrijn en Lek River (Rhine delta) in the Netherlands at Hagestein lock 277
8.33 Projected 2020 flood stage reductions by hydrogeomorphic province, Rhine River 278
8.34 Projected flood hydrograph change after management implemented, upper Rhine 278
8.35 Location of different measures for the Room for the River program in the Rhine delta 280
8.36 Closure of Maeslantkering flood control structure, mouth of Rhine River, North Sea 282
9.1 Vision of the Room for the River program in the Netherlands 285
9.2 Integrated approach to flood management within a sinking delta city, Rotterdam, NL 290

Tables

1.1 Comparative physical and human indices of lower Mississippi and lower Rhine rivers *page* 6
2.1 Drivers and processes associated with drainage basin zones for large rivers and deltas 11
2.2 Characteristics of different types of fluvial response to different types of human disturbances 18
2.3 Characteristics of channel patterns 21
3.1 Rainfall totals for the Great Flood of 1993, Missouri and upper Mississippi basins 38
3.2 Characteristics of major atmospheric teleconnections, linkages to streamflow variability 39
3.3 Flooded areas along the Amazon River for varying ENSO conditions 41
3.4 Atmospheric rivers: characteristics and hydrologic influences 43
3.5 Hydrologic influences on bank erosion along alluvial rivers 47
3.6 Sedimentary structures and pedogenic indices for Holocene fluvial deposits 52
3.7 Classification schema for spillage sedimentation along lowland rivers 56
3.8 Soil properties developed atop different channel belts, Mississippi River 60
3.9 Drainage area and delta area for a range of larger rivers 68
3.10 Factors influencing the morphology and sedimentary characteristics of deltas 70
3.11 Characteristics of deltaic depositional systems 70
3.12 Characteristics of the delta cycle 74
4.1 Irrigation projects in the Mekong basin 82
4.2 Dam types 83
4.3 Drainage basin distribution and ownership of large dams in Spain 86
4.4 Significant impacts of dams and reservoirs on sediment trapping along larger rivers 93
4.5 Comparison of global fluvial sediment budget and modification by human activity 93
4.6 Comparison of sediment yields and runoff before and after dam impoundment 94
4.7 Environmental issues related to dams and reservoir sediment storage 95
4.8 Characteristics of main-stem reservoirs along Huanghe River, China 97
4.9 Functional fluvial geomorphic surfaces and ecological relevance 106
4.10 Global retention of phosphorus by river reservoirs, 1970, 2000, and 2030 112
4.11 Ranking of river basins by reservoir storage of phosphorus for 2000, projected for 2030 112
4.12 Reservoir flows and dissolved oxygen levels for rivers in the Tennessee Valley Authority 115
4.13 Main drivers of dam removal 116
4.14 Cost to rehabilitate US dams 117
4.15 Comparison of submerged and subaerial weight of reservoir sedimentary deposits 123
4.16 Reservoir sediment eroded after dam removal for US case studies 124
4.17 Storage capacity and estimated life spans of reservoirs in Missouri River Basin 132
4.18 Reservoir management approaches: sediment sluicing and sediment flushing 134
4.19 Reservoir management strategies of Sanmenxia Reservoir and sedimentary response, Huanghe River 136
5.1 Common structural and nonstructural measures of channel engineering for flood control and navigation with implications for river management 139
5.2 Annual volume of sand and gravel mining from the lower Mekong River by nation and grain size 141
5.3 Importance of sedimentology in river channel engineering 143
5.4 Characteristics of appropriate riprap in the design of alluvial channel bank protection 149
5.5 Channel adjustment of the lower Mississippi to meander bend cut-offs in association with Quaternary floodplain deposits 157
5.6 Development of Giles Cut-off (1933) along the lower Mississippi including the progressive increase in flow and sediment dredged 158

6.1 Common riparian land-based structural measures associated with flood control 169
6.2 Common types of flood control dikes along rivers 169
6.3 Summary characteristics of US National Levee Database 171
6.4 Important factors to consider in dike design 173
6.5 Geophysical indices of floodplain deposits in relation to dike design to prevent problems with dike failure (underseepage, piping, sliding, erosion, river erosion) 174
6.6 Deficiencies in dikes caused by construction activities 174
6.7 Natural and anthropogenic water bodies within embanked floodplains 183
6.8 Comparison of embanked oxbow lake infilling along the Lower Mississippi 192
6.9 Crevasse (dike breach) events associated with the 1893 flood along the lower Mississippi in Louisiana 197
6.10 Floodplain erosion and sedimentation from the dike breach complex from the Great Flood of 1993 along the middle Mississippi River, Miller City, Illinois 200
6.11 Continuum of lowland channel and floodplain adjustment in response to the sequence of flood control engineering 203
7.1 Types of flood diversion structures along large lowland rivers 207
7.2 Annual subsidence rates along major lowland rivers and deltas impacted by human activities 219
7.3 Geophysical properties associated with peat classification 221
7.4 Factors to consider in the management of channel avulsions and bifurcations 232
7.5 Issues associated with flooding in urban areas within flood basins 235
7.6 Geophysical characteristics of Standard Project Hurricane for the 1965 Lake Pontchartrain and Vicinity Project in comparison with Category-I,-II, -III, -IV, and -V hurricane parameters 243
7.7 Causes and consequences of flood management failures associated with the New Orleans Hurricane Katrina flood disaster 247
8.1 Strategies and approaches common to integrated river basin management 250
8.2 Techniques and benefits of utilizing environmentally friendly "soft" bank protection rather than "hard" measures 255
8.3 Management steps for design of channel reconnection projects 257
8.4 Guidelines for design of floodplain side channels 259
8.5 Ecosystem service benefits associated with channel and lake reconnection 262
8.6 Characteristics of four subdeltas within the modern Mississippi delta lobe 267
8.7 Recommendations for operation strategy of large sediment diversion structures 272
8.8 Design of freshwater and sediment diversions to offset deltaic land loss 274
8.9 Primary measures to lower flood stage levels and expected minimum reduction 281
9.1 Overview of "lessons learned" in relation to management options 286

Preface and Acknowledgments

The importance of writing a book about the geomorphic and environmental impacts of hydraulic engineering to lowland rivers has long seemed obvious. I've mainly studied and resided in lowland coastal plain settings, other than an early stint in the Midwest having been born in Wisconsin. This includes north Florida and the lower St. Johns River valley, the Holocene Mississippi floodplain in Baton Rouge, Louisiana, and Austin, Texas at the upper fringe of the Gulf Coastal Plain. And now, atop late-Holocene sand dunes at the terminus of the Rhine delta in the Netherlands.

Since the early 1990s I've conducted research on the theme of human impacts to lowland rivers, particularly topics related to hydraulic engineering and flooding. The concept for the book really began to take shape over a decade ago after my return to Austin from Utrecht in late summer 2008. At Utrecht we focused on an explicit comparison of management approaches between the Mississippi and Rhine, a timely theme given the 2005 flooding of New Orleans and new developments in flood management in the Netherlands. Colleagues in the Department of Geography and the Environment at UT Austin encouraged me to embark on the sinuous journey of writing a monograph. I was keen. Well, life is rich – and complex – and in 2010 it involved relocating to the Netherlands and two universities that resulted in a series of start and stop attempts to write the book. A semester research sabbatical from Leiden University College in autumn 2015 finally provided the watershed moment to hunker down and tackle the book, although I didn't realize it would take another six years.

In some ways this work is a historical treatise in that it draws upon prominent themes and topics of where I've studied, taught, and researched. I was fortunate to complete my doctorate at Louisiana State University (LSU), immersed in the rich tradition of scholarship on fluvial–deltaic processes and engineering geomorphology of the lower Mississippi River prominently developed by R. J. Russell and H. N. Fisk, among others. At LSU my doctoral supervisor, R. H. Kesel, embraced these themes and also emphasized the importance of archival and historic research as complementary and essential to scholarship in geomorphology. This is necessary when working on large rivers and can be particularly rewarding when working on rivers that have long been utilized for human settlement. And I became acquainted with several themes herein examined during my MSc thesis work at the University of Florida with Joann Mossa, who also completed her doctorate at LSU. Joann introduced me to the lower Mississippi during a "Friends of the Pleistocene" field trip (actually focused on the Red River valley) in Louisiana. With repeated trips between Baton Rouge and Vicksburg along Scenic Highway (U.S. 61), I became fully ensconced in the lore of the lower Mississippi alluvial valley. And many trips to New Orleans, that great sinking delta city, provided a unique perspective on the difficulty of flood management in a complex urban environment undergoing high rates of ground subsidence. At LSU I had courses and attended lectures by esteemed scholars, including H. H. Roberts, J. M. Coleman, G. W. Stone, among others, that oozed of Mississippi delta science.

I've been fortunate to have an academic career over the past twenty-five years that enabled me to conduct research and annually teach courses closely aligned with themes contained herein, including at the University of Texas at Austin, and in the Netherlands at the University of Amsterdam and Leiden University. At UT Austin, I had great scholars as colleagues, including Karl Butzer, who was always supportive and instilled a measured perspective on current fluvial disasters that can only be appreciated when working across longer timescales ("we've seen this before..."). Receipt of a US Fulbright Fellowship spent in residence (2007–2008) with Hans Middelkoop's group at the Institute of Physical Geography at Utrecht University was invaluable for shedding light on several themes herein reviewed, and for gaining an appreciation for the high level of scholarship in geomorphology and the precision to which flood management science is practiced in the Netherlands. I've also been fortunate to have great students who enquire about topics we might have thought were worked out but were not, thereby unwittingly helping to elucidate gaps in knowledge and stimulate new research ideas. Teaching and research really are bidirectional.

It's important to acknowledge that this treatise could not have been written without the tremendous body of scholarly materials from which it draws. This includes materials produced by government agencies in the form of reports and data sets invaluable to academic research. I'm particularly grateful for a range of materials provided by the U.S. Army Corps of Engineers (esp. Potamology Program) and Rijkswaterstaat. Additionally, digital archives and databases from a wide range of organizations were important for providing historic photographs, imagery, maps, and data. These especially included Actueel Hoogtebestand Nederland (lidar DEM), American Rivers, Bangkok Metropolitan Agency, Bank Swallow Technical Advisory Committee (California), Australian National Committee on Large Dams, DamRemoval.eu, Delfland Water Authority, Delta Regional Authority (Mississippi), German Federal Waterways and Shipping Administration, Google Earth Pro, International Commission on Large Dams, International Commission for the Protection of the Danube, International Commission for the Protection of the Rhine, Louisiana Coastal Protection and Restoration Authority, Louisiana State University, Department of Geography & Anthropology, ATLAS lidar data, Mekong Eye, National Archives of the Netherlands, Murray-Darling Basin Authority, National Oceanic Atmospheric Administration, NASA Earth Observatory, New Orleans Historic Collection, Occidental College Special Collections, Regional Archives Dordrecht, Rijksmuseum Amsterdam, Spanish Ministry of Environment, State Library of Louisiana Historic Collection, Tennessee State Library and Archives, United Nations Food and Agriculture Organization, U.S. Department of Agriculture (NRCS), U.S. Fish and Wildlife Service, U.S. AID (Mekong ARCC Program), U.S. Geological Survey, U.S. Library of Congress, and Utrecht University Digital Historic Map Collection. An exhaustive attempt was made to properly secure rights and permissions for all figures and tables herein utilized. I'm grateful to the following individuals who kindly granted permission to use their personal photographs or figures, including Prof. S. Darby (Mekong River erosion), Prof. L. Fitcher (point bar diagram), L. Lefort (New Orleans area subsidence), N. Olsen (Bayou Goula towhead island, Mississippi River), and J. Rusky (Head of Passes, Mississippi River).

The quality and quantity of new published academic research along the themes herein examined is especially impressive and requires considerable effort to keep pace. The Mississippi and Rhine have long served as hearths of fluvial scholarship, and recent research increasingly has direct environmental, interdisciplinary, and societal relevance. An impressive flow of new large river research concerns long neglected regions, providing exceptions to the general body of knowledge of fluvial geomorphology developed in mid-latitude North America and Western Europe. This especially includes rivers in Asia and South America.

I'm keenly aware of esteemed scholars who have passed over the last decade or so, and from whom I've learned so much and whose ideas permeate this treatise, including Leal Mertes (2005), Luna Leopold (2006), Gilbert White (2006), Henk Berendsen (2007), Gordon Wolman (2010), Stan Schumm (2011), Jim Knox (2012), Jess Walker (2015), Karl Butzer (2016), Wil Graf (2019), and Ken Gregory (2020). I'm grateful to and remain in awe of their extensive contributions and dedication to river science.

I'm very appreciative of Cambridge University Press, specifically editors M. Lloyd, S. Lambert, Z. Pruce, S. Duveau, and F. Mathews Jebaraj for prompt feedback and, especially, the time and space to complete the monograph, which extended long after my relocation from Texas to the Netherlands. Critical reviews and sharp insights provided by E. M. Latrubesse, J. M. Daniels, F. T. Heitmuller, M. K. Steinberg, M. C. LaFevor, among others, and my students, who were unsuspecting critics for several years, allowed me to explore and develop a range of materials. And last – but certainly not least – this work could not have been completed without support and sacrifice from my family, who endured far too many "fragmented" weekends and "cut-off" holidays while I toiled. Thanks for your patience over all these years.

1 Fluvial Lowlands and the Environment: Cause for Concern

1.1 INTRODUCTION AND SCOPE OF WORK

Fluvial lowlands are fascinating landscapes – big, flat, wet, and inherently unstable. Large fluvial lowlands are among Earth's most unique and productive settings, rich in resources and opportunities. For millennia humans have coexisted within fluvial lowlands but with variable successes. From the standpoint of human settlement, safety from flooding is the most important issue. Flooding is not simple however, and in lowland floodplains it occurs by several distinct processes, such as overbank flooding from upstream sources, groundwater-induced flooding, intense local rainfall events, or flooding generated by large coastal storm events. In the most extreme scenarios flooding is simultaneously caused by several mechanisms, exacerbated by headwater land use change and lowland engineering that drives ground subsidence.

Many of Earth's larger rivers have been transformed by hydraulic infrastructure. Through efforts to make fluvial lowlands habitable, regulated rivers are a fundamental characteristic of the Anthropocene (Gregory, 2006; Hudson and Middelkoop, 2015). From the standpoint of fluvial classification, diking of rivers for flood control has created an entirely new anthropogenic fluvial typology: the embanked floodplain. The hydrology, sedimentology, and geomorphology of embanked floodplains have characteristics that fundamentally differ from natural rivers (Middelkoop, 1997; Hesselink et al., 2003), with consequences to the geodiversity and biodiversity of fluvial lowlands (Nienhuis and Leuven, 2001; Schramm et al., 2009).

Over centuries the Netherlands has developed a variety of approaches to manage large flood events derived from upstream and coastal hydrologic processes (van de Ven, 2004; de Bruin, 2006; van Heezik, 2008). Large floods in 1993 and 1995 (Figure 1.1), however, provided valuable lessons in regard to flood vulnerability in relation to global and local environmental change. The two high-water events were a stimulus in the development of a new strategy to manage flooding within the Netherlands, and in fostering a more integrated river management approach within Europe (European Council, 2000, 2007). A decade later, in North America, the 2005 flooding of New Orleans from Hurricanes Katrina and Rita served to shine the spotlight on the issue of flood vulnerability of large urban flood basin environments (Figure 1.2). The event revealed – in stark definition – that the might of massive hydraulic infrastructure is dependent upon an adequate understanding of the flood basin sedimentology in which it is constructed, as well as the governmental framework in which flood management is implemented. In contrast to the response in Europe, however, the flood disaster in New Orleans has resulted in incremental rather than sweeping changes in the philosophy of flood management (Hudson, 2018).

This treatise examines environmental impacts to large fluvial lowlands and deltas caused by hydraulic infrastructure and flood management from a geomorphic, hydrologic, and sedimentary perspective. A range of hydraulic infrastructure types and their impacts to different fluvial environments are systematically examined, including upstream dams, main-stem channel engineering, floodplain embankment by dikes, and flood and sediment diversion structures within flood basins and deltas.

The scope of the book concerns larger alluvial valleys and deltas and how their processes over the past decades and centuries have been altered by human impacts. Where appropriate a distinction is made for pure fluvial or coastal features and processes (channel incision, storm surge, overbank deposition, etc.), but otherwise fluvial–deltaic and fluvial lowlands are used interchangeably. The book draws upon a range of fluvial geomorphic topics, including concepts and processes related to hydroclimatology, floodplain hydrology, channel erosion, overbank sedimentation, channel dynamics, channel avulsion, ground subsidence, and fluvial–deltaic sedimentology. Additionally, because construction and operation of hydraulic infrastructure is implemented by specific governmental institutions, the book also concerns how management and hydraulic engineering is manifest as different types of environmental impacts to fluvial lowlands.

Figure 1.1. Flood protection in a densely populated region. Waal River (Rhine delta) above flood stage in 1993, the largest high-water (flood) event since the devastating 1953 flood, only to be surpassed two years later by the 1995 event. The two events stimulated a new paradigm of flood protection in the Netherlands and the Rhine basin, with broader implications across the European Union. (Photo source: Rijkswaterstaat archives.)

Figure 1.2. Failure of 17th Street Canal flood wall and catastrophic flooding of New Orleans, Louisiana associated with Hurricanes Katrina and Rita in 2005. The event received global attention and resulted in swift upgrades to New Orleans' long neglected flood control system. The broader influence of the event on US flood management is more incremental than paradigm changing. (Photo source: U.S. Army Corp of Engineers.)

1.2 EXTENT AND SCALE OF FLUVIAL LOWLANDS

Large fluvial lowlands are among Earth's most distinctive environments. Their importance to humans and biophysical processes is vastly disproportionate to their area, which range from about 5% to 0.5% of Earth's land surface (Overeem and Syvitski, 2009; Science, 2021). Fluvial lowlands have broad low-gradient valleys, multiple channel belts, clayey backswamp prone to extended periods of inundation, and are fed by rivers with low specific stream power (Nanson and Croke, 1992; Phillips and Slattery, 2007; Latrubesse, 2008). The concept of space is important to understand fluvial lowlands. River valleys need to be sufficiently wide to preserve older channel belts and cohesive backswamp deposits, which results in complex topographic and sedimentary environments that influence flood pulse dynamics and related ecosystems (Saucier, 1994; Tockner et al., 2000; Stouthamer and Berendsen, 2001; Hudson and Colditz, 2003; Toonen et al., 2016). Additionally, broad river valleys are associated with abrupt lateral changes in energy from high-velocity conditions adjacent to channel banks to tranquil distant flood basins, which influences sediment dispersal. The low gradient of large valleys reduces drainage, contributing to the distinctive hydrology of fluvial lowlands. The border between delta and alluvial valley is not always obvious, although neotectonic faulting often represents a structural border (Saucier, 1994; Cohen, 2003). In lowland rivers neotectonics is also responsible for altering river courses and warping valleys, which can exacerbate flooding and alter floodplain water bodies (Burnett and Schumm, 1983; Guccione et al., 2000). Above the delta apex lowland rivers typically extend upstream to where there is an abrupt increase in slope (gradient) and reduction in valley width. This often, but not always, occurs at a significant change in geologic structure, lithology, and/or the confluence of a major tributary.

Deltas are not very old, from a geologic perspective. Most major deltas have formed since the early mid-Holocene, roughly 7,500 BP. This corresponds to the cessation of high rates of Holocene sea level rise (Stanley and Warne, 1994; Blum and Törnqvist, 2000). The importance of sea level rise is that it reduces river slope and induces sedimentation. Rivers with large sediment loads deposit a thick sedimentary wedge of deltaic deposits associated with river mouth progradation, advancing land over sea. Large river deltas have extensive subaqueous deposits that can extend to the continental shelf edge (Coleman and Wright, 1971; Nittrouer et al., 1986). Sustained sea level rise resulted in continued fluvial–deltaic deposition but also reduced river gradient. This caused the river mouth to periodically switch course (an avulsion), perhaps about every thousand years or so, thereby forming a new channel belt and a new delta lobe. Abandoned channel belts subside and become buried and infilled with organic and fine-grained deposits. Subsequent erosion by marine processes drives sequential coastline change, including barrier island migration (Penland et al., 1988; Blum and Törnqvist, 2000).

Because fluvial–deltaic lowlands are young surfaces it has important implications to physical processes, as well as related human activities. A key point is that Holocene fluvial–deltaic deposits are porous and soft, and therefore prone to compaction and subsidence (Dokka, 2006; Stouthamer and van Asselen, 2015). This is particularly true in clayey backswamp and low-lying flood basins, which contrasts with higher and denser sandy-silty deposits of natural levees and point bars within the active channel belt. By comparison with upstream channel belts, delta lobes are deposited atop thick units of porous clay and organic material, which rapidly subside upon dewatering and compaction (Yuill et al., 2009). Additionally, the high organic content of flood basin environments, especially peat basins, drives subsidence from oxidation at the surface, particularly following drainage and diking for flood protection (van Asselen et al., 2009; Erkens et al., 2015).

1.3 HUMAN IMPACTS TO FLUVIAL LOWLANDS

Fluvial lowlands are sensitive to changes in climate, neotectonics, ground subsidence, and marine processes at the coast. These factors are exacerbated by different forms of human activities and directly influence geomorphic adjustment and the flood regime of fluvial lowlands (Figure 1.3). Despite constraints imposed by the physical environment many of Earth's large fluvial lowlands support high population densities and large cities (Komori et al., 2012; Best and Darby, 2020). This necessitates management integrated within a regional- and local-scale geomorphic framework that does not adversely impact hydrologic and sedimentologic processes, particularly in the context of global environmental change scenarios (Middelkoop et al., 2002; WMO, 2004).

The proliferation of dams has fragmented Earth's riparian landscapes and resulted in upstream trapping of fluvial sediments within numerous reservoirs. The streamflow regime of many rivers has been fundamentally altered by dams, which reduces downstream flow variability and drives riparian environmental change (Graf, 2006; Grill et al., 2015). This especially includes terrestrialization of formerly dynamic fluvial environments, reducing aquatic habitat and degrading riparian fisheries. With sediment trap efficiencies of upstream dams often exceeding 80%, the coastal sediment flux of many rivers is greatly diminished. This has dramatically decreased downstream sediment loads vital to lowland floodplains related to wetland protection and ecosystem services, including flood safety. Foremost, sediment reduction is linked to sinking deltas and loss of coastal wetlands (Blum and Roberts, 2009; Syvitski et al., 2009; Allison et al., 2016) (Figure 1.4).

Many deltas are supplied by rivers that drain humanized landscapes impacted by upstream dams, with downstream sediment loads insufficient to counter projected sea level rise of 0.4–1.2 m by 2100. The Nile delta, for example, is rapidly eroding because its sediment load has drastically declined due to upstream reservoir storage (Stanley and Warne, 1998; Becker and Sultan, 2009). Immediately downstream of the Aswan High Dam the Nile River transports less than 1% of its upstream sediment load, as all of the sediment is trapped in Lakes Nasser and Nubia (Ahmed and Ismail, 2008). Other large rivers commonly transport less than 10% of their upstream sediment loads immediately downstream of dams. While further downstream some of the sediment load is replenished because of tributary inputs and fluvial reworking of older alluvial deposits, global sediment flux to the ocean is only about half of pre-dam sediment loads. Since the construction of a cascade of dams along the upper Mekong (Lancang) in China beginning in the early 1990s, the sediment flux to the delta has declined by 74%, with many more dams planned. Many large rivers, therefore, no longer have adequate sediment loads to counter sea level rise, resulting in drowning of deltaic landscapes and greater flood risk to increasingly larger urban populations. More significant than absolute sea level rise is relative sea level rise, which is directly linked to land subsidence and can only be understood through local analysis of sedimentology and human activities (Ibàñez et al., 1997; Minderhoud et al., 2020).

Balancing the needs of flood safety with environmental sustainability is the art of good science for river management (Figure 1.5). Flood basins along large fluvial lowlands with high populations serve a variety of functions and are frequently associated with conflict between competing stakeholders. Large flood basins include vast natural resources, including wetlands that support high biodiversity. Large downstream populations, however, rely on these flood basins to temporarily store flood waters, abating downstream flood crests. Flood basin sedimentation infills floodplain water bodies that drives an ecological transition, with direct consequences to flood basin flora and fauna. Increased flood basin sedimentation also reduces floodwater storage space, resulting in less room for water and accelerating downstream flood wave transmission.

Embanked floodplains are highly anthropogenic fluvial settings (Middelkoop, 1997; Nienhuis, 2008, pg. 116). For many decades embanked floodplains were considered a sort of biophysical wasteland, but they are increasingly appreciated for the

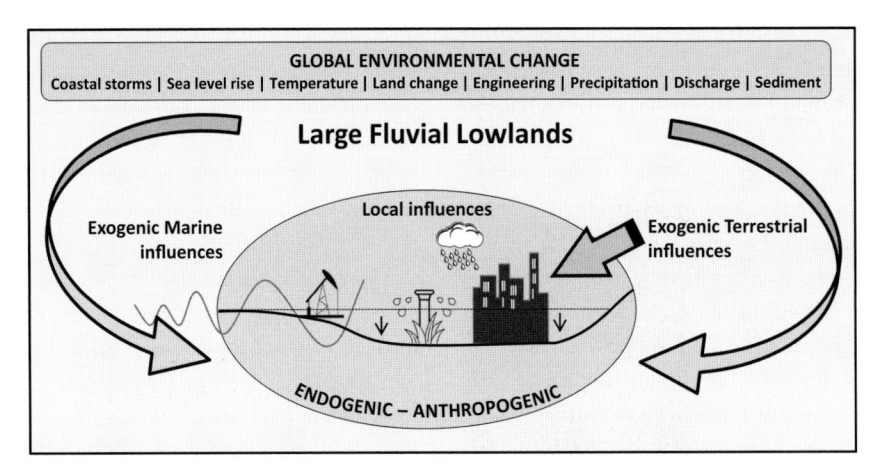

Figure 1.3. Exogenic and endogenic influences on fluvial–deltaic vulnerability, emphasizing upstream (drainage basin), marine, and local influences on flooding by natural and anthropogenic drivers. Extraction of fossil fuels, groundwater withdrawal for agriculture, and urbanization are major drivers of ground subsidence across fluvial lowlands.

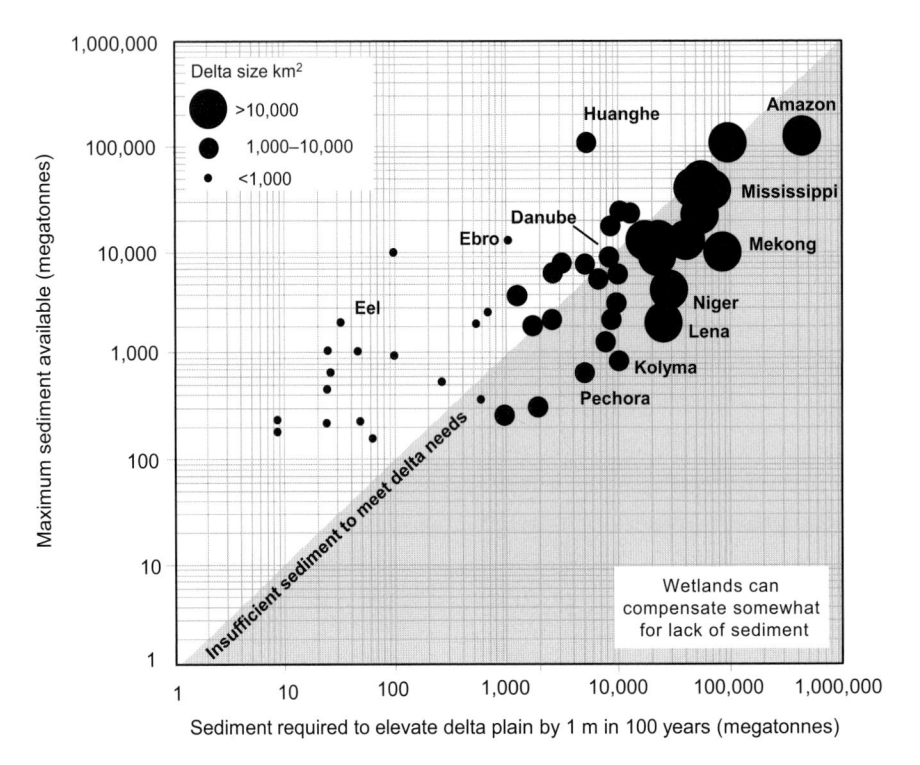

Figure 1.4. Delta size in relation to sediment required to raise delta surfaces by 1 m for protection against sea level rise. Reduction in sediment delivery is caused by upstream dams and land change. (Source: Giosan et al., 2014.)

provision of valuable ecological services in an otherwise humanized landscape (Hudson et al., 2008; Biedenharn et al., 2018). This is because flood basins outside of the embanked floodplain are most often utilized for intensive agriculture or urban settlement (White, 1945; Pinter, 2005). And the use of flood basins for agriculture and settlement requires pumping and drainage that trigger oxidation of organic-rich soil and peat. This initiates a positive feedback that requires further drainage and pumping, accelerating subsidence and increasing flood risk (FAO, 1988; Galloway et al., 1999; Erkens et al., 2015).

The dire condition of humanly altered flood basins is epitomized in several Asian mega-deltas, which includes eleven or so large rivers that drain east and southeastern Asia. The Intergovernmental Panel on Climate Change (IPCC) identified Asian mega-deltas as particularly vulnerable to global environmental change (Cruz et al., 2007). Asian mega-deltas

Figure 1.5. The Atchafalaya basin, a vital component of the flood control system for the lower Mississippi River. The largest flood basin in North America supports a unique freshwater ecosystem undergoing high rates of sedimentation and terrestrialization. (Author photo, October 2003.)

Figure 1.6. Rice farming and aquaculture (shrimp) in the Mekong River delta, Thuan Hoa Commune, Vietnam. High rates of groundwater withdrawal associated with such activities drive ground subsidence. (Photo source: Donald Bason/DAI, USAID, used with permission.)

produce enormous amounts of food for local consumption by megacity residents, and for global consumers who increasingly demand protein. Fish and shellfish production has proliferated across several of the Asian mega-deltas (Figure 1.6), although it is not sustainable because of the industry being heavily reliant upon large amounts of freshwater supplied by groundwater pumping, which accelerates subsidence and increases relative sea level rise (Higgins et al., 2013; Minderhoud et al., 2020).

1.4 A TALE OF TWO RIVERS, AND BEYOND

This treatise especially utilizes the Rhine and lower Mississippi Rivers to review key processes and concepts related to historic and modern flood management strategies and their environmental impacts to fluvial lowlands. And where relevant, specific case studies for an international range of rivers are provided – including several of the Asian mega-deltas. There are multiple reasons for this approach. The first is that the Rhine and Mississippi are

Table 1.1 *Comparative physical and human indices of lower Mississippi and lower Rhine rivers*

Indices	Rhine	Mississippi
Drainage basin area (km^2)	185,000	3,210,000
Average discharge (m^3/s)	2,150	18,400
Diked floodplain width (km)*	1.8	10.8
Delta plain area (km^2)	3,100[a]	36,480
Suspended sediment discharge to coast (tons/yr)	1,250,000[b]	145,000,000[c]
Natural areas (%): Alluvial valley*, delta plain	5, <5[d]	35, 60[e]
Population	7,200,000[f]	2,000,000[g]
Main urban areas	Rotterdam, Amsterdam, The Hague, Utrecht, Arnhem, Dordrecht	New Orleans, Baton Rouge, Lafayette
Ranking of port size (2016): # container ships / tonnage[h]	Rotterdam: 12 / 6	South Louisiana: not ranked (<100) / 13

Notes: *between main flood control dikes along main-stem river valley upstream of delta apex and distributary branches;
[a]Middelkoop and van Haselen (1999); [b]Frings et al., 2019 (Figure 5); [c]Meade and Moody (2010); [d]based on Nienhuis (2008) and Middelkoop and van Haselen (1999); [e]lower Mississippi River Conservation Committee 2015; [f]limited to delta; 2012; [g]Twilley et al. (2016); [h]2016 American Association of Port Authorities: World Port Ranking for 2016 – Top 100 Ports (www.aapa-ports.org/unifying/content.aspx?ItemNumber=21048, accessed July 2, 2020).
After Hudson (2018).

of great societal and scientific value. Importantly, both systems have had considerable scientific analysis and represent mature bodies of research that can be drawn upon to elucidate common themes (de Bruin, 2006; Hudson et al., 2008), providing familiarity to the reader and continuity in narrative. Additionally, the Rhine and lower Mississippi differ greatly in terms of size, geomorphology, and human activities (Table 1.1), providing different types of lessons as related to lowland river management. The Rhine and the Mississippi are representative, as much as any, of the general situation across Europe and the United States. The Rhine is representative of a new European approach to river management and was actually a stimulus for the diffusion of concepts developed more broadly across Europe, and beyond. The Mississippi dominates the continental drainage of North America. And as 99% of its basin lies within the United States, the Mississippi is very much representative of a US approach to river and flood management (White, 1945). Finally, while the Rhine and Mississippi are both well-known there are many contrasts in management approaches and especially their history of engineering that makes for interesting comparisons to elucidate key concepts in flood management and lowland floodplain geomorphology (Pinter et al., 2006a; Hudson et al., 2008).

In addition to the Rhine and Mississippi, case studies from an international assortment of rivers are reviewed (Figure 1.7), including the Mekong, Huanghe, Nile, and Sacramento Rivers, among others, to arrive at general conclusions about the environmental impacts of hydraulic engineering and flood management to lowland fluvial systems.

1.5 OUTLINE AND NARRATIVES

Utilizing a drainage basin framework, the contents are logically ordered and chapters are discrete, and generally organized in downstream order. Three main narratives are interwoven throughout the text, including (1) the importance of the older sedimentological framework as a control on modern fluvial dynamics and hydraulic engineering, (2) the unintended geomorphic and environmental consequences of land use and main-stem hydraulic engineering – which may require many decades to unfold – and, subsequently, (3) the need to develop further management options that weren't initially planned, or possibly even conceived. This third narrative can be succinctly summarized as "managing management." Because of the scale and complexity of large rivers, such management options can require decades to operationalize. Ultimately, because each large river is unique, management must be tailor adapted to specific physical and ecological parameters, in addition to the societal values of riparian stakeholders.

A further theme that emerges is the importance of agriculture and food production as a driver of lowland fluvial degradation, including by dams to provide stable supplies of freshwater to irrigation networks, for floodplain embankment of agricultural lands, as a reason to stabilize river channels to ensure their navigability, and as an obstacle to wetland management and reconstruction in deltaic wetlands. The importance of agriculture in driving river regulation is likely to become more important as agricultural expansion occurs within uncultivated lands until about 2050. Thus, with increasing

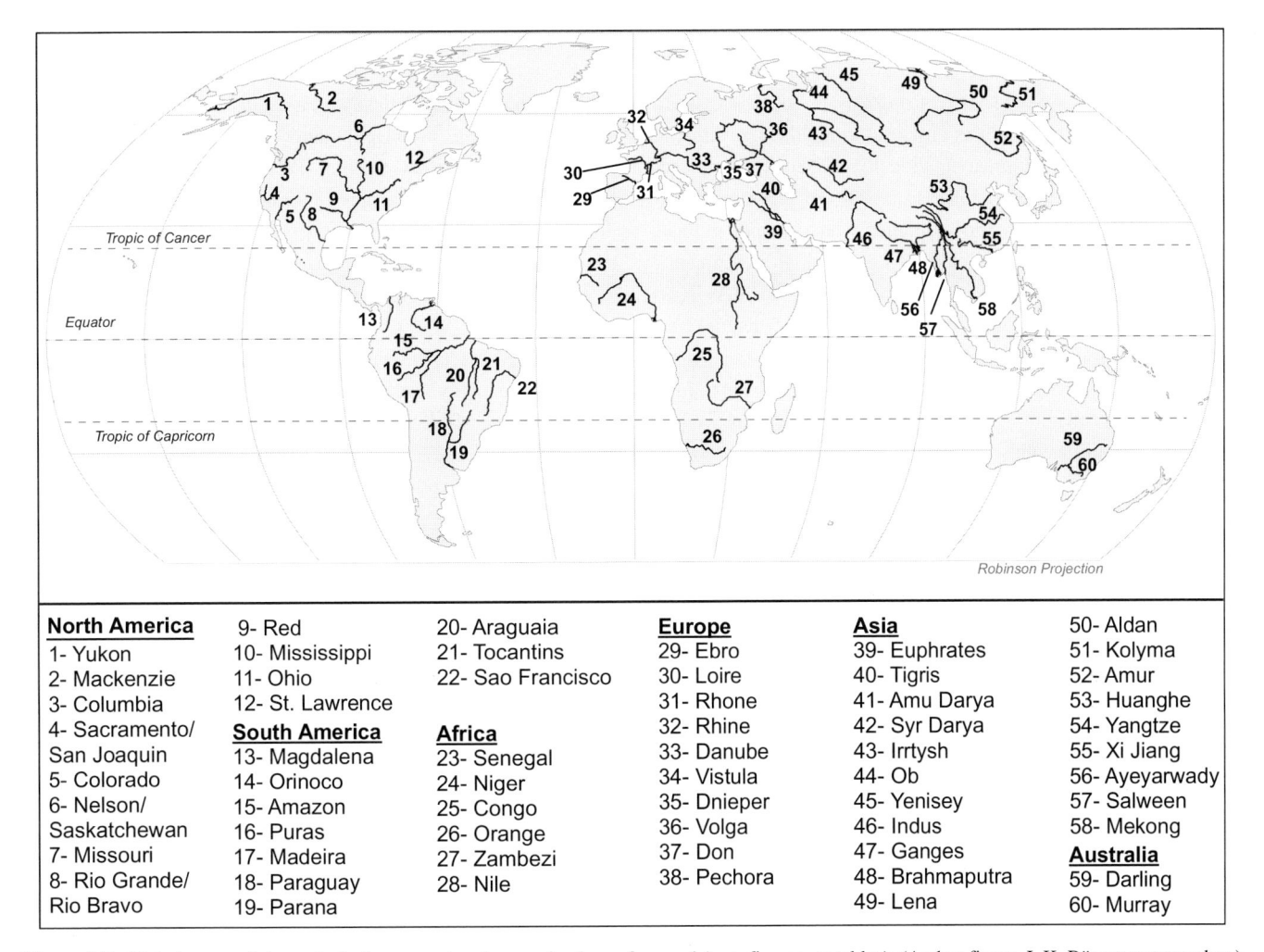

North America			Europe	Asia	
1- Yukon	9- Red	20- Araguaia	29- Ebro	39- Euphrates	50- Aldan
2- Mackenzie	10- Mississippi	21- Tocantins	30- Loire	40- Tigris	51- Kolyma
3- Columbia	11- Ohio	22- Sao Francisco	31- Rhone	41- Amu Darya	52- Amur
4- Sacramento/ San Joaquin	12- St. Lawrence	**South America**	32- Rhine	42- Syr Darya	53- Huanghe
5- Colorado	13- Magdalena	**Africa**	33- Danube	43- Irrtysh	54- Yangtze
6- Nelson/ Saskatchewan	14- Orinoco	23- Senegal	34- Vistula	44- Ob	55- Xi Jiang
7- Missouri	15- Amazon	24- Niger	35- Dnieper	45- Yenisey	56- Ayeyarwady
8- Rio Grande/ Rio Bravo	16- Puras	25- Congo	36- Volga	46- Indus	57- Salween
	17- Madeira	26- Orange	37- Don	47- Ganges	58- Mekong
	18- Paraguay	27- Zambezi	38- Pechora	48- Brahmaputra	**Australia**
	19- Parana	28- Nile		49- Lena	59- Darling
					60- Murray

Figure 1.7. Global range of rivers, including many herein examined or referenced (text, figures, or tables). (Author figure, I. K. Bürger cartographer.)

regulation many of the management issues herein observed for intensively regulated mid-latitude rivers will need to be observed for large tropical rivers.

1.5.1 Contents

Chapter contents are organized thematically rather than regionally. The book consists of three sections consisting of nine chapters, and ranges from headwater processes to lowland rivers and deltas, the main thrust of the book. The study concludes by reviewing a range of environmentally oriented engineering approaches increasingly employed under the framework of "integrated river basin management."

The first section provides a rationale and approach for the book (Chapter 1), as well as a brief overview of upland environmental change and how it relates to lowland floodplains and deltas (Chapter 2). Here the focus is on the impacts of land degradation to downstream fluvial processes, as well as briefly reviewing fundamental hydrologic and geomorphic processes concerning fluvial adjustment. Following, hydroclimatic, river bank erosion, and fluvial sedimentary processes associated with lowland floodplains and deltas are reviewed (Chapter 3) to provide a baseline foundation of concepts and terminology for material covered in subsequent chapters.

The second section consists of four chapters that systematically examine the environmental impacts of conventional hydraulic infrastructure and flood management on different segments of rivers and deltas. This includes the impacts of dams, main-stem channel engineering, flood control systems, embanked floodplain processes, and flood basins and deltas. The section begins with a review on the impacts of dams on rivers (Chapter 4), including dam removal and reservoir sediment management. While many hundreds of small dams are being removed in Europe and North America, there are millions of dams across Earth. Dams and reservoirs are essentially a permanent facet of Earth's riparian landscapes. Because they degrade riparian environments it is essential that effective dam and reservoir management strategies be designed so that associated ecosystems are not adversely impacted by changes to downstream streamflow and sediment loads.

Regulating rivers for navigation and flood control to meet societal demands requires a variety of engineering modifications to river channels (Chapter 5). Here the focus is on classic "hard engineering" structures, such as groynes (wing dikes), revetment, cut-offs, and dredging, approaches globally utilized. The emplacement of hydraulic infrastructure is associated with direct modification to channel morphology and hydraulics, which abruptly initiates fluvial adjustment, including channel aggradation, incision, and widening. While the science of river engineering has developed sophisticated hydraulic models, large rivers ultimately require considerable time to fully adjust (decades), and do so in unintended ways. This suggests the importance of a historical approach to fully comprehend the magnitude and direction of change, as well as subsequent feedbacks with other fluvial environments.

Following examination of river channels, the book explores the characteristics of embanked floodplains, the riparian lands between dikes, with a focus on the lower Mississippi and the Rhine (Chapter 6). History illustrates that the evolution of a specific dike system is a complex enterprise that involves different governmental scales of operation, policy, and requires many years to complete. As embanked floodplains are synonymous with the Anthropocene it behooves scientists to understand their processes, and especially how they vary from "natural" floodplains that represent the foundation of knowledge. As embankment results in overbank deposits being constrained to a narrow corridor, the sediments fill up the embanked floodplain over time. In the process, the sediment buries wetlands and shallow floodplain water bodies, resulting in very different – anthropogenic – floodplain styles with reduced capacity to support riparian ecosystems. Following the review of embanked floodplains, we then turn our focus to flood basins (Chapter 7), large expansive settings beyond higher channel belts that are comprised of cohesive sediment and organics. In contrast to the active dynamics of channel belts, expansive flood basins are much more quiescent. Despite their overall stability a key geomorphic process in flood basins is subsidence, which is much greater than the adjacent sandy channel belts. This occurs because flood basins are drained for agriculture, which sets off a positive feedback – and further subsidence – requiring further drainage and pumping of groundwater to be utilized for agriculture or human settlement. Ground subsidence increases the potential for river avulsion, which can result in the complete switching of a channel belt. In regard to the latter, we especially focus on deltaic flood basins with large urban populations that provide unique challenges to flood management. Perhaps the most illustrative example of these challenges is found in the Asian megacities located atop sinking mega-deltas (Overeem and Syvitski, 2009; Hanson et al., 2011), characterized by rapidly growing populations, landscape degradation, and increased vulnerability (Tellman et al., 2021). In this context it is also important to consider marine processes, especially storm surge events. The latter is especially well illustrated by reviewing the infamous 2005 flooding of New Orleans by Hurricanes Katrina and Rita. With New Orleans as a final example we bring together many aspects of the preceding chapters that highlight issues related to the design of urban flood control systems in relation to fluvial-deltaic sedimentology and flood basin hydrology.

The third section looks ahead to modern styles of floodplain management that focus on sustainable design of rivers and floodplains that seeks to balance nature and flood safety. It does this by providing a systematic review of "integrated river basin management" approaches, which in some ways is facilitated by measures adapted to the Rhine, and then diffused more broadly across the European Union, and beyond. The specific flood management approach recently employed in the Rhine delta is "room for the river," – a hybrid strategy that incorporates soft and hard engineering, a detailed understanding of the local floodplain geomorphology, sedimentology, and ecology, and stakeholder involvement. The procedures being employed in the Mississippi delta are very different from the Rhine, in part because of the contrasting physical and human environment (Table 1.1). The key measure to be employed to restore and preserve coastal deltaic wetlands in the Mississippi delta is sediment diversion (Science, 2021, pg. 334). The design of sediment diversion strategies draws upon historic sedimentologic lessons learned from the study of subdeltas that formed in the Mississippi delta over about the past two centuries. Lastly, Chapter 9 provides a synthesis of key themes and "lessons learned" from earlier chapters, and, with regard to more sustainable approaches to the management of fluvial lowlands, reasons for concern and reasons for optimism.

2 Rivers and Landscapes: A Drainage Basin Framework

> The most important environmental change that influenced fluvial activity ... during last 10,000 years involved the conversion of a late Holocene mosaic of prairie and forest to a landscape dominated by cropland and pastureland ...
>
> —J. C. Knox (2006)

2.1 THE FLUVIAL SYSTEM: A LARGE RIVER PERSPECTIVE

Large rivers are supplied by many smaller rivers that drain varied landscapes. The tripartite division of drainage basins into supply, transport, and deposition zones provides an organizational framework to consider spatial variability in fluvial form and process. Headwater processes influence large rivers and deltas, suggesting the importance of employing a drainage basin perspective in examining human impacts to fluvial lowlands (Schumm, 1977; Syvitski et al., 2009). From upstream to downstream, fluvial systems are distinctively nonlinear with increasing drainage area. The nonlinearity pertains to key hydraulic, hydrologic, topographic, and sedimentary processes and features that drive change to lowland floodplains and deltas at the terminus of large river basins (Figure 2.1). Nevertheless, because of their vast size, larger fluvial lowlands have unique processes and features that can seem to be independent of basin headwaters (Phillips, 2003; Lewin and Ashworth, 2014a,b).

2.1.1 Headwater Zone

Primary controls in the headwater zone are geology, topography, climate, vegetation, and humans (Table 2.1). Parent material minerology and sedimentology under specific climatic regimes influence rates of rock weathering, infiltration, and runoff, and ultimately downstream discharge and sediment regimes (Milliman and Meade, 1983). Fundamental distinctions usually occur between igneous crystalline, clastic sedimentary, and carbonate lithologies that influence upper basin soil development and erosional and hydrologic processes. Sediment sources in basin headwaters include mass wasting, particularly in steep terrain, and surface erosion by runoff. The latter includes sheet flow, rills, gullying as well as fluvial incision of first-order streams.

The streamflow of first-order tributaries is strongly coupled to hillslope hydrologic processes, with slope and infiltration being key governors. The high gradients of larger headwater streams results in high shear stress[1] being competent in transporting all but the largest-sized clasts. Tectonics influences uplift rates that renew episodes of fluvial incision in addition to triggering mass wasting events that provide pulses of coarse sediment. Large rivers draining the tectonically active high Himalayan–Tibetan Plateau and the Andean Mountains, for example, have the highest sediment yields globally (Milliman and Syvitski, 1992; Latrubesse and Restrepo, 2014).

In addition to natural biophysical processes, human land use has a tremendous influence on drainage basin processes across the headwaters of many large watersheds, and has an extensive historic legacy (Knox, 1977; Dotterweich, 2008; Houben, 2008; Middelkoop et al., 2010; Coe et al., 2011; James, 2013; Trimble, 2013; Kidder and Zhuang, 2015). Knowledge of human impacts on headwater soil and hydrologic processes can aid in the development of effective management that greatly reduces downstream sediment loads to the lower reaches of major fluvial systems. The immense Chinese Loess Plateau (640,000 km^2), for example, headwaters for the middle Huanghe River, underwent accelerated soil erosion caused by improper land use practices for millennia, which formed a network of erosional gullies and incised canyons (Figure 2.2). This resulted in the Huanghe River formerly having the highest sediment concentrations and sediment loads of any major river in the world (Wang et al., 2019). The ochre-colored suspended sediment derived from eroded silt resulted

---

1 Mean boundary shear stress (τ_0) in Newtons per square meter (N/m^2) of channel bed, and is calculated as $\tau_0 = \rho g Rs/w$, where ρ = fluid density (1,000 kg/m^3); g = gravitational acceleration (9.81 m/s^2); R = hydraulic radius (m), R = A/W$_p$ (where A = w × d, W$_p$ = w + 2d); d = mean channel depth; w = channel width (m, commonly bankfull width); and s = water surface slope (m/m).

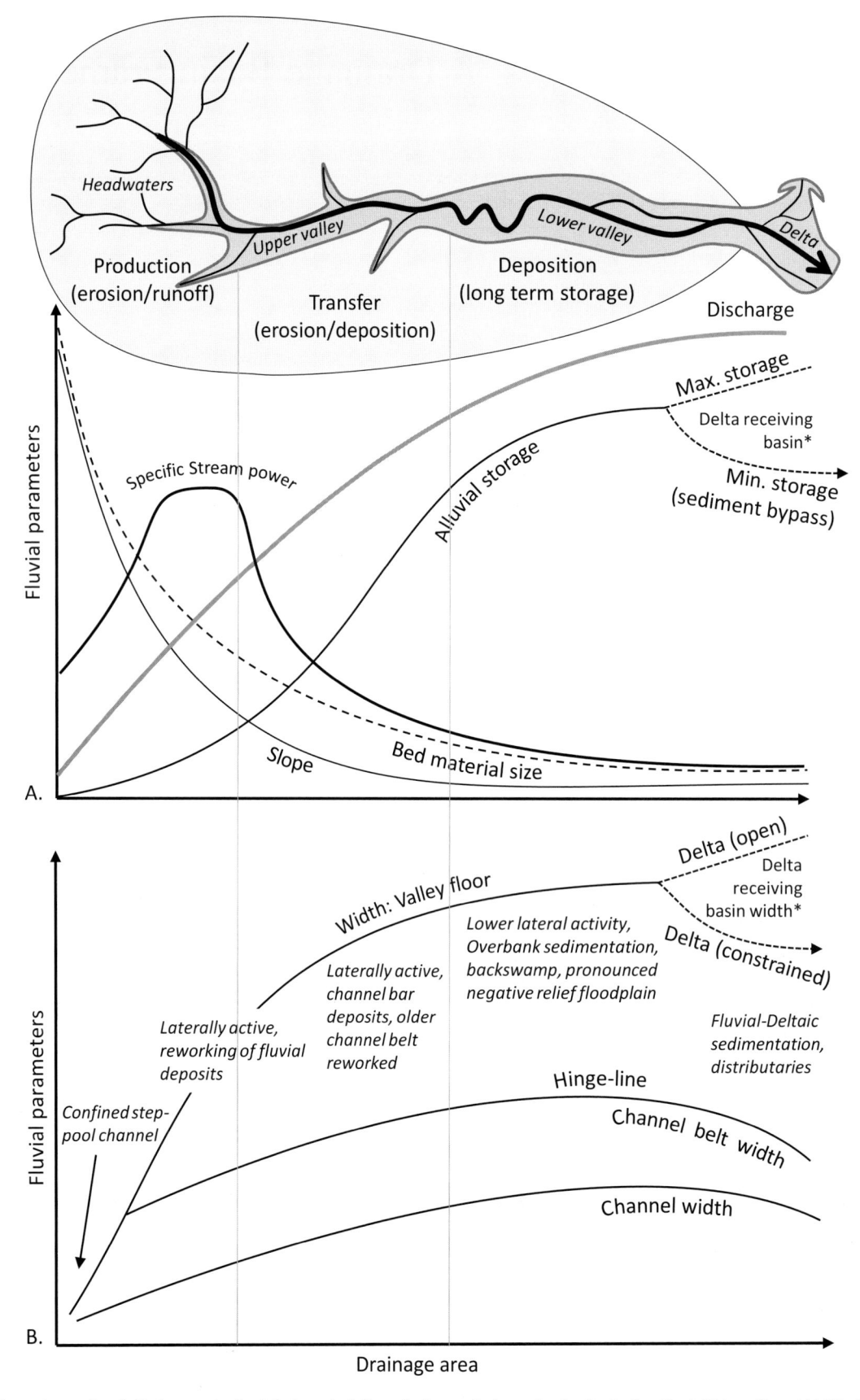

Figure 2.1. Downstream (spatial) changes in fluvial characteristics of a larger drainage basin displaying fluvial hinge-line. (A) Discharge, slope, specific stream power, bed material size (sand), and alluvial storage. (B) Width of valley and deltaic receiving basin, channel belt, and channel. Note (A) delta receiving basin width and accommodation space vary according to whether older subaqueous deltaic deposits constrain active delta progradation, with possible sediment bypass to continental shelf, and (B) hinge-line and reduction in channel belt width in lower reaches of alluvial valley and delta. (After Church, 2002; Schumm, 1977; Macklin et al., 2012; and modified by author for large rivers and deltas.) See Table 2.1.

Table 2.1 *Drivers and processes associated with drainage basin zones for large rivers and deltas. See Figure 2.1.*

Zone	Location	Dominant processes	Drivers
Production zone	Headwaters	Erosion and runoff, tectonic activity	Climate, vegetation, geology (tectonics, structure, topography, lithology), soil, humans (land use, water extraction)
Transfer zone	Upper–middle main valley	Streamflow and sediment transport, lateral migration and/or channel bar development, floodplain stripping (erosion)	Valley morphology (slope, width), discharge regime (stage range, flood frequency, magnitude, duration, seasonality), water surface slope, location of tributary inputs, sediment load (volume, size), neotectonics, hydraulic engineering
Deposition zone	Lower main valley	Overbank sedimentation and inundation, lateral migration and channel bar development, crevasse and avulsion, surface–groundwater exchange, warping and subsidence	Valley morphology (slope, width, topography), discharge regime (stage range, flood frequency, magnitude, duration, seasonality), sediment load (volume, size, ratio of bed to suspended load), older geomorphic and sedimentary deposits (bank material), local/regional precipitation regime, neotectonics, hydraulic infrastructure
	Delta	Flood regime, overbank sedimentation, crevasse and avulsion, storm surge, subsidence, vegetation change	Geometry of receiving basin (depth, area, shape), fluid density differences between river flow and receiving basin (hyperpycnal or hypopycnal), local/regional precipitation regime, groundwater fluctuation, vegetation, humans (land use and hydraulic infrastructure), marine dynamics (tidal, wave, wind, and storm surge regime), neotectonics, sea level fluctuation, salinity

Figure 2.2. Land degradation by soil erosion in the Chinese Loess Plateau, with lower Daxia River valley (left), which flows into the Middle Huanghe River in Gansu Province. (Photo date July 2009, Licensed CC BY-SA 3.0.)

in the Huanghe being referred to as the Yellow River. The excessive sediment loads resulted in high rates of downstream sedimentation that drove dynamic fluvial adjustments in the lower reaches of the basin (see Chapters 6 and 7).

2.1.2 Transfer Zone

The middle reaches of drainage basins are characterized by streamflow and sediment transport (Table 2.1). Here, a wider valley provides limited space for the temporary storage of fluvial deposits. But steep valley slopes in combination with increased discharge results in much higher specific stream power (Figure 2.1). As alluvium is mainly comprised of non-cohesive channel bar deposits, higher rates of bank erosion and lateral migration result in the efficient reworking and subsequent downstream transfer of alluvial deposits. Important influences on the transport zone include upstream sediment and discharge regime, including the location of major tributary inputs. Larger tributaries disrupt smooth downstream trends by debauching sediment at different sizes and rates because of variable upstream geology and land use. The timing of sedimentary inputs is often asynchronous with the main-stem hydrograph, as hydroclimatic mechanisms often vary between larger tributary catchments. Additionally, floodplain reworking often continues for decades after headwater land degradation and upstream sediment loads have abated (Knox, 2006). Thus, the middle reaches of a basin work as a kind of "sputtering" conveyer belt, coupled to upstream controls, albeit with lapses, lags, and accelerations that result in variable residence times for alluvial deposits.

2.1.3 Deposition Zone

The boundary between the transfer zone and deposition zone occurs downstream of the confluence with larger tributaries and a pronounced reduction in valley gradient (Figure 2.1). As the river enters the deposition zone it has already received the vast majority of its sediment and discharge from upstream sources. The deposition zone is especially characterized by the long-term storage of cohesive deposits in flood basins and backswamp. Floodplain topography exerts a strong control on the connectivity and size of flood basin ecosystems, which include lotic and lentic water bodies, the latter somewhat disconnected from active main-stem hydrologic processes (Miranda, 2005; Paira and Drago, 2007; Korponai et al., 2016). The deposition zone is also characterized by increased valley width, reduction in bed material size, as well as a pronounced reduction in specific stream power[2] (Lecce, 1997a; Phillips and Slattery, 2007). The increased accommodation space provides an opportunity for floodplain sediment storage over millennial timescales, which includes moribund channel belts – paleochannels – abandoned following an avulsion. Despite the scale of large basins, however, rivers within the deposition zone can efficiently respond – within decades – to major fluctuations in sediment and discharge caused by upstream land use change, dams, and main-stem hydraulic engineering (Hudson and Kesel, 2006; Latrubesse et al., 2009; Wang et al., 2019).

---

[2] Specific stream power (Ω) in watts per square meter (W/m^2) of channel bed. Stream power is calculated as $\Omega = \rho g Q s/w$, where ρ = fluid density (1,000 kg/m^3); g = gravitational acceleration (9.81 m/s^2); Q = discharge (m^3/s); s = slope (m/m); w = channel width (m, commonly bankfull width). See Lecce (1997) for discussion of fluvial stream power calculation and downstream spatial variability.

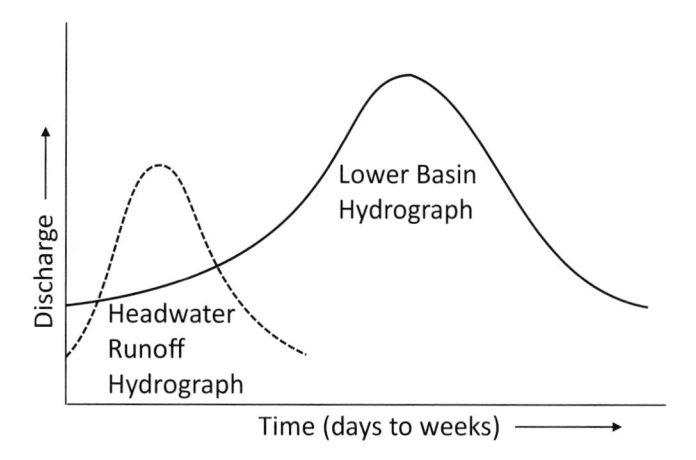

Figure 2.3. Downstream translation and attenuation of flood wave over hours to weeks, illustrating differences in hydrograph shape in relation to (i) increased travel distance, (ii) gradient reduction, (iii) streamflow input from tributaries, and (iv) water transmission and storage into alluvium and floodplain water bodies. (After Jones, 1989.)

The hydroclimatology of the lower reaches of large basins is often distinct from the upper basin, such that flooding of broad alluvial valleys and deltas may occur under sunny skies. While discharge increases, the hydrograph undergoes a distinctive change in shape – becoming smoother – as it translates downstream (Figure 2.3). Additionally, intense pluvial events can directly generate inundation and flooding of local water bodies and flood basins in lowland floodplains, while the main-stem river is well below flood stage. Flood basin inundation is exacerbated because of the inverted topography and low permeability of cohesive sedimentary deposits (Mertes, 1997; Lewin and Ashworth, 2014a). Ground subsidence is a fundamental control on flood basin inundation (see Section 7.3), within alluvial valleys and especially deltas (Dokka, 2006; Syvitski et al., 2009; Giosan et al., 2014). Land sinking occurs naturally, primarily driven by compaction of soft sedimentary and organic layers, but is greatly exacerbated by human activities, such as land drainage and groundwater pumping (Morton et al., 2006; Erkens et al., 2015), resulting in substantial flood risk across many of Earth's major deltas (Figure 2.4).

Deltas extend seaward from alluvial valleys in the lowermost reaches of the deposition zone (Figure 2.4), which includes subaqueous extension into a receiving basin (Nittrouer et al., 1986; Coleman et al., 1998). The apex of deltas, the upstream border with the alluvial valley, is usually the avulsion node of a major distributary branch and its location often controlled by neotectonics as well as the upstream limit of sea level influence (Berendsen and Stouthamer, 2001; Phillips, 2011; Ganti et al., 2016). The dynamics and morphology of deltas are a product of upstream fluvial inputs (discharge and sediment) relative to the marine energy provided by waves and tides, in addition to the accommodation space provided by the receiving basin. Indeed, portions of a delta plain may be actively accreting and prograding

Figure 2.4. Subsiding Sacramento–San Joaquin River Delta in California, with Sacramento at top and San Joaquin at bottom, separated by Sherman Island. Intensive agriculture extends to channels with flood control dikes constructed along river banks. Sherman Island has subsided ~8 m since the late 1800s and average elevation is ~3 m below sea level, increasing flood risk. (Photo date: 23.9.2004 by World Island. Licensed under CC BY 2.0.)

due to fluvial inputs while other portions are being eroded and reworked by waves into sandy beach ridges (Panin, 1999). This truism alone implies we need to carefully apply – if not rethink – management approaches that utilize the concept of equilibrium as a conservation target or restoration goal, certainly across such large spatial scales as delta plains (e.g., Renwick, 1992).

2.2 DOWNSTREAM IMPLICATIONS OF LAND COVER CHANGE

2.2.1 General Principles

Landscape transformation for agriculture is a major form of global environmental change (James and Marcus, 2006; Turner et al., 2007). Human domination of Earth's landscapes is apparent when considering that some 40% of natural lands have been converted for agricultural purposes. Over 5 billion hectares of land is utilized for grazing and cultivation, with a vast acceleration since the Industrial Revolution (Figure 2.5). Earth's largest biome are agricultural landscapes (Foley et al., 2005). Agricultural transformation of Earth's surface drives changes to soils and water resources, and agriculture accounts for 70% of global freshwater withdrawal. By changing ground cover, soil infiltration, and runoff, agricultural land use change drives land degradation and soil erosion across basin headwaters, which often has downstream hydrologic, ecologic, and geomorphic consequences for riparian lands and deltas (Nienhuis et al., 2020).

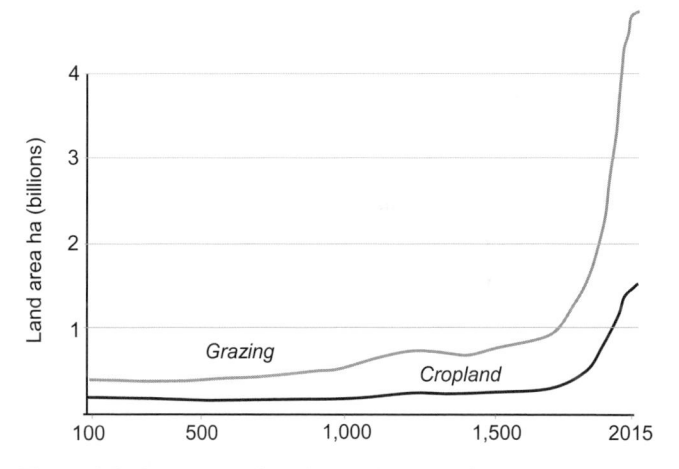

Figure 2.5. Global area of cropland and grazing since 100 AD. "Cropland" refers to FAO category: "arable land and permanent crops." "Grazing" refers to FAO category: "permanent meadows and pastures." (Source: History Database of the Global Environment [HYDE]. Klein Goldewijk et al., 2017.)

Decades of process-based research in government-managed field experiment stations using approaches such as standard runoff plots and paired basins, among others, established fundamental empirical relationships between runoff and soil erosion for different agricultural practices. Such studies commonly utilize specific soil and vegetation assemblages as a reference baseline, and compare with variation in agricultural practices such as crop type, root depth, crop rotation, plant density, plowing, and

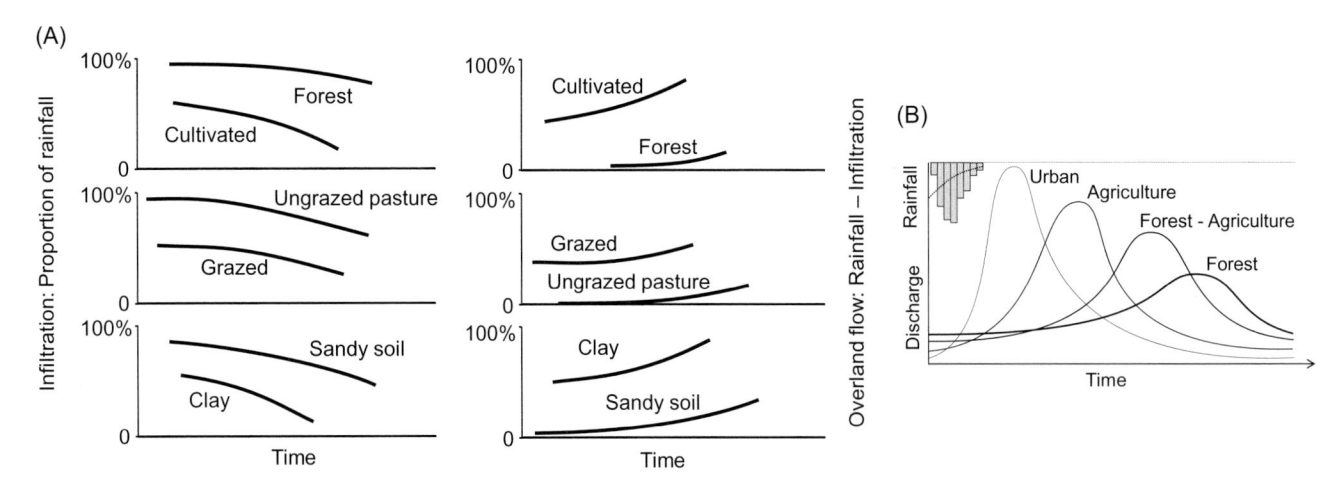

Figure 2.6. Influence of varying land surface conditions on surface hydrologic processes, including (A) land cover and soil types on infiltration and runoff and (B) continuum of streamflow response. (Sources: (A) U.S. National Resources Conservation Service (B) unknown source.)

tillage, among other factors. An important consideration in landscape response to land change is soil infiltration, which under natural conditions exhibits variability according to amounts of soil humus, texture, structure, vegetation, and slope. After pore saturation, most soils attain steady-state infiltration capacity within about one to three hours. But when natural landscapes are transformed to agricultural landscapes, infiltration rates often decrease and runoff proportionately increases (Figure 2.6A). The lowest rates of soil erosion under natural land are generally forest cover and grassland, particularly in humid temperate climates. Natural landscapes buffer the influence of runoff events because of surface roughness associated with microtopography, roots, and litter. Soil aggregates with high clay–humus complex resist shear stress during runoff events. Additionally, soil aggregates with higher amounts of organic matter store water after a rainfall event, resulting in slower rates of percolation and evaporation. This is beneficial to cultivation but also to maintaining channel baseflow (groundwater) between rainfall events. In the downstream reaches of larger basins this headwater influence manifests as reduced flow variability, reducing flood peaks and increasing low flows during dry periods.

An enormous collective body of work involving process-based experimental study and regionally focused historic geomorphic research has unequivocally established that modern mechanized agriculture produces soil erosion rates that range from several times to several orders of magnitude greater than natural soil erosion rates and new soil production (Happ et al., 1940; Knox, 1972; Trimble, 1976; Montgomery, 2007; Dotterweich, 2008).

The natural inherent buffer (shear strength) to increased runoff provided by healthy soil is diminished with modern mechanized agriculture, which over time degrades soil structure and soil chemistry (loss of cations), thereby increasing landscape sensitivity to increased shear stress during runoff events (Trimble, 2013, p. 55). Reduced soil organic matter results in individual soil particles being more easily dislodged by kinetic energy from rain splash, especially atop bare fields between plantings. With ensuing rainfall events, the dislodged particles are easily flushed downslope, and the runoff is eventually able to generate sufficient shear stress to selectively remove topsoil rich in silt and clay, organic matter, and nutrients. Topsoil loss, therefore, subsequently results in increased flow variability to downslope channels, characterized by higher peak flows, reduced lag times, and reduced baseflow (Figure 2.6B).

2.2.2 Land Degradation: From Case Study to Concept

Regionally, extensive analyses of land degradation and drainage basin response were conducted for many characteristic landscapes representing a range of environmental conditions and agricultural practices. These include central Germany (Houben, 2008; Hoffmann et al., 2009, 2013; Driebrodt et al., 2009), the Belgium loess belt (Poesen and Govers, 1990; Verstraeten et al., 2017), the upper US Midwest (Happ et al., 1940; Knox, 1972, 1977, 2006; Trimble, 1976, 2013; Beach, 1994; Renwick and Andereck, 2006), the US deep south (Happ et al., 1940; Trimble, 1974), the Mayan lowlands of central America (Beach et al., 2008), southern Africa (Meadows and Hoffman, 2010), the Mediterranean (Vita-Finzi, 1969; Butzer, 1971; Koulouri and Giourga, 2007; García-Ruiz, 2010), New South Wales (Terry, 1945; Graham, 1992; Butzer and Helgren, 2005), and the Chinese Loess Plateau (Shi and Shao, 2000; Sun et al., 2014), among many notable others.

As an example, a major era of land degradation and soil erosion in the Midwestern United States was initiated after the mid-1800s, after European settlement converted natural forests and grasslands to grazing and croplands, especially cereals (Happ et al., 1940; Knox, 1977, 2006; Baker et al., 1993; Beach, 1994; Faulkner, 1998; James and Lecce, 2013;

Figure 2.7. Post-settlement alluvium in Driftless Area of Wisconsin, overlying organic rich topsoil in tributary of Coon Creek, Vernon County, WI. (Source: Happ et al., 1940.)

Trimble, 2013, figure 7.20). Resulting land mismanagement caused abrupt and deep soil horizon truncation manifested by sheet wash, rills, and extensive gullying. Soil truncation significantly degraded one of the world's great soil resources, reducing topsoil depths across some regions by about half. While some of the eroded soil was transported downstream through the fluvial system, much of the eroded topsoil was redistributed to lower reaches of small headwater basins and continues to be stored as post-settlement alluvium or legacy sediment atop buried topsoil horizons (Knox, 1972; James, 2013; Trimble, 2013). The phenomenon was meticulously documented by Happ et al. (1940) in an extensive survey by the U.S. Soil Conservation Service upon recognition of the severe problem of soil erosion in the early 1900s (Figure 2.7).

While case studies have to be properly contextualized before being applied elsewhere, many provide valuable insights into how landscapes in other regions may respond. The concepts embedded in Knox's observation (2006, quoted earlier) regarding the substantial environmental disturbance caused by conversion of natural forests and grasslands to intensive agricultural land use in the mid-1800s are of hemispheric relevance – if not global – as much as it pertains to southwestern Wisconsin.

Great progress in reducing land degradation and soil erosion has occurred in some regions, which when combined with sediment trapping behind dams (Chapter 4) has resulted in diminished downstream suspended sediment loads for a number of major rivers. Soil erosion rates in the Midwestern United States greatly declined after about the 1940s because of improved land management. Common management measures adapted included contour plowing, crop rotation, shelter belts, and other structural modifications to reduce soil erosion, particularly by overland flow (Renwick and Andereck, 2006; Trimble, 2013).

In Europe, considerable land has been taken out of cultivation and afforestation is the general pattern, resulting in reduced soil erosion for some regions (Panagos et al., 2016), as well as new management challenges concerning vegetative disturbance to agricultural landscapes highly valued as cultural heritage.

Specific motivations for rural communities to forego agricultural practice vary regionally and include demographic change, regional and national economic policies, depopulation associated with rural to urban migration, and sociological change (Bucala-Hrabia, 2017). In former European Eastern Bloc nations, post-communist changes in land use planning and shifts to market-based agricultural systems have been a disincentive for local farmers to continue practicing agriculture, resulting in fallow fields and increased forest cover. Soil erosion rates in the Western Carpathians of Poland have declined considerably since the collapse of communism (Bucala, 2014), which in some cases serves as a longer historical continuum over the past century and a half. In the Homerka catchment (19.3 km^2) within the Vistula River basin headwaters, for example, forest cover increased by 68% and cultivated lands decreased by 92%. Based on the widely utilized RUSLE model, soil erosion rates show a marked decline. Since peak agricultural intensity in the mid-1800s, soil erosion rates have declined by 77%, from 18.13 t/ha in 1860 to 4.11 t/ha by 2010 (Kijowska-Strugała et al., 2018).

2.2.2.1 EROSION REDUCTION OF THE CHINESE LOESS PLATEAU

A developing success story with major downstream implications for a densely populated region concerns episodic soil erosion in the Chinese Loess Plateau. For decades, Chinese agencies and international development organizations attempted to reduce soil erosion and restore the degraded landscape. And soil erosion has greatly reduced in the Chinese Loess Plateau (Chen et al., 2007; Shi et al., 2017). Effective measures include terracing, vegetation planting, changes to crop types, and construction of numerous check-dams to store eroded soil within small valleys and reduce inputs into the middle Huanghe River. Additionally, government organizations have effectively incentivized local stakeholders to manage soil erosion (Fu et al., 2017). Indeed, sediment inputs into the middle Huanghe River have declined annually by some 500 million tons (Figure 2.8) and downstream sediment loads have reduced by as much as 90% over past decades (Wang et al., 2019). The change is so dramatic that it is already revealed in reservoir sedimentation studies in main-stem dams along the Huanghe River (Chapter 4). Ironically, effective management in the upper Huanghe basin is now contributing to downstream channel degradation and associated fluvial riparian problems, requiring new management approaches in the alluvial valley and delta to address the issues of declining sediment loads (Syvitski et al., 2009; Wang et al., 2019).

Conceptually, numerous regional historic land degradation case studies have been synthesized to arrive at general models that link with downstream riparian response across a variety of landscape scales (e.g., Wolman, 1967; Knox, 1977; Schumm, 1977; Faulkner, 1998; James, 2011; Hoffmann et al., 2013; Trimble, 2013, figure 6.8; Verstraeten et al., 2017). Such models have great value in conceptualizing the landscape response to two major global environmental change phenomena: urbanization and expansion of agricultural lands. The global demographic shift toward urbanization, while some rural areas are in demise, has direct relevance to the conceptual model put forward by Wolman (1967) over five decades ago (Figure 2.9). While the downstream streamflow response to urbanization is well established in engineering hydrology, channel and ecosystem adjustment caused by erosion and aggradation are less often considered. And they are more complex because of being dependent upon the sensitivity of hydraulic and sedimentary indices that require detailed field data to adequately characterize (water surface profiles, bed material size, depth of active layer, bank material, critical aquatic habitat, etc.).

The magnitude and character of downstream fluvial and environmental response is often out of accord with the severity of upstream land degradation (Butzer et al., 2008; Verstraeten et al., 2017). This occurs because of inherent buffers, feedbacks, and thresholds within the fluvial system, which vary considerably from basin to basin due to inherent geomorphic controls (Figure 2.10 and Table 2.2). Thus, extending results from Western Europe or the Midwestern United States to other settings should be done with care, and with specific information about the physical and human setting (Houben, 2008). The first-order control on downstream response is the hillslope to valley coupling (or connectivity), which ultimately influences whether extensive land degradation occurring upstream is detectable downstream. Other physical basin characteristics are also important and include basin size, tributary locations, valley width, relief, slope, and vegetation regime (Faulkner, 1998). Human factors are perhaps the most complicated to generalize, in part because of extensive landscape histories that precondition soil to erosion or resulted in substantial soil loss in prior periods of occupation (Verstraeten et al., 2017). It is now well established that European colonists did not cultivate pristine landscapes in the Americas (Denevan, 2000; Doolittle, 2000; Mann, 2005), as most soilscapes already exhibited some degree of disturbance that conditioned their sensitivity of response (Butzer et al., 2008; James, 2011). Thus, it remains challenging to predict downstream changes in sedimentation and environmental disturbance following upslope land degradation, particularly with increasing drainage area (Knox and Daniels, 2002).

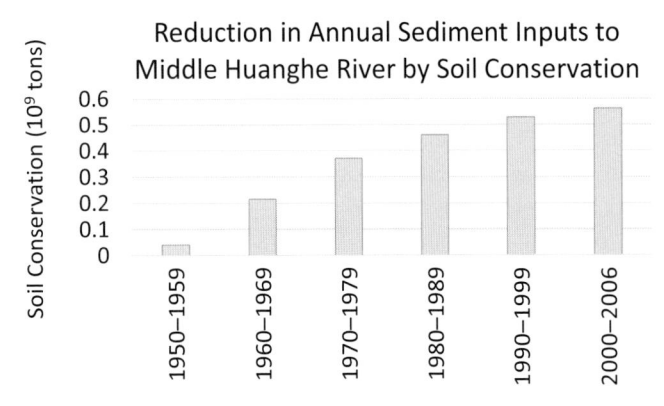

Figure 2.8. Reduction in annual sediment inputs (billions of tons) to middle Huanghe River due to soil conservation. (Source: Shi et al., 2017; reproduced with permission.)

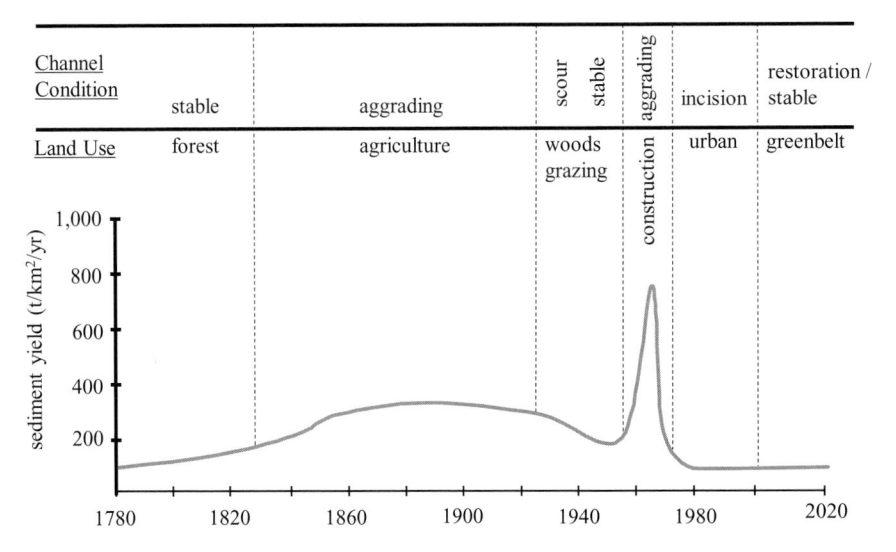

Figure 2.9. Channel adjustment to changes in sediment yield driven by land use change, from forest to urbanization. Based on small streams in the vicinity of Baltimore, Maryland, USA. (Source: Wolman, 1967, with author modification.)

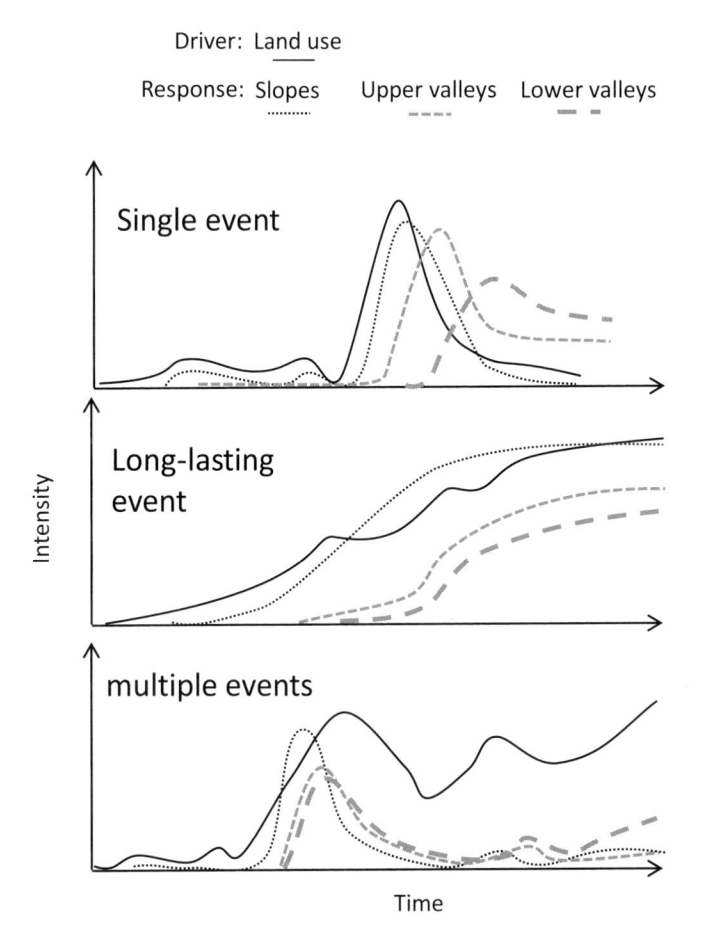

Figure 2.10. Fluvial response for three different settings impacted by intensive agriculture, including upper Midwestern United States (upper), Belgium loess belt (middle), and eastern Mediterranean (lower). Variation in nonlinearity of erosion response due to period and duration of disturbance, condition at time of disturbance, and post-disturbance management. See Table 2.2. (Source: Verstraeten et al., 2017.)

2.2.2.2 HISTORIC HYDRAULIC GOLD MINING IN CALIFORNIA

While misaligned agricultural practices can result in adverse consequences to landscapes and downstream riparian environments, a more extreme form of land degradation with long-term consequences are inappropriate mining practices (James, 1989, 2006). Many historic and contemporary case studies have documented the adverse impacts of mining on fluvial systems and the downstream transport and storage of polluted sediments to the deposition zone (Lecce and Pavlowsky, 2001), including large rivers such as the Rhine (Middelkoop, 2000). Historic gold mining in California from the mid-1800s to the early 1900s represents among the most extreme examples (Figure 2.11).

Hydraulic mining impacts from the Sierra Nevada foothills provides useful management lessons as regards the long-term consequences of such activities to downstream geomorphology, hydrology, ecology, and the design of hydraulic infrastructure that are still being comprehended. The hydraulic mining activities in California were particularly problematic because of the intensity and scale of operations within the Sierra Nevada foothills, and the lack of effective management for the copious volume of debris (Gilbert, 1917). New methods of hydraulic mining developed onsite were tailor made for the hydrogeomorphic setting, and efficiently reduced vast placer deposits to debris that was funneled into Yuba, Bear, American, and smaller tributaries of the Sacramento River (Mount, 1995, pg. 204). The activities resulted in 1.1 billion m^3 of mining debris being transported to a range of depositional settings between the Sierra Nevada foothills and San Francisco Bay (James, 1989; James et al., 2017). Much of the sediment was stored where narrow valleys widened and today exist as anthropogenic fan deposits and terraces. Considerable sediment was transported downstream as a channel bed wave (James, 2006), ultimately resulting in 876 million m^3 of debris being deposited in the San Francisco Bay system between 1849 and 1914 (Gilbert, 1917, table 5). While Gilbert (1917) characterized the downstream movement of hydraulic mining debris as a continuous sediment wave (Figure 2.12), it was actually more of a channel bed wave that was episodic, with considerable lag times (James, 2006). The arrival of the bed wave resulted in increased channel bed elevations, driving higher river stage levels that increased flooding (Mount, 1995, pg. 208).

In addition to debris produced from hydraulic gold mining, the processing activities used considerable amounts of mercury (Hg), which was ultimately released directly into the environment (Mount, 1995, pg. 215; Bouse et al., 2010; Singer et al., 2013). Deposition zones along the Yuba River valley, a tributary to the Sacramento River that drains the core mining areas, store some 6,700 kg of mercury (Nakamura et al., 2018). Alarmingly, however, this estimate is far less than what was produced during the mining period (by orders of magnitude). Where did it go? The discrepancy can likely be explained by enormous amounts of mercury already transported to the lower Sacramento and San Francisco Bay (Nakamura et al., 2018). The downstream significance of the sediment, and particularly the mercury contamination, has continuing ramifications to management of riparian ecosystems, including for Chinook salmon (*Oncorhynchus tshawytscha*), a federally endangered species that utilizes the hydraulic infrastructure of the Sacramento flood control system during a critical phase of its lifecycle (Section 7.2.3).

The case studies and syntheses noted above are essential to contextualize riparian environmental impacts driven by different land use practices and climate change in other regions, if prudently applied. The value of historical-based models (i.e., Figure 2.10) of river basin response to environmental disturbance are especially important to consider given varying projections of land cover change (Figure 2.13), and where it will occur. Most of

Table 2.2 *Characteristics of different types of fluvial response to different types of human disturbances (see Figure 2.10)*

Human disturbance	Fluvial reaction	Region and case study
Single event: Major peak in disturbance (decades to centuries) Low-intensity events before and after peak	**Sediment wave**: Slopes and upstream valleys respond early Downstream valleys may respond after peak in human pressure due to sediment reworking of legacy sediment (result is attention of flood wave)	Midwestern United States (Driftless region, Wisconsin; Knox, 1972; Trimble, 1974)
Long-lasting event: human impact gradually increasing through time short-term fluctuations are possible but not important	**Continuous fluvial aggregation**: slope-channel coupling when threshold in human pressure is crossed delayed response of the floodplain compared to the slopes	European loess belt (e.g., Dijle, Belgium; Notebaert et al., 2011; Broothaerts et al., 2013)
Multiple events: Human pressure fluctuations strongly through time Alternation of events with intense human disturbance and periods of low activity	**Sediment wave**: First major event results in a sediment wave Soil exhaustion generates negative feedback: subsequent events have lower impact on fluvial system (sediment wave peaks before land use) Reworking of legacy sediment especially in downstream valleys	Mediterranean environments (e.g., Bugdoz, SW Turkey; Dusar et al., 2012; D'Haen et al., 2013; van Loo et al., 2017)

Source: Verstraeten et al. (2017).

Figure 2.11. Hydraulic gold mining in the 1870s at Malakoff Diggings, Sierra Nevada foothills, California. (Photographer: C. E. Watkins, figure source: U.S. Geological Survey.)

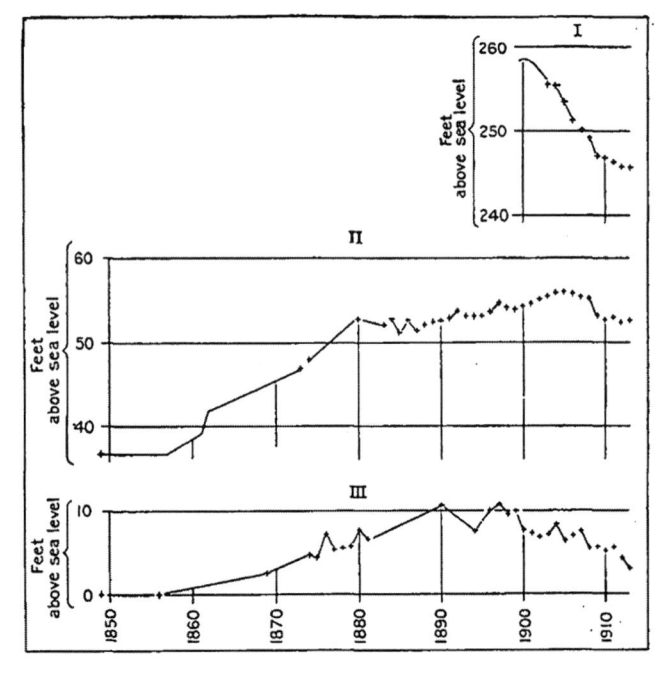

Figure 2.12. Changes in low flow levels in the Sacramento basin in relation to channel bed aggradation. (I) Yuba River near Smartsville; (II) Yuba River at Marysville; (III) Sacramento River at Sacramento. Variation in response to downstream moving channel bed wave. This was a key figure in Gilbert's 1917 study because of related low-water-level fluctuations caused by deposition of mining debris on the channel bed, and subsequent incision. Note: y-axis in feet. (Source: Gilbert, 1917, figure 4, p. 30.)

the tremendous increase in land cover change in support of agriculture and urbanization is occurring within semiarid landscapes of Africa, South America, and Asia, with Asian coastal lowlands and mega-deltas increasingly being utilized for intensive protein production (Chan et al., 2013; Higgins et al., 2013; Ritchie and Roser, 2013).

Increasing reliance upon irrigation-based agriculture will have highly variable benefits that depend in part on the configuration and orientation of the watershed in relation to changes in

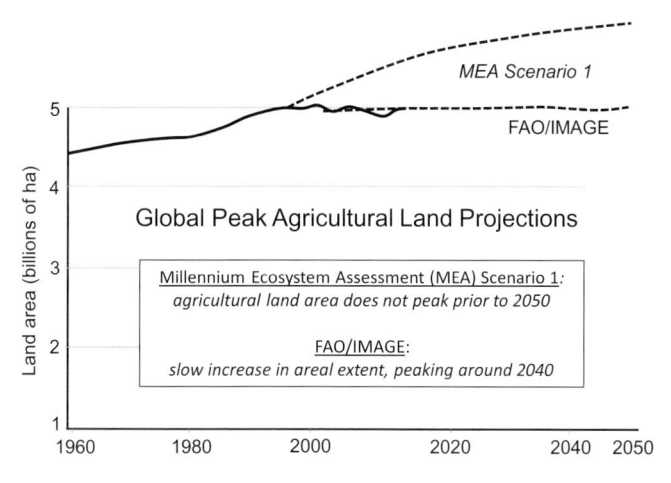

Figure 2.13. Projections in global agricultural land until 2050, since 1960. (Source: UN FAO Database, Ritchie and Roser, 2013.)

precipitation. The Indus River basin, for example, is projected to receive lower amounts of precipitation in the downstream reaches of the basin where most agriculture occurs within the Indus Basin Project, an extensive irrigation system that heavily withdraws freshwater from the Indus River. But higher precipitation in mountainous headwaters are expected to increase streamflow totals overall, which potentially could benefit downstream irrigation networks (Okada et al., 2018). But this is highly dependent upon construction of new mega dams in the Himalayan headwaters (Section 4.2.2.1), which are expected to reduce downstream flow levels. Thus, it is essential that upstream reservoir outflow release schedules are timed to critical plant growth stages of downstream crop cultivation.

2.3 RIVER CHANNEL PATTERNS

2.3.1 Channel Classification

Because headwater land cover and climate change can drive adjustment to downstream riparian environments its essential to have foundational knowledge of river form and process. River channel patterns are classified according to a variety of criteria and typologies. Although the topic has received considerable attention by fluvial geomorphologists, the search for a universal classification scheme remains incomplete (Carling et al., 2014).[3] Leopold and Wolman (1957) classified river channel patterns as straight, braided, meandering, or anastomosing. Their classification schema, however, did not adequately consider very large rivers (Figure 2.14).

Additionally, the typology was too dependent upon sinuosity as a distinguishing criterion, which is directly related to channel slope. Reliance upon slope as a discriminating criterion is particularly problematic at the meandering–braided threshold. Straight channels are seldom considered, and few larger rivers have straight alluvial channels for very long distances, with exception of deltaic distributaries with low specific stream power and cohesive channel banks.

A broad typology of classification considers (1) whether rivers are single channel or multichannel (anabranching) and (2) whether rivers are laterally stable or laterally active (Figure 2.15). A practical advantage of this typology is that it squarely relates to management issues, such as navigation and channel bank stabilization. Thus, herein the categories of Nanson and Knighton (1996) are utilized (Table 2.3), with the addition of mega rivers (Latrubesse, 2008), because of their size and unique characteristics. While each class of channel patterns has dominant processes and features that make it unique, they also have common processes and morphologic features (pools, bar accretion, cut-offs, etc.), and in some instances are transitional.

2.3.2 Single-Channel Pattern

A first step to developing sustainable rivers is to understand the characteristics and controls on river channel processes and adjustment. Alluvial rivers have three modes of adjustment, specifically planform, cross-sectional, and longitudinal. Each mode of adjustment provides a different profile to contextualize and quantify river channel change with specific morphologic indices. Meandering rivers have the most distinctive and symmetrical morphology (Figure 2.16), and are herein emphasized to review several basic geomorphic concepts and processes, some of which applies to multiple channel pattern typologies.

2.3.2.1 MEANDERING RIVERS: MORPHOLOGY AND POINT BAR SEDIMENTOLOGY

The sinusoidal pattern of meandering rivers is among Earth's most unique landscapes. Meandering river flow patterns result in distinctive erosional and depositional processes associated with indices of channel geometry scaled to specific discharge levels.

The streamflow index considered most important to controlling channel morphology is the bankfull discharge,[4] which occurs at a frequency of about every one to two years. The importance of bankfull discharge is that it is also (often) considered the dominant discharge,[5] the streamflow responsible for

---

3 The topic of channel pattern classification has received considerable attention from fluvial geomorphologists. Thorough and insightful reviews that address issues related to theory, terminology, scale, and parameterization are provided by Ferguson (1987), Nanson and Knighton (1996), Latrubesse (2008), and Kleinhans and van den Berg (2011).

4 Bankfull discharge is the amount of streamflow a channel can accommodate before flowing overbank as floodwater (Wolman and Leopold, 1957; Woodyer, 1968; Williams, 1978a; Knighton, 1998).

5 Dominant discharge is the discharge magnitude responsible for the most geomorphic work (i.e., Wolman and Miller, 1960). For further discussion on the concept and limitations related to utilization of bankfull discharge, see Benson and Thomas (1966) and Church (2015).

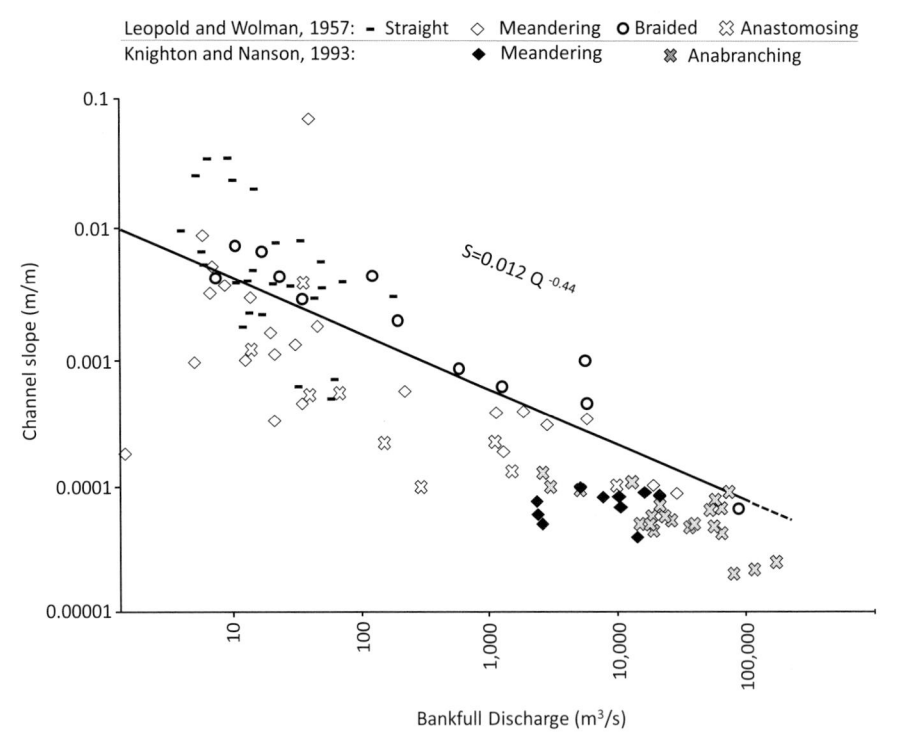

Figure 2.14. Channel pattern distinguished by slope-discharge relation after Leopold and Wolman (1957). Extension for very large rivers by Knighton and Nanson (1993), as plotted by Latrubesse (2008, figure 10), includes Amazon basin rivers and Yangtze River.

the most physical work (e.g., erosion and sediment transport) (Wolman and Miller, 1960). In addition to being geomorphically significant, because of inherent linkages between geodiversity and biodiversity the bankfull discharge is also an important control on riparian ecosystems. For some rivers, however, rather than bankfull discharge indices of streamflow at different frequencies of occurrence, such as mean annual flood or a specific discharge duration (10%, 5% flow duration) are considered the dominant discharge, particularly for rivers in dry climatic settings (i.e., Wolman and Gerson, 1978).

Fundamental to understanding river adjustment are inherent interrelations between different indices of channel morphology (Figure 2.16). Meandering river morphologic indices are dependent upon discharge, which is considered the "master variable" as concerns river channel adjustment. Thus, channel width and meander wavelength can then be used as a surrogate for discharge (~energy).

A number of specific channel morphologic features are interrelated. The planform geometry of meandering rivers is intricately related to the longitudinal profile, or the thalweg geometry. In particular, the spacing of pools and riffles is generally five to seven times channel width (Figure 2.17), or half of the meander wavelength. This is inherent to meandering reaches spanning from the smallest of tidal creeks to large rivers, such as the lower Mississippi (Keller and Melhorn, 1978; Hudson, 2002; Gibson et al., 2019; Wu and Mossa, 2019). Indeed, strong interrelations between the planform and channel bed morphology is an indication that meandering rivers, by systematically altering their meso-scale morphologic features, also meander in the vertical dimension (Keller and Melhorn, 1978; Thorne, 1991; Hudson, 2002).

Meander wavelength and radius of curvature range from about ten to fourteen and two to three times the channel width (Figure 2.18), respectively (Williams, 1986). Such relations are interesting from a theoretical perspective and have practical relevance to river management authorities. This is because it provides a quantitative framework to assess channel adjustment to human impacts, and to guide river restoration projects. And the morphometric perspective provides a formal means to contextualize the scope of past environmental change. In this case, the relationship between channel geometry and streamflow can be utilized to develop formulae to estimate the paleo-discharge of older river courses from morphologic dimensions of abandoned channels (Schumm, 1968; Williams, 1984). Such analysis provides insight into the importance and implications of fluvial adjustment to climate change, especially when combined with sedimentary indices (Knox, 2000).

In addition to scale-dependent morphologic indices, indices that are independent of scale are useful to assess channel adjustment, such as w/d and sinuosity. The ratio between channel width and depth (w/d) provides a quantitative index that expresses the shape of the channel cross section (i.e., shallow and wide vs. narrow and deep). The w/d index can be used to

Table 2.3 *Characteristics of channel patterns, including single-channel patterns (meandering and braided) and multichannel (anabranching) patterns, which include anastomosing, wandering, and mega rivers*

Characteristics	Anabranching			Single channel	
	Anastomosing (laterally stable)	Wandering (laterally active)	Mega rivers**	Meandering	Braided
Gradient	Low	Moderate	Very low	Low-Moderate	High
Sinuosity	High	Low	Low	High	Low
Width/Depth	Low	Moderate	Moderate	Moderate	High
% silt/clay in bank material	High	Low	Moderate	Moderate	Low
Lateral migration	Very low	High	Moderate	Low to high	High, variable
Vegetated islands	Yes	Yes (less mature)	Yes	Yes (less mature)	Possible (temporary and less complex)
Island overtopped* at bankfull stage?	No	No	No	No	Yes (bars dominant)

* Overtopped is used rather than inundated, because irregular channel bank topography allows seepage into low depressions below bankfull stage. Additionally, this assumes a "mature" older island rather than recently formed.
** A unique category of anabranching channels for Earth's largest rivers (see Latrubesse, 2008).
After Nanson and Knighton (1996).

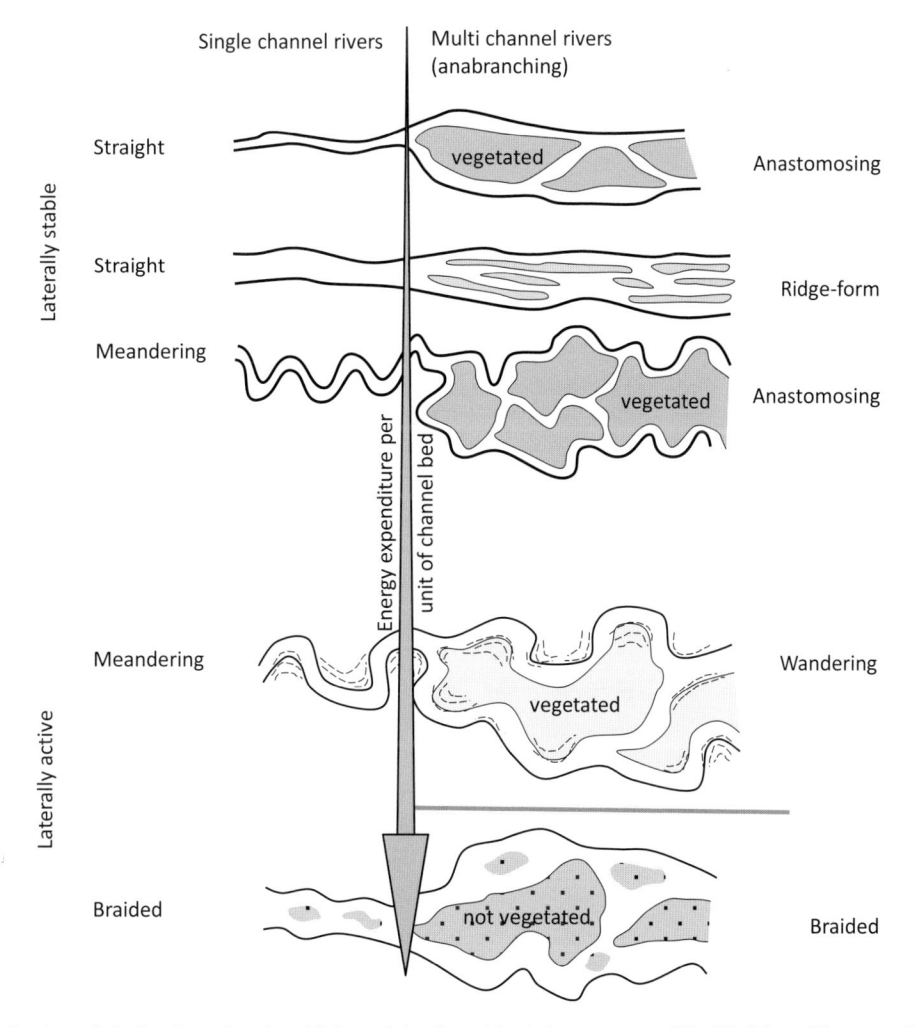

Figure 2.15. Characterization of single-channel and multichannel (anabranching) river patterns. (Modified from Nanson and Knighton, 1996 and Dunne and Aalto, 2013.)

Planform profile

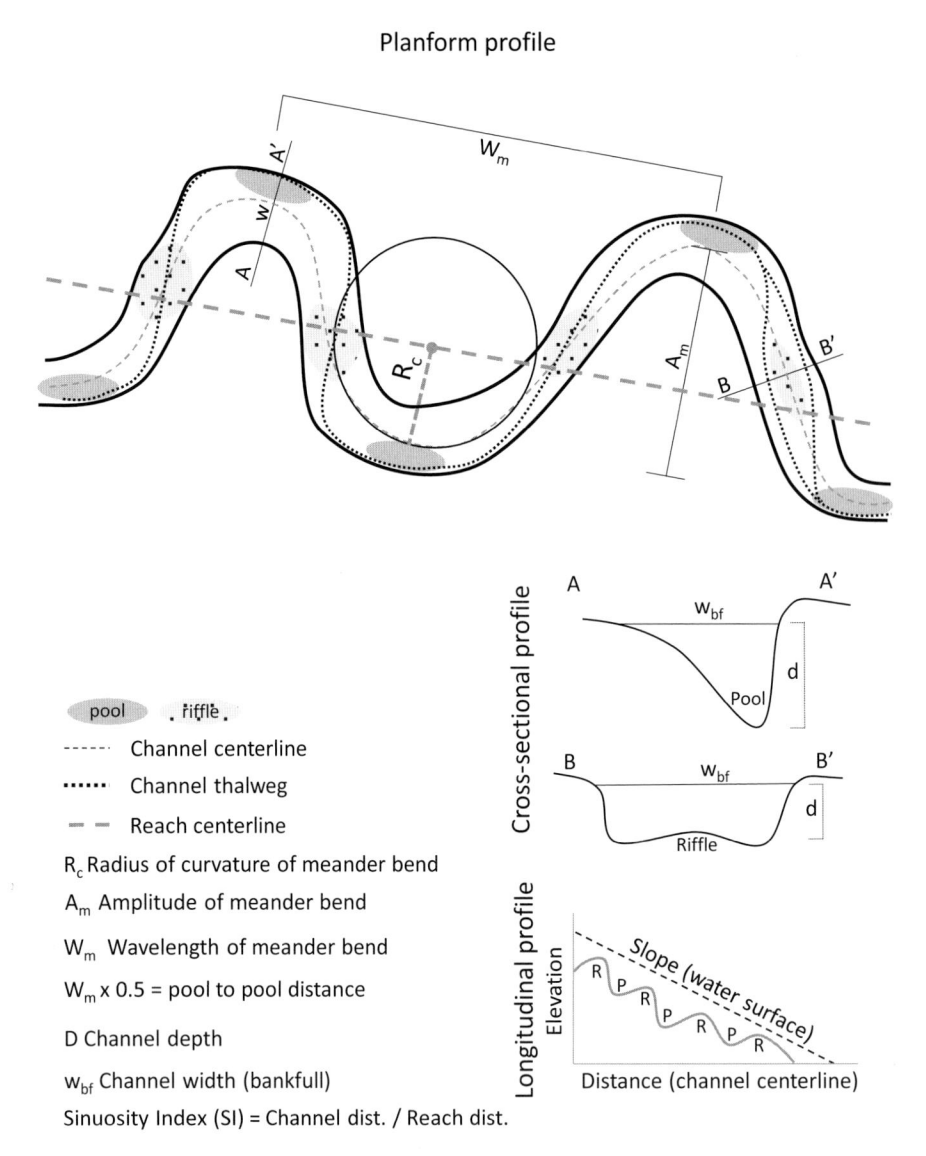

Figure 2.16. Morphologic indices of meandering river channels: planform, cross-sectional, and longitudinal profiles. Sinuosity index: ≥1.5; meander wavelength: 10–14 × channel width; radius of curvature: 2–3 × channel width. Pool to pool spacing: = half meander wavelength or 5–7 × channel width. (Author figure.)

Legend:
- pool, riffle
- Channel centerline
- Channel thalweg
- Reach centerline
- R_c Radius of curvature of meander bend
- A_m Amplitude of meander bend
- W_m Wavelength of meander bend
- $W_m \times 0.5$ = pool to pool distance
- D Channel depth
- w_{bf} Channel width (bankfull)
- Sinuosity Index (SI) = Channel dist. / Reach dist.

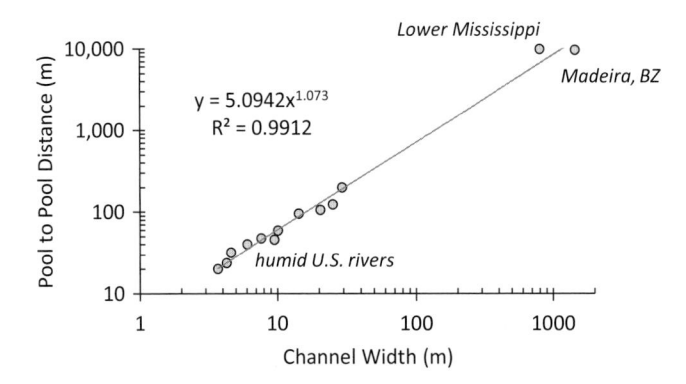

Figure 2.17. Relation between pool-to-pool spacing and channel width illustrating dearth of research on larger rivers. (Source: lower Mississippi [Hudson, 2002], Madeira [Gibson et al., 2019], humid US rivers [Keller and Melhorn, 1978].)

Plot: $y = 5.0942 x^{1.073}$, $R^2 = 0.9912$

directly compare the shape of channel cross sections for rivers that vary in size. Additionally, because channel cross sections are sensitive indicators of disturbances caused by upstream sediment loads or engineering (e.g., groynes, dredging, revetment) that drive channel erosion and aggradation, w/d is often utilized to assess temporal changes in cross-sectional channel morphology.

In comparison to anastomosing and braided rivers, indices of channel morphology and sediment load of meandering rivers are intermediate (Table 2.3). The pattern of deposition and erosion, however, is very distinctive along meandering rivers and is spatially dependent upon the location of point bars and cutbanks. Larger meandering rivers are dominated by suspended sediment transport, which may account for some 90% of the total sediment load (Knighton, 1998). Meandering rivers have an intermediate slope and w/d, which systematically varies in accordance with

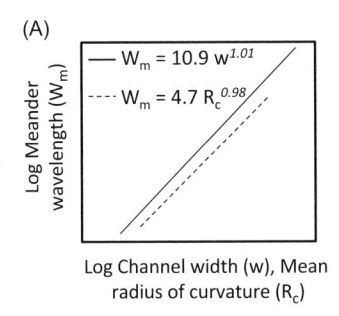

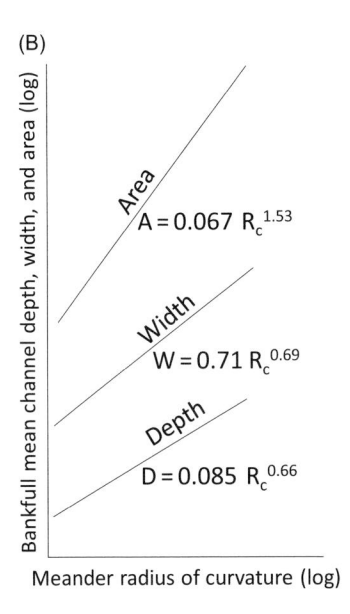

Figure 2.18. Relations between planform and cross-sectional channel indices. (A) Meander wavelength and channel width (~discharge surrogate). (Source: Summerfield, 1991.) (B). Radius of curvature and channel depth, width, and area. Equations developed from rivers varying in climate, geology, and size (flumes to Mississippi). (See Williams, 1986, figure 4, for specific river data.)

Plot A equations: $W_m = 10.9\, w^{1.01}$ and $W_m = 4.7\, R_c^{0.98}$.

Plot B equations: $A = 0.067\, R_c^{1.53}$, $W = 0.71\, R_c^{0.69}$, $D = 0.085\, R_c^{0.66}$.

pools and riffles. Reach-scale variations in channel slope are related to variation in channel sinuosity. Channel slope increases by meander neck or chute cut-offs whereas slope reduction occurs by lateral migration, especially extension of meander bends and increase in meander amplitude (Figure 2.16). In addition to systematic variation in cross-sectional channel shape (w/d) between successive meander bends, variation in w/d occurs because of differences in bank material cohesion (% silt/clay), flow variability, and lateral activity.

In addition to the maintenance of river channel stability, it is essential that management organizations consider the interrelations between morphologic dimensions of the planform and longitudinal channel profiles. This is because of linkages between fluvial geodiversity and aquatic biodiversity. Pools and riffles provide distinct forms of aquatic habitat related to differences in dissolved oxygen, temperature, velocity, sediment, organic matter, and nutrients (Langeani et al., 2005; McCabe, 2011).

The sedimentary signature of a meandering river is a distinctive fining-up grain-size trend (Figure 2.19). A vertical profile through a meandering river channel belt reveals a complete range of channel sedimentary, geomorphic, and hydraulic and hydrologic processes associated with point bar deposition. The sedimentology of a vertically accreting point bar is related to flow regime, which can be expressed by the Froude number (F),[6] a dimensionless index computed between the ratio of lateral flow velocity to vertical forces (Bridge, 2003). The base of a vertical floodplain profile is commonly coarse channel bar sediment deposited atop an erosional surface during high flow regime (Folk and Ward, 1957; Saucier, 1994). Such deposits are commonly gravel, pebbles, and coarse sands, and may also include rip-up clasts. In rivers with distinctive pool–riffle sequences basal lag deposits are indicative of deep thalweg scour. During high flow events pools undergo the highest velocity and shear stress along a channel reach (Hannan, 1969; de Almeida and Rodríguez, 2011). The pattern of scour and deposition is related to flow divergence at the pool in association with helical flow around the bar. Moving up the profile, the vertical fining trend reveals sandy cross-bedding and current laminations associated with dune migration (bed load transport) along the flanks of the channel bars during high discharge events. The fining trend is indicative of hydraulic sorting as flow velocity (and boundary shear stress) decreases toward the upper flank of the point bar. These sedimentary deposits form what are commonly described as scroll bars (Folk and Ward, 1957), a fundamental topographic component of meandering river point bars (e.g., Hickin and Nanson, 1975), although some authors liken their sedimentology and topography to incipient (and ephemeral) natural levee deposits (Mason and Mohrig, 2019).

A distinctive environmental attribute of point bars is their time-transgressive ecological succession (Hickin, 1984). This is associated with older and more distant point bar ridges having larger and more complex vegetation assemblages, whereas ridges closer to rivers edge commonly have one or two dominant vegetation types. Distant swales are less frequently inundated by river water, and develop aquatic ecosystems with hydrophytic vegetation attuned to the local climate. Additionally, thick clayey swale bottoms enable near perennial water bodies, whereas swales close to rivers edge rapidly drain and dry out upon recession of high water, resulting in deep mud cracks.

---

[6] Froude number (F), a dimensionless index of flow regime defined as the ratio of inertial to gravitational force, which is a comparison of water velocity relative to the velocity of surface waves. It is calculated as $F = U/\sqrt{gd}$, where U = stream velocity, g = gravitational acceleration, and d = flow depth. Three states of flow regime are $F < 1$ = subcritical flow (lower flow regime), $F \cong 1$ = critical flow, and $F > 1$ = supercritical flow (upper flow regime). See Bridge (2003) for further discussion as related to fluvial sedimentology.

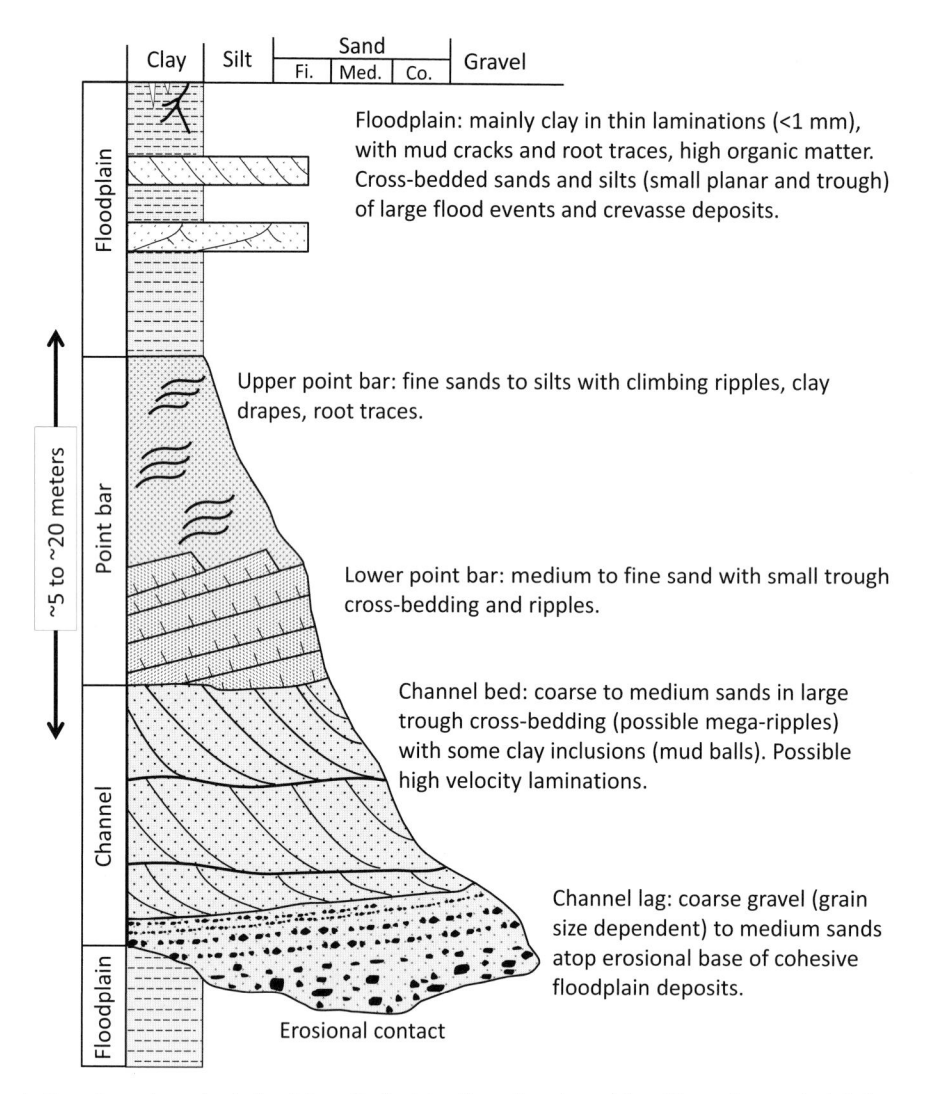

Figure 2.19. Point bar stratigraphy and overbank flood deposits for laterally active channel bar, illustrating vertical fining-up trend. (Redrawn with permission from source: L. Fitcher.)

The theoretical explanation for why rivers meander has practical value to river management. A river increases its sinuosity to reduce its unit rate of energy expenditure (Leopold and Wolman, 1960; Langbein and Leopold, 1966; Huang and Nanson, 2007). The development of a meandering channel pattern, therefore, results in a spatially symmetrical distribution of boundary resistance along the river course, and thus a more uniform rate of energy expenditure. This is essential to channel pattern maintenance. A local increase in valley gradient results in a local increase in energy (stream power), which can result in adjustment of channel sinuosity to compensate for higher energy. Along some river reaches, the higher valley gradient exceeds a threshold by which the river can maintain a meandering pattern (Schumm and Khan, 1972). This can result in channel pattern adjustment – from meandering to braided – associated with a lower sinuosity and higher channel slope, but which is compensated for with an increase in channel bed width to increase flow resistance (Leopold and Wolman, 1957).

2.3.2.2 BRAIDED RIVERS

Braided rivers have multiple channels separated by active sedimentary bars during low stage. Channel bars are generally submerged at bankfull stage, resulting in a dominant single channel and a reduction in sinuosity. The wide channel (high w/d) and steep slope of braided rivers is necessary for sediment transport, which is bed load dominated. Channel bars are seasonally active and reworked (erosion and deposition), although some are partially vegetated with pioneer species (willow) and can remain stable for decades.

Four conditions are associated with development of a braided river pattern (Knighton, 1998, p. 231), including (1) sediment load dominated by coarse (i.e., noncohesive)

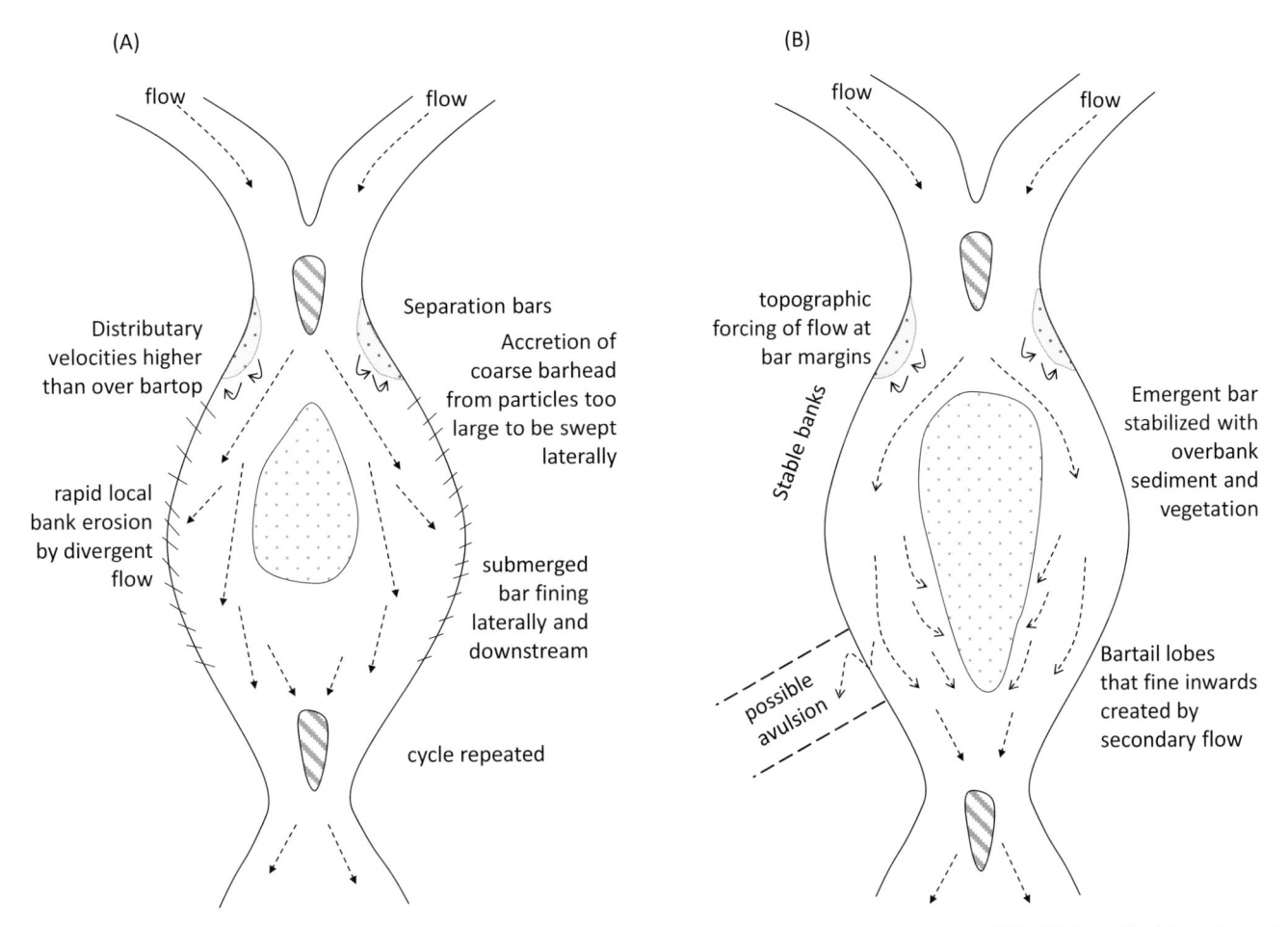

Figure 2.20. Geomorphic features of mid-channel bars for unstable (A) and stable (B) banks. (Source: Ashmore, 1991, 1996, modified from Leopold and Wolman, 1957.)

bed material, (2) high seasonal flow variability, (3) high bank erodibility (low cohesion ~ % silt/clay), and (4) high valley slope. The latter is associated with high specific stream power necessary for particle entrainment that drives bed load transport and bank erosion processes in braided rivers.

The depositional mechanism of braided river pattern development is oriented toward understanding the formation of mid-channel bars (Figure 2.20). While mid-channel bars are also associated with meandering (Hooke, 1986) and anabranching rivers (Nanson and Knighton, 1996; Best et al., 2007; Latrubesse, 2008), they are especially associated with braided rivers (Leopold and Wolman, 1957).

Mid-channel bars are formed where sediment transport competence decreases below particle specific velocity or shear stress thresholds required for transport (Figure 2.20), resulting in deposition of coarse channel lag deposits. The location of channel lag deposition is often just downstream a zone of flow convergence and associated channel bed scour. Further, development of an incipient central bar from channel lag deposits drives flow separation during low flow and deflection of higher velocity flow toward the channel bank, away from the channel bar. Reduced velocity along incipient channel bar sides and downstream velocity shelter induces further deposition along bar margins, resulting in the channel bar increasing in size laterally and longitudinally (Leopold and Wolman, 1957). A critical stage in the evolution of mid-channel bars is whether the deflected flow is sufficiently concentrated to cause bank erosion. This is key because it influences the overall shape of the channel bar, and specifically whether the channel bar becomes laterally or longitudinally oriented (Figure 2.20).

In comparison with point bars, the sedimentology of braided river floodplains often lacks a distinctive fining-up grain-size trend. Rather, the vertical particle size profile is dominated by abrupt changes from sand to gravel because of variable flow conditions manifest by coarse-grained trough cross-strata with occasional planar sandy deposits.

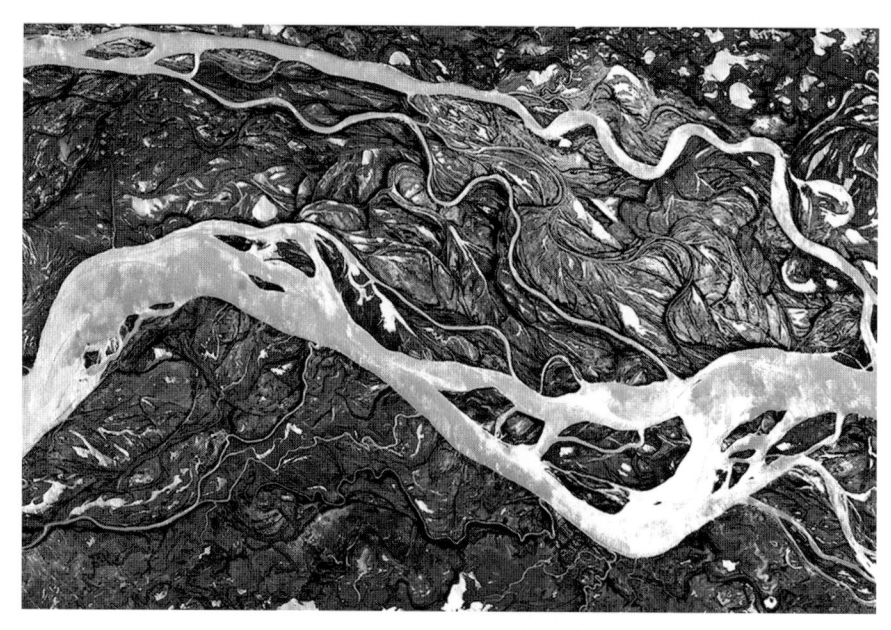

Figure 2.21. Dynamic anabranching middle Rio Paraná, northern Argentina. (Source: US Astronaut photograph, NASA. Date: April 9, 2011.)

2.3.3 Multichannel: Anabranching Rivers

Anabranching rivers are dominated by two or more channels for much of their distances (Figure 2.21), and includes wandering, anastomosing, and mega-rivers. Two primary processes important to anabranching river patterns are avulsion and channel bar accretion for anastomosing and wandering channels, respectively. A key criterion for anabranching rivers, in particular to distinguish from braided rivers, is that they have multiple channels separated by vegetated islands that are about the same height as the adjacent floodplain. It is important to note that anabranching rivers – at a reach scale – include "general" geomorphic and sedimentary processes common to meandering and braided rivers. This may include meander bend migration and mid-channel bar formation (Figure 2.20). But the dominant geomorphic expression of anabranching rivers is multiple channels separated by vegetated channel islands (Figure 2.21) not overtopped at bankfull stage.

2.3.3.1 ANASTOMOSING

Anastomosing rivers have multiple sinuous channels separated by vegetated islands emergent at bankfull stage (Table 2.3) that reconnect downstream (Leopold and Wolman, 1957; Makaske, 2001). Sediment transport is wash load dominated. Channel banks are stable because of cohesive sedimentary material and low specific stream power (≤ 2.5 W/m^2). In comparison with braided and meandering rivers, anastomosing river channels have lower w/d ratios and therefore larger hydraulic radii. The importance of narrow and deep channels is that it enables anastomosing rivers to reduce flow resistance, thereby maximizing the specific stream power (and boundary shear stress) necessary for bed load transport over low slope gradients (Nanson, 2013; Huang and Nanson, 2007). From the perspective of at-a-station hydraulic geometry, the cohesive channel banks and low specific stream power result in the width (b) exponent being much lower than that reported by Leopold and Maddock (1953).

Avulsion is the dominant process of channel splitting and occurs during long-duration flood events by hydrologic reconnection with former channels (paleochannels) and formation of new flood-scoured channels within cohesive flood basins by avulsion (Makaske, 2001; Stouthamer and Berendsen, 2001; Slingerland and Smith, 2004). Smaller channels often represent an unsuccessful avulsion because of cohesive flood basin deposits and thick peat being sufficiently resistant to fluvial incision (Aslan et al., 2005; Makaske et al., 2007; Stouthamer and van Asselen, 2015). Anastomosing rivers are common to lowland rivers and deltas (Nanson, 2013). While anastomosing rivers are characterized by interconnecting channels within alluvial valleys, anastomosing rivers in deltas function as a coastal draining distributary network.

2.3.3.2 WANDERING

Wandering anabranching rivers are distinct from anastomosing. While wandering channel pattern has been proposed as transitory between anastomosing and braided, the importance of channel bar dynamics to the overall pattern suggests that wandering channels are indeed a pure channel pattern (Church, 1983; Rice et al., 2009). The bed material of wandering channels is predominantly coarse, and may be dominated by sand, gravel, or cobbles. Wandering channels are identified across a range of geographic locations where there is sufficient bed load, perennial streamflow, and moderate channel bed slope, such as downslope of mountainous British Columbia (Church, 1983; Desloges and Church, 1989). Channel adjustment in wandering channels primarily occurs as bar accretion forces the channel thalweg to shift laterally, resulting in an irregular pattern of erosion of noncohesive channel banks (Church, 1983).

Islands within wandering channels, in comparison with anastomosing, are subject to higher rates of reworking and are less likely to have mature vegetation assemblages. Vegetated islands along the lower Fraser River, Canada, for example, attain a scale-dependent size within about a century before being reworked (Rice et al., 2009). Along some wandering channel reaches, lateral mobility and ample coarse bed material result in appreciable scroll-bar deposits, although the absence of oxbow lakes within the floodplains informs that meandering does not fully develop.

2.3.3.3 MEGA RIVERS

A unique category of anabranching are mega rivers, which includes Earth's largest rivers having an average annual discharge >17,000 m³/s (Latrubesse, 2008). Mega rivers are worthy of their own classification because they differentially discriminate on slope-discharge plots (Figure 2.14), and because they transport most of the world's runoff and sediment loads. The nine largest rivers on Earth (ranked by discharge) are anabranching mega rivers, with the tenth largest river being the meandering lower Mississippi. Despite their global presence and overall importance, mega rivers do not fall neatly into other river pattern classifications. Aside from having large average discharge values, mega rivers have very low specific stream power, very low channel slope, fine bed material (D_{50} = medium sand), and multiple low sinuosity channels separated by vegetated islands (Latrubesse, 2008). The sinuosity of mega rivers is mainly less than 1.3 for the primary and secondary channels. Specific reaches may have higher sinuosity values, but are restricted to relatively short segments. Additionally, over short segments some individual reaches include prominent meander bends and lateral migration, but do not typically develop meander neck cut-offs. Mega river channel slopes are very low (< 0.00007) and can approach 0.0 m/m, and may extend for distances > ~1,000 km. For such low slopes and low specific stream power, the only way to transport sediment over great distances is to concentrate stream power within narrower reaches, which is provided by the high islands that separate flow into multiple channels. Indeed, width (b) exponents for at-a-station hydraulic geometry for mega rivers is less than 0.1 (Latrubesse, 2008, 2015).

2.4 FLUVIAL ADJUSTMENT TO ENVIRONMENTAL DISTURBANCE

An important outcome of the search for a "general" method to precisely discriminate alluvial river channel patterns is increased understanding of fluvial controls on river adjustment, which has enormous practical application to river management.

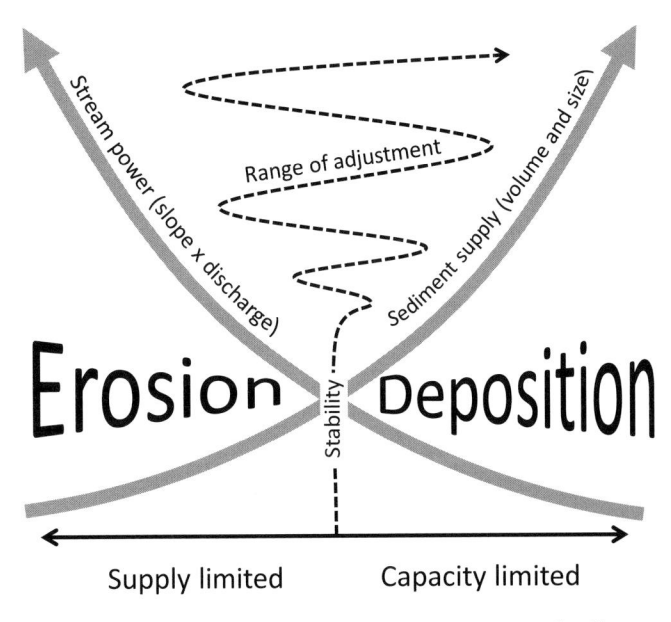

Figure 2.22. Conceptual relation between stream power and sediment supply with implications to erosion and deposition for supply and capacity limited scenarios. (Modified from Julien, 1995.)

A framework of hydraulic versus sedimentary controls provides river management organizations with general signposts as to the direction of change that can be expected should the dominant fluvial controls be modified (Lane, 1957; Ferguson, 1987; Julien, 1995) (Figure 2.22), as well as insights into the magnitude and character of related riparian environmental changes. This is commonly considered from changes to discharge and sediment load caused by headwater landscape change, such as agriculture and mining, as well as direct channel engineering associated with reduced sinuosity, increased slope, and knickpoint formation (Figures 2.23 and 2.24), the latter being a common management approach for navigation and flood control (Chapter 5).

Because changes in streamflow and sediment drive channel pattern adjustment, environmental change occurring in basin headwaters should be considered as to whether it impacts downstream discharge variability and sediment transport, particularly bed load (Schumm, 1968; Leigh, 2008). If inherent geomorphic thresholds are exceeded such changes can be expected to result in one of four scenarios (Figure 2.25), with the direction (+, −) of morphologic response contextualized through a series of conceptual equations (Schumm, 1968). Specific fluvial responses to changes in discharge or bed load, whether triggered by anthropogenic or natural drivers, are manifest as changes to fluvial indices such as channel width, depth, w/d ratio, meander wavelength, sinuosity, or a change in channel slope (Yodis and Kesel, 1992; Leigh, 2008; Latrubesse et al., 2009). Utilizing such an explicit framework provides a means to formally consider riverine response to drainage basin disturbances caused by different

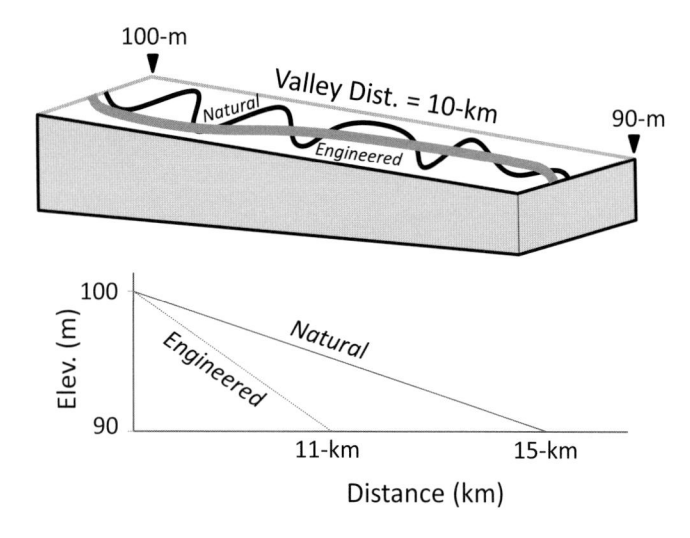

Figure 2.23. Adjustment of channel sinuosity and slope with channel straightening by a sequence of engineered meander bend cut-offs. Slope increase represented as a channel knickpoint. (Author figure.)

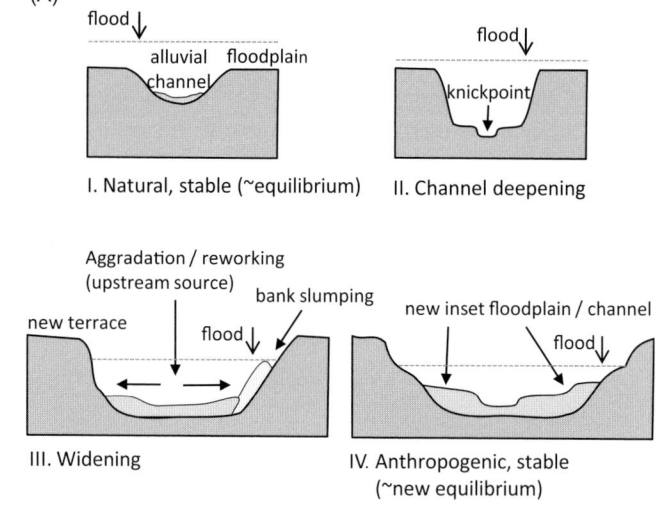

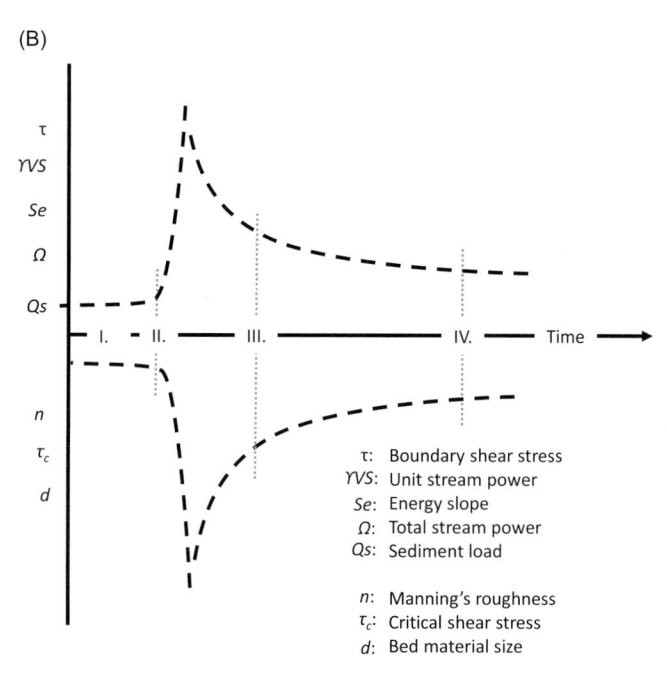

facets of global environmental change, particularly land cover change and climate change, both of which directly impact sediment loads and streamflow variability.

Much attention is justifiably oriented to understanding geomorphic adjustment in response to changes in discharge, and particularly changes to flood regime, for different climate change scenarios. River channels are complex entities comprised of a range of morphologic sedimentary features, each created by different processes (dunes, lateral bars, meander bends, pools, riffles, etc.). The sedimentary character of alluvial channels is adjusted to a range of discharge magnitudes that have a certain frequency of occurrence. Thus, increased flooding can drive geomorphic change that results in a channel pattern change, which could be environmentally detrimental and disrupt human activities for extensive periods of time. For rivers close to a channel pattern threshold (e.g., Figure 2.14), increased stream power may temporarily (perhaps for decades) or permanently initiate a change to channel pattern, such as a transition from meandering to braided, while the size of the channel adapts to a new discharge regime. This could be manifest in several ways,

Figure 2.24. Sequence of fluvial response to knickpoint incision caused by channelization for flood control. (A) Channel and flood hydrology and (B) hydraulic and sedimentary parameters. (After Yodis and Kesel, 1992; Downs and Gregory, 2004; Simon and Rinaldi, 2006.)

including combinations of increased erosion of meander bends to accommodate larger flow levels, channel bed degradation that disturbs benthic aquatic habitat, and disruption to navigation and related commerce. From a management perspective, the altered pattern of erosion and deposition will temporarily require – for intensively utilized rivers – increased channel bed dredging, placement of groynes, or channel bed and bank protection measures to reduce erosion that threatens floodplain activities

$$Q^+, Q_{sb}^+ \rightarrow w^+, d^\pm, (w/d)^+, \lambda^+, S^-, s^\pm$$

$$Q^-, Q_{sb}^- \rightarrow w^-, d^\pm, (w/d)^-, \lambda^-, S^+, s^\pm$$

$$Q^+, Q_{sb}^- \rightarrow w^\pm, d^+, (w/d)^\pm, \lambda^\pm, S^+, s^-$$

$$Q^-, Q_{sb}^+ \rightarrow w^\pm, d^-, (w/d)^\pm, \lambda^\pm, S^-, s^+$$

Q: streamflow (discharge); Q_{sb}: sediment bed load; w: channel width; d: channel depth; (w/d): channel width to depth ratio; λ: meander wavelength; S: channel sinuosity; s: channel slope

Figure 2.25. Channel morphologic adjustment in response to changes to streamflow and sediment load. (After Schumm, 1968.)

(settlement, industry, agriculture, flood control). Ideally, such measures would not utilize conventional hard engineering approaches (Section 5.1), and instead use a more integrated approach less environmentally abrasive to aquatic and riparian ecosystems (Section 8.2).

A sedimentary framework provides guidance to river management authorities about expected geomorphic adjustments and environmental disturbances caused by changes to sediment load. An abrupt increase in sediment load driven by conversion of natural land cover to agriculture, for example, may change bed material size and increase bed load transport. A common geomorphic adjustment is the development of sedimentary bars and increase in channel width, which occurs because of increased bank erosion. Ultimately, such changes can lead to a transition to a braided river channel pattern with a reduced sinuosity and higher channel bed slope. Such a scenario is a common geomorphic response, often at about the scale of a channel segment, and commonly requires bank protection measures to mitigate the erosion problem.

To comprehensively consider broader fluvial environmental implications of a transition in geomorphic regime driven by large sediment pulses, we need only look back to the epic work of Gilbert (1917) that chronicled such adjustments in the Sacramento basin as triggered by hydraulic gold mining in the Sierra Nevada foothills.

Extensive downstream disturbance occurred to the geomorphology, hydrology, and engineering structures of the Sacramento basin as a consequence of historic hydraulic gold-mining activities. The enormous volume of sediment (as noted above) produced by the mining activities drove geomorphic changes within the tributaries and main-stem valley of the Sacramento River that has been well chronicled (see James, 1989, 2006), including channel bed aggradation, increase in channel width, reduced sinuosity, and creation of alluvial terraces that remain as legacy sediment (James et al., 2009; James et al., 2017). For example, channel bed aggradation in the lower Yuba River ranged from ~20 m at the upstream location to 4.8 m in downstream locations (James et al., 2009). This is complementary to that reported by G. K. Gilbert, who noted a maximum debris thickness within the adjacent floodplain of 12.8 m (Gilbert, 1917, at Daguerre Pt., plate XX photo A).

The signature feature of the hydraulic mining debris reported by Gilbert (1917) was the downstream migrating sediment pulse that drove geomorphic and hydrologic changes with long-term consequences to hydraulic infrastructure. Gilbert's sediment wave concept provides river managers with a sense of the timing, magnitude, and style of downstream impacts that can be expected for decades after cessation of mining activities. While Gilbert (1917) underestimated the persistence of the hydraulic mining debris and the sediment wave is more accurately portrayed as a right-skewed channel bed wave (James, 2006), the observations provided by Gilbert (1917) as relates to changes to hydrology and geomorphology in response to an episodic sediment pulse remain prophetic.

Much of the hydraulic infrastructure emplaced within the Sacramento River basin to manage the sediment pulse from hydraulic gold mining was demolished within a few years by the volume of debris, including sediment retention dams breached in floods. In addition to influences on stream hydrology, and flood hydrology in particular, the copious amount of headwater-derived mining debris had direct consequences to existing flood control infrastructure, as well as future flood risk. A contemporary problem associated with the historic mining is that flood control dikes (levees) were largely constructed by dredging coarse debris directly out of the adjacent channel bed, and the dikes (levees) were located too close to the channel banks. In part, this was designed to hydraulically funnel sediment downstream and to protect floodplain lands from being buried by sedimentary debris. For dike stability and flood protection, however, this was less than ideal. Additionally, the dikes were emplaced atop coarse-grained flood deposits of mining debris, rather than thick cohesive top-stratum, as currently recommended (USACE, 1998b), resulting in excessive underseepage and dike failure. This increased flood risk behind the dikes and stimulated development of an enormous flood control system, which includes the Yolo basin flood bypass structure (see Section 7.2.3). Gilbert (1917) vividly documented linkages between excessive channel bed aggradation, hydrology, flooding, and flood control that directly link with current riparian struggles in the Sacramento basin, including environmental management.

In addition to the downstream fluvial geomorphic and engineering impacts caused by the pulse of mining debris, the riparian ecosystems would have been decimated by decades of excessive sedimentation. Quality biologic baseline data prior to the gold-mining era is sparse, but pre-disturbance aquatic and floodplain ecosystems are considered to have been highly diverse and teeming with native Chinook salmon (*Oncorhynchus tshawytscha*), among other species, that are now federally endangered (Yoshiyama et al., 2001). The current severely degraded status of several Sacramento basin rivers is considered to be directly related to the extensive hydraulic mining that occurred between the 1850s and the late 1800s, including mercury (Hg) pollution, and is the focus of major federal restoration efforts (USACE, 2018b).

Changes to the lower Sacramento River basin caused by anthropogenic headwater erosion and downstream sediment pulse concerns more than the topic of channel pattern adjustment, but also changes to flood regime in lowland rivers, hydraulic engineering, riparian ecosystems, and flood control infrastructure – core foci of this treatise.

Chapter 3 transitions from the drainage basin scale to riparian processes, with an explicit focus on lowland rivers and deltas, including hydroclimatology, and erosional and depositional processes.

3 Hydrologic and Geomorphic Processes in Fluvial Lowlands and Deltas

[D]ata from eastern and western United States and the examples from India indicate a remarkable uniformity in the recurrence interval of overbank flooding.

—Wolman and Leopold (1957)

3.1 CONTEXT OF LARGE RIVERS

Each lowland river is unique, having a distinctive combination of geomorphic, human, and environmental features and processes that require tailor-made approaches for sustainable river management. This chapter examines flooding and geomorphic processes across larger lowland rivers and deltas. This is approached primarily from a "natural" perspective, and with attention to those processes and features that relate to human activities and lowland hydraulic engineering examined in subsequent chapters. Specifically, this includes a review of hydroclimatology, river bank erosion, flooding and overbank sedimentary processes, and deltaic geomorphic processes.

By virtue of being coastal draining rivers that support large populations and economic activities, most of the lowland rivers examined in this treatise are larger rivers. While there are multiple definitions for large rivers, some of the criteria utilized include geology, hydroclimatology, morphology, and discharge and sediment flux (i.e., Potter, 1978; Schumm and Winkley, 1994; Gupta, 2007; Latrubesse, 2008).

Geologically, large rivers usually have drainage basins that link two or more physiographic provinces (Potter, 1978; Dunne and Aalto, 2013). This often includes headwaters in orogenic belts or cratons that drain to axial trunk valleys controlled by regional-scale structural lows with a seaward extension. And large river valleys and deltas require large amounts of sediment (Figure 3.1). This is usually supplied by high mountains, and tectonically active mountain systems in particular provide energy for high rates of headwater erosion and sediment production (Milliman and Meade, 1983; Milliman and Syvitski, 1992; Hori and Saito, 2007; Latrubesse and Restrepo, 2014).

Hydroclimatology is also distinctive within large basins, as their extensive spatial extent results in larger tributary watersheds being influenced by different climatic processes. Additionally, the flood regime is usually influenced by precipitation variability driven by larger-scale atmospheric mechanisms with distinctive seasonality strongly related to Earth's general circulation. This results in large rivers having a characteristic annual flood pulse (i.e., Junk et al., 1989; Tockner et al., 2000) that provides a "rhythm" to geomorphic and ecosystem processes, as well as related human activities (Figure 3.2). This is particularly the case within rural areas in tropical developing nations because of human settlements being directly dependent upon riparian natural resources (i.e., WinklerPrins, 2002).

Synoptic scale circulation, including midlatitude cyclones (fronts), anticyclones, tropical cyclones, and monsoonal precipitation mechanisms, is influenced by larger hemispheric-scale circulation, such as the seasonal position of the intertropical convergence zone (ITCZ) and the strength and position of polar and subtropical jet streams (Hirschboeck, 1988). Additionally, global patterns of precipitation and flooding are influenced by teleconnections such as the North Atlantic Oscillation (NAO) or the El Niño Southern Oscillation (ENSO) (Wang et al., 2017; Muñoz et al., 2018).

Additional criteria for identifying large rivers include specific indices such as drainage area, channel length, channel width, or average annual discharge (e.g., Potter, 1978; Schumm and Winkley, 1994; Gupta, 2007; Hori and Saito, 2007; Latrubesse, 2008; Dunne and Aalto, 2013; Lewin et al., 2016). With exception of very large rivers (Amazon, Yangtze, Congo, etc.) such indices inevitably result in different rankings and in omission of societally important lowland rivers with large deltas heavily utilized for navigation and flood control (Sacramento, Rhine, Red, etc.). The Global Commission on Continental Paleohydrology (GLOCOPH) Large Rivers Working Group, for example, considers large rivers to have a mean annual discharge $\geq 1,000$ m^3/s. Such a strict criterion would include the Rhine River (185,000 km^2, 2,150 m^3/s) but ignore the enormous Murray–Darling River basin (1,072,000 km^2, 550 m^3/s), of vital importance to southeast Australia.

The approach herein adapted, therefore, is to include larger rivers that also have societal importance. In general, common physical characteristics of large(r) fluvial lowlands examined in this treatise include (1) a large amount of space (i.e., valley

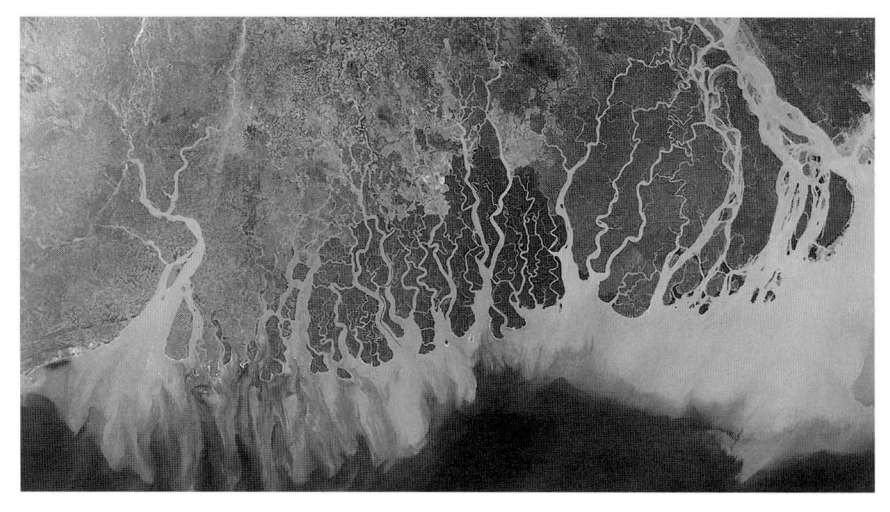

Figure 3.1. Ganges–Brahmaputra delta – the largest subaerial delta dominated by high sediment loads and large tidal flux. (Source: November 9, 2011, MODIS image. NASA Earth Observatory.)

Figure 3.2. Settlement along the lower Amazon, with elevated house to cope with annual flooding. (Author photo, January 2001.)

width) to store relict fluvial deposits not connected to the modern hydrologic regime, (2) floodplain inundation influenced by both local- and continental-scale precipitation mechanisms, (3) and floodplains and deltas dominated by cohesive sedimentary deposits and overbank processes.

3.2 HYDROCLIMATOLOGY OF LARGE RIVERS

3.2.1 Hydrologic Regime

Flooding of large lowland rivers is among Earth's most vital natural processes and is driven by atmospheric and surficial controls. While large-scale precipitation mechanisms in upper tributaries are essential to inundation of lowland rivers, cohesive floodplain deposits and older floodplain topography in the lower reaches of large drainage basins result in local-scale precipitation events also being important to floodplain inundation.

Climatic mechanisms that drive streamflow variability and flooding vary temporally and spatially, and are regionally specific (Hirschboeck, 1988; Lecce, 2000; Wohl, 2007). Important atmospheric drivers to the flood hydrology of lowland river valleys range from intense convectional storms to tropical cyclones to global-scale circulation mechanisms (Figure 3.3).

Annual flood pulses along large rivers are driven by seasonal alterations of Earth's general circulation, manifest by alterations to the timing, strength, and position of discreet atmospheric phenomena (Figure 3.4). These include the Intertropical Convergence Zone (ITCZ), the polar and subtropical jet streams, trade winds, monsoons, tropical cyclones, midlatitude cyclones, polar lows, and high-pressure cells (anticyclones) that influence precipitation and streamflow variability across large basins (Knox, 1983, 2000; Hirschboeck, 1988; Mechoso et al., 2001; Muñoz et al., 2018). The importance of these mechanisms in producing runoff is augmented by the size and orientation of large land masses and mountain systems relative to their proximity to ocean basins (i.e., moisture source). Periodic changes in jet stream orientation results in a steep north–south dip in westerly flow across the midlatitudes, from zonal to meridional (Figure 3.5), with implications to flood hydrology and water resource management (Knox, 1993). Additionally, atmospheric rivers (discussed below) associated with persistent midlatitude cyclones transfer corridors of water vapor thousands of kilometers from tropical oceans to midlatitude lands (Newell et al., 1992; Zhu and Newell, 1998; Dettinger, 2011).

The flood pulse of large tropical and neotropical rivers (Figure 3.6) is primarily influenced by the seasonal shift in the

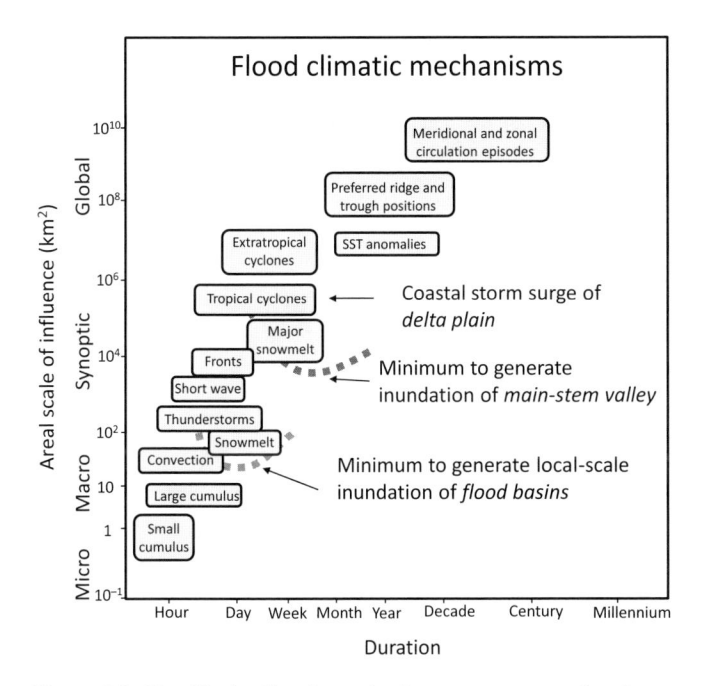

Figure 3.3. Flood hydroclimatic mechanisms across temporal and spatial scales, including minimum required to generate local-scale inundation, overbank flooding, and delta plain inundation by coastal storm surge. (Source: Hirschboeck, 1988, with author modifications.)

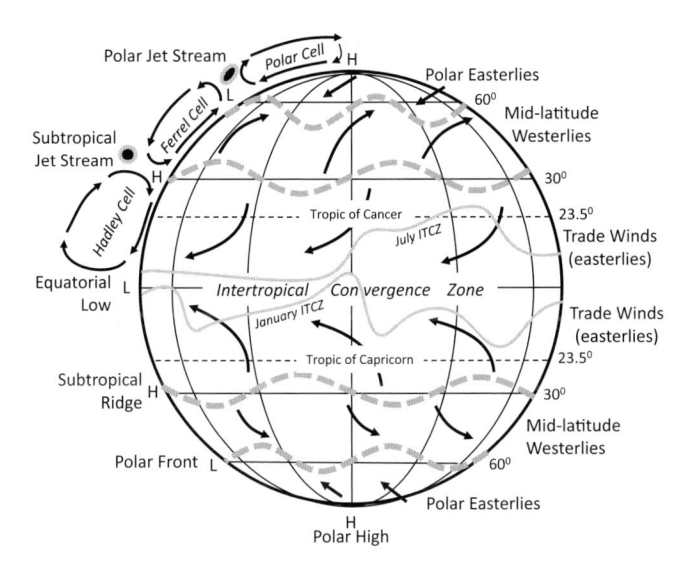

Figure 3.4. Earth's atmospheric circulation showing major winds and pressure systems.

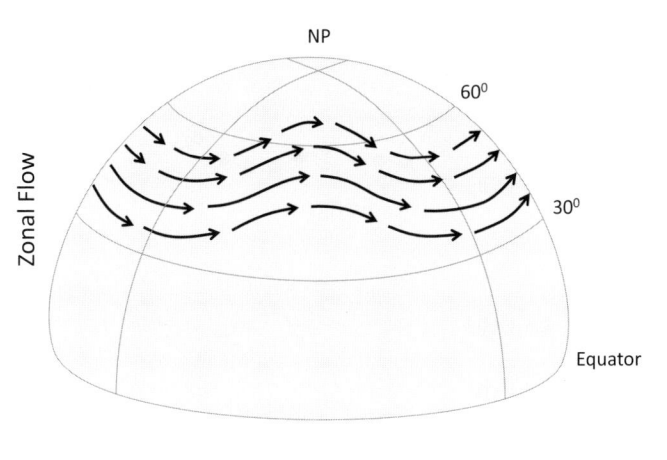

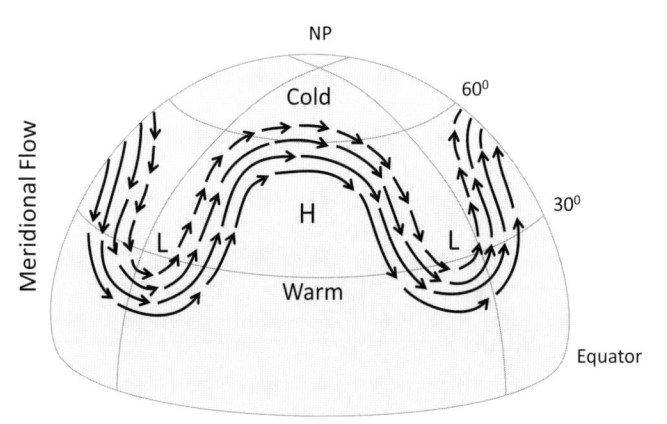

Figure 3.5. Zonal (top) and meridional (bottom) flow over northern hemisphere for different jet stream paths.

position of the ITCZ in relation to humid trade winds and the position of high-pressure cells (Hayden, 1988), which have migrated significantly over the past couple of millennia (Lechleitner et al., 2017). Annually, these mechanisms are globally the most consistent and result in large tropical rivers having a distinctive rhythmic "flood pulse" (i.e., Junk et al., 1989). The main flood pulse for the lower Nile River occurs in September (Figure 3.6), and is driven by summer monsoonal rainfall. This occurs with ascension of easterly trade winds over the Blue Nile headwaters in the Ethiopian highlands (~10° north latitude), associated with a northerly shift in the ITCZ (Wohl, 2007; Williams et al., 2012). Additionally, an important moisture source to the Nile headwaters derives from evapotranspiration of rainforest over lush West Africa, which is then supplied to the crucial Ethiopian highlands. While comprising just 10% of the total basin area, the Ethiopian highlands are vital to Nile waters in Egypt, supplying about 85% of the total streamflow that arrives at Aswan, Egypt (Gebrehiwot et al., 2019). With migration of the ITCZ south of the equator, humid easterly trade winds from the Indian Ocean deliver large amounts of rainfall to the Congo basin headwaters in central Africa. The annual peak discharge for the Congo River occurs over November–December (Figure 3.6), while a distinctive secondary peak occurs over March–April (Wohl, 2007).

Large floods along large rivers usually develop by a combination of factors. And while floods may be triggered by single large atmospheric events, extended wet conditions (e.g., high antecedent soil moisture) is usually necessary for large storms to generate large floods (Muñoz et al., 2018; Khanal et al., 2019). This is exemplified in Southeast Asia because of its distinctive

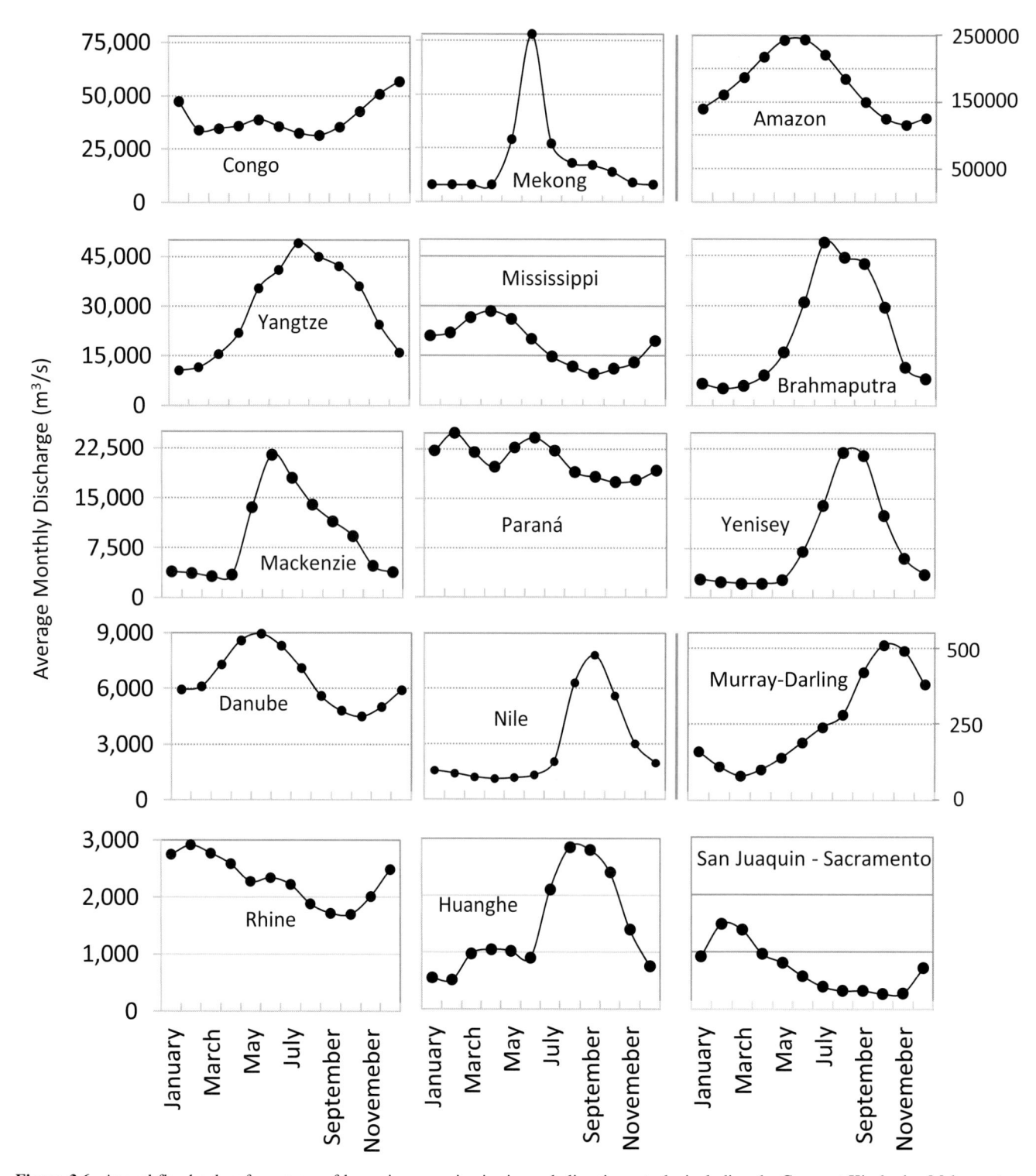

Figure 3.6. Annual flood pulses for a range of large rivers varying in size and climatic controls, including the Congo at Kinshasha, Mekong at Mukdahan, Amazon at Obidos, Huanghe at Huayuankou, Yangtze at Datong, Brahmaputra at Bahadurabad, Mackenzie at Norman Wells, Yenisey at Igarka, Danube at Ceatal Izmail, Nile at Aswan, and Murray–Darling at Lock 9 (data from Wohl, 2007). Data sources for Rhine at Rees, DE (Uehlinger et al.,2009), Rio Paraná at Corrientes (Mechoso et al., 2001), lower Mississippi at Vicksburg, MS (USGS 07289000), and combined discharge of San Juaquin near Vernalis, CA (USGS 11303500) and Sacramento at Rio Vista, CA (USGS 11455420).

monsoonal wet season that occurs with the seasonal shift of the ITCZ and development of low-pressure troughs over the enormous Asian land mass (Mariotti, 2007). Additionally, the Himalayan "water towers" serves as headwaters for a number of large Asian river basins, with orographic influences exacerbating the impacts of tropical cyclones. Thus, combined with its high population density Southeast Asia is particularly vulnerable to extensive flood disasters (Okada et al., 2018).

The enormous flood event along the lower Chao Phraya River basin (160,500 km^2) in 2011, the largest flood of record, serves as a key example. The 2011 monsoon season started early in March and persisted into October, and the wet season between May and October recorded 1,440 mm of precipitation, 143% above the annual average (Komori et al., 2012). The trigger for the flooding was the July landfall of tropical cyclone Nock-ten, which dumped copious amounts of rainfall in the mountainous basin headwaters. The combination of high antecedent soil moisture, basin morphometry, reservoir mismanagement, and poor land management decisions in the lower reaches of the basin resulted in flooding that persisted over five months (Komori et al., 2012), from August through December, devastating metropolitan Bangkok at the delta apex (see Section 7.5.2).

Annual flow regime is determined by average variability in monthly streamflow, which defines the duration of both high flow and low flow seasons. Two key controls on flow variability are climate and drainage area. Dry zone rivers display greater variability in annual peak discharge, with humid tropical zone rivers displaying lower variability (Wohl, 2007), and the largest rivers are tropical rivers (Latrubesse, 2008). Thus, for very large river basins the Murray–Darling (1,070,000 km^2) has the highest coefficient of variation (ratio of standard deviation to mean) in annual peak discharge whereas the Amazon (6,915,000 km^2) has the lowest (Figure 3.7). Flow variability is an important streamflow index because it directly relates to the duration and proportion of the riparian zone that is subaerial over the hydrologic year, with rivers having greater flow variability being especially vulnerable to emplacement of upstream dams (Graf, 1999, 2006) (Section 4.3.1).

Midlatitude rivers often display considerable streamflow variability because of strong seasonality in precipitation and temperature. This often includes streamflow influenced by both rainfall- and snowmelt-derived runoff processes. The streamflow regime for the lower Mississippi reveals the dominance of the Ohio basin, which drains humid east-central portions of the United States, and smaller contributions from the much larger Missouri basin, which drains drier west-central United States (Figure 3.8). The discharge regime of the lower main-stem Rhine River (Figure 3.9) is bimodal because of the streamflow sources of its main tributaries (Middelkoop et al., 2001; Uehlinger et al., 2009; Khanal et al., 2019). A June peak occurs along the High and Alpine Rhine associated with snowmelt and rainfall in the Alps, which is fed by the Aare River (Figure 3.9).

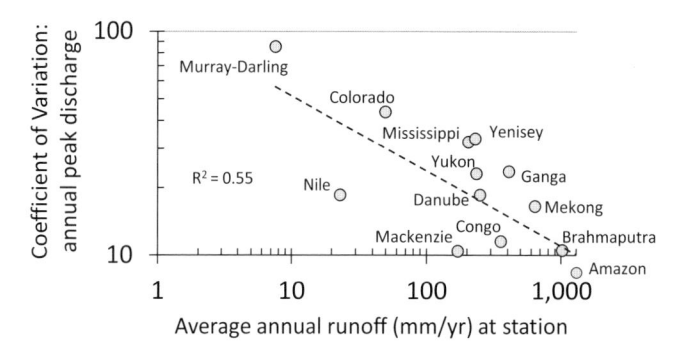

Figure 3.7. Mean annual runoff for large rivers in relation to annual peak discharge variability expressed as coefficient of variation. (From Wohl, 2007.)

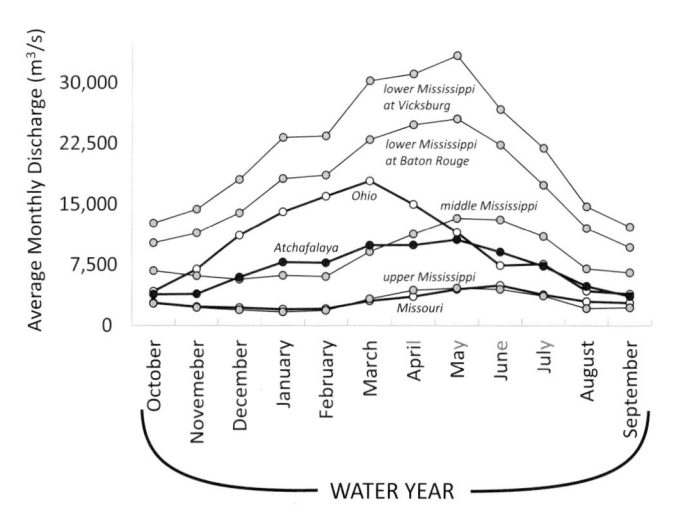

Figure 3.8. Discharge regime for the Mississippi basin (2008–2020). The maximum average monthly discharge in Vicksburg occurs in May at 33,405 (m^3/s). The large reduction in discharge at Baton Rouge is due to the Atchafalaya distributary capturing ~30% of the main-stem Mississippi discharge. Of note is that this data captures a period of high discharge flows, with large floods in 2008, 2011, 2016, 2018, 2019, and 2020 requiring opening of the Bonnet Carré Spillway in New Orleans. Gauging stations and period of record include the upper Mississippi River at Keokuk, IA (USGS 05474500, yrs 2008–2020), Ohio River at Olmsted, IL (USGS 03612600, yrs 2013–2020), Missouri River at Hermann, MO (USGS 06934500, yrs 2008–2020), lower Mississippi River at Vicksburg, MS (USGS 07289000, yrs 2008–2020), lower Mississippi River at Baton Rouge, LA (USGS 07374000, yrs 2008–2020), Atchafalaya River at Simmesport, LA (USGS 07381490, yrs 2010–2019). (Author figure. Data source: USGS as noted.)

The main peak discharge event in the lower Rhine (and delta) occurs in February, and is associated with rainfall- and snowmelt-derived runoff primarily from central and southern Germany. While the overall contribution of the Aare River to the lower Rhine discharge is minimal, it is vital to sustaining an intensive navigation system along the main-stem Rhine during drier summer months. Indeed, the severe European summer drought of 2018 resulted in historic low flows and greatly

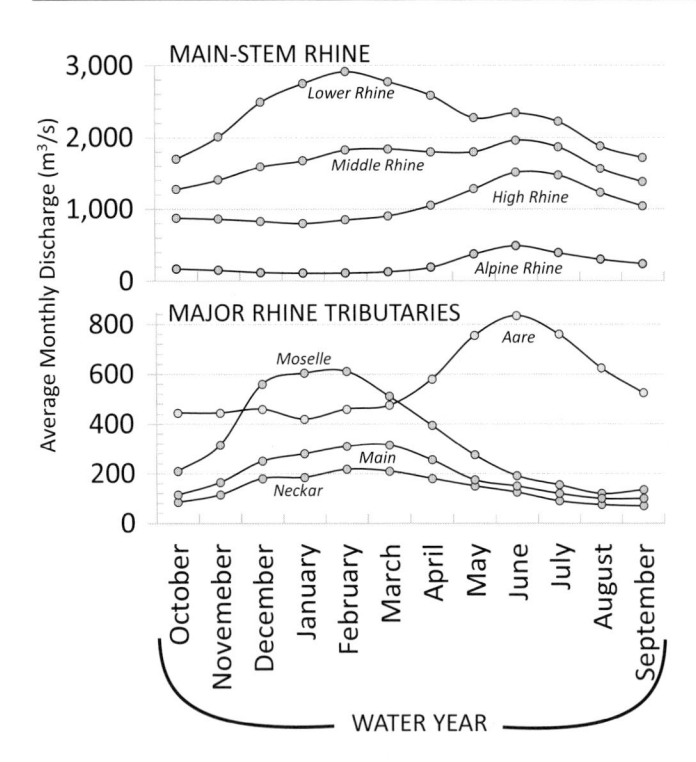

Figure 3.9. Discharge regime for Rhine River and tributaries (Aare, Main, Neckar, Moselle Rivers). Rhine data: monthly averages from 1931 to 2003 for lower Rhine at Rees, DE (20 km upstream of NL-DE border), middle Rhine at Kaub, DE (45 km upstream of Moselle confluence), high Rhine at Basel, CH, and alpine Rhine at Diepoldsau, CH. Rhine tributary data: monthly averages from 1963 to 2003. See Rhine basin map for locations. (Author figure. Data source: Uehlinger et al., 2009.)

Figure 3.10. Historic hunger stone exposed along Elbe River bed during the severe 2018 European drought. Hunger stones with carved markings record very low flow levels and were associated with severe drought, reduced agricultural productivity, and famine during medieval times. (Photo source: Dr. B. Gross, August 21, 2018. Hungerstein in Dolní Žleb-Niedergrund Nr. 1.)

reduced navigation and associated commerce along the main-stem Rhine (Hanel et al., 2018). The severity of the continental-scale drought was revealed in exposure of historic "hunger stones" along several larger central European rivers, such as the Rhine, Weser, and Elbe Rivers (Figure 3.10).

Decadal-scale alterations to Earth's general circulation influence the flood regime of large rivers (Knox, 1983; Berri et al., 2002; Naik and Jay, 2011; Muñoz and Dee, 2017). Over mid-latitudes this can be seen in changes to the position of the subtropical and polar jet streams, which influence continental airflow and precipitation patterns. In North America, zonal circulation is associated with westerly flow across the midlatitudes, and is generally associated with dry conditions (Knox, 2000). Meridional circulation, however, results in the jet stream extending into lower latitudes, which effectively serves as a conveyer belt to pull in moisture from the warm Gulf of Mexico and deliver it to higher latitudes. Meridional circulation, therefore, is associated with higher precipitation in central North America, and larger floods in the upper Mississippi basin (Knox, 1993; Dirmeyer and Kinter, 2009; Muñoz and Dee, 2017).

Knowledge of linkages between atmospheric drivers and hydrologic responses provides important insights into understanding implications of changes to Earth's general circulation associated with future climate change scenarios, particularly for those that have undergone historic fluctuation (Toonen et al., 2016; McGregor, 2017; Daniels et al., 2019). The shift from zonal to meridional flow in North America has decadal-scale oscillations that result in regional changes to climate and flood regime, with important consequences to water resources and flood management. Between 1895 and 1950, for example, west-to-east zonal flow conditions prevailed across middle portions of North America and were associated with moderate precipitation and streamflow conditions (Figure 3.11). After 1950, however, the jet stream developed a southerly loop into lower latitudes and meridional-scale conditions prevailed, resulting in increased flood magnitudes in the upper Mississippi basin (Knox, 1993, 2000).

In higher latitude rivers, annual temperature variability across river basins strongly influences the timing of discharge pulses produced by snow-melt runoff, as well as ice dam breakup (e.g., Brooks, 2000; Rokaya et al., 2018). The annual streamflow regime of rivers that seasonally freeze records a steep rise in discharge in late spring and early summer months associated with snowmelt runoff and breakup of river ice that focuses the majority of geomorphic processes within a reduced hydrologic period (Walker et al., 1987).

For Arctic flowing rivers hydrologic processes that control the annual flow regime are somewhat scale dependent (Walker and Hudson, 2003). Very large Arctic rivers, such as the Ob, Yenisei, Lena, Yana, Kolyma, and Mackenzie Rivers, have their headwaters south of the zone of continuous permafrost. While much of the lower reaches of the rivers are seasonally frozen, they sustain baseflow discharge during winter months, which is

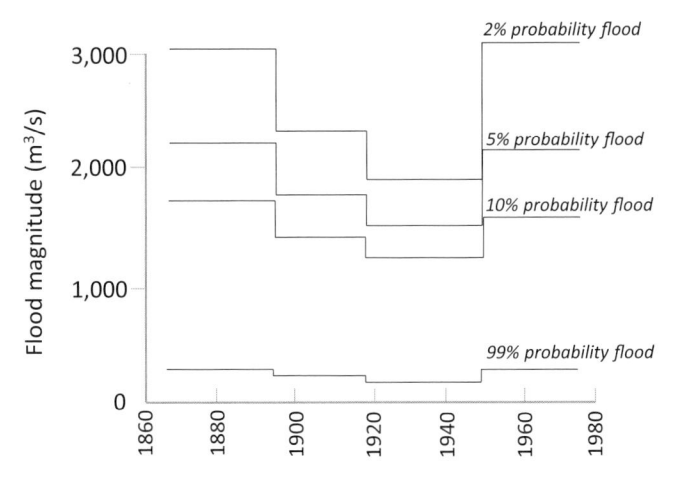

Figure 3.11. Historic hydroclimatology of the upper Mississippi River at St. Paul, MN. Flood magnitude for different probabilities (based on log-Pearson Type III) calculated for about a 100-year time span. Periods are statistically different and associated with changes in global circulation. Zonal circulation results in lower flood magnitudes between 1895 and 1950. Higher magnitude flooding associated with meridional circulation prior to 1895 and after 1950. (Source: Knox, 1983.)

partially supplied by precipitation from westerly migrating fronts (cyclones). The annual flow regime of the enormous Mackenzie River basin (1,870,000 km^2) (Figure 3.6), for example, reveals a sharply rising curve in the early summer with a peak discharge in June of 22,000 m^3/s, and has a winter baseflow discharge of 4,000 m^3/s where it discharges into the Arctic Ocean (Brooks, 2000).

Arctic rivers having headwaters within continuous permafrost completely freeze, resulting in a substantially reduced hydrologic year. The Colville River (53,000 km^2), for example, freezes solid to the channel bed in the long winter, and has but a four-month hydrologic year over which all streamflow and associated geomorphic processes occur (Walker et al., 1987; Walker and Hudson, 2003). Hydrologically, spring is flood season, and associated with a distinctive ice breakup phase consisting of three short periods over about four weeks. Flood season begins at the end of May and includes pre-breakup flooding, breakup flooding, and post-breakup flooding. The summer streamflow period then lasts for three months, ending in late September at the onset of the Arctic winter.

Large flood events along the lower reaches of large rivers require that all tributaries are contributing substantial discharge. And the largest flood events require that the timing of peak discharge events in the upper and lower tributaries are synchronized, and preceded by wet conditions (Hirschboeck, 1991; Muñoz and Dee, 2017).

3.2.1.1 GREAT FLOOD OF 1993: MISSOURI AND MISSISSIPPI BASINS

The Great Mississippi Flood of 1993 had an approximately 500-year recurrence interval across the upper and middle Mississippi basins, which includes the Missouri basin (Wahl et al., 1993; Pitlick, 1997). The flooding was generated by a persistent low-pressure trough that efficiently transferred moisture via the "Maya Express," an atmospheric river from the warm Caribbean and Gulf of Mexico, to the upper Mississippi and Missouri basins. The trouble had actually started almost a year earlier, as higher than average rainfall and lower than average temperatures in summer and autumn 1992 resulted in high antecedent soil moisture across the Midwestern United States (Lott, 1993; Wahl et al., 1993; Pitlick, 1997). Additionally, high reservoir storage in the Missouri and upper Mississippi basins reduced further runoff storage capacity, which was needed because of the high snowfall and persistent high magnitude rainfall events that occurred into spring and summer. Large portions of Iowa had more than 100 cm of rainfall between April and August. The June–July rainfall totals across a large swath of the upper Mississippi and Missouri River basins were the highest in over a century (Table 3.1), with most stations recording more than 200% above average rainfall (Lott, 1993). The event resulted in numerous dike breach events that flooded rural populations and farmland, and resulted in considerable cropland damage by floodplain stripping and deposition of coarse flood deposits (see Section 6.6.3.3). The summer 1993 Missouri and Mississippi River flooding remains among the most expensive flood disasters in US history (Pinter, 2005; NOAA, 2019).

But downstream, in the lower Mississippi, the summer 1993 event was but a moderate-sized pulse. This is because tributaries within the southern and eastern portions of the Mississippi basin, particularly the Ohio basin, were simultaneously receiving below average precipitation (Lott, 1993). Parched conditions were driven by warm and dry anticyclonic flow caused by an Atlantic subtropical ridge (the Bermuda High) settling over the southeastern United States (Wahl et al., 1993; Dirmeyer and Kinter, 2009). As the Ohio River supplies nearly half (Meade, 1995) of the annual discharge to the lower Mississippi, the lower Mississippi is geomorphically adjusted to the Ohio River's discharge, and thus was able to accommodate the streamflow delivered from the middle Mississippi basins. Most gauging stations in the lower Mississippi did not flood in summer 1993. In Memphis, TN, for example, the discharge crest for the summer 1993 event peaked August 7 at 65.1 masl, 1.3 m below the official flood stage at 66.42 masl (Rivergages.com).

3.2.2 Teleconnections and Streamflow Variability

In addition to discrete atmospheric phenomena, teleconnections – large-scale atmospheric linkages of climatic variability across great distances (thousands of kilometers) – influence streamflow variability in faraway watersheds (Richey et al., 1989; Ward et al., 2010; Kundzewicz et al., 2019). Teleconnections are defined by anomalies from average temperature or air pressure

Table 3.1 *Rainfall totals in June–July, 1993 at National Weather Service offices within Missouri and upper Mississippi basins*

Location	New record (cm)	Old record (cm) (year)	Normal (cm)	% > normal
Chicago, IL	36.6	28.5 (1970)	18.8	195
Springfield, IL	42.3	42.1 (1981)	17.5	243
Moline, IL	48.3	46.8 (1969)	23.4	207
Dubuque, IA	41.9	40.4 (1969)	20.8	201
Concordia, KS	59.7	47.8 (1967)	20.6	290
Omaha, NE	45.0	36.1 (1967)	18.3	246
Fargo, ND	30.5	30.0 (1975)	14.0	218
Williston, ND	25.7	23.3 (1963)	11.2	230
Huron, SD	35.1	33.8 (1984)	15.2	230
Green Bay, WI	34.5	33.5 (1990)	16.5	209

IA = Iowa, WI = Wisconsin, SD = South Dakota, ND = North Dakota, NE = Nebraska, IL = Illinois, KS = Kansas.
Data from Lott (1993).

conditions over specific reference regions. This results in negative or positive anomalies that can be statistically linked to somewhat predictable changes in terrestrial variability in temperature, wind, and precipitation patterns. Statistical indices that define teleconnections are constantly reviewed and refined by scientists and government agencies to better understand the periodicities and duration of teleconnections, and to develop robust understandings of their influences to water resources and surficial environments. The main teleconnections that influence terrestrial precipitation and riverine streamflow variability are the North Atlantic Oscillation (NAO), Northern Annular Mode (NAM), Southern Annular Mode (SAM), Indian Ocean Dipole (IOD), Pacific Decadal Oscillation (PDO), Atlantic Multidecadal Oscillation (AMO), and the El Niño Southern Oscillation (ENSO). Their geographic and spatial range of influence varies greatly, and periodicities of teleconnections range from days to decades (Table 3.2). The timing and influence of some teleconnections to water resources are somewhat consistent, while others are highly dynamic and unpredictable. Failure to understand linkages between teleconnections and precipitation variability can result in substantial degradation to natural and anthropogenic water resources with considerable societal consequences (Caviedes, 2001; Dettinger, 2011; van Dijk et al., 2013; Mehta, 2017).

Although there are challenges to identifying explicit linkages between teleconnections and streamflow variability (Kundzewicz et al., 2019), some teleconnections are recognized as important drivers of flood hydrology (Table 3.2). In northern Europe flooding is largely driven by the NAO and the NAM. Over the past six centuries, for example, all major flood events along the Rhine River were associated with winter positive phases of NAO, with some flooding also coincident with the AMO (Toonen et al., 2016). In southwestern Europe it is the opposite, with the negative phase of NAO driving flooding for the Douro, Guadalquivir, and Tagus Rivers draining the Iberian Peninsula (Benito and Machado, 2012). The IOD is important to east and central Africa, including streamflow variability in the Congo basin (Ficchì and Stephens, 2019). In North America, the negative phase of NOA is important to streamflow variability in the east – and especially rivers in the northeast (Coleman and Budikova, 2013). ENSO is most important to flooding in south, central, and western North America, particularly positive (warm) phases during winter months. The decadal periodicity to the PDO is also important to Pacific draining rivers in North America (Naik and Jay, 2011). By comparison to other major teleconnections much less is known about the SAM. But studies over the past decade have shown it influences streamflow variability within its core region, which logically includes southern portions of Africa, Australia, and South America (Table 3.2). This includes influencing low and high flows of the Negro (Argentina) and Zambezi Rivers (Milliman and Farnsworth, 2011).

Nevertheless, the most important teleconnection to streamflow variability is ENSO (Ward et al., 2010; Kundzewicz et al., 2019) from the standpoint of (1) its global extent, (2) its synergetic influence with other teleconnections, and (3) because it directly modifies the highest proportion of Earth's runoff.

3.2.3 El Niño Southern Oscillation and Hydrologic Regime

Driven by sea surface temperature anomalies along a 10° latitude × 50° longitude swath of the equatorial Pacific,[1] ENSO has a global influence that can be detected with historic and modern

---

[1] Niño 3.4 sea surface temperature (SST) anomalies are averaged over three months ending with the current month, which defines the Oceanic Niño

Table 3.2 *Major teleconnections and their characteristics, including linkages to streamflow variability for specific river basins*

Teleconnections*	Index characteristic	Periodicity (duration)	Terrestrial core region of influence	Significant influences on streamflow and/or flooding (river basin)
Atlantic Multidecadal Oscillation (AMO)	10-year running-mean of detrended sea surface temperature anomalies in the North Atlantic (0°–70°N, 80°W–10°E).	40–80 years (10–20 years)	North America and Europe	Enfield et al., 2001 (lower Mississippi); Hodgkins et al., 2017 (range of small to medium "natural" rivers in Europe and North America); Toonen et al., 2016 (Rhine)
El Niño Southern Oscillation (ENSO)	Oceanic Nino Index – sea surface temperature anomalies in equatorial Pacific (Niño region 3.4, 5°N–5°S × 170°W–120°W); Southern Oscillation Index – normalized air pressure differences between Darwin and Tahiti	2–7 years (9–18 months)	Eastern Australasia, central-western South America	Richey et al., 1989 (Amazon); Eltahir and Wang, 1999 (Nile); Hudson, 2004 (Pánuco-MX); Ward et al., 2010 (Amazon, Murray–Darling); Naik and Jay, 2011 (Columbia); Räsänen and Kummu, 2013 (Mekong); Wei et al., 2014 (Yangtze); Muñoz and Dee, 2017 (lower Mississippi); Hardiman et al., 2018 (Yangtze)
Indian Ocean Dipole (IOD)	Dipole Mode Index (DMI): Sea surface temperature anomalies between western (50°E–70°E and 10°S–10°N) and eastern (90°E–110°E and 10°S–0°N) Indian Ocean basin	2–4 years (6 months)	East Africa, southern India, Indonesia, northern Australia	van Dijk et al., 2013 (Murray–Darling); Ficchì and Stephens, 2019 (Congo, Rovuma-MZ)
North Atlantic Oscillation (NAO)	Air pressure differences at sea level between the subpolar low (Reykjavík, Iceland) and the subtropical high (at Lisbon or Azores)	4–10 years (with high variability)	Northeastern North America, Western Europe	Coleman and Budikova, 2013 (eastern US rivers); Toonen et al., 2016 (Rhine); Benito and Machado, 2012 (Douro, Guadalquivir, Tagus)
Northern Annular Mode (NAM)[1]	Air pressure anomalies in the north Pacific and north Atlantic (20°N) and the Arctic (90°N)	Days to annual (high variability)	Eurasia, North America, eastern Canada, North Africa, Middle East	Savelieva et al., 2000 (Ob, Yenisey, Lena, Indigirka, Kolyma); Ionita et al., 2011 (Elbe); Kryjov and Gorelits, 2019 (Onega, Northern Dvina, Mezen', Pechora –RU)
Pacific Decadal Oscillation (PDO)	Sea surface temperature anomalies in north Pacific (20°N to Gulf of Alaska)	20–30 years	Western North America	Pavelsky and Smith, 2004 (Lena, Mackenzie); Naik and Jay, 2011 (Columbia)
Southern Annular Mode (SAM)	Zonal air pressure anomalies at sea level between 40°S and 65°S	Weekly to annual (1–3 months)	Southern Africa, Australia, New Zealand, southern South America	Areneo and Villalba, 2014 (Atuel-CL); Milliman and Farnsworth, 2011, figure 3.20 (Zambezi, Rio Negro-AR); Lara et al., 2015 (Baker-CL and AR); Muñoz et al., 2016 (Biobío, Puelo-CL)

Notes: [1]NAM is considered to include the Arctic Oscillation (AO) (see Thompson and Wallace, 1998; 2000; Riviere and Drouard, 2015; Dai and Tan, 2017).

*Sources: **AMO** – Kerr, 2000; Enfield et al., 2001; Alexander et al., 2014; Hodgkins et al., 2017; **ENSO** – Trenberth, 1997; Ward et al., 2010; Wolters, 2010; NOAA (accessed 2.12.2020); **IOD** – Saji et al., 1999; Endris et al., 2019; **NAO** – Hurrell, 1995; Kingston et al., 2006; Coleman and Budikova, 2013; Guimarães Nobre et al., 2017; Wang et al., 2017; **NAM** –Thompson and Wallace, 1998, 2000; Polyakov and Johnson, 2000; Riviere and Drouard, 2015; Dai and Tan, 2017; He et al., 2017; **PDO** – Mantua et al., 1997; NOAA (accessed 2.12.2020); **SAM** – BoM (accessed 2.16.2020); Hendon et al., 2007; Dieppois et al., 2016; Lim et al., 2016.
(Author table.)

streamflow records and paleoenvironmental data far beyond its terrestrial hearth along coastal Peru and northern Chile (Caviedes, 2001, p. 15). The warm and cool phases of ENSO, El Niño and La Niña, respectively, influence precipitation, wind, and temperature patterns far across the globe, with strong seasonal and regional variation. El Niño occurs at a frequency of about every two to seven years and has a duration from nine to eighteen months. La Niña events are somewhat less frequent and usually persist over a shorter duration (Trenberth, 1997; Wolters, 2010).

For river basins located under ENSO impact zones, streamflow variability can be influenced by eight scenarios. These include variations of warmer, cooler, drier, or wetter that are – depending on location – driven by either El Niño or La Niña. The strongest influence of ENSO on streamflow variability occurs with either cool and wet or warm and dry conditions because soil moisture and runoff are both directly influenced by precipitation and temperature (i.e., evaporation), and some river basins experience both conditions (Ward et al., 2010).

The ENSO drives a crucial component to the flow regime of many of Earth's rivers, which over long periods increases flow variability by either making low flow periods lower, high flow periods higher, or both (Zorn and Waylen, 1997; Ward et al., 2010; Naik and Jay, 2011). Riparian aquatic environments along many rivers are dependent upon ENSO-driven flood pulses to provide sustenance to riparian water resources (Florsheim and Dettinger, 2015). A phase of ENSO may provide a counter influence to the opposite phase. Along the lower Guadalupe River on the Texas Gulf Coast, for example, El Niño results in increased winter rainfall that drives flood pulses, which is important for riparian and aquatic environments to recover from dry years caused by La Niña (Hudson et al., 2012).

The magnitude to which ENSO influences streamflow variability and associated geomorphic processes within a particular watershed also depends on how El Niño or La Niña impact zones overlay the drainage network (Foley et al., 2002; Hudson, 2004; Naik and Jay, 2011). For large river basins, major tributaries may be differentially influenced by negative and positive ENSO phases, resulting in a "buffering" effect (i.e., Ronchail et al., 2005). Additionally, the sensitivity of water resources to ENSO also depends upon environmental history as well as contemporary water management strategies (Caviedes, 2001; van Dijk et al., 2013). Knowledge of the timing and intensity of ENSO periods, therefore, provides crucial information to water resource managers as regards reservoir impoundment for hydropower and irrigation during floods and droughts (Naik and Jay, 2011; van Dijk et al., 2013).

3.2.3.1 ENSO AND SOUTH AMERICAN RIVERS

The streamflow of South American rivers differentially reveals the influence of both El Niño and La Niña. La Niña is associated with the largest streamflow totals, including for the massive Amazon, Orinoco, and Sao Francisco basins (García and Mechoso, 2005; Ronchail et al., 2005; Jiménez-Muñoz et al., 2016). Increased Amazon streamflow during La Niña conditions is driven by trade wind intensification from the warm tropical Atlantic, increasing moisture flow and precipitation in northern headwaters (Ronchail et al., 2005). The actual influence of La Niña on streamflow in the lower Amazon (at Ōbidos) is lagged by about half a year because of the delay in precipitation response to changes in the Southern Oscillation Index (SOI), soil moisture residence time, and discharge transmission through the extensive drainage system (Richey et al., 1989; Foley et al., 2002). La Niña conditions are associated with a 7% higher average annual discharge along the lower main-stem Amazon at Ōbidos (Foley et al., 2002). But as the greater influence of ENSO to streamflow occurs during the high flow period (i.e., Ward et al., 2010), it is the Amazon flood peaks that are most influenced. Indeed, La Niña conditions were associated with the enormous floods of 2009 and 2012, the two largest recorded floods on Earth (Satyamurty et al., 2013; Filizola et al., 2014).

Because streamflow variability driven by ENSO manifests differently in the northern and southern portions of the Amazon basin (Richey et al., 1989; Foley et al., 2002; García and Mechoso, 2005), its influence on floodplain inundation varies between the dry season and flood season (Table 3.3). In comparison to neutral conditions, floodplain inundation during El Niño is reduced by 12% (~20,000 km^2). For La Niña, an additional 35% of floodplain inundation occurs relative to neutral conditions, increasing from 170,079 km^2 to 184,058 km^2. The manner in which the floodplain inundation occurs involves an intricate network of floodplain connectivity that defines the distinctive Amazonian flood pulse (Dunne et al., 1998; Lewin et al., 2016; Latrubesse and Park, 2019).

The strongest overall influence of ENSO to Amazon streamflow occurs during El Niño, which is associated with lower precipitation totals caused by a reduction in convection over the northern, eastern, and western headwaters. Such conditions are associated with the severe droughts of 1925/26, 1982/83, 1997/98, and 2015/16 (Richey et al., 1989; Foley et al., 2002; Jiménez-Muñoz et al., 2016). Interestingly, southern reaches of the Amazon basin experience increased streamflow during El Niño, which is likely caused by the influence of warmer sea surface temperatures in the south Atlantic (Foley et al., 2002; Ronchail et al., 2005). Indeed, across the southern Amazon drainage divide, into the headwaters of the massive La Plata

Index (ONI). An El Niño or La Niña event occurs when the ONI exhibits positive (warm) or negative (cool) anomalies (defined as $> \pm 0.5°C$) for a minimum of five consecutive values. Source: U.S. National Oceanic and Atmospheric Administration (www.ncdc.noaa.gov/teleconnections/enso/enso-tech.php).

Table 3.3 *Flooded areas along the Amazon River for Neutral, El Niño, La Niña conditions**

	Minimum flooded area (km^2)	Maximum flooded area (km^2)
Neutral	20,378	170,079
El Niño relative to Neutral conditions	+2,483 (dry season)	−20,203 (flood season)
La Niña relative to Neutral conditions	+7,049 (dry season)	+13,979 (flood season)

* Based on model simulations for flood (April, May, June) and dry (November, December, January) seasons.
Source: Foley et al. (2002).

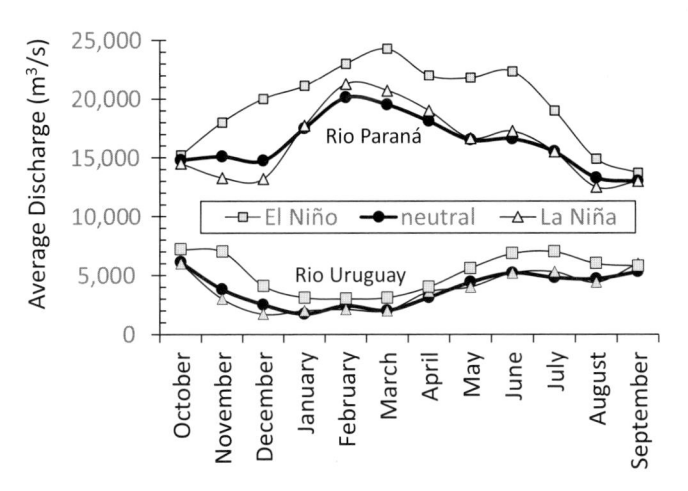

Figure 3.12. Long-term average monthly discharge for ENSO phases for La Plata basin, including Rio Uruguay (Paso de los Libres, 1909–2000) and lower Rio Paraná (Corrientes, 1904–2001). Rio Uruguay is a large tributary to the Rio Paraná. Differences in streamflow are statistically significant. (Source: García and Mechoso, 2005.)

Figure 3.13. Nilometer on Isle of Rhoda, Cairo, Egypt. Nilometers provide the world's longest continuous streamflow data record and extend Nile flow levels back five millennia. (Artist: David Roberts, 1838; licensed by CC.)

(Rio Paraná) basin, El Niño conditions are associated with the largest streamflow totals whereas La Niña conditions vary little from neutral conditions (Figure 3.12). Of sixteen large floods on the Rio Paraná between 1904 and 2000, eleven occurred during El Niño events (Prieto, 2007).

3.2.3.2 ENSO AND OTHER LARGE RIVERS

ENSO has a global imprint on the hydrologic signature of many rivers, often in association with other components of Earth's general circulation (Ward et al., 2010; Kundzewicz et al., 2019). El Niño is key to the flood regime of several of the large rivers in Asia, such as the Mekong (Räsänen and Kummu, 2013) and the Yangtze (Wei et al., 2014), particularly in combination with a westerly (landward) shift in position of the subtropical high-pressure cell that modifies summer monsoons and tropical cyclone tracks. In Africa, El Niño is responsible for some 30% of the annual variability in Nile River discharge, and has become more important. Long-term flow records for the Nile River at Cairo, Egypt, obtained with the Nilometer – the longest continuous flow record in the world (Figure 3.13) – establishes that contemporary El Niño events are more important to streamflow variability than over the prior two millennia (Eltahir, 1996; Eltahir and Wang, 1999).

The influence of ENSO to streamflow variability for river basins within temperate zone continental interiors is also related to temperature variability, because of its influence on snow melt runoff and evaporation, which influences future runoff events because of its effect on antecedent soil moisture (Kundzewicz et al., 2019). The influence of ENSO on lower Mississippi River streamflow is complex, and displays a six-to-twelve-month lag with positive precipitation anomalies because of high antecedent soil moisture that builds up during wet periods in autumn and winter – preceding the main spring flood season. While human impacts are also responsible, the largest lower Mississippi floods are associated with El Niño events, including the massive floods of 1973, 1983, and 2011, in addition to the infamous 1927 flood disaster (Muñoz and Dee, 2017). In the Pacific Northwest it is La Niña that is most strongly correlated with higher streamflow totals, in association with the PDO (Whitfield et al., 2010). The average annual discharge of the enormous Columbia River is higher when La Niña occurs during cold (positive) PDO phases (114% of neutral conditions) and lower when El Niño occurs during warm (negative) PDO phases (87% of

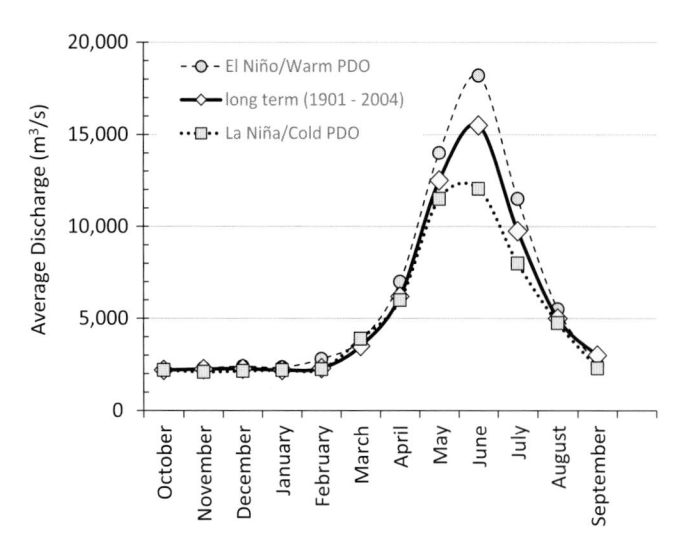

Figure 3.14. Influence of ENSO and PDO on average monthly discharge for Columbia River (USGS 14105700 at The Dalles, OR, USGS 14105700). Drainage area at station: 613,827 km^2 (upstream of confluence with Willamette River). (Source: Naik and Jay, 2011.)

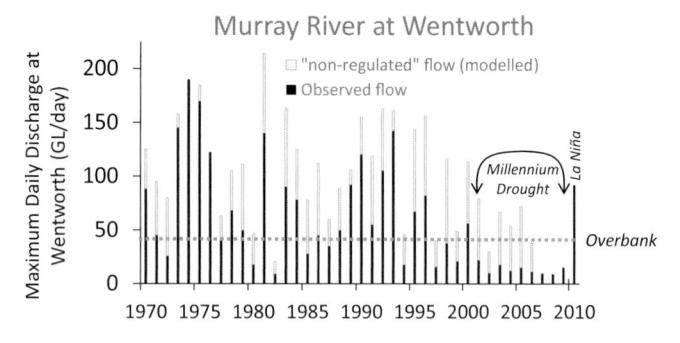

Figure 3.15. Maximum daily discharge for the Murray River at Wentworth (lower basin), revealing flow reduction during the great Millennium Drought (2001–2009) with El Niño conditions in southeastern Australia. The drought ended with an abrupt shift to flooding caused by a shift to La Niña conditions. Large irrigation withdrawals worsened the drought for riparian ecosystems and stakeholders (compare observed to non-regulated flow). (Sources: Van Dijk et al., 2013, with 2009 and 2010 data from the Murray-Darling Basin Authority for the Murray River at Wentworth Weir at Lock 10).

neutral). And average monthly variability (Figure 3.14) reveals considerably greater differences, particularly during high flow months of May, June, and July (Naik and Jay, 2011).

The great Murray–Darling basin (1,072,000 km^2) of southeastern Australia is the most strongly influenced large river by ENSO. Streamflow variability in the Murray basin alternates like a metronome between positive and negative phases of ENSO. This is particularly the case when amplified Walker Circulation in the western tropical Pacific occurs in association with the SAM and IOD in a positive phase, which drive extreme drought conditions during El Niño (van Dijk et al., 2013; BoM, 2016). Flooding occurs during La Niña conditions with SAM and IOD in opposite phases, as occurred following the severe Millennium Drought in Southeast Australia (Figure 3.15).

3.2.4 Atmospheric Rivers

Atmospheric rivers – essentially lower tropospheric conveyer belts of water vapor (Newell et al., 1992) – are extensive meandering plumes of moisture that link warm tropical oceans with midlatitude lands (Table 3.4). Atmospheric rivers may extend for several thousand kilometers and be several hundred kilometers in width and 2–3 km in thickness (Dettinger, 2011; Guan and Waliser, 2015). They usually travel in advance of westerly cold fronts associated with extratropical cyclones (Ralph et al., 2013), and are substantial precipitation-makers during winter months (Gimeno et al., 2014).

The size and frequency of atmospheric rivers makes them responsible for redistributing copious volumes of tropical water. The volume of water vapor transported by a single atmospheric river may exceed the flow of the Amazon River! (Zhu and Newell, 1998).

Three to five atmospheric rivers are located across the globe at any given moment. While only covering about 10% of Earth's surface atmospheric rivers transport 90% of Earth's meridional water vapor (Zhu and Newell, 1998; Guan and Waliser, 2015). Where atmospheric rivers are forced over tall mountains they produce considerable orographic snowfall that, with spring and summer snowmelt, serves as a vital water resource to westerly draining river basins (Dettinger, 2011; Lavers and Villarini, 2015). Some atmospheric rivers are sufficiently frequent that they're formally named, such as Australia's Northwest Cloud Bands (Telcik and Pattiaratchi, 2014), Maya Express, and the Pineapple Express – the latter provides substantial Pacific water to western and central North America (Dirmeyer and Kinter, 2009; Dettinger, 2011; Ralph et al., 2013).

Atmospheric rivers are a global phenomenon that influence precipitation variability to all continents (Guan and Waliser, 2015; Paltan et al., 2017). Key regions of influence include western North America (SE Alaska, British Columbia, Washington, Oregon, California) (Dettinger, 2011; Guirguis et al., 2019), South America (southern Chile) (Viale et al., 2018), Australia (Telcik and Pattiaratchi, 2014), South Africa (Blamey et al., 2017), and northwestern Europe (Lavers and Villarini, 2015). Atmospheric rivers are responsible for as much as 20–30% of average precipitation along portions of western North America and Europe (Lavers and Villarini, 2015), and the annual arrival of atmospheric rivers is especially important to regions vulnerable to annual water stress. Atmospheric rivers provide 25–50% of California's annual water supply (Guan et al., 2010; Dettinger, 2011; Ralph et al., 2013). Southern Chile receives 45–60% of its annual precipitation from atmospheric rivers (Viale et al., 2018).

Extension of atmospheric rivers into the central and south-central United States contributes to substantial precipitation and

Table 3.4 *Some characteristics and features of atmospheric rivers and hydrologic influence*

Dimensions	Length: ~1 to ~10 × 103 km; Width: ~1 to ~6 × 102 km; Thickness: ~2 to 3 km	Newell et al., 1992; Dettinger, 2011; Guan and Waliser, 2015; Lavars and Villarini, 2015
Thickness of integrated water vapor (IWV)	>2 cm	Dettinger, 2011; Ralph et al., 2013; Guan and Waliser, 2015
# per day	3–5 (in either hemisphere)	Zhu and Newell, 1998; Guan and Waliser, 2015; Blamey et al., 2017
Global areal extent	<10% of Earth's circumference	Zhu and Newell, 1998; Guan and Waliser, 2015
Seasonal frequency	Highest # in winter (DJF), also autumn (SON) and spring (MAM), with few in summer (JJA)	Dettinger, 2011; Lavers and Villarini, 2013; Ralph et al., 2013; Blamey et al., 2018
Water vapor transport	90% of extratropical meridional water vapor flow	Zhu and Newell, 1998
Average duration of associated precipitation event at landfall	20 h	Ralph et al., 2013
Runoff	22% of annual global runoff	Paltan et al., 2017
Key regions where ARs influence flooding	Pacific Northwest and California, Canada and United States (Dettinger, 2011; Ralph et al., 2013; Lavars and Villarini, 2015; Konrad and Dettinger, 2017; Paltan et al., 2017), Midwest and south-central United States (Dirmeyer and Kinter, 2009; Paltan et al., 2017), western Europe (Lavers et al., 2011; Paltan et al., 2017), South Australia (Paltan et al., 2017), western South Africa (Paltan et al., 2017)	

flooding, particularly with moisture deriving from the Caribbean and western Gulf of Mexico delivered by the "Maya Express" during enhanced meridional circulation.

Similar to the 1993 flooding, the 2008 Mississippi basin flood season was preceded by considerable snowfall and months of wet conditions that resulted in high antecedent soil moisture. Additional high precipitation totals supplied by the Maya Express in spring 2008 contributed to extensive flooding in the US Midwest (Dirmeyer and Kinter, 2009). And in comparison to the 1993 event, rainfall totals were much higher to the east of the Mississippi, in the Illinois and Wabash River basins (Holmes et al., 2010). The result was an enormous flood crest along the lower Mississippi, which for several gauging stations was among the highest. At Natchez, MS, for example, the 2008 flood crest on the lower Mississippi ranked as the second highest flood of record, higher even than the infamous 1927 flood. And downstream, toward New Orleans, the threat of levees being overtopped resulted in the U.S. Army Corps of Engineers opening the Bonnet Carré Spillway for thirty-one days, the first time the massive flood diversion structure had been opened in eleven years (See Section 7.2.2.3).

3.3 CHANNEL BANK EROSION ALONG LOWLAND RIVERS

3.3.1 Bank Erosion and the River Environment

River bank erosion is a natural geomorphic process essential to maintenance of riparian ecosystems. Because river banks are transitional between water and land they serve as a vital riparian

Figure 3.16. River bank erosion (bank height ~20 m, including the submerged portion) on a heavily mined section of the Mekong River near Phnom Penh, Cambodia. (Photo courtesy of Prof. S. Darby.)

ecotone (Florsheim et al., 2008; Gurnell et al., 2016). Active channel bank heterogeneity is ecologically significant to specific flora and fauna, which utilize sedimentary deposits and channel bank micro-morphologic features as habitat. Bank erosion drives vegetation succession, creating varied riparian habitat niches. In addition to macro invertebrates, larger organisms, including riparian avian species such as bank swallows (*Riparia riparia*), preferentially identify coarse friable channel bank deposits for burrowing and nesting (Moffatt et al., 2005; Silver and Griffin, 2009).

Bank erosion is of considerable concern to river management because it threatens hydraulic infrastructure, property, and disrupts navigation (Figure 3.16). Flood control dike systems should be designed with knowledge of bank erosion processes and historic channel pattern adjustment. Accelerated bank erosion results in a large supply of riverine sediment and causes channel bed aggradation (Kesel et al., 1992) that can require dredging to maintain a navigable channel (Smith and Winkley, 1996). Concerns with bank erosion along large rivers drive major hydraulic infrastructure projects, particularly riprap emplacement and concrete revetment construction to arrest bank erosion (see Section 5.3). Such projects need to consider the sedimentology of channel banks, as well as the geometry of channel bends (Thorne, 1989). Recognition of geodiversity and biodiversity associated with natural river bank processes are a key motivation in "river rewilding" strategies (Brown et al., 2018), which results in removal of some channel bank protection measures (See Section 8.2.4).

As with much of the base of knowledge in fluvial geomorphology, because of the difficulty in conducting fieldwork in large rivers with complex channel bank stratigraphy, bank erosion studies primarily derive from field investigations of smaller rivers, laboratory flume studies, and mathematical modeling. River managers should be prudent, therefore, before extending such outcomes to larger lowland rivers. This is because (1) bank erosion along larger lowland rivers is dominated by mass failure of fine-grained cohesive bank material rather than hydraulic entrainment of coarse sediment, (2) is dependent upon flood duration more than flood magnitude, and (3) in comparison to smaller rivers, vegetation is less important to bank erosion along large rivers because plant root density is negligible in lower bank strata.

3.3.2 Process of Bank Erosion

Two sequential processes primarily drive channel bank erosion along lowland rivers: hydraulic entrainment and mass wasting. These two erosional processes are often "set up" by prior weathering activities and antecedent moisture conditions (Thorne, 1989). Hydraulic entrainment occurs at the scale of individual particles, mainly along the channel bed. The result is channel bed scour or incision. Hydraulic entrainment of particles occurs when boundary shear stress exceeds the critical shear stress of individual particles comprising the channel bed.[2] This occurs with increases in stage, which increases the hydraulic radius and water surface slope (hydraulic gradient) between the rising limb and crest of the discharge hydrograph, thereby increasing rates of sediment entrainment and bed scour (Richards, 1982; Knighton, 1998).

Channel bed scour by hydraulic entrainment during high discharge events is characteristic along cutbanks, particularly in deeper pools at the outside of bends. Hydraulic entrainment occurs as the momentum of streamflow is directed toward outer banks during higher stages with high boundary shear stress along the banks and at the channel bed, accentuated by super-elevation of streamflow. Additionally, flow patterns at cutbanks have large eddies and secondary divergent currents that sweep to the channel bed, increasing particle entrainment (Hickin, 1977; Richards, 1982; Clifford, 1993).

Mass wasting of heterogeneous floodplain deposits results in the largest amount of bank erosion along lowland rivers (Thorne, 1991; Bull, 1997). The importance of hydraulic entrainment to the process of bank erosion is that channel bed scour at the base (toe) of the channel bank results in over-steepening of the potential failure plane (i.e., higher shear stress) within the bank material. A linear or arcuate potential failure plane extends obliquely from the floodplain surface to perhaps the maximum scour depth at the cutbank. As scour depth increases at the cutbank, the failure plane (slope) gradient also increases, increasing boundary shear stress and bank failure potential (Thorne and Abt, 1993). Imminent bank failure is often revealed by the presence of linear tension cracks within the floodplain, perhaps ~0.5 to 5-m from the channel bank.

3.3.2.1 SEDIMENTARY CONTROLS

Sedimentology is a key control on the process of channel bank erosion along lowland rivers (Turnbull et al., 1966; Hey and Tovey, 1989; Thorne, 1991; Parker et al., 2008). At a local scale most channel banks along lowland rivers are comprised of coarse-grained bottom-stratum and fine-grained top-stratum deposits (Figure 3.17). Bottom-stratum deposits are comprised of noncohesive sandy channel bed and bar deposits and erode by direct hydraulic entrainment or gravity flow (slip flow). Top-stratum deposits are comprised of cohesive fine-grained overbank deposits, including clayey backswamp and channel fill deposits, and erode by mass wasting after being saturated and undercut by erosion of underlying bottom-stratum deposits. At channel reach scales the sedimentology of bank material may range from sand to clay, resulting in varying resistance to hydrologic and hydraulic processes. At watershed scales bank material becomes increasingly cohesive downstream with increasing drainage area while specific stream power decreases (Lawler et al., 1997). This often results in lower bank erosion rates toward the coast, particularly within deltaic distributary channels (Kolb, 1963; Hudson and Kesel, 2000).

---

[2] Critical entrainment: $\tau_c = kg(p_s - p)D_{50}$, where k = Shields entrainment parameter for dimensionless critical shear stress (0.045 = common value); g = gravitational constant (9.81 m/s^2), p_s = density of sediment (2.65 g/cm^3), p = density of water (1.0 kg/m^3), d_{50} (median particle size of bed material, m). Boundary shear stress in N/m^2, $\tau = (gp)RS$ 122.5 RS, where R = channel hydraulic radius (A/Wp, where A = w × d, Wp = 2d + w), S = water surface slope (m/m). While entrainment should theoretically occur when $\tau > \tau_c$, particle entrainment often does not occur until the ratio of τ_c to $\tau \geq 2$, because of particle packing, shape, bed armoring, and other factors at the fluid–channel bed interface (for discussion of assumptions, see Richards, 1982; Ferguson, 1987; Wilcock, 1998; Gomez, 2006).

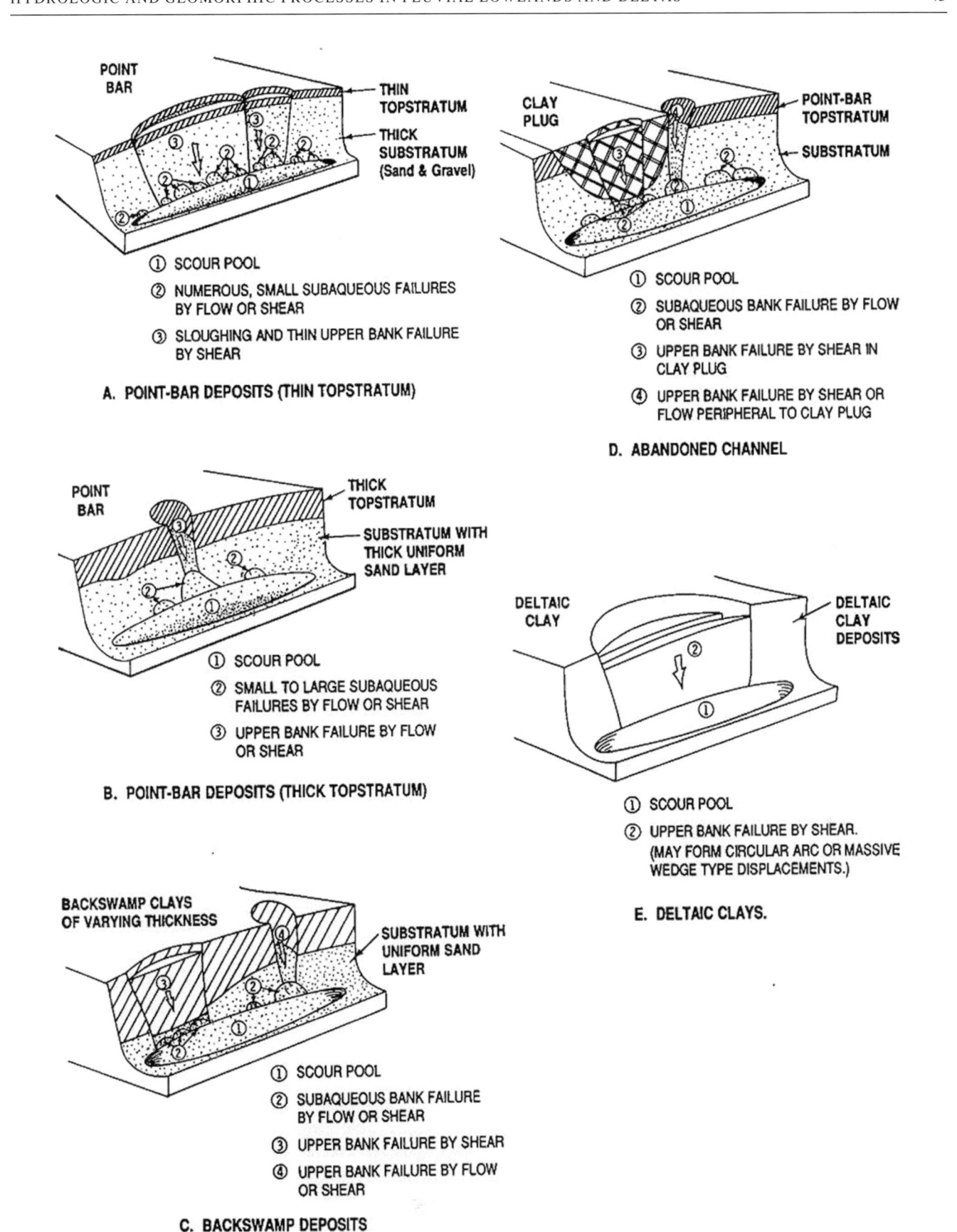

Figure 3.17. Channel bank erosion mechanisms along large alluvial rivers for a range of floodplain depositional settings – particularly contrast between sandy bottom-stratum and clayey cohesive top-stratum deposits. The shape of bank failure is related to the thickness of the top-stratum and bottom-stratum, which influences scour and shear failure. Five scenarios are provided, with particular contrasts between sandy point bar deposits, cohesive clay plugs, and clayey deltaic deposits. (From Saucier, 1994, based on Turnbull et al., 1966.)

Figure 3.18. Bank caving by arcuate slumping along the lower Amazon River, Brazil, downstream of confluence with Tapajos River. Note detachment along channel bank and large failure block in water. For scale, height of boat ~3 m from water surface. (Author photo, January 2001.)

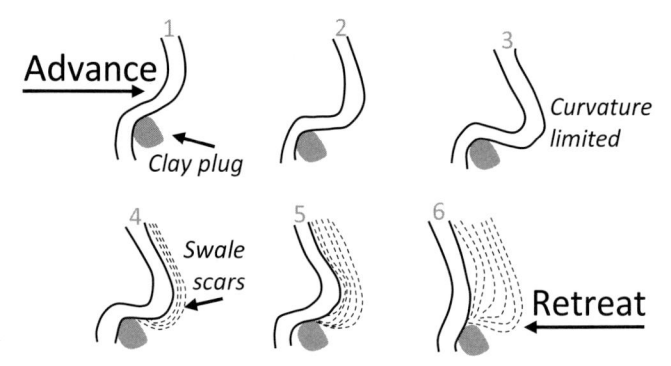

Figure 3.19. Channel planform adjustment to hard points (i.e., clay plugs, backswamp) in floodplain deposits. Advance: Local hard point forces higher rates of bank erosion upstream and the formation of a promontory, which induces deep scour adjacent to the hard point and higher rates of erosion of adjacent sandy bank material, resulting in meander bend extension. Retreat: In the case of particularly sharp (i.e., low rc/w) bends, flow separation at bend apex can result in reduced velocity along the outside of the bend, inducing deposition and lateral bar formation that results in an opposite pattern of adjustment (retreat). (Modified from Thorne, 1998.)

Noncohesive sandy and silty bank material often fails by sloughing (Figure 3.17), a mass wasting mechanism associated with elongated sheets (or ribbons) of bank material collapsing into the river (Turnbull et al., 1966; Thorne, 1991). Such deposits are rapidly reworked and assimilated into the active sediment load (Hickin and Nanson, 1984). Clayey cohesive bank material, however, often erodes by slab failure at slopes >60° (Thorne, 1989, 1991). In comparison to noncohesive deposits, cohesive slabs are generally thicker shorter blocks (or wedges) of alluvium (Figure 3.17). In the case of particularly cohesive deposits, failure blocks may remain along the channel bed for a couple of years or so before being assimilated into the active sediment load as fine-grained wash load, and perhaps reduce bank erosion during subsequent discharge events. Arcuate slumps are a distinctive type of bank failure associated with cohesive deposits, which are rotational and slide along a curvilinear failure surface with slopes <60° (Thorne, 1991). Upon failure initiation (Figure 3.18), such blocks of alluvium may slide down beyond the bank toe, requiring several years before completely sliding into the river. In such cases they provide a distinctive pattern, and can result in trees at oblique angles relative to in situ growth position.

Considerable reach-scale variability in bank erosion rates occurs because of spatial variability in channel bank material, including local cohesive "hard points," and is manifest in the channel planform morphology (Figures 3.19 and 3.20). Older sedimentary hard points exposed in the active channel bank material, including clay plugs or backswamp, are effectively nonerodable and result in channel pattern adjustment of the reach upstream and downstream of the hard point (Fisk, 1944, 1952; Thorne, 1998). Resistant bank material has important implications to channel engineering for flood control and navigation, requiring considerably more excavation and dredging to engineer meander bend cut-offs (Winkley, 1977) (see Section 5.4.1).

3.3.2.2 HYDROLOGIC CONTROLS

Hydrologic influences on channel bank erosion include discharge magnitude and duration as well as the seasonality of events (Table 3.5). Channel bank erosion is mainly related to high flow periods because of the influence of discharge on stream power (Turnbull et al., 1966; Bull, 1997; Darby et al., 2013) and channel bank moisture (Hooke, 1979; Thorne, 1991; Rinaldi and Casagli, 1999; Simon and Collison, 2001). As river stage rises, water is driven (infiltrates) into alluvium, saturating channel bank material. This influences the process of bank erosion in two ways. First, saturation of sandy material negates particle cohesion (pore suction) associated with capillary soil moisture. The resulting positive pore pressure reduces bank shear strength. Secondly, the greater weight of the saturated sedimentary matrix increases shear stress along the failure plane, increasing erosion potential by mass wasting.

While erosion of sandy channel bank material is primarily dependent upon increasing discharge (stream power), erosion of cohesive channel banks is also dependent upon channel bank moisture, and is not necessarily related to discharge magnitude (Wolman, 1959; Hooke, 1979; Thorne, 1991). Field observations have confirmed that large discharge events can result in little channel bank erosion while, alternatively, moderate-sized discharge events can be associated with high rates of bank erosion (Hooke, 1979). This occurs because of flood pulse sequence

Table 3.5 *Hydrologic influences on bank erosion along larger alluvial rivers*

Streamflow quality/ index	Importance / role
Event magnitude (m^3/s)	Directly increases stream power and boundary shear stress: (1) Channel bed: entrainment of coarse bed material, scour (deepening), undercutting of channel bank; (2) Channel bank: limited entrainment of coarse (noncohesive) bank sediment and cohesive aggregates.
Seasonality of events (wet/dry, cool/warm, freezing)	Warm and dry summer allows channel bank drainage and evaporation of soil moisture. Cool and wet winters result in low evaporation and saturated channel bank material, increasing vulnerability to sequential flood events.
	High water table from alluvial aquifer: high antecedent bank moisture, return flow results in piping and rills along banks.
	Needle ice and freeze–thaw of bank material results in physical weathering (prepares sediment for erosion). Ice-jam events: gouging/abrasion of channel banks by ice debris/rafting, scour from flooding.
	Cantilever failure (frozen banks).
Timing (frequency, sequence)	Flood frequency during a wet season, time (days) between sequential flood events relative to channel bank drainage (higher bank erosion from second event).
Duration (# days)	Time over which high channel bed shear stress and bank saturation occurs.

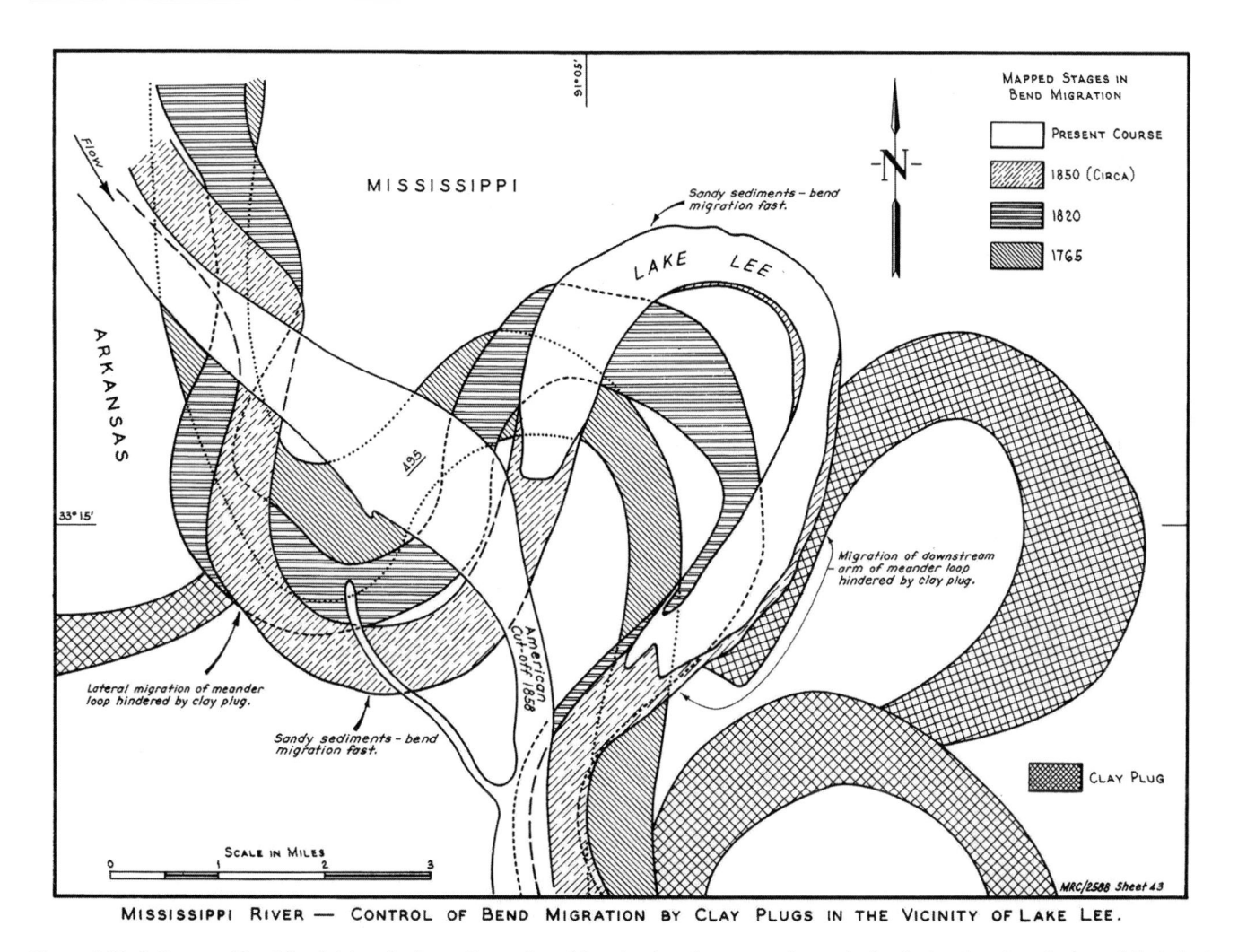

Figure 3.20. Influence of local floodplain cohesive sedimentology (clay plugs) on the pattern of meander bend migration along the lower Mississippi from 1765 to early 1900s. Two key influences are illustrated, including (i) the arrest of meander bend rotation associated with exposure of hard clay along the channel bank and (ii) downstream deflection of bend migration because of clay plug in contact with the active channel. (From Fisk, 1944, plate 4.)

and seasonality in relation to antecedent channel bank (soil) moisture.

The timing of bank failure events does not necessarily coincide with flood crest (i.e., peak discharge). In addition to flood peak, other important hydrologic influences on bank erosion include (1) the sequence of discharge events within a wet season, (2) the seasonality of discharge events, particularly related to saturated versus dry bank material, and (3) the flood duration (Table 3.5). Much bank erosion occurs just after the flood crest, on the receding limb of the flood wave because heavy saturated bank material is not hydraulically supported by the water column. Additionally, return flow as groundwater seepage through saturated alluvium can initiate erosion by creating pipes and rills where it exfiltrates (van Balen et al., 2008), particularly at the contact between sedimentary layers. The sequence of discharge events during a wet season is important, as high rates of bank erosion can be generated by moderate-sized events because of saturated bank material. Bank material remains saturated longer during cool winter conditions, increasing vulnerability to sequential events. Finally, in high-latitude regions ground freezing and needle ice are important physical weathering mechanisms that prepare sediment for erosion by subsequent discharge pulses. Rafting ice sheets can cause bank erosion by gouging and scour during seasonal ice-jam breakup.

3.4 FLOODING ACROSS FLUVIAL LOWLANDS

3.4.1 Hydrologic Pathways

Lowland river flooding is generated by watershed-scale and local-scale sources, and the most severe flooding occurs when both mechanisms are synchronous.

Watershed-scale flood mechanisms include (1) overtopping of natural levees, (2) overbank flow supplied via a network of conduits and low surfaces, and (3) exfiltration of groundwater flow through the sedimentary matrix (Figure 3.21). Larger floods overtop natural levees along the main-stem channel and generally exceed about a five-year recurrence interval, although this certainly varies according to the age and development of natural levees (Cazanacli and Smith, 1997; Hudson et al., 2013; Lewin and Ashworth, 2014a). Indeed, whilst the sedimentology and geomorphology of natural levees are directly associated with the process of flooding more than any depositional landform, natural levees are actually the least inundated component of the floodplain. Compared to clayey backswamp basins, natural levees of coarse sandy–silty deposits are moderately permeable, resulting in rapid drainage after flood crest recession.

The channel–floodplain boundary is exceptionally "leaky." Hydrologic connectivity along lowland rivers initially occurs at levels below official government flood stage levels (Hudson et al., 2013; Phillips, 2013). Seepage occurs by low sloughs, batture (tie) channels, crevasses, small tributaries, and old channels that function as conduits between the main-stem river and floodplain water bodies (Aalto et al., 2008). Thus, more common than floodwaters overtopping river banks is that discharge is distributed across the floodplain by a variety of discrete topographically controlled hydrologic pathways (Figures 3.21 and 3.22). Crevasses are of particular importance to local flooding and the identification of crevasse splays from satellite imagery and high-resolution digital elevation models (DEMs) provides an indication of flood activity, and the possibility of an avulsion (Shen et al., 2015). Such local-scale mechanisms often escape the attention of agencies charged with floodplain management, although local residents, and farmers in particular, are intricately knowledgeable of their functioning because of being detrimental to grazing and cultivation.

An additional mechanism in which upper basin streamflow sources inundate floodplain lowlands is by groundwater. The seasonal discharge pulse transfers main-stem streamflow into floodplain alluvium, driving a rise in the height of the water table of the alluvial aquifer (Krinitzsky and Wire, 1964; Sharp, 1988). The seepage of streamflow through the floodplain alluvium occurs by two types of flow, conduit and matrix (Whiting and Pomeranets, 1997; Burt et al., 2002). Conduit flow results in the highest groundwater flow rates, and is essentially through elongated pathways of coarse sediment, mainly old channel belts. Such pathways can result in groundwater flow from 10 to 100 m/day. Many floodplain water bodies no longer connected by surface conduits are connected to the active channel by such seasonal groundwater pulses (Poole et al., 2002). In contrast to conduit flow, matrix flow through fine-grained clayey deposits, such as clayey backswamp, has extremely low flow rates and effectively serves as an aquiclude (Krinitzsky and Wire, 1964; Heath, 1983). The subaqueous zone where alluvial groundwater derived from main-stem river sources mixes with older sources of alluvial groundwater is referred to as the hyporheic zone, a buried ecotone with unique geochemistry and microbial communities (Stanford and Ward, 1993). The surficial location where such groundwater exfiltrates to the floodplain is often indicated by springs and bogs, which supports wetland vegetation that can be detected with satellite imagery.

3.4.2 Local-Scale Controls on Lowland Flooding

Because of the topographic compartmentalization of large floodplains by older geomorphic deposits, such as relict channel belts, inundation on large floodplains can be driven by local-scale climatic mechanisms such as intense convectional thunderstorms (i.e., Figure 3.21). Smaller flood basins and topographic

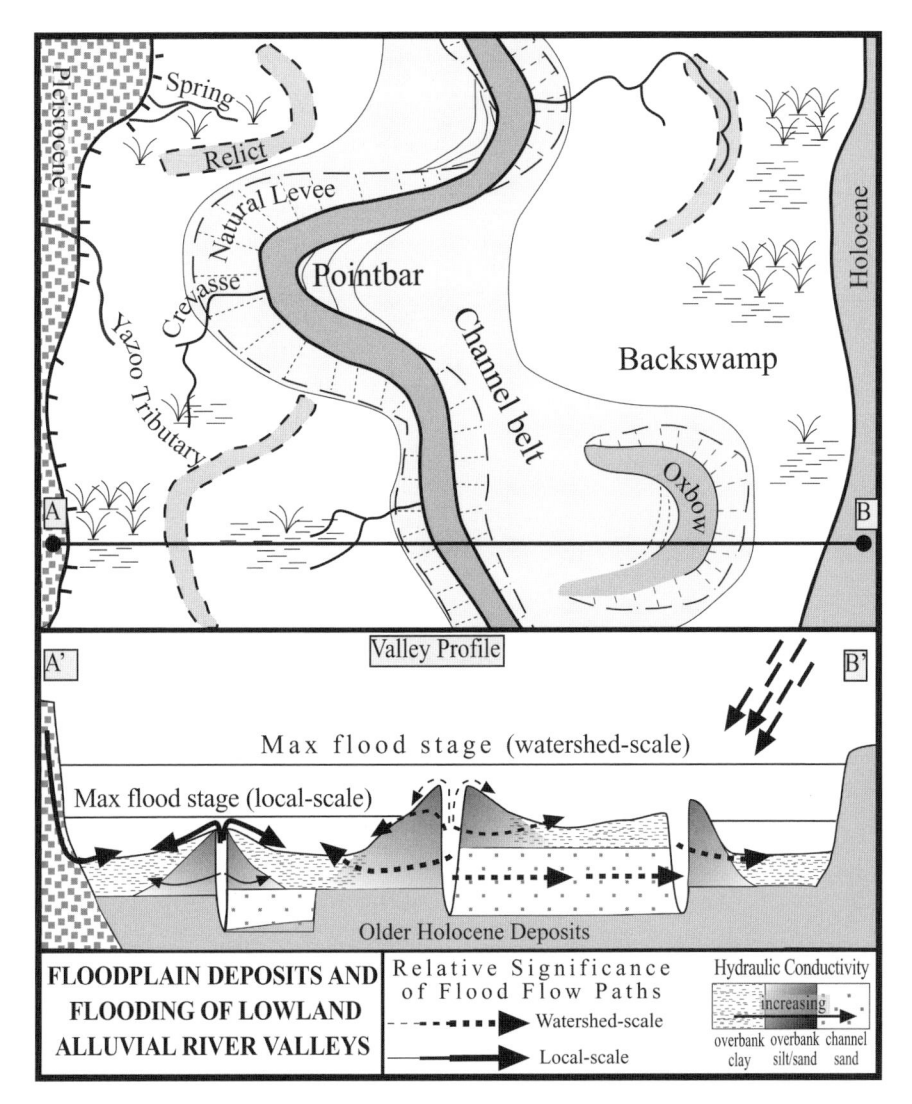

Figure 3.21. Hydrologic pathways for a lowland floodplain with meander belt topography. Specific pathways are distinguished between watershed-scale and local-scale mechanisms and also related to the sedimentology (hydraulic conductivity). (Author figure, modified from Hudson and Colditz, 2003.)

depressions can be inundated during different periods than the main-stem (upper basin) sources. Floodplain patches can be inundated by runoff generated by pluvial (overhead) precipitation events, spring sapping along valley bluffs, or runoff supplied by local yazoo-style tributaries (Figure 3.23A and B). These local mechanisms produce runoff that is topographically confined to lower-lying depressions within flood basins. The waters are unable to flow over higher main-stem natural levees and are not able to infiltrate into the clayey alluvium. Management of these local-scale mechanisms usually involves direct modification to the floodplain, including straightening (shortening) of yazoo streams, drainage ditches, culverts, sluice gates, and pumping. Indeed, the enormous Yazoo River flood basin along the lower Mississippi is a textbook model for the challenges and controversies associated with local-scale flooding within a major alluvial valley (Section 7.2.2.1).

3.5 LOWLAND FLOODPLAIN GEOMORPHOLOGY

3.5.1 Negative Relief Floodplains

Lowland floodplains are dominated by two broad sets of fluvial sedimentary processes: overbank flooding and lateral accretion (Wolman and Leopold, 1957). These processes create two distinctive morpho-sedimentary units within a river valley, specifically coarse-grained channel belts raised above fine-grained flood basins (Figure 3.24).

Negative relief floodplains are the dominant topographic expression (Figure 3.24) within the deposition zone (Russell, 1939), and impose an active control on hydrologic, sedimentary, and ecosystem processes (Lewin and Ashworth, 2014a; Park and Latrubesse, 2019). Geomorphically, a channel belt includes a

Long Profile of Floodplain Topography and Hydrologic Connectivity: Lower Mississippi River, Vicinity of Natchez, MS

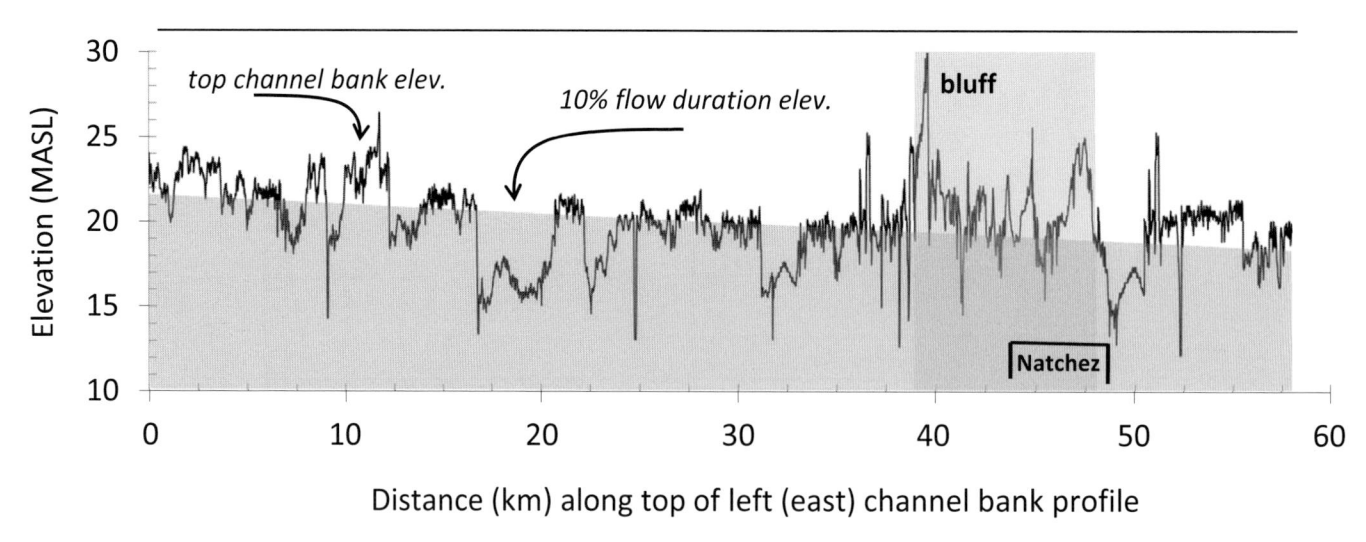

Figure 3.22. Channel bank elevation profile in relation to the 10% flow duration along the lower Mississippi River, near Natchez, revealing numerous hydrologic pathways for a moderate magnitude event. For comparison, the official flood level at Natchez is 19.9 masl, which is a 6% flow duration and nearly 1-m higher than the flow level at 10% duration. (Author figure, after van der Most and Hudson, 2018.)

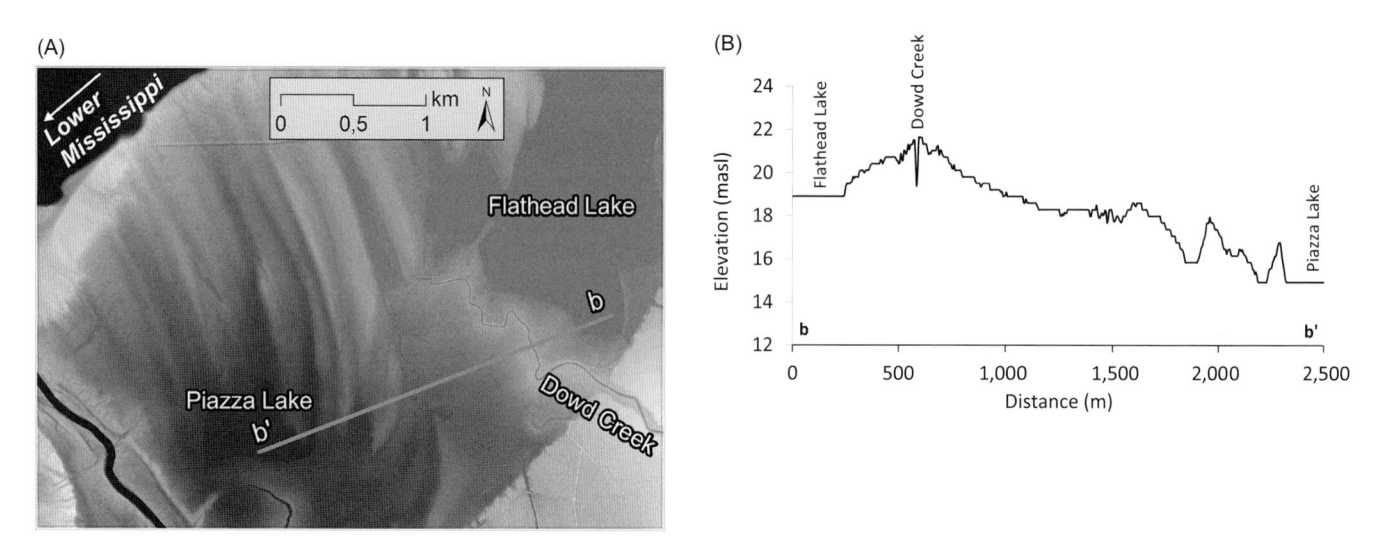

Figure 3.23. Local-scale floodplain water bodies (depression lakes) along the lower Mississippi valley (upstream of Natchez) portrayed with a lidar digital elevation model (A) and a topographic profile (B), illustrating different sources of local-scale hydrologic connectivity on the floodplain with distinctly different water surface levels. Additionally, the local-scale mechanism can be seen as influencing the floodplain topography of a major river, as sedimentation from Dowd Creek has created negative relief atop a flood basin within the Mississippi valley. (Author figure. Data source, lidar digital elevation model, Louisiana State University, Atlas/LSU.)

combination of sedimentary deposits having distinctive topographic signatures, including natural levees, meander scroll, oxbow lakes, and crevasse splay deposits. Large lowland rivers often include multiple Holocene channel belts, which may be associated with different discharge fluvial regimes. Distinguishing specific geomorphic features to reconstruct the fluvial landscape requires careful consideration of the floodplain topography, as well as a variety of criteria pertaining to sedimentary and pedogenic characteristics of the fluvial landform (Table 3.6). Such indices are related to sedimentary structure (i.e., thickness, morphology), pedogenic development, and presence of biological organisms (plant, shells), and are somewhat

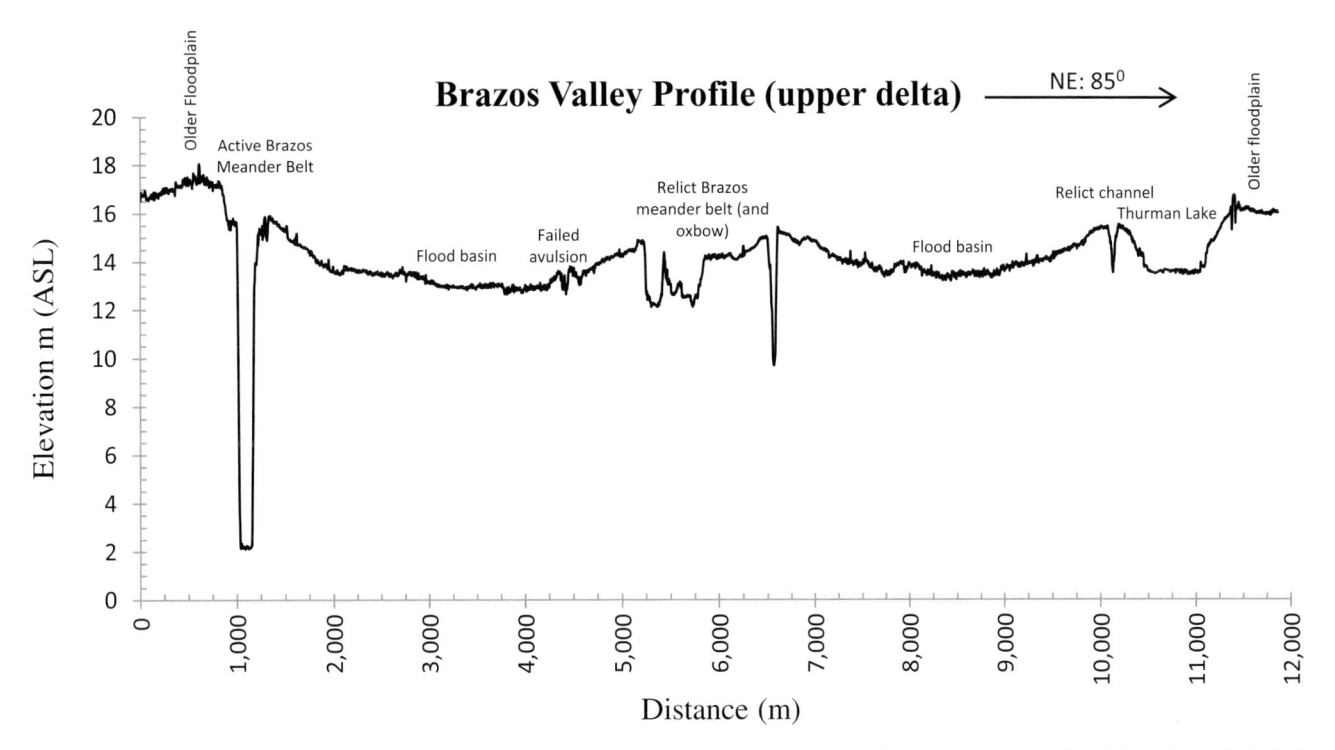

Figure 3.24. Valley profile across that lower Brazos River floodplain, near apex of delta. The profile shows the active floodplain has a relief of ~3 m because of active channel belts and two other relict channel belts being higher than adjacent flood basins. (Data source: lidar from Texas Natural Resource Information Systems, Austin, TX.)

distinct for each depositional environment (point bar, natural levee, backswamp, crevasse splay, abandoned courses).

3.5.2 Overbank Sedimentary Deposits

The topographic complexity of large lowland river valleys results in considerable spatial heterogeneity in sedimentation patterns (Asselman and Middelkoop, 1998; Aalto et al., 2008; Lewin et al., 2016; Heitmuller et al., 2017). In association with channel bank irregularities (i.e., Figure 3.22) that enable floodwater seepage across a range of flow magnitudes (Hudson et al., 2013), floodplain sedimentation within large lowland river valleys can be envisioned as a complex depositional web, with each floodplain comprised of a unique combination of sub-basins and internal linkages (Aalto et al., 2003; Shen et al., 2015; Park and Latrubesse, 2017). In addition to main-stem sedimentary deposits, lowland floodplain depositional systems also include secondary linear and prior-form components unique to each lowland river (Figure 3.25).

Floodplain depositional features along the main-stem river include natural levees, channel bars and islands, banktop splays, and swales and chutes (Figure 3.25, Table 3.6). With the exception of fine-grained wash load (clay/silt) deposited within swales or as clay drapes, the majority of these overbank sedimentary deposits along main-stem rivers are coarse (sand, coarse silt).

Sediment is transported across the floodplain by suspension and saltation as floodwaters overtop channel banks. Although often depicted as a floodwater sheet across a smooth floodplain surface, floodwaters are quickly disrupted by floodplain topography, vegetation, and human structures (roads, fields, fences, etc.), which induces deposition and disrupts smooth spatial patterns of sedimentation (i.e., Pizzuto, 1987; Asselman and Middelkoop, 1998).

3.5.2.1 NATURAL LEVEES

Natural levees are relatively coarse-grained ridges of overbank flood deposits aligned along channel banks and are the key structural form along lowland rivers. The size of natural levees (width, height) is related to the range in flood stage and sediment load. The signature pattern of natural levee particle size and thickness (Table 3.6) exhibits distinctive lateral fining and is nonlinear. Individual sandy flood strata commonly range from 1 cm to ~20 cm in thickness near the levee crest, with abrupt lateral reduction to ~1 mm along the backslope as the natural levee transitions into clayey flood basins (Brierley et al., 1997; Cazanacli and Smith, 1997; Hudson and Heitmuller, 2003; Leigh, 2018). Hydrologically, natural levees are topographically the most significant component of negative relief floodplains, and natural levees associated with abandoned channels influence the pattern of floodwaters and floodplain water bodies.

The process of sedimentation along natural levees is associated with an abrupt velocity reduction and sediment sorting. Coarse sediment is deposited adjacent to the channel bank, which

Table 3.6 *Occurrence of minor sedimentary structures and pedogenic indices in Holocene fluvial depositional environments*

Sedimentary structures		Natural levee	Point bar accretion	Backswamp	Abandoned channels*	Abandoned courses	Crevasse splay
Bedding	Thick (>15 cm)					D	
	Medium (5–15 cm)	C			C	C	C
	Thin <5 cm	C–A		C	A		A
Morphology	Parallel laminations (texture)	C	A	C	C–D	A	C
	Parallel laminations (color)			A	C		
	Lenticular laminations			R			
	Wavy laminations		C	R	C		
Cross-laminations	Simple					C	
	Planar		C			A	
	Trough		A				
Ripple laminations	Current		D–A			C	A
	Ripple-drift		A			A	
	Wave		A				
Scour and fill			D			C	
Plant remains	Distinct particles	C	C	A	R	R	
	Finely divided		A	C	A	C	R
	Bedded			C	R		
Shell fragments				R			
Clay inclusions			C			A	
Load casts						C	
Distorted laminations		R		C	C		
Burrows		A		D	R		D
Nodules	Ferruginous	A		C			C
	Calcareous	C		A	R		
Desiccation features		C		C	R		C
Oxidation mottling		D–A		C–R			A

D = dominant; A = abundant; C = common; R = rare * upper, fine-grained "clay plug" portion only.
Modified from Saucier (1994).

includes saltating bed load deposited as ripples and dunes (Figure 3.26A). Flood sedimentation is especially variable adjacent to channel banks (Kesel et al., 1974; Heitmuller et al., 2017; Leigh, 2018). Along floodplain reaches that lack adequate boundary resistance (cohesive soil, vegetation), high velocity floodwaters can result in local zones of floodplain stripping (Magilligan et al., 1998) immediately adjacent to thick banktop splay deposits (sand sheets). Dense floodplain vegetation results in high flow resistance, and high rates of deposition. The influence of individual (isolated) trees can often result in highly variable spatial patterns of sedimentation and erosion, characterized by proximal thick deposits, floodplain stripping by flow concentration around the tree (and tree root exposure), and a velocity shelter on the lee side characterized by a near absence

Spillage Model for Large River Floodplain

MS: Mainstream sediments
MSa: natural levee
MSc: Channel bars & islands
MSb: Bank-top splays

SL: Secondary linear systems
SLd: Accessory channels
SLf: Crevasse with delta
SLe Crevasse splays
SLg Internal drainages

PF: Prior-form following
PFh: Cutoff & paleochannel fills
PFj: Point bar swales & chutes
PFi: ponded lake filling
PFk: Diffuse overbank spreads

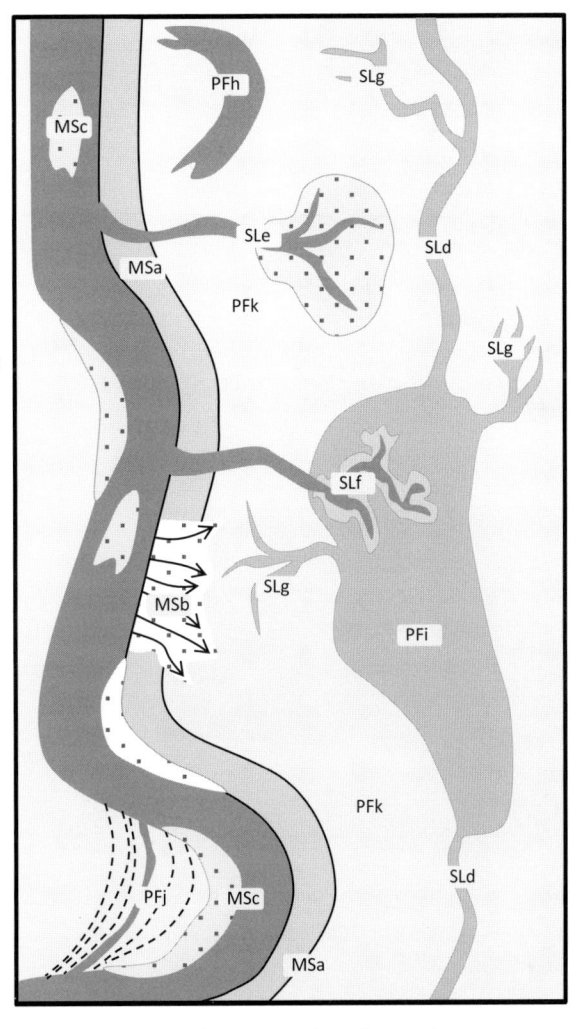

Figure 3.25. Spillage sedimentation model for large river floodplains, emphasizing three main sedimentation zones, secondary linear systems, and prior-form (relict) pathways linking sediment movement from the main-stem channel to various depositional environments. See Table 3.7 for further definition and supporting citations. (After Lewin et al., 2016, with author modification.)

of sedimentation (Figure 3.26B). Over time, the topography associated with banktop splays is smoothed by subsequent floods and suspended sediment deposition, which accumulate as part of the overall natural levee deposit. The accumulation of natural levee sediments is related to rates of lateral migration and floodplain reworking (Figure 3.27), as laterally stable channels provide more time for flood sediments to accumulate, and thus result in larger (higher and wider) natural levees (Hudson and Heitmuller, 2003; Hudson, 2004).

Natural levee width increases downstream, with increasing drainage area and sediment load, until an interesting reversal of scale dependence occurs, resulting in a decrease in natural levee size (Figure 3.28). As rivers approach the coast there is usually a reduction in the coarse sediment available for natural levee construction, as well as a reduction in flow variability (between low and high stage). Thus, downstream of the hinge-line the height and width of natural levees decrease (Russell, 1936; Kolb, 1963; Hudson and Heitmuller, 2003). Along the lowermost Mississippi (Figure 3.29), for example, the width of natural levees decreases from about 6.5 km to about 0.4 km toward the river mouth (Kolb, 1963).

3.5.2.2 CREVASSE SPLAY

Crevasse splays are high-energy coarse-grained hydro-geomorphic landforms that extend beyond natural levees into flood basins. Crevasse splays form fan-like deposits extending outward from a river bank, especially on the outside of bends. Crevasse channels represent a discrete mechanism for delivering coarse

Figure 3.26. Thick sand sheet deposition adjacent to channel bank from record duration 2018–2019 flooding along the lower Mississippi River, ~5 km downstream of Natchez, MS, which persisted for 286 days (above official flood stage). (A) Exposure of cross-bedding (in trench) associated with dune (bed load) migration (10 cm rod increments). (B) Sand sheet deposition with disruption due to tree. Note tree root exposure caused by scour, and absence of sedimentation behind tree due to velocity shelter. High water mark on tree. (Author photos, October 2019.)

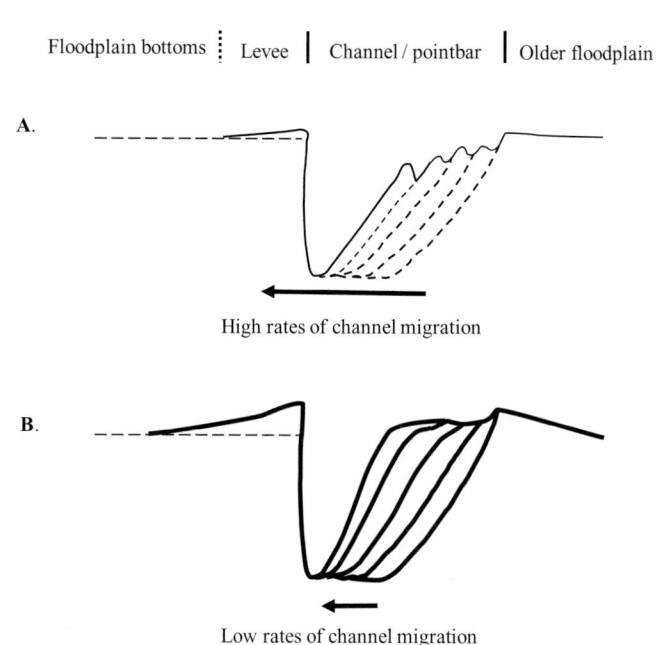

Figure 3.27. Relation between variable lateral migration rates (A. High rates, B. Low rates) with the development of natural levees and meander scroll topography. (After Hudson, 2004.)

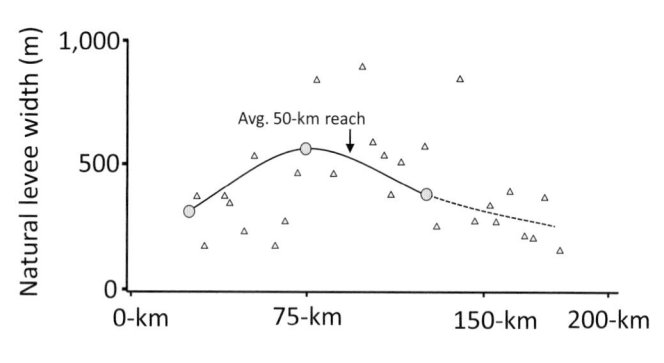

Figure 3.28. Downstream pattern in natural levee width along the lower Rio Pánuco, a large gulf coastal plain river in eastern Mexico. The pattern displays nonlinearity in levee width associated with a distinctive fluvial hinge-line caused by reduction in gradient, flow variability, and particle size. (After Hudson and Heitmuller, 2003.)

suspended sediment from high energy to low energy environments (Fisk, 1944; Saucier, 1994).

Whereas the crevasse splay is depositional, the main crevasse channel is erosional. Beyond the natural levee the crevasse channel network is classically dendritic in pattern because of secondary crevasse formation and bifurcation. Disruptions to dendritic patterns often occur because of older floodplain topography or human structures. Individual crevasse channels also include minor natural levees, which can be subaqueous where crevasse channels form lacustrine deltas (Cazanacli and Smith, 1997; Lewin and Ashworth, 2014b). Crevasse channels are oriented tangent to the channel cutbank, or somewhat down valley because of the flood wave influence. Channel breach and crevasse formation is especially initiated with high-stage and long-duration flooding, which provides sufficient energy to breach the channel bank and incise into flood basin deposits. Subsequent discharge pulses can reactivate the crevasse below flood stage so that it functions for

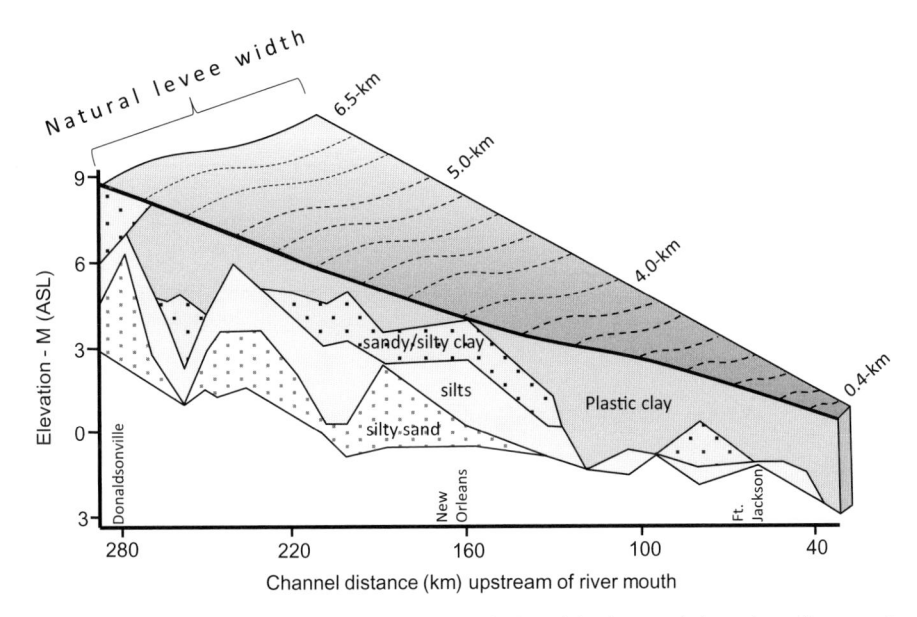

Figure 3.29. Longitudinal profile of natural levee width, thickness, and vertical particle size trend along the main-stem channel of the Mississippi delta, between Donaldsonville, Louisiana to 38 km upstream of the river mouth (Head of Passes, LA). The profile is downstream of the fluvial hinge-line for the lower Mississippi River. Data from twenty-four borings made on or adjacent to natural levee crest. (From Kolb, 1963.)

decades, resulting in continued coarse sediment transport into flood basins (Nienhuis et al., 2018). Providing sufficient boundary conditions, crevasse channels can continue to enlarge and ultimately result in an avulsion, which results in main-stem channel abandonment because of formation of a new channel belt (see Section 7.4).

Crevasse splay dimensions are dependent upon several factors, including sediment load, grain size, frequency of hydrologic activity, flood duration, as well as the geometry and hydrology of the flood basin (Millard et al., 2017; Nienhuis et al., 2018). The slope (energy gradient) between the channel bank and flood basin is an important control on the depth of crevasse scour and distance of sediment transport.

The splay deposit includes sediment reworked (eroded) from the flank of the natural levee (Shen et al., 2015). Coarse sediment is transported by saltation and is deposited within dunes and ripples, with finer suspended sediment deposited along the margins of fan deposits (Farrell, 1987). The sedimentary structure and pedogenic characteristics of splay deposits have some similarity to natural levee deposits (Table 3.6), although displaying higher energy bedding and a stronger presence of bioturbation because of being deposited over backswamp (Figure 3.25, Table 3.7). The thickness of crevasse splay varies spatially with distance from the main channel, where it may range from ~10 cm to 100 cm. At the distal margins of the splay, thicknesses may range from 1 cm to 10 cm. Older crevasse deposits are buried by fine-grained flood basin deposits and are easily recognizable by abrupt change in particle size and color within vertical sedimentary profiles (Farrell, 1987).

3.5.2.3 FINE-GRAINED FLOOD SEDIMENTATION AND LOW-ENERGY ENVIRONMENTS

Fine-grained overbank sedimentation along large lowland rivers primarily occurs within flood basins or old channels. The depositional environments include backswamp and floodplain water bodies, such as sloughs and paleochannels (Farrell, 1987; Bridge, 2003). Sediment pathways into low-energy floodplain environments include diffuse overbank clayey spreads as well as funneling of coarser flood sediments by sloughs, swales, and crevasse channels (Figure 3.25). Sedimentary deposits within quiescent backswamp basins include massive thick near-featureless clay, infrequently interrupted by large flood events resulting in thin (~1 mm) "stringers" of coarser sediments (~silts) (Table 3.6). By comparison, the sedimentary deposits of abandoned channels are less homogeneous because of the variety of connectivity pathways and proximity to higher-energy (river) conditions, which can result in alternating layers of sands, silts, and clays (Krinitzsky, 1970; Toonen et al., 2012, 2016).

3.5.2.4 FLOOD BASIN AND BACKSWAMP DEPOSITS

Beyond the limits of coarse sediment delivery, overbank flooding over millennia results in the thick accumulation of fine-grained (clay/fine-silt) flood deposits in low-energy slack water environments, such as backswamp. Annual sedimentation rates are low, perhaps <1 mm/yr, and are initially associated with thin-bedded parallel laminations (Table 3.6).

The slow accumulation of cohesive deposits within backswamp environments is also associated with high amounts of

Table 3.7 *Classification schema for spillage sedimentation forms along large lowland river floodplains, including the relict floodplain and valley topography (see Figure 3.25 for visualization of the sedimentation web)*

Spillage types	Subtypes	Characterization	Sources
MS: Mainstem sediments	Natural levees (MSa)	Floodwater sheet sedimentation (not channelized)	Kesel et al., 1974; Smith, 1986; Saucier, 1994; Brierley et al., 1997; Cazanacli and Smith, 1997; Ferguson and Brierley, 1999; Hudson and Colditz, 2003; Hudson and Heitmuller, 2003; Aalto et al., 2008; Adams et al., 2004; Klasz et al., 2014; Park and Latrubesse, 2015; Heitmuller et al., 2017; Leigh, 2018
	Banktop splays (MSb)	Coarse sediment deposited within bed load sheets	Coleman, 1969; Alexander et al., 1999; van Dijk et al., 2013; Shen et al., 2015; Leigh, 2018
	Channel bars and islands (MSc)	Active depositional channel features. Larger vegetated surface includes overbank deposition	Kesel et al., 1992; Gurnell et al., 2002; Benedetti, 2003; Rozo et al., 2014; Mardhiah et al., 2015; Wintenberger et al., 2015; Wang and Xu, 2015; Leli et al., 2017; Hudson et al., 2019
	Point bar swales and chutes (MSd)	Arcuate depressions along active point bars, lacustrine at low flow, activated and flushed at > bankfull flows	Hickin and Nanson, 1975; Latrubesse and Franzinelli, 2002; Grenfell et al., 2012, 2014; Rozo et al., 2014; Harrison et al., 2015; Heitmuller et al., 2017
SL: Secondary linear	Accessory channels / yazoo rivers (SLe)	Yazoo channels and moribund channels draining local runoff sources	Dunne et al., 1998; Lewin and Ashworth, 2014a, 2014b; Heitmuller et al., 2017
	Crevasse splay (SLf)	Channel breach, extending beyond the natural levee	Farrell, 1987; Smith et al., 1989; Smith and Pérez-Arlucea, 1994; Bristow et al., 1999; Toonen et al., 2016; Millard et al., 2017
	Crevasse deltas (SLg)	Channelized flow into lake basins, with subaqueous levees	Tye and Coleman, 1989; Latrubesse and Franzinelli, 2002; Rowland et al., 2014
	Internal drainages (SLh)	Smaller flood basin drainage channels, not related to main-stem topography	Day et al., 2009; Trigg et al., 2012; Lewin and Ashworth, 2014b
PF: Prior / relict elements	Cut-off and paleochannel fills (PFi)	May include oxbow lakes from active and abandoned channel courses	Toonen et al., 2012; Dieras et al., 2013; Constantine et al., 2014; Muñoz et al., 2018
	Ponded lake filling (PFj)	Depression lakes between topographic features, basins created by neotectonics, and valley side depressions	Maurice-Bourgoin et al., 2007; Paira and Drago, 2007; Citterio and Piégay, 2009; Park and Latrubesse, 2019
	Diffuse overbank spreads (PFk)	Main sedimentation into backswamp and flood basin lowlands, off of the natural levee	Pizzuto, 1987; Magilligan, 1992; Allison et al., 1998; Asselman and Middelkoop, 1998; Aalto et al., 2003, 2008; Benedetti, 2003; Törnqvist and Bridge, 2003; Heitmuller et al., 2017; Park and Latrubesse, 2019

From Lewin et al. (2016) with author modification.

organic material, and core samples often include considerable plant macrofossils. Ferruginous and calcareous nodules are common and abundant, with some oxidation and mottling according to water table fluctuation (Table 3.6). Spatially, backswamp thickness varies mainly according to valley size (depth and width) (Magilligan, 1992; Saucier, 1994). In coastal plain rivers, backswamp often exceeds several meters up-valley and may be tens of meters deep down-valley, toward the coast. Along the lower Mississippi valley, backswamp deposits are ~10 m in the northern portions of the alluvial valley to about ~35 m in the southern portions of the valley (Fisk, 1952; Krinitzsky and Wire, 1964; Saucier, 1994).

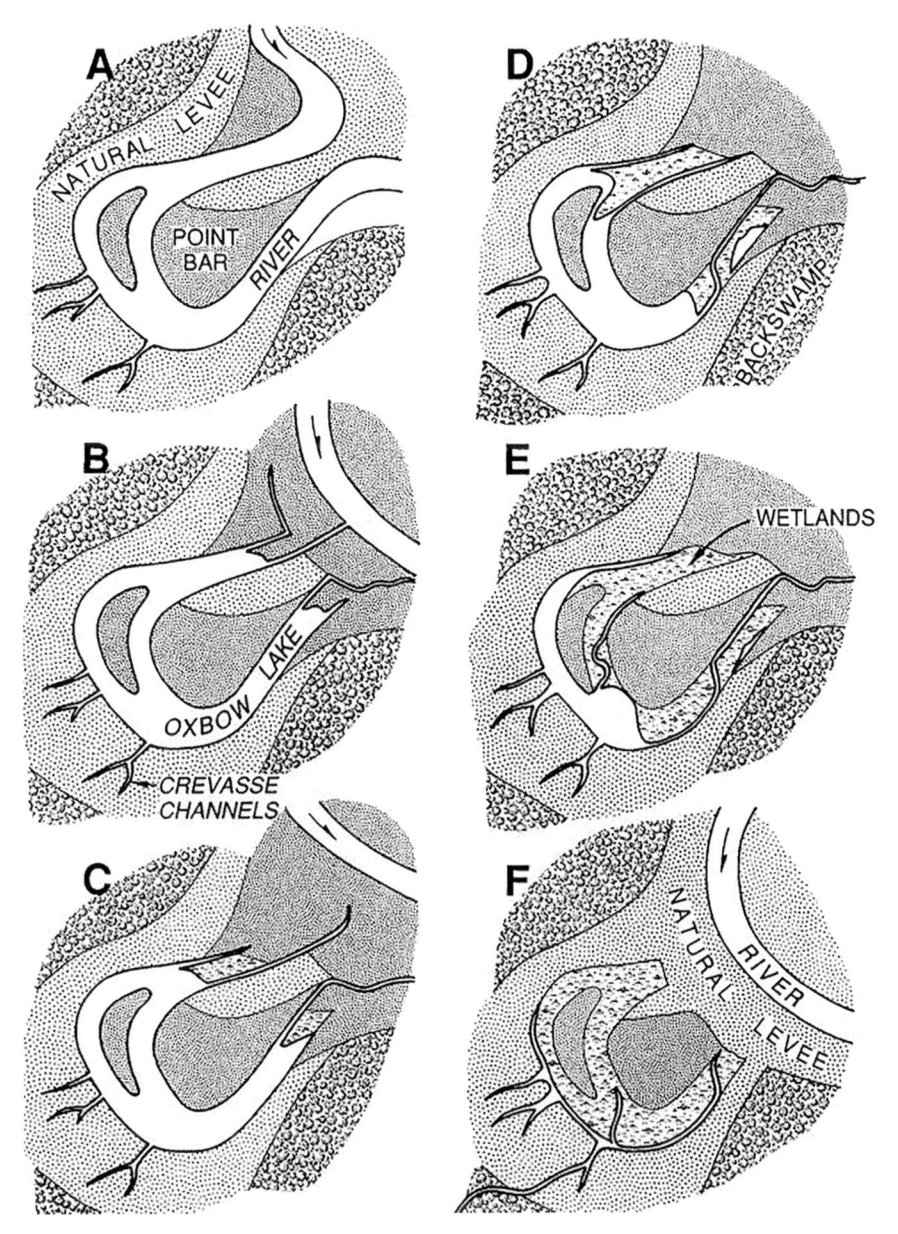

Figure 3.30. Evolution of oxbow lake environments associated with sedimentary infilling of a meander neck cut-off. Different stages include: A. pre cut-off, B. cut-off and initial coarse channel plug deposits, C. main-stem migration and fine-grained sedimentation, D. pronounced batture (tie) channel with lacustrine deltaic wetlands, E. further fine-grained sedimentation and wetland extension, F. main-stem migration towards oxbow with natural levee burial, fine-grained sedimentation and wetland extension across oxbow lake.

3.5.2.5 INFILLED CHANNEL DEPOSITS

As regards channel management and engineering, an important sedimentologic component of lowland river valleys is infilled channel deposits (Fisk, 1947). There are three types of channel cut-offs, including (1) meander neck cut-offs, (2) chute cut-offs, and (3) elongated cut-offs. The signature cut-off associated with active meander belts form oxbow lakes (Figure 3.30), arcuate water bodies created by either a meander neck cut-off or a chute cut-off. The former is primarily associated with erosion by lateral migration, whereas the latter is associated more with erosion by flood scour. Additionally, elongated cut-offs created by local avulsions within the active channel belt result in floodplain water bodies subject to infilling (Gagliano and Howard, 1984; Toonen et al., 2012). Recent cut-offs within the active channel belt retain their morphologic dimensions and serve as floodplain lakes and wetlands (Figure 3.30), water bodies important to contemporary riparian ecosystem dynamics and in particular aquatic habitat for fish (Winemiller et al., 2000).

There are two key types of deposits in channel cut-offs; coarse-grained channel plugs and fine-grained lacustrine deposits

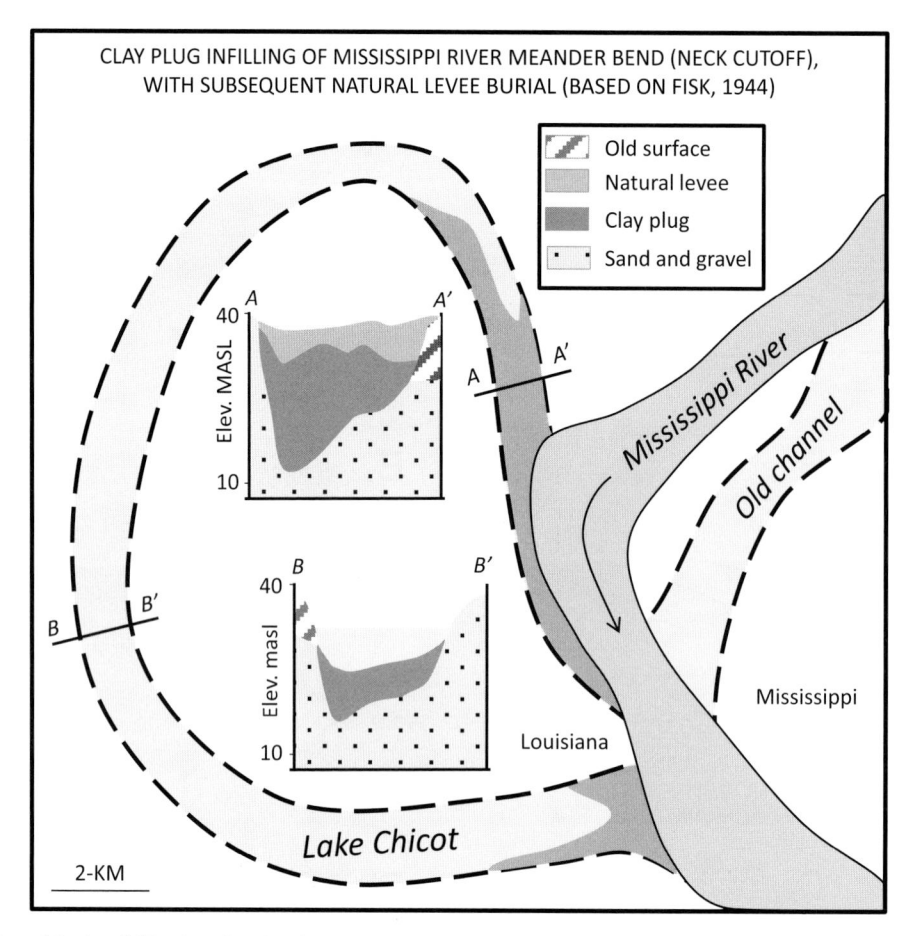

Figure 3.31. Simplied model of variable clay plug development and subsequent natural levee burial in meander neck (oxbow lake) cut-off along the lower Mississippi River. (Based on Fisk, 1944.)

within the interior. Immediately after the cut-off, coarse bed load sediments are deposited at the entrance and exit of the cut-off, resulting in formation of a channel plug several meters in thickness. With continued aggradation, the crest of the channel plug becomes subaerial and the vigorous growth of hygrophytic vegetation further induces deposition of bed load and suspended sediments (Figure 3.31). Because the channel plug forms a topographic barrier between the active river and cut-off, rates of lacustrine sedimentation within the cut-off interior abruptly decline (Toonen et al., 2012). Hereafter, infilling of channel cut-offs varies somewhat between meander, neck, and elongated cut-offs.

The dominance of either clayey lacustrine or sandy bed load deposits within old channels depends on the type of cut-off and particularly the evolution of flow conditions over ensuing centuries (Toonen et al., 2012). The interior of meander neck cut-offs is dominated by massive homogeneous clayey lacustrine deposits, with a sharp demarcation with the coarse-grained channel bed. Such features are likely to eventually form a "clay plug" (Fisk, 1944; Krinitzsky and Wire, 1964). Chute cut-offs and elongated cut-offs, however, are frequently reoccupied by energetic flows after the cut-off, perhaps for decades, which results in sandier deposits within the lake interior (Fisk, 1952; Krinitzsky, 1970, figures 5.7 and 5.8; Saucier, 1994; Toonen et al., 2012). Lacustrine sedimentation rates are influenced by the proximity of the active channel to the cut-off, and whether it has remained stationary or migrated away from the cut-off. Other factors that influence sedimentation include the main-stem sediment and flood regime, and whether the lake basin is connected to the main-stem channel by a batture channel. Importantly, batture (tie) channels result in connectivity and sediment delivery with the channel basin interior for discharge events below flood stage (Shields and Abt, 1989; Rowland et al., 2014).

Infilled channels are prime locations to reconstruct paleo-flood histories along lowland rivers, which may reveal climatic change and various types of human impacts (Knox, 2000; Muñoz et al., 2018; Daniels et al., 2019). Along the Rhine River in the Netherlands, for example, coring and dating of channel infill deposits provided key sedimentologic data to reconstruct the paleo-flood history. The largest paleo-floods in the Rhine delta were estimated to be comfortably below the current dike height design for a flood magnitude of 16,000 m^3/s (Toonen et al.,

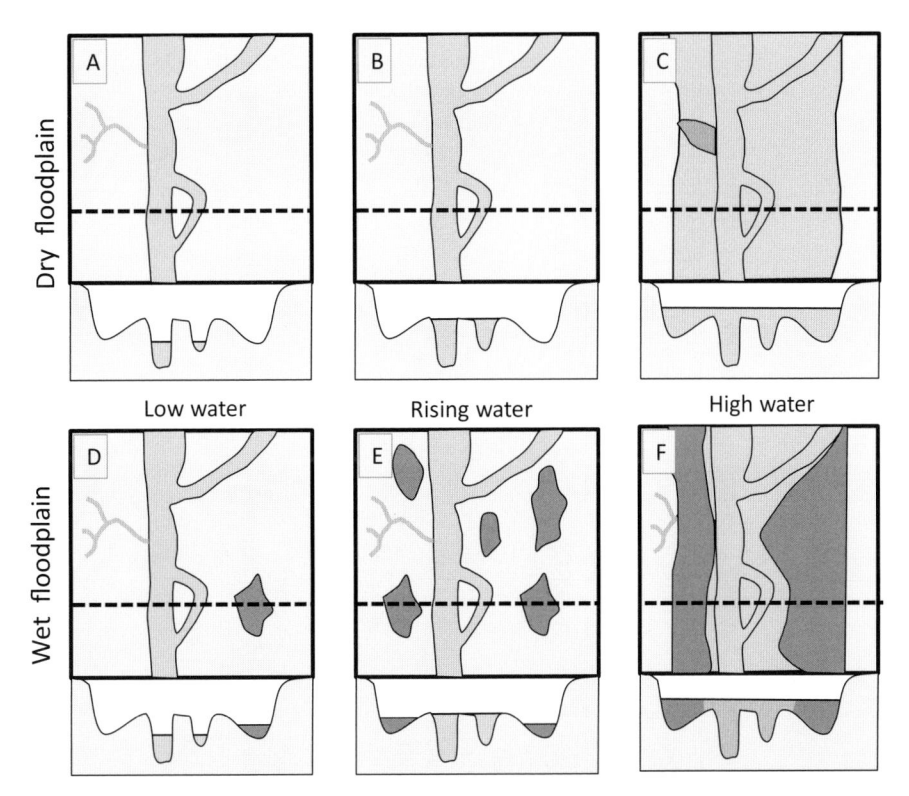

Figure 3.32. Perirheic zone floodplain hydrology, emphasizing two general mechanisms for floodplain inundation and distinction between overbank flooding and water/sediment mixing associated with dry floodplains relative to floodplains already inundated by local sources. The different model stages represent specific combinations of low, rising, and high water in relation to either a dry floodplain (A–C) or a wet floodplain (D–F). The perirheic zone limits lateral sediment dispersal. (Author figure from Mertes, 1997, modified by Lewin and Ashworth, 2014a.)

2016). Along the lower Mississippi, the sedimentologic record from oxbow lakes spanning the past five centuries were instrumental in revealing that the magnitude of the 100-year flood has increased by 20%, with the vast majority of the increase caused by hydraulic engineering (Muñoz et al., 2018).

3.5.2.6 PERIRHEIC ZONE

A unique aspect of large lowland floodplain hydrology that influences sedimentation is the perirheic zone (Mertes, 1997), which occurs at the location where main-stem floodwaters meet floodplain water supplied by local-scale sources (Figure 3.32). The perirheic zone influences floodwater mixing and sediment dispersal across the floodplain. Briefly, flooding of a dry floodplain results in a broad pattern of floodwater mixing and sediment dispersal by turbulent diffusion (i.e., Pizzuto, 1987). The lateral dispersal of flood sediment is limited when main-stem flooding occurs after a floodplain is already inundated by local sources, such as by small tributaries or pluvial events. In such cases, sediment dispersal and deposition is limited to a narrow zone along the riparian edge, and sedimentation occurs by convection.

In the case of floodwaters flowing into an already inundated floodplain (Figure 3.32), contact between sediment-rich overbank floodwaters with sediment-poor flood basin water creates a steep (lateral) concentration gradient (Pizzuto, 1987; Mertes, 1997). In this case, lateral transport of sediment across the floodplain is limited, and as a result the majority of sediment is deposited immediately adjacent to the channel bank. The location and extent of the perirheic zone can be documented along large river floodplains by remote sensing analysis of satellite imagery, which can be used to estimate the suspended sediment concentration of floodwaters. The perirheic zone during the epic Great Flood of 1993 on the Missouri and middle and upper Mississippi Rivers, for example, was identified where floodwaters with high suspended sediment values (>200 mg/L) were immediately adjacent to low suspended sediment values (<60 mg/L) (Mertes, 1997).

3.5.3 Floodplain Pedogenesis

Fluvial sedimentary deposits eventually evolve into floodplain soil, with pedogenic development varying temporally and spatially. Alluvial soils differentially evolve according to floodplain depositional environment, particularly due to differences in sedimentation, biological activity, and floodplain hydroclimatology (Table 3.8). Additionally, transformation of fresh alluvium into soil requires time, as the conversion of organic matter into humus and chemical weathering of

Table 3.8 Soil properties developed along active Mississippi River meander belt depositional environments, depicting a spatial transect (from left to right)

Depositional setting	Channel bank, natural level crest	Natural levee crest	Natural levee crest and backslope	Natural levee back-slope	Natural levee back-slope, backswamp	Natural levee back-slope, backswamp	Backswamp	Backswamp	Backswamp
Typical horizon sequence	A–C	A–C	A–B–C	A–Bg–Cg	A–Bg–Cg	A–Bg–Cg	O–A–Bg–Cg	A–Cg	Oa1–Oa2
Clay content (%)	0–18	0–18	18–35	18–35	35–60	>60	>60	>60	<30
Key characteristic	Minimal soil development	Minimal soil development	Mineral soil, translocation and hums formation	Water logged (gleyed soil)	Water logged (gleyed soil)	Water logged (gleyed soil)	Water logged (gleyed soil)	Water logged (gleyed soil)	Organic soil
Solum thickness range (cm)	10–25	10–25	50–100	50–125	75–150	90–150	60–125	0–25	125+
Soil classification*	Typic Udipsamment	Aeric Fluvaquent	Aeric Fluvaquent	Typic Fluvaquent	Vertic Haplaquept	Vertic Haplaquept	Typic Fluvaquent	Typic Hydraquent	Typic Medisaprist

Active lower Mississippi meander belt (Hmb1; Saucier, 1994). *U.S. NRCS soil series designation; see Autin et al. (1991).
Modified from B. J. Miller in Autin et al. (1991).

minerals minimally require many decades and centuries. Pedogenic characteristics of floodplain soils are important to recognize as they provide insight into floodplain environmental processes, and to more accurately reconstruct fluvial chronologies (Blum et al., 1994; Aslan and Autin, 1998; Leigh, 2008). Pedogenic criteria are also employed to quantify riparian ecosystem services, such as biogeochemical cycling associated with phosphorus, nitrogen, and carbon sequestration (Zehetner et al., 2008; Hoffman et al., 2013). Pedogenic criteria are especially important to map wetland soils (Tiner, 2016), crucial for restoration of riparian lands heavily modified for agriculture and flood control.

A key formative step in floodplain soil development is the reworking of original sedimentary structure by physical weathering processes (Farrell, 1987). The internal sedimentary structure (e.g., cross-bedding, laminations) of surficial alluvial deposits can be reworked within several years – if not sooner – with vigorous bioturbation by plant root growth and burrowing organisms (Figure 3.33). Bioturbation results in assimilation of fine-grained and thin-bedded flood sedimentary layers into topsoil within months, especially in warm lush climates. Seasonal freezing and thawing of the floodplain surface, including needle ice, churns and destroys sedimentary structure of recent overbank deposits. Additionally, floodplain agricultural activities, such as cultivation and grazing, disrupt sedimentary layers. Fieldwork to sample flood deposits – particularly fine-grained sediments – therefore, should ideally occur within several months of floodwater recession.

Spatially, soil properties systematically change from natural levee crest to floodplain bottoms due to differences in soil hydrology and vegetation (Autin et al., 1991). Such changes are imposed by the topography and represent a soil catena (Table 3.8). Bioturbation and soil development increases with distance from the active channel. In essence, there is a transition from coarse-grained and well-drained oxidized soils at the natural levee crest to fine-grained and poorly drained soils along the natural levee tail, where it merges into poorly drained flood basins. Extending along a natural levee profile from crest to tail, an idealized sequence of soil horizons may range from A–C to A–B–C to O–A–B–C (Table 3.8). The dominant chemical weathering process changes from oxidation at the natural levee crest to reduction at the more frequently inundated natural levee tail, resulting in redoximorphic and gleyed features becoming more prominent (Tables 3.6, 3.8). Additionally, with distance from the natural levee crest, translocation of clay and humus results in soil horizonation. Soils of natural levee backslope and backswamp are organic rich and the high-water table results in gleyed soil color (Table 3.8).

Temporally, conversion of sedimentary deposits into distinct alluvial soil requires centuries. Over millennia, time becomes an important driver of alluvial soil pedogenesis because of its influence on chemical weathering and in the development of distinct soil horizons by the translocation of clay, organic matter, and nutrients (Aslan and Autin, 1998), which results in distinct alluvial soils within similar depositional environments (Figure 3.34). In the lower Mississippi valley, for example, older complementary alluvial deposits varying in age by ~2,000 years displayed a stronger presence of iron nodules, greater reworking of sedimentary structure, development of cutans (clay film on soil peds), and a stronger presence of illuviation. Adjacent backswamp soil displayed stronger pedogenic mineralization, including iron and calcite nodules, and soil structure with distinctive slickensides.

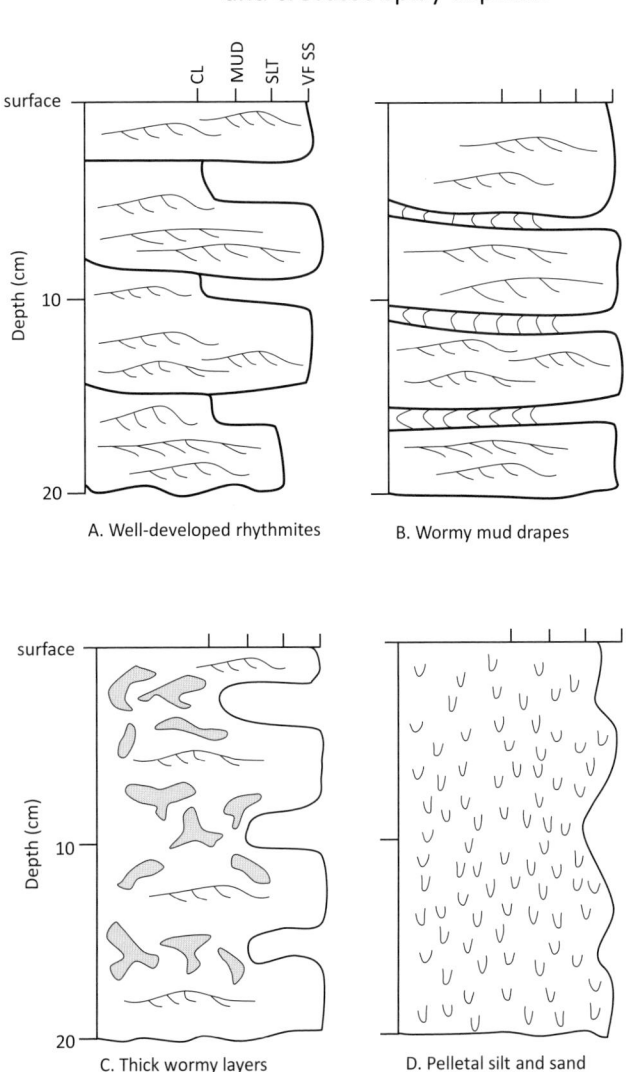

Figure 3.33. Progressive alteration to thin-bedded sedimentary structure (cross-bedding) of surficial natural levee and crevasse deposits by bioturbation. (Based on Farrell, 1987.)

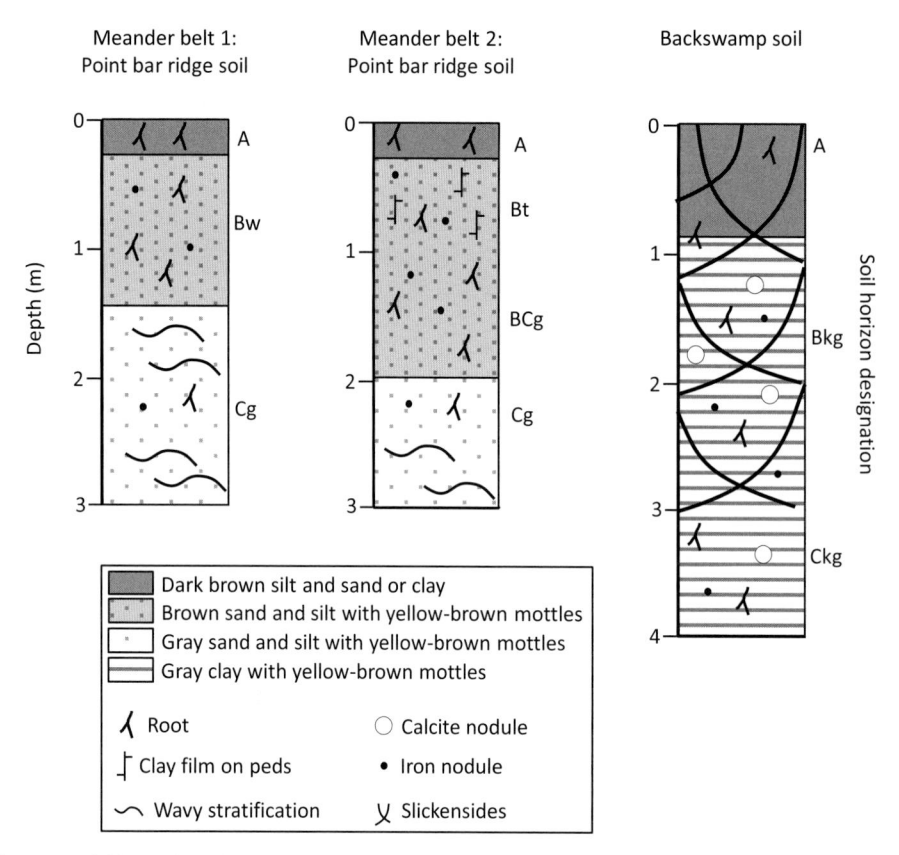

Figure 3.34. Soil development within sandy point bar deposits for two meander belts and adjacent clayey backswamp environments in lower Mississippi valley. (Source: Aslan and Autin, 1998.)

3.6 NEOTECTONIC INFLUENCES ON LOWLAND RIVERS

3.6.1 Faults and Uplift Domes

Neotectonics drives fluvial adjustment over timescales that concern river and flood management (Schumm et al., 2002, pp. 117, 124; Dokka, 2006; Mertes and Dunne, 2007). Recent and ongoing crustal deformations imposed upon historic and modern rivers alter relict fluvial landforms as well as active fluvial processes. Two key structures that drive neotectonics along alluvial rivers are faults and uplift domes. Such structures commonly move from ~1 to 10 mm/yr and can result in differential vertical adjustment on either side of the axis (Figures 3.35 and 3.36), in addition to lateral warping of valley surfaces. Among a variety of responses to neotectonics that influence alluvial rivers includes vertical displacement of channel belts, floodplains and terrace profiles, changes to channel pattern and sinuosity, changes to local flood regime and overbank sedimentation, changes to floodplain water bodies (Figure 7.37), channel incision (entrenchment), channel bank failure, sediment liquefaction and sand blows and, possibly, channel avulsion (Guccione et al., 2000; Schumm et al., 2002; Gupta et al., 2014; Gasparini et al., 2016). Neotectonic influences on the lower Indus River drove alternating changes to channel pattern and sinuosity (Figure 3.37B). Neotectonic influences on the active lower Mississippi meander belt are revealed as changes in slope (gradient) to the longitudinal profile (Figure 3.38).

Neotectonic rates of movement are highly variable, and structures may effectively remain dormant for decades to centuries before undergoing higher rates of displacement (Schumm et al., 2002). Additionally, earthquakes can result in abrupt adjustment of alluvial surfaces of several meters or more during a single event, such as occurred with the 1811–1812 New Madrid earthquakes along the lower Mississippi valley. The event altered floodplain slope profiles and resulted in lateral warping that created Reelfoot Lake (Guccione et al., 2000; Carlson and Guccione, 2010).

3.6.2 Influence of Neotecontics and Sea Level Rise on the Rhine Delta

The transition from lowland floodplain to delta plain is often influenced by both neotectonics and sea level rise. The delta apex often occurs along an active neotectonic hinge-line (Stanley and Hait, 2000; Cohen, 2003; Gouw, 2007), which

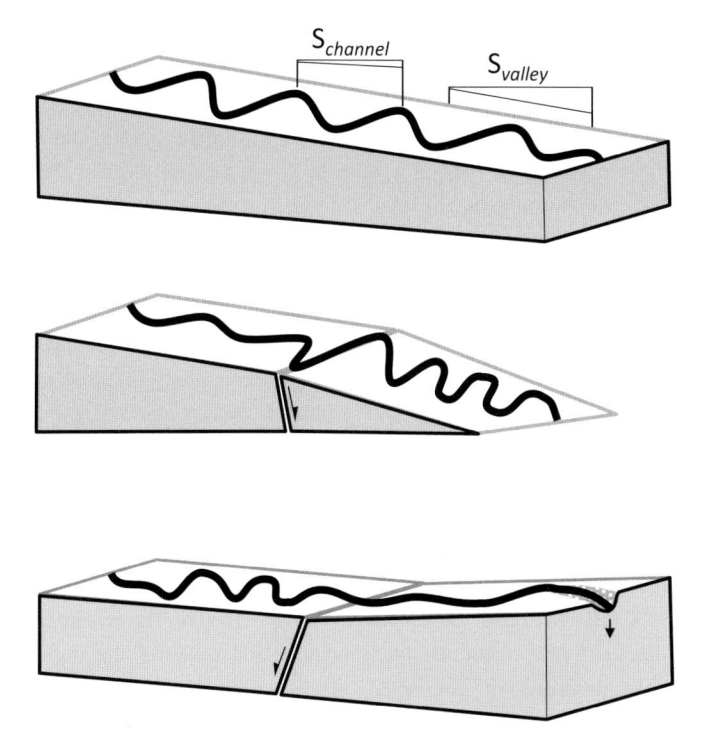

Figure 3.35. Channel sinuosity response to changes in valley slope imposed by neotectonic faulting. (Source: Ouchi, 1985 and Schumm et al., 2002.)

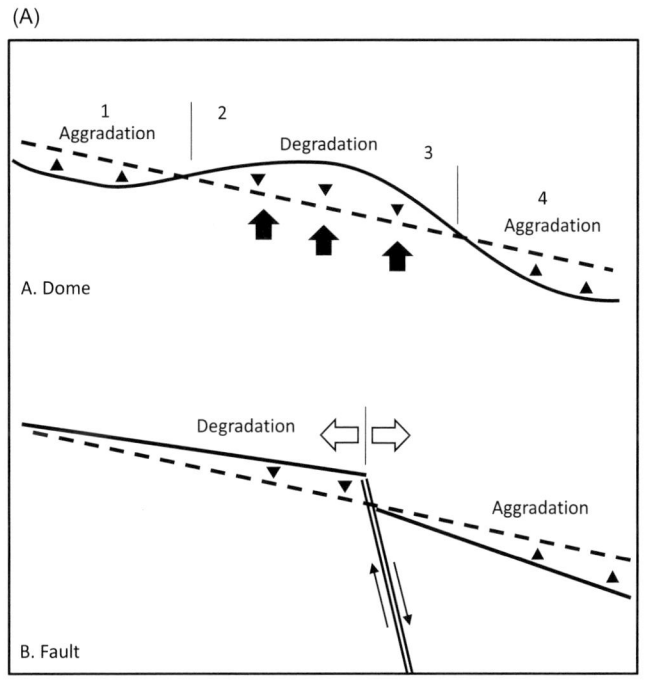

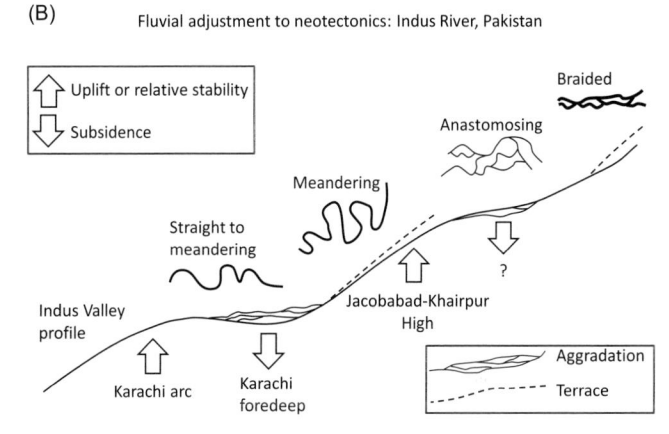

Figure 3.36. Fluvial adjustment to uplift domes and faulting: (A) General model of aggradation and degradation (Holbrook and Schumm, 1999). (B) Channel pattern adjustment of lower Indus River, Pakistan (Harbor et al., 1994).

often coincides with the upstream-most distributary node. The upper delta surface is shouldered by Pleistocene terraces that grade down valley and are buried by Holocene deposits toward the coast. Nevertheless, apart from the channel bifurcation node the actual transition from floodplain to delta plain includes considerable overlap in contemporary fluvial processes, at least within the upper delta. This is especially the case for overbank hydrologic and sedimentary processes related to natural levee, crevasse, and flood basin processes.

The apex of the Rhine delta occurs at about the border of Germany and the Netherlands (Figures 3.40 and 3.41), which is associated with a definitive flexure in the longitudinal profile (slope gradient) (Cohen, 2003). Upstream of the delta apex the floodplain slope is 0.0002 m/m, and thereafter reduces to 0.0001–0.00015 m/m in the upper and central delta, before flattening to nearly ~0 m/m in the tidal-lagoon portions of the coastal delta (Cohen et al., 2002; Stouthamer et al., 2011).

Important structural controls on channel and flood basin dynamics occur within the lower Rhine embayment (Figure 3.41), which includes several fault zones associated with large SE to NW aligned horst and graben structures (van Balen et al., 2005). These structures include the subsiding Venlo Graben and Roer Valley Graben, which are separated by the slightly elevated Peel Horst (Cohen et al., 2002; van Balen et al., 2005). Structural controls also delineate the upper and middle portions of the Rhine delta, with the delineation of the lower, central, and upper delta influenced by the Peel Boundary Fault zone (Figure 3.41).

Late Pleistocene and early Holocene channel belts were displaced from 1 m to 2 m by tectonics, although little deformation has occurred since the mid-Holocene (Stouthamer and Berendsen, 2001; Cohen et al., 2002). This older deformation, however, results in segments of late Holocene channels being aligned along terrace scarps, with older bank material likely influencing historic rates and patterns of channel migration (Cohen, 2003).

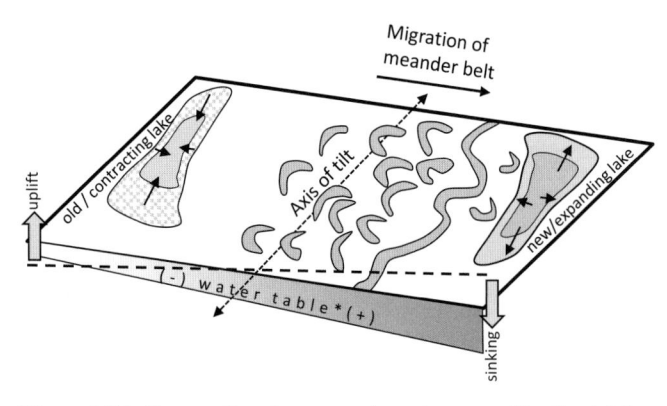

Figure 3.37. Cross-valley changes to channel, water table, floodplain water bodies to lateral warping. (After Adams, 1980, and Charlton, 2008, with author modifications.)

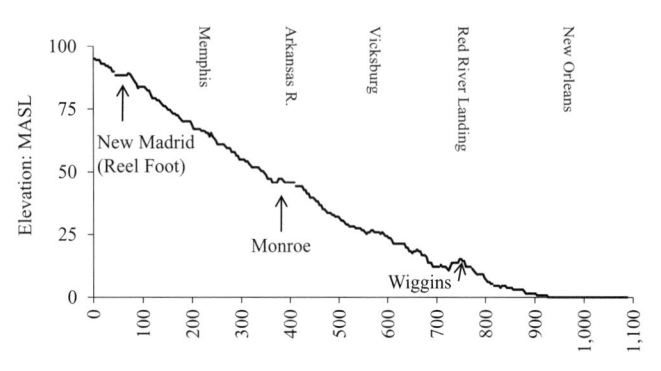

Figure 3.38. Longitudinal profile of lower Mississippi active channel belt (Hmb-1) from Ohio River to Gulf of Mexico (before cut-off program of 1930s). Three distinctive structural controls are revealed, including Reelfoot Fault (New Madrid Seismic Zone), Monroe Uplift Dome, and Wiggins Arch. Arkansas River confluence at about location of Monroe Uplift, whereas Red River and delta apex at about Wiggins Arch. See Figure 3.39 for map of Mississippi alluvial valley and delta (After Hudson et al., 2008.)

A key approach to understanding sea level rise influences on deltaic processes is to analyze sedimentary profiles to discern crevasse and avulsion[3] activity. By analyzing the characteristics and chronology of clastic sedimentary and organic (peat) deposits, the overbank flood regime and paleo-groundwater levels were reconstructed for the Rhine delta (Berendsen and Stouthamer, 2001; Cohen, 2003; Stouthamer et al., 2011; van Asselen et al., 2017) (Figure 3.42). Such analyses revealed that, following a period of rapid sea level rise and backfilling (8500 BP to 5500 BP), avulsion nodes shifted eastward, upstream (Figure 3.42). Between 5500 BP and 3000 BP the great majority of channel avulsions in the Rhine delta occurred within the central delta at sites along the Peel Horst Fault Zone, as well as near the delta apex along the hinge-line. The location of the highest number of avulsions occurred (at Wijk bij Duurstede) where the Rhine River was topographically trapped between the Peel Horst Fault Zone and the Pleistocene ice-pushed ridge (Berendsen and Stouthamer, 2001; Cohen, 2003). Upstream of the Peel Boundary Fault Zone avulsion activity resulted in formation of smaller secondary channels, rather than an entire main-stem avulsion (Figure 3.42). Because smaller channels were unable to cross the fault zone topography, they eventually rejoined the main-stem river upstream of the fault zone (Stouthamer and Berendsen, 2001; Cohen et al., 2002; Stouthamer et al., 2011). This changed between about 3000 BP to 1000 BP when avulsion activity significantly increased and channel switching occurred all over the delta plain because of increased discharge and sedimentation (Figure 3.42). After about 1000 BP, the only avulsions that occurred were triggered by human influences (Stouthamer et al., 2011, figure 5).

3.7 DELTAIC GEOMORPHOLOGY

3.7.1 Primary Controls

Deltas consist of a network of distributaries in different stages of evolution. The delta plain framework is constructed of multiple delta lobes, individually prograding, subsiding, or eroding at variable rates and over different timescales (Coleman and Wright, 1971; Penland et al., 1988; Panin, 2015). This ultimately results in a mosaic of fluvial–deltaic and coastal–deltaic environments, with some deltas simultaneously being dominated by fluvial, lacustrine, or marine processes that support a diverse range of aquatic and terrestrial ecosystems (Suanez and Provensal, 1996). The Danube River delta (2,950 km^2), for example, is primarily dominated by fluvial controls at its main active distributary while less active distributaries are simultaneously dominated by wave processes (Panin, 1999, 2015).

In general, fluvial processes and freshwater conditions are predominant in the upper delta plain, which have large flood basins, lakes, and forested swamps with thick peat deposits. A transition from freshwater to brackish to saline conditions occurs in the lower delta plain, toward the coast. Between delta lobes, interdistributary bays undergo tidal flux and are subject to salt water intrusion, with the predominant environments consisting of lagoons and mudflats. Vegetation predominantly consists of salt marsh and/or mangroves.

---

[3] Avulsion is the switching of a river channel belt, usually driven by flooding and crevasse development (Berendsen and Stouthamer, 2001; Slingerland and Smith, 2004). See Section 7.4 for further review.

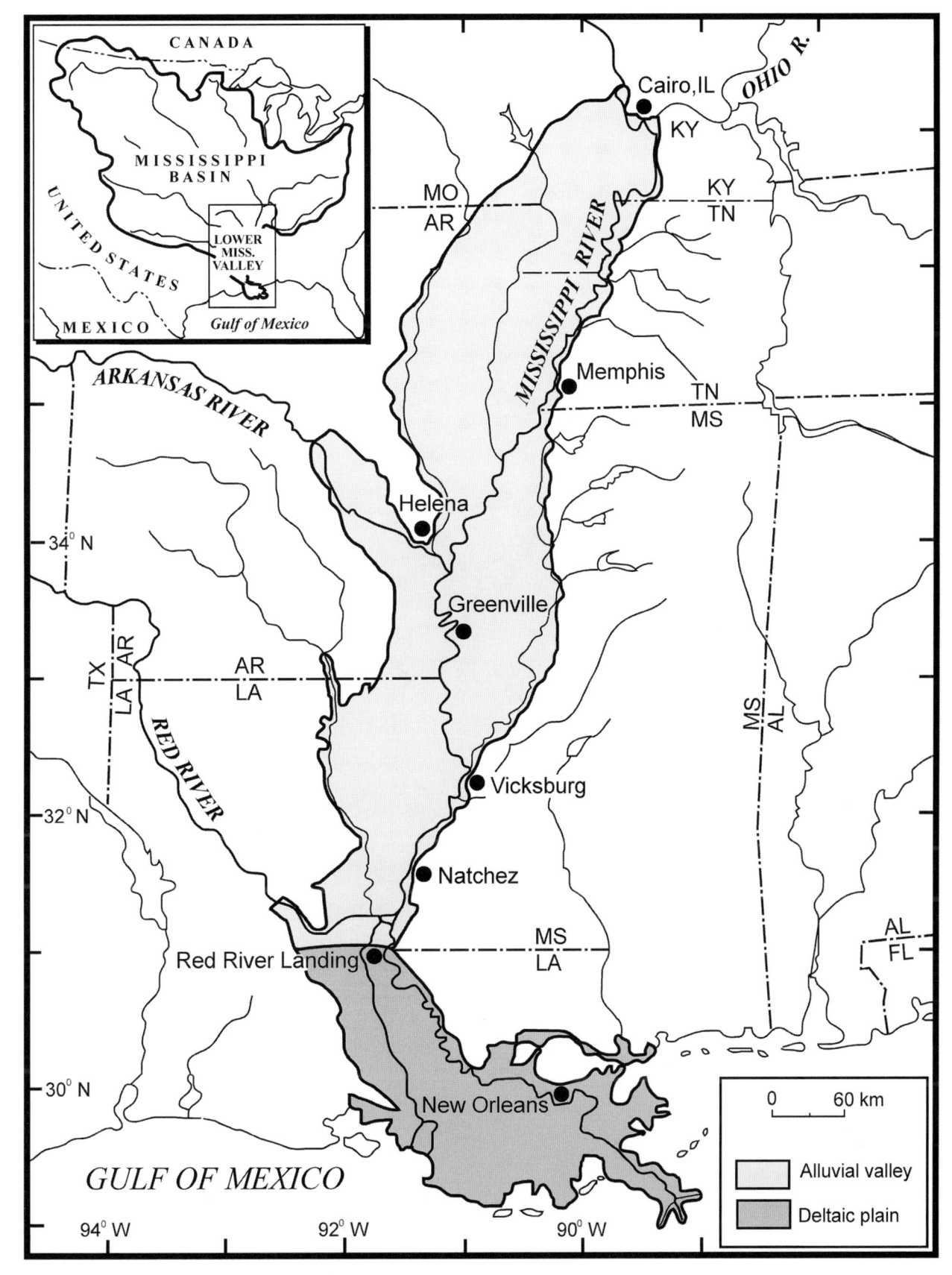

Figure 3.39. Lower Mississippi alluvial valley and delta plain. (Author figure.)

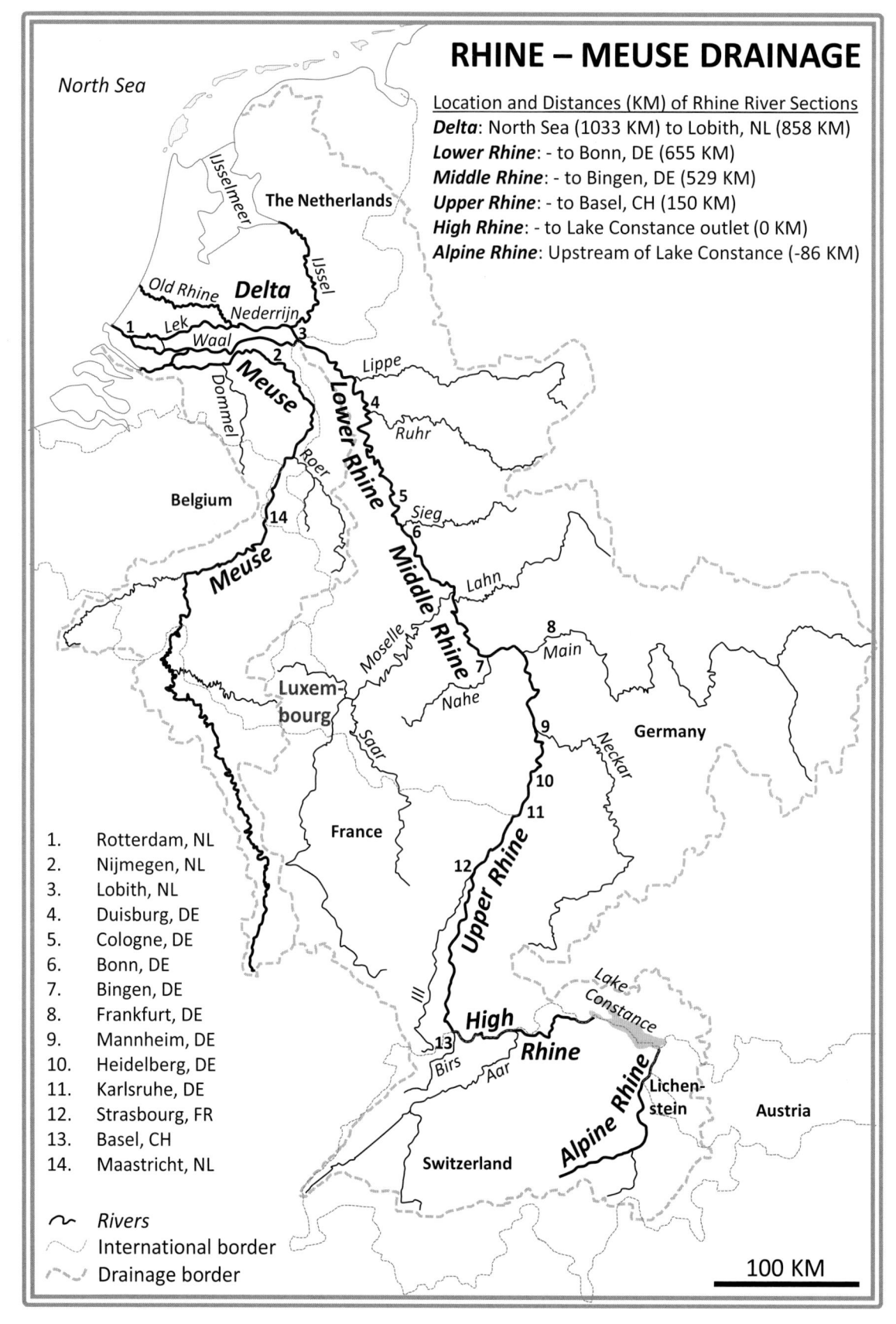

Figure 3.40. Rhine–Meuse drainage basin, including specific sections and locations noted in text. (Author figure.)

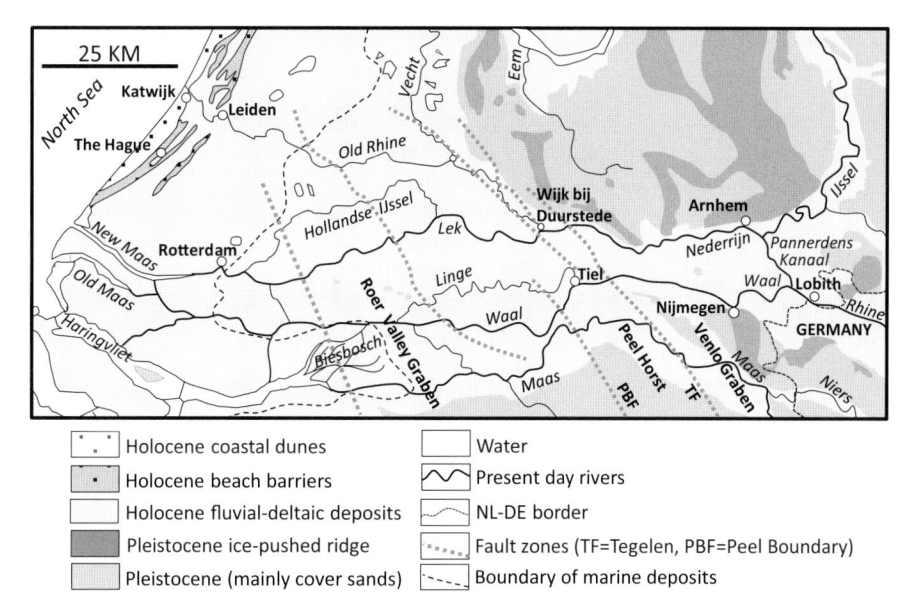

Figure 3.41. Quaternary geomorphology of Rhine delta region in vicinity of major rivers. (Source: Cohen et al., 2002 and Stouthamer et al., 2011.)

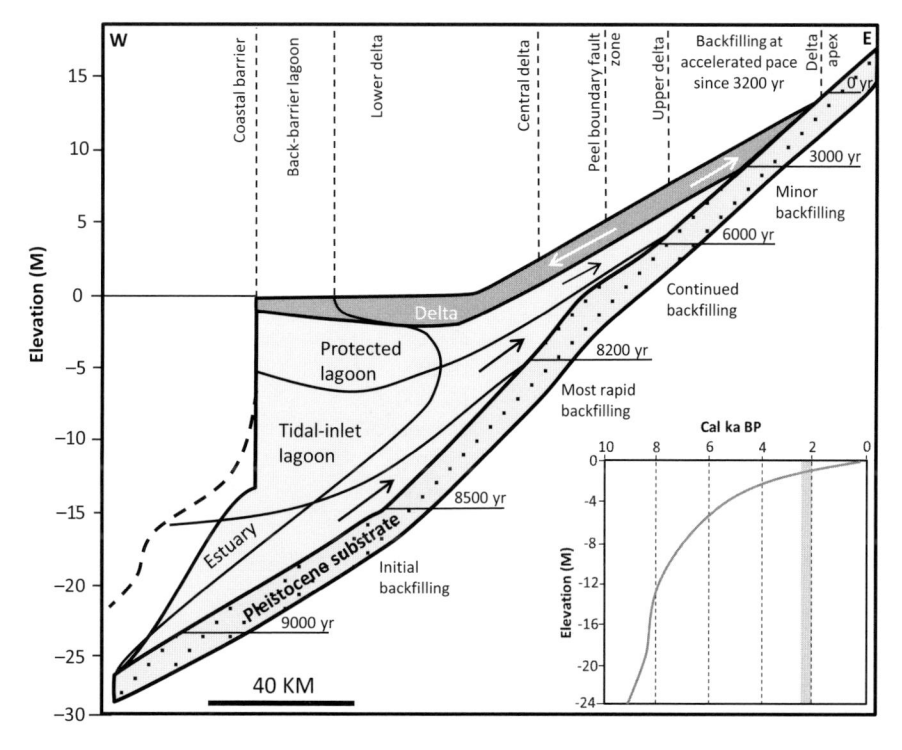

Figure 3.42. Influence of sea level rise on Holocene backfilling of Rhine delta, with timing and development of major depositional environments as regards Holocene sea level rise (inset). (From Stouthamer et al., 2011.)

The primary control on delta size is upstream drainage area (Coleman and Wright, 1971; Galloway, 1975), as the largest river basins generally have the largest deltas (Table 3.9 and Figure 3.43). The Amazon River has the largest delta (467,078 km^2).[4] Other important natural controls on delta size are streamflow and sediment load driven by upstream climate, geology, and drainage basin relief. Earth's larger deltas are

---

[4] Note on Figure 3.43: The extremely large size (area) for the Amazon refers largely to the subaqueous delta, which extends to the shelf edge. With exception of the Pororoca (extensive tidal bore), little marine water enters the Amazon River mouth and it is generally not considered estuarine. A large proportion (~20%) of the Amazon sediment flux is resuspended and transported northeasterly along the Atlantic coast of South America. See Nittrouer et al. (1986; Dunne et al., 1998; Meade, 2007).

Table 3.9 *Drainage area and delta area for a range of larger rivers*

River basin (continent)	Drainage area (km^2)	Delta area (km^2)*
Amazon (SA)	6,915,000	467,078
Nile (AF)	3,349,000	12,512
Mississippi (NA)	3,230,000	28,568
Ob (AS)	2,990,000	5,593[1]
Parana (SA)	2,966,900	17,500[2]
Yenisey (AS)	2,580,000	5,485[1]
Lena (AS)	2,490,000	43,563
Niger (AF)	2,273,946	19,000[3]
Yangtze (AS)	1,940,000	66,669
Mackenzie (NA)	1,787,000	7,400[3]
Ganga–Brahmaputra (AS)	1,712,700	105,641
Volga (EU)	1,400,000	27,224
Zambezi (AF)	1,331,000	12,000[4]
Indus (AS)	1,165,000	29,524
Niger (AF)	1,112,000	19,135
Murray–Darling (AU)	1,072,000	942[1]
Orinoco (SA)	989,000	20,642
Yukon (NA)	850,000	129,000[5]
Danube (EU)	816,000	2,950[3]
Mekong (AS)	811,000	93,781
Tigris–Euphrates (AS)	792,000	18,497
Huanghe (AS)	752,443	36,272
Colorado (NA)	640,000	7,700[6]
Pearl (AS)	425,000	9,100[7]
Irrawaddy (AS)	411,000	20,571
Senegal (AF)	337,000	4,200[3]
Godavari (AS)	295,000	6,500[3]
Magdalena (SA)	257,438	1,800[3]
Rhine–Meuse (EU)	185,000	3,100[8]
Red (AS)	160,000	11,908
Chao Phraya (AS)	159,000	11,329
Burdekin (AU)	130,100	2,100[3]
Sacramento–San Juaquin (NA)	109,000	2,978[9]
Grijalva–Usumacinta (NA)	108,100	17,028
Rhone (EU)	95,500	1,740[10]
Ebro (EU)	85,550	625[3]
Mahakam (AS)	77,100	1,800[11]
Po (EU)	71,700	13,398
Colville (NA)	69,000	1,750[3]
Tana (AF)	65,000	3,800[3]
Ord (AU)	46,000	3,950[3]

*Delta area (km^2) from Coleman and Wright (1975) unless otherwise indicated. Other delta area sources: 1 = subaerial delta measured with Google Earth Pro imagery by author, 2 = Kandus and Quintana (2018), 3 = Coleman and Roberts (1989), 4 = Beilfuss (2018), 5 = Thorsteinson et al. (1989), 6 = Coleman et al. (1986), 7 = Wu et al. (2018), 8 = Middelkoop et al. (2010), 9 = Ingebritsen et al. (2000), 10 = Provansal (2004), 11 = Gastaldo et al. (1995).

located along coastal margins with broad continental shelves. This includes the large mega-deltas of Southeast Asia, which, although being located within tectonically active zones, have continental shelves of sufficient width for the storage of considerable deltaic deposits (Goodbred and Saito, 2012). The enormous deltas of the Indus and Ganges–Brahmaputra Rivers, for example, are formed within structural troughs adjacent to major transform fault systems (Grall et al., 2018).

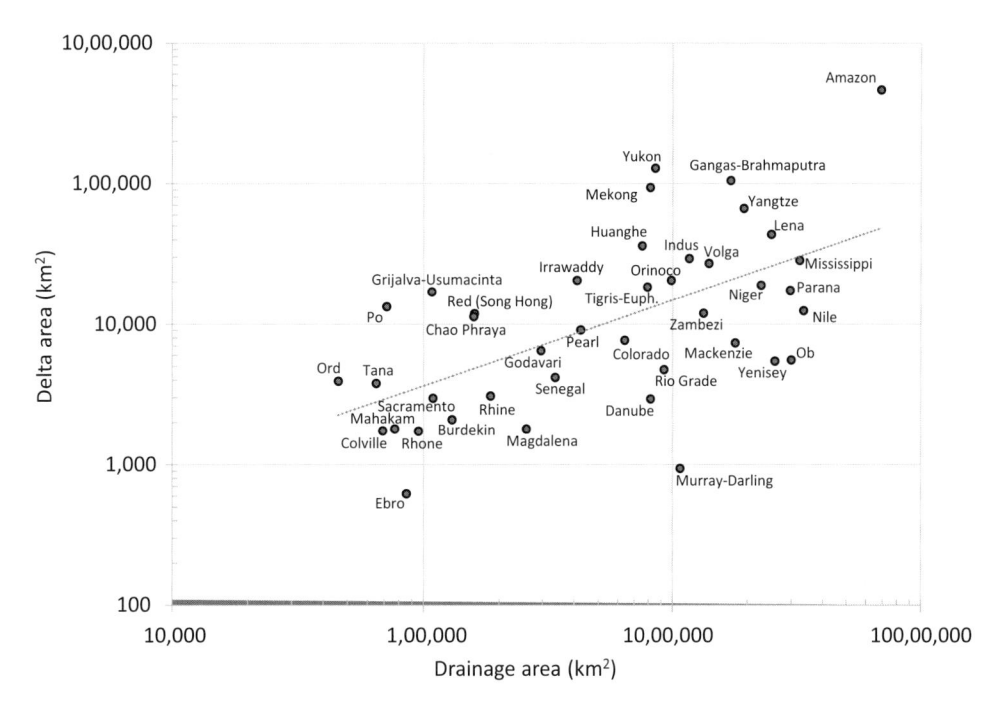

Figure 3.43. Delta area in relation to drainage area for major rivers. (Data: Coleman and Wright, 1975; Coleman and Roberts, 1989; and others. Table 3.9 for sources.)

Deltas are not located at the terminus of all drainage basins. Rivers flowing into narrow continental shelf along convergent plate boundaries often do not have deltas. This includes the Columbia River (North America), which empties into a tectonically active narrow coastal zone bordered by a deep trench. The Congo River, which is bedrock controlled in the lowermost reaches, is the largest river without a delta.

Additional influences on deltas include climatic, tectonic, and human factors (Table 3.10). The geometry (depth, shape, area) of the coastal receiving basin is an important influence on delta pattern because of the control on the accommodation space available to store deltaic deposits and influences on near-shore coastal dynamics (Cohen et al., 2002). The receiving basin is a cast infilled with deltaic deposits. The geometry of the receiving basin is influenced by arrangement of older delta lobes, neotectonics, and subsidence, which can be influenced by sediment loading (Roberts, 1997; Törnqvist et al., 2008). Shallow receiving basins result in broad deltaic sediment dispersal. Narrow receiving basins result in deltas prograding further out to sea, whereas deep receiving basins result in smaller subaerial deltas.

Human activities are also an important influence on delta morphology (Table 3.10). Upstream land degradation, particularly deforestation and intensive agricultural activities, influences delta evolution by increasing sediment loads (Middelkoop et al., 2010; Nienhuis et al., 2020). The late-Holocene and historic evolution of the Huanghe, Danube, and Ebro deltas, for example, are both noted for being influenced by high upstream sediment loads generated by intensive anthropogenic activities, which accelerated delta progradation and extension of the distributary network (Fan et al., 2006; Somoza and Rodríguez-Santalla, 2014; Zheng et al., 2017; Nienhuis et al., 2020). Indeed, despite a great reduction in global sediment loads due to upstream dams, especially for large rivers (Section 4.3.2), land cover change across basin headwaters is driving delta land gain for smaller rivers at a rate of 54 ± 12 square kilometers annually, which is not expected to continue due to sea level rise (Nienhuis et al., 2020).

The tremendous global reduction in sediment flux to the coast is an indication of the problems facing the world's large deltas. The reduction in sediment supply to deltas principally occurs because of sediment being trapped by dams, improved land management and afforestation in the upper basin, and construction of flood control dikes. Additionally, lowland agriculture and petroleum activities alter the delta surficial and groundwater hydrology, driving delta subsidence. These changes reduce sedimentation and accelerate delta drowning as the delta enters a destructive phase, resulting in wetland erosion and shoreline retreat (Stanley and Warne, 1998; Syvitski et al., 2005, 2009; Nienhuis et al., 2020).

3.7.2 Delta Form and Process: Fluvial, Marine, and Tidal Controls

While delta size (area) is scale dependent, the pattern and morphology of deltaic deposits is largely determined by the relative dominance of fluvial, tidal, or wave processes (Figure 3.44 and Table 3.11), which may then be further

Table 3.10 *Factors influencing the morphology and sedimentary characteristics of deltas*

River regime	Flood stage	Sediment load	Quantity of suspended load and bed load (stream capacity) increases during flood
		Particle size	Particle size of suspended load and bed load (stream competence) increases during flood
	Low river stage	Sediment load	Stream capacity diminishes during low river stage
		Particle size	Stream competence diminishes during low river stage
Coastal processes	Wave energy		High wave energy with resulting turbulence and currents erode, rework, and winnow deltaic sediments
	Tidal range		High tidal range distributes wave energy across an extended littoral zone and creates tidal currents
	Current strength		Strong littoral currents, generated by waves and tides, transport sediment alongshore, offshore, and inshore
Tectonic factors	Stable area		Rigid basement precludes delta subsidence and forces deltaic plain to build upward as it progrades
	Subsiding area		Subsidence through down-warping coupled with sediment compaction allows delta to construct overlapping sedimentary lobes as it progrades
	Rising area		Uplift of land (or lowering of sea level) causes river distributaries to cut downward and rework their sedimentary deposits
Climatic factors	Wet area	Hot or warm	High temperature and humidity yield dense vegetative cover, which aids in trapping sediment transported by fluvial or tidal currents
		Cool or cold	Seasonal character of vegetative growth is less effective in sediment trapping; cool winter temperature allows seasonal accumulation of plant debris to form delta plain peats
	Dry area	Hot or warm	Sparse vegetative cover plays minor role in sediment trapping and allows significant aeolian processes in deltaic plain
		Cool or cold	Sparse vegetative cover plays a minor role in sediment rapping; winter ice interrupts fluvial processes: seasonal thaws and aeolian processes influence sediment transportation and deposition
Human factors	Upstream: Dams Land use change Downstream: Hydraulic infrastructure Agriculture		Trap sediment, reducing coastal sediment flux Increased runoff and soil erosion (degradation), possible reduction in sediment loads (improved land management, afforestation) Fossil fuel facilities (including canals, pipelines, and related physical changes to delta topography) Water extraction (particularly groundwater)

After Morgan (1970, table 1), with author modification (human factors).

Table 3.11 *Characteristics of deltaic depositional systems*

	Fluvial dominated	Wave dominated	Tide dominated
Geometry	Elongate to lobate	Arcuate	Estuarine to irregular
Channel type	Straight to sinuous distributaries	Meandering distributaries, abrupt change in direction with sand ridges	Flaring straight to sinuous distributaries
Bulk composition	Muddy to mixed	Sandy	Variable
Framework facies	Distributary mouth bar and channel fill sands, delta margin sand sheet	Coastal barrier and beach ridge sands	Estuary fill and tidal sand ridges
Framework orientation	Parallels depositional slope	Parallels depositional strike	Parallels depositional slope

From Galloway (1975).

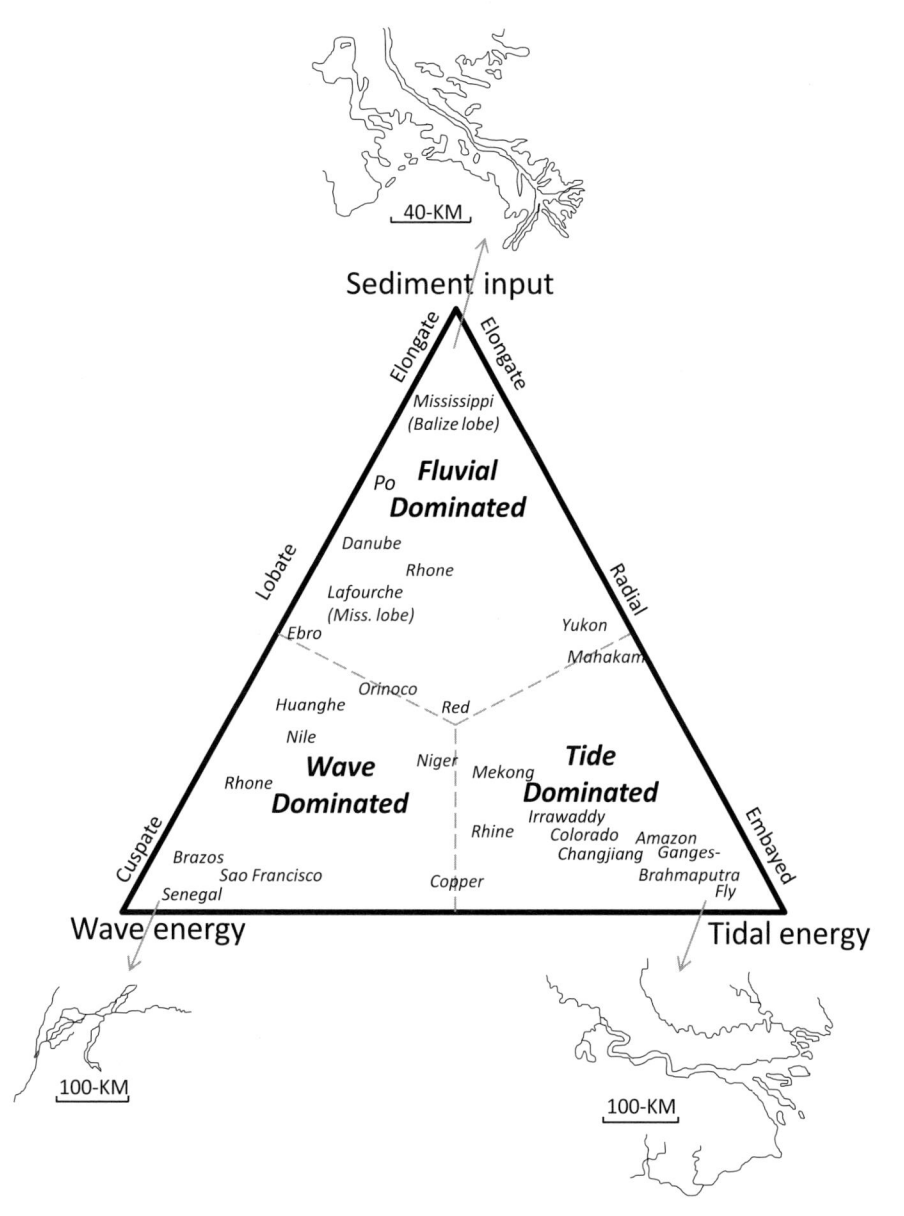

Figure 3.44. Delta classification according to fluvial, wave, or tidal influences. Plotting of deltas based on Galloway (1975), with modification according to Elliott (1986), Nienhuis et al. (2015), and author. (Table 3.12 for further characterization.)

characterized.[5] The relative importance of these three key drivers is illustrated by considering delta position within a ternary plot, referred to as the "delta triangle," which developed from systematic inventories of delta characteristics in the 1960s and 1970s (e.g., Scruton, 1960; Fisher et al., 1969; Coleman and Wright, 1975) and formally put forward by Galloway (1975). In this approach, the ultimate end-member river deltas of the three axes for fluvial, wave, and tidal are the Mississippi, Senegal, and Fly Rivers, respectively (Figure 3.44). The general framework provided by the delta triangle has management implications because of providing insights into the direction of change that could occur were one or more of the drivers changed (i.e., reduction in sediment loads). While modifications have been proposed based upon further analysis of geophysical controls (e.g., Dominquez, 1996; Nienhuis et al., 2015), the general conceptual framework of the delta triangle remains useful for comparisons across a range of delta types.

---

[5] Each of the three drivers (fluvial, wave, tidal) may be further characterized, with fluvial including discharge (average annual, flood magnitude) and sediment (volume, size) regime, wave energy (amplitude, wavelength, period, and angle of approach), as well as tidal amplitude (daily, spring tide) and tide frequency (diurnal or semidiurnal).

3.7.2.1 TIDAL-DOMINATED DELTAS

Tidal-dominated deltas have an embayed morphology, with wide estuarine river mouths that accommodate a large amplitude semidiurnal tidal regime (two highs and the two lows daily). Most large tidal deltas have >2 m average daily tidal amplitudes and >3 m spring tidal amplitudes (Goodbred and Saito, 2012). The high vertical tidal range drives large horizontal exchanges of water and sediment, resulting in the reworking and deposition of coarse sediments into elongated sand bodies, with sand bars and islands oriented perpendicular to the coastline (Galloway, 1975). Large quantities of fine-grained deposits are stored within marsh and mudflats that are hydrologically connected by a dense network of smaller tidal channels.

Tidal-dominated deltas are associated with some of Earth's largest rivers with the highest sediment loads (Goodbred and Saito, 2012; Nienhuis et al., 2020). This includes the Asian "mega-deltas," especially the Yangtze (Changjiang), Indus, Mekong, Ayeyarwady, and Ganges–Brahmaputra. An important reason for tidal deltas being so large is that several drain tectonically active high mountainous regions, including the Himalayan–Tibetan plateau and the Indonesian archipelago, associated with the world's highest sediment yields (Milliman and Meade, 1983). Unfortunately, a number of tidal-dominated deltas are also heavily degraded by human activities, which has made them increasingly vulnerable to sea level rise, climate change, and land use change (Xue, 1993; Xiquing, 1998; Syvitski et al., 2013; Giosan et al., 2014). While the Asian mega-deltas directly support a population of some 200 million and an intensive globally important food production system, several are seriously threatened by a range of human activities.

3.7.2.2 WAVE-DOMINATED DELTAS

Wave-dominated deltas are characterized by littoral drift processes reworking fluvial sands into linear beach ridges, with the wash load (silt/clay) transported offshore – beyond the deltaic zone – by coastal currents (Galloway, 1975; Nienhuis et al., 2015). The sedimentary deposits of wave-dominated deltas, therefore, are coarser grained than either tidal- or fluvial-dominated deltas. The removal of fine-grained sediments from wave-dominated deltas results in steeper shoreface slopes, in comparison with fluvial or tidal deltas. The planform shape of wave-dominated deltas is cuspate, formed by successive accretionary sandy beach ridges along the coastline. Successive beach ridges are separated by elongated swales and shallow water bodies that provide but limited space for storage of fine-grained fluvial sediments. The overall high wave energy relative to fluvial inputs results in active reworking of distributary mouth bar deposits. Thus, wave-dominated deltas tend to have smooth coastlines with few abandoned distributaries emerging at the coast.

There are several key examples of wave-dominated deltas. The São Francisco (608,770 km^2) is one of the great rivers of Brazil and has long been considered the epitome of a large (~800 km^2) wave-dominated delta (Galloway, 1975; Dominquez, 1996). While the São Francisco has a large average annual discharge (2,980 m^3/s), coastal littoral processes efficiently rework mouth bar deposits into an extensive strand plain. The upper delta includes freshwater swamps and linear marshy swales. The river mouth provides downdrift shelter (or groyne effect), providing habitat for extensive mangrove. The Senegal River has among the most intensive wave-dominated deltas of a large river, which is revealed in the abrupt redirection of its main course at a right angle to its predominant orientation. Other large rivers with deltas commonly classified as wave dominated include the Grijalva, Huanghe, and Nile Rivers, with the geometry of the Nile also being (formerly) influenced by high fluvial sediment loads.

The transition from wave-dominated to fluvial-dominated delta can be quantitatively characterized by formally considering indices of fluvial sand supply relative to the ability of waves to remove fluvial sand (Figure 3.45), which relates to wave energy (McBride et al., 2007; Nienhuis et al., 2015). Such indices provide a useful framework for characterizing delta morphology in explicit reference to physical processes. Additionally, such indices provides insights into modeling changes in delta morphology in relation to environmental change. In this context, a reduction in sediment supply, such as by upstream dams and reservoir storage, implies that fluvial-dominated deltas will transition to wave-dominated deltas, manifest by coastal land loss and shoreline retreat. This is well exhibited along the modern Nile delta, which effectively no longer discharges sediment to the coast and has entered a destructional (transgressive) phase, with distributary mouth bar deposits being reworked into linear beach ridges (Stanley and Warne, 1998).

3.7.2.3 FLUVIAL DELTAS

Fluvial-dominated deltas have substantial discharge and sediment loads relative to the marine energy of the receiving basin. Rather than sediment being reworked by marine processes, coarse sediment is deposited in a distributary mouth bar. The high discharge enables finer sediments to be transported beyond the littoral (wave) zone, where it is deposited as a broad subaqueous deltaic deposit referred to as the prodelta base (Figures 3.46 and 3.47). Fluvial deltas, therefore, prograde into their receiving basin perpendicular to the coastline. The ultimate end member of a fluvial-dominated delta is the Mississippi, and the active (Balize) delta lobe in particular (Figure 3.44). The modern Mississippi River has an enormous discharge (18,400 m^3/s) and large sediment load (145 × 10^6 tons/yr), although the sediment load has substantially declined – by more than half – since the early 1950s (Meade and Moody, 2010). An important control on the geometry of the active Mississippi delta lobe is

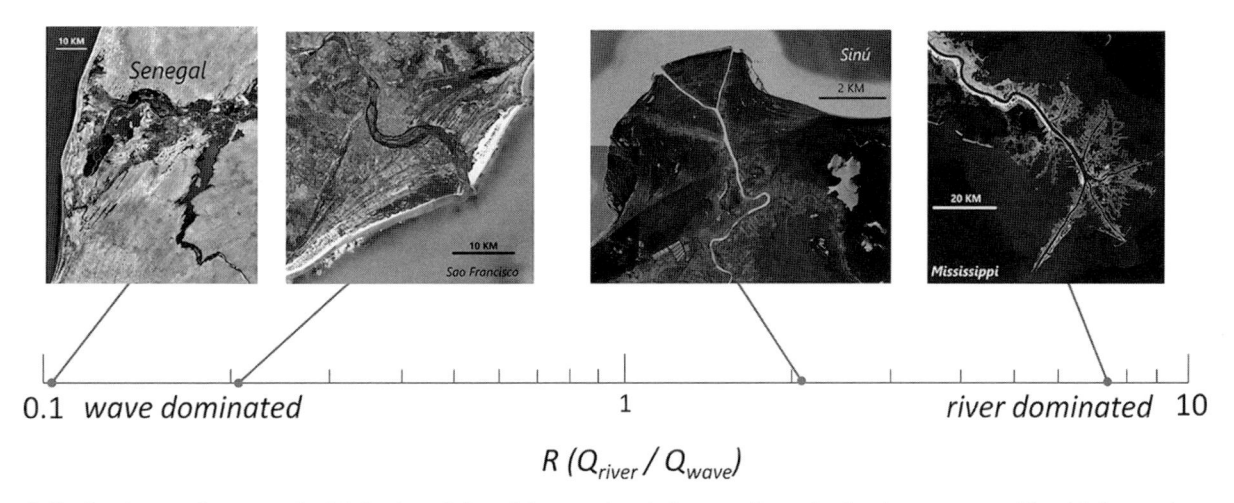

Figure 3.45. Continuum of wave- to fluvial-dominated river delta types in relation to sediment load and wave energy (R), with 0 = maximum wave influence and 10 = maximum fluvial dominance. Deltas (left to right): Senegal, Sao Francisco, Sinú, and Mississippi. $R = Q_{river}/Q_{wave}$; Q_{river} (kg/s) = sand load (kg/s); Q_{wave} (kg/s) = $90 * H_s^{12/5} * T_p^{1/5}$, with H_s = significant wave height (m) and T_p = wave period (s). (After Nienhuis et al., 2015, imagery from Google Earth Pro.)

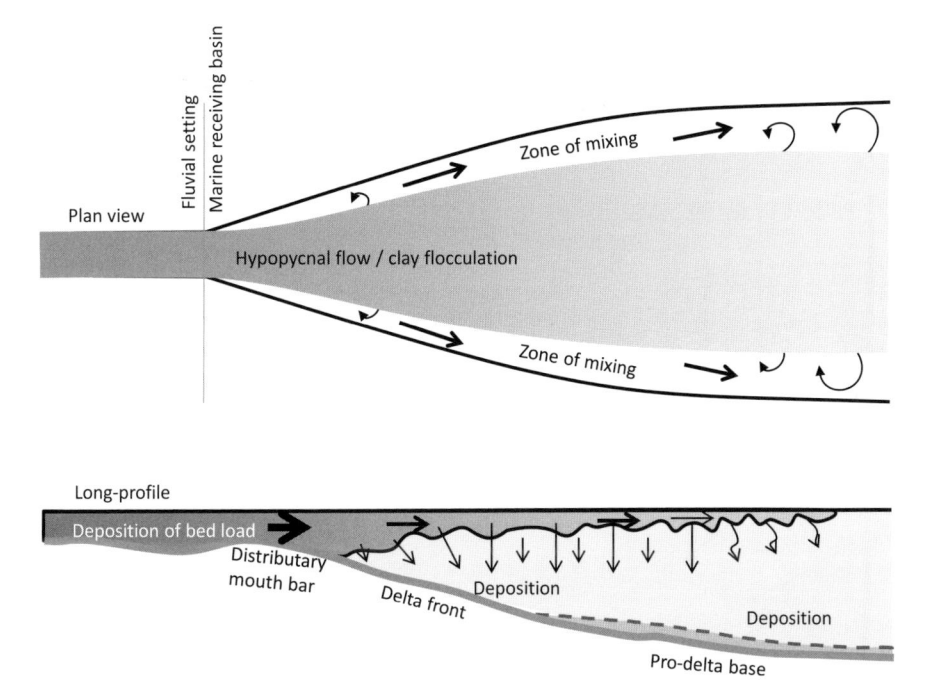

Figure 3.46. Fluvial–deltaic sedimentation in marine receiving basin in relation to major deltaic sedimentary units with clay flocculation in a hypopycnal condition (marine water denser than fluvial water). (After Frazier, 1967.)

that the river discharges into the Gulf of Mexico, a shallow microtidal receiving basin, and is laterally confined by adjacent older delta lobes (Coleman et al., 1998).

3.7.3 The Delta Cycle

Fluvial-dominated deltas are well characterized by the delta cycle, a conceptual model that systematically considers each type of fluvial–deltaic and coastal–deltaic process in relation to delta landforms and sedimentary deposits (Figure 3.48). The delta cycle formally relates time (years) to delta land area (km²) and is conceptualized in a curvilinear pattern of delta growth and loss over about a 2,000-year period (Coleman and Gagliano, 1964; Roberts, 1997). The delta cycle includes the following four stages: (1) initial progradation, (2) enlargement of delta lobes, (3) abandonment and transgression, and (4) reoccupation and growth of a new delta lobe (Table 3.12). A major contribution of the delta cycle is that it provides a lens

Table 3.12 *Characteristics of the delta cycle (see Figure 3.48).*

Phase	Stage	Timescale (years)	Primary processes	Depositional environments and/or landforms
Regressive	1. Initial progradation	50–200	Bed load deposition, channel bifurcation and river mouth progradation, clay flocculation	Distributary mouth bar, subaqueous delta front and prodelta base, riparian ribbon forest along low natural levees
	2. Enlargement of delta lobe	200–1,000	Flooding and overbank deposition, natural levee breach and crevasse formation, channel mouth bifurcation and distributary network development, continued progradation	Natural levees heightening and extension, crevasse splay deposits, subdelta formation, flood basins and interdistributary bay sedimentation, mature swamp forests and marsh wetlands in flood basins and interdistributary bays, hardwood forest along natural levees
Transgressive	3. Abandonment and transgression	200–500	Reduced stream gradient, Compaction and subsidence, Avulsion or stream capture	Flanking recurved spits, barrier island formation (attached), tidal inlets
	4. Avulsion / reoccupation and growth of new delta	200–500	Avulsion, marine processes reworking coarse distributary mouth bar deposits	Detached barrier island arc, salt water intrusion, wetland dieback and coastal land loss; rapid deltaic development in new delta lobe

Based on multiple sources, including Penland et al. (1988); Roberts (1997); Coleman et al. (1998).

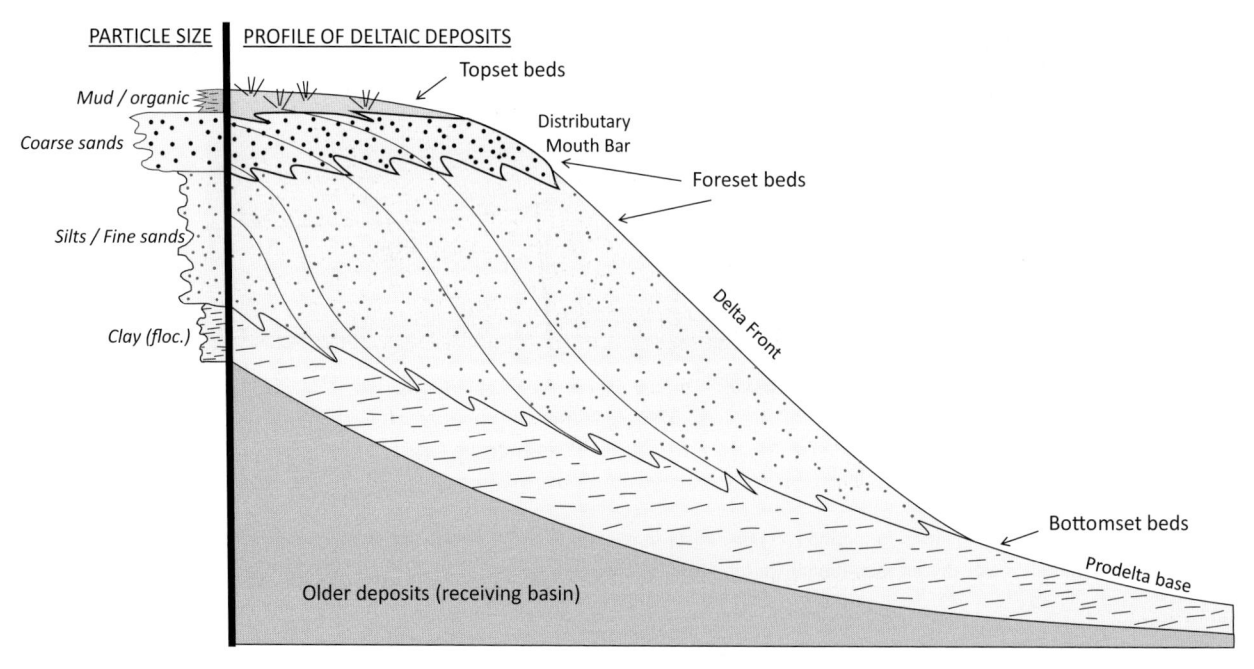

Figure 3.47. Prograding fluvial deltaic deposits, emphasizing coarsening-up vertical particle size trend. Progradation of distributary mouth bar deposits over pro-delta base. (After Elliott, 1986.)

in which to view all processes and environments within the enormous and complex Mississippi delta plain (Figure 3.49), an invaluable conceptual framework to formulate new management strategies during a period of sediment decline and rapid land loss (see Section 8.5).

3.7.3.1 DELTA CYCLE: STAGE I – INITIAL PROGRADATION

A new delta lobe is initiated with debauching of sediment at the river mouth into an open body of water (stage I). This results in immediate sediment sorting, and the deposition of coarse bed material into a

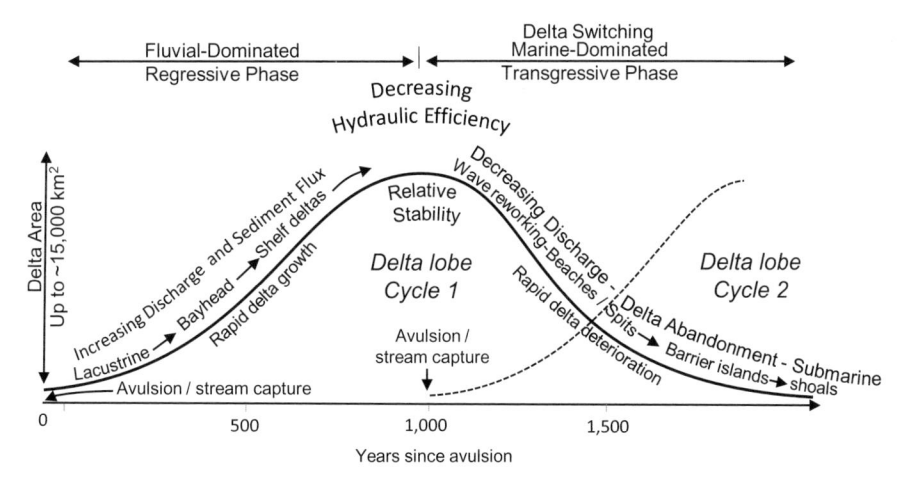

Figure 3.48. Delta cycle, with delta lobe initiation and abandonment in relation to changes in fluvial–deltaic environments. (Table 3.12 for description.) (After Coleman et al., 1998.)

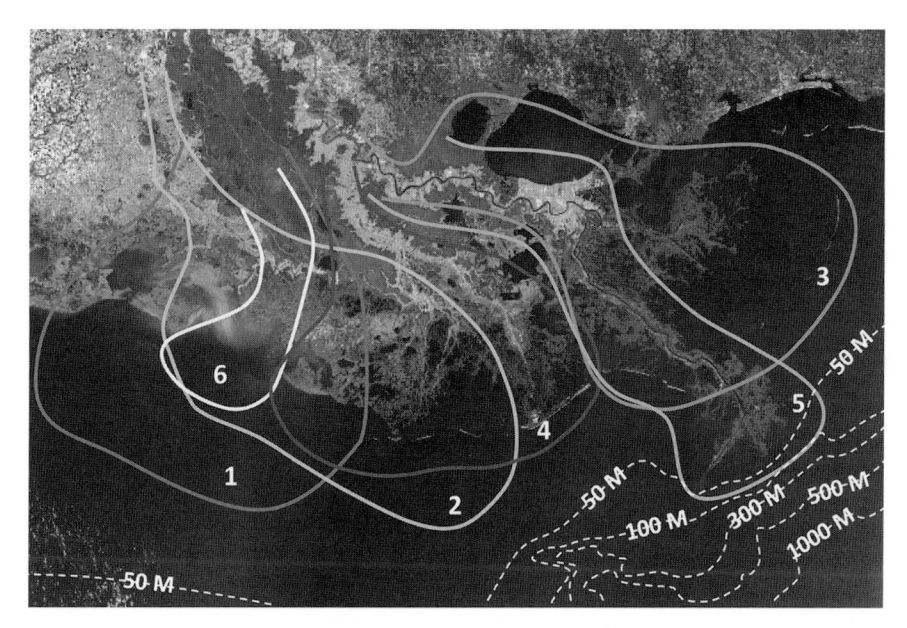

Figure 3.49. Mississippi delta plain with individual delta lobes. Chronology and area (km^2) of delta lobes: (1) Maringouin / Sale Cypremort: ~7500–5000 years BP (15,030 km^2); (2) Teche: ~5500–3800 years BP (13,570 km^2); (3) St. Bernard: 3569 years BP (15,470 km^2); (4) LaFourche: ~1491 years BP (11,310 km^2); (5) Plaquemines – Balize (Modern): 1322 years BP – present (9,930 km^2); (6) Atchafalaya – Wax Lake: 400 years BP – present (2,800 km^2). (After Wells et al., 1983; with chronology from Frazier, 1967 and Törnqvist et al., 1996.)

distributary mouth bar. This stage is associated with rapid land construction (Figure 3.47). Continued deposition of coarse bed material results in growth and advance of the distributary mouth bar. This is important because it results in channel bifurcation, and the initial creation of the deltaic distributary network. The transport of streamflow and sediment within two channels reduces the hydraulic efficiency of the delta, which drives further deposition and growth of the distributary network by channel bifurcation. Deposition of fine-grained (silt/clay) suspended sediment results in formation of two major types of deltaic deposits (Figures 3.46 and 3.47). This includes the delta front, which primarily consists of silt, and the prodelta base, which largely consists of flocculated clay, and ultimately represents the key "natural" driver of delta subsidence (Törnqvist et al., 2008).

3.7.3.2 DELTA CYCLE: STAGE II – ENLARGEMENT OF DELTA LOBE

As the progradational river mouth and channel advances further into the receiving basin, annual flooding builds up coarse-grained natural levees by the process of overbank sedimentation (stage II). Thus, a new delta finger forms, which consists of dense and coarse-grained (sandy/silty) distributary mouth bar

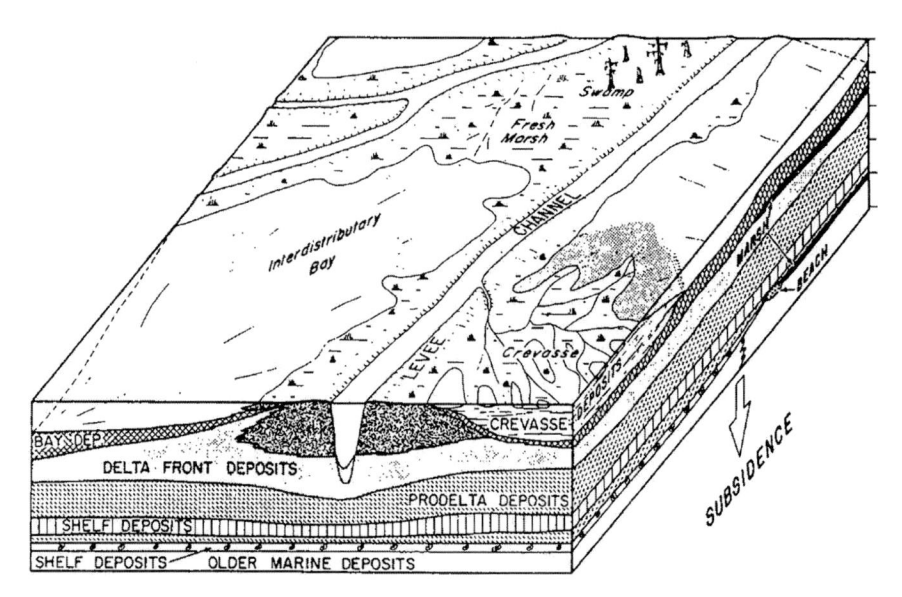

Figure 3.50. Fluvial–deltaic environments associated with a subsiding delta lobe. Flood events initiate crevasses that create subdeltas. (Saucier, 1994.)

deposits and natural levee deposits. Further, river mouth progradation (lateral advance) and sediment deposition within the receiving basin results in a vertically stratified regressive sequence of deltaic deposits that includes (1) a distributary mouth bar of sands and coarse silts, (2) low-angle delta front deposits of silts, and (3) a broad prodelta base of thinly laminated clay deposited atop older marine sediments. Thus, in contrast with fluvial deposits that have a fining-up vertical sequence, deltaic sedimentation and river mouth progradation result in a distinctive coarsening-up vertical sequence (Figure 3.47). The deposition of denser sandy/silty deposits atop fine-grained cohesive clay deposits initiates delta lobe subsidence, a fundamental environmental phenomenon that is greatly exacerbated by human impacts on deltaic processes.

3.7.3.3 DELTA CYCLE: STAGE III – ABANDONMENT AND TRANSGRESSION

Following development of an initial delta finger, the delta lobe rapidly enlarges and becomes more complex (stage III). In this stage, the initial distributaries created at the main channel mouth bifurcation rapidly prograde, and then again bifurcate. As the distributaries advance, overbank flooding simultaneously builds up the natural levees, disconnecting flood basins from riverine processes except during floods. Within flood basins and interdistributary bays a continuum of wetland environments develop, ranging from freshwater forested swamps, brackish marsh, to saline open bay environments (Figure 3.50). Seasonal river flooding delivers clastic silt and clay to low energy swamp and marsh environments by sheet flow, resulting in delta plain accretion. The submerged swampy conditions result in accumulation of thick peatlands. Closer to the coast, the salinity and energy regime increases, from the salt marsh to open bay environments, which undergo considerable disturbance by coastal storm surge events. Large river flood events breach emerging natural levees and form crevasse channels, which deliver coarser sediments (silt and fine sand) to fine-grained subsiding marsh environments (Figure 3.50).

An important component of stage III of the delta cycle is development of crevasses and subdeltas, which form from flood breaches of natural levees (Roberts, 1997). Subdeltas replicate the form and processes of delta lobes on a smaller spatial and temporal scale. Subdeltas are about an order of magnitude smaller than main-stem delta lobes and function for about one or two centuries. Collectively, subdeltas result in a large expansion of overall land area associated with a delta lobe. Importantly, subdeltas increase the overall sedimentary and geomorphic complexity of delta lobes (Roberts, 1997), enhancing environmental resilience to storm surge events and providing a valuables management tool in the context of modern coastal land loss and the design of sediment diversion structures (see Section 8.5).

Continued progradation and bifurcation of distributaries results in a broad network of channels that extend further into the receiving basin, but which individually transport an increasingly smaller proportion of the main-stem river's discharge and sediment load. Combined with reduction in gradient this reduces the overall hydraulic efficiency of the delta lobe, a key factor that contributes to its demise.

The third stage in the delta cycle is characterized by abandonment and transgression, as marine processes begin to dominate over declining fluvial inputs (Figure 3.48). This occurs for several reasons, including reduced downstream hydraulic efficiency driving upstream flooding, leading to avulsion or stream capture. Second, progradation of coarser and denser channel bed and natural levee deposits over soft prodelta base clay deposits results in compaction and subsidence (Stanley and Clemente, 2014). Compaction of the prodelta base is enhanced because of dewatering of porous flocculated clay, driving further delta subsidence. While some subsidence is annually checked by accumulation of new flood sediments (Hensel et al., 1999), accretion is substantially reduced upon completion of the upstream avulsion or stream capture. Thus, subsidence becomes dominant across the interdistributary environments, which enables sea level – and storm surge events – to advance further inland across the delta plain.

The transgressive phase of the delta model is associated with a distinctive evolutionary sequence of coastal environments (Roberts, 1997; Penland et al., 1988). At the mouth of the abandoned delta lobe, the large reduction in discharge and fluvial sediment results in marine processes beginning to rework coarser distributary mouth bar sediments. As marsh dieback and subsidence accelerate within interdistributary bays, wave processes rework the distributary mouth bar sands into an erosional headland with flanking barrier islands and recurved spits (Penland et al., 1988). Large storm surge events are able to breach sand dunes, depositing coarse sediment within the back barrier salt marsh environment. Dune breach scars eventually enlarge into tidal inlets, further increasing marine influence within the marsh and mudflats of intra-deltaic lagoons and leading to increased plant dieback and land loss. Further erosional enlargement of tidal inlets results in permanent detachment of spits from the headland, forming a transgressive barrier island arc. Tidal inlets and dune breach events by storms result in "rollover" of barrier island sands. Barrier island rollover is a distinctive mechanism that results in the landward migration of the barrier island arc, but in its wake, it also results in the winnowing (loss) of sand and formation of a subaqueous sand sheet. Sinking and erosion of the original delta lobe occurs at a greater rate relative to landward advance of the barrier island arc, resulting in coastal lagoon expansion as the barrier island arc becomes further isolated from mainland (Penland et al., 1988).

3.7.3.4 DELTA CYCLE: STAGE IV – AVULSION OR REOCCUPATION AND NEW DELTA GROWTH

The final stage of the delta cycle is formation of a new delta lobe. This is initiated by upstream avulsion forming a new river course, or by reoccupation of an old river course. In either case, the historic geomorphology of the delta plain is an important control on the location, development, and morphology of the new delta lobe (Aslan et al., 2005). In the case of an avulsion, the shape and morphology of the new delta lobe is considerably influenced by the gradient and topography of the receiving basin. High flanking natural levees along old river courses serve as topographic barriers to the spreading of fluvial–deltaic sediments across flood basins, representing a control on the orientation and width of the new delta complex. Additionally, older deltaic deposits may reduce the volume of the receiving basin. From the standpoint of an old channel being reoccupied, this may occur where there was formerly a partial avulsion. Thus, the channel, although initially smaller, is essentially ready to accommodate discharge and transport sediment. As the discharge increases, the older infilled sediments are flushed and the channel enlarged to accommodate increasing streamflow of the new delta lobe. In other cases, crevasse and smaller distributary channels are unable to enlarge because of resistant peat and cohesive sedimentary deposits of flood basins (Makaske et al., 2007).

3.7.3.5 MISSISSIPPI DELTA WETLAND LOSS

The Mississippi delta plain is in a transgressive (destructive) phase, and is rapidly declining due to subsidence and erosion by marine processes. The coarse deposits of older delta lobes as well as the active Balize delta lobe are being reworked into sandy spits and barrier islands (e.g., Penland et al., 1988; Roberts, 1998) while swamp and marsh environments associated with flood basins and interdistributary bays have undergone a precipitous loss of wetlands. The rate of wetland loss averaged 45 km^2/yr between 1932 and 2016, the highest rate of land loss in the world (Couvillion et al., 2017).

The land loss in the Mississippi delta over the past century is caused by both natural and human factors. Human factors are especially driven by a precipitous decline in sediment loads, the construction of hydraulic engineering structures to support flood control and industrial activities, and pumping of water and petroleum (fluid withdrawal). Over the latter half of the twentieth century the annual suspended sediment load of the Mississippi precipitously declined – from 450 million tons/yr to 145 million tons/yr – because of sediment trapping behind large dams and improved land management in the upper basin (Meade and Moody, 2010). Substantial hydraulic infrastructure, which includes a network of flood control dikes – hundreds of kilometers long – align the Mississippi River and its distributaries (USACE; Rogers, 2008). Activities associated with petroleum exploration and industry have resulted in fluid extraction, including excessive industrial groundwater pumping, and an extensive network of canals and pipelines that drives further ground subsidence and land loss (Ko and Day, 2004; Morton et al., 2006; Barras et al., 2008; Morton and Bernier, 2010; Couvillion et al., 2017). High rates of wetland loss between 1932 and 2016 resulted in 4,833 km^2 of coastal–deltaic wetlands being converted to open water, about 25% of the total wetland area of 1932 (Couvillion et al., 2017). The greatest period of land loss occurred between 1956 and 1978, which squarely coincides with the peak of petroleum extraction and groundwater withdrawal (Figure 3.51).

Reduction in land loss since the mid-1970s and particularly since the 1990s is a welcome trend, but it is likely too late.

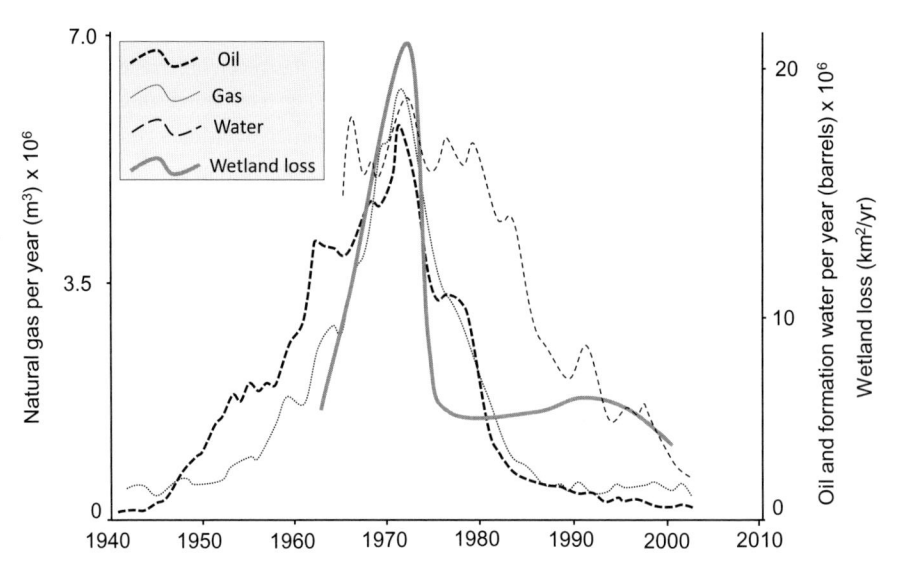

Figure 3.51. Composite annual fluid production (oil, natural gas, and formation water) from delta plain oil and gas fields with wetland loss rates measured at five representative hot-spots. (Source: US Geological Survey, 2013.)

Reduced rates of wetland loss are likely related to a decline in petroleum withdrawal and industrial water pumping (Figure 3.51). Formation of new deltaic lands associated with rapid infilling of the Atchafalaya basin and subaerial growth of the Atchafalaya and Wax Lake Outlet deltas represents an exception to the overall trend of high land loss (Coleman et al., 1998; Barras et al., 2008), and introduces additional management considerations. Aside from this minor positive exception to the overall trend, the future of the Mississippi delta plain is bleak. Even modest sea level rise projections in relation to the sediment loss problem will result in an unprecedented amount of coastal land loss by 2100 (Rutherford et al., 2018), effectively drowning all but a skeletal framework of distributaries across the delta plain (Blum and Roberts, 2009).

The importance of understanding drivers of wetland loss caused by sea level rise requires consideration of possible neotectonic controls on coastal subsidence (Dokka, 2006). While the Mississippi delta includes a number of older (primarily) east–west trending growth faults, unlike in the Rhine delta growth faults in the Mississippi delta have historically not influenced channel belt processes (Saucier, 1994; Armstrong et al., 2013; Chamberlain et al., 2018). And deeper structural features are not responsible for deltaic subsidence over the Holocene (Törnqvist et al., 2008). The underlying Pleistocene foundation of the Mississippi delta is relatively stable, with rates of vertical displacement at 0.1 mm/yr (Törnqvist et al., 2006). Recent fluid withdrawal, however, is activating growth faults that cause delta subsidence and wetland loss (Morton et al., 2006; Morton and Bernier, 2010).

Evolution of the Mississippi delta plain fits the framework of the "delta cycle," a general model that characterizes the spatial–temporal dynamics and adjustments to deltaic processes and depositional environments. The inherent historic framework of the delta cycle reveals the consequences of changes to fluvial adjustments associated with, for example, changes to sedimentation, avulsion and delta lobe switching, ground subsidence, and sea level rise. Such adjustments are common in river deltas heavily impacted by humans, and can be expected to occur in future decades owing to different manifestations of global environmental change. Thus, application of the delta cycle provides insights into how to implement a variety of deltaic and coastal restoration strategies, such as sediment diversion structures (see Section 8.5.3) to build new wetlands (Chamberlain et al., 2018). Because the delta cycle was developed from study of the Mississippi, the textbook example of a large fluvial dominated delta, its extension to other delta types (i.e., wave, tidal) as a management concept should proceed with care and with modification as appropriate (e.g., Dominquez, 1996; Hori and Saito, 2007).

3.8 CLOSURE: LOOKING FORWARD WITH AN EYE ON THE PAST

Knowledge of fundamental fluvial processes and dynamics – from basin headwaters to the deposition zone – is key to understanding human impacts to lowland rivers as well as the development of effective management and restoration strategies. While management is an inherently forward-looking enterprise, it is just as important to look back, as large lowland rivers and deltas are inherently historic systems that include both a Quaternary and historic human context.

There are two key reasons why a historic perspective is essential to the future sustainability of fluvial lowlands. First, modern processes are literally situated within – and influenced by – a historic framework that alters contemporary spatial and temporal dynamics of lowland fluvial systems. Older floodplain units, for example, have sedimentological and physical properties that differ from modern sedimentary deposits, possibly because of being deposited under a different climatic regime. Sedimentary units such as dense clay plugs and backswamp require greater shear stress to generate cutbank erosion, but also buffer and condition fluvial response to unintended human impacts by hydraulic engineering. Variable erosion rates associated with heterogeneous bank material results in asymmetric channel pattern adjustment. In the case of rivers that are intensively utilized for navigation and settlement it requires tailor-made engineering and management on a reach-by-reach basis, dependent upon detailed sedimentologic information that requires detailed field investigation. Flood processes across broad fluvial lowlands are heavily influenced by older fluvial deposits, with the sedimentary matrix controlling groundwater flux and the surficial topographic expression of negative relief floodplains influencing patterns of floodwater and sediment dispersal as well as nutrient sequestration. The careful study of historic floodplain deposits is essential to disentangling fluvial response to natural versus human impacts, and hopefully provides some clarity as to drivers of degradation of fluvial environments.

Secondly, the historic system informs future management strategies. This is especially important because massive hydraulic infrastructure of large rivers requires decades to implement, and enormous investment. Reconstruction of historic channel migration rates provides quantitative insights into the consequences of "rewilding" strategies, whereby hard engineering structures to prevent erosion – such as groynes and revetment – are altered or removed. Such data inform whether flood control dikes will be threatened by bank erosion, how much additional property should be acquired to accommodate a freely meandering river, and vulnerability of alluvial cultural heritage. In the delta, reconstructing the chronology and timescales of activity for specific fluvial–deltaic processes and environments across the broader delta plain (rather than just the active delta) provides crucial data to assess which areas are most vulnerable. And to realistically prioritize areas that can be saved, and identifying those that will be lost. This is especially necessary in the design of sediment diversion structures, which will be a vital tool to preserve and construct new patches of wetlands because of delta drowning by subsidence and sea level rise (Section 8.5.3). Finally, a thorough knowledge of past fluvial adjustment provides insights into the direction and range of possible fluvial and hydrologic responses to climate change, with future global warming projections (2°C) being somewhat similar to warming that occurred over the mid-Holocene Altithermal (Knox, 1993, 2000).

The perilous condition of many of the world's rivers and deltas raises the fundamental question: How to best implement effective strategies to sustain and restore such complex and sensitive environments that support a range of biophysical and human activities? The answer is largely under our feet, contained within the muds and sands comprising older fluvial and deltaic deposits.

4 Dams, Rivers, and the Environment

... their effects far exceed any effects likely to occur from global climate change over a period of several centuries.
—W. L. Graf (2003)

As noted earlier, a variety of conditions will control the response on different rivers.
—Williams and Wolman (1984, p. 14)

4.1 HYDRAULIC INFRASTRUCTURE AND SOCIETY

4.1.1 Context and Focus

Many rivers are substantially degraded because of human reliance upon water resources to support intensive agriculture, industry, and settlement. This includes emplacement of "hard engineering" structures that modify landscapes and water supplies, such as dams, dikes (levees), groynes, river bank protection, and irrigation networks, among other forms.

Dams and reservoirs are a specific type of hydraulic infrastructure constructed to meet a variety of societal demands and provide stable water supplies for irrigation, navigation, industry, flood control, enabling agricultural activities and human settlement to expand upon marginal lands (Williams and Wolman, 1984; Graf, 2006). Dams and reservoirs, however, have many unintended adverse consequences. These include altered hydrologic regimes and sediment trapping that impact downstream environments, and in some cases increase human vulnerability to global environmental change. Additionally, constructing dams to reduce flood risk is a false justification. This is because flood control results in reduced downstream flow variability for the purpose of promoting floodplain development for settlement and agriculture within a predictable physical environment, which is the actual justification.

This chapter singularly focuses on dams, the major form of hydraulic infrastructure in the upper and middle reaches of drainage basins, and which impact downstream fluvial lowlands and deltas. The chapter begins by examining the motivation for dam construction and establishes their near ubiquitous presence upon the global riparian landscape. Next, the environmental impacts of dams are systematically reviewed, including hydrologic, sedimentary, geomorphic, and ecological impacts. While some nations are experiencing a surge in dam construction, others are practicing dam removal, which holds great potential to restoring downstream environments. Many dams, however, are effectively permanent components of the riparian environment (Figure 4.1), which behooves government organizations to develop strategies for sustainable management of reservoir sediments. While some dams are constructed in the lower portions of drainage basins, hydraulic infrastructure in lowland floodplains and deltas primarily consists of flood control dikes and hard structures to stabilize rivers in support of navigation. These forms of hydraulic infrastructure are systematically examined in subsequent chapters (Chapters 5, 6, and 7), including environmentally progressive approaches within the framework of integrated river basin management (Chapter 8).

4.1.2 Dams, Agriculture, and Water Resources

Hydraulic societies were ancient agrarian civilizations organized around water resource management by robust government institutions, as originally described by Wittfogel (1957). Historically, the concept applied to lowland rivers in Mesopotamia, eastern China, the lower Indus, the Nile in Egypt (see Butzer, 1976; Wolman and Giegengack, 2007), as well as pre-Columbian societies in the Americas (see Denevan, 2000; Doolittle, 2000; Mann, 2005). The concept of hydraulic societies reinforces the point that – because of their extensive size and sophistication of design – hydraulic infrastructure along large rivers require well-functioning government institutions for planning, construction, and management and maintenance. This was true for ancient societies, and as we're constantly reminded, remains valid today.

Global freshwater withdrawals have increased greatly since 1900 and especially after the 1950s (Figure 4.2). Some 70% of global freshwater withdrawals are to support agriculture. And because of reservoir evaporation, the actual percentage of global freshwater withdrawals could be closer to ~80%, as many

Figure 4.1. Red Bluff Diversion Dam on Sacramento River, California, constructed from 1962 to 1964 (closure) to support 61,000 ha of irrigated agriculture for massive Central Valley Project (decommissioned in 2013 for environmental reasons), supplied by two large irrigation canals (photo). (Source: U.S. Bureau of Reclamation, 2004. Licensed under CC BY 2.0.)

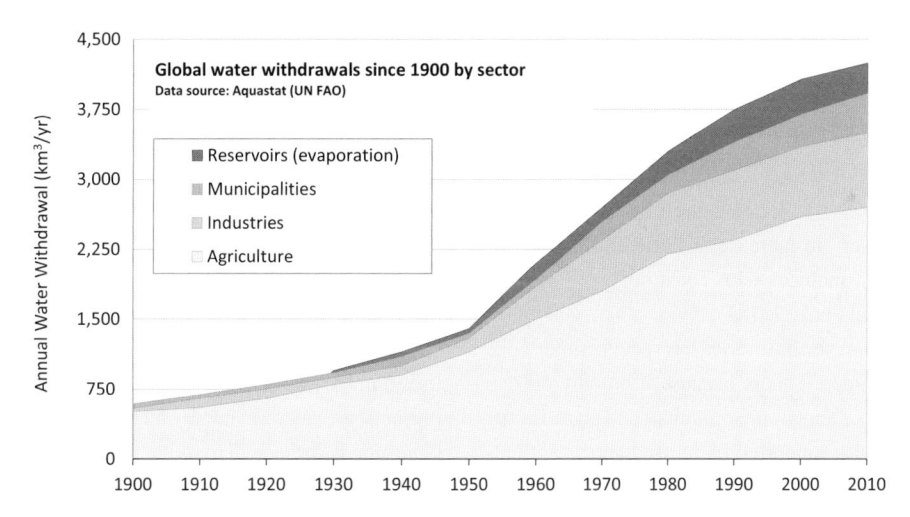

Figure 4.2. Global water withdrawals by sector since 1900, showing steep increase after the 1950s. (Source: U.N. FAO data in Koohafkan et al., 2011.)

reservoirs in dry regions were historically constructed to support agriculture and have high rates of evaporation. For societies heavily dependent upon riverine freshwater resources, evaporative losses can be a substantial component of the annual water budget, increasing vulnerability to water scarcity caused by climatic fluctuations or upstream political rivals. The evaporation rate at Lake Nasser and Lake Nubia, the massive 5,275 km^2 reservoir that impounds the Nile River upstream of the Aswan High Dam in Egypt and Sudan, ranges from an average daily maximum in June of 10.9 mm/day to an average daily minimum in January of 3.8 mm/day. The amount of Nile River water lost by evaporation in Lake Nasser ranges from 10 to 16 billion m^3/yr. While the Nile River supplies Lake Nasser with 55 billion m^3/yr, evaporation results in a loss of 18–29% of annual totals (El Bakry, 1994; FAO, 2016; El-Mahdy et al., 2019). Within such a hyperarid region this constitutes a substantial loss of freshwater, and increases downstream water stress for riparian agricultural stakeholders.

Table 4.1 *Hydraulic infrastructure associated with irrigation projects in the Mekong basin**

Stage	Main components of irrigation systems							Total
	Reservoirs	Weirs	Traditional weirs	Sluices	Pumps	Colmatage canals**	Others, or unknown	
Cambodia	25	47		1	23	133	94	323
Lao PDR	175	1,014		83	1,264		9	2,545
Thailand	5,116	1,347	63	37	1,182		1,004	8,749
Viet Nam	50	46		157		4	4	261
Total	5,366	2,454	63	278	2,469	137	1,111	11,878

* As registered on the DMFPF project of the Mekong River Commission's Agriculture, Irrigation and Forestry Project in 2003.
** Built by the French under colonial system for the purpose of transporting silty river water over natural levees into backswamp to build up agricultural lands (see Dennis, 1990; FAO, 2011a).
Data source: Hortle and Nam (2017).

Many dams exist as part of a larger agricultural system that requires stable freshwater supplies from extensive irrigation networks. Indeed, irrigation networks necessitate additional forms of engineering structures, including weirs, sluices, pumps, ditches and canals to distribute water between reservoirs and fields (Figure 4.1). Thus, the riparian impact of dams is very much part of a broader impact of agriculture on the environment. The Mekong basin, for example, receives considerable attention because of the recent boom in dam construction for hydropower. But already the impact of agriculture on water resources across the Mekong basin is extensive (Hortle and Nam, 2017). Irrigation projects in the Mekong basin include an extensive amount of hydraulic infrastructure for support, including 5,366 dams and reservoirs, 2,454 weirs (low-flow dams), 2,469 pumps, and 278 sluices (Table 4.1). Paradoxically, to ensure regional food security because of projected losses to inland Mekong River fisheries caused by intensive dam construction for hydropower, an increase in agricultural lands of 19–63% is required, which includes dam construction to support irrigation networks (Pokhrel et al., 2018).

The hundredth meridian in North America has special importance as it relates to water resources. This is because it broadly corresponds to the 50 cm isohyet of annual precipitation, and signifies the boundary between agriculture supported by rainfall versus agriculture supported by water supplied by irrigation networks (Powell, 1895; Stegner, 1954). The climatic significance of the hundredth meridian is that it represents the approximate westerly limit of humid tropical moisture supplied from the Gulf of Mexico (Seager et al., 2018). Much of the land beyond the hundredth meridian effectively lies in the rain shadow of the mountain west and basin and range region, requiring considerable hydraulic infrastructure to support agriculture. Of course, beginning in the late nineteenth century, and prior to the development of numerous dams and irrigation networks west of the hundredth meridian (Reisner, 1993 [1986]), the western Great Plains were already being developed and cultivated based on a flimsy scientific posit known as "rain follows the plow," which was especially pushed by land speculators to encourage westward development (Smith, 1947; McLeman et al., 2014). The idea was that, following the onset of agriculture in dry regions, rainfall would increase because high evapotranspiration rates from crops would produce summer cloud bursts to satiate thirsty lands. This nonsensical and dangerous contention placed agriculture and human settlement at an environmental tipping point, one that was dramatically exceeded in the dry decade of the 1930s, the era of the American Dust Bowl. Unfortunately, the concept was exported to dry southeastern Australia, within the Murray River basin as farmers crossed into drylands north of Goyder's Line, beyond the 25 cm isohyet (Andrews, 1938).

With a growing population requiring a 70% increase in food production by 2050 and potentially doubling or tripling by 2100 (Crist et al., 2017), large increases in cultivated lands are projected to drive enormous increases in global freshwater withdrawals (Mulligan et al., 2020). Because much of the cultivation will be for water-intensive crops and include new agricultural lands developed in semiarid regions, future agricultural expansion will heavily rely upon water supplied by dams and irrigation networks (UNWWDR, 2020). Additionally, many irrigation networks established in the 1960s and 1970s as part of international development projects are in dire need of repair, maintenance, and updating. Combined, these will require great investment by governments. The U.N. Food & Agricultural Organization (FAO) estimates that US$960 billion is required to improve and expand irrigation networks across ninety-three developing nations by 2050 to address projected climate change, including increased water loss due to higher rates of evaporation (Koohafkan et al., 2011).

4.2 GLOBAL EXTENT OF DAMS

4.2.1 First Step: Accounting of Dams and Reservoirs

To formally assess the global imprint of dams on the environment requires firm numbers of dam size and distribution. An accurate global estimate of dams, however, is far from complete.

Table 4.2 *Dam types*

Arch dam	A concrete or masonry dam that is curved upstream to transmit the major part of the water load to the abutments. Double curvature arch dam: An arch dam, which is curved vertically as well as horizontally.
Buttress dam	A dam consisting of a watertight part supported at intervals on the downstream side by a series of buttresses. A buttress dam can take many forms, such as a flat slab or a massive head buttress. Ambursen dam: A buttress dam in which the upstream part is a relatively thin flat slab usually made of reinforced concrete. Multiple arch dam: A buttress dam composed of a series of arches for the upstream face.
Coffer dam	A temporary structure enclosing all or part of the construction area so that construction can proceed in the dry. A diversion cofferdam diverts a stream into a pipe, channel, tunnel, or other watercourse.
Diversion dam	A dam built to divert water from a waterway or stream into a different watercourse, often to supply irrigation networks.
Embankment dam	Any dam constructed of excavated natural materials or of industrial waste materials. Earthen dam: An embankment dam in which more than 50% of the total volume is formed of compacted earthen material generally smaller than 3 in. in size. Hydraulic fill dam: An embankment dam constructed of materials, often dredged, which are conveyed and placed by suspension in flowing water. Rockfill dam: An embankment dam in which more than 50% of the total volume is composed of compacted or dumped cobbles, boulders, rock fragments, or quarried rock generally larger than 3 in. in size.
Gravity dam	A dam constructed of concrete and/or masonry, which relies on its weight and internal strength for stability. Hollow gravity dam: A dam constructed of concrete and/or masonry on the outside but having a hollow interior relying on its weight for stability. Crib dam: A gravity dam built up of boxes, crossed timbers, or gabions filled with earth or rock. Roller-compacted concrete dam: A concrete gravity dam constructed by the use of a dry mix concrete transported by conventional construction equipment and compacted by rolling, usually with vibratory rollers.
Hydropower dam	A dam that uses the difference in water level between the reservoir pool elevation and the tailwater elevation to turn a turbine to generate electricity.
Industrial waste dam	An embankment dam, usually built in stages, to create storage for the disposal of waste products from an industrial process. The waste products are conveyed as fine material suspended in water to the reservoir impounded by the embankment. The embankment may be built of conventional materials but sometimes incorporates suitable waste products. Mine tailings dam. An industrial waste dam in which the waste materials come from mining operations or mineral processing.
Overflow dam.	A dam designed to be overtopped.
Regulating (Afterbay) Dam	A dam impounding a reservoir from which water is released to regulate the flow downstream.
Saddle dam (or dike)	A subsidiary dam of any type constructed across a saddle or low point on the perimeter of a reservoir.

Source: U.S. Society on Dams, *Types of Dams*. www.ussdams.org/dam-levee-education/overview/types-of-dams/ (accessed May 2020).

The large range in the size of dams, differences in national accounting procedures and criteria, differences in ownership (public or private), and the extensive history of dams make arriving at a global tally of contemporary dams a challenging task. Additionally, there is no consensus on how to categorize dams. The British Dam Society identifies four types of dams while the U.S. Society on Dams identifies eleven types of dams (Table 4.2). Part of the issue is that dams are characterized in several ways, including their usage (purpose), construction materials, and design (shape).

While accounting of smaller dams will continue to be a work in progress, great improvements have been made in tallying larger dams so that a clearer global picture has emerged with regard to their distribution and characteristics. These data are essential to assess the broader environmental imprint of dams upon the landscape and water resources. Over the past decade or so, national and international stakeholder organizations have systematically accumulated information pertaining to a variety of dam criteria. Increasingly, these include important descriptive data such as dam height, type of dam, ownership, date of construction, purpose of dam, date of closure, reservoir capacity, river basin, upstream impounded area, streamflow volume, etc. The most authoritative and scholarly recognized organization for large dams is compiled by the International Commission on Large Dams (ICOLD), a nongovernmental organization (trade industry) that promotes standards and guidelines pertaining to best practices regarding construction and management of dams and reservoirs. ICOLD defines large dams as having a height of 15 m or greater (from foundation to dam crest) or a dam between 5 and 15 m that impounds greater than 3 million m^3.

Data from the most recent ICOLD tally identifies 57,985 large dams (as of September 2019) across ninety-six member nations,

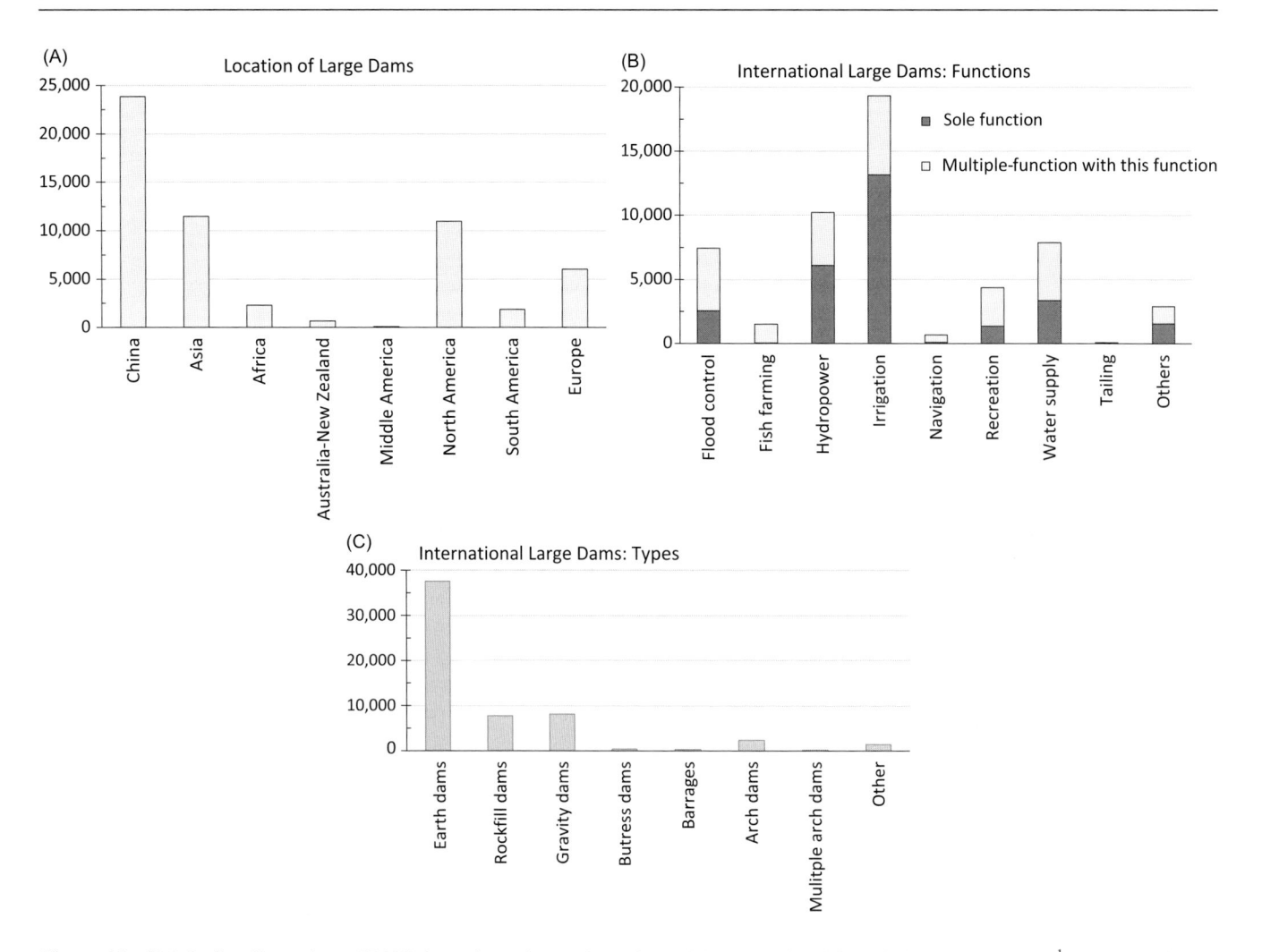

Figure 4.3. Global tally of large dams (57,985) from ninety-six member nations of the International Commission on Large Dams[1] (September 2019). Large dams are classified by (A) location (continent or region), (B) functions (sole function and multiple functions), and (C) types of large dams. Location (continent or region): China plotted separate from Asia because of very high number of large dams; Asia included Middle Eastern nations and Turkey because dams mainly in Tigris-Euphrates drainage to the Persian / Arabian Gulf; Australia and New Zealand classified together; North America included Canada, United States, and Mexico; Europe included Russia. Author figure. Data source: International Commission on Large Dams (ICOLD).

providing a view as to their location and function (Figure 4.3A–C). China stands out as the global leader in large dams. About half (47%, 13,142) of the world's large dams are constructed for the sole purpose of irrigation in support of agriculture, although an additional 6,180 large dams primarily built for irrigation have multiple purposes (flood control, hydropower, etc.). Large dams constructed for irrigation are especially concentrated in India, Spain, South Africa, Southeast Asia, and in the United States west of the hundredth meridian (Biemans et al., 2011). Most significant large dam construction occurred over a four-decade period spanning the 1950s and 1980s, and earthen dam structures dominate (Figure 4.3). The total volume of reservoir impoundment exceeds 6,500 km^3 (Figure 4.4), and will rapidly increase over the next several decades.

Advances in geospatial technologies such as high-resolution satellite imagery and topographic data (lidar) hold great promise for characterizing Earth's impounded riparian landscapes (Walter and Merritts, 2008; Poeppl et al., 2015; Mulligan et al., 2020). Such information can then be coupled with important landscape matrices, such as land cover, precipitation, streamflow volumes, slope, sediment loads, and other river basin indices. The Global Reservoir and Dam Database (GRanD), for example, assimilated a global geospatial database of dams mapped from 1800 to 2010 (Lehner et al., 2011). Although providing detailed attribute data on dams it includes only 6,862 records (<10% of global large dams). A recent effort at a global tally utilized several types of satellite imagery at varying spatial scales (i.e.,

---

[1] International Commission on Large Dams (www.icold-cigb.org/).

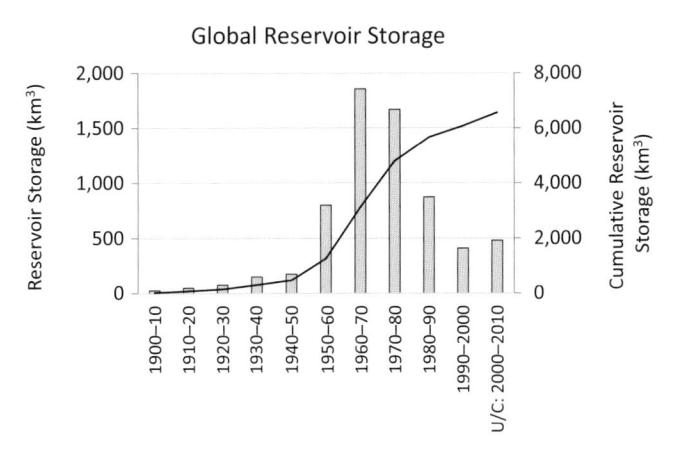

Figure 4.4. Global growth of reservoir storage capacity between 1900 and about 2010, including dams under construction and planned (U/C) between 2000 and 2010. (Source: Walling, 2006.)

Landsat, Ikonos, Spot imagery) to identify dams with concrete structures, including large dams from GRanD and ICOLD databases, in addition to some medium-sized dams. The GlObal geOreferenced Database of Dams (GOODD) contains >38,000 dams. Such approaches are also essential to identify long decommissioned dams – particularly numerous small dams – buried in historic (legacy) reservoir deposits, but which continue to influence contemporary fluvial riparian processes (i.e., Walter and Merritts, 2008; Poeppl et al., 2015; Brown et al., 2018).

If considering a larger range of impoundment types, including mill dams, low weirs, and primitive "check dams," there are surely millions of dams on Earth. And if including impoundments with a minimum surface area of 100 m^2, it is estimated there could be about 16 million dams on Earth. This potentially increases Earth's freshwater surface area by 306,000 km^2, a global increase of 7% (Lehner et al., 2011; Mulligan et al., 2020).

To further appreciate the scope of river disturbance by dams it is useful to provide data that pertain to specific basins and nations. In the United States, there are some 2.5 million dams impacting every basin larger than 2,000 km^2 (NRC, 1992; Heinz Center, 2002; U.S. EPA, 2016). Most of China's dam construction is in the Huanghe and Yangtze basins, with the latter having over 50,000 dams (Lehner et al., 2011; Yang et al., 2011). India has 5,100 large dams (International Commission on Large Dams, www.icold-cigb.org/) and is heavily investing in dams and reservoirs, with 313 large dams under construction (NRDI, 2016). An important concern is that many of India's dams under construction and proposed will be located in active seismic zones within the Himalayas, including regions that are intensely politically contested with Pakistan and China. Dams in India were historically overwhelmingly constructed for irrigation in support of agriculture and food security. India's new mega dams, however, are being constructed primarily for hydropower. Indeed, India is now the world's fifth largest hydropower producer (IWPDC, 2020). With 1,538 large dams (MITECO, 2020) Spain is the most impounded nation in Europe. This represents 2.5% of the world's dams, and Spain's dams have a total storage capacity of 54.6 km^3 (Batalla, 2003). Although Spain also has Earth's oldest functioning dam (Figure 4.5), most of Spain's functional dams were constructed in the 1970s and 1980s and are privately owned (Table 4.3), with some 5,500 being less than 15 m in height (Figure 4.6).

Figure 4.5. (A) World's oldest dam. Roman era Cornalvo dam in Extremadura, Spain impounds the Rio Albarregas since 130 AD. The dam is part of a system of hydraulic infrastructure that supplies water to Emerita Augusta (Mérida) by aqueduct, similar to the adjacent (B) Roman era Acueducto de los Milagros. The gravity dam has a concrete core and an outer earthen structure for support, with stone cladding. Size: 28 m high, 194 m long. (Source: Licensed under CC BY 2.0.)

Australia is the driest inhabited continent, and internationally has the highest per capita surface water impoundment of any nation (AWA). Australia has 557 large dams, according to the Australian National Committee on Large Dams (Figure 4.7). Most of Australia's dams are in Queensland (120 dams), Victoria (113 dams), and New South Wales (135 dams), although much smaller and wet Tasmania has 100 large dams. Unfortunately, new large dams are proposed for tropical northern Australia, including the

Table 4.3 *Drainage basin distribution and ownership of large dams in mainland Spain*

River basin	Public-owned dams	Privately owned dams
Andalusian Mediterranean Basins	3	44
Douro	38	107
Ebro	75	224
Eastern Cantabrian	0	14
Galician Coast	0	24
Guadalete & Barbete	0	27
Guadalquivir	51	71
Guadiana	39	151
International Basins of Catalonia	0	16
Jucar	32	22
Mino-Sil	6	70
Tagus	66	218
Tinto, Odiel & Piedras	0	45
Western Cantabrian	3	68
Total	313	1,101

Data source: Spanish Ministry of Environment. MITECO (2020).

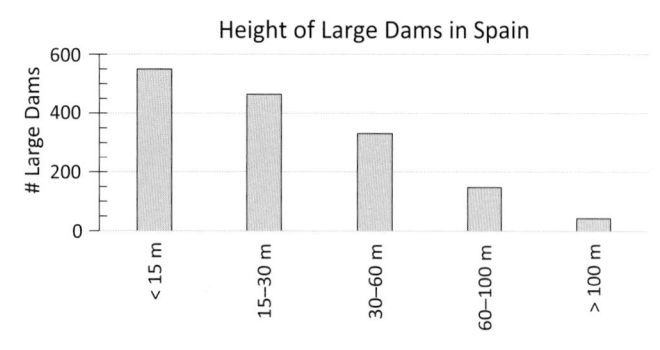

Figure 4.6. Height of large dams in Spain (total = 1,538). (Source: Spanish Ministry of Environment, MITECO, 2020.)

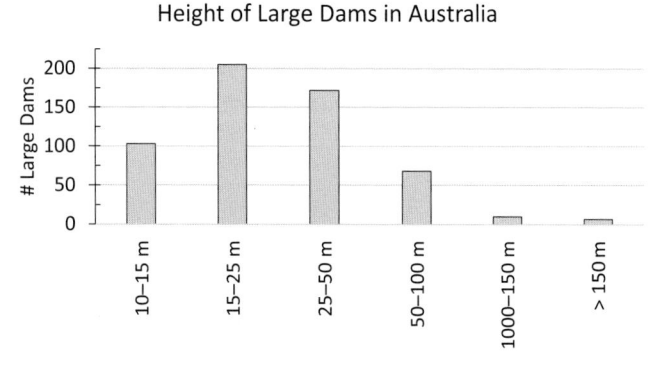

Figure 4.7. Large dams in Australia organized by height (m). Large dams in Australia are >15 m, or >10 m if reservoir >1,000,000 m^3 and maximum discharge >2,000 m^3/s. (Data: ANCOLD, 2020.)

Fitzroy (Western Australia), Darwin (Northern Territory), and Mitchell (Queensland) basins. The main driver behind proposed dam construction is to provide water for irrigation to support massive investment in agricultural development.

Canada has over 14,000 dams with 1,157 considered large dams (15 m high, or >5 m high and impounding >3,000,000 m^3) and 49 dams considered very large (>60 m high) (Canadian Dam Association, 2019). Dam construction in Canada is overwhelmingly for hydropower, including 860 of its large dams. Dams constructed for irrigation are primarily in central and western provinces, including Saskatchewan, Alberta, Manitoba, and eastern portions of British Columbia. Most water demands for agriculture and industry are in the south, while Canada's rivers mainly flow north. There are also many Canadian dams constructed to store mining tailings. Unlike in the United States, which has not constructed a large dam in decades, Canada has completed several large dam projects within the last decade and is finalizing works on a major dam to the Peace River in British Columbia.

The United States has a conflicted history with dams over about the past century, being a global leader in both dam construction and dam removal. The United States has 91,457 large dams (>1.8 m high) on its national registry (Figure 4.8), with 137 very large dams that each store more than 1.2 km^3 of water (U.S. National Inventory of Dams, 2020[2]). Of these dams, the average age is fifty-seven years, 7% are built for hydropower, 74% have a high hazard potential, and 69% are regulated at the state level (as of April 2020). Additionally, as with many nations, in the United States there are numerous small dams locally operated to manage small fishing ponds, farm ponds, amenity lakes, and old mill dams not in the NID. Many old mill dams, in particular, are obsolete and relict, having long lost their original purpose because of upstream land degradation and soil erosion driving reservoir infilling, particularly those constructed in the eighteenth and nineteenth centuries (Trimble, 1974; Walter and Merritts, 2008). As an example, for the Apalachicola-Chattahoochee-Flint basin (52,720 km^2) in the US southeast, a GIS analyst identified over 25,000 reservoirs, although only 6% were included in the National Inventory of Dams (NID) database (Figure 4.8; U.S. EPA, 2016).

4.2.2 River Fragmentation by Dams and Reservoirs

The environmental impacts caused by dams and reservoirs can be appreciated by considering changes to freshwater resources. From the standpoint of the volume of freshwater, in the beginning of the twentieth century, reservoirs increased global freshwater supplies over natural supplies by 5% (18 km^3/yr). By the end of the

---

[2] The U.S. National Inventory of Dams (NID) includes dams ≥25-feet (7.6-m) high and >15 acre-feet of storage, ≥50 acre-feet of storage and >6 feet high (1.8-m), or are hazardous (could cause potential or actual loss of life and significant property or environmental damage) (https://nid.sec.usace.army.mil/).

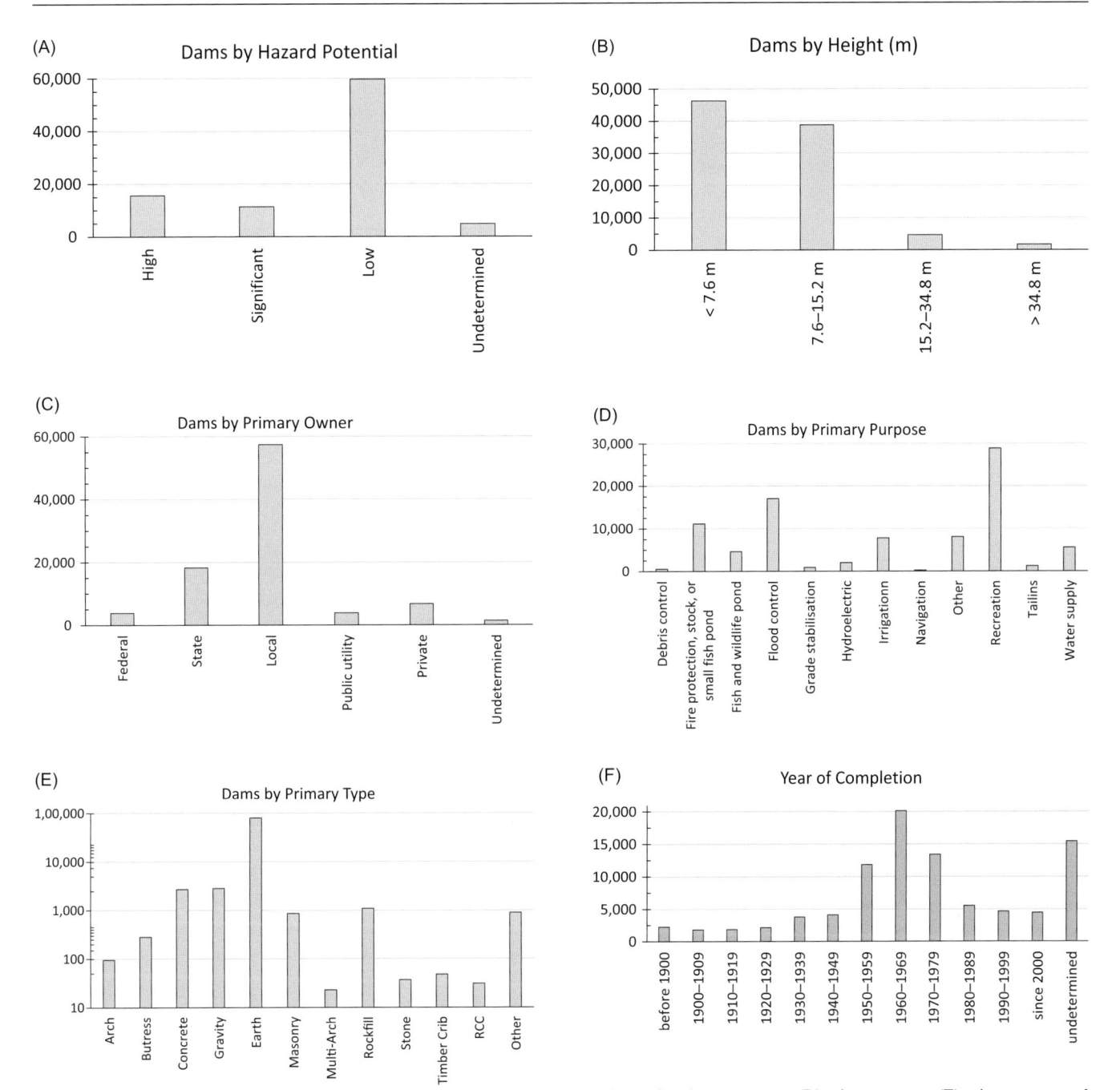

Figure 4.8. The United States has 91,457 dams (A) decade of completion, (B) height, (C) primary purpose, (D) primary owner, (E) primary type, and (F) hazard potential. Note: y-axis for dam type (E) is logarithmic. (Author figure. Data source: U.S. Army Corps of Engineers National Inventory of Dams.)

twentieth century, the volume of surface water supplies had increased by 40% (460 km^3/yr) over natural supplies (Biemans et al., 2011). From the perspective of the global surficial footprint (area) of the impact of reservoirs, the cumulative area of irrigation increased from 40 Mha in 1900 to 215 Mha in 2000. Over half of Earth's global runoff is regulated by dams, with most dams located in basins with heavy agriculture. The total cumulative storage capacity of reservoirs is ~20% of Earth's annual runoff (Biemans et al., 2011). The streamflow of all larger US rivers is impacted by dams, and a volume of water equal to 75% of the average annual runoff is stored in reservoirs (Graf, 1999).

An important principle of impounded river management – and a cruel irony for riparian ecosystems and stakeholders – is that river basins in dry environments are more adversely impacted by

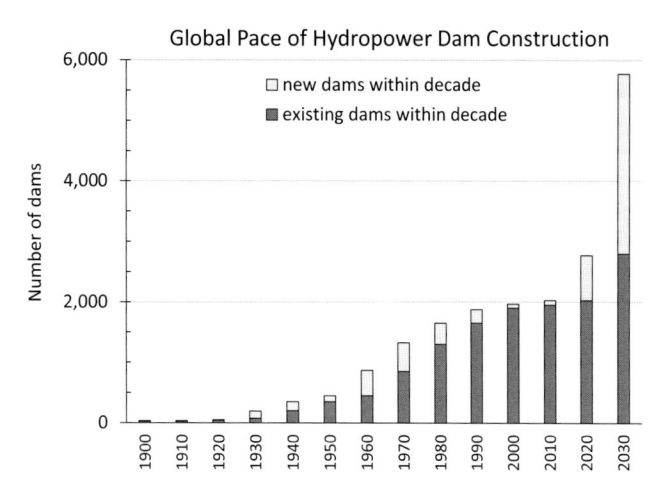

Figure 4.9. Global acceleration of dam construction for hydropower per decade since 1900, including dams under construction and planned through the 2030s. (Source: Zarfl et al., 2015.)

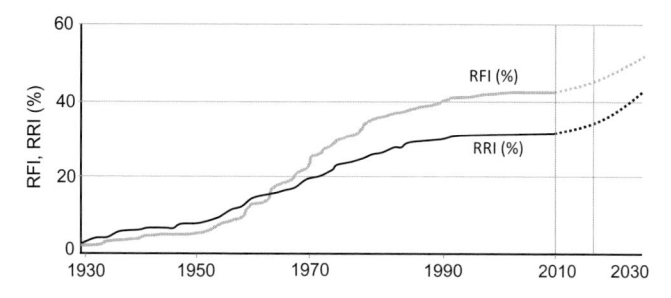

Figure 4.10. Global trend and future trajectory (to 2030) for the river fragmentation index (RFI) and river regulation index (RRI). Indices derived between 1930 and 2010 and estimated until 2030 (based on GRanD) for hydropower scenarios (dotted lines). Values reflect area-weighted means of indices across all basins. (Source: Grill et al., 2015, with data from Zarfl et al., 2015.)

dams than those in wet environments (Graf, 1999; Magilligan et al., 2008).

Dam construction is charging forward in Asia, South America, Africa, and the Balkans in Europe, despite vital environmental and management lessons learned from experiences across an international array of physical and economic settings. Of great concern is that an additional 3,700 large dams are either under construction or planned for construction. Incredibly, within about a decade the number of large hydroelectric dams on Earth will double (Figure 4.9). While this would provide an additional 1,700 GW of hydropower (Zarfl et al., 2015) it will come at a substantial environmental cost (e.g., Winemiller et al., 2016; Zarfl et al., 2019). Should the planned dam construction be completed, a total of 89% of Earth's flow volume would be impacted by dams, particularly in view of forthcoming impoundment of the Amazon (Grill et al., 2015).

Unfortunately, the massive increase in dam construction is located within the headwater supply and transfer zones of some of Earth's largest river basins, including the Yangtze and Amur Rivers in China. And new dams are especially being constructed within large tropical basins, such as the Amazon (368 new dams), Brahmaputra (396 new dams), Mekong (120 new dams), and Congo (35 new dams) Rivers (Zarfl et al., 2019). Damming tropical rivers is especially unfortunate because the riparian corridors representing among Earth's highest biodiversity and the habitat of numerous riparian species – including charismatic aquatic megafauna as well as endemic and yet to be inventoried species – could be lost (Winemiller et al., 2016; Latrubesse et al., 2017).

The impacts of dams on drainage basin hydrology can be characterized by indices that quantify the degree of fragmentation (disruption to longitudinal connectivity) to the drainage network and also by the degree to which annual flow volumes are impounded (Graf, 1999, 2001, 2006; Magilligan et al., 2003; Zarfl et al., 2014; Grill et al., 2015). Two approaches to assess the impact of dams on a river basin is the river fragmentation index (RFI) and the river regulation index (RRI). The RFI expresses the degree that riverine longitudinal connectivity is physically obstructed by dams (Figure 4.10), and strongly relates to the degree that fish migration is directly impacted by dams. The RRI expresses the degree that annual runoff of a basin is impounded by dams, and is important to issues related to downstream flow competence, water scarcity, and riparian habitat. Changes to the RRI and RII mirrored global dam construction over the twentieth century (Figure 4.10), and are expected to undergo sharp increases between 2010 and 2030 with the new boom in dam construction.

While global changes to the RRI and RII over most of the twentieth century were somewhat similar, they often vary considerably for specific drainage basins because of differences in the types of dam constructed, and particularly because of basin hydroclimatology. The annual streamflow regime of the Danube, Europe's second largest river, is minimally impacted by mainstem dams (RRI < 20%) because the reservoirs are essentially large narrow run-of-the-river structures that do not store much water relative to the large volume of annual runoff supplied from the humid headwaters. But the lower Danube is substantially fragmented by dams with an RFI of ~90%, which effectively means that fish migration no longer occurs. And Danube basin fragmentation is projected to become much worse if some 2,700–3,000 (small to large) hydropower dams in the Balkan headwaters are completed (Hockenos, 2018; Zarfl et al., 2019).

A key concern as regards the Balkan dam boom is that they're being hastily constructed with very little oversight, including very little baseline or post-dam environmental data. A total of 636 new hydropower dams are already operational in the Balkans. Alarmingly, a scant 2.2% of the dams had environmental monitoring stations within the upstream or downstream vicinity. This includes only fourteen of the dams with hydrologic monitoring stations, only six with fish monitoring stations, and only four had macroinvertebrate monitoring stations. The capability to conduct environmental impact assessments of the Balkan dam construction boom is severely hampered. And for Balkan nations in the EU it will be difficult to comply with the Water Framework Directive (Huđek et al., 2020).

The RFI for the Amazon is expected to increase from near zero in 2010 to ~20% by 2030 because of the large number of dams being constructed (Grill et al., 2015; Latrubesse et al., 2017). But because of the substantial annual runoff volumes of the Amazon, over the same period the RRI is only expected to increase from about zero to ~5% (Grill et al., 2015).

In contrast to humid regions, rivers impounded by reservoirs in semiarid settings tend to have higher RRI values, particularly with broad reservoirs in flat terrain. Rivers draining the US Great Plains and western interior regions, including the Colorado River basin, are the most heavily impacted in the United States because a large proportion of the basin's total annual water yield is impounded within reservoirs (Graf, 1999, 2001). The high proportion of annual runoff impounded by dams produces considerable downstream water stress in dry regions, which is challenging when managing water for both irrigation and riparian ecosystems. While dry impounded regions in the United States are increasingly introducing dam and reservoir management strategies to mitigate the impact of altered streamflow regimes, in other dryland regions the rapid pace of dam construction is causing additional water stress.

4.2.2.1 INDUS BASIN PROJECT

The Indus Basin Project is the world's largest irrigation system and includes some 26.3 Mha of cultivated lands within a dry region fed by one of the world's largest rivers. The irrigation project was born from the Indus Water Treaty signed in 1960, which has been a model of trans-boundary cooperation between India and Pakistan for seven decades. To manage water supply to the irrigation district three large reservoirs were constructed (Chashma, Mangla, and Tarbela). Additional hydraulic infrastructure comprising the irrigation project includes twelve inter-river canals, forty-five canals extending 60,800 km, twenty-three barrages and siphon structures, communal watercourses, farm channels, and field ditches that extend some 1.6 million km to provide water to over 90,000 farmer-organized water courses (FAO, 2016). The irrigated lands provide some 95% of Pakistan's food production. But the system is fragile, needing repairs, and under stress because of increasing competition from hydropower dams in the Himalayan headwaters of the Indus basin (Raza et al., 2019).

Despite the long-term effectiveness of the Indus Water Treaty, the race to build mega dams in the Indus basin headwaters are straining relations between India and Pakistan, as well as China (Laghari et al., 2012; Raza et al., 2019). India is building several large hydropower plants in the Sutlej, Beas, and Ravi River basins that will impound streamflow that drains to Pakistan (FAO, 2016). China is building dams on the upper Sutlej River in China, which then drains to India before eventually joining the Indus River in Pakistan as a major tributary. Pakistan is constructing the Diamer-Bhasha Dam (2028 completion), which at 272 m high will be the tallest concrete filled roller dam in the world and have a massive storage capacity of 10 km^3. Pakistan is also constructing the Bunji Dam (190 m high) on the upper Indus River main-stem as well as five additional dams to form a North Indus River cascade.

Collectively, the new Indus River basin dams will reduce water and sediment flow to the largest irrigation project in the world, particularly in Pakistan's Punjab and Sindh provinces, the latter of which includes much of the heavily degraded Indus delta (Qureshi, 2011; FAO, 2016). Already some 95% of Indus River water is extracted for irrigation before reaching the coast, with zero flow in prolonged dry periods (FAO, 2011). This means that the extensive irrigation districts in the dry lower basin and delta will experience additional water stress (Qureshi, 2011), with expected declines in riparian and delta fisheries (Laghari et al., 2012).

Overall, new dam construction in the Indus Himalayan headwaters is projected to result in a modest increase in river fragmentation (RFI from ~60% to ~75% over 2010–2030), but because the structures are mega dams the degree of impoundment will be staggering, with the RRI increasing from ~30% to ~180% over 2010–2030 (Grill et al., 2015).

4.3 ENVIRONMENTAL IMPACT OF DAMS AND RESERVOIRS

4.3.1 Streamflow Regime

Upon impoundment, the impact of dams to streamflow is immediate and can be assessed by analyzing a time series of streamflow before and after dam installation (Figure 4.11). Changes to streamflow regime principally occurs in two ways: (i) a reduction in flow variability and (ii) a shift in seasonality of high- and low-flow periods (Graf, 2001; Magilligan et al., 2008). The former is manifest by a reduction in maximum (annual peak events) discharge and an increase in minimum (low) flows. The latter results in a change to the seasonality of high- and low-flow periods, potentially being out-of-phase with critical downstream ecological processes. The shift to a more monotonous flow regime is especially concerning because reduced flow variability directly reduces riverine geodiversity (i.e., physical integrity), to which riparian biodiversity is tightly dependent (Ward and Stanford, 1995; Nislow et al., 2002; Pegg et al., 2003; NRC, 2005). Channel morphologic features and geomorphic processes such as

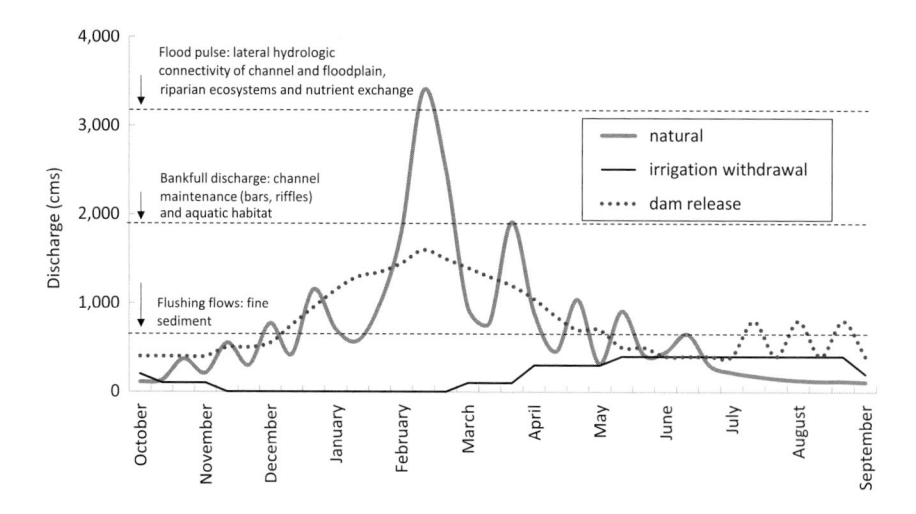

Figure 4.11. Conceptual model of annual flow regime for natural and impounded rivers, including "sawtooth" pattern from storage and release schedule for hydropower generation as well as withdrawal for irrigation to support agriculture during dry periods. (Author figure.)

pools and riffles, sediment flushing, lateral migration, and planform geometry are controlled by moderate-sized discharge pulses at about bankfull or flood stage. Channel and floodplain (lateral) connectivity is crucial to providing riparian wetland ecosystems with nutrients and water, and for migration of aquatic organisms. Benthic riparian ecosystems require periodic "flushing flows" to cleanse algae from gravel, remove fine sediment from interstitial pores, reorganize meso-scale bed topography, and resupply spawning gravel (Power et al., 1996). Because many dams are designed to provide water for crop cultivation, withdrawals for irrigated agriculture impose water stress on downstream riparian environments, particularly during vulnerable low-flow periods (Batalla et al., 2004; van Dijk et al., 2013).

4.3.1.1 UNITED STATES

The wide physical geographic variability across the United States elucidates regional differentiation on the downstream impacts of dams to streamflow regime. And the greatest impacts occur along rivers with a high ratio of reservoir storage capacity to mean annual water yield (~dry regions). At a national scale, very large dams in the United States have reduced annual peak discharge pulses by 67% and low-flow discharge has increased by 52%. The ratio between the annual maximum and mean flow has decreased by 60% (Graf, 2006). These hydrologic changes are epitomized along the middle Missouri River, for example, where the post-impoundment discharge during the spring high-flow period is significantly lower, a critical issue because of coinciding during periods of fish spawning (Pegg et al., 2003).

An additional important streamflow alteration with ecological implications is changes to the timing (seasonality) of high- and low-flow periods. This often occurs because of dam management in relation to stakeholder interests (e.g., irrigation schedules, electricity demands, recreation, flood management). In such cases, critical high-flow pulses may be offset by weeks to months. Rather than coinciding with the "natural" wet period, higher flows occur during low-flow periods (Graf, 2006). The most problematic aspect of changes to the timing of high- and low-flow periods occurs when they are out of accord with the environmental rhythms of the riparian corridor, such as seed germination, avian water fowl nesting, or the dependence of fisheries lifecycles for feeding and spawning (Graf, 2002; NRC, 2005; Górski et al., 2012), as discussed below.

The reduced streamflow variability of US rivers is mainly consistent with international trends in changes to downstream streamflow regime following impoundment and dam closure, with variation due to basin-specific physical and stakeholder influences, particularly in support of agriculture.

4.3.1.2 EUROPE: EBRO AND VOLGA RIVERS

Extensive national and regional analyses of the impacts of dams on flow regimes similar to that conducted in the United States (i.e., Graf, 2006) have not been conducted for Europe. It is well acknowledged, however, that streamflow regimes of European rivers are heavily impacted by dams. Indeed, 90% of Earth's most obstructed rivers are located in Europe (Kornei, 2020). An extensive analysis of pre- and post-dam streamflow trends for the Ebro basin (85,530 km^2) in Spain identified reduced flow variability (Batalla et al., 2004). This was manifest by higher low flows due to reservoir outflow during dry periods (summer) to support irrigation while the storage of winter streamflow reduced maximum flow values. The two-year and ten-year flood magnitudes were reduced by over 30% on average for most gauging stations, including in the lower Ebro River (at Tortosa) downstream of the Mequinenza and Ribarroja reservoirs (Figure 4.12).

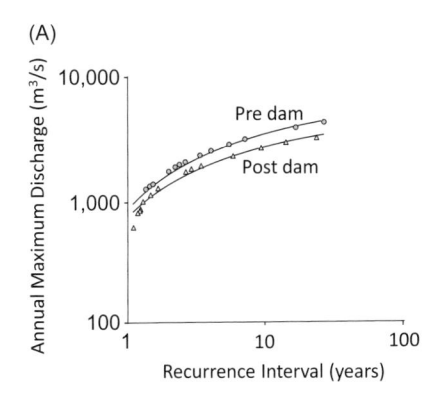

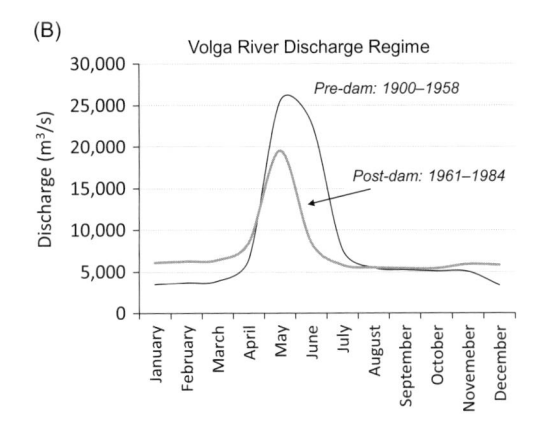

Figure 4.12. Changes to flow regime for two large European rivers, before and after impoundment. (A) Reduction in annual maximum floods, lower Ebro River (at Tortosa) downstream of the Mequinenza and Ribarroja reservoirs. (B) Changes to lower Volga River discharge regime after dam closure over 1959–1960. (Sources: Ebro River, Batalla, 2003; Volga River, Schmutz and Moog, 2018 with data from Górski et al., 2012.)

Like many large northern midlatitude rivers, the Volga (1,400,000 km^2) serves as a major economic corridor. The Volga River is navigable for some 3,200 km through a series of eleven dams and canals that link northern ports along the White and Baltic to southern ports in the Caspian and Black Seas (Avakyan, 1998; Middelkoop et al., 2015; Rodell et al., 2018). The Volga annually handles two-thirds of inland Russian navigation. Following impoundment of the lower Volga River behind the Volgograd Hydroelectric station in 1959 (at closure the largest hydroelectric dam in the world, remains largest in Europe), annual flow variability declined overall, including an increase in low flows by over ~1,000 m^3/s (Figure 4.12B). The average maximum discharge reduced from 34,500 (m^3/s) prior to dam closure (1959) to 26,800 (m^3/s) after impoundment between 1959 and 1999 (Korotaev et al., 2004). Of key importance to the hydrology of the lower Volga River is the vast amount of water withdrawn for industry, domestic use, and especially irrigation for agriculture, with the latter annually accounting for some 200 Mha of cultivated lands. Water consumption greatly increased following dam construction, increasing from 6 km^3/yr pre-dam to 24 km^3/yr post-dam, an amount that is about 10% of the water discharged into the Caspian Sea (Middelkoop et al., 2015). The Volga contributes 80% of the annual runoff supplied to the Caspian Sea, and annual water withdrawals for agriculture are contributing to Caspian Sea levels lowering (Rodell et al., 2018).

4.3.1.3 YANGTZE AND THREE GORGES DAM

The streamflow regime downstream of the world's largest dam, Three Gorges Dam (closure in 2003) along the middle Yangtze River has also changed downstream streamflow patterns (Zhang et al., 2016; Tian et al., 2019). Reduced flow variability extends some ~1,600 downstream, with reductions ranging from about 11% below the dam to 4% near the delta (Wang et al., 2013). These modest changes in river stage, however, mask hydrologic problems associated with management of Three Gorges Dam, in particular the abrupt up and down ramping and flow release schedule that produces frequent out-of-season flow variability. The flow release results in an oscillating pattern of drying out and backwater flooding of the massive floodplain lake system, including Poyang Lake (Zhang et al., 2016). Not surprisingly, and with fair warning (i.e., Leopold, 1998), the downstream riparian environment has substantially degraded since closure of Three Gorges Dam, in association with sediment reduction and geomorphic adjustment (Zhang et al., 2016; Cheng et al., 2018).

4.3.1.4 NILE, ASWAN HIGH DAM (LAKE NASSER AND LAKE NUBIA), AND WATER POLITICS

Because of its link to Egyptian antiquity, the historic Nile streamflow regime is perhaps the world's most iconic coupled agro–hydro system, with irrigated agriculture extending along the entire Nile valley and across the delta plain (Figure 4.13). For over six millennia the Nile annually delivered nutrient-rich floodwater and silts to irrigated agricultural lands. But this abruptly ceased upon closure of the Aswan High Dam over 1965–1967 (Figure 4.14). The prior dynamic flow regime characterized by a seasonal fluctuation in river levels of 5–8 m has been greatly reduced to 1–2 m. As is common to downstream flow variability, Nile River low flows are moderately higher while the large annual flood pulses are greatly reduced, from ~8,000 m^3/s to less than ~3,000 m^3/s (Figure 4.15). A system of weirs downstream of Aswan High Dam maintains moderate flow stage levels to increase the river's irrigation potential. Nile impoundment has greatly reduced annual runoff volumes, from 80 km^3 to 30 km^3 from pre-dam to post-dam, respectively (Liu et al., 2017). Although agricultural productivity increased with impoundment, the high demands on the irrigation system results

Figure 4.13. The Nile extends ~1,000 km from Aswan High Dam (bottom of image) to the delta without tributary inputs. Riparian agriculture effectively covers the entire valley and delta and contrasts with the hyper-arid Sahara. (Licensed by CC.)

Figure 4.14. Lake Nasser and Aswan High Dam, and Nile River in Egypt. Lake Nasser formed with dam closure over 1965–1967. The reservoir covers 5,250 km^2 and extends for ~500 km, with 350 km as Lake Nasser in Egypt and 150 km as Lake Nubia in Sudan. Storage capacity for the reservoir is 162 billion m^3. Reservoir sedimentation is measured regularly by bathymetric surveys along twenty-one fixed cross sections (Ahmed and Ismail, 2008). Scale: main axis of reservoir is ~3 km in width. North is to lower right of image. (Source: NASA photo, April 12, 2015.)

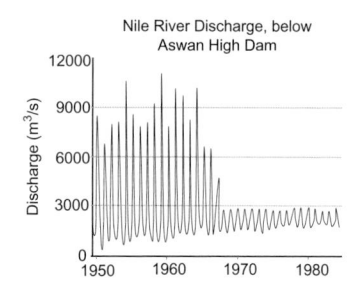

Figure 4.15. Streamflow variability of Nile River below Aswan Dam between 1950 and 1985, revealing impacts to annual flood pulse (higher low flows, much lower peak flows) after dam closure over 1965–1967. (Figure Source: Vörösmarty and Sahagian, 2000).

in frequent water stress for some 95% of Egypt's riparian population dependent upon the Nile River for 90% of its freshwater.

Today the main issue concerning Nile River management is located some 3,200 km upstream of Cairo and concerns impoundment of the Blue Nile by the Grand Ethiopian Renaissance Dam. As the largest reservoir in Africa, the Grand Ethiopian Renaissance Dam will store 1.5 times the average annual flow of the Blue Nile (Wheeler et al., 2016). This will greatly increase the index of river impoundment (i.e., Grill et al., 2015) and, based on lessons learned from dam impacts on dryland rivers in the United States, the downstream riparian zone will be substantially degraded because of reduced flow variability (e.g., Graf, 2006).

For well over a decade, imminent impoundment of the Blue Nile in Ethiopia has increased water stress and political tensions between Egypt and Ethiopia. Aside from the longer-term management scheme of the dam and reservoir – such as changes in water quantity, quality, and the outflow release schedule – a pressing issue receiving considerable international attention is the period of reservoir infilling (Wheeler et al., 2016; El-Nashar and Elyamany, 2018). If the reservoir is rapidly infilled, it requires that the reservoir impounds all upstream Blue Nile waters (zero outflow). Even this extreme measure will require about three years to fill the reservoir to a level where hydropower turbines can be activated. And this is dependent upon the rainfall patterns over the Ethiopian highlands (El-Nashar and Elyamany, 2018). Such a rapid rate of reservoir infilling would effectively desiccate downstream riparian lands in Sudan, upstream of Egypt's Aswan High Dam. International agreements between Nile basin nations with regard to water withdrawals have been especially ineffective but are particularly critical in view of imminent closure of the Grand Ethiopian Renaissance Dam (Wheeler et al., 2016).

Table 4.4 *Impacts of reservoir sediment trapping along several large rivers*

River (nation)	Reduction in sediment loads (%) below dam
Nile (Egypt)	100
Orange (South Africa)	81
Volta (Ghana)	92
Indus (Pakistan)	76
Don (Russia)	64
Krishna (India)	75
Ebro (Spain)	92
Kizil Irmak (Turkey)	98
Colorado (USA)	100
Rio Grande / Rio Bravo (USA / Mexico)	96

Source: Walling (2006), with data reported from Vörösmarty et al. (2003).

Table 4.5 *Comparison of two estimates of the global fluvial sediment budget and its modification by human activity*

Component	Syvitski et al. (2005)	Walling (2006)
Prehuman land-ocean flux (Gt/yr)	14.0	14.0
Contemporary land-ocean sediment flux (Gt/yr)	12.6	12.6
Reduction in flux associated with reservoir trapping (Gt/yr)	3.6	24.0
Contemporary flux in the absence of reservoir trapping (Gt/yr)	16.2	36.6
Increase in prehuman flux due to human activity (%)	22	160
Reduction in contemporary gross flux due to reservoir trapping (%)	16	66

Source: Walling (2006).

4.3.2 Impact of Dams to Fluvial Sediment Flux

4.3.2.1 GLOBAL PERSPECTIVE

The global proliferation of dams means that copious amounts of Earth's riverine sediments are trapped in upstream reservoirs, small and large (e.g., Morris and Fan, 1998; Walling, 2006; UNESCO-IHP, 2011; Tockner et al., 2016). Trap efficiencies of main-stem dams along large rivers tend to be very high (Table 4.4), with many reservoirs trapping greater than 80% of upstream sediment flux (Vörösmarty et al., 2003). Many US dams, including large dams in the Mississippi basin, have trap efficiencies greater than 95%. The large main-stem Colorado dams trap up to 100% of upstream sediment loads (except when intentionally flushed), as does the Nile River upstream of the Aswan High Dam (Williams and Wolman, 1984).

Fluvial sediment delivery to the coast has precipitously declined since the mid-1900s (Syvitski et al., 2005; Walling, 2006; UNESCO-IHP, 2011). Globally, some 12.6 GT of sediment loads are annually discharged into the oceans (Table 4.5). While estimates vary widely, the global decline in sediment flux ranges from 16% to 66% of annual riverine sediment loads (Syvitski et al., 2005; Walling, 2006). Much of the decline in sediment flux is attributed to the impact of dams, although improved land management, channel engineering, and climate change also reduces sediment loads (Batalla et al., 2006; Wang et al., 2007; Walling, 2008b; Meade and Moody, 2010; Liu et al., 2017). Because dam construction coincides with different periods of human activities and climatic fluctuation, changes to sediment loads after dam closure can vary greatly from river basin to river basin (Table 4.6).

Reservoir sediment storage has diverse implications to fluvial environments. This includes upstream, within, and downstream of the reservoir (Table 4.7). Reservoir sedimentation can result in upstream flooding and reduced navigation, and aquatic habitat degradation. Problems associated with sediment storage in reservoirs includes reduced hydropower potential because of reduced storage capacity, storage of contaminated sediments, and reservoir stratification impacting aquatic habitat within the reservoir and downstream (discussed further, below). The high sediment trap efficiency of many dams substantially reduces downstream fluvial sediment loads, with consequences to geomorphic adjustments and riparian habitat. In addition to the quantity of sediment being reduced, the quality of sediment is often also reduced. In combination with the altered flow regime fine bed material is removed, resulting in downstream channel bed material becoming coarser, and in some cases leading to channel bed armoring.

4.3.2.2 SEDIMENT DECLINE TO NILE, LINK TO AGRICULTURE

For millennia, rains in the Ethiopian Highlands generated runoff that transported some 95% of the Nile River's sediment load. Sediment was principally supplied to the main-stem Nile by the Blue Nile (324,530 km^2) and Atbara (166,875 km^2) Rivers, both of which enter the main-stem Nile upstream of Aswan High Dam. But after impoundment behind the Aswan High Dam all upstream Nile sediment is stored in Lake Nasser (Ahmed and Ismail, 2008; Liu et al., 2017). Suspended silt concentrations averaged 3,800 mg/L between 1929 and 1963 but abruptly declined to only 129 mg/L by 1965, two years after the start of infilling Lake Nassar reservoir (Raslan and Salama, 2015). The average annual sediment load in the pre-dam period was 120 million tons, but was reduced to 0.2 million tons after dam construction (Liu et al., 2017). Further downstream, sediment loads in the

Table 4.6 *Comparison of sediment yields and runoff before and after dam impoundment for selected rivers*

River basin		Mean annual sediment load ($\times 10^6$/t)	Mean annual specific sediment yield (t/km^2)	Mean annual runoff (volume – km^3)	Mean annual runoff (depth – mm)
Liaohe	Pre-dam	46.4	384	5.8	48
	Post-dam	7.9	65	1.7	14
Huanghe	Pre-dam	1,243	1,653	50	66
	Post-dam	149	198	10	13
Mississippi	Pre-dam	400	134.2	490	158
	Post-dam	145	48.6		
Nile	Pre-dam	120	38.6	80	26
	Post-dam	0.2	0.1	30	10
Volga	Pre-dam	26	18.8	254	179
	Post-dam	8	5.8		

Data: Liu et al. (2017).

Nile only marginally recover, as there are no appreciable tributaries for about the final 1,000 km (Figure 4.16). A minor amount of sediment is supplied from aeolian inputs, bank erosion and channel bed scour. Low flow structures (weirs) control stage levels in support of navigation and floodplain irrigation (Morris and Fan, 1998; Ahmed and Ismail, 2008). At the coast the Nile discharges very little sediment (~5 × 10^6 tons/yr) to the Mediterranean, with the substantial reduction in sediment loads contributing to coastal erosion and retreat of the Nile Delta (Stanley and Warne, 1998).

Impacts to the Nile sediment and streamflow regime caused by the Aswan High Dam aren't merely significant, they're symbolic. The lower Nile is the premier example of a millennia-old, intensive floodplain-agricultural coupled system, especially dependent upon annual floodwaters and sediment for irrigation and fertility. As an early hearth of irrigation science, Nile farmers utilized an ingenious system of sluice gates, water wheels, dams, and buckets to maximize the potential of furrow-and-field irrigation methods (Butzer, 1976). But the iconic floodwater irrigation system was all but eliminated with construction of the Aswan High Dam, and initially reduced downstream productivity. Crop yields have increased since the dam was completed, but at a high cost. The Nile delta has continued to degrade because of a lack of fluvial sediment (Becker and Sultan, 2009; Stanley and Clemente, 2014), and Nile delta agriculture now requires significantly larger "hard engineering" structures, including larger canals, ditches, and larger mechanical pumps for irrigation and drainage (Molle, 2018). Additionally, the Nile agricultural system is dependent upon high amounts of chemical fertilizer to maintain high agricultural yields, an issue that results in interesting and unintended impacts to coastal fisheries production (Nixon, 2003), as noted further in text. Of course, the artificially high water table is problematic with regard to the annual deficit in rainfall (165 mm/yr) relative to evapotranspiration (1,500 mm/yr), and results in land degradation through soil salinization (Ahmed and Ismail, 2008; Molle, 2018).

4.3.2.3 SEDIMENT DECLINE TO YANGTZE AND HUANGHE RIVERS

Some of the large rivers in Asia have more recently experienced sharp reductions in sediment loads following dam construction, and the changes continue to rapidly unfold because of many new dam construction projects (Syvitski et al., 2005; van Binh et al., 2019). This is particularly true of the Yangtze and Huanghe Rivers in China. Suspended sediment loads along the Yangtze River downstream of Three Gorges Dam at the lowermost sediment station (Datong, 565 km upstream of the delta) were already in decline. In the 1950s and 1960s, suspended sediment load averaged 507 Mt/yr but had declined in the period before dam construction to 320 Mt/yr (1993–2003). After closure of Three Gorges Dam (TGD), the sediment loads between 2003 and 2012 steeply dropped (Figure 4.17), to 145 Mt/yr. Between 2003 and 2012, an astonishing 182 Mt/yr (80% of the total sediment load) was trapped behind Three Gorges Dam (Yang, et al., 2015). Between 2002 and 2008, the median particle size (d_{50}) of the channel bed material increased from 0.36 mm to 25 mm at Yichang (44 km downstream of TGD) but only increased from 0.18 mm to 0.19 mm at Chenglinji (420 km downstream of TGD) (Zheng et al., 2018). Between these two stations (at 195 km downstream of TGD), the d_{50} of bed material increased from 0.19 mm (pre–TGD) to 0.251 mm by 2010, seven years after closure (Zhang et al., 2016).

The Huanghe River is also referred to as the Yellow River because of its ochre colored sediment that derives from erosion of the Chinese Loess Plateau. The Huanghe previously transported 1.08 Gt of sediment per year to the coast (at Lijin, 40 km from delta) – Earth's highest annual sediment load. By 2005 this

Table 4.7 *Environmental and functional issues related to dam and reservoir sediment storage*

Issue	Characteristics
Upstream Reservoir Impacts	
Deposition above pool elevation	Upstream deposition increases flood levels, reduces navigational clearance below bridges, possible water table rise, inducing soil waterlogging and possible salinization, and can increase evaporative losses. Deposition upstream of reservoir pool level reduces reservoir storage capacity.
Within Reservoir Impacts	
Loss of storage capacity	Reduction in firm yield causing water rationing, reduced hydropower, and flood control benefits.
Reservoir stratification	Temperature and oxygen stratification in deep reservoirs can result in degraded water quality for biota within reservoir, and outflow release. Changes include temperature, water chemistry, and dissolved oxygen levels.
Contaminated sediment	Reservoir can trap and bury contaminated sediment, effectively removing it from the active biotic environment. Contaminated sediments can cause lake quality to deteriorate. Sediment removal can remobilize contaminated sediment and are very costly.
Organic sediment deposition	Oxygen demand exerted by organic sediment contributed from upstream or primary producers within a lake can result in anaerobic water in lower depths of reservoir.
Turbidity	Turbid water reduces depth of photic zone, decreases primary productivity. Reduced visibility interferes with fish feeding. Reduced clarity makes lakes aesthetically unpleasant for recreation.
Navigation	Sediment fills navigation channels and locks. Interferes with boating, fishing, and marina access.
Wildlife	If depositional areas become wetlands, significant wildlife benefits may occur. Fine sediment deposition can degrade and reduce fish habitat.
Downstream Reservoir Impacts	
Air pollution	Exposure of fine sediment to winds during reservoir drawdown can result in dust storms. Greenhouse gases: methane (CH_4) and carbon dioxide (CO_2) produced by decomposition of organic materials (esp. deep reservoirs).
Reduced coarse sediment load	River bed may incise, accelerate bank erosion. Lowered base level may initiate erosion along tributary channels, and desiccate wetlands. River bed will coarsen and may become unsuitable for spawning. Bridge pier and river training works may be undermined, river bank structures may be threatened, and deposits of contaminated river sediment may be remobilized.
Reduced flow variability	Reduced peak discharge can lead to reduction in channel size and vegetative encroachment, reducing downstream channel conveyance. Increased low flows remove functional surfaces that represents habitat.
Reduced fine sediment load	Reduce sedimentation and dredging in navigational channels. Increased erosion along riparian lands, loss of sediment-dependent wetlands, reduced nutrient and sediment inputs to floodplains and riparian wetlands. Downstream water clarity benefits water-based recreation and sediment-sensitive species, including coastal ecosystems.
Sediment release	Timing and magnitude of sediment release may influence downstream environmental and economic activities associated with the river.
Water quality	Reservoir outflow may include water that is lower in dissolved oxygen, lower temperature, and water chemistry that is harmful to riparian ecosystems.
Coastal	Reduced sediment flux to estuarine, deltaic and coastal zone can drive salinity increases, loss of wetlands, coastal retreat.

Sources: Morris and Fan (1998), Scheueklein (1990), Randle et al. (2019), Winton et al. (2019), Morris (2020), and author.

had declined to 0.15 Gt/yr, which is only 14% of its historic sediment load (Wang et al., 2007).

A key reason for the decline in sediment loads of the Huanghe River is construction of seven main-stem dams between 1968 and 1998 (Table 4.8). It should be noted, however, that in addition to main-stem dam emplacement several other explanations are also responsible for the large decline in sediment loads over the past several decades, for both the Huanghe and Yangtze Rivers. These include improved land management (noted in Chapter 2), increased water withdrawals for agriculture and industry, and reduced precipitation due to climate change (Walling, 2006; Wang et al., 2007; Yang et al., 2015; Shi et al., 2017). For the Huanghe, the impact of main-stem dam construction and improved management of the Loess Plateau has been so

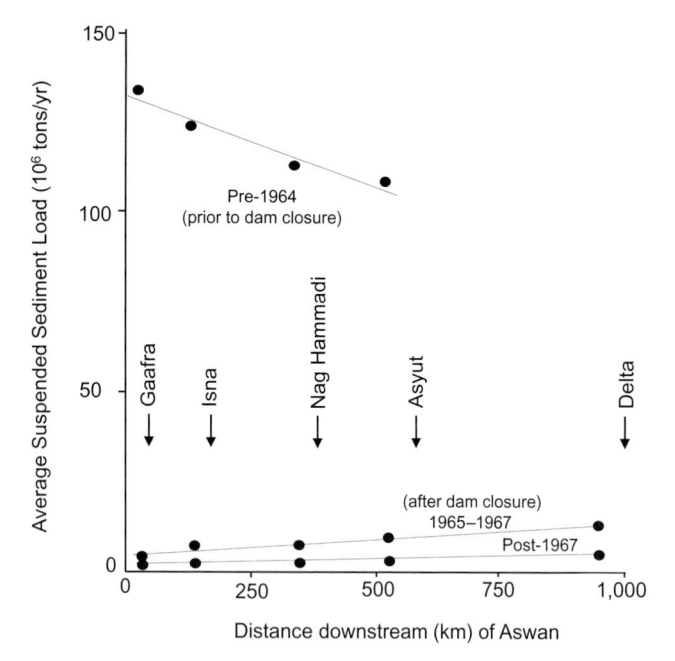

Figure 4.16. Downstream changes to suspended sediment loads along the Nile River before and after closure of Aswan High Dam in 1964. Suspended sediment comprises 30% clay, 40% silt, and 30% fine sand (Ahmed and Ismail, 2008). (Source: Morris and Fan, 1998.)

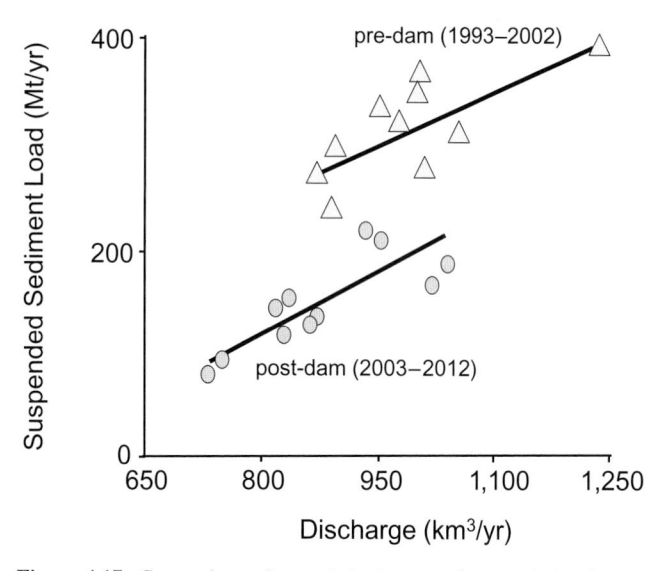

Figure 4.17. Comparison of annual discharge and suspended sediment loads before and after impoundment of the Yangtze River by Three Gorges Dam: Pre-dam (1993–2002) and post-dam (2003–2012). Datong is 565 km upstream of the river mouth. (Source: Yang et al., 2015.)

effective at reducing downstream sediment loads that downstream channel engineering measures were implemented to manage channel bed incision. Between the 1960s and 2000, soil conservation in the Chinese Loess Plateau reduced sediment inputs into the Huanghe River by some $\sim 300 \times 10^6$ tons/yr. On the Yangtze, 65% of the reduction in downstream sediment loads since 2003 is attributed to closure of Three Gorges Dam, 5–14% is attributed to climate change (reduced precipitation), and the remaining cause of sediment load reduction is attributed to upstream dams and soil conservation (Chen et al., 2007; Yang et al., 2015).

4.3.2.4 SEDIMENT DECLINE FOLLOWING MEKONG DAM CONSTRUCTION

The Mekong delta receives much less sediment than before construction of a cascade of dams along the upper Mekong (Lancang) in China, which started with closure of Manwan Dam in 1993. The most recent analysis of "before and after" comparison of sediment loads shows a large sediment flux reduction to the Mekong delta, by some 74%. The pre-dam (before 1993) sediment flux to the Mekong delta was 166.7 million tons/yr (± 33.3) but has now declined to 43.1 million tons/yr for 2012–2015 (van Binh et al., 2019), which varies by ± 20 million tons per year depending on whether ENSO is in a negative or positive phase, respectively (Ha et al., 2018).

Closure of the Manwan Dam on the upper Mekong main-stem abruptly reduced downstream suspended sediment loads by >60% at Gajiu, 2 km below the dam, with modest sediment recovery further downstream (Fu et al., 2008). Between 1993 and 2003, 26.9–28.5 million tons/yr were trapped in the Manwan reservoir, resulting in reservoir sediment storage of 295.9–313.5 million tons over an eleven-year period (Fu et al., 2008). The large amount of reservoir sedimentation is concerning, as 21.5–22.8% of the total storage capacity of the Manwan reservoir was lost over an eleven-year period following impoundment. As a whole, the trap efficiency of six large main-stem dams along the upper Mekong in China ranges from 61% to 92% (Figure 4.18). As regards downstream sediment loss this is alarming when considering that some thirteen additional main-stem dams are planned or under construction upstream, toward the basin headwaters (Kummu and Varis, 2007; Fan et al., 2015) (Figure 4.19). And already thirty-seven more dams exist in the lower basin in Laos, Cambodia, Thailand, and Vietnam.

Detecting changes to the Mekong delta sediment flux highlights the importance of having continuous sediment records before and after impoundment, having sediment gauging stations in key upstream and downstream locations within the drainage basin, and having access to the data (Walling, 2008a, 2009). A study by van Binh et al. (2019) provides the longest period of analysis – spanning five decades – that envelops the pre- and post-dam periods (1961–2015). Crucially, the analysis utilizes sediment load data for years 1965–2003 at the Jiuzhou station (China), above the upper Mekong cascade of dams.

The post-dam period for the Mekong River is a story that continues to unfold because of construction and planning of numerous large dams (Figure 4.19) throughout the basin (Walling, 2008a,b; Pokhrel et al., 2018). Because of the large amount of drainage area in the lower Mekong, there is concern

Table 4.8 *Characteristics of main-stem reservoirs along the Huanghe River, China*

Reservoir	Upstream drainage area (10^3 km^2)	Reservoir capacity (10^9 m^3)	Date of commissioning
Longyangxia	131.4	24.7	October 1986
Liujiaxia	181.8	5.7	October 1968
Qingtongxia	275	0.606	April 1967
Sanshenggong	314	0.08	April 1961
Wanjiazhai	395	0.896	October 1998
Sanmenxia	688.4	9.64	October 1960
Xiaolangdi	694.5	12.65	October 1999

Source: Shi et al. (2017).

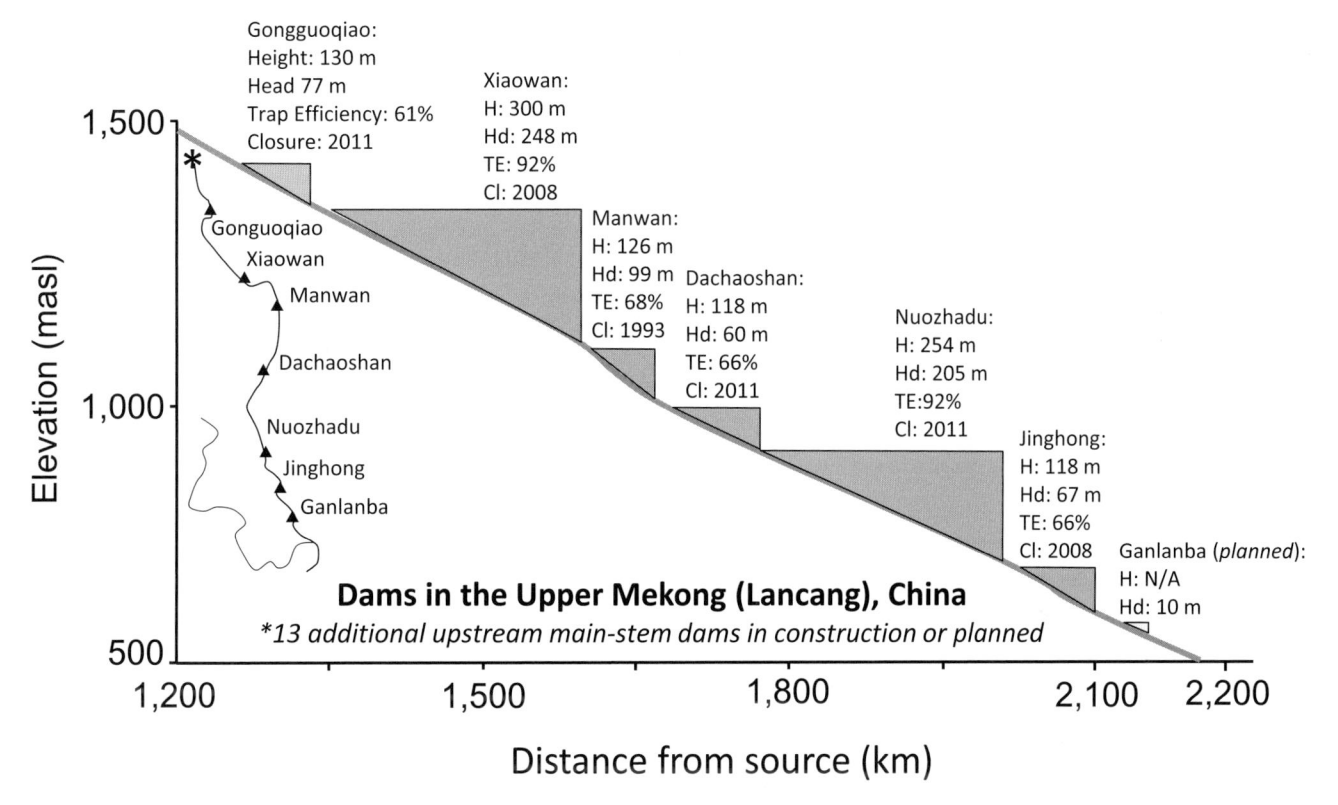

Figure 4.18. The "stair-step" profile of a ~1,000 km segment of the upper Mekong (Lancang) River in China, upstream of Laos border, including trap efficiency (%), hydraulic head, dam height, and year of closure. *Upstream of this reach, seven large main-stem dams are under construction, including Miaowei (140 m high), Dahuaqiao (106 m), Huangdeng (202 m), Tuoba (158 m), Lidi (74 m), Wunonglong (136.5 m), and Gushui (220 m); and further upstream, six additional main-stem hydroelectric dams are planned or in construction, extending the stair-step profile toward the Himalayan headwaters. (Source: InternationalRivers.org Fact Sheet on upper Mekong (Lancang) Dams, May 2013, accessed April 12, 2020. Figure source: Kummu and Varis, 2007, with author modifications and updates.)

over the ongoing and future development of forty-two dams in the Se San, Srepok, and Se Kon Rivers in the lower basin, which join together to form the largest tributary to the Mekong. If the 133 dams planned for the Mekong are constructed only 4% of the Mekong River sediment load will be discharged to the delta (Kondolf et al., 2014a, 2014b; van Binh et al., 2019), which would be devastating to the deltaic geomorphology and associated aquatic environments (Piman and Manish, 2017).

4.3.2.5 SEDIMENT DECLINE FOR SEVERAL EUROPEAN RIVERS: RHINE, EBRO, VOLGA

The sediment loads of European rivers have been impacted by humans for centuries and millennia. Several rivers are representative of the impacts of dams to European rivers, including the Ebro, Volga, and Rhine Rivers.

A longitudinal profile of the Rhine River basin reveals the artificial stair-step pattern in the Upper and High Rhine because

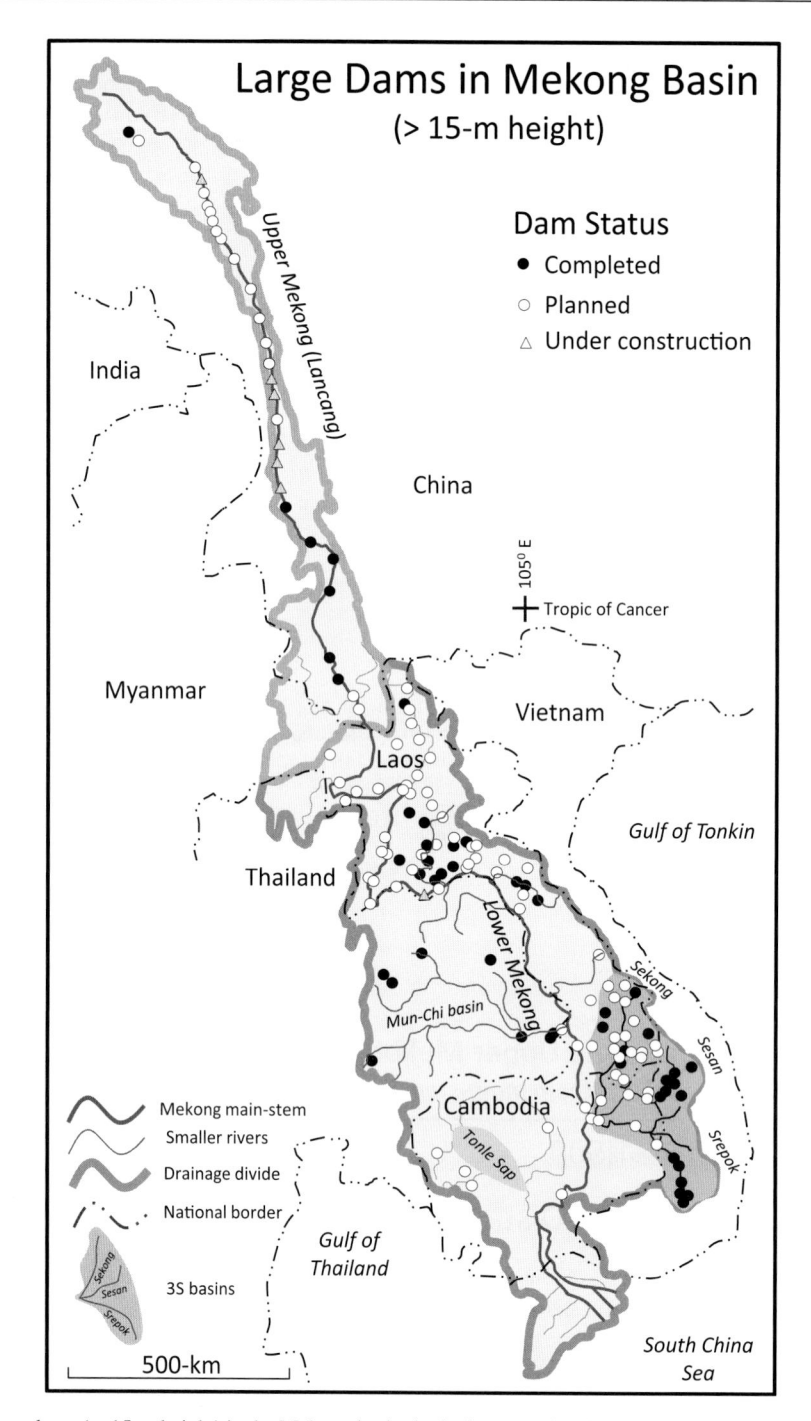

Figure 4.19. Location of large dams (> 15-m height) in the Mekong basin, including completed, under construction, and planned. The border between the Upper Mekong (Lancang) and Lower Mekong divide is indicated. The "3S" river basins refer to the Sekong, Sesan, and Srepok Rivers, which join to form the largest tributary in the lower Mekong. (Author figure. Data sources include Mekong River Commission and Räsänen et al., 2017 (for large dams).)

of intensive impoundment. This includes the Iffezheim Dam, the lowermost and last dam to be emplaced along the main-stem Rhine River (Figure 4.20). Suspended sediment loads along the Rhine River (185,000 km^2) has increased and decreased with episodes of land cover change and hydraulic engineering (Vollmer and Goelz, 2006; Spreafico and Lehmann, 2009; Middelkoop et al., 2010; van der Perk, 2019). The more recent episode, since about the 1950s, reveals about a 70% decline, which is dramatic for a river that had already been intensively impacted by human activities. Annual suspended sediment loads declined from 4×10^6 tons to 1.2×10^6 tons by 2016, and have remained relatively stable over about the past ten years (van der

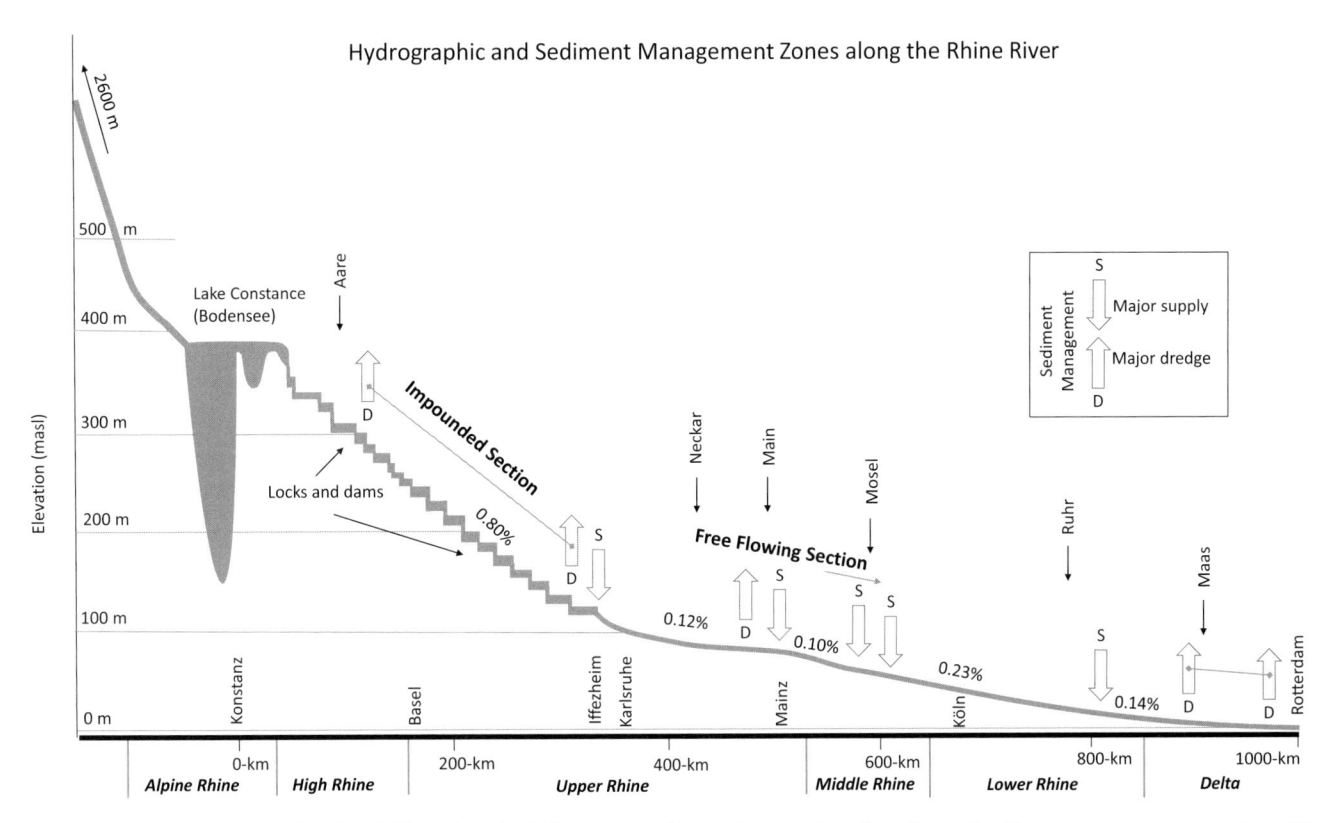

Figure 4.20. Longitudinal profile of the Rhine River, including geomorphic province, major tributaries, and sediment management operations. The stair-step pattern in the Upper and High Rhine is due to intensive impoundment. Gambsheim hydroelectric dam is located at 309-km, 25 km upstream of Iffezheim dam. (After Götz, 2008 and Frings et al., 2019.)

Perk, 2019). The primary source of the decline since the early 1950s is likely the completion of two main-stem dams in the Upper Rhine, specifically Gambsheim and Iffezheim dams (Figure 4.20) completed in 1974 and 1977, respectively. As a whole there are twenty-one main-stem dams on the Upper and High Rhine. The sediment starvation problem along the Rhine has resulted in channel bed incision along large reaches of the main-stem Rhine in the delta and alluvial valley (Quick et al., 2020) that threatens navigation and associated infrastructure, requiring innovative sediment management approaches (see Section 8.2.3).

The sediment flux of the lower Ebro River (85,530 km^2) was historically high, annually discharging 20–30 million tons/yr into the Mediterranean (late nineteenth century). But the Ebro suspended sediment loads abruptly declined to 3.3 million tons/yr following a period of dam building in the early 1960s. By the early 2000s, and ~299 dams later, the sediment load transported to the delta had declined to a meager 0.29 million tons/yr – less than 2% of pre-dam sediment loads (Rovira and Ibàñez, 2007). The dramatic reduction in Ebro River sediment load has had substantial adverse consequences to its delta, and ground subsidence is of particular concern (Tena and Batalla, 2013). About 45% of the emergent Ebro delta will be drowned by 2100 because of subsidence and rising sea levels. The strategy to mitigate against higher sea levels includes building new lands by sediment diversion structures within the delta and managing reservoir sediment by flushing operations to mobilize stored reservoir deposits for transport through a cascade of reservoirs (Rovira and Ibàñez, 2007; Tena and Batalla, 2013).

Impoundment of the lower Volga River behind the Volgograd Dam trapped sediments and resulted in a large decline in annual suspended sediment loads. Prior to dam closure (1934–1953), the suspended sediment loads of the lower Volga River were 12–18.5 million tons/yr, which declined to 7.4 million tons/yr in the post-dam period (1961–1982). The annual suspended sediment load at the delta apex substantially declined from the pre-dam to the post-dam period, being 26 million tons/yr and 7.9 million tons/yr, respectively (Middelkoop et al., 2015).

4.3.2.6 SEDIMENT DECLINE TO MISSOURI AND MISSISSIPPI RIVERS

The fact that upstream dam efficiencies are so high and that a lot of fluvial sediment still makes it to the coast implies that sediment loads somewhat recover (Williams and Wolman, 1984; Phillips et al., 2004). This occurs largely because of downstream channel scour and erosion (discussed subsequently) and because impacts of upstream dam trapping can be buffered by sedimentary inputs from downstream tributaries. The downstream sediment load recovery is often less than 50% of upstream sediment loads.

Figure 4.21. Gavins Point Dam and Lewis and Clarke Lake, the lowermost dam on the Missouri River. The structure is an embankment dam of earthen and chalk-fill materials. Dimensions: closure in 1955, upstream drainage: 723,825 km^2, 23 m high, 2,650 m long, total reservoir capacity is 606,873 m^3, surface area of 12,700 ha, depth of 14 m, and maximum length of 40 km. (Photo and data source: U.S. Army Corps of Engineers.)

Following closure of the downstream-most main-stem dam in 1955 on the Missouri River, Gavins Point Dam in North Dakota and South Dakota (Figure 4.21), downstream suspended sediment loads immediately declined by 99%. At Hermann, Missouri, a distance of 1,047 km downstream of Gavins Point Dam, suspended sediment loads in the post-dam period (1957–1980) only recovered to ~30% of pre-dam amounts (Figure 4.22), which includes sediment inputs from the Platte River basin (10.1 × 10^6 tons/yr) (Heimann, 2016). The precipitous decline in suspended sediment loads is detected further downstream in the middle Mississippi River (at St. Louis, Missouri), and all the way downstream to the Mississippi delta (Figure 4.23).

As the case with many larger rivers, the timing of dam construction in the Mississippi basin occurred while other forms of hydraulic engineering and land cover change were also occurring, which further reduced downstream suspended sediment loads (Figure 4.24). Additional factors that reduced downstream suspended sediment loads in the Mississippi include improved land management in the upper basin (as with the Huanghe and Yangtze Rivers) and channel engineering in the lower basin. As channel bank erosion was previously a source of sediment for the lower Mississippi River, hydraulic engineering works such as groynes (wing dikes) and concrete revetment further reduce downstream sediment loads (Kesel et al., 1992; Kesel, 2003; Meade and Moody, 2010).

4.3.3 Impacts to River Channels

It is difficult to conceive of anything that could more fundamentally disturb an alluvial river than a large dam obstructing streamflow and trapping the vast majority of its sediment. And considering the sophistication of the science it is somewhat surprising that in 2021 we're unable to accurately predict the direction, style, magnitude, spatial extent, and duration over which downstream geomorphic adjustment to a river will occur upon impoundment – at a level of precision useful to direct management decisions. More than anything, this serves as a reminder as regards the complexity and uniqueness of each fluvial system, and that subtle – but significant - differences in slope, particle size, flood regime, etc. result in varying forms of fluvial adjustment to flow impoundment (Williams and Wolman, 1984; Brandt, 2000; Phillips, 2003). And as much as any subject that concerns lowland rivers, the inability to precisely and accurately predict downstream fluvial adjustment to dams highlights the importance of a local – case-by-case approach – that requires copious field and ground data.

A fundamental change that occurs to initiate channel bed incision is the "hungry water" phenomenon. This develops because sediment loss due to upstream reservoir trapping results in excess stream power and sediment transport capacity downstream relative to the sediment load available for transport (Kondolf, 1997; Habersack et al., 2013). Provided the channel bed is close to the threshold of erosion (i.e., Figure 2.22), the excess stream power is then able to incise the channel bed (Williams and Wolman, 1984; Wilcock, 1998; Chin et al., 2002; Smith et al., 2016). Channel bed incision can then destabilize channel banks and lead to lateral widening, resulting in increased channel width (i.e., Figure 2.24). In the case of bedrock controls, channel incision may be limited, although lateral migration and increased channel widening may be substantial.

The benchmark works by Williams and Wolman (1984) and Graf (1999, 2006) analyzed large databases spanning considerable geomorphic and climatic variability, and provide signposts to make several general statements about the downstream impacts of dams to rivers because of altered streamflow patterns and "hungry water." Following impoundment, rivers are likely to undergo incision, especially immediately downstream of the dam, with some lateral adjustment, especially if the channel banks are noncohesive and if the streamflow regime is not significantly altered. The magnitude of incision diminishes with distance downstream. Reduced flow variability results in channel narrowing, and a formerly active floodplain is likely to become stabilized by encroaching woody vegetation to the detriment of aquatic and riparian habitat (Marston et al., 2005). Most of the main-stem channel adjustment occurs within about the first decade, and continues for decades although it is difficult to generalize how much, and for how far downstream the geomorphic adjustment will occur.

Temporally, the pattern of channel bed incision for rivers spanning a range of geologic and climatic settings mainly follows a negative exponential pattern. A number of channel cross sections along the Missouri, Colorado, Red, and

DAMS, RIVERS, AND THE ENVIRONMENT   101

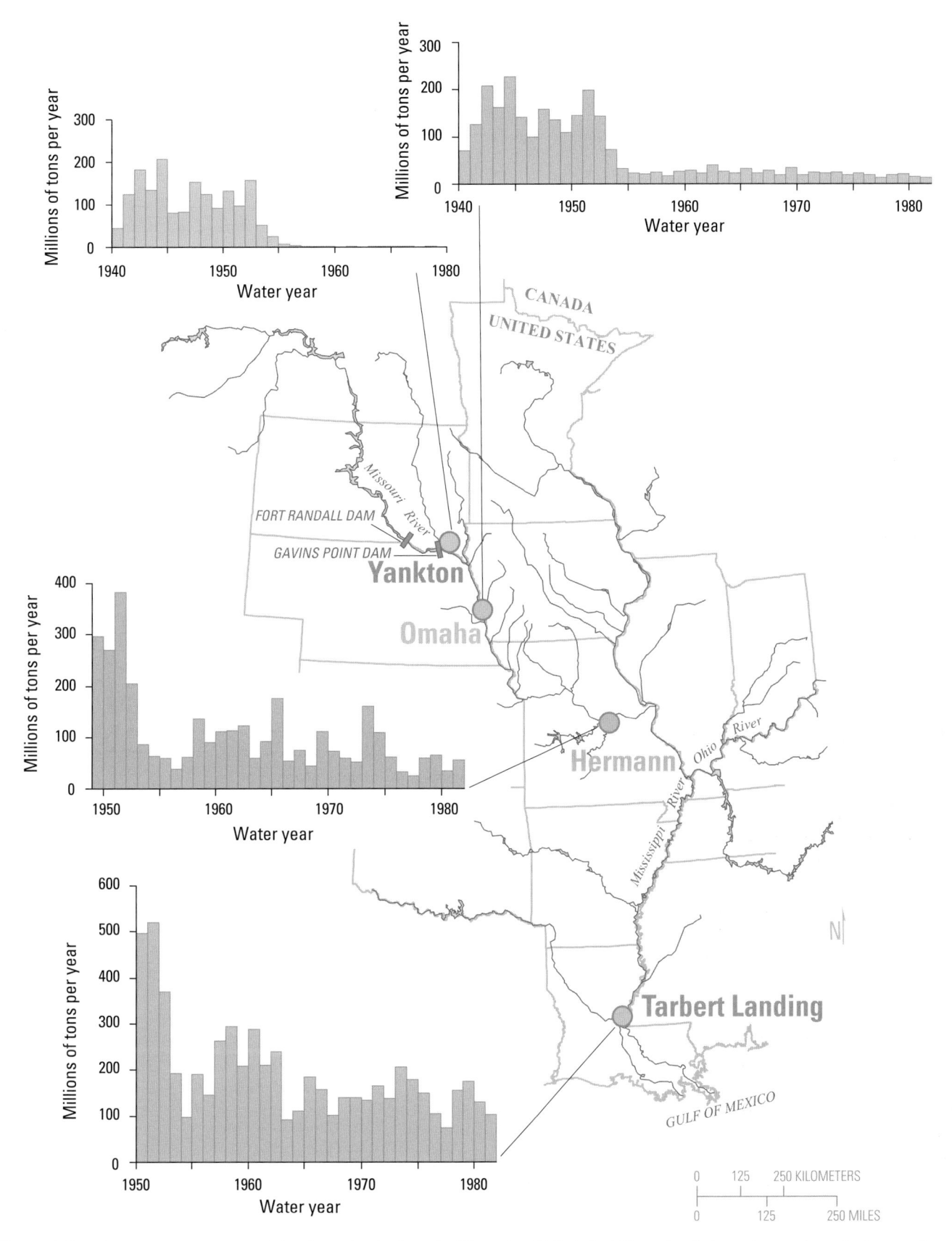

Figure 4.22. Historic changes to suspended sediment loads along the Missouri and Mississippi Rivers in relation to impoundment by large mainstem dams. (Source: Alexander et al., 2012.)

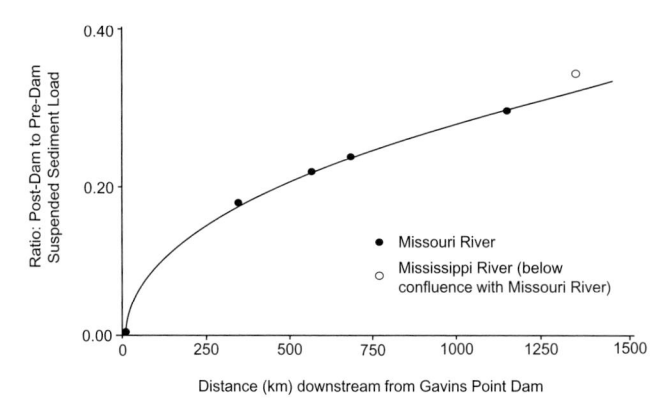

Figure 4.23. Ratio of pre-dam and post-dam suspended sediment loads along the Missouri River downstream from Gavins Point Dam (closure in 1955) in South Dakota to below the confluence with the Mississippi River at St. Louis, Missouri. Stations and pre- and post-year dataset include Yankton (1940–1952, 1957–1969), Omaha (1940–1952, 1957–1973), St. Joseph, Kansas City, and Hermann (1949–1952, 1957–1976) located 8, 314, 584, 716, 1,147 km downstream from Gavins Point dam. (Source: Williams and Wolman, 1984.)

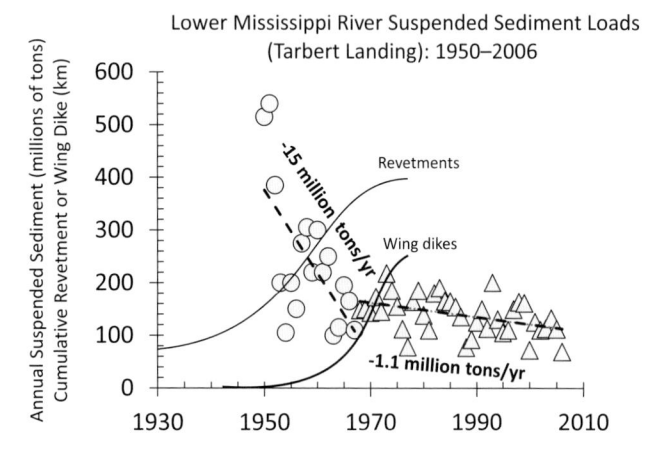

Figure 4.24. Historic change to suspended sediment loads for the lower Mississippi River at Tarbert Landing, Louisiana. Trend lines coincide with years 1950–1967 (−15 million tons/yr) and 1968–2006 (−1.1 million tons/yr). Cumulative distance (km) for revetment and wing dike (groyne) construction included for USACE Memphis and Vicksburg Districts. (Source: Meade and Moody, 2010.)

Chattahoochee Rivers in the United States reveal channel bed incision ranging from about 0.5 m to 6.0 m (Figure 4.25). Most of the channel bed incision occurs within about the first decade, and then decreases or ceases over subsequent decades. The temporal pattern of adjustment may also be somewhat linear or stepwise, and Williams and Wolman (1984) were careful to emphasize both "regular" and "irregular" styles of channel bed degradation, including incision and channel widening by lateral adjustment, following flow impoundment.

Europe's two largest rivers exhibit clear downstream channel bed incision following impoundment (Figure 4.26). The lower Volga River underwent some ~1.5 m of incision following impoundment behind Volgograd Dam in 1960, and continued to incise (Figure 4.26A). A consequence of channel bed incision is that floodplain connectivity occurs by less frequent higher-discharge magnitudes (Middelkoop et al., 2015).

Following extensive upstream impoundment in the 1950s the Danube River in Austria, for example, illustrated a pattern of adjustment that is somewhat opposite many US rivers (Figure 4.26B). The Danube basin is intensively fragmented by dams (Grill et al., 2015). Sixty-nine dams were constructed in the upper Danube basin between the 1950s and 1990s that resulted in some 90% of the upper Danube being impounded (Habersack et al., 2016). Effectively, no coarse sediment makes it over the main-stem dams along the Danube. Despite some 200,000 tons of coarse sediment annually being dumped downstream of the Freudenau hydropower dam in Vienna, higher shear stress downstream of the dam results in channel bed incision (Habersack et al., 2013). The annual rates of channel bed incision average ~2.0 cm/yr, and channel bed lowering continues (Habersack et al., 2016). An additional influence on channel bed incision of the Danube is that, as with many rivers intensively utilized for navigation and settlement in Europe and North America (Hudson and Middelkoop, 2015; Quick et al., 2020), channel straightening and other hydraulic engineering works have reduced lateral sediment inputs and altered flow patterns, further driving channel bed incision (Habersack et al., 2016; Schmutz and Moog, 2018).

4.3.3.1 DOWNSTREAM PROPAGATION OF CHANNEL DEGRADATION

Spatially, along large rivers, channel bed degradation following impoundment may extend for tens to hundreds of kilometers downstream of the dam (Ma et al., 2012; Smith et al., 2016; Smith and Mohrig, 2017; Quick et al., 2020). The amount of incision usually declines with downstream distance, being dependent upon the amount of change to the sediment and streamflow regime, the location and character of downstream tributary inputs, as well as the sedimentology and lithology of channel bed and bank material, valley slope, and human impacts (Brandt, 2000; Chin et al., 2002; Phillips et al., 2004; Smith et al., 2016; Lai et al., 2017).

The closure of Three Gorges Dam in 2003 was a watershed moment in environmental sciences. Because of being the world's largest dam on a very large river it has provided riparian sciences an unprecedented opportunity to examine the impacts of dam construction on a major river across a variety of fields. After closure of Three Gorges Dam, the Yangtze River abruptly transitioned from a depositional phase into an erosional phase. From the mid-1950s to mid-1980s the rate of channel bed deposition was about +90 Mt/yr of aggradation, but this reversed to about −50 Mt/yr of erosion following closure of Three Gorges Dam. Annual rates of average channel bed incision (estimated from

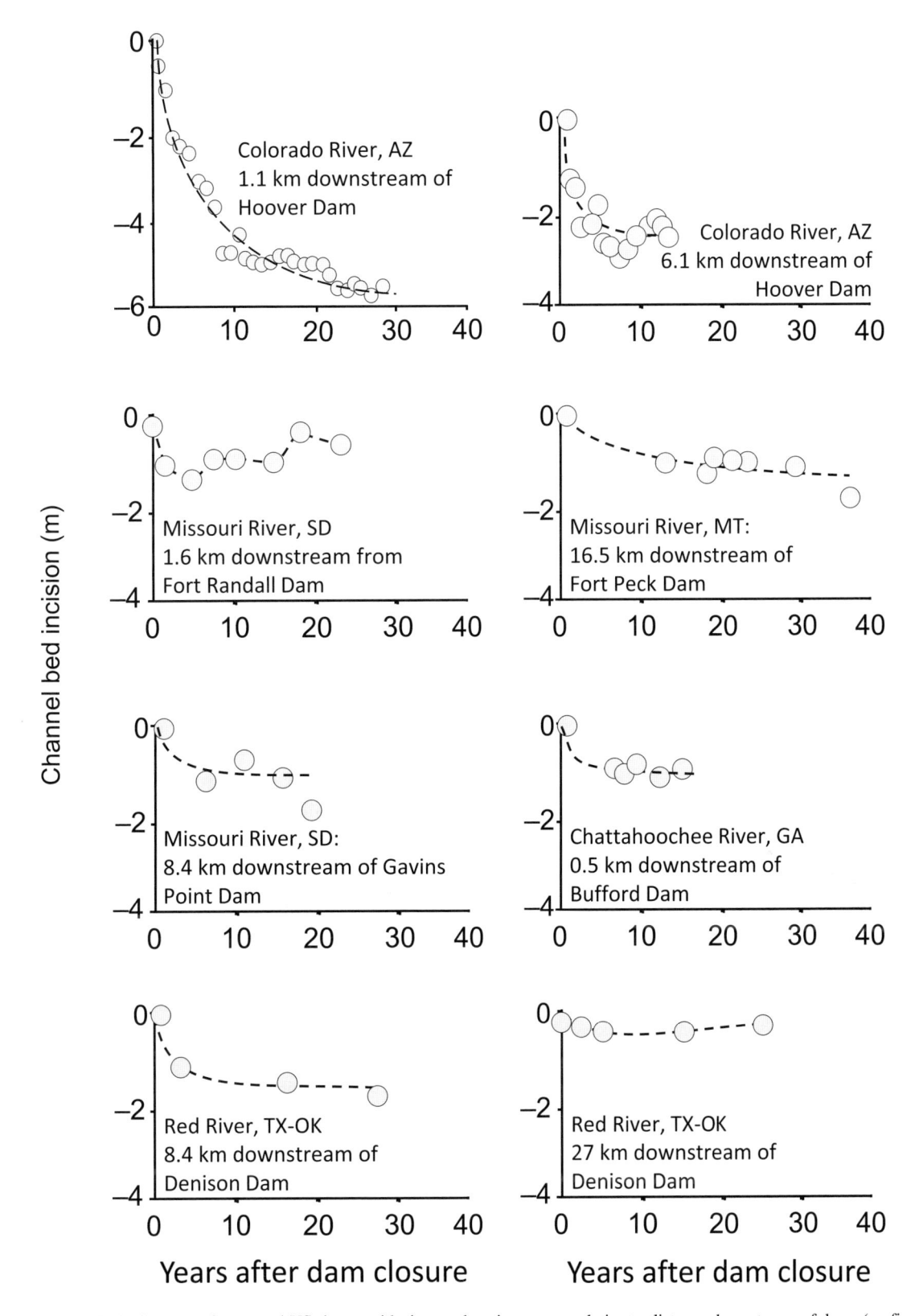

Figure 4.25. Channel bed adjustment for several US rivers, with river and station names relative to distance downstream of dams (on figure). (Data: Williams and Wolman, 1984.)

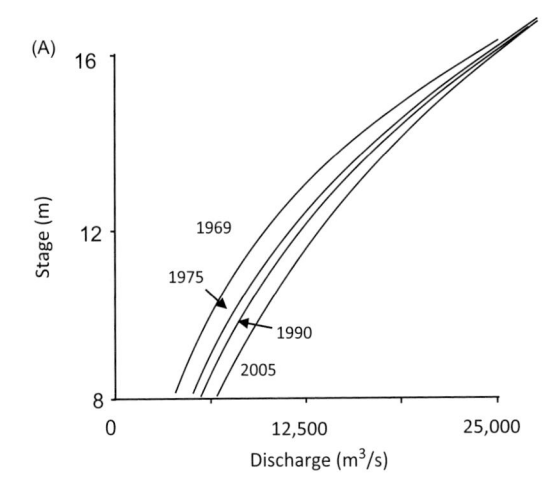

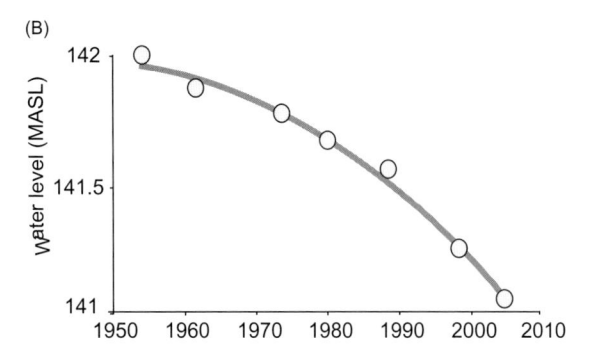

Figure 4.26. Channel bed incision for Europe's two largest rivers revealed with hydrologic data. (A) Lower Volga River at Volgograd, Russia inferred by changes to stage-discharge relationships. Channel bed incision likely prior to 1969 data (dam closure over 1959–1960). (Source: Górski et al., 2012.) (B) Incision of upper Danube River in Austria inferred with low water level data between 1950 and 2003 (at Wildungsmauer) following intensive impoundment from the 1950s to 1990s. (Source: Habersack et al., 2016.)

low-flow stage data) between 2004 and 2012 decreased downstream, from 10.2 to 3.4 cm/yr at Jingjiang (50–420 km downstream TGD), 4.9 to 2.5 cm/yr in the middle reach (420–910 km downstream TGD), to 1.7 to 0.9 cm/yr in the lower reach (910–1,550 km downstream TGD) (Wang et al., 2013). Averages across the channel bed mask local variability, and along a 70-km long segment of the middle Yangtze (Shashi reach, ~190 km downstream of TGD), channel thalweg incision averaged 2 m between 2002 and 2010, with a maximum thalweg incision of 8 m (Zhang et al., 2016). Channel scour moved progressively downstream, and it effectively ceases where the channel slope abruptly decreases. Sediment starvation is detected within the delta, manifest as reduced coastal marsh accretion and net erosion of the subaqueous delta front (Yang et al., 2011; Yang et al., 2015; Zheng et al., 2018). The well-documented case of the Yangtze River and Three Gorges Dam illustrates that following impoundment the geomorphology of large rivers can rapidly adjust.

Coastal draining rivers in Texas exhibit very different downstream responses to impoundment and illustrate the importance of drainage basin controls on the style of post-dam geomorphic adjustment (Williams and Wolman, 1984). Even within a region as homogeneous as the Texas coastal plain, fluvial geomorphic responses to dam impoundment can vary widely, complicating the development of general statements regarding the impacts of dams to rivers (Phillips, 2003).

From east to west Texas, the downstream riparian impacts of dam closure vary greatly, emphasizing the importance of streamflow variability to geomorphic response. In contrast to considerable downstream incision of the lower Trinity River (Phillips et al., 2005; Smith et al., 2016), the Sabine River along the Texas–Louisiana border revealed very little geomorphic adjustment to installation of Toledo Bend Reservoir in 1967. Sediment transport to the coastal zone continues, bars continue to accrete, and bends continue to migrate. This is likely because of sediment decoupling of the upper and lower reaches of larger basins, which buffers the downstream impacts of dams (Phillips, 2003). Additionally, the dam is a flow-over structure with little change to hydrologic regime (Phillips, 2003; Heitmuller and Greene, 2009). But to the west, along the lower Rio Grande/Bravo at the Texas–Mexico border, the impact of dam closure has been the opposite, and geomorphic and hydrologic processes have effectively been arrested.

The natural high-flow variability of the Rio Grande/Bravo along the US–Mexico border made the river more sensitive to change after impoundment (e.g., Graf, 2006). A series of mainstem dams, beginning in 1916 with Elephant Butte dam in southern New Mexico, tremendously altered a once dynamic riparian environment that drains 557,722 km^2 of southwestern North America. Along the lower Rio Grande/Bravo, Falcon Dam and Reservoir (443 km upstream of Gulf of Mexico) was closed in 1953. This greatly reduced downstream stream power such that persistent sediment transport no longer occurs. The median and maximum recorded discharge in the pre-dam period (1934–1951) was 48.7 m^3/s and 872 m^3/s, respectively. But in the post-dam period (1954–2011) median and maximum discharge greatly reduced to 5.6 m^3/s and 459 m^3/s, respectively (Swartz et al., 2020). This is significant because suspended sand in the lower Rio Grande/Bravo is not transported until a discharge of ~6 m^3/s (at median discharge). The majority of suspended sediment transport occurs by relatively infrequent events, specifically at discharge magnitudes ranging between 300 m^3/s and 400 m^3/s that have a flow duration of <5% (Hudson and Mossa, 1997). Rates of meander bend migration declined from 11.6 m/yr to 0.9 m/yr between the pre-dam and post-dam periods, respectively (Swartz et al., 2020). The once great dynamic lower Rio Grande/Bravo is effectively locked in place.

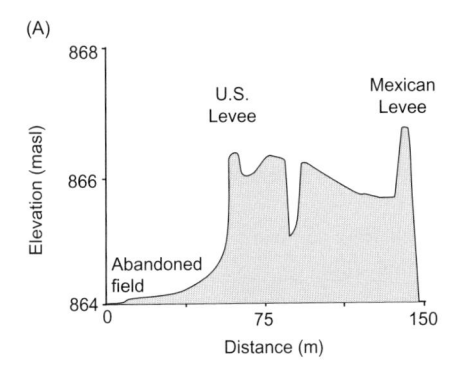

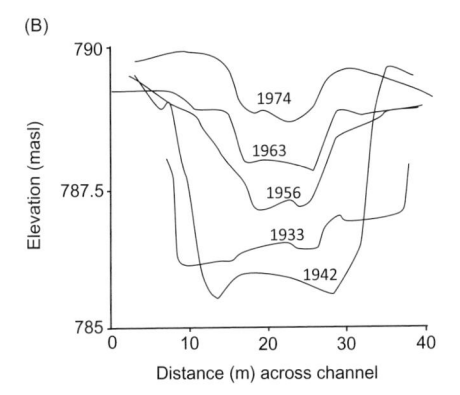

Figure 4.27. Changes to Rio Grande/Bravo at Presidio, Texas/Ojinaga, Chihuahua. (A) Deposition between the levees (dikes) resulting in embanked floodplain aggradation. (B) +3 m of channel bed aggradation and narrowing due to reduced flow conveyance caused by upstream impoundment and water withdrawal. (Source: Collier et al., 1996, data from Everitt, 1993.)

Flow impoundment of the Rio Grande at Elephant Butte Reservoir (New Mexico) and downstream water withdrawals for irrigation have reduced streamflow to such a degree that aggradation, rather than incision, is the dominant form of geomorphic adjustment (Figure 4.27). At Presidio, Texas/Ojinaga, Chihuahua, Mexico, the Rio Grande/Bravo channel underwent some ~3 m of channel bed aggradation between 1933 and 1974, with about 20 m of channel narrowing. The reduced streamflow results in aggradation from smaller valley-side downstream tributaries because of a lack of flow competence of the mainstem Rio Grande/Bravo (Everitt, 1993; Collier et al., 1996).

4.3.4 Impacts to Riparian Landscapes

4.3.4.1 TERRESTRIALIZATION OF RIPARIAN ENVIRONMENTS

Reduced hydrologic variability and channel stabilization following upstream flow impoundment drives downstream riparian environmental change. Woody vegetation encroachment over formerly active channel bars and siltation of sloughs and backwater areas directly reduces critical aquatic habitat linked to specific biologic functions, such as fish feeding and spawning. Thus, strongly related to hydrologic, sedimentary, and geomorphologic changes noted earlier are changes to riparian ecosystems, including the types of flora and fauna associated with floodplain and channels (Ligon et al., 1995; Grams and Schmidt, 2002; Schumm, 2007).

For a large range of regulated rivers in the United States, a comparison with upstream unregulated reaches found that downstream regulated reaches have low-flow channels that are 32% wider and high-flow channels that are 50% narrower (Graf, 2006). As relates to the riparian zone, floodplains of regulated rivers are 79% less active than upstream river reaches not influenced by dams. The functional portion of the river, which includes channel bar surfaces and floodplain environments in contact with streamflow at different frequencies and durations (Table 4.9), is 72% smaller for impounded reaches than upstream reaches not influenced by dams. And, in agreement with fragmentation indices of streamflow disruption, rivers within semiarid and arid regions with high-flow variability are especially heavily impacted because woody vegetation takes advantage of reduced flood pulse disturbances (Johnson, 1994; Marston et al., 2005; Graf, 2006). This scenario played out along the upper Snake River (4,510 km^2) in Wyoming and Idaho (United States) following the closure of Lake Jackson Dam in 1906 (Marston et al., 2005). In the case of the Snake River, species-poor forest vegetation assemblages, including Blue spruce, Cottonwood, and other mixed forest varieties degraded the riparian environment by replacing a rich mosaic of willow-alder and shrub-swampland and dynamic unvegetated surfaces that supported a diverse riparian ecosystem. Growth of woody vegetation reduced local channel avulsions and associated structural riparian diversity (geodiversity), further degrading lateral hydrologic connectivity. In the case of the upper Snake River, this degraded the native fishery habitat by reducing the spawning area for the Snake River fine-spotted cutthroat trout (*Oncorhynchus clarkii*).

4.3.4.2 CHANGES TO THE PLATTE RIVER, UNITED STATES

The Platte River drains 219,900 km^2 of the US Midwest and Rocky Mountains and is a prime example of a large braided river impacted by impoundment. The Platte is a major sediment source to the Missouri River, and annually discharges 10.1 million tons of suspended sediment into the Missouri River at Plattsmouth, below Omaha, Nebraska (Heimann, 2016). Its dynamically adjusting channel and wetlands within a semiarid environment represent a crucial recharge station to migratory birds, including federally endangered Sandhill Cranes (*Antigone canadensis*). And the abundant low sandy channel bars are prime nesting

Table 4.9 *Functional fluvial geomorphic surfaces and ecological relevance*

Functional surface	Definition	Ecological significance
Low-flow channel	Channel along thalweg, occupied by mean annual low flow	Aquatic habitat with longest annual duration of inundation
High-flow channel	Channel occupied by high flows, between floodplain and low-flow channel	Aquatic flora and fauna adapted to high flood potential with fast flowing water and erosion potential
Low bar	Sediment accumulation at margin of the low-flow channel, materials mobilized regularly	Location for unstable communities, frequent instability, often the site of pioneer or invasive vegetation
High bar	Sediment accumulation at margins of the high-flow channel or valley side, materials mobilized infrequently	Location for moderately stable communities, often for pioneer species
Island	Sediment accumulation with surface above mean annual low-flow level and below floodplain level, not attached to channel margins	Similar to low bar surfaces, with higher surface having flora communities similar to floodplain
Active floodplain	Nearly level surface adjacent to the low-flow channel, separated from channel by banks, inundated regularly	Stable community adjusted to frequent inundation, complex patches of vegetation
Inactive floodplain	Nearly level surface next to the low-flow channel, separated from the channel by banks, seldom inundated	Stable mature community not adjusted to frequent inundation, less complex patches of vegetation than active floodplain.
Engineered surface	Surface constructed, builtup, or excavated by human activities	Often bare, or with planted communities

Frequency definitions for the modern river adjustment to the contemporary geomorphic regime. Specific frequencies vary regionally. From Graf (2006).

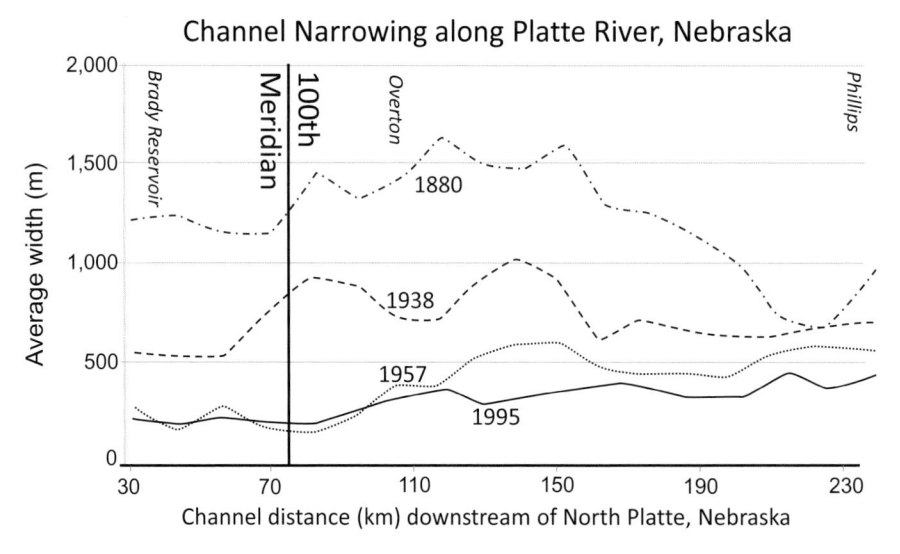

Figure 4.28. Changes to average channel width (m) along Platte River between Brady and Phillips, NE between 1880 and 1995. Kingsly dam is 118 km upstream of Brady Reservoir. (Source: Schumm, 2007.)

habitat for interior least terns (*Sterna antillarum*) and piping plovers (*Charadrius melodus*), of concern because of historic degradation to this dynamic habitat (NRC, 2005).

The Platte River channel has considerably narrowed since the mid-1800s, which is primarily attributed to flow impoundment and water withdrawals for irrigation (Williams, 1978b; Johnson, 1994; Schumm, 2007). Between 1880 and 1995, average channel width decreased from ~1,250 m to ~300 m (Figure 4.28). The main culprit is the 1941 closure of Kingsley Dam (and Lake McConaughy reservoir), the world's second largest hydraulic fill dam. And before completion of the reservoir the Platte River had experienced considerable riparian change because of irrigation withdrawals for agriculture, and closure of dams in Wyoming in the early 1900s (Johnson,

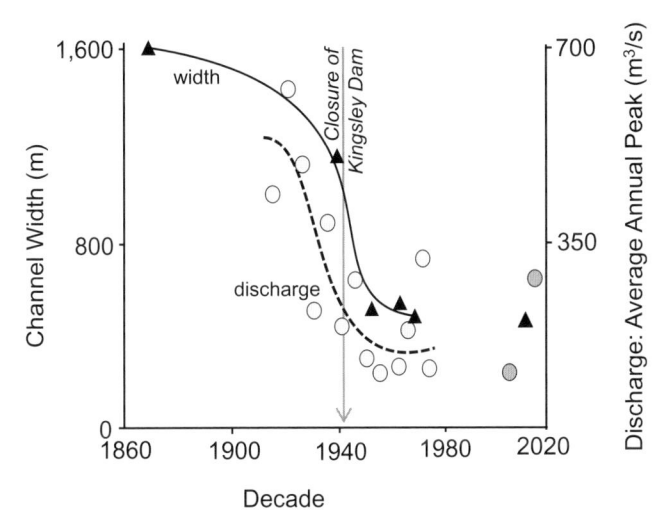

Figure 4.29. Historic changes in channel width and annual maximum peak discharge for the Platte River near Overton, Nebraska (USGS 06768000) in relation to upstream closure of Kingsley Dam in 1941 along North Platte River (~120 km downstream). Recent annual maximum discharge values are 104 (m^3/s) and 297 (m^3/s) for water years 2009 and 2019, respectively. Recent channel width value is 517 m for 2015 (September 27), measured between vegetated banks at 1 km intervals along a 5-km channel reach at Overton, NE in Google Earth Pro by author. (Data source: Williams, 1978b, with author modifications.)

1994). This initiated a dramatic change to the riparian environment, as the 1938 channel width was about half its 1880 width (Williams, 1978b; Schumm, 2007). The trend in narrowing and stabilization continued until the mid-1990s. For the Platte River the increase in low flows is particularly problematic, and results in much of the formerly sandy braid plain and active channel bars being terrestrialized by woody vegetation.

Changes to the Platte River riparian corridor vividly illustrate the importance of maintaining a historic flow regime with regard to functional geomorphic surfaces (Table 4.9). Specifically, this includes flow variability (with higher high flows and lower low flows) as well as maintaining the seasonality of low- and high-flow periods (NRC, 2005).

After impoundment behind Kingsley Dam, the annual maximum discharge declined from about 450 m^3/s to 175 m^3/s, which has remained constant through 2019 (Figure 4.29). The historic flow regime of the Platte River annually had numerous no-flow days, with an average of 78 per year at Overton, NE (Schumm, 2007). The local alluvial water table was too deep to support woody vegetation growth across the broad exposed dry channel bed surface. The inability of woody vegetation to colonize channel bars helped maintain a dynamic channel bed, with sedimentary bars frequently mobilized and reworked during higher flow periods (Johnson, 1994). Since impoundment there are zero no-flow days (at Overton, NE), as the flow regime has changed from intermittent to perennial (Schumm, 2007).

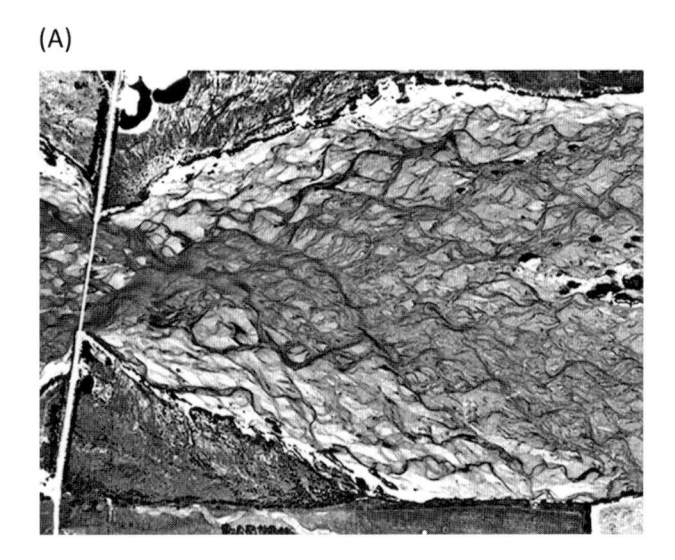

Figure 4.30. Comparison of historic and modern Platte River riparian landscape imagery (same area). (A) Historic (1938) vertical aerial photograph of the Platte River at Kearney. Numerous active sedimentary bars and few vegetated bars (islands). Distance along bridge (between vegetated banks) is 386 m. (B) Platte River at Kearney, NE September 27, 2015. Compare with 1938 photo. Distance along bridge between vegetated channel banks is 264 m. (Sources: (A) U.S. Geological Survey vertical air photo, 9″ × 9″ (23 cm × 23 cm) at 1:10,000 scale, Platte River Program – Historical 1938 Aerial Photography 47_53, (B) similar area as A from Google Earth Pro.)

The seasonality of the high and low flow periods is also important, as prior to impoundment the timing of the premodified flood pulse coincided with the period of seed germination for *Populus-Salix* in June. Importantly, the high-flow disturbance period in late spring and early summer limited root growth and sedimentary bar colonization, as revealed by comparison of historic air photo with modern imagery (Figure 4.30). After impoundment and substantial water withdrawals for agriculture resulted in the lowering of seasonal flood pulses, *Populus-Salix* vegetation rapidly colonized formerly active channel bars and transformed the riparian environment (Johnson, 1994; NRC, 2005) (Figure 4.30).

The scenario reviewed for the Platte River has played out across numerous riparian lands after upstream flow impoundment, particularly in dryland settings. In addition to changes to riparian sedimentary and vegetation environments, changes to fisheries are acute.

4.3.5 Impacts to Aquatic Organisms and Fisheries

The physical obstruction posed by dams combined with downstream terrestrialization of channel bed and riparian aquatic habitat often results in dramatic declines to riverine fisheries, with variability due to basin specific natural and human conditions, including dam and reservoir management. Aquatic megafauna are especially vulnerable to post-dam riparian changes; because of their size they are more likely to be damaged by turbines, and also because they require more space to accommodate different phases of their life cycle (Winemiller et al., 2016).

4.3.5.1 YANGTZE FISHERY DECLINE DOWNSTREAM OF THREE GORGES DAM

Because of pollution and hydraulic engineering projects, the Yangtze River fishery was already in trouble prior to construction of Three Gorges Dam. Since closure of Three Gorges Dam in 2003, however, the fishery has continued to decline. The number of carp eggs and carp larvae has plummeted from about 3.5 billion in 1997 to less than 500 million by 2003 (Figure 4.31A). The situation has improved in the past decade with more environmentally friendly approaches to reservoir management, increasing to almost 1.5 billion carp larvae and eggs by 2015 (Cheng et al., 2018). An important management approach is to coordinate the reservoir release schedule with critical life cycle stages of key aquatic organisms. In the case of the Yangtze River it involves reintroducing lateral hydrologic connectivity between the river and the massive floodplain lakes, such as Dongting Lake, which serves as vital carp habitat (Yia et al., 2010; Ru and Liu, 2013).

4.3.5.2 DOWNSTREAM IMPACTS TO VOLGA FISHERY

Changes to the lower Volga River imposed by closure of Volgograd Dam (1959) reduced riparian spawning areas by ~80%, and is associated with large declines in fish populations, particularly the iconic Russian sturgeon (*Acipenser gueldenstaedtii*) (Secor et al., 2000; Maltsev, 2009). The annual fish catch in both the channel and the floodplain since impoundment of the lower Volga has plummeted (Figure 4.31B). In addition to woody vegetation encroachment, numerous side channels that served as prime aquatic habitat for fish spawning infilled with sediment (Middelkoop et al., 2015). In addition to impoundment, other factors related to the decline of the lower Volga fishery includes water pollution, water stress caused by extensive withdrawal for agriculture, and illegal fishing (Ruban et al., 2019).

4.3.5.3 NILE COASTAL FISHERY AND INDUSTRIAL AGRICULTURE

The impact of impoundment on downstream fisheries is quite different for the Nile River. Because of a 90% reduction in Nile flooding after closure of the Aswan High Dam (Oczkowski et al., 2009), nutrients historically associated with the Nile flood pulse were impounded in Lake Nasser, resulting in a collapse of coastal primary productivity (Dorozynski, 1975). The loss of the "Nile bloom" resulted in an abrupt decline of the Nile delta fishery. The annual sardine catches, for example, declined from 37,000 tons (1962–1965) to 6,500 tons (1966–1970) (Nixon, 2003).

An unintended consequence of Egypt's moves to industrialized agriculture and its booming delta urban population, however, caused the fishery to rebound in the 1980s. This occurred because of large increases in chemical fertilizers (phosphate and nitrogen) to support agriculture, which is consumed by a much larger urban coastal population. Annual fertilizer use increased almost fourfold, from about 340,000 tons to 1,300,000 tons (Figure 4.31C). The sequence is fairly straightforward. The fertilized agricultural products are consumed, and the raw sewage is discharged into the lagoons and marine environments, increasing primary productively. A substantial proportion (60–100%) of the Nile delta fishery is supported by nutrients from fertilizer that is largely supplied by sewage disposal (Oczkowski et al., 2009). The Aswan High Dam is not the cause of the nutrification of the Nile delta waters, but it is part of a larger industrial agricultural system that utilizes large amounts of chemical fertilizers (Dorozynski, 1975). Unexpectedly, annual fish catch totals now far exceed historic levels (Nixon, 2003; Oczkowski et al., 2009).

4.3.5.4 MEKONG FISHERY AND AQUATIC MEGAFAUNA

Fluvial riparian environments along the Mekong are being strangled by dams, while the delta is drowning because of sediment starvation (Syvitski et al., 2005; Kummu and Varis, 2007; Kondolf et al., 2014b; Lu et al., 2014; Winemiller et al., 2016; Best and Darby, 2020; Minderhoud et al., 2020; Yoshida et al., 2020). With 60 million living within 15 km of the lower Mekong, much of the population depends upon riparian services provided by a free-flowing river, including transportation, navigation, agriculture, commerce, and sustenance (Dugan et al., 2010). The Mekong inland fishery is the world's largest and produces some 2.0–2.6 million tons of fish per year, accounting for 49–82% of animal protein consumed by the Mekong population (Yoshida et al., 2020). Proposed fish ladders and gates to facilitate fish migration around dams are inadequate to cope with the diversity of behavior exhibited by the numerous fish species (Dugan et al., 2010; Ziv et al., 2012; Yoshida et al., 2020). As 40–70% of the Mekong fish are migratory, dam construction along the main-stem and tributaries blocks migratory fish, and

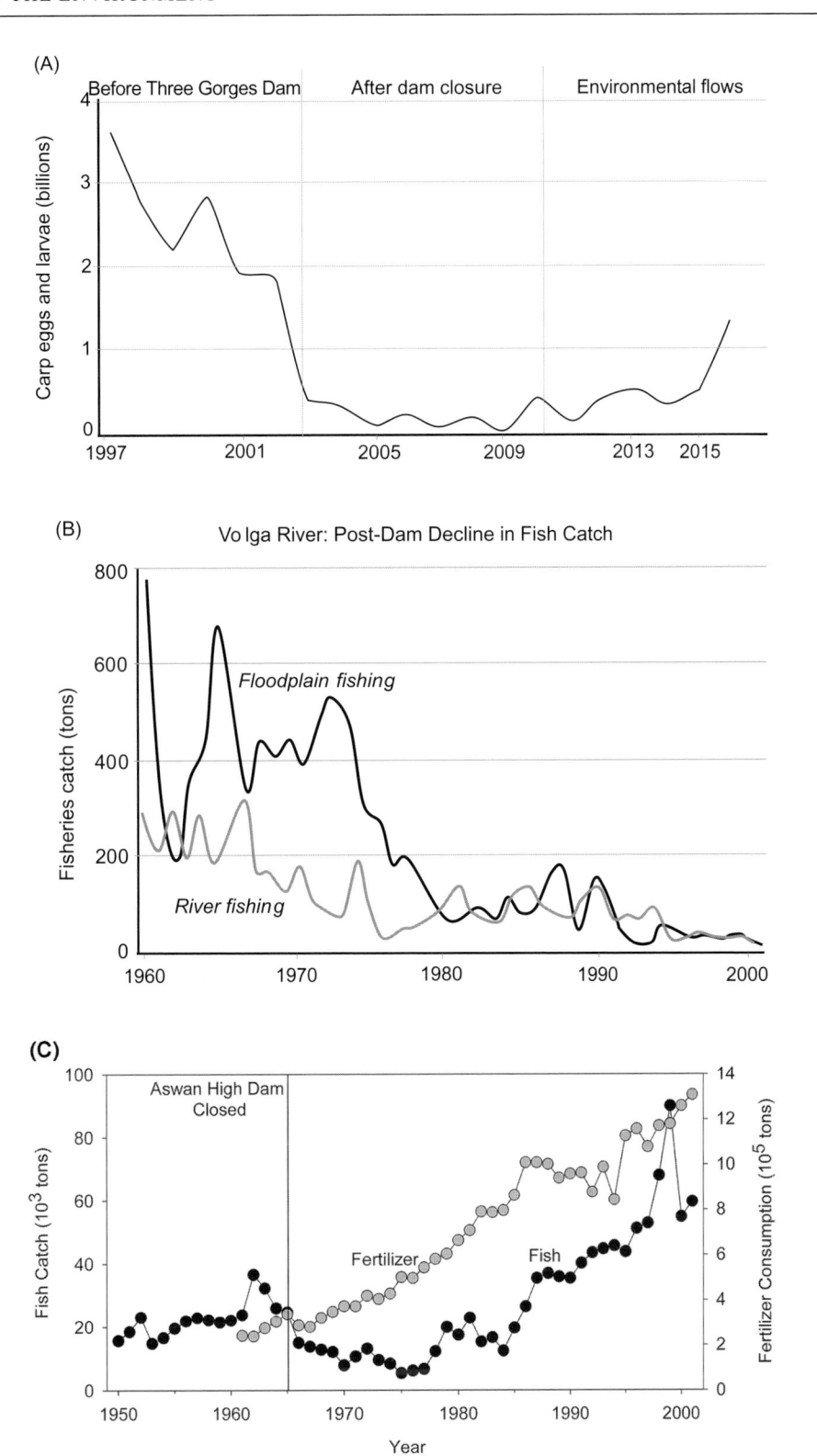

Figure 4.31. Three different downstream responses of riverine fisheries to dam construction, including (A) decline in carp eggs and larvae on Yangtze River before and after closure (2003) of Three Gorges Dam, with subsequent environmental flows program after 2011; (B) Volga River decline in riverine and floodplain production after closure of Volgograd Dam in 1959; and (C) coastal fish catch before and after Aswan High Dam impoundment (1964) of the Nile in relation to increased fertilizer use for intensive agriculture. (Sources: Volga River, Schmutz and Moog, 2018, with data from Górski et al., 2012; Yangtze, Cheng et al., 2018; Nile, Oczkowski et al., 2009.)

threatens livelihoods. In addition to the main-stem channel, the large tributary channels in the middle and lower Mekong basin are also impounded.

A special aquatic megafauna that inhabits the lower Mekong is the Irrawaddy Dolphin (*Orcaella brevirostris*), which migrates between the estuarine delta and middle reaches of the Mekong. As with river dolphins in many impounded rivers, the Irrawaddy dolphin is heavily threatened (Winemiller et al., 2016). Up to half of the remaining Mekong population of Irrawaddy Dolphin, estimated at only eighty individuals (Krützen et al., 2018), were expected to be lost because of being in the vicinity of the construction process (e.g., earth works, explosions, dredging) of the Don Sahong hydroelectric dam (commissioned 2020), located in Laos 2 km from the Laotian–Cambodian border (Dugan et al., 2010). A sliver of hope occurred in March 2020 when the Cambodian government abandoned large dams planned for the lower Mekong, and put a ten-year moratorium on any new dams on the Mekong. This is crucial because additional large main-stem dams in the lower Mekong would block upstream fish migration with the rich delta nurseries, as well as the Tonle Sap inland lake and wetland system, which in itself represents 60% of Cambodia's annual fish catch (Johnstone and Sithirith, 2018).

4.3.6 Reservoirs and Water Quality

Reservoirs not only change streamflow and sediment regimes, they also change water quality in ways that are significant to aquatic organisms and human health (Petts, 1986; Hortle and Nam, 2017; Jiang et al., 2018). While simple run-of-the-river and flow-over dams have less influence on water quality, large and deep impoundment reservoirs potentially influence temperature, chemistry, and biota of reservoir water over extensive areas, with potential downstream implications (Figure 4.32).

Reservoirs capture what is supplied by rivers draining upstream landscapes. This includes runoff, clastic sediment, organic sediment, as well as nutrients and pollutants. Many rivers draining agricultural landscapes have high nutrient loads, especially nitrogen and phosphorus, which results in eutrophic reservoir waters ideal for algae blooms. Algae blooms reduce oxygen levels, and if extreme can result in hypoxic conditions (oxygen <2 mg/L), which represents a direct form of aquatic habitat degradation (Brandao et al., 2017). Blue-green algae (cyanobacteria) blooms are of particular concern where reservoir waters are used for consumption, as toxins can be poisonous to humans and animals (Mur et al., 1999). Fine-grained sedimentary deposits change reservoir chemistry by adsorbing nutrients and metals, resulting in potentially toxic sludge that can be discharged into rivers as reservoir outflow (Fovet et al., 2020; Palanques et al., 2020). Outflow of deep reservoir water often smells of hydrogen sulfide (H_2S), as the gas is produced in oxygen-poor waters with microbial decomposition of organic materials (Young et al., 1976; Rashid, 1995; Winton et al., 2019).

4.3.6.1 THERMAL STRATIFICATION

Thermal stratification in large reservoirs varies with changes in riverine inputs and temperature (Rashid, 1995; Jiang et al., 2018). Dissolved oxygen levels in Lake Nasser and Nubia, for example, vary seasonally and spatially along the ~500-km long reservoir. Reservoir stratification is pronounced in July and

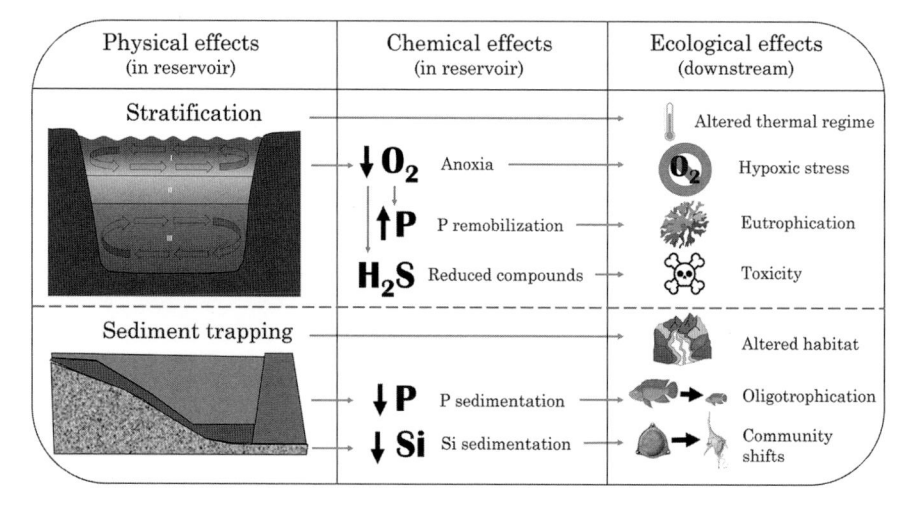

Figure 4.32. Conceptual synthesis of chemical and physical characteristics of reservoir water quality relevant to aquatic ecology, particularly for large and deep reservoirs. Reservoir stratification occurs at depths where water temperature and dissolved oxygen levels abruptly change. Warmer temperatures in summer months lead to thermal stratification with several defined layers, including the epilimnion (upper layer) sensitive to mixing by wind and with higher dissolved oxygen, metalimnion (middle layer) at the thermocline where temperature abruptly changes with depth, and hypolimnion (bottom layer), the coolest (10–14°C) layer that is relatively stagnant with low levels of dissolved oxygen. (Source: Winton et al., 2019. Licensed under CC 4.0.)

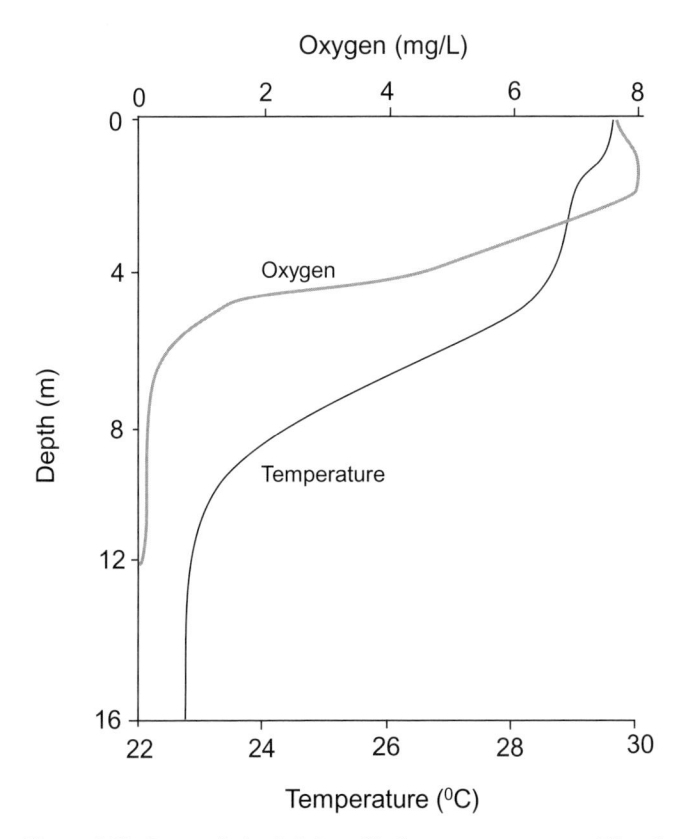

Figure 4.33. Reservoir depth (m) profile for water temperature (C) and dissolved oxygen (mg/L) in 2004 for stratified Nam Leuk Reservoir, Laos, Mekong basin. (Source: Hortle and Nam, 2017.)

August with dissolved oxygen levels averaging between 9.5 mg/L at the surface and 0.0 mg/L at a depth of 15 m. In the late autumn and winter, the water is mixed and does not exhibit stratification and the reservoir is oxygen saturated from the surface to bottom (Rashid, 1995). Large reservoirs in the lower Mekong basin have higher temperature layers (Figure 4.33) that commonly range from 28°C to 30°C and somewhat "float" upon denser cooler layers (Hortle and Nam, 2017). In cooler climatic settings, head-of-reservoir spring and summer water temperatures can be too warm and detrimental to the life cycle of aquatic organisms, particularly concerning spawning habitat for fish accustomed to cooler water temperatures.

For decades these issues were documented for a range of reservoir types across an international range of environments (Young et al., 1976; Rashid, 1995; Kunz et al., 2011), and are of pressing concern considering the current boom in large dam construction (Winton et al., 2019).

4.3.6.2 RESERVOIRS AND PHOSPHORUS STORAGE

Because many dams are built to support irrigation and agriculture it is not surprising that increases in agriculture over the twentieth century resulted in increased nutrient loading of reservoirs. Globally, 12% of total phosphorus loads transported by rivers were trapped in reservoirs in 2000, a 5% increase over 1970 levels (Table 4.10). While total river phosphorus loads modestly increased between 1970 and 2000, the amount of phosphorus retained in reservoirs effectively doubled, as total phosphorus retained in reservoirs increased from 22 to 42 Gmol/yr between 1970 and 2000. Thus, the much higher amount of retention is due to the increase in reservoir storage capacity that occurred over this period (e.g., Figure 4.4). And the ongoing massive dam construction boom that will result in hundreds of new large reservoirs is projected to result in further increases in phosphorus storage, with some scenarios projecting an increase from 12% to 17% by 2030 (Maavara et al., 2015). It should be noted that riverine phosphorus budgets (Table 4.10) also reveals an alarming increase in phosphorus bypassing reservoirs, which contributes to coastal eutrophication and the proliferation of hypoxic dead zones at the mouths of rivers around the world (Breitburg et al., 2018).

The Mississippi basin has been the global leader in reactive phosphorus retained in reservoirs since 1970, with about 50% of its phosphorus load retained in its 700 odd reservoirs (Table 4.11). The amount of reactive phosphorus retained in reservoirs in 2000 was 920×10^6 mol/yr and by 2030 will undergo a modest increase. But, by 2030 the Mississippi will be replaced by the Yangtze River as the highest reservoir retention of reactive phosphorus, which at $2,898 \times 10^6$ mol/yr will represent only 35% of its total load (Table 4.11). Other rivers projected to enter into the top 10 rankings include the Paraná, Huanghe, Zaire, and Mekong Rivers. These rivers are not only undergoing increased dam construction, but also are located in regions undergoing increased land cover change for mechanized agriculture that is reliant upon artificial fertilizers, specifically high amounts of phosphorus. The Yangtze River, for example, is projected to see an enormous increase in reactive phosphorus from $3,758 \times 10^6$ mol/yr in 2000 to $8,327 \times 10^6$ mol/yr by 2030. Interestingly, the Zambezi River basin will remain among the top ten for phosphorus retention, although it has by far the lowest phosphorus loads. This is because the high trap efficiency of its reservoirs, increasing from 62% to 73% between 2000 and 2030, which includes Kariba Reservoir, the world's largest reservoir by volume (Kunz et al., 2011; Giosan et al., 2014; Maavara et al., 2015; Winton et al., 2019).

4.3.6.3 DOWNSTREAM CHANGES TO TEMPERATURE AND OXYGEN

Dam and reservoir management strategies influence water quality within reservoirs and downstream of dams. In this regard, a fundamental consideration is the depth of the reservoir thermocline/oxycline relative to the spillway intake. Many large reservoirs have outflow levels deeper than ~10 m, which is usually

Table 4.10 *Global retentions of total phosphorus and reactive phosphorus* by reservoirs for 1970, 2000, and 2030*

Global estimates	1970	2000	2030GO**
Global river TP load, Gmol/yr	312	349	384
Global river RP load, Gmol/yr	113	133	175
TP retained, Gmol/yr	22	42	67
RP retained, Gmol/yr	9	18	36
Fraction of global TP load retained (%)	7	12	17
Fraction of global RP load retained (%)	8	14	21

* Reactive phosphorus defined as sum of total dissolved phosphorus, exchangeable phosphorus, and particulate organic phosphorus and is considered the fraction of total phosphorus potentially bioavailable.
** GO = Global Orchestration scenario. See Alcamo et al. (2006) for description of millennium ecosystem assessment scenarios.
Source: Maavara et al. (2015).

Table 4.11 *Ranking of basins by reservoir storage of reactive phosphorus for 2000 and a projected scenario for 2030*

Rank	Watershed	No. of reservoirs	Reactive phosphorus load, 10^6 mol/yr	Reactive phosphorus load retained, 10^6 mol/yr	Retention (%)
1	Mississippi	700	1,880	920	48.9
2	Zambezi	50	863	531	61.5
3	Volga	17	1,320	500	37.9
4	Yangtze	358	3,758	480	12.8
5	Paraná	70	2,410	357	14.8
6	Ganges–Brahmaputra	83	8,961	322	3.6
7	Yenisei	6	840	267	31.8
8	Niger	52	687	262	38.1
9	Nile	10	624	239	38.3
10	Dnepr	6	438	202	46.1
YRR 2030 (GO scenario)					
1	Yangtze	500	8,327	2,898	34.8
2	Mississippi	700	2,294	1,124	49.0
3	Paraná	418	3,912	676	17.3
4	Mekong	140	3,283	650	19.8
5	Zambezi	65	884	649	73.4
6	Ganges–Brahmaputra	483	10,006	621	6.2
7	Niger	74	1,422	568	39.9
8	Volga	17	1,334	506	37.9
9	Zaire	20	2,462	417	16.9
10	Huang He	51	1,033	402	38.9

Source: Maavara et al. (2015).

below the thermocline, even during warm and dry conditions (Young et al., 1976; Winton et al., 2019). The release of potentially cold, hypoxic, and nutrient-enriched waters can have detrimental consequences to downstream riparian environments. Cooler dam outflow water along the Yangtze River resulted in a delay of fish spawning by several weeks (Zhong and Power, 2015). Macroinvertebrate communities along Guadalupe River exhibited qualities consistent with ecological stress, with reduced diversity and higher numbers of fewer species downstream of Canyon Dam, a deep storage reservoir in the Texas Hill Country (Young et al., 1976).

Outflow from large reservoirs is clear and cold with lower oxygen levels. Because of the depth and size of reservoirs, the volume of outflow potentially alters downstream aquatic environments for tens of kilometers. This is shown for two of the larger tributaries to the lower Mekong (i.e., Figure 4.19). Streamflow temperatures varied by four and six degrees (Celcius) 56 km and 78 km downstream of recently constructed dams along the Sesan

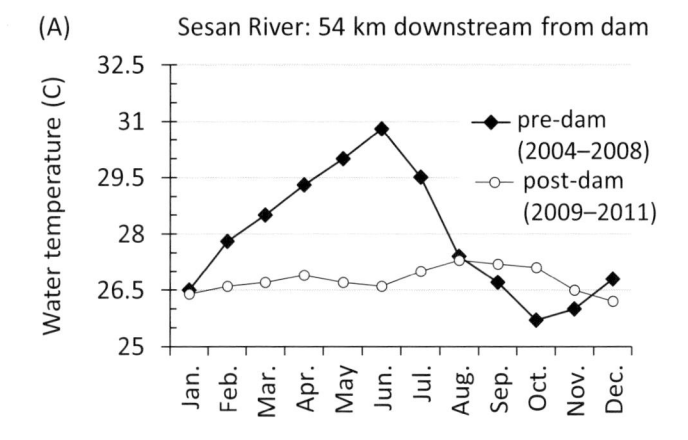

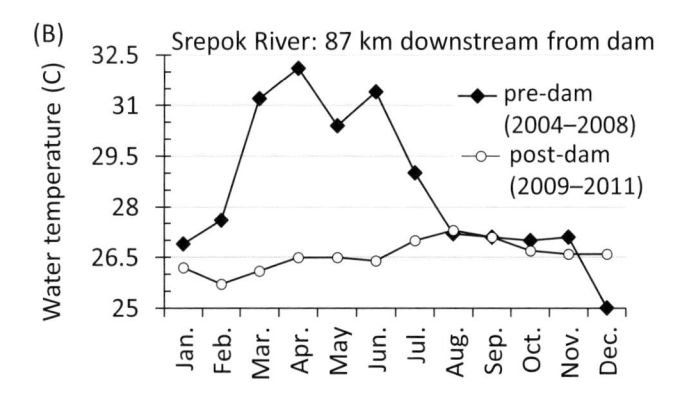

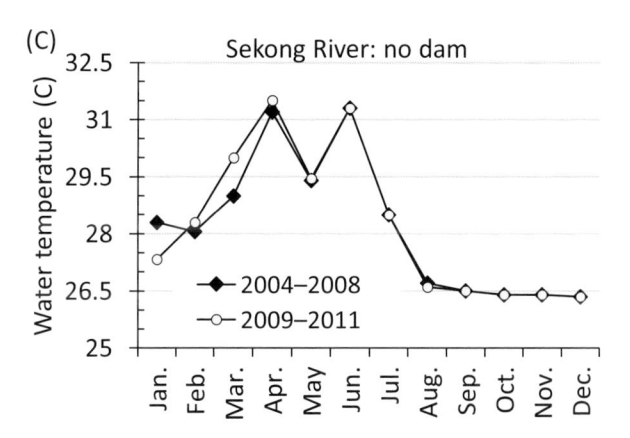

Figure 4.34. Downstream impact of lower Mekong basin dams on water temperature. Comparison of pre-dam (2004–2008) and post-dam (2009–2011) average monthly streamflow temperatures for the (A) Sesan and (B) Srepok Rivers, with dam construction in 2008 and 2009. No major dams were constructed on the (C) Sekong River over the study period, which provides a "control" for comparison. Temperature monitoring stations along the Sesan and Srepok Rivers were 56 km and 78 km downstream of dams, respectively. See Figure 4.19 for location of rivers within the lower Mekong basin. (Source: Bonnema et al., 2020.)

and Srepok Rivers, respectively (Figure 4.34). The streamflow for both rivers underwent substantial cooling during the dry season, which persisted into the first half of the wet season (May, June, July). Thus, reduced streamflow variability caused by dams occurs with temperature, as it does with water level (stage). The importance of this is highly seasonal and related to the life cycles of aquatic organisms, including fish behavior related to feeding and spawning (Bonnema et al., 2020).

4.3.6.4 RESERVOIR WATER QUALITY MANAGEMENT: TENNESSEE VALLEY AUTHORITY CASE STUDY

The Tennessee Valley Authority (TVA) is a massive federal economic development project in the southeast United States that was initiated during the Great Depression of the 1930s (Figure 4.35). The main focus of the TVA was to build dams for hydroelectricity, flood control, and navigation. The TVA operates forty-nine dams, with twenty-nine built for hydroelectricity and twenty non-power-generating dams with a primary purpose of flood control. The dams are located within the Tennessee River basin (105,900 km^2), which drains the southwestern Appalachian Mountains and flows westerly, joining the Ohio River ~40 km upstream of its confluence with the Middle Mississippi to form the lower Mississippi. The Tennessee River is heavily impounded and includes nine main-stem dams and locks that create a 1,050-km long navigable corridor. The approach to reservoir management upon completion of the dams was to maximize hydropower, which resulted in storage and release schedules that did not coincide with riparian ecosystems; essentially the opposite of environmental flows.

Reservoir outflow for many dams built by the TVA had poor water quality that did not meet established environmental standards (Table 4.12). A study of twenty dams found that over a thirty-six-year period, downstream flow volumes were too low to support critical riparian functions of aquatic habitat (Higgins and Brock, 1999). The downstream distance where streamflow volumes were too low exceeded 20 km for ten dams. The Holston River downstream of Cherokee Dam has an average discharge of 130 m^3/s, but for a distance of 76 km below, the dam only averaged a minimum flow of 2 m^3/s. This effectively resulted in a dry channel bed, although the region has among the highest annual rainfall totals in the continental United States. Additionally, sixteen of the TVA dams had outflow with dissolved oxygen levels below environmental standards (Table 4.12). The French Broad River, downstream of Douglas Dam, had average minimum dissolved oxygen levels of just 0.9 mg/L for an average of 113 days per year over a distance of 129 km downstream of the dam (Higgins and Brock, 1999). Such low levels of dissolved oxygen represent hypoxic conditions that can be fatal to immobile benthic macroinvertebrates while fish abandon the segment. The river effectively becomes a riparian dead zone.

Fortunately, in 1991 the TVA initiated changes to its reservoir management strategy to improve downstream ecological conditions (Bednarek and Hart, 2005). To increase the wetted surface

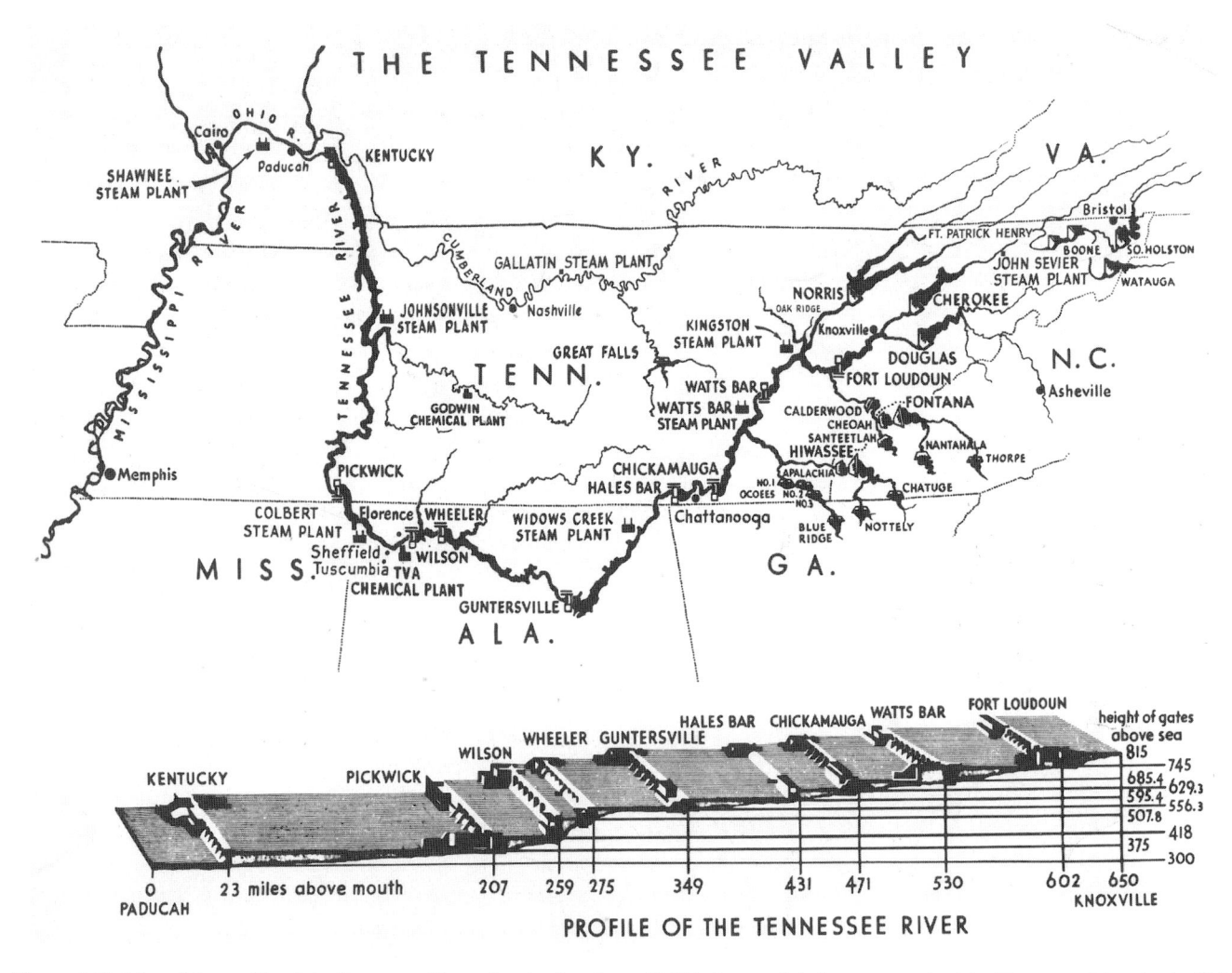

Figure 4.35. Map of the profile of the Tennessee River, showing locations of TVA dams and drainage network of the Tennessee River basin. Unit conversion: 1 mile = 1.6094 km, 1 ft = 0.3048 m. (Source: Tennessee State Library and Archives.)

area of the channel bed, measures included turbine pulsing to create a steady outflow, increased minimum flow volumes, and construction of low weirs downstream to maintain water levels at critical heights. To increase dissolved oxygen levels, modifications were made that involved injecting oxygen into reservoir water as well as structures immediately below the dam to create turbulence and aeration of reservoir outflow water. The measures were largely deemed effective, and both abiotic and biotic conditions improved downstream of the dams. Average dissolved oxygen levels increased by ~34%. The average increase in minimum velocity and discharge was 59% and 528%, respectively. While it was difficult to discern the importance of increased flow velocity relative to increased dissolved oxygen, macroinvertebrate assemblages significantly improved across the study segments. The number of macroinvertebrate families increased by 36% and there was a 13% reduction in the number of macroinvertebrates tolerant of low water quality conditions (Bednarek and Hart, 2005).

Downstream rehabilitation of impounded rivers by modifying dam and reservoir operations is critically important because of increased stressors of land cover change and climate change, and because many dams will effectively remain a permanent component of the riparian landscape. But designing reservoir outflow for environmental flows remains a work in progress (Bednarek and Hart, 2005; Rader et al., 2008; Kunz et al., 2013). While the efforts of the TVA to improve the downstream environmental integrity of riparian environments were deemed somewhat successful, the TVA's highly regulated rivers far from resemble a natural stream, and restoration was not the goal. Riparian ecologists have developed an intricate knowledge of critical life cycle phases of stream biota and their relation to abiotic components (e.g., bed material, flow velocity, depth, temperature). Such knowledge has resulted in conceptual models that provide useful signposts to guide river managers in environmentally effective dam and reservoir management, including lateral connectivity with floodplains (Ward and Stanford, 1995). The actual success of such mitigation

Table 4.12 *Reservoir release flows and dissolved oxygen (DO) levels for selected reservoirs in the Tennessee Valley Authority project*

Dam, River	Mean discharge (m^3/s)	Mean minimum daily flow (m^3/s)	Stream length impacted by low flow (km)	Mean minimum DO (mg/L)	Mean # days DO below target*	Stream length impacted by low DO (km)
Appalachia, Hiwassee	62	4	24	5.0	64	3
Blue Ridge, Toccoa	17	0	21	3.4	83	24
Boone**, South Fork Holston River	72	6	0	3.9	46	16
Chatuge, Hiwassee	13	0	29	1.3	91	11
Cherokee, Holston	130	2	76	0.2	122	80
Douglas, French Broad	197	4	40	0.9	113	129
Fontana**, Little Tennessee	114	1	2	2.7	54	8
Fort Loudoun**, Tennessee	466	41	0	3.7	17	68
Fort Patrick Henry, South Fork Holston River	75	20	53	3.8	59	8
Hiwassee**, Hiwassee	60	1	0	4.1	82	5
Norris, Clinch	119	2	21	1.0	120	21
Nottely, Nottely	12	0	23	1.0	81	5
South Holston, South Fork Holston	28	0	23	0.8	122	10
Tims Ford, Elk	27	1	69	0.4	199	64
Watauga, Watauga	20	0	13	3.8	66	3
Watts Bar, Tennessee	798	124	0	4.5	27	48
Chickamauga, Tennessee	984	192	0	5.3	0	0
Kentucky**, Tennessee	1,797	376	0	6.1	0	0
Ocoee No. 1, Ocoee	40	5	19	6.9	0	0
Pickwick**, Tennessee	1,614	312	0	6.5	0	0
Total	–	–	413	–	1,346	503

Note: Based on daily streamflow (m^3/s) and weekly DO (mg/L) data from 1960 through 1996. Data affected by release improvements were omitted.
* DO target is 6.0 mg/L for nine dams with cold water downstream fishery, and 4.0 mg/L for eleven dams with warmwater downstream fishery.
** Tailwaters not impacted by low flows due to backwater from downstream reservoirs.
Source: Higgins and Brock (1999).

measures, however, varies greatly. In view of each river and reservoir representing a unique combination of natural and human controls, such management measures are only likely to be effective with a detailed knowledge of the local fluvial environment (Robinson et al., 2003).

4.4 DAM REMOVAL

The dam paradigm in North America, western Europe, and some of Australia is dam removal.

4.4.1 Drivers of Dam Removal

Dam removal is increasingly viewed as a vital tool in river restoration and is initiated by a variety of drivers (Pohl, 2002; Lejon et al., 2009; Magilligan et al., 2016; Ding et al., 2018). The main driver to remove dams is environmental, initiated by recognition of adverse impacts on rivers and associated environments (Table 4.13). The most important environmental justification for dam removal is to restore connectivity to support and restore fish habitat. Despite the strong interest in increasing aquatic ecosystem habitat, however, ecological justifications are not always singularly sufficient to bring down a dam. Other drivers of dam removal include safety,

Table 4.13 *Main drivers* of dam removal*

Drivers	Characterization	Case study example: River name, country (author)
Economic	Dam removal less expensive than maintenance and/or repairs; Loss of income from lack of fishing and or recreation may be greater than revenue from electricity generation; Hydropower no longer viable; Environmental costs	Kennebec River (Edwards Dam), United States (Lewis et al., 2008); Katsunai River, Japan (Noda et al., 2018)
Environmental	Restrict migration of aquatic organisms, especially fish, with decline and possible extinction; Reduction in biodiversity; Changes to dynamics, timing, and quality of streamflow, reduced sediment, and geomorphic adjustment; Reservoir impoundment causes algae blooms, greenhouse gas emissions, reservoir stratification, and degraded outflow	Chichiawan Stream, Taiwan (Chang et al., 2017); Elwha River, United States (Lierman et al., 2017); Hedströmmen River, Sweden (Törnblom et al., 2017)
Safety	Requires regular maintenance; Liability cost; Degradation of structure; Structure not adapted to changes in upstream flow and sediment loads (climate and land cover); Sedimentation threatens structural integrity; Impoundment causes upstream hazards to infrastructure and navigation	Carmel River, United States (Mussetter and Trabant, 2005)
Cultural heritage and aesthetics	Dam and reservoirs displace sacred spaces and degrade vital ecosystems; Material culture drowned by impoundment; Displacement of communities; Value of free-flowing rivers based solely on appearance	Elwha River, United States (Guarino, 2013); Penobscot River, United States (Opperman et al., 2011)

* Most dam removal projects include a combination of drivers. (Author table)

Figure 4.36. Removal of the 64-m high Glines Canyon Dam on the Elwha River, Washington in February 2012. (Photo source: U.S. Geological Survey.)

economics, and cultural heritage and aesthetics (Table 4.13). Free-flowing rivers are valued by society for their own sake, and thus aesthetics are also deemed to be an important motivation for dam removal. Most dam removal projects involve at least a couple of drivers. The high-profile case of the Elwha Dam project (Figure 4.36), for example, was driven by a combination of environmental and cultural heritage issues related to indigenous people and their connection to fish migration, specifically Pacific salmon.

4.4.1.1 LEGISLATION AND POLICY INSTRUMENTS

Government institutions across international, federal, state/provincial, and local scales have varying types of policy instruments that pertain to specific drivers of dam removal. The U.S. Clean Water Act (Sec. 404) and Endangered Species Acts, for example, are federal legislative tools that can be used in environmentally motivated dam removal 'projects', although they are seldom employed (Bowman, 2002; Doyle et al., 2003b; Opperman et al., 2011). The relicensing of a dam requires a review of the Endangered Species Act as it pertains to adverse environmental impacts as well as measures to manage and mitigate adverse impacts to riparian environments. Sweden has the federal Environmental Objective, which identified sixteen ecosystem goals to be attained by 2020. This included a stated priority of maintaining healthy rivers, lakes, groundwater, and wetlands, among others. A specific facet of the objective was to restore 25% of valuable rivers by 2010, a stimulus for dam removal efforts in Sweden based on environmental justifications (Lejon et al., 2009). For rivers that intersect international borders it is important to have international agreements and institutions to specify cooperation in support of healthy rivers.

The European Union's Water Framework Directive (WFD) requires that all EU nations manage water in the context of sustainability. Indeed, Article 4 of the WFD includes specific ecologic and hydrologic targets that nations are supposed to attain by 2027 (Directive 2000/60/EC). To adapt to the WFD, the EU developed the River Basin Management Plan as a common strategy with specific deadlines to meet each of the WFD requirements. While the requirements were in accord with the concept of dam removal to improve riparian health, a number of nations did not view the WFD in these terms and were slower to pursue dam removal projects. The implementation and enforcement of the EU's WFD in the context of dam removal was recently strengthened (May 2020) by the EU Biodiversity Strategy for 2030. The Biodiversity Strategy requires restoration of at least 25,000 linear km of rivers to a free-flowing state through the removal of barriers (dams and weirs), in addition to floodplain and wetland restoration.

4.4.1.2 ECONOMICS AND DAM REMOVAL

Economics is a vital driver of dam removal and includes multiple facets. While revenue associated with electricity generation is often an important primary consideration for hydropower dams, other economic considerations include revenue associated with tourism and recreation, land use and associated income generation, property values, taxes, fisheries production, sediment management, and dam maintenance and repair (Lawson, 2016). Dam removal projects include formal cost–benefit analyses, which compare the cost of removal to the economic benefits of removal as well as the cost of alternatives to dam removal. Alternatives to dam removal often include the cost of upgrading to meet new standards as regards safety, mechanical and electrical generation, and environmental mitigation, particularly fish passages. The alternatives to dam removal are often twofold higher than the cost of dam removal (Lawson, 2016). Sediment management (discussed below) is usually the most expensive part of dam removal. For some dams, changes to residential property values can be a larger consideration. In some instances, property values increased after dam removal for rivers in Maine and Wisconsin (United States), although it depends on the proximity to the river (Lewis et al., 2008; Provencher et al., 2008; Auerbach et al., 2014; Lawson, 2016). And each economic factor may be used to make a case for or against dam removal depending on the competing stakeholders (Morris and Fan, 1998; WCD, 2000; Heinz Center, 2002).

After a decades-long dispute about removing four dams along a 225 km stretch of the lower Snake River, Washington (Ice Harbor, Lower Monumental, Little Goose, and Lower Granite dams) that provide power, flood control, water for irrigation, and navigation, but only about 5% of the regional electricity, the U.S. EPA in 2020 decided not to remove the four dams. The stated reason was concern that a loss of power generation would increase electricity rates for consumers. Earlier economic analyses had shown a net benefit of $310 million, and that the river recreation gains actually exceeded lost revenue from reservoir recreation (Loomis, 2002). The dams ensure ~450 km of navigation from the Pacific Ocean at the mouth of the Columbia River to Lewiston, Idaho. But the dams are also associated with steep decline in Pacific salmon stock as well as declines in species that heavily rely upon salmon for feeding, which includes Southern Resident killer whales.

Table 4.14 *Cost to rehabilitate US dams*

Category	No. of dams	Cost
Nonfederal	87,199	$60.7 billion
Federal	3,381	$4.2 billion

Source: ASDSO (2016).

Globally, numerous dams are older than five decades (e.g., Figure 4.4) and in need of costly repairs, maintenance, and upgrading (Morris and Fan, 1998; Heinz Center, 2002; Doyle et al., 2008). The average age of US dams is fifty-seven years (as of 2020), and over 10,000 dams (>2 m) are older than seventy years (e.g., Figure 4.8). The cost of upgrading and repairing such infrastructure is enormous, and was estimated in 2016 at about $65 billion (Table 4.14). Because there are so many old dams with questionable structural integrity, an additional strong motivation for dam removal is liability, which ultimately relates to economics (Heinz Center, 2002). Most dams are locally owned and many owners are increasingly opting for dam removal to avoid expenses associated with liability or repairs. Small dam removal projects often cost less than the cost of safety repairs (Born et al., 1998; Bowman, 2002). Thus, the same rationale for the original construction of dams – economics – is now being utilized to justify their removal.

4.4.2 Stakeholders and Dam Removal

Getting different stakeholders to work together to accomplish a common goal is always challenging, and is a core element of "integrated water resource management" (Lejon et al., 2009; Opperman et al., 2011). Stakeholder types involved in dam removal projects potentially include energy companies, recreational and commercial fishing organizations, environmental organizations, indigenous peoples, cultural heritage preservationists, property owners, developers and the business community, and governmental organizations at local, state/provincial, federal, and international scales. In North America, indigenous peoples are important stakeholders in dam removal efforts, particularly coastal communities whose livelihoods are linked to fisheries impacted by dams. Some environmental organizations are explicitly oriented to dam removal, such as International Rivers, American Rivers, and DamRemoval.eu.

Stakeholder involvement in the dam removal process should occur at all stages of the project. Success requires that all stakeholders be immersed in plans and progress as the project evolves.

As much as general "lessons learned" are important, the reality is that all dam removal projects are different (Stanley and Doyle, 2003). Each project has unique cultural, political, and historical dimensions that impact the outcome. Although government agencies and scientists may be in support of dam removal, opposition from local stakeholders can result in the failure or extensive delay of a seemingly well-intended project (Born et al., 1998; Lejon et al., 2009; Fox et al., 2016; Magilligan et al., 2017). Regionally, there can be great differences in the representation of different stakeholder types, and whether they have favorable or negative views of dams. In western portions of North America, a mill dam over a hundred years old may be considered valuable "cultural heritage" by some stakeholders, while in Europe structures of that age may seem less unique.

4.4.3 International Extent of Dam Removal

4.4.3.1 DAM REMOVAL IN NORTH AMERICA

Dam removal is occurring at an accelerated pace and being guided by increasingly sophisticated procedures and monitoring conventions, although there is great regional and national and international variability across the United States, Canada, and Mexico.

In total, 1,699 dams were removed in the United States between 1900 and 2019 (Figure 4.37). And 806 dams were removed over the past decade, nearly double the previous decade. Most dams removed are less than 5 m in height. In the United States, the hotspots for dam removal are in the northeast, northwest, and the northern Midwest (American Rivers, 2020).

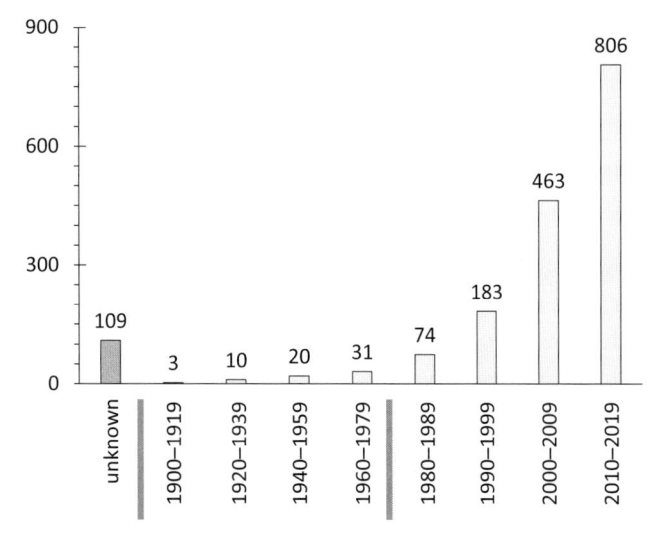

Figure 4.37. US dam removal between 1900 and 2019. Total dams removed are 1,699 (May 22, 2020). Dam removal between 1900 and 1979 grouped in twenty-year periods and from 1980 to 2019 by decade. (Author figure. Data source: American Rivers, 2020.)

Numerous mill dams in New England have been removed. Many of these are well over a hundred years old and represent the region of the most intense and oldest river impoundment in the United States (Walter and Merritts, 2008). The largest dam removal project in New England is the Great Works Dam in 2012 and the Veazie Dam in 2013 along the lower Penobscot River (22,196 km^2) in Maine. The project also included modification to dam structures to create fish passages, and was driven to restore some 3,200 km of riparian habitat to anadromous fish, such as Atlantic salmon (*Salmo salar*), alewife (*Alosa pseudoharengus*), American shad (*Alosa sapidissima*), and Atlantic sturgeon (*Acipenser oxyrinchus oxyrinchus*). And it worked, as fish populations quickly returned within a few years, with anadromous fish species increasing while slower lacustrine species declined (Watson et al., 2018). California had twenty-three dams removed in 2019, and Pennsylvania had fourteen. Texas had one dam removed in 2019, and has only had eight in total removed. The paucity of dam removal projects in Texas is concerning because it has the highest number of dams and the largest proportion of its surface water impounded of any state in the United States (Graf, 1999).

In terms of height, 349 dams that are higher than 5 m have been removed in the United States. In the Pacific Northwest, dams are being removed in smaller coastal draining rivers where it is mainly about restoration of Pacific salmon habitat (all five species), which quickly return to spawn in upstream rivers for the first time after many decades (Lierman et al., 2017). The largest dam removal project in the world – to date – occurred along the Elwha River (820 km^2) in the Olympic Peninsula of coastal Washington over a couple of years (see discussion below). The removal project occurred in two phases, the 33-m Elwha River Dam in 2011 followed in 2012 by a phased removal of the 64-m Glines Canyon Dam (Ritchie et al., 2018).

Dams are also being removed in Canada, although at a slower pace than the United States. The removal of small dams in Canada is tempered by the fact that large dams have recently been completed and are still being constructed along major waterways. Since 2006 twenty-one dams have been removed in British Columbia, including four large dams >9 m in height. Dam removal is proceeding within established removal procedures and guidelines. A positive sign that dam removal is becoming a mature environmental industry in Canada is that sophisticated guidelines are being developed within clear policy frameworks at the scale of individual provinces. Ontario, for example, has the Lakes and Rivers Improvement Act (LRIA) that specifically includes a range of fluvial geomorphic criteria to develop a dam removal plan, including sedimentary and morphologic adjustments (Ontario Ministry of Natural Resources, 2011).

In Mexico, dam removal is not occurring with any documented frequency. Indeed, the trend is somewhat the opposite and several large dams were constructed in the 2000s. In 2020, Mexico confirmed it will go forward with the massive Chicoasén II hydroelectric dam (240 MW) in the deep canyons of the Grijalva River in the mountainous southern state of Chiapas. The dam is the second phase of a larger hydroelectric site that includes the 261-m high Manuel Moreno Torres (Chicoasén) Dam, the tallest dam in North America. Several other proposed Mexican dams, fortunately, have been stalled or canceled over the past decade because of substantial opposition from environmentalists, human rights supporters, and indigenous peoples.

4.4.3.2 DAM REMOVAL IN THE EUROPEAN UNION

Dam removal in the European Union[3] is occurring with much greater frequency over the past decade, with France leading the way (Fernández Garrido, 2018; Schiermeier, 2018). The increasing pace of dam removal is occurring for several key reasons, including: (1) The European Union Water Framework Directive (WFD), which requires free-flowing rivers to attain good ecological status by 2027 (noted above); (2) environmentally oriented national strategies supported with science that view river networks as linked to landscapes; and (3) the development of effective nongovernmental organizations that champion the cause of dam removal. Nevertheless, like the United States there is considerable regional variability in the pace of dam removal. Many of the debates and the roles of stakeholders with regard to dam removal projects are similar to the United States (Born et al., 1998; Lejon et al., 2009; Jorgensen and Renofalt, 2013; Magilligan et al., 2017). To obtain a clearer picture of dam removal across the EU it is crucial that a standard system of data reporting is developed.

A recent "fitness check" of the effectiveness of implementation of the EU Water Framework Directive revealed that the quality of some 40% of European rivers is under "hydromorphological" stress caused by dams and hydraulic engineering (EEA, 2018; EC, 2019). This recognition further increased attention to the issue of dam removal, as well as passage in 2020 of the EU Biodiversity Strategy for 2030 that includes a stated purpose to greatly increase the length of free-flowing rivers in Europe. Dam Removal Europe (damremoval.edu) is the key NGO, and has singularly focused on removing barriers (dams and small weirs) that are obstacles to fish migration, which like North America, is the primary driver of dam removal. A recent development (as of 2018) is the formation of AMBER (Adaptive Management of Barriers in European Rivers), a consortium of twenty key private and public specialists that represent a range of stakeholders involved with dam removal in Europe. AMBER includes partners from the hydropower industry, river authorities, nongovernmental organizations, and university scientists with the intention to enhance collaboration and knowledge transfer about the process of dam removal. An important task being undertaken by AMBER is to arrive at a tally and site-specific characterization of dam removal projects, which is difficult because of different protocols for reporting, as noted for Spain. Additionally, an important element of AMBER is "citizen science," which directly involves the public in data collection and dissemination. Combined, these two thrusts of European dam removal have increased the visibility of the topic to a range of public and private stakeholders, increasing public participation in the process of dam removal.

According to the DamRemoval.eu NGO, some 5,000 dams (and weirs) have been removed across five European nations as of 2018. This includes France with 2,425 dam removals (partially removed obstacles: 5,728), Sweden with >1,600, Finland with >450, Spain with >250, and Great Britain with 156 dam removals (Fernández Garrido, 2018). Many EU nations have ambitious plans for dam removal that link national strategies with the EU WFD. Among the several significant dams removed in Europe are the Retuerta dam (14 m) and the La Gotera dam (8 m) in the Duoro basin (Spain) in 2013 and 2011, respectively. The latter was initiated as part of the Spanish National Strategy of River Restoration to increase hydrologic continuity through the Alto Bernesga Biosphere Reserve of the Man and Biosphere Programme. In France's nationally symbolic Loire basin, the Saint Etienne du Vigan dam (14 m) on the Allier tributary and the Maisons Rouges dam (4 m) on the Vienne tributary were both removed in 1998 as part of the French governments "Plan Loire Grandeur Nature." The main focus is to increase fish migration to the upper basin headwaters, particularly for spawning of Atlantic salmon (*Salmo salar*). The largest dam removal project in Europe occurred along the Sélune River in Normandy, France. The 36-m high Vezins Dam was the first of two dams to be removed (2019), with the second being the 15-m high La Roche Qui Boit removed in 2021. Combined, the removal of the two dams reopen 91 km of a coastal draining river to migrating Atlantic salmon, eels, and other riparian wildlife, which empties into the bay of Mont-Saint-Michel – a UNESCO world heritage site and key stakeholder related to cultural heritage.

Although progress is being championed in scientific media (Schiermeier, 2018), dam removal varies greatly across the European Union. And it often varies within individual nations between states/provinces or basin, as with the issue of data reporting in Spain. Data reporting procedures in Spain are not centrally coordinated and instead are organized at the scale of the larger drainage basins, which vary considerably in format and in public accessibility (Rincón Sanz and Gortázar Rubial, 2016). In the Douro River basin (98,400 km^2), for example, 116 dams and barriers were removed between 2009 and 2016, the highest in any basin in Spain. Of these dams, 31 were <2 m height, 16 were 2–5 m height, 2 were >10 m height, but 67 of the dams removed

---

[3] Includes the United Kingdom at time of preparation.

in the Duoro basin did not include a recorded dam height or descriptive information (Ministry for Ecological Transition, MITECO, 2020). In comparison to the Duoro Basin, which has 145 large dams, the Ebro River basin (80,093 km^2) is the most impounded in Spain, and has 299 large dams. But data from the Ebro River basin authority only reported 5 dams removed (Rincón Sanz and Gortázar Rubial, 2016).

4.4.4 Science of Dam Removal

Effective dam removal is the essence of multi- and interdisciplinary science, and commonly includes consultation with engineers, earth scientists, biologists, economists, archaeologists and historians, social scientists, among others. While the science is important, however, specific scientific activities need to be intricately synchronized with the legal and policy framework in which dam removal projects exist (Aspen Institute, 2002; Bowman, 2002; Heinz Center, 2002; Doyle et al., 2003b; USSD, 2015; Randell and Bountry, 2017).

Depending on the size and sensitivity of the riparian area, the time required to remove a dam can vary greatly, with even modest-sized dams (~5 m) frequently taking a decade or two to remove. This is largely because of requirements to obtain a range of permits and the importance of having conducted research on the potential impacts of the prospective dam removal project (USSD, 2015). This is particularly the case when dam removal projects may degrade wetlands, if the riparian area includes endangered species or other important ecosystems, or if valuable cultural materials are associated with downstream riparian area as well as the dam and reservoir structure. Smaller dams and weirs without such critical habitat concerns are often (but not always) less complicated to remove, particularly smaller run-of-river structures that store little sediment. In such cases, the removal process is often regulated at the state or provincial level. Indeed, many states and provinces have detailed guidelines and "checklists" for how landowners can remove smaller structures (i.e., Ontario Ministry of Natural Resources, 2011). The removal of small dams and structures as a community-based program has also proven effective at restoring aquatic habitat, particularly if effectively coordinated by a larger government agency (Lenhart, 2003). Regardless of reservoir size, however, legal disputes between different stakeholders can significantly delay (by decades) or completely stymie a dam removal project.

Dam decommissioning is expensive. Dam removal cost widely varies according to the scope and sensitivity of the project (Aspen Institute, 2002; Randle et al., 2019). Funding may come from a variety of sources, such as private power companies, environmental organizations, and government agencies at international, national, state/provincial, and local scales. It is essential to establish stakeholder responsibility for cost early in the exploratory phase, as cost can escalate. Morris and Fan (1998) report that the largest proportion of dam removal cost is associated with sediment management – at 48%. The actual removal of the hydraulic infrastructure and environmental engineering are 30% and 22%, respectively. And if the dam removal project includes substantial polluted reservoir deposits, the sediment management cost significantly increases (Evans, 2015).

The science of dam removal is rapidly developing into a sophisticated and diverse enterprise. This is largely because of (1) being able to draw upon a rich theoretical foundation of knowledge in fluvial dynamics and aquatic ecosystem processes developed over past decades, (2) increasing availability of historic datasets, and (3) sophisticated monitoring and data collection equipment.

The specific research activities associated with a dam removal project should be organized within a temporal–spatial framework that includes a pre-dam-removal phase, removal phase, and post-dam-removal phase that spans upstream (control site), dam and reservoir, and downstream reaches. The latter may also include shallow marine and beach environments for river basins tightly coupled to coastal dynamics.

Most dam removal projects begin by establishing specific hydrogeomorphic and biological reference targets, which may be historic prehuman disturbance "baseline" conditions (if possible to establish) or other environmental targets mutually agreed to by relevant stakeholders (Aspen Institute, 2002; Heinz Center, 2002; Stanley and Doyle, 2003; USSD, 2015). In practice, many datasets for reference conditions (prior to dam construction) may extend back only about a decade. In such cases, it is ideal that this includes both a downstream and upstream control reach (above reservoir backwater).

Although a return to a natural riparian regime and the replenishment (reactivation) of stored reservoir sediments to downstream reaches is a common goal, the most important task is to assess the potential impacts of an abrupt sediment pulse to downstream physical and biological habitat. Thus, a key element to dam removal planning is to develop a sediment management strategy (Figure 4.38) that outlines the sequence of procedures according to varying qualities of reservoir deposits (Randle et al., 2019).

4.4.4.1 MANAGING RESERVOIR SEDIMENT REMOVAL

Reservoir drawdown is a crucial step in dam removal, and management of sedimentary deposits during dam removal follows a sequence of specific steps (Figure 4.38). The method of drawdown is of key importance to sediment mobilization during and after reservoir drawdown. Rapid reservoir drawdown results in considerable hydraulic flushing and scour of reservoir deposits (Scheueklein, 1990), and produces a large sediment pulse that can be detrimental to downstream riparian environments (Grimardias et al., 2017). A slower drawdown is more benign to downstream riparian environments, although it can

RESERVOIR SEDIMENT MANAGEMENT STRATEGY

1. Identify sediment concerns

2. Collect reservoir and river data

3. Evaluate potential for contaminated sediment
   - *Concerns?*
   - NO → (continue to 4)
   - YES OR UNKNOWN → Sediment sampling and analysis: Do containment concentrations exceed sediment quality criteria?
     - NO → (return to 4)
     - YES → Conduct biological analysis and estimate sensitivity: Can sediment be relocated?
       - YES → (return to 4)
       - NO → Assume cap/isolate, stabilize, and/or sediment removal → Assess risk of future contaminant release → 10. Develop monitoring and adaptive management plan

4. Determine relative reservoir sediment volume and probability of impact
   - *Negligible?*
   - YES → **DONE** with sediment analysis
   - NO → Small, medium, or large relative reserve volume

5. Refine potential sediment consequences and estimate risk

6. Develop dam removal and sediment management alternative

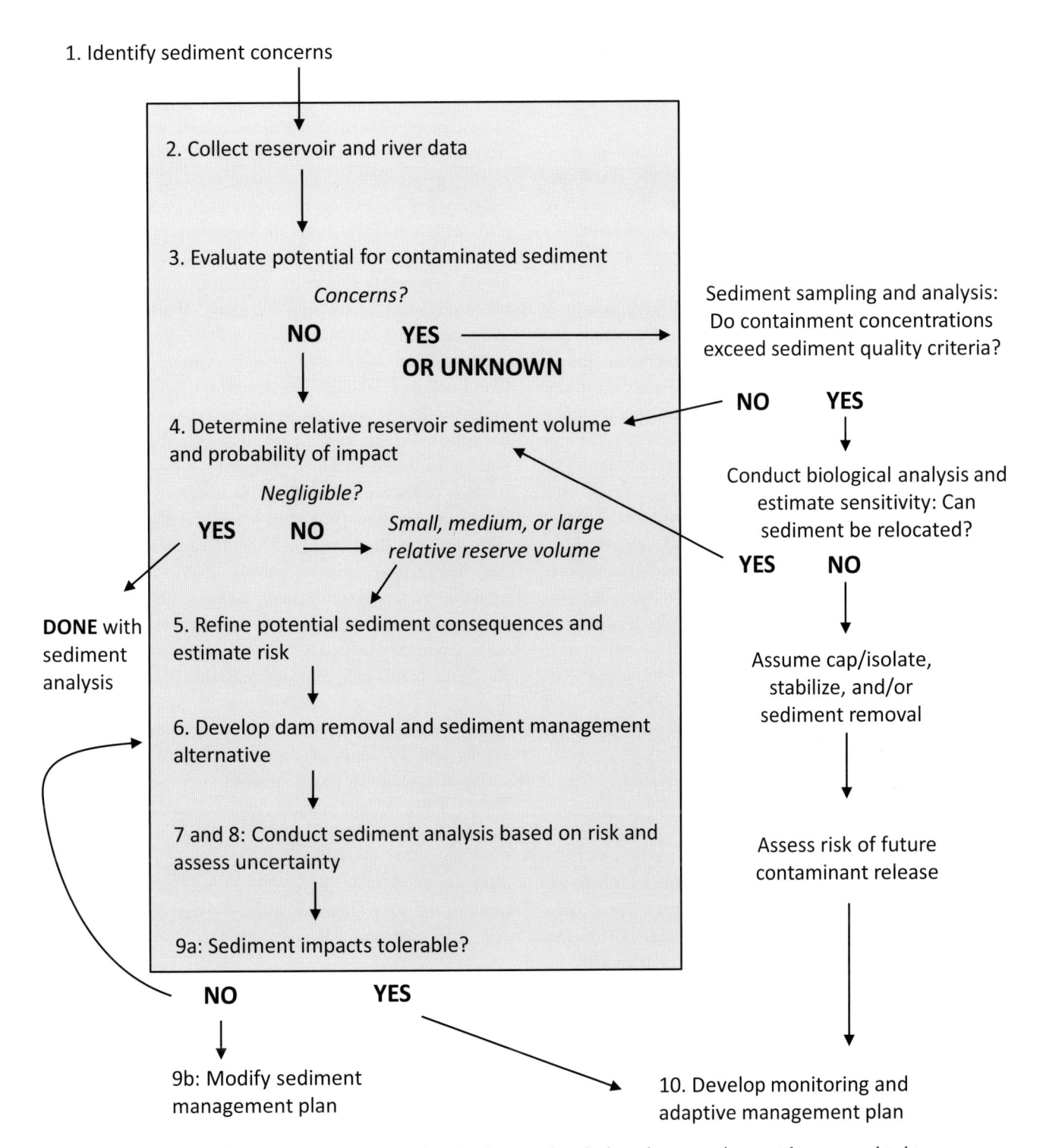

7 and 8: Conduct sediment analysis based on risk and assess uncertainty

9a: Sediment impacts tolerable?
   - NO → 9b: Modify sediment management plan → (return to 6)
   - YES → 10. Develop monitoring and adaptive management plan

Figure 4.38. Steps for removal of reservoir sedimentary deposits. Sequence largely depends upon environmental concerns related to reservoir deposits. (Source: Randle et al., 2019.)

result in a larger volume of reservoir deposits that are disconnected from active fluvial processes. Additionally, upon drawdown a sharp knickpoint forms near the dam base (Neave et al., 2009). Subsequent upstream headcut migration of the knickpoint through reservoir deposits further removes sediment by triggering mass wasting and erosion of noncohesive deposits (Morris and Fan, 1998; USSD, 2015).

Because reservoir deposits sequester large amounts of phosphorus and nitrogen from upstream agricultural landscapes (Maavara et al., 2015), the release of nutrients into the active fluvial system and eventually into downstream coastal waters is a concern during the dewatering phase of dam removal (Riggsbee et al., 2012).

Many dams impound polluted reservoir deposits and it is important that dam removal does not result in contaminated sediment pulses impacting downstream environments (Evans, 2015). Sediment pollution from dam removal is of particular concern in reservoirs downstream of old industrial and mining areas (Tullos et al., 2016). In such cases, removal of contaminated deposits must precede dam removal, or dam removal may be untenable. Along the Hudson River in New York, the 1973 removal of Fort Edward Dam resulted in several hundred thousand cubic meters of sediment contaminated with polychlorinated biphenyls (PCBs) to be transported downstream. After decades of legal disputes with industry about whether the polluted sediment should be capped or removed, some 2 million m^3 of sediment was eventually dredged from the river bed, in addition to floodplain restoration, at a cost of $561 million (Evans, 2015). And legal disputes between the state of New York and industry are ongoing as to whether the polluted sediment has been satisfactory removed.

4.4.4.2 REMOVAL OF TOXIC RESERVOIR DEPOSITS

The Milltown Dam removal project along Clark Fork (59,320 km^2), the largest river in Montana, included reservoir deposits polluted with arsenic, copper, and other metals associated with a legacy of mining (Sando and Vecchia, 2016). The dam had been constructed in 1907 and within months an enormous flood (~500-year RI) deposited tons of polluted sediment into the reservoir (Evans, 2015). The reservoir is part of Clark Fork Basin Superfund Complex, the largest superfund site in the U.S. (U.S. EPA, 2016). The environmental and health problems associated with the contaminated deposits had been known since the early 1980s when residents started to become ill from polluted drinking water. The issue came to a head in the winter of 1996 when a large ice jam threatened to breach the dam. To prevent Milltown dam from being damaged by the ice jam the reservoir water level was lowered. But the reservoir drawdown was too abrupt and resulted in hydraulic scour of polluted reservoir deposits, releasing a toxic pulse that resulted in a massive fish kill downstream (Diamond, 2006; Evans, 2015). Prior to finally dismantling the 12.8 m tall dam in spring 2008 (initial breach March 28), some 2.2 million m^3 of polluted deposits (40% of total reservoir volume) were hauled away by train and buried outside of the riparian zone (U.S. EPA, 2016). By far the largest proportion of the Milltown Dam removal cost was managing reservoir sediment removal ($120 million).

The environmental disaster associated with the abrupt 1996 drawdown of the Milltown dam on Clark Fork River provided a useful lesson employed in the dam removal strategy. Rather than an abrupt drawdown, reservoir drawdown occurred very slowly over nearly two years, extending from June 1, 2006 to the dam breach, on March 28, 2008. While arsenic levels increased during the drawdown, it increased much less than copper (Figure 4.39). Nevertheless, despite the attempt to limit pollutant mobilization during the drawdown, the increased suspended sediment concentrations during the drawdown period (June 2006–March 2008) flushed 140,000 metric tons of particulates, which included 44 tons of copper and 6.4 tons of arsenic eroded from the reservoir. This is interesting because it shows that even when exercising great caution with reservoir drawdown, it should reasonably be expected that fine-grained deposits will be hydraulically excavated from the reservoir. Additionally, higher sediment loads during the drawdown phase were also attributed to structural landscape modifications (roads, facilities, excavation, etc.) associated with preparation of dam removal. Following the dam breach, an additional 420,000 tons of suspended sediment were transported downstream, which included 15.8 tons of arsenic and 169 tons of copper. A discharge of 250 m^3/s during the removal phase resulted in a sediment load of 20,000 while several months later the same discharge produced a sediment load of about 2,000 tons (Sando and Lambing, 2011). Overall, copper, arsenic, and suspended sediment decreased by 53%, 29%, and 22%, respectively, between the start of water year 1996 and the end of water year 2015 (Sando and Vecchia, 2016).

4.4.4.3 REWORKING OF RESERVOIR DEPOSITS

To evaluate changes to fluvial sediment budgets associated with dam removal it is necessary to calculate downstream sediment loads associated with the reworking of reservoir deposits. Sediment mass (tons) is commonly estimated from the volume (m^3) of reservoir deposits (Randle and Bountry, 2017). Accurate sediment load estimates from reservoir deposits require data on the weight of specific particle size classes, and whether they are subaqueous (submerged) or subaerial (above the water surface) deposits. This data is obtained from reservoir sedimentation surveys that measure the bathymetry and topography of reservoir deposits, as well as the particle size distribution obtained from

Table 4.15 *Comparison of submerged and subaerial weight of reservoir sediment*

Dominant particle size	Always submerged (g/cm^3)	Above water (g/cm^3)
Clay	0.64–0.96	0.96–1.28
Silt	0.88–1.20	1.20–1.36
Clay-silt	0.64–1.04	1.04–1.36
Sand-silt	1.20–1.52	1.52–1.76
Clay-silt-sand	0.80–1.28	1.28–1.60
Sand	1.36–1.60	1.36–1.60
Gravel	1.36–2.00	1.36–2.00
Sand-gravel	1.52–2.08	1.52–2.08

Source: Randle and Bountry (2017), based on Morris and Fan (1998).

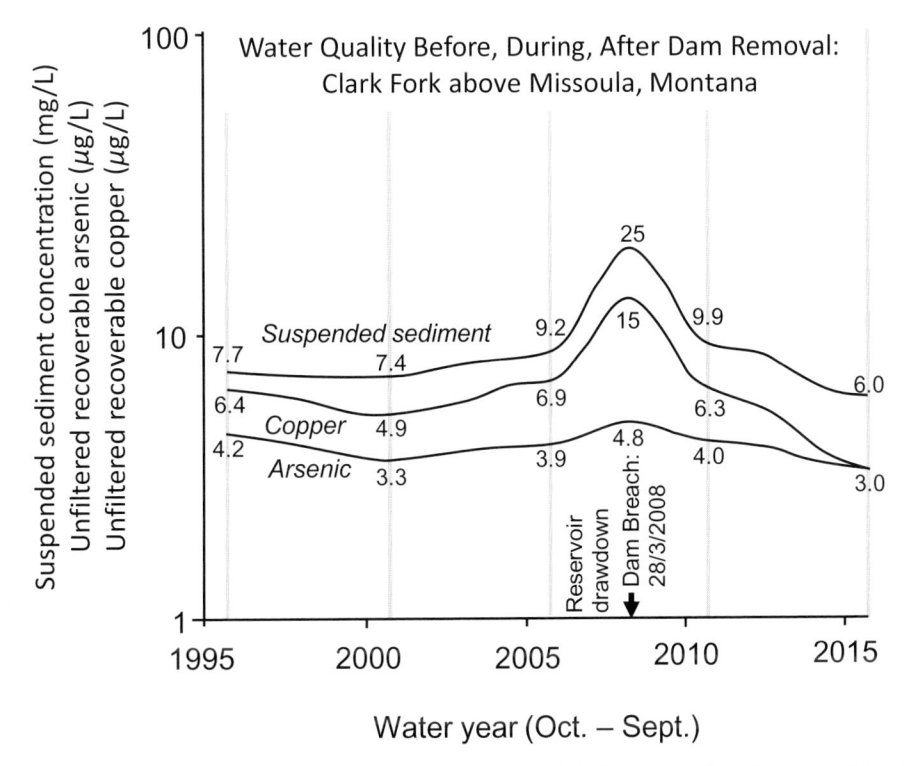

Figure 4.39. Arsenic, copper, and suspended sediment between 1996 and 2015 for Clark Fork River. Sampling was 4.8 km downstream of Milltown Dam. Note: y-axis is logarithmic (USGS station i.d.: 12340500). (Source: Sando and Vecchia, 2016.)

borings and geophysical mapping (Yang, 2006).[4] The weight of fine-grained deposits varies considerably according to whether it is submerged or above water (Table 4.15). Additionally, cohesive deposits, such as clay, substantially compact over time, which reduces pore space and increases sample weight. For example, after fifty years of compaction clay loses 35% of its original volume (Figure 4.40), which also increases reservoir storage capacity.

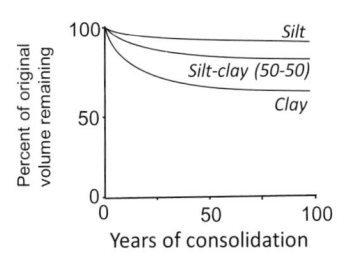

Figure 4.40. Reservoir sediment compaction over time for different particle size classes over one hundred years. Values are for continuously submerged deposits. Sand not shown because of minimal consolidation. (Source: Annandale et al., 2016.)

[4] See chapter 9 in Yang (2006) for a thorough review of methods to estimate reservoir storage capacity.

Table 4.16 *Reservoir sediment eroded after dam removal for US case studies*

Dam, river, state	Sediment type	Dam height (m)	Reservoir sediment volume (m^3)	Short term (<1 yr)		Long term (>1 yr)	
				Years after dam removal	Sediment erosion volume (%)	Years after dam removal	Sediment erosion volume (%)
Condit, White Salmon, Washington	60% sand, 35% mud, 5% gravel	38	1.8 million	0.7	72		
Glines Canyon, Elwha, Washington	44% mud, 56% sand and gravel	64	16.1 million	1	37	5	72
Elwha, Elwha, Washington	47% mud, 53% sand and gravel	32	4.9 million	1	20	5	50
Rockdale, Koshkonong, Wisconsin	35% sand, 45% silt, 20% clay	3.3	287,000	0.8	17		
Ivex, Chagrin, Ohio	Mud	7.4	236,000	0.2	13		
La Valle, Baraboo, Wisconsin	45% sand, 40% silt, 15% clay	~2	10,000–15,000	1	8		
Brewster, Brewster Creek, Illinois	70–99% silt and clay	2.4	18,000	1	8	3.7	13
Milltown, Clark Fork, Montana	Sand*	12.8	5.5 million	0.4	77		
Simkins, Patapsco, Maryland	Sand, fine gravel	3	67,000	1	73	3.6	94
Merrimack Village, Souhegan, New Hampshire	95% sand	4	62,000	1	63	1.5	79
Marmot, Sandy, Oregon	~50% gravel, 50% sand	15	750,000	1	53	1.8	58
Savage Rapids, Rogue, Oregon	70% sand, 30% gravel	12	150,000	0.4	50		
Lost Man, Lost Man Creek, California	predominantly sand and gravel	2.1**	2,000	0.6	30		
Brownville, Calapooia, Oregon	Gravel	2–3	10,000–15,000	1	30	1.9	38
Secor, Ottawa, Ohio	Sand	2.5	5,000–9,000	0.4	10		
Stronach, Pine, Michigan	70% sand, 30% gravel	3.6–5	800,000	1	1	2.8	3

*Approximately 40% of sediment excavated prior to dam breaching was part of Superfund remediation efforts.
**Dam height from Sacklin and Ozaki (1988).
Data: Randle and Bountry (2017).

Most reservoir deposits reassimilated into the active fluvial system are eroded within the first year or two of dam removal, particularly for smaller elongated reservoirs with coarse deposits (Major et al., 2017). After the first couple of years, the amount of reservoir sediment reworking abruptly decreases as the channel stabilizes and as reservoir deposits become anchored with vegetation. A review of twenty-one dam removal projects in the United States (Table 4.16) revealed that over 40% of reservoir deposits were reworked and resupplied to the active fluvial system within the first year of dam removal by moderated streamflow events (Major et al., 2017). The proportion of reservoir deposits that eventually become assimilated back into the active sediment regime varies according to several factors, including sediment texture (fine or coarse), lateral channel dynamics, flood regime, and the size of the reservoir relative to the river channel. The amount of reservoir reworking appears to be less for broader reservoirs with finer-grained deposits and less lateral channel activity (Doyle et al., 2003a), although more documentation of dam removal projects is needed across lowland settings with fine-grained reservoir deposits.

An additional factor that influences reservoir sediment reworking includes the biogeographic regime relative to the seasonality of reservoir drawdown (Cannatelli and Crowe-Curran, 2012). If vegetation quickly colonizes reservoir deposits following reservoir drawdown, sedimentary deposits can effectively remain "locked in place" within the old reservoir site. This can be a desirable outcome. If not, vegetation may have to be removed to increase the possibility of sediment reworking (Bountry and Randle et al., 2017). Channel incision within the reservoir deposits forms anthropogenic terraces, reducing the possibility of erosion and reworking during flood events. Additionally, some reservoir sediments deposited atop natural terraces are unlikely to be hydrologically connected during subsequent flood events. In some cases, temporary training dikes can steer laterally active channels toward reservoir deposits to increase their potential for erosion. Fine-grained reservoir deposits distributed across wide valleys are less likely to be reassimilated into the active fluvial system.

4.4.5 Post Dam Removal Response

Comparisons of before, during, and after field data from biological sampling, sediment sampling, and morphologic measurements illustrate the influence of dam removal to riparian environments. Downstream environments adjust and recover to sediment pulses produced by dam removal usually within a few years, particularly fish populations. While there are few adverse consequences to aquatic ecosystems after dam removal, benthic organisms may require additional time to become reestablished (Doyle et al., 2003a). Biological assemblages may be different than pre-dam assemblages because of environmental changes having occurred since impoundment (Foley et al., 2017).

Bed material usually becomes finer following dam removal. This occurs for two reasons. Firstly, the influence of impoundment usually results in winnowing of fine deposits so that the channel bed material becomes coarser. Secondly, finer reservoir deposits are more easily mobilized and flushed downstream than the coarser deposits, which moves downstream as a slower wave of coarse deposits. The interstitial pore spaces of the channel bed active layer may initially become clogged with fine sediments (the embeddedness problem) immediately downstream of the former dam site, decreasing with distance to perhaps tens of kilometers downstream. This is a key consideration and represents a direct form of aquatic habitat degradation that can be important to benthic macroinvertebrates and fish spawning. Even several centimeters of embeddedness over a few months can be detrimental to the benthic habitat of aquatic organisms, and of crucial concern if the species is endangered. This again highlights the importance of the timing of dam removal, as the streamflow variability following dam breach is critical to mobilization and downstream assimilation of reservoir deposits into the active sediment load. Downstream fining of bed material is not always obvious. Downstream of two small dams removed in Wisconsin, the reworking of fine-grained reservoir deposits (muddy sand) did not result in appreciable change in downstream bed material size (Doyle et al., 2003b).

The downstream channel morphology adjusts rapidly to dam removal, although it is dependent upon receipt of coarse sediment pulses. Channel bed aggradation is the initial morphologic response after dam removal, with some infilling of pools and aggradation of bars. Minor channel bed incision often occurs after several years as the initial coarse sediment pulse is transported downstream by higher flows. Overall, the particle size of bars increases, and grain size distributions become less well sorted. Some channel cross sections increase width-to-depth ratios, becoming shallower and wider, particularly braided rivers. Increased flow variability results in more frequent lateral connectivity between channel and riparian wetlands, including renewed overbank sedimentation.

4.4.5.1 ELWHA RIVER DAM REMOVAL CASE STUDY

The Elwha River dam removal project in the US Pacific Northwest is an ideal "laboratory" to study the impacts of dam removal on geomorphic and related ecosystem processes for a coarse-grained system. This is because it is the world's tallest dam removal project and occurred on a modest-sized (833 km^2, $Q_{-avg.}$ 42.7 m^3/s) mountainous (peak 1,452 m elev.) gravel bed river that drains directly to the coast. And Elwha River and Glines Canyon Dams were only 7.4 and 21.6 km upstream of the river mouth at the Juan de Fuca Strait, respectively. Thus, detectable geomorphic and ecological changes to downstream riparian and coastal environments was all but assured. The study includes a range of morphologic, biological, and hydraulic, and sedimentologic sampling within a conventional research framework to capture the disturbance (spatially: upstream, at reservoir, downstream; temporally: before, during, and after).

The Elwha River Dam removal project is an example of stakeholders joining forces to dismantle dams for river restoration. The S'Klallam people of the Salish Indigenous First Nations historically heavily relied upon annual salmon runs in the Elwha basin for prosperity and spiritual wellness. They fought the dams for many decades. By joining with several environmental stakeholders they were able to finally breach the bureaucracy and power of the hydroelectric industry, and bring down the dams. The Elwha River Ecosystem and Fisheries Restoration Act of 1992 was the key legislative piece to trigger the removal of the Elwha River Dams. The Act specified that the Secretary of Interior should purchase the dams to initiate complete restoration of the Elwha River, including its anadromous fisheries.

Removal of the Elwha River dams occurred in stages to minimize the downstream disturbance of the sediment pulse, with reservoir drawdown starting in September 2011. The two dams were removed in phases, with bed material first moving past the breached 32-m Elwha Dam site in spring 2011 and across the breached 64-m Glines Canyon Dam site autumn 2011, although removal of the structures extended to 2014 (East et al., 2018). Over a five-year period (2011–2016) ~20 million tons of reservoir deposits were reworked by lateral channel migration and incision. This was about two-thirds of the total sediment volume stored in the reservoir, which resulted in downstream sediment pulses being some twenty to forty times greater than "natural" annual maximum sediment loads (East et al., 2018).

By 2015, already 90% of the reservoir deposits had been discharged to the coastal zone (Ritchie et al., 2018). Only minor amounts of long-term storage of reservoir sediment will occur as overbank deposits, although some sediment storage evidently occurred within side channels (East et al., 2018). The median particle size decreased from cobbles (144–174 mm) before dam removal to gravel (18 mm) by 2013. By summer 2017, particle size had become somewhat coarser to 40 mm (gravel), but was nevertheless considerably finer than before dam removal (Figure 4.41). Because the particle size did not change upstream of the dam (in the control reach), it is concluded that downstream fining of the channel bed occurred because of the reworking of reservoir deposits.

Morphologically, the downstream channel reaches adjusted to the pulses by changing from a riffle morphology to a plane-bed morphology. Surveys for downstream channel cross sections showed variable geomorphic response and recovery (East et al., 2018). By spring 2013, the reach averaged bed elevation change ranged from about +0.38 m to +0.82 m of aggradation (Figure 4.42). But already within a few years most of the channel bed aggradation from the initial sediment pulse had been reworked, as channel incision ensued. By summer 2017, the average bed elevation changes at downstream research sites showed a net aggradation, but it had decreased to +0.15 to +0.35 m. These average values mask local-scale geomorphic variability, and channel bars are indeed now a more prominent component of the channel bed environment than prior to dam removal, with some bars +1.0 m. Additionally, the channel thalweg displays greater lateral mobility than previously, with one reach having laterally migrated by about 30 m.

Ecological changes followed geomorphic changes after removal of Elwha River dams. Tributary streams above the Elwha River Dam were being recolonized for spawning within a couple years by the river's iconic fish, Coho salmon (Liermann et al., 2017). At the coast, the copious sediment volumes resulted in about 1 m of vertical aggradation and some 400 m of lateral (deltaic) aggradation (Ritchie et al., 2018), which are already being colonized by pioneer vegetation (Foley et al., 2017). The prograding river mouth is associated with a salinity shift from coastal to more freshwater environments, including fish and macroinvertebrates communities. A number of activities are being conducted by government agencies and environmental organizations, including monitoring of sediment, vegetation, benthic invertebrates, including radio telemetry to map the movements of anadromous fish. The Elwha River dam removal project specifically applies to humid braided rivers (Figure 4.43), although a number of the findings are in accord with dam removal trajectories observed in other settings (Foley et al., 2017).

4.4.6 Synthesis of Dam Removal Results

While returning to a pristine riparian status is not widely championed by river managers and scientists, dam removal is now a key component of the "river restoration" enterprise (Magilligan et al., 2017). The detailed inventory and analysis of morphologic, sedimentary, hydrologic, and ecosystem data of numerous dam removal case studies provide sufficient resolution for development of a general model of river recovery to dam removal (Figure 4.44). Important differences to consider include the type of reservoir deposits (fine or coarse grained) and the shape and size of the reservoir (long and narrow or wide and shallow). This sets up a key distinction in the response because of the vegetation regime, and the potential to anchor fine-grained deposits "permanently" after dam removal by woody vegetation. Additionally, the reservoir drawdown in relation to the type and environmental condition of reservoir sediments is a key consideration, particularly with regard to the release of stored nutrients with the dewatering phase as well as the release of contaminated sediments into the active system. While fish are a key component of riparian ecosystems, other types of biota such as benthic invertebrates are also important to monitor, and have a slower recovery time than fish. Some benthic organisms may not recover because of broader watershed scale environmental changes to the system. Woody riparian vegetation that quickly colonizes stabilized sedimentary bars and wetlands after reservoir drawdown requires substantial efforts to remove, and may become a permanent condition of the riparian environment.

Drawing upon numerous case studies it can be concluded that dam removal improves downstream riparian geodiversity (i.e., physical integrity) and biodiversity. A couple of decades of research can be summarized with the following points regarding the impact of dam removal on rivers (Aspen Institute, 2002; Heinz Center, 2002; Doyle et al., 2003a; Tullos et al., 2016; Foley et al., 2017; Magilligan et al., 2017; Major et al., 2017).

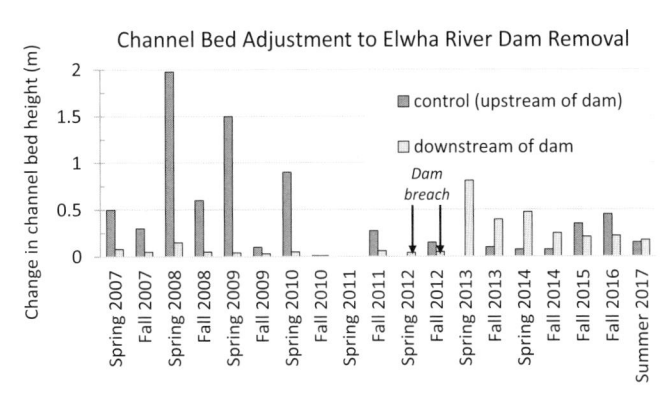

Figure 4.42. Reach-averaged channel-bed elevation changes between seasonal topographic surveys, referenced to earlier baseline survey data (not shown). Control reach averaged six cross sections spaced along a ~100-m channel segment located 1.5 km upstream of the upper end of Lake Mills (Glines Canyon Dam reservoir). The downstream reach was centered at 1.9 km below Elwha Dam and included six cross sections over a 172 m long channel segment. The two arrows indicate the approximate date when reservoir deposits were first transported past the two breached dam sites (Elwha, left and Glines, right). (Source: East et al., 2018.)

1. Rivers are resilient and recover (somewhat) quickly from decades of impoundment, as well as the disturbance from the dam removal sediment pulse and post-dam higher flow variability. Recovery mainly takes from one to several years, not decades (Foley et al., 2017; East et al., 2018).
2. The streamflow regime (variability, timing) quickly returns to a condition similar to the upstream streamflow regime (above the dam), but not necessarily to the pre-impoundment regime because of the possibility of other changes (e.g., land use) since impoundment (Sando and Vecchia, 2016; Ritchie et al., 2018).
3. Reservoir deposits are efficiently reworked and transported downstream within a few years by moderate-sized flow events. Two main controls on reservoir sediment reworking is the dynamics of channel head-cutting after breach and whether the deposits are coarse or fine-grained, with the former most efficiently reworked. Fine-grained deposits in wide reservoirs may effectively be locked in place because of rapid vegetation growth after reservoir drawdown and effectively removed from the active fluvial system (Riggsbee et al., 2007; Major et al., 2017).

Figure 4.41. Downstream changes to Elwha River channel before and after dam removal (photos looking upstream). Note changes to particle size of bed material and presence of wood: (A) One week prior to dam removal

Figure 4.41 (*cont.*) (September 2011). (B) Eleven months after dam removal, but prior to arrival of main sediment pulse (August 2012). (C) Shortly after sediment pulse (September 2014). (D) Almost six years after dam removal, four years after sediment pulse (July 2017). (Source: East et al., 2018.)

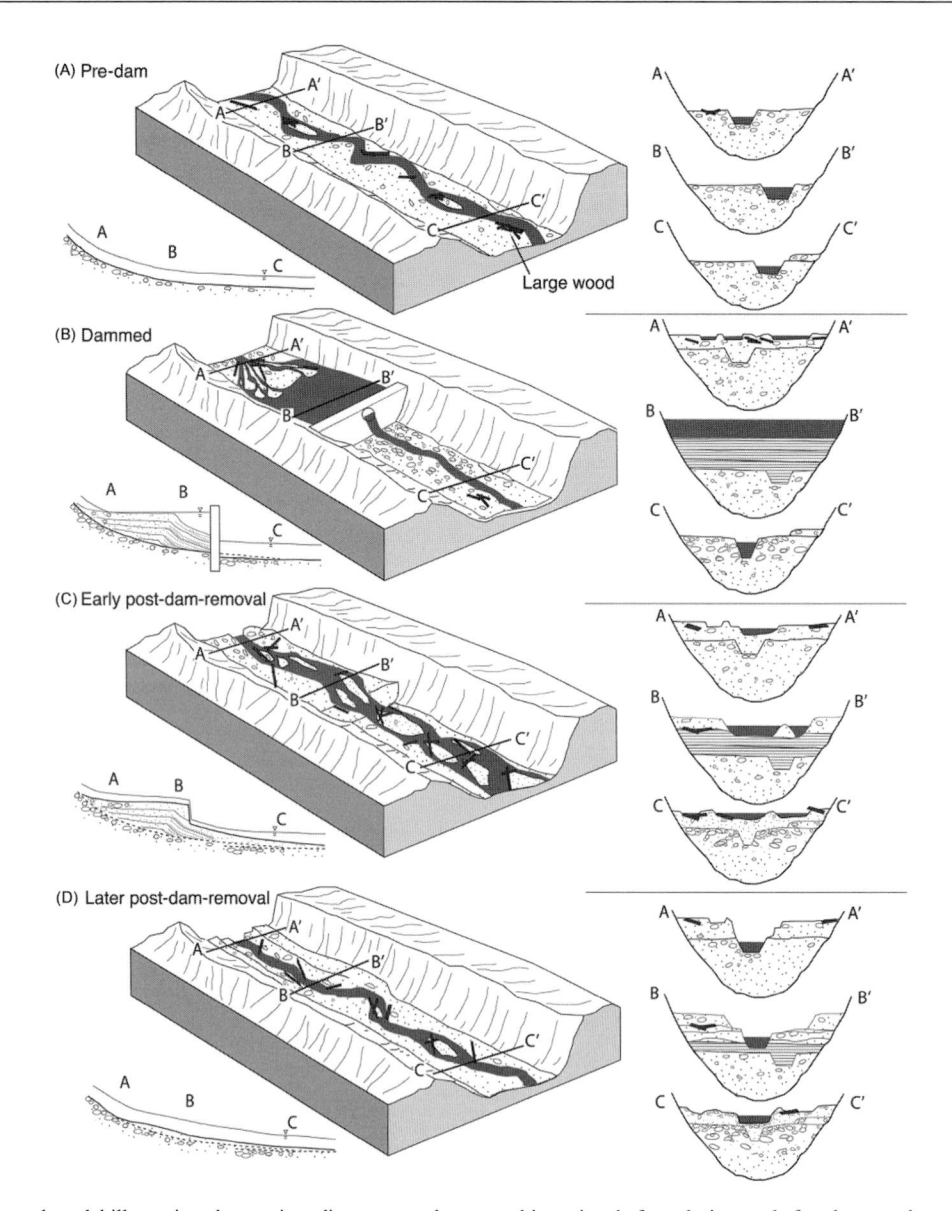

Figure 4.43. Conceptual model illustrating changes in sedimentary and geomorphic regime before, during, and after dam emplacement and dam removal based on the Elwha River dam removal project in the Olympic Peninsula of Washington (United States). (A) Prior to dam emplacement. (B) Dammed river, showing deltaic reservoir sedimentation. Downstream of the dam the river is incising, narrowing, and channel bed material is becoming coarser. (C) Initial post-dam removal phase (weeks to one to two years), showing braided channel above and below the former dam site with exposed reservoir deposits. Longitudinal profile of water surface and reservoir sediment includes migrating knickpoint. Downstream of the dam, new deposition of finer-grained sediments buries coarse-grained armored channel bed sediment that characterized the impounded system. (D) Later in the post-dam removal phase (two to ten years), channel incision occurs through reservoir deposits and into initial post-dam removal deposits downstream of the dam. (Figure source: East et al., 2018.)

4. Channel bed material downstream of a dam removal project initially becomes finer, and then somewhat coarser, but less coarse than the impoundment phase. Changes to channel bed sediment texture are greatest near the former dam site and diminish downstream (East et al., 2018).
5. Riparian – and linked coastal – ecosystems recover quickly, although there are differences in the trajectories of recovery between fish, benthic organisms, and vegetation (Foley et al., 2017; Lierman et al., 2017; East et al., 2018).
6. Environmental drivers are the stimulus for most dam removal projects, especially to increase fish passage. Other important drivers include dam safety, economics, and cultural heritage (Aspen Institute, 2002; Heinz Center, 2002; Stanley and Doyle, 2003; Lejon et al., 2009).
7. Reservoir geometry is an important first-order control on how efficiently – and if at all – reservoir deposits are reworked and reassimilated into downstream fluvial

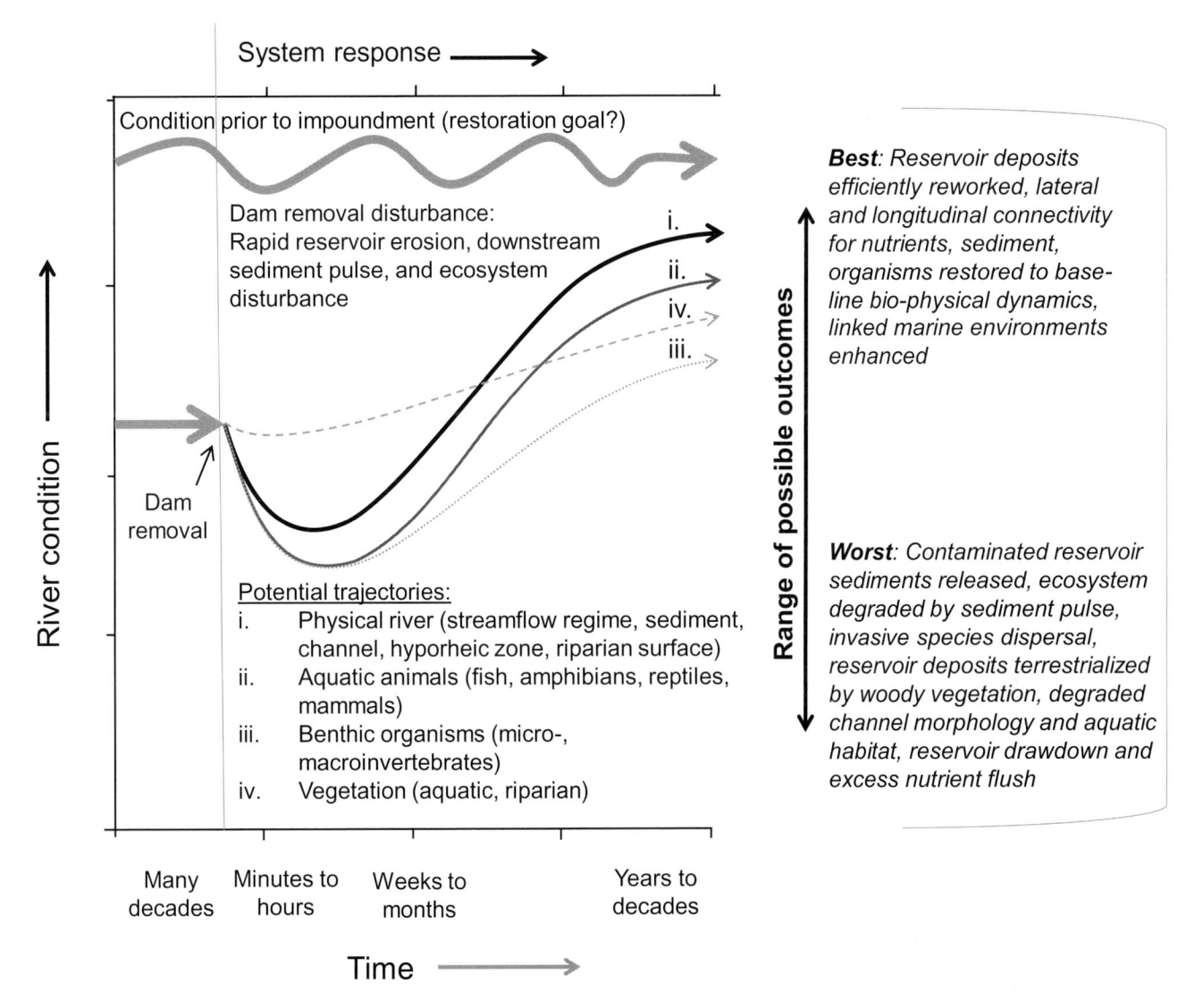

Figure 4.44. Conceptual model of river response to dam removal, including abiotic and biotic trajectories. The endpoints of possible outcomes, "best" and "worst," range from full recovery of riverine geodiversity (physical integrity) and biodiversity to a completely degraded and polluted river associated with contaminated sediment release and a sediment pulse that obliterates channel aquatic habitat. (Figure developed after Foley et al., 2017; Doyle et al., 2003b; and others.)

system (Riggsbee et al., 2007; Major et al., 2017). More documentation is needed of dam removal in lowland settings with fine-grained deposits in larger and wider reservoirs.

8. Reservoir drawdown is a key phase of dam removal and should be timed with hydroclimatic and biogeographic regimes, including spawning, feeding, migration, seed germination (Cannatelli and Crowe-Curran, 2012; Lierman et al., 2017).

9. More research is needed to understand biogeochemical changes caused by dam removal, as dewatering of reservoir deposits results in sequestered nutrients being released into active fluvial and marine environments (e.g., Riggsbee et al., 2007).

10. Despite safeguards, reservoir deposits with polluted sediments are likely to be released into the active fluvial system because of the process and activities associated with dam removal (reservoir drawdown, land preparation activities) (Sando and Vecchia, 2016).

11. The time required for the bureaucracy of dam removal takes much longer than the actual recovery of the river, and often takes decades (e.g., Heinz Center, 2002).

12. While dam removal has been largely successful, the large gap in knowledge between small and large rivers should not be underestimated. Before proceeding with large lowland dam removal dominated by muddy reservoir deposits, much more research is needed on the physical processes of sediment mobilization in large reservoirs along lowland rivers.

4.4.6.1 DAM CONSTRUCTION AND DAM REMOVAL: CONTRASTING INTERNATIONAL TRENDS

Globally there are two contrasting international paradigms with regard to the environmental impact of dams on rivers: dam construction and dam removal. In Asia, South America, the Balkans in Europe, parts of Africa, and northern Australia, the main trend is dam construction – primarily large hydroelectric dams – but also for new irrigation projects to support agriculture and food production. In most of Europe, North America, and southern Australia, where "contemporary" dams have been installed over a longer period of time, the *dominant* paradigm is dam removal. While mainly limited to smaller structures, dam removal is a vital step toward improved riparian health. The large number of case studies, varying by physical geography, history, engineering, and socioeconomic context, collectively provide an invaluable database of "lessons learned" to inform removal of larger dams in the future.

The benefits of dam removal on a large river would extend far beyond the downstream limits of the river basin, and would be symbolic of a fundamental shift in human–nature relationships, and homage to the legacy of scholars that established a foundation of knowledge regarding dam impacts to riparian environments.

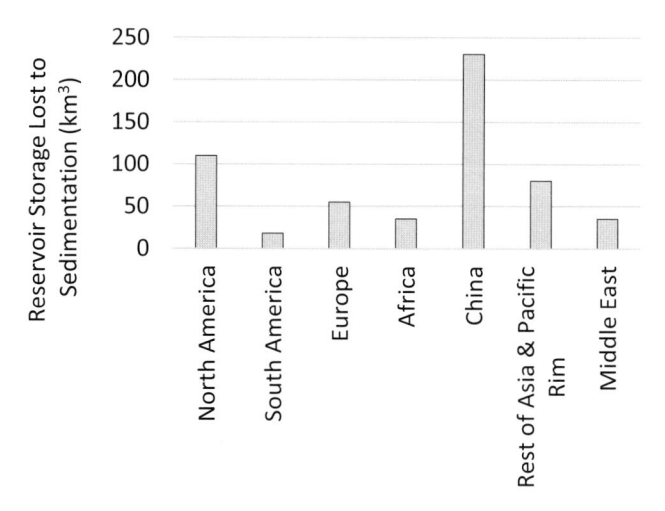

Figure 4.45. The global distribution of reservoir storage lost to sedimentation. (Figure source: Walling, 2006.)

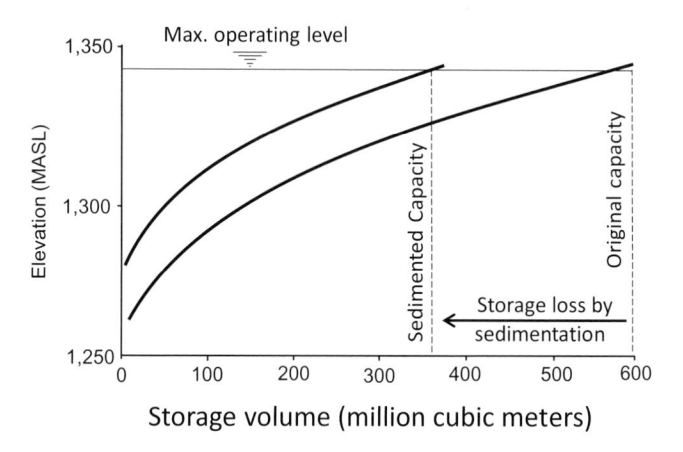

Figure 4.46. Reservoir elevation-storage: Comparison of original relationship and curve shift and storage volume lost due to reservoir sedimentation. (Source: Annandale et al., 2016.)

4.5 MANAGING RESERVOIR DEPOSITS

4.5.1 Changing Reservoir Storage

While some dam and reservoir management strategies have already been noted above, in this section we focus specifically on reservoir sediment management.

The increasing pace of dam removal provides some reason to be optimistic about increased number of barrier-free river segments. But the reality is that most dams and reservoirs will remain a permanent component of the global riparian environment. Because of a high sediment trap efficiency all dams will *eventually* infill with sediment. Annually there is about a 1% global loss in reservoir capacity (Morris and Fan, 1998; Syvitski et al., 2005; Kondolf et al., 2014b). China has had the greatest loss of reservoir storage capacity (Figure 4.45), much of which is due to high rates of historic soil erosion in the loess plateau. The global loss in reservoir capacity averages ~45 km^3/yr, which is equivalent to having to replace about 300 large dams annually (Wang et al., 2005).

Earth's historic riverine landscapes are littered with small infilled reservoirs, especially headwater dams and mill dams that have been buried for decades and centuries (Trimble, 1974; Morris and Fan, 1998; Walter and Merritts, 2008; James, 2013; Brown et al., 2018).

Trapping fluvial sediments in reservoirs has implications to the functioning of reservoirs and dams, and has environmental consequences to associated riparian lands upstream and downstream of the dam as well as within the reservoir. Unfortunately, information about the status of reservoir storage volumes and the projected life span of reservoirs is woefully insufficient (Morris and Fan, 1998; Graf et al., 2011; Patterson et al., 2018; Randle et al., 2019). This is important because it limits our ability to make general statements about water resource planning and reservoir storage capacity (Figure 4.46), or to develop coordinated strategies for downstream riparian management to address climate change.

Large variation in monitoring periods and monitoring procedures complicate estimating the life span of existing reservoirs (Morris and Fan, 1998; Graf et al., 2011; Randle et al., 2019). For example, the 537 reservoirs owned and operated by the U.S. Army Corps of Engineers are organized across 7 divisions and

37 districts. But because of different monitoring and data reporting procedures across the divisions, only 65% of historic reservoir data are available in an accessible online format (Patterson et al., 2018).

Although reservoir storage capacity is directly influenced by rates of sedimentary infilling and sediment type, dam and reservoir construction occurred with little information (data) about future sedimentary inputs. During the peak of dam construction, national sediment monitoring networks were only in the process of being established. Indeed, government agencies were still developing protocols for suspended sediment sampling (e.g., Graf et al., 2011; Gray and Landers, 2014). Many large dams were installed with less than a decade of sediment monitoring (Mossa et al., 1993), which limits effective pre- and post-dam assessment of dam impacts to downstream sediment loads and associated riparian environments. This issue is particularly acute in developing nations. The absence of long-term sediment records is especially daunting in view of the current surge in dam construction in regions that transport large proportions of Earth's water and sediment (i.e., South America, Southeast Asia, Africa) and that are also undergoing considerable changes to climate and land cover (e.g., Walling, 2008b).

Regional and local controls on sediment production and reservoir management result in great variability in annual storage capacity loss and the estimated life span of reservoirs. Obtaining global-scale estimates of reservoir storage capacity loss is important, but only at a global scale. It's essential that losses in reservoir storage capacity be worked out at a "river basin by river basin" basis, because the numbers are highly sensitive to a variety of factors related to the fluvial system (soil erosion, topography, hydrologic regime) as well as the reservoir configuration and dam operation. In the Missouri River basin, for example, the annual storage capacity loss for larger reservoirs ranges from 0.1% to 2.9% (Graf et al., 2011). The storage capacity for reservoirs in the Missouri basin is projected to decrease by 15% between 2000 and 2100 (from about 82% to 67%). The life span of large reservoirs in the Missouri basin ranges from about 100 years to over 1,000 years (Table 4.17).

Large reservoirs can represent sustainable infrastructure for the long term, and dam and reservoir operations need to be incorporated into future water resource and sediment management strategies (Graf et al., 2011).

4.5.2 Reservoir Management Strategies

Effective management of reservoir sediment is crucially important in view of (1) the decreasing life span of reservoirs, (2) limited suitable locations to develop new dams and reservoirs, (3) the need to reduce downstream impacts to riparian environments caused by sediment starvation, (4) public opposition to new dam construction, and (5) to mitigate varying climate change scenarios (Kondolf et al., 2014b).

Utilizing dams and reservoirs to sustainably manage fluvial sediment enhances the physical and biological integrity of rivers downstream of dams, and extends the life span of essential water resource infrastructure. Several main strategies exist to manage reservoir sediment (Wang and Hu, 2009; Morris, 2020). These include (1) sediment yield reduction, (2) sediment routing, and (3) sediment removal (Figure 4.47). A fourth category includes adaptive strategies that primarily require modifications to the dam and reservoir structure or changes to the usage of the reservoir.

The two main types of sediment routing include sediment bypass and sediment pass-through, which is also commonly referred to as sediment sluicing (Figure 4.47). Sediment bypass requires large channels or pipes to transport coarse sediment upstream of the reservoir – before the sediment enters the reservoir – to below the reservoir. The procedure is often utilized for coarse sediment, and is expensive to implement because of the massive physical infrastructure. The strategy for sediment bypass is mainly timed with discharge pulses so that coarser sediment is transported around the reservoir without deposition.

Sediment sluicing is designed to prevent reservoir sedimentation during discharge pulses and allow suspended sediment to pass through the reservoir (Table 4.18). Sluicing transports sediment through the reservoir by opening passages in the lower dam structure during the rising limb of the discharge hydrograph (Figure 4.48) when suspended sediment concentration is highest (Morris, 2020). This option works well for wash load (silt/clay) and fine sand. A benefit of sluicing is that it does not require modification to the existing dam structure and works with the natural flood cycle, which helps to make it more environmentally friendly (Kondolf et al., 2014b).

Sluicing is most effective in long and narrow reservoirs that are common to hilly or mountainous terrain. Three Gorges Dam along the middle Yangtze River was engineered to sluice sediment over the long flood season, which usually also results in mobilization of some existing reservoir deposits (Figure 4.49). Three Gorges Reservoir is ideal for sluicing as it is ~600 km long and less than 1.5 km in width (Kondolf et al., 2014a). While sluicing is considered more environmentally friendly than sediment flushing (Table 4.18), however, aquatic environments downstream of Three Gorges Dam underwent substantial degradation following impoundment, with a near collapse of some fish species (i.e., Figure 4.31). Sediment sluicing is also used on the Sanmenxia Reservoir along the Huanghe (Yellow) River (Wang et al., 2005), which has long had an enormous sedimentation problem. The concept of sluicing is in accord with the Chinese strategy to release muddy flow and store clear water (Wang and Hu, 2009; Espa et al., 2019).

Table 4.17 *Changes in storage capacity and estimated life spans for U.S. Army Corps of Engineers reservoirs in the Missouri River Basin*

Dam	River sub-basin	Drainage area upstream (km^3)	Total storage capacity (km^3)	Year closed	Year last sedimentation survey	Total lost capacity (%)	Mean annual capacity loss (%)	Estimated life (yr)	End date
1. Gavins Point	Main stem	723,825	0.6664	1955	2007	21.7	0.4	240	2195
2. Fort Randall	Main stem	682,387	7.7742	1952	1996	1.27	0.3	346	2298
3. Big Bend	Main stem	645,740	2.3446	1963	1997	9.1	0.3	374	2337
4. Oahe	Main stem	630,615	29.1224	1958	1989	2.6	0.1	1,192	3150
5. Garrison	Main stem	468,617	30.2330	1953	1988	3.7	0.1	946	2899
6. Fort Peck	Main stem	149,502	23.5694	1937	2007	5.6	0.1	1,250	3187
7. Perry	Kansas	2,893	0.8910	1969	2001	31.8	1.0	100	2069
8. Tuttle Creek	Kansas	24,936	2.6542	1962	2000	5.9	0.2	644	2606
9. Milford Lake	Kansas	9,831	1.3885	1967	1994	27.7	1.0	97	2064
10. Harlan County	Kansas	19,776	1.0046	1952	2000	15.4	0.3	311	2263
11. Wilson	Kansas	18	0.9421	1964	1995	35.5	1.1	87	2051
12. Kanopolis	Kansas	20,357	0.5517	1948	1993	6.9	0.2	652	2600
13. Clinton	Kansas	950	0.4906	1977	1991	11.9	0.8	118	2095
14. Pomona	Osage	834	0.3053	1963	1989	24.0	0.9	108	2071
15. Melvern	Osage	904	0.4477	1972	1985	15.7	1.2	83	2055
16. Hillsdale	Osage	373	0.1972	1981	1993	17.8	1.5	67	2048
17. Stockton	Osage	3,004	2.0567	1969	1987	39.5	2.2	46	2015
18. Pomme de Terre	Osage	1,582	0.8005	1961	1974	37.3	2.9	35	1996
19. Harry S Truman	Osage	29,784	6.4283	1979	1992	8.9	0.7	146	2125
20. Long Branch	L. Chariton	282	0.0802	1978	1988	12.1	1.2	83	2061
21. Rathbun	Chariton	1,422	0.6807	1969	1999	ND	ND	100	2069
22. Smithville	L. Platte	552	0.3041	1979	1993	9.5	0.7	147	2126
23. Blue Springs	L. Blue	85	0.0328	1988	ND	ND	ND	100	2088
24. Longview	L. Blue	129	0.0579	1985	ND	ND	ND	100	2085

ND indicates no data.
Table source: Graf et al. (2011), based on data from U.S. Army Corps of Engineers.

Sediment flushing (or drawdown flushing) is a management approach designed to remove sediment that is already deposited within the reservoir (Morris and Fan, 1998), and must be carefully implemented (Grimardias et al., 2017; Espa et al., 2019). The strategy works by opening low gates near the bottom of the dam structure to lower reservoir levels (Figure 4.50). This increases the hydraulic gradient of the reservoir, and initially scours fine-grained deposits in the lower reservoir, resulting in high suspended sediment concentrations. This is followed by scour of coarser delta deposits higher up in the reservoir that produces bed load transport (Figure 4.50). The timing of reservoir drawdown flushing often occurs during low or moderate flow conditions, rather than during the seasonal flood pulse, which can therefore result in downstream channel bed aggradation. Environmentally, reservoir flushing can be devastating to downstream aquatic habitat, particularly if the channel bed is utilized for critical biological functions – such as fish spawning – and is buried by coarse sandy deposits (Peteuil et al., 2013). The

Classification of Sediment Management Strategies for Reservoirs

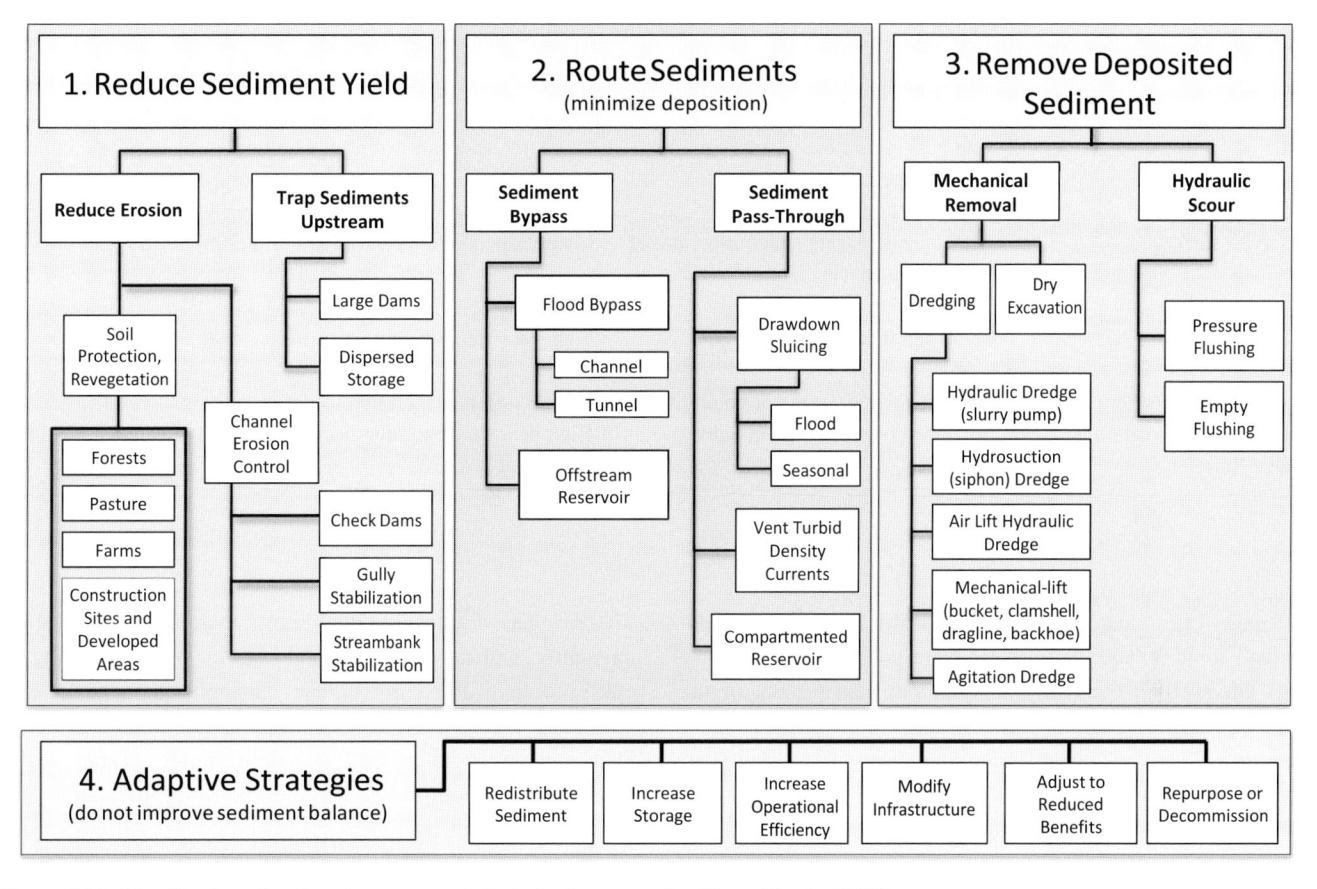

Figure 4.47. Classification of sediment management strategies for reservoirs. (From Morris, 2020.)

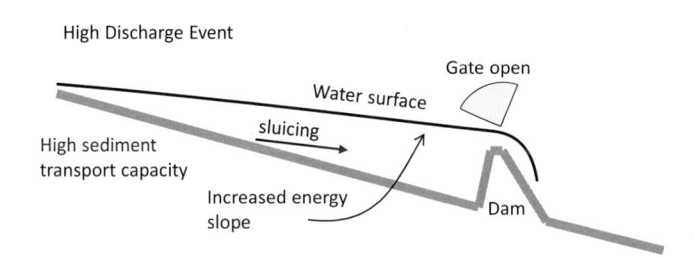

Figure 4.48. Profile of dam and reservoir operation during a sediment sluicing event, with gate opening timed with flood pulse. (Source: Kondolf et al., 2014a.)

timing of reservoir drawdown flushing events should be synchronized with crucial ecological functions occurring in the downstream riparian environments, and ideally should occur during the seasonal high-flow period (Wang and Hu, 2009; Kondolf et al., 2014a; Espa et al., 2019).

An important consideration within the reservoir concerns the scour and downstream mobilization of polluted reservoir sediments (Evans, 2015). The Flix Reservoir along the lower Ebro River in Spain, for example, is only 110 km from Ebro delta

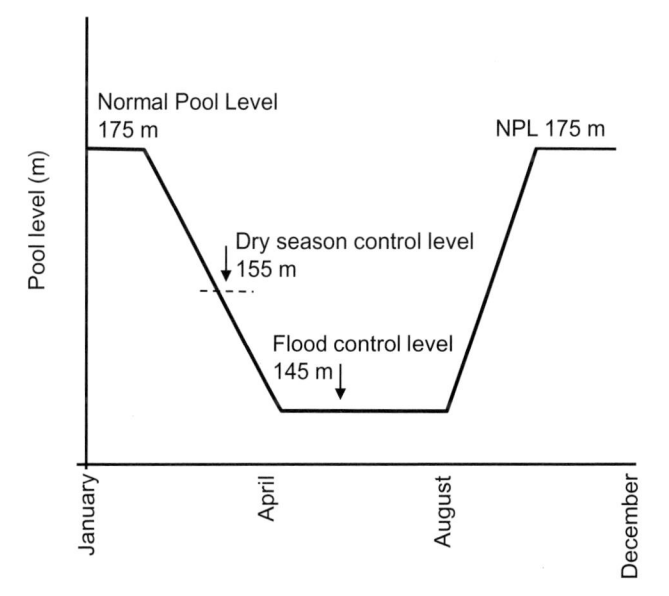

Figure 4.49. Seasonal pool operation at Three Gorges Dam, Yangtze River, China. Drawdown procedure results in pool lowering by some 30 m during the flood season relative to the normal pool level. (Source: Kondolf et al., 2014a.)

Table 4.18 *Comparison of reservoir management approaches: sediment sluicing and sediment flushing*

Parameter	Sluicing	Flushing
Timing	Always coincides with natural flood events	May not coincide with natural flood events, may have predetermined (fixed) dates
Outlet capacity	Can pass large floods with minimum backwater	Discharge and drawdown may be limited by low-level outlet capacity
Sediment discharge	Sediment outflow $\approx$ inflow	Sediment outflow $\gg$ inflow
Reservoir intakes	May operate during sluicing periods if so designed	Cannot operate (concentration too high, water level too low)
Recover lost capacity	Primarily intra-annual deposits	Yes
Redeposition in downstream channel	Little significance due to high discharge (flood)	Significant post-flushing clear water release may be needed
Typical erosion pattern	Retrogressive erosion not common	Retrogressive erosion may occur
Extreme spike in sediment concentration	Extreme concentrations avoided	High peak concentration occurs when full drawdown level is reached

From Morris (2020).

wetlands. Unfortunately, adjacent industries historically discharged some 300,000 tons of sediment polluted with mercury (Hg) directly into the reservoir. Improper management of the reservoir flushing regime results in erosion and transport of mercury-polluted sediments from the reservoir to the Ebro Estuary (Palanques et al., 2020).

Finally, along some rivers the main sediment management option is to dredge sediment from reservoirs or other storage sites and transport the sediment downstream by barge, and then dump the sediment into the river for assimilation into the active bed material layer. This approach is utilized along the Rhine Elbe, and Danube Rivers (see Section 8.2.3), for example, to address the sediment starvation problem that has occurred because of numerous dams, hydraulic infrastructure, and land management, which has reduced upstream sediment inputs (Habersack et al., 2016; Frings et al., 2019).

4.5.2.1 HUANGHE RIVER RESERVOIR SEDIMENTATION: CHANGING STRATEGIES

Options for managing reservoir sediment need to be adaptable to changing conditions, including climate change and land use change (Wang and Hu, 2009; Graf et al., 2011). This can require changes to the dam structure and reservoir management strategies that differ from the original design. Such is the case with the Huanghe River basin. Reservoir sedimentation problems along the Huanghe are well known, and some 21% of the total storage capacity for its seven main-stem reservoirs (i.e., Table 4.8) was lost to reservoir sedimentation by 1989, only about two decades after the first dam was commissioned (Liu et al., 2017).

Sanmenxia Dam was completed in 1960 (closure) as the first main-stem dam on the Huanghe River (i.e., Table 4.8). The original design was to supply hydropower, irrigation, navigation and downstream flood control, the latter being an epic problem along the lower Huanghe associated with numerous avulsions (see Chapter 7). The initial reservoir management strategy was to store water (Table 4.19). Unfortunately, Sanmenxia was not designed to address the enormous Huanghe sediment loads, the largest sediment loads on Earth, supplied by a legacy of land mismanagement in the upstream Chinese Loess Plateau. Copious sedimentation immediately ensued after dam closure, as only a meagre ~7% of incoming sediment was transported downstream (Table 4.19). A staggering 93% of sediment was stored within the reservoir. Incredibly, by 1965 Sanmenxia reservoir had already lost 41.5% of its reservoir capacity below 335 masl and 62.9% of its storage capacity below 330 masl (Figure 4.51). The sedimentation problems resulted in having the dam redesigned to accommodate sediment sluicing beginning in the early 1970s (Table 4.19), so that the muddy water was released during high-flow events. By the mid-1970s the reservoir had already attained a more consistent storage capacity (Wang et al., 2005).

The revised sediment sluicing strategy for Sanmenxia Dam is now part of a larger sediment management approach for the Huanghe basin (Figure 4.52). Since 2002 the Yellow River Conservancy Commission regulates river flow and sediment transport to manage channel bed degradation along a cascade of dams and reservoirs (Liu et al., 2017; Ma et al., 2017; Kong et al., 2020). The importance of developing an appropriate reservoir sediment management strategy for the Huanghe is as important as any river in the world. Flood disasters along the lower Huanghe River are linked to channel bed aggradation attributed to landscape and sediment mismanagement. The substantial

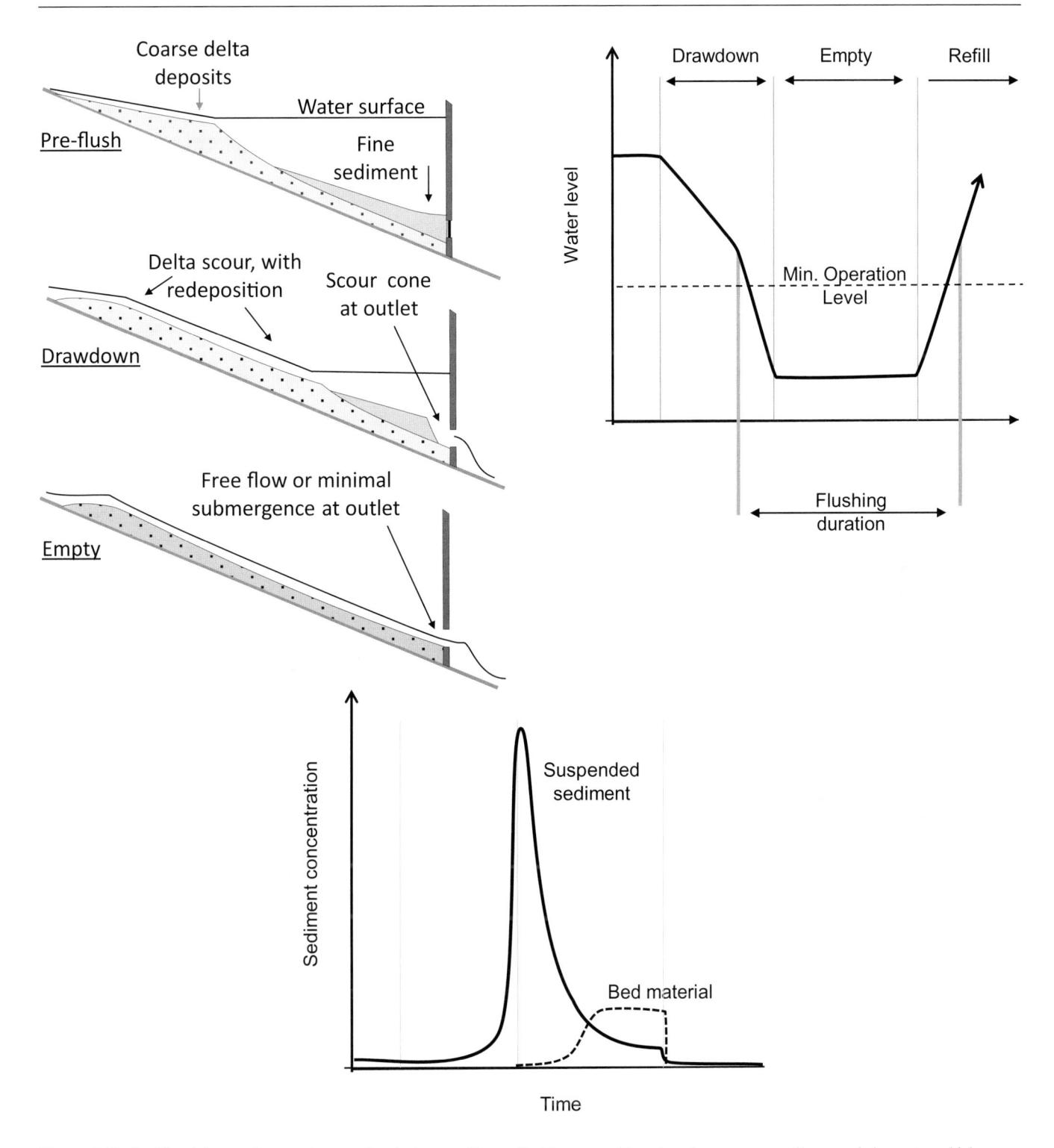

Figure 4.50. Profile of dam and reservoir operation during a sediment flushing event. Note drawdown occurs until reservoir is empty, which scours both fine-grained and coarser-grained reservoir deposits. High suspended sediment peaks occur prior to bed material mobilization. (Source: Morris, 2020.)

Table 4.19 *Changes in reservoir management strategies of Sanmenxia Reservoir and sedimentary response*

Date	Management approach	Pool level (m) Max	Pool level (m) Average	Quantity of reservoir sedimentation and erosion (billions of m^3) Upstream Tongguan	Quantity of reservoir sedimentation and erosion (billions of m^3) Downstream Tongguan	Sediment inflow and outflow Total outflow (billions tons)	Sediment inflow and outflow Total outflow (as % of inflow)	Bed elevation at Tongguan (m)
9.1960–3.1962	Storing water	332.58	342.02	+0.32	+1.43	0.11	6.8	+4.5
4.1962–6.1966	Before reconstruction	325.9	312.81	+0.36	+1.61	3.39	58.0	+4.4
7.1966–6.1970	Initial reconstruction	320.13	320.13	+1.58	+0.01	7.38	82.5	+5.0
7.1970–10.1973	Further reconstruction	313.31	298.03	+0.38	−0.28	5.93	105	+3.1
11.1973–10.1978	Control operation	317.18	305.60	−0.07	+0.12	6.68	100	+3.1

Tongguan located ~100 km upstream of Sanmenxia Reservoir.

Data source: Wang et al. (2005).

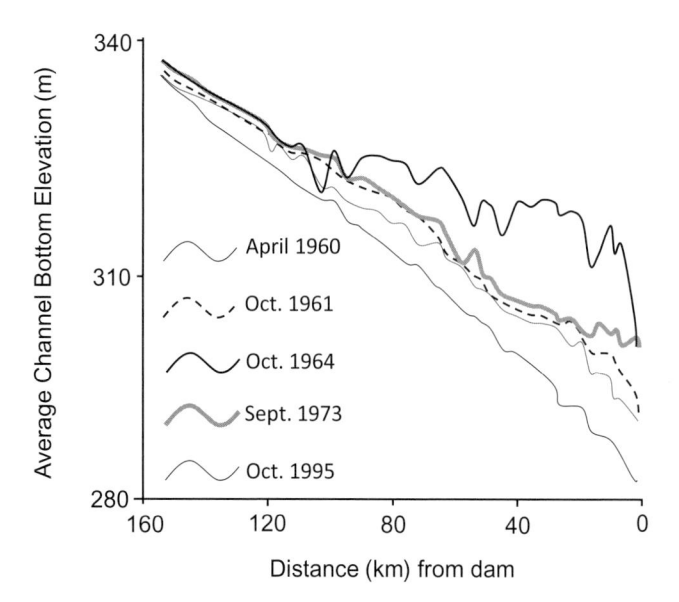

Figure 4.51. Longitudinal profile of reservoir deposits for Sanmenxia Dam, Huanghe River. Abrupt aggradation occurred soon after closure, and cessation of aggradation after effective reservoir management strategies implemented. (Source: Wang et al., 2005.)

Figure 4.52. Sediment flushing event of Sanmenxia Dam, Huanghe River. Sanmenxia Dam closed in 1960 and is the first main-stem dam on the Huanghe River (contributing drainage area 688,400 km^2). (Photo date: July 26, 2013. Source: R. Mueller, licensed by CC.)

population and economic activities associated with the lower Huanghe riparian environments means that the issue is not only a vital societal concern, but it also influences fluvial dynamics in the lower basin related to channel avulsion, and has implications to the global economy (Section 7.5.2).

4.5.2.2 CLOSURE: RESERVOIR MANAGEMENT AND SUSTAINING RIVERS

The appropriate sediment management approach for a particular dam and river basin is dependent upon a variety of factors including sediment volume and sediment type (sand/gravel or silt/clay), reservoir geometry (size and shape), streamflow variability, downstream geomorphic and environmental conditions, and expenditures. Adapting the appropriate sediment management strategy can extend the life span of reservoirs beyond the conventional design-life paradigm (Figure 4.53). And, as noted along the Huanghe River, sediment management approaches should be adaptive to accommodate changing boundary conditions and societal priorities because of its impact on the supply of upstream sediment loads (Graf et al., 2011; Kondolf et al., 2014a). Identification of an appropriate reservoir sediment management strategy should be seen as a fundamental component of integrated river basin management to support navigation and associated economic activities, mitigate against climate and land use change, and to sustain downstream riparian and linked coastal environments.

It is interesting to ponder a sediment management strategy to Gavins Point Dam (Figure 4.21) on the Missouri River in view of the need to provide sediment to support physical riparian habitat and related biodiversity (NRC, 2011), and the urgent need to supply sediment to construct wetlands in the Mississippi delta (Ahn et al., 2013; Kemp et al., 2014). Proposed drawdown flushing strategies designed to scour reservoir deposits and mobilize stored sediment for downstream transport, however, are complicated by the need to consider both the sediment budget of the lower Mississippi as well as specific habitat needs of aquatic species along the lower Missouri, including the endangered Pallid Sturgeon (*Scaphirhynchus albus*) (Coker et al., 2009; Elliot et al., 2020; Jacobson et al., 2009). Nevertheless, identifying a "sustainable" approach to mobilize reservoir sedimentary deposits stored behind Gavins Point Dam, and other Missouri basin dams (Table 4.11), should be of highest priority to responsible government organizations.

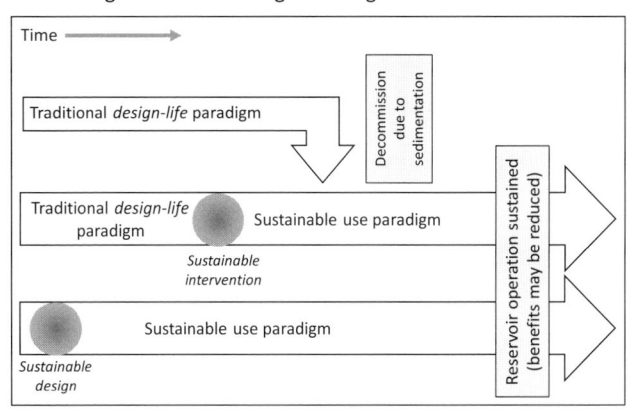

Figure 4.53. Models of dam and reservoir design related to sediment management. Conventional "design-life" paradigm for dams and reservoirs can be modified toward a sustainable use paradigm. New dam construction should include features adaptable to changing sediment management strategies. (Source: Annandale et al., 2016.)

5 Hydraulic Engineering for Channel Stability and Flood Control: The Sequence of Geomorphic Adjustment

[A] host of broader problems of channel adjustment to external controls ... might be thought of as physiographic problems.
— L. B. Leopold and G. M. Wolman (1960)

[T]his large alluvial river is not monotonous in appearance ... and not completely controlled by hydrology and hydraulics.
— S. A. Schumm and W. J. Spitz (1996)

5.1 OVERVIEW OF APPROACHES TO CHANNEL ENGINEERING

River channel engineering has a long tradition in the science of flood control. In Chapter 4 we examined the influence of dams on rivers, the most massive form of hydraulic engineering. While many dams are constructed in basin headwaters and transfer zones, other forms of hydraulic engineering are imposed directly within main-stem channels of lowland rivers. Dikes, groynes, dredging, bank protection, and channelization approach flood control and channel stabilization with channel engineering (Schumm and Winkley, 1994; Mosselman et al., 2000; Mosselman et al., 2004; Gregory, 2006; Pinter et al., 2006a; Hudson et al., 2008; Habersack et al., 2016; Hudson, 2017). Channel engineering influences channel hydraulics and sediment transport processes and drives geomorphic adjustment of alluvial rivers. Such measures, however, have consequences to a variety of riparian processes that require time to unfold, and often requires further engineering to manage the unintended disturbance (Smith and Winkley, 1996; Hudson et al., 2008; Vorogushyn and Merz, 2013; Bravard and Gaydou, 2015; Habersack et al., 2016).

This chapter reviews geomorphic impacts of discrete types of hydraulic engineering imposed upon lowland rivers. The focus is primarily on structural measures (Table 5.1), which result in emplacement of hydraulic infrastructure within the channel. A case study of the lower Mississippi is utilized to contextualize the sequence of channel engineering activities as part of an overall flood control and channel management program, including unintended geomorphic consequences.

Channel engineering along lowland rivers primarily occurs for two purposes: (1) flood conveyance and (2) channel stability. The motivation for the latter approach is often to protect engineering structures, especially to create a navigable channel to support shipping, as well as to protect dike systems from being undermined because of channel bank erosion (Figure 5.1). Specific procedures, such as meander neck cut-offs, dredging, tree snag removal, groynes, and channel cross-section modification, are designed to increase discharge conveyance to reduce flood stage and flood duration (Table 5.1). These procedures directly influence hydraulics, and commonly reduce channel roughness (resistance) and increase channel velocity. Combined, this results in higher flow velocities and a faster moving discharge pulse that lowers flood stage levels and reduces the duration of flood events (Smith and Winkley, 1996; Gregory, 2006; Pinter et al., 2006b; Singer and Aalto, 2009).

Structural engineering approaches are commonly utilized for flood control and river "improvement," especially channel cut-offs and straightening. This may occur over the scale of a single meander neck, a chute cut-off, or an elongated channel reach (discussed below). Critically, cut-offs reduce channel sinuosity and increase channel slope at the location of the knickpoint (reviewed in Chapter 2).

To increase flow conveyance, channel re-sectioning is often completed to modify the cross-sectional channel bank geometry. This is accomplished by reducing the angle of channel banks and increasing channel depth, thereby increasing the channel cross-sectional area of streamflow during higher discharge magnitudes. Channel re-sectioning is frequently implemented in association with cut-offs as part of a larger plan of channelization (Mosselman et al., 2004; Singer et al., 2008).

Groynes (wing dikes) reorient the cross-sectional pattern of stream velocity to a specific channel section, with the goal to create a self-scouring channel. Combined with deposition of bed material within the groyne field the active channel is narrowed and, if correctly aligned, greatly reduces the amount of annual channel dredging (Echevarria-Doyle et al., 2020).

Revetment and riprap is constructed to prevent channel bank erosion, while channel bed incision is arrested by installation of

Table 5.1 Common structural and nonstructural measures of channel engineering for flood control and navigation with implications to river management

Engineering action	Primary justification	Intended effect	Potential adverse impact	Hydraulic/ morphologic location	Hydrologic events (low, moderate, high)	Time-scale for impact	Spatial impact (upstream, local, downstream)
Structural measures							
Groynes (wing dikes)	Navigable and stable channel	Narrow channel, increase depth	Disrupt sediment budget, trap coarse bed material, channel bed incision	Channel bed, banks	Low, moderate, high	10 years to permanent	Local, downstream
Cut-offs and straightening	Flood risk	Reduce frequency and duration of flooding	Knickpoints and channel incision, channel widening, disconnect floodplain habitat (chute channels have lower impact than neck)	Meander bend neck (or chute)	High	5 years to decades	Upstream, local, downstream
Dams*	Flood reduction	Reduce downstream flood stage and duration	Sediment trapping, reduced flow variability, degraded water quality of outflow	Valley and channel	High and low (dependent upon reservoir release strategy)	Instantaneous to decades	upstream, reservoir, and downstream
Revetment and riprap, concrete weir	Channel stability	Prevent bank erosion, prevent channel bed incision	Thalweg scour, disrupt sediment budget	Channel banks (cutbanks), channel bed	Moderate, high	1 year to decades	Local, downstream
*Nonstructural** measures*							
Sediment dredging	Navigation	Increase depth	Disrupt sediment budget	Bars, riffles, floodplain (if dredge spoil)	Low to moderate, high (if dredge spoil)	Instantaneous	Local, downstream
Sediment mining	Aggregate for construction and industry	None	Bank erosion, channel bed scour, knickpoint development, local avulsion, disrupt sediment budget	Channel bed and bars, floodplain	Low, moderate	Instantaneous to decades	Local, downstream, upstream
Sediment replenishment	Navigation	Reduce channel incision	Disrupt sediment budget, channel bar formation	Thalweg, pools	Moderate, high	Instantaneous, 10 years	Local
Tree snag removal	Navigation	Increase depth, reduce flow resistance	erosion	Channel bars	Low, moderate	Instantaneous, 10 years	Local, upstream

* Only applies to flood control dams, as dams and reservoir comprehensively reviewed in Chapter 4. **Refers to measures not permanently embedded in channel as hard structures.
Source: Author table.

Figure 5.1. The highly engineered Danube River, east of Vienna, Austria, trained with groynes and straightening by meander bend cut-offs. (Source: International Commission for the Protection of the Danube River.)

concrete weirs in the channel bed. In the upper Rhine delta the installation of concrete weirs in the channel bed of the Waal River have locally reduced incision and serve as grade control, although they have also caused channel bed incision downstream of the structures (Mosselman et al., 2004; Ylla Arbos et al., 2020).

Nonstructural channel engineering that does not include emplacement of hydraulic infrastructure also impacts the fluvial system. This occurs primarily by channel bed dredging and mining of aggregate, in addition to sediment replenishment (Section 8.2.3). Channel dredging is often required as part of the overall construction activities for hard engineering. More common, however, is channel bed dredging to accommodate navigation during low-flow conditions (Figure 5.2). This occurs along shallow riffles, crossings, and at river mouth bars. Most dredging does not alter the overall sediment budget because the sediment is dumped in downstream sections where it is reassembled within the active sediment load (Frings et al., 2019; Echevarria-Doyle et al., 2020).

But along some rivers considerable dredge spoil is dumped along the adjacent floodplain. This effectively removes coarse sediment from the active system and creates dredge spoil splay or spoil mounds on the floodplain surface (Figure 5.3), which may be reworked by channel bank erosion or become semipermanent anthropogenic fluvial landforms (Mossa et al., 2017). Along the lower Mississippi delta, for several decades sediment dredged to maintain a navigable channel at the river mouth (Head of Passes, LA) has been rerouted over the channel bank into subsiding wetlands to create new lands (Thompson et al., 2019; USACE, 2020) (Figure 5.4). The discharge of dredge sediments into wetlands is heavily regulated (in some nations) because of concerns with damaging marsh environments, particularly habitat for endangered species. Rerouting dredge sediment in the Mississippi delta has had a limited effect in rebuilding channel banks and adjacent wetlands (USACE, 2020) and is being championed by the dredging industry as "sediment recycling" (rather than dredge spoil) (Powers, 2018). But, because dredge material is predominantly sand and coarse silt it is not transported far beyond the channel bank and is less effective than sediment diversion structures that transport higher

Figure 5.2. Cutterhead of dredge Bill Holman during dredging of Ohio River below McAlpine Locks, which accumulates sediment during high-flow periods, requiring annual dredging to maintain suitable navigation channel. The rotating cutterhead churns up the river bed, which is ultimately discharged into a deposit area. Obliteration of the channel bed benthic layer is part of the "cost" of hydraulic dredging. (Photo date: July 2002, Source: USACE. Licensed under CC 2.0.)

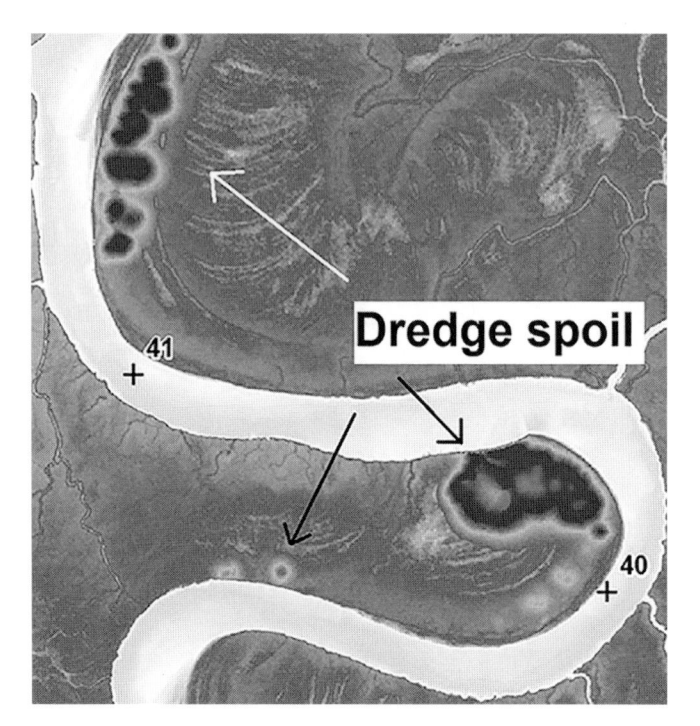

Figure 5.3. Dredge spoil landforms atop the channel floodplain of the Apalachicola River, Florida depicted in a lidar DEM. The two larger landforms are 7 m (top) and 12 m (middle) above the floodplain surface. Additional smaller round features along the channel bank. For scale, the river channel width is about ~150 m. Channel numbers refer to distance in miles upstream of the Gulf of Mexico (40 miles = 64.4 km). (Source: Mossa et al., 2017.)

Table 5.2 *Annual volume of sand and gravel mining from the lower Mekong River by nation and grain size*

Country	Extraction (m³/yr)			
	Sand	Gravel	Pebbles	Total
Laos	904,100	10,000	454,500	1,368,600
Thailand	3,677,200	857,740	0	4,534,940
Cambodia	18,748,503	2,044,940	0	20,793,443
Vietnam	7,750,000	0	0	7,750,000
Total	31,079,803	2,912,680	454,500	34,446,983
%	90	8	1	100

Source: Bravard et al. (2013).

Figure 5.4. Dredge spoil and new land creation at retention area near Head of Passes, Louisiana at the mouth of the lower Mississippi. Dredge spoil of sandy deposits increases the delta framework. For scale, width of Mississippi River at Head of Passes (river mouth) in background is about 1,800 m. (Photo source: USACE, 2020.)

Figure 5.5. Sand mining on the Mekong River, Vietnam. Illegal sand mining is resulting in extensive channel bed degradation in the lower Mekong. (Photo source: Mekongeye, Licensed under CC 2.0.)

proportions of silt and clay to rebuild deltaic wetlands across much broader areas (Kemp et al., 2014) (See Section 8.5.3).

Tree snag removal reduces channel roughness and increases stream velocity (e.g., Hickin, 1984), although having some adverse environmental consequences. Historically, tree snag removal was a major focus of nonstructural channel engineering to reduce log jams (discussed below) and improve river navigability (Wohl, 2014).

A troubling international trend is the explosive growth of coarse channel bed sediment removal by aggregate mining, which is largely in support of rapid urbanization and expansion of transportation networks (Bravard et al., 2013; Jordan et al., 2019; Best and Darby, 2020; Hackney et al., 2020). Fluvial aggregate mining occurs directly in the channel bed and also within coarse floodplain deposits (point bars). Sand mining directly modifies channel morphology and alters related hydraulic processes. The consequences of excessive aggregate mining include the development of channel bed knickpoints and subsequent incision, channel bank destabilization and erosion, and local channel avulsion (Mossa and McLean, 1997; Mossa and Marks, 2011).

The lower Mekong River represents a premier example of fluvial degradation (Figure 5.5) caused by excessive aggregate mining (Bravard et al., 2013; Best and Darby, 2020), which is primarily for sand extraction. Sediment mining in the lower Mekong in Laos, Thailand, Vietnam, and especially Cambodia removes an average of 34.5 million tons/yr (Bravard et al., 2013) to 50 million tons/yr (Hackney et al., 2020), with 90% of the aggregate removal being sand (Table 5.2). This amount far exceeds the annual amount of sand supplied to the Mekong delta, which is estimated at 6.2 ($\pm$2) million tons/yr (Hackney et al., 2020). The negative sediment budget for coarse sediment in the lower Mekong is driving channel bed degradation, with bank erosion destabilizing settlements and infrastructure (i.e., Figure 3.16). Precise quantification of sediment budgets is essential to implement effective management strategies, which along some sediment-starved rivers includes the dumping of sediment into scour zones to reduce channel bed incision and improve channel stability (Götz, 2008; Kondolf et al., 2014b; Habersack et al., 2016; Frings et al., 2019).

For rivers that have been impacted by a range of human activities for many decades to centuries, channel degradation is driven by multiple causes. In such cases individual responses to discreet disturbances can be difficult to discern, as adjustment to a specific disturbance may require decades to occur while others are simultaneously being imposed upon the system (Schumm, 1991; Middelkoop and van Haselen, 1999; Mosselman et al., 2004; Gregory, 2006; Hudson and Kesel, 2006; Pinter et al., 2006a; Quick et al., 2020). Such influences may have contrasting impacts, such as land degradation increasing downstream sediment loads while dam construction reduces downstream sediment loads.

The above issue is well illustrated across a range of heavily engineered European rivers. The once dynamic meandering Elbe River in Germany is intensively controlled and modified with groynes, revetment, and straightened with cut-offs, in addition to being impounded by 292 upstream dams (IKSE, 2005). This has resulted in a large reduction in sediment inputs and higher channel bed shear stress, which has caused an extensive ~200 km long "erosion zone" in the middle Elbe (120–320 km). Some reaches have incised by up to ~2 m since end of the nineteenth century, and incision continues at rates of 1–2 cm/yr (König et al., 2012; Vollmer, 2020). This prolonged disturbance has resulted in a range of riparian environmental impacts, including reduced floodplain connectivity and degradation of channel bed habitat for fish spawning. The Austrian Danube is sediment starved because of numerous upstream dams (Chapter 4), and is straightened and controlled with revetment and groynes. The Danube River between Vienna and Bratislava, SK had a dynamic and sinuous anabranching channel, but is now canalized for navigation and flood control with a sinuosity of only 1.3. Channel bed incision of 1–2 cm/yr requires artificial inputs of gravel as part of the regular management regime (Habersack et al., 2016). The heavily engineered lower Arno River in Tuscany, Italy has deeply incised over the past 150 years (Cencetti and Tacconi, 2005) by an average of ~7 m (Figure 5.6A and B), with channel bank widening of 38 m (Rinaldi and Simon, 1998).

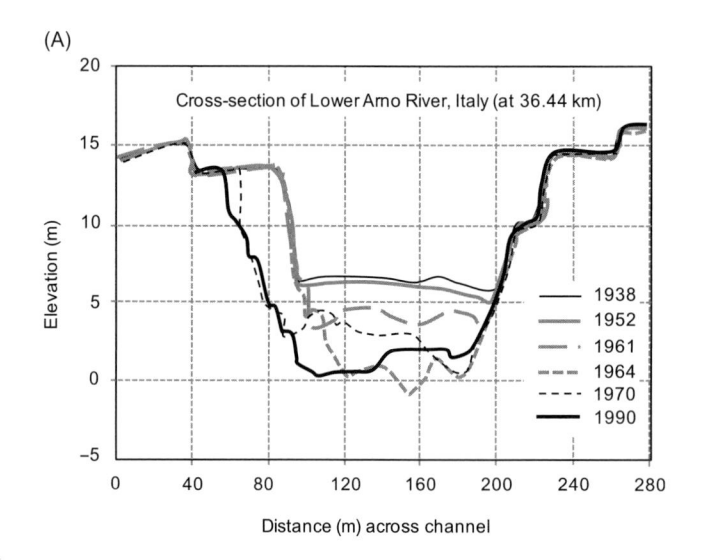

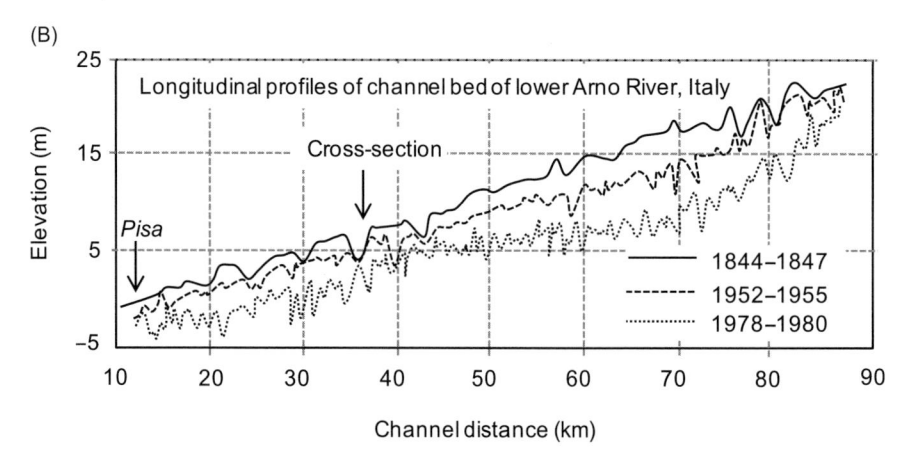

Figure 5.6. Historic degradation of the lower Arno River (Valdarno valley) in Tuscany, Italy over about a 150-year period following upstream land use change and engineering works. (A) Longitudinal profiles between 1844 and 1980. (B) Channel bed incision and bank widening. (Source: modified from Rinaldi and Simon, 1998.)

Table 5.3 *Importance of sedimentology in river channel engineering*

Sedimentary unit	Cohesive / noncohesive	Sedimentological index or feature	Management/engineering issues
Floodplain deposits			
Point bar	Noncohesive	Particle size, thickness of unit, sorting	Cut-offs, influences rate of adjustment, orientation of cut-off to channel, erodable bank material, high permeability
Channel bar	Noncohesive	Particle size, thickness of unit, sorting	Erodable bank material, high permeability
Clay plugs (infilled channels)	Cohesive	Particle size, thickness of unit, plasticity, organic content	Resistant to bank erosion, requires substantial equipment to excavate, low permeability
Crevasse splay	Noncohesive	Particle size, thickness of unit, sorting	Erodable bank material (minor role), provide stable foundation (surface), high permeability
Natural levee	Noncohesive (along channel bank)	Particle size, thickness of unit	Erodable bank material, provide stable foundation, moderate permeability
Backswamp	Cohesive	Particle size, thickness of unit, organic content, plasticity	Resistant to bank erosion, requires substantial equipment to excavate, subsidence, poor drainage (low permeability)
Channel bed sediments			
Bed material	Noncohesive	Particle size, shape and sorting, imbrication, depth of active layer, presence of armor layer	Incision and/or aggradation, buried structures
Bed load	Noncohesive	Rate of transport, timing of transport, bed waves	Scour (incision) of channel bed, loss of support for structures, destabilizes channel banks, higher stage levels

Source: Author table.

5.2 INFLUENCE OF SEDIMENTOLOGY ON CHANNEL ENGINEERING

5.2.1 Channel Bank Material: Cohesive versus Noncohesive

The design and functioning of channel engineering activities is dependent upon channel and floodplain sedimentology (Table 5.3), which influences the emplacement of structures, its long term effectiveness, and the magnitude and style of channel adjustment (Fisk, 1947; Kolb, 1963; Saucier, 1994; Smith and Winkley, 1996; Thorne, 1998). A major distinction occurs in regards to whether floodplain deposits and bank material are cohesive or noncohesive. Along large alluvial valleys and fluvial dominated deltas it is common that engineered structures are in contact with heterogeneous bank material (see Section 3.3.2), both vertically and at the reach scale (i.e., along a cutbank). This is often manifest as coarser-grained (sandy, gravelly) bottom-stratum deposits along the toe of the channel bank being capped with fine-grained cohesive overbank top-stratum deposits. Noncohesive deposits are usually associated with former channel bars, created by bed load deposition, and sandy crevasse or natural levee deposits formed by the accumulation of flood deposits. The high erodibility of coarser deposits facilitates excavation during the construction phase of engineering. Additionally, coarse deposits are easily eroded because of hydraulic changes imposed by the engineered channel, resulting in considerable increase in channel width. This can be a concern to the integrity of the engineered structure (e.g., Dunbar et al., 1999; Downs and Gregory, 2004), and influence the geomorphic evolution of the river.

An important engineering quality of cohesive bank material is that it effectively serves as bedrock control (Fisk, 1947; Thorne, 1998; Dunbar et al., 1999). Cohesive channel bank material is difficult to excavate during the construction phase of channel engineering, potentially requiring other types of excavation equipment (tailor designed), and along larger rivers the size of required equipment results in large increases in project expenditure. The key geomorphic significance of cohesive bank sediments is that it is inherently stable after initial channel bed incision, as channel banks maintain steep profiles without failure and subsequent widening. Clay plugs and backswamp deposits (Figure 5.7) are the two main cohesive Holocene sedimentary units likely to be encountered along lowland rivers, and backswamp becomes increasingly thick in the lower reaches of alluvial valleys.

Figure 5.7. Buried Holocene backswamp deposits exposed along active channel bank of the lower Mississippi River, ~10 km downstream of Natchez, MS. The cohesive deposits restrict lateral channel migration along this reach. Sediment texture of overbank deposits was light brownish gray (10 YR 6/2) fine sand (d_{50} 0.15 mm) whereas the backswamp deposits were dark blueish-gray (Gley 2 3/5 BG) clay (d_{50} < 0.004 mm). Note in situ cypress stumps within backswamp deposits (below line), consistent with historic maps depicting land cover. For scale, height of vertical dashed line is 2.8 m. (Author photo: October 2017.)

5.2.2 Importance of Bed Load

An additional major sedimentary distinction in regard to channel engineering is whether sediment load is dominated by bed load or suspended load (Schumm and Winkley, 1994). In the lower reaches of large rivers and deltas the dominant mode of sediment transport is suspension, with coarser-grained sediment entrained from the channel bed. A general rule-of-thumb is that about 10% of the total sediment load is bed load (Knighton, 1998), although this can vary widely because of sand being transported both in suspension and as bed load (Nittouer et al., 2008; Knox and Latrubesse, 2016; Frings et al., 2019). Nevertheless, although bed load is a low proportion of total sediment load it is the major control on river channel adjustment. Rivers transporting higher volumes of bed load require a wider channel for sediment transport (Ferguson, 1987; van den Berg, 1995). This implies a "critical" width is maintained to ensure channel stability. Thus, engineering modifications that increase bed load, such as cut-offs and knickpoint development can drive downstream channel degradation (Winkley, 1977; Schumm and Winkley, 1994). The dynamics and timing of bed load transport is also of interest. The particle size and volume of bed material influences the depth of channel bed scour, which is important to the design and long-term stability of hydraulic infrastructure. Large waves of bed material can require many decades to move through a fluvial system, with delayed consequences to hard engineering structures such as groynes and bridges and river stage levels (Gilbert, 1917; James, 2006; Knox and Latrubesse, 2016).

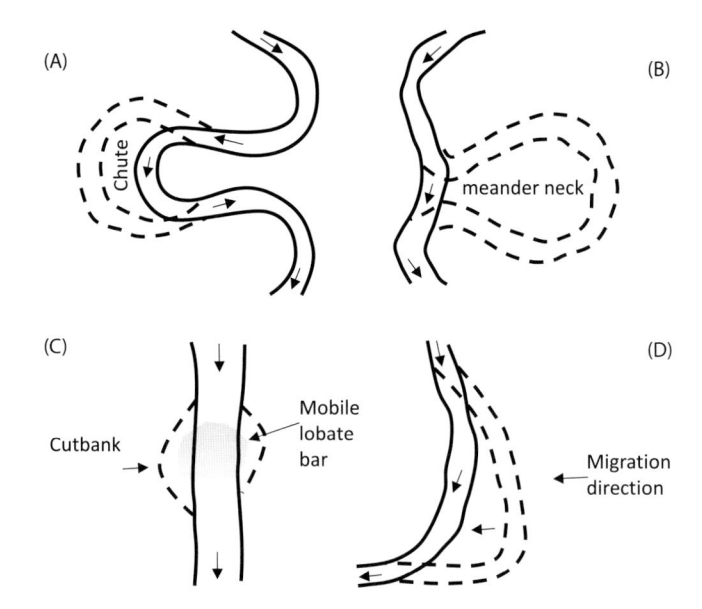

Figure 5.8. Types of cutoffs and lateral adjustment. (A) Chute cut-off; (B) Meander neck cut-off; (C) Mobile bar cut-off; (D) Relocation of river course.

5.3 ENGINEERING FOR FLOOD CONTROL AND NAVIGATION

5.3.1 Channelization

River straightening by meander bend cut-offs is an effective way to initially reduce flood stage along large lowland rivers (Smith and Winkley, 1996; Downs and Gregory, 2004; Shankman and Smith, 2004; de Bruin, 2006). There are several approaches to channel straightening, including bend straightening, chute cut-off, and meander neck cut-offs (Figure 5.8). In some instances an entire river reach is moved, resulting in abandonment of multiple bends.

Although the conceptual framework for designing a channel cut-off is not complex, getting the engineering correct is more challenging, particularly upon larger rivers. Straightening a river channel reduces local flow resistance because of streamflow not having to be routed around bends and because the overall area of channel bed in contact with the flow is reduced. Most importantly, cut-offs increase channel slope, a direct control on velocity and boundary shear stress (Section 2.1). Thus, increased velocity changes the hydraulic geometry with compensatory reductions in flow depth, which reduces flood stage. Flood stage is also geomorphically lowered because increased velocity and channel slope drives channel bed incision. In addition to meander neck cut-off, other types of channel cut-offs include chute cut-offs and bend flattening (Figure 5.8).

Successful engineering of meander neck cutoffs requires a number of considerations (e.g., Table 5.1). It is important to note that, in addition to pilot channel excavation across the point bar, additional engineering modifications are necessary immediately

HYDRAULIC ENGINEERING FOR CHANNEL STABILITY AND FLOOD CONTROL 145

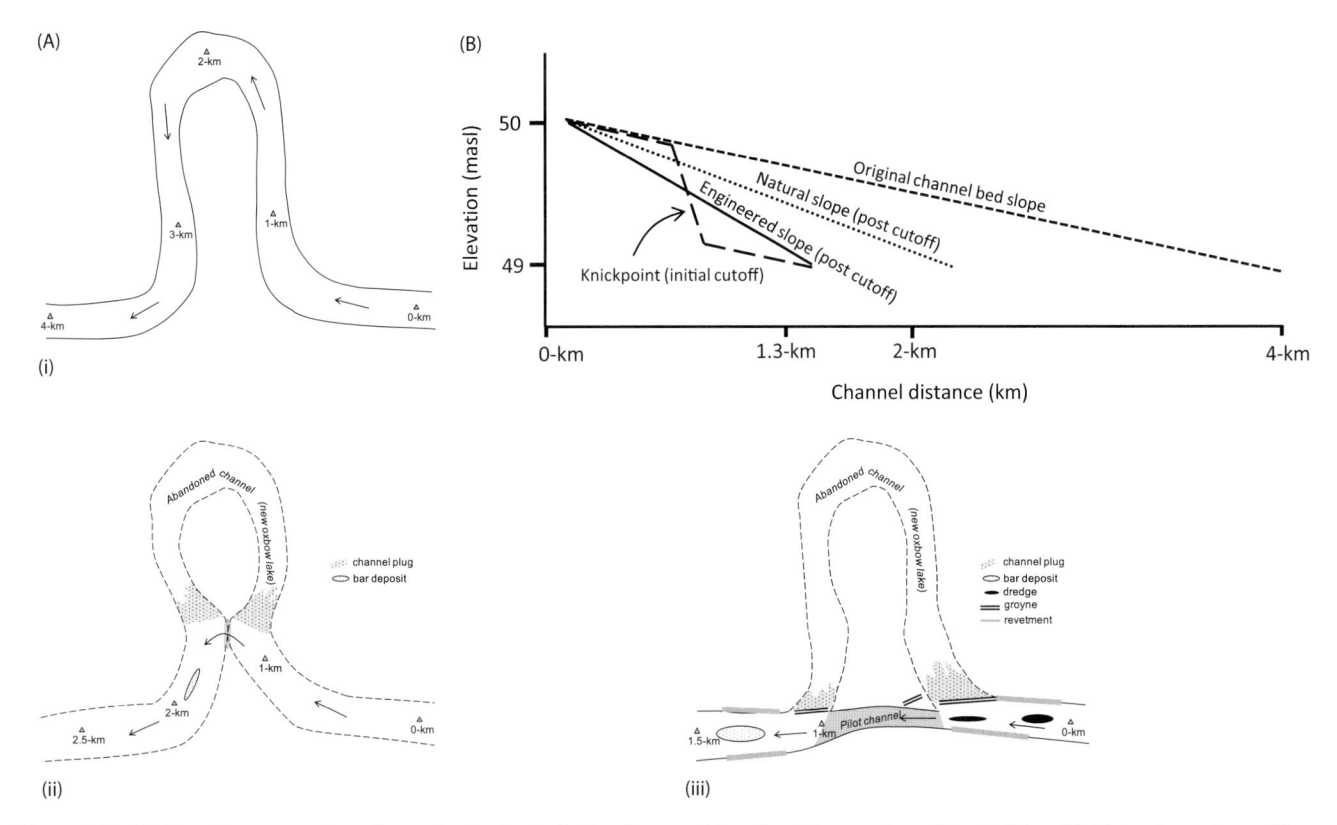

Figure 5.9. (A) Planview perspective of meander bends, including (i) natural baseline (ii) natural cutoff, and (iii) artificial (engineered) cutoff. Artificial cut-off includes a pilot channel and upstream and downstream channel alignment procedures. (B) Subsequent knickpoint development in channel long profile (distance × elevation) for natural and artificial cut-offs. Note that location of pilot channel results in artificial cutoffs having steeper post-cutoff slopes than natural. Author figure.

upstream and downstream of the meander cut-off. These activities are oriented around several issues, including (1) ensuring proper downstream transfer of coarse sediment after the cut-off, (2) ensuring proper alignment of upstream and downstream channel reach, and (3) protection of channel banks from erosion by sloughing (Figure 5.9).

After deciding on a location for the meander bend cut-off, which is usually the shortest distance across the meander neck aligned with upstream and downstream channel segments, the right-of-way on either side of the proposed pilot channel is cleared (Figure 5.10). For large rivers, this action in itself is a considerable disturbance to local land cover as the activities result in the removal of much floodplain forest and riparian wetlands. For the Mississippi cut-off program, land was cleared along a ~300 m corridor along the axis of the pilot channel (Ferguson, 1940). The pilot channel is aligned tangent to the axis of the meander neck, connecting upstream and downstream bends. A pilot channel is excavated to a depth from 3 m to 6 m below very low flow stage, such that the top bank width is ~200 m and the bottom width is ~100 m. Although the substratum is generally coarse grained, in some instances cohesive clayey deposits are encountered, and thus dredging should extend to a depth below the clay layer. This ensures that scour and enlargement of the pilot channel rapidly ensues upon breach of the cut during rising stage levels. Excavation of the pilot channel begins at the downstream bank and works toward the upstream bank. In most cases the excavated sediment is simply dumped directly in

Figure 5.10. Pilot channel creation for an engineered meander neck cutoff along the lower Mississippi River in the 1930s as part of the 1928 Mississippi Rivers & Tributaries Project. (Photo from USACE.)

the river channel immediately on the downstream side of the pilot channel. A thin plug of bank material is initially left intact at the upstream side of the pilot channel.

From the standpoint of the timing of cut-off activities, the pilot channel should be excavated during dry season (low flow), prior to the next discharge pulse. Upon completion of the narrow pilot channel across the sandy point bar, the narrow lip of bank material is breached as the flood wave rises. The physical processes by which the cut-off proceeds include fluvial entrainment and incision into coarse-grained channel bed (point bar) deposits resulting in over steepening of banks along the pilot channel. Mass wasting (sloughing) and enlargement of pilot channel banks rapidly ensues, particularly with saturation of noncohesive coarse materials during rising river stage levels. To ensure that the pilot channel is properly oriented with the thalweg of the upstream and downstream channel reaches requires channel bed dredging and channel bank excavation. This also requires reshaping of channel cross sections (bank angles) to smoothly convey flow events. The active formation of the new cut-off and erosion of the old meander bend produces large amounts of bed material, with much of it deposited immediately downstream of the new cut-off in riffles and crossings (e.g., Zinger et al., 2011). In the case of fine-grained deposits (silt and clay), much of this will be entrained and become part of the suspended load. In the case of improperly aligned channels, subsequent flow events result in considerable deposition of coarse sediment (shoaling). Revetments and groynes, therefore, are often installed immediately upstream, downstream, and along the new cut-off to prevent excessive bank erosion and to "train" the alignment. Along large rivers with distinctive seasonality to the flow regime, complete cut-off of coarse-grained meander bends usually occurs within several years, although bends with a complex planform geometry may require up to a decade (Winkley, 1977; Gagliano and Howard, 1984).

Geomorphically, upon the final breach of the pilot channel, a knickpoint is formed in the channel long-profile (e.g., Figure 5.9). A knickpoint occurs where there is a large increase in the local channel over a relatively short segment (Winkley, 1977). Channel knickpoints form by meander bend cut-offs, whether occurring naturally or by humans. By comparison with natural cut-offs, however, engineered cut-offs are associated with a greater reduction in channel length and, therefore, a steeper channel slope (Echevarria-Doyle et al., 2020). The initial knickpoint (pilot channel) has a much higher slope (Figure 5.9). While natural meander bends initially develop a knickpoint, this is compensated by slope reduction upstream or downstream of the cut-off by lateral migration (increased in sinuosity). Engineered knickpoints, however, are abruptly forced to efficiently complete the cut-off.

While the pilot channel can reasonably be expected to incise and widen within coarse-grained point bar deposits, in some instances pilot channel incision and completion of the new cut-off is slowed because of contact with resistant sedimentary deposits in the channel bed, such as tertiary bedrock or buried backswamp deposits (e.g., Figure 5.7). This can require a large expenditure of engineering effort to further excavate the pilot channel so that it is able to enlarge and accommodate the entire streamflow.

Channelization for flood control by meander neck cut-off has existed for centuries. The concept and practice of channelization within government institutions would have evolved with knowledge exchanges, especially China, the Mediterranean, northern Europe, and, later, North America. It is unlikely, therefore, that the origin of the concept can be attributed to a single entity as the approach was likely individually developed. Support for this position can be appreciated based on an increasing amount of evidence of sophisticated hydraulic engineering by pre-Columbian societies in the Americas, including construction of large drainage structures, canals, and cut-off channel modifications (Denevan, 2000; Mann, 2005; Erickson and Walker, 2009).

Support for prehistoric hydraulic engineering in the Amazon basin is provided by contemporary analogues (Raffles and WinklerPrins, 2003). Along the Ucayali River in the Peruvian Amazon, for example, local rural communities manually engineered an extensive cut-off of a sinuous channel reach to reduce local flooding (Abizaid, 2005; Coomes et al., 2009). In this case, using simple basic hand tools local people were able to initiate a cut-off through sandy point bar deposits by digging a 7-km long ditch that was initially only 1-m deep and from 1 to 2-m wide (Figure 5.11). Because of the length of the channel, however, the slope of the "pilot channel" had a gradient advantage eighteen times greater (steeper) than the much longer natural channel. By completing their excavation ahead of the flood wave the "pilot channel" took advantage of natural hydraulic processes to accelerate scour, resulting in deepening and elongation of the ditch. After several flood cycles (years) a 71-km sinuous channel reach had cut-off, reducing the channel length by 64 km (Figure 5.11). Hydrologic changes due to the cut-off substantially impacted livelihoods of local populations that depend upon riparian landscapes and waterways for agriculture and transportation. Upstream of the cut-off the river no longer annually floods, which indicates typical channel bed incision, whereas previously the river flooded annually. The downstream reach, however, has experienced channel bed aggradation and an increase in flood frequency and duration. The channel bed has aggraded by some 2 m and the flood stage has increased by 0.5–1.0 m. Whereas previously the high natural levees along the downstream reach flooded only every five to ten years, they now flood every year.

While the comprehensive characterization of indigenous and local peoples use of hydraulic engineering is increasingly being

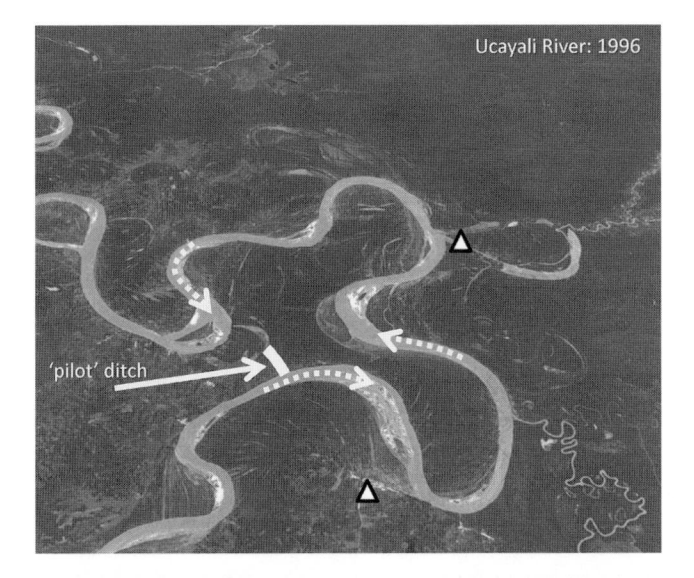

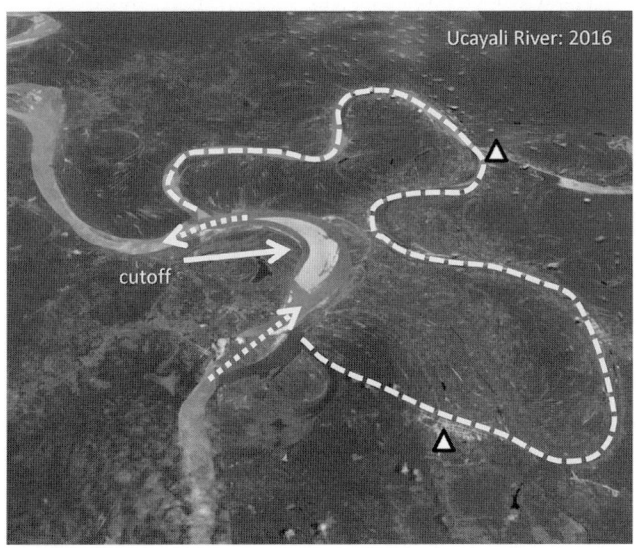

Figure 5.11. Extensive cut-off of meandering channel segment along the Ucayali River at Masisea, in the Peruvian Amazon. Before the cut-off (top, 1996) and about two decades after the cut-off (bottom, 2016). Triangles mark the same point in space. For scale, in the 2016 image the channel width (foreground) is 823 m. (Author figure. Image source: Google Earth Pro.)

appreciated, a number of engineering programs along large industrialized rivers have been extensively documented. Among the most well-known channelization projects are the Tulla cut-offs and straightening of the middle-upper Rhine River in southern Germany near Worms (de Bruin, 2006; Uehlinger et al., 2009). This extensive channelization project was completed over about a six-decade period between 1817 and 1880, and was based on a master plan by General J. G. Tulla. While the overall plan included flood control, a key motivation for the Tulla cut-offs was to improve navigation. The efforts were seen as a major public works project, which also had the effect of nation building by enhancing coordination between neighboring riparian states.

Along a particularly sinuous and anabranching segment of the upper-middle Rhine the Tulla cut-offs reduced the channel length by 81 km (Figure 5.12). In addition to channelization the project included dikes, revetment, and groynes to stabilize the channel and embank the floodplain. The channel width was stabilized and narrowed, from a variable width ranging from 325 m to 450 m to a uniform width averaging 240 m. The removal of hundreds of sand bars and channel islands resulted in a gain of over 100 km^2 of reclaimed land, and a tremendous loss in geodiversity and degradation to biodiversity. The increase in channel slope and constriction of flow within a narrower channel drove channel bed incision of up to 7 m in the upper Rhine. The channel bed incision decreased river stage (water level), which lowered the water table of the alluvial aquifer (Frings et al., 2014). Thus, riparian wetlands degraded and were terrestrialized by encroaching vegetation as well as being converted into arable land, which requires irrigation. Channel bed incision along some reaches was halted by the presence of resistant bedrock that resulted in formation of rapids that, somewhat ironically, impede navigation (Uehlinger et al., 2009).

The Tulla channelization program was hailed as a major success and the scope and visibility of the works assured that they received public support and wide international acclaim. Unfortunately, the cut-off program occurred before there was much scientific understanding of the importance of hydrologic connectivity between rivers and floodplains (i.e., Junk et al., 1989; Hudson et al., 2008).

The *assumed* success of the Tulla cut-offs stimulated development of similar channelization programs along other large industrialized rivers, especially in Europe, Australia, North America, and China. Among many, additional important examples of heavily engineered rivers straightened for flood control include the Yangtze, Ebro, Rhône, Danube, Volga, Missouri, Tisza, Ohio, and the Elbe (Bravard and Gaydou, 2015; Constantinescu et al., 2015; Middelkoop et al., 2015; Ollero et al., 2015), among others. These activities frequently resulted in tens of kilometers of channel being removed from the active river within a limited overall distance, resulting in abrupt reduction in sinuosity and increase in slope. Cut-off projects were also carried out across numerous smaller rivers. While it is not possible to tally the total reduction in channel length, in the United Kingdom a total of 450 km of channel distance was removed from smaller alluvial rivers (Downs and Gregory, 2004). In essence, the Tulla cut-offs set forth a wave of "hard" engineering of large alluvial rivers over about a seventy-five-year period in the twentieth century that resulted in substantial degradation of lowland river environments, which continue to adjust, and require further management.

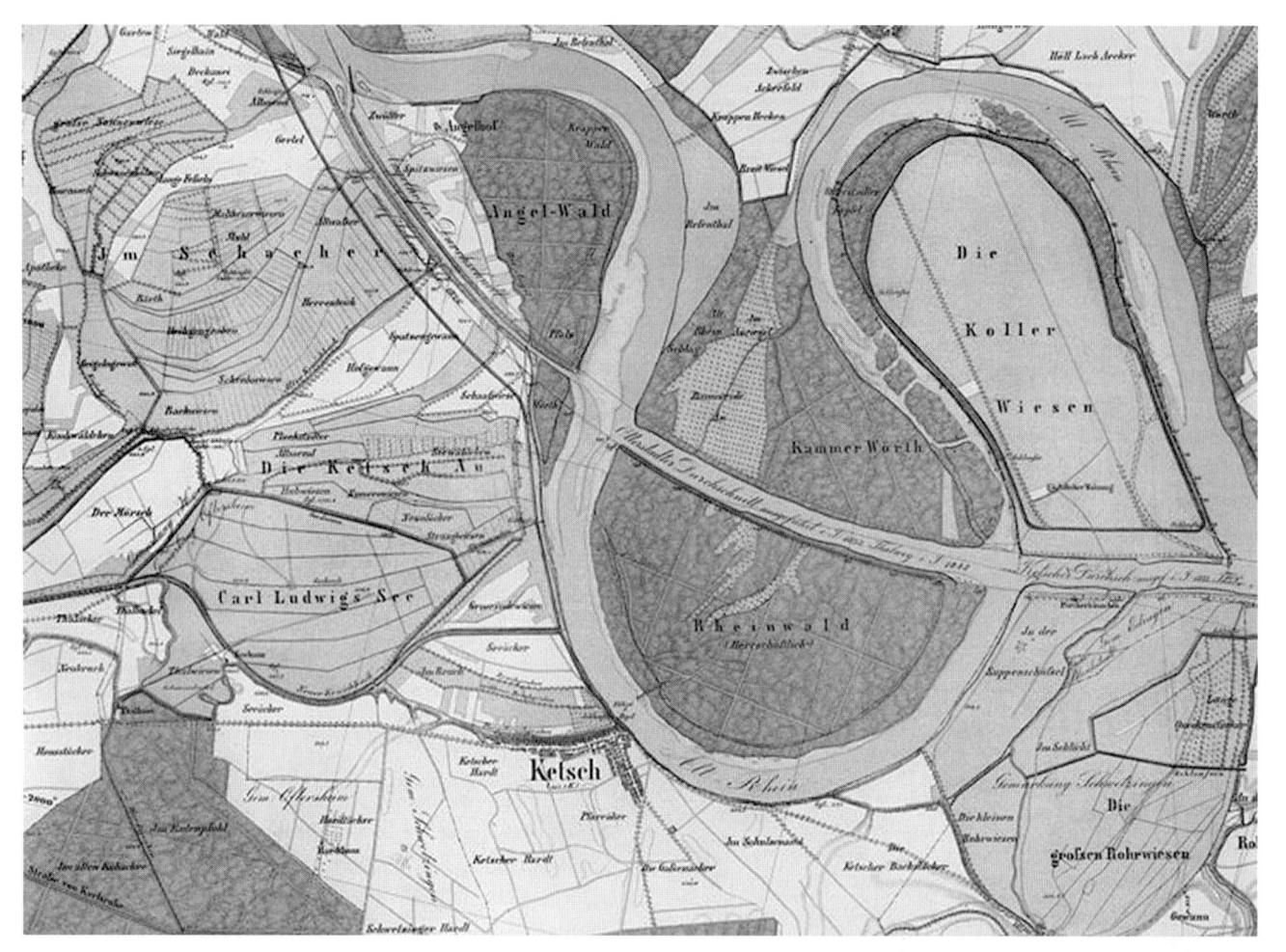

Figure 5.12. Channel straightening as part of the Tulla cut-off plan (1833) of the upper-middle Rhine River, ~5 km upstream of Mannheim, Germany. Overall channel sinuosity was reduced from 2.8 to 1.2, effectively becoming a straight channel. (Licensed by CC.)

5.3.2 Bank Protection: Revetments and Riprap

Revetments have likely existed for as long as bank erosion has represented a problem to humans. The persistent problem of bank erosion has resulted in considerable experimentation to protect river channels, including use of materials such as tree root wads, tree trunks, steel, concrete, rocks, and planting of trees. For centuries the most common type of revetment material by government agencies has been the simple river willow (*Salix*). Willow is supplied from local riparian sources and its low cost, rapid growth and extensive availability ensures that it is utilized across an international range of rivers. Along smaller alluvial rivers willow is utilized for bank protection by inserting live willow stakes directly into the channel bank, whereby they rapidly grow. This mainly protects from hydraulic abrasion, which is not the dominant process of erosion along large alluvial rivers (Section 3.3.2). Willow is utilized for bank protection along larger rivers by emplacement of trunks within a large wooden frame. The willow is fastened to the frame with small flexible branches, creating a highly permeable and somewhat flexible structure. The willow fascine mattress is usually installed by securing one side against the bank, and then extending it into the channel and ballasted with rock or concrete (Elliot, 1932).

An additional method of bank protection involves the use of common riprap, large aggregate installed directly atop eroding river bank. While willow revetments are usually installed at the top and mid channel bank, riprap is also installed along the channel base to stabilize the river bank toe (Fischenich, 2003; Pepper and Rickard, 2008). The design of riprap bank protection should be related to the physical dimensions of the channel, specifically the slope angle and the flow velocity at the channel base. The high internal friction of riprap enables it to remain stable atop steep banks (Table 5.4). Additionally, the inherent roughness reduces flow velocity at the channel bed.

While riprap and willow are common along moderate-sized lowland rivers, concrete is the favored material for revetment construction along larger rivers. The most effective revetment type is the articulated concrete revetment, developed and refined by the U.S. Army Corps of Engineers in the 1950s after decades of trial and error (e.g., Markham, 1916; Elliot, 1932).

Table 5.4 *Characteristics of appropriate riprap in the design of alluvial channel bank protection*

Parameter	Characteristic
Bank grading	W_{85}/W_{15} = 3.4–16 (for smooth, well-graded curve)
Shape	Blocky
Angle of repose	35°–42°
Angle of internal friction	40°–45°
Density	2.5–2.7
Revetment thickness	$\geq 2 \times D_{50}$, or 1–1.5 × stone diameter
Porosity	25–40%

Source: Based on Pepper and Rickard (2008) (U.K. Environment Agency; Escarameia, 1998).

Figure 5.13. Historic willow revetment construction along the lower Mississippi River from the 1920s. Photo shows ballasting of a wooden fascine mattress. (Source: USACE.)

Articulated concrete revetment consists of large concrete pads linked together with steel cable. Concrete revetment slabs are manufactured in casting fields located along the floodplain (Figure 5.14). The concrete slabs are then transported on specially designed barges where the steel cable is affixed. The revetment is constructed along the graded river bank, with the barge moving outward as the flexible revetment is "draped" along the river bank (Smith and Winkley, 1996). The flexibility of the articulated concrete revetment enables it to adjust to channel erosion, ensuring its stability for modest amounts of channel change (Figure 5.15).

Hydraulically, because of the smooth surface of concrete, articulated revetment reduces channel bank roughness and results in increased near-bank channel velocities. The stabilized reach increases channel shear stress at the channel bank toe, resulting in channel bed scour and deepening (i.e., Mosselman et al., 2000; Kiss et al., 2019). To ensure the physical integrity of the revetment structure and that it is not undermined by bank erosion, revetment should be extended from the top of the channel bank to the scour pool at the toe of the channel bank or channel thalweg (Figure 5.16), a lesson learned by the US Army Corps of Engineers after decades of trial and error (Markham, 1916; Elliot, 1932).

An outcome of the success of articulated revetments is that concrete covers the channel banks of many larger alluvial rivers utilized for navigation and flood control, Revetment is often constructed to manage unintended consequences of other forms of channel engineering (Hudson et al., 2008). To address problems with flooding along the lower Tisza River (Poland), an extensive engineering transformation was undertaken in the late 1800s, including construction of 2,900 km of dikes (levees) to increase agricultural productivity (Amissah et al., 2018). Channel straightening occurred with twelve meander bend cut-offs, shortening a 131 km segment by 40 km, which increased channel slope from 2.2 cm/km to 2.9 cm/km. Channel incision and lateral channel bank migration ensued. Channel revetment was installed over 25% of the eroding bends, mainly between the 1930s and 1960s, to reduce bank erosion initiated by channelization for flood control for the

Figure 5.14. Concrete revetment casting field along the lower Mississippi River at St. Francisville, Louisiana (45 km north of Baton Rouge). The scale of the operation is massive, and annually produces 120,000 concrete pads. Individual pads are woven together with steel wire into 42.7-m-wide sections. (Author photo, 2003.)

Figure 5.15. Installation of modern articulated concrete revetment along lower Mississippi River. (Photo source: USACE.)

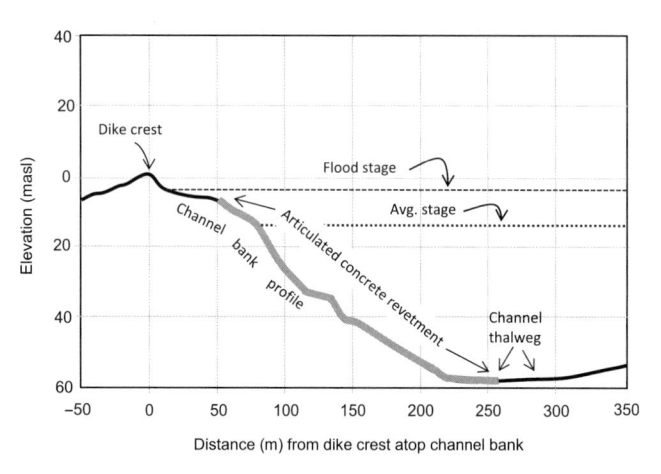

Figure 5.16. Placement of articulated concrete revetment along channel bank profile for the lower Mississippi River at New Orleans. (Based on USACE.)

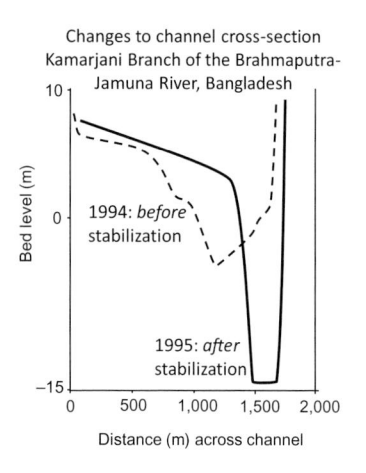

Figure 5.17. Changes to channel cross section before (1994) and after (1995) bank stabilization (along right side of figure) for the Kamarjani Branch of the Brahmaputra–Jamuna Rivers, Bangladesh. Note large increase in channel depth. (Source: Mosselman et al., 2000.)

purpose of protecting the dikes. The revetment was associated with the development of scour pools and an increase in channel depth by 1–2 m (Amissah et al., 2018; Kiss et al., 2019). Along the highly dynamic braided Kamarjani branch of the massive Brahmaputra–Jamuna Rivers in Bangladesh, lateral channel stabilization by hard structures and groynes was immediately effective in stopping bank erosion, but also resulted in channel bed scour and deepening (Figure 5.17). Before bank stabilization, the measured channel width, outer bend depth, and bank migration rate was 1,640 m, 11.5 m, and 200 m/yr, respectively. After bank stabilization in 1995, the measured channel width, outer bend depth, and bank migration rate was 1,850 m, 20.5 m, and 0 m/yr, respectively. In addition to the influence of the hard structure on flow processes, the reduction in sediment supply from reduced rates of lateral migration (i.e., bank erosion) also contributes to channel bed deepening at the toe of hard structures (Mosselman et al., 2000).

Channel stabilization often reduces overall riparian species diversity. The concrete surface of revetment degrades aquatic and terrestrial riparian environment of channel banks by changing channel bed velocity distributions, channel bed deposits, and by physical obliteration or sealing over channel banks. The inherent roughness and irregular height of natural channel banks are critical to healthy riparian ecosystems, providing refuge for fish, microhabitat, and linkages to backwater habitat (Newson and Newson, 2000; Florsheim et al., 2008; Massey et al., 2017). The terrestrial portion of the river bank is important to flora and fauna adjusted to the dynamic physical environment provided by river banks. A major bank protection program along the Sacramento River to protect flood control infrastructure was controversial because of the environmental impact of concrete revetment to the riparian ecosystem. Specifically, concrete revetment paved over burrowing riparian bank swallow (*Riparia riparia*) habitat, which nest directly within sandy channel bank alluvium (Moffatt et al., 2005; Givetz, 2010; BSTAC, 2013).

While concrete is seemingly permanent, within a few decades concrete revetment often degrades and requires repairs (Elliot, 1932; Kiss et al., 2019), and ubiquitous stone riprap is then often necessary to keep the revetment from collapsing into the river (Figure 5.18). For concrete revetment to remain stable requires that it is properly installed along river banks, and in particular extends to the channel thalweg (e.g., Figure 5.16) where scour and boundary shear stress is usually highest. Revetment that does not extend to the thalweg can be undermined by scour and bank collapse, with variation influenced by channel bank sedimentology (Figure 5.19). Bank material with higher cohesion (% clay) better withstands moderate amounts of channel bed scour whereas sandy channel banks easily collapse with increased channel bed scour following revetment installation (Dunbar et al., 1999).

5.3.3 Groynes

A dynamic and shifting channel bed is problematic for flood control and navigation, as bank caving and channel migration undermine dikes and associated hydraulic infrastructure (Echevarria-Doyle et al., 2020). As with other engineering measures, humans have been trying to control the position of channel lines for many centuries. A principal way to accomplish this is to construct channel groynes (wing dikes), which directly influence the hydraulics of erosion and deposition (Figure 5.20). Groynes are elongated rigid structures that extend from the edge of a river bank toward the center of the channel (e.g., Figure 5.1). Groynes have historically been constructed of a range of materials, including tree branches, root wads, steel fencing, iron beams, concrete, riprap, and basalt. Because groynes directly interact with flow, the construction of groynes influences channel flow patterns and results in local-scale erosion and deposition, primarily of coarser bed sediments (Mosselman et al., 2004). Along large rivers, groynes are primarily constructed for the purpose of maintaining navigation and shipping. While groynes were often constructed with the idea that they only influenced low- and moderate-magnitude events, increasingly groynes are viewed from the broader perspective of the overall flood control system, including their role in influencing flooding (TenBrinke et al., 2004; Pinter et al., 2010; Remo et al., 2018).

Figure 5.18. Degradation of modern articulated concrete revetment along lower Mississippi River (downstream of St. Francisville, Louisiana), now patched up by miscellaneous placed riprap. (Author photo: October 2017.)

Figure 5.20. Groynes along the Dutch Rhine. Note successive trapping of bed material and channel narrowing (left bank), including new (right bank) "longitudinal dam" design (upper, looking downstream) and vegetation growth and new floodplain formation indicating channel stability following aggradation (lower, looking upstream). (Source: Rijkswaterstaat.)

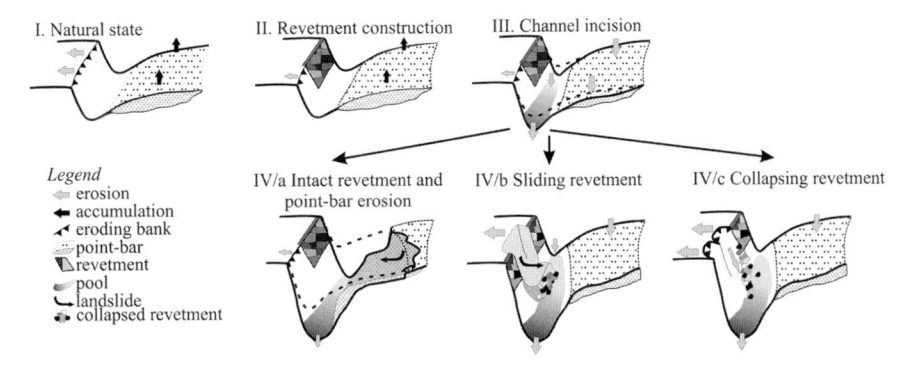

Figure 5.19. Sequence of channel changes after concrete revetment construction, including incision by scour pool development at revetment toe, and failure due to channel bank oversteepening. (Source: Kiss et al., 2019.)

Groynes are usually installed perpendicular to the thalweg, but this depends somewhat upon the angle of the thalweg within the larger channel and the flow pattern (Przedwojski et al., 1995; Smith and Winkley, 1996; Mosselman et al., 2004; Ten Brinke, 2005). Groyne placement is often along the inside of a bend, or along shallow bars along straighter reaches. In this way the groynes take advantage of the inherent helical flow patterns associated with a river bend, especially along a meandering reach. Helical flow assures that coarse sediment is transported laterally, across the channel, from the deeper portion of the thalweg to the inside of point bars. Along straighter reaches the accumulation of large amounts of coarse bed material occurs within riffles, crossings, or alternate bars, which is also an ideal placement for groyne fields (Smith and Winkley, 1996). Additionally, individual groynes are often installed at the upstream entrance to side channels to prevent flushing flows, thereby resulting in sediment infilling and side channel closure, ultimately increasing flow concentration within a single-thread channel (Smith and Winkley, 1996; Alexander et al., 2012).

Groynes considerably modify local hydraulics and sedimentation patterns (Ten Brinke et al., 2004; Malik and Pal, 2019). Hydraulic roughness and boundary resistance increase with groyne emplacement, thereby reducing local flow competence to transport coarse sediment. Bed material, therefore, is deposited and stored within groyne fields, at least in initial decades after their emplacement (Mosselman et al., 2004; Alexander et al., 2012). Where groynes are spaced closer together, a large primary eddy develops between adjacent groynes (Figure 5.21A). In combination with minor eddies adjacent to the channel bank, the reduced velocity results in a scalloped pattern of coarse sediment deposition along the inside of the channel (Przedwojski et al., 1995). Importantly, the pronounced development of the primary eddy forces the main flow toward the channel center (or alternate bank), such that higher velocities result in a self-flushing channel bed, which is ultimately the goal of groyne emplacement. This is especially desirable for navigation because of the channel maintaining sufficient depth for barge traffic, reducing the need for channel dredging.

The construction of groynes often occurs within clusters – three to ten, referred to as groyne fields. The spacing of groynes is scale dependent, meaning that along larger rivers individual groynes are installed further apart. Where groynes are spaced further apart, the flow pattern from the main channel enters the groyne field and prevents appreciable sediment deposition along the downstream side of the groyne field, while lower flow velocities along the upstream portions of the groyne field enable coarse sediment deposition (Figure 5.21B).

Groyne spacing also depends upon fluvial dynamics and the importance of the channel segment to navigation. Some rivers effectively have groynes placed along their entire length. Along the lower Rhine, from Bonn, Germany to the North Sea, groynes are spaced ~200 m apart and maintain a stable channel. Prior to the intensive construction of groynes, however, the Rhine had a much wider and dynamic channel that included a number of shallow crossings, channel bars, and islands (Middelkoop and van Haselen, 1999; Nienhuis, 2008; Uehlinger et al., 2009). The first groynes were constructed along the Rhine many centuries ago (even by Roman engineers) by simple measures (van Veen, 1948; Middelkoop, 1997; van de Ven, 2004). In the sixteenth and eighteenth centuries, the Waal River became the main distributary for the Rhine, and land was "reclaimed" by groyne constructed of riprap and willow (*salix*), which grew from stems inserted directly into the channel bed. While such measures were initially limited to specific "problem" reaches, they were effective at reducing channel velocities and entrapping coarse bed material. Thus, the channel banks and bars were stabilized and the width decreased from an average of 340 m to 165 m. With subsequent addition of fine-grained sedimentation deposited during the waning stage of flood events, the deposits effectively became "welded" to the channel banks. This new low anthropogenic floodplain surface provided additional space for human activities, especially agriculture.

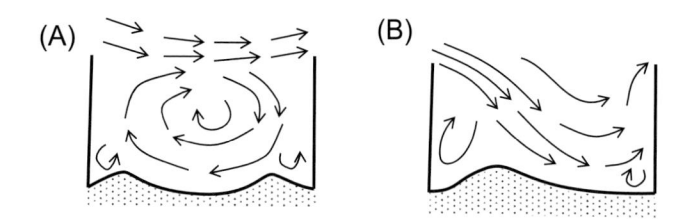

Figure 5.21. Patterns of groyne field flow circulation and sediment deposition in relation to the distance (spacing) between sucessive groynes. Groynes spaced close together (A) and groynes spaced further apart (B). Not to scale. (Source: Przedwojski et al., 1995.)

Floodplain expansion in association with channel stabilization and narrowing caused by groyne construction is a scenario that has played out along many heavily engineered rivers and floodplains, or is in the process of occurring. Examples of rivers in Europe where the process has effectively reached its final stage includes the Elbe, Rhine, and Danube Rivers, which are also heavily utilized for shipping (Mosselman et al., 2004). Several large North American rivers, especially the Missouri River, have effectively reached a mature stage (Alexander et al., 2012). The lower Missouri River is of interest because of being heavily utilized for shipping but also because it previously had a highly dynamic sand bed channel. Groyne construction (in addition to upstream flow impoundment) has contributed to stabilization and narrowing of the lower Missouri, reducing the channel width from 635 m to 325 m (upstream of St. Louis). Groynes along the Missouri River are effective from the standpoint of the channel becoming self-flushing, which has reduced the amount of dredging required to maintain a navigable channel (Alexander et al., 2012).

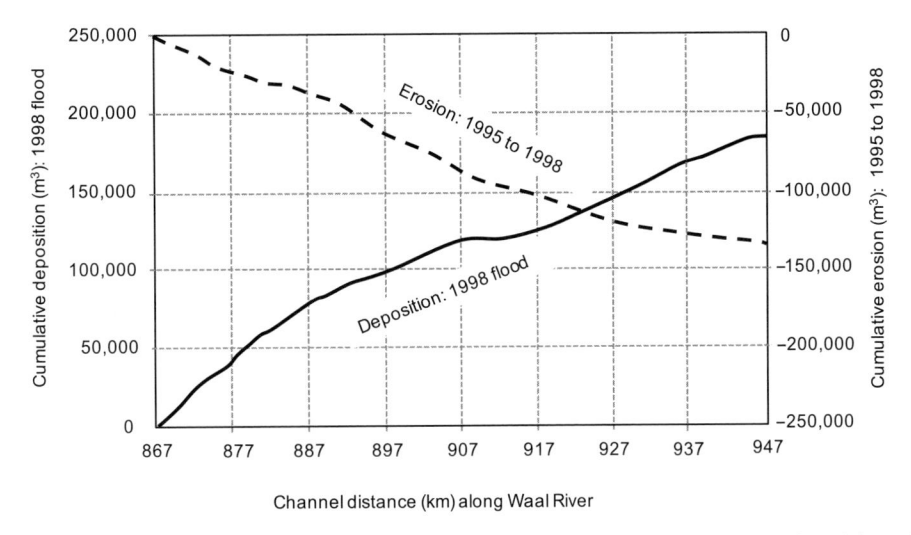

Figure 5.22. Cumulative incremental sand erosion between 1995 and 1998 and subsequent abrupt deposition of sand from the 1998 flood within groyne fields along the Waal River, main distributary of Rhine delta. (Modified from Ten Brinke et al., 2004.)

The pattern of sedimentation along rivers with intensive groyne fields varies from natural channels. From the perspective of the bed load sediment regime, 34% of the Rhine's sediment is associated with deposition and reworking within the groyne fields (Frings et al., 2014, 2019). The amount of time for groyne field sedimentary deposits to be assimilated back into the active channel bed load is about a decade, but it also depends on the discharge and sediment regime and geometry of the groyne field. Erosion occurs during low and moderate discharge years, manifest by incremental erosion and downstream transport of coarse sediment. Deposition of sand within groyne fields primarily occurs during years with large discharge events, resulting in pulsed sand deposition within groyne fields and sediment trapping that is somewhat in balance with the erosion (Figure 5.22).

In addition to hydrologic and hydraulic processes influencing sedimentation within groyne fields, an increasing amount of research has established that the wake from ship traffic substantially influences groyne field sedimentation (Ten Brinke et al., 2004). The amount of erosion caused by ship wake along the Waal River in the Netherlands, for example, was significantly greater along the south river bank in comparison to the north river bank. This is because flow along the south river bank has greater shear stress, which is caused by heavy cargo-laden ships going upstream from Rotterdam to central Germany and beyond. Much of the ship traffic along the north river bank includes empty vessels that induce much lower shear stress and wake, resulting in lower rates of sediment reworking.

5.3.4 Tree Snag Removal and Log Jams

Rivers utilized for navigation and flood control are engineered for stability. While modern environmentally oriented channel engineering approaches seek a nexus between channel stability and conserving riparian geodiversity and biodiversity, historically the latter was not a management priority. Engineering for channel stability historically included systematic removal of tree snags. Large woody debris is resistance to flow, and trees embedded in the river channel alter local sedimentation and flow patterns (Gurnell et al., 2002; Wohl, 2014). Thus, an additional type of river "improvement" is tree snag removal, an activity that was intensively imposed along many coastal plain rivers from the mid-1800s to early 1900s.

The problems that woody debris presented to navigation and river management in North America and Australia between the mid-1800s and early 1900s is somewhat unique because of the combination of rapid industrialization, agriculture, and expanding settlement from coastal areas to continental interiors (Montgomery et al., 2003). Long-settled regions such as Europe and parts of Asia, however, had already removed much forest cover. Afforestation in Europe over the past century has resulted in a significant increase in forest cover, and some 43% of the European Union is now forested, and expected to increase because of projected reductions in agricultural land cover (Nègre, 2019). The regrowth of riparian forests, particularly where river rewilding occurs (e.g., Brown et al., 2018), can be expected to increase woody debris load to rivers, presenting management challenges to downstream rivers that remain intensively utilized (Piégay and Marston, 1998).

In addition to their geomorphic, hydrologic, and ecological influences, log jams pose a substantial challenge to managed rivers. While the removal of tree snags has long been a government strategy to increase flow conveyance and improve navigation, it is at the peril of associated ecosystems because of its role in reducing aquatic biodiversity (Piégay and Gurnell, 1997; Benke and Wallace, 2003; Erskine and Webb, 2003; NSW DPS, 2007; Wohl, 2014). Along US Gulf Coastal Plain rivers, snags and log jams were a persistent problem to navigation that required great expenditures by the federal government to manage under the guise of river "improvement" from the mid-1800s to the early 1900s (Annual Report of the Chief of Engineers, 1890, 1904 Wohl, 2014). Historic navigation of the Murray River basin, Australia was considerably hampered by enormous amounts of woody debris, requiring intensive snag removal campaigns between the 1850s and early 1920s (Colloff, 2014, pg. 179).

Woody debris was intensively and systematically removed from US rivers during the late 1800s and early 1900s (Wohl, 2014). The amount of tree snags removed from rivers varied widely and averaged about ~1,000 tree snags per river, with some rivers, such as the Guadalupe and Tombigbee having more than ~10,000 snags annually removed (Figure 5.23). The reason for the five decade or so period of intensive tree snag removal in the United States is twofold. First, conversion of bottomlands to agriculture, encouraged with the U.S. Swamp Lands Acts of the mid-1800s and other policy initiatives, resulted in extensive deforestation of riparian forest for timber and agricultural expansion, often under the guise of "improvement." In the southern United States, cut trees were harnessed into log rafts and floated downstream to timber mills during higher flow periods. But the erratic flow of many lowland coastal plain rivers, particularly those along the central and western Gulf Coastal Plain, resulted in the loss of many trees and log rafts were frequently stuck atop shallow bars (Wohl, 2014). The second reason is that many coastal plain rivers were actively migrating and had not yet been intensively engineered for navigation and bank erosion (i.e., revetment and groyne construction), which occurred several decades later. Thus, seasonally, large amounts of woody debris were fed into rivers by eroding riparian lands. The period for intensive snag removal for other rivers was somewhat later, and timed with the era of hard channel engineering for navigation in the mid-twentieth century. The Apalachicola River (50,945 km^2) in Florida, for example, annually had more than 10,000 snag removals between the 1950s and 1970s (Mossa et al., 2017).

5.3.4.1 GREAT RIVER RAFT: RED RIVER, LOUISIANA

The most notorious log jam is the historic Great River Raft, which formed along the Red River (174,824 km^2) valley in Louisiana, upstream of its confluence with the lower Mississippi and Atchafalaya Rivers (Figure 5.24). Additionally, at different periods portions of the raft extended upstream into southwestern Arkansas and northeast Texas. While the precise age of the "Great Raft" is unknown, some estimates are that it initiated as long as 2000 years BP (Torres and Harrelson, 2012; Mossa, 2013), if not earlier. It is very likely that log jams have always been a component of the riparian environment, particularly along the Red River because of its erodable banks and highly variable flow levels. The log rafts along the Red River were such a dominant component of the riparian environment

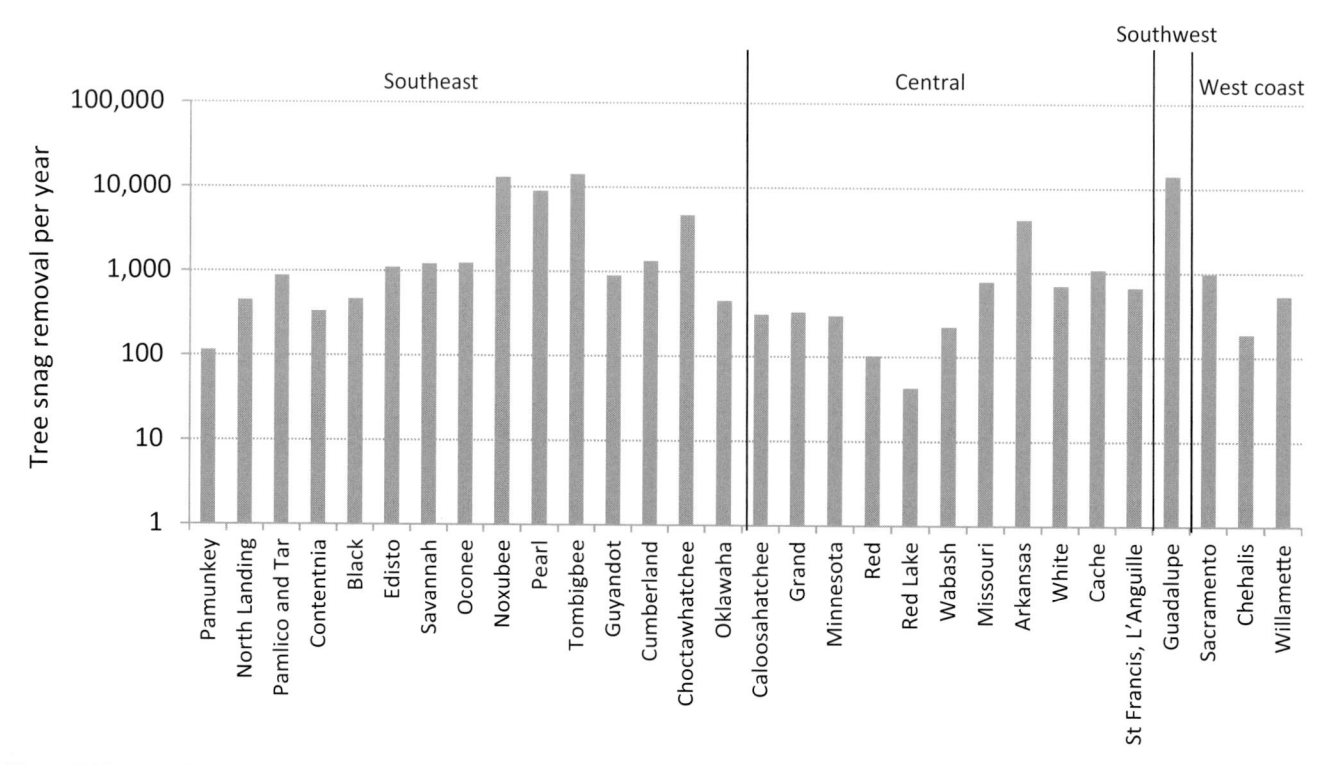

Figure 5.23. Annual number of snags (trees in river bed) removed from rivers in the United States for navigation improvement over about a twenty-five-year period (on average) between late 1800s to early 1900s. Note: Y-axis is logarithmic. (Data source: Wohl, 2014.)

HYDRAULIC ENGINEERING FOR CHANNEL STABILITY AND FLOOD CONTROL

Figure 5.24. View of Red River Raft from the 1870s. (Photo source: State Library of Louisiana, Historic Photograph Collection.)

(Figure 5.25A) that indigenous Native Americans (Caddo) organized portions of their livelihoods around the distinctive floodplain environments created by the backwater flooding. The Red River log jams were described by early French and Spanish explorers, as well as scientists including Lyell (1837) following his travels to Louisiana. The Red River raft grew rapidly over the seventeenth and eighteenth centuries. By the early nineteenth century, when navigation became an increasingly important management priority along the Red River, the raft had grown to its largest extent, being some 260 km in length. The Red River Raft was finally cleared in the 1830s by the federal government, although by the 1870s continued log jam formation required further removal.

Because of its extensive size and duration, the Red River raft substantially altered the floodplain geomorphology, hydrology, and ecology of portions of the Red River valley, with potential downstream consequences to avulsion of the Mississippi (Aslan et al., 2005). The backwater effect resulted in increased flood duration and frequency. These changes included new side channels and lakes revealed in historic cartography (Figure 5.25A and B), including Caddo Lake and Cross Lake (Report of the Secretary of War, 1855). The backwater effect and altered flood regime resulted in considerable channel bed aggradation and floodplain accretion.

The eventual removal of the Red River raft and backwater conditions also had substantial impacts to the geomorphology and riparian environments of the Red River valley. The lowering of river stage caused riparian wetlands and lakes to drain within several decades, which created new lands that could be cleared for settlement and agriculture. The improved hydraulics also resulted in higher streamflow velocities, which drove channel bed incision, essentially the removal of backwater sediment that

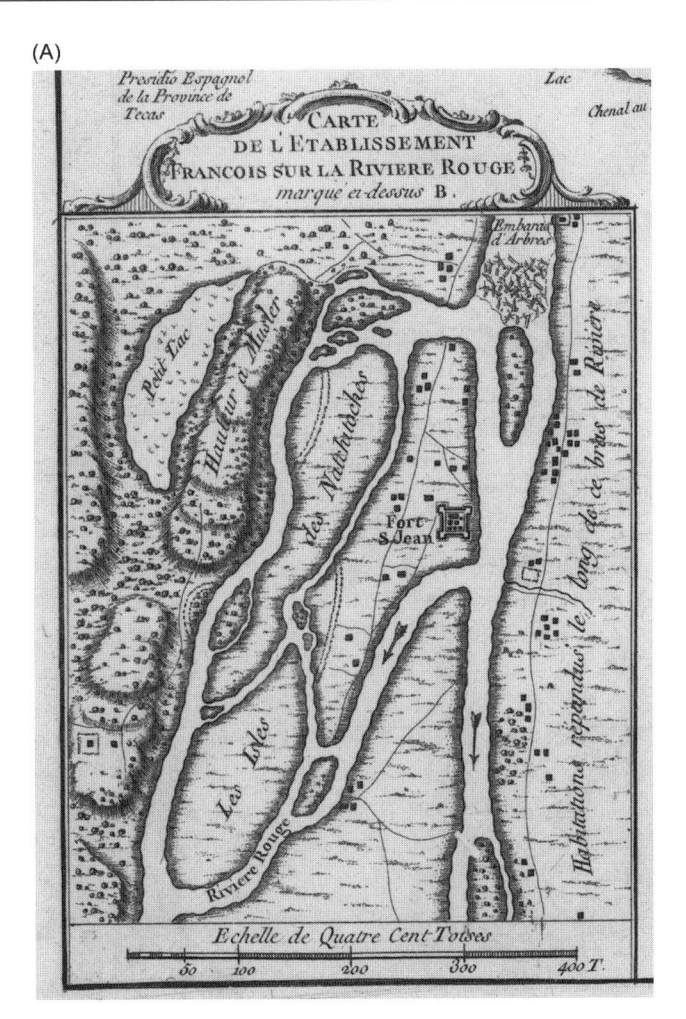

Figure 5.25. (A) Historic map (1764) displaying altered channel pattern and floodplain hydrography of the Red River in northwestern Louisiana in vicinity of Natchitoches, Louisiana. (B) Inset, log jam (*Embaras d' Arbres*), anabranching pattern, and islands. (Source: Bellin, 1764, from U.S. Library of Congress, Geography and Map Division, Washington, DC.)

accumulated during the log jam period. Finally, removal of the Great Raft is considered to have accelerated the capture of the Mississippi River discharge by the Atchafalaya River (see Section 7.4.3.1), which historically also experienced persistent log jams (Anonymous, 1828; Fisk, 1952; Comeaux, 1970; Reuss, 1998; Mossa, 2013; Barnett, 2017).

5.4 IMPACTS OF A MASSIVE FLOOD CONTROL SYSTEM: LOWER MISSISSIPPI CASE STUDY

5.4.1 The Cut-off Program

Isolating cause-and-effect in fluvial systems impacted by human activities for extended periods is complex, and difficult (Schumm, 1991; Thorne, 2002; Gregory, 2006; Pinter et al., 2006a; Hudson and Middelkoop, 2015). Multiple forms of human activities are superimposed across headwaters and lowlands in time and space. Analyzing such impacts to rivers, especially large lowland rivers, requires a variety of detailed data types and a tight chronology of when disturbances were imposed upon specific components of the system (Schumm and Winkley, 1994; Middelkoop and van Haselen, 1999; Biedenharn et al., 2000; Thorne, 2002; Harmar et al., 2005; Alexander et al., 2012; Remo et al., 2018). Because the impacts of discrete engineering activities are not limited to the channel reach where they are constructed, additional forms of channel engineering are required to manage subsequent channel adjustments. That is, the impacts of prior (older) channel engineering results in a suite of unintended geomorphic consequences that requires further management (Hudson et al., 2008). Thus, along most large rivers modified for flood control, engineering activities often occur within a specific sequence. While some measures are indeed planned, other engineering measures had not been anticipated.

The scale of large lowland fluvial systems necessitates a great deal of experimentation by government organizations to develop new technologies, such as concrete revetment. And such strategies require decades to design, implement and perfect (i.e., Markham, 1916; Elliott, 1932; Ferguson, 1940; van Veen, 1948; Winkley, 1977).

Until an engineering structure is actually tested with real hydrologic events its effectiveness is only theoretical. The lessons learned from numerous engineered lowland rivers is that it requires decades to assess how various infrastructure operate together (Winkley, 1977; Kiss et al., 2019), as well as comprehending the geomorphic, hydrologic, and associated environmental impacts (Schumm and Winkley, 1994; Harmar et al., 2005; Hudson and Kesel, 2006; Pinter et al., 2006b; Remo et al., 2018). This is especially the case with large lowland rivers because of inherent lag-times associated with scale and geomorphic response. An ideal case study that illustrates these issues occurs along the lower Mississippi, and specifically implementation of its 1928 Mississippi River & Tributaries Project.

The channelization program associated with the 1928 Mississippi River & Tributaries Project was focused to meander neck cut-offs within the Vicksburg District in the 1930s, and in the latter 1940s to chute cut-offs between Arkansas City and Memphis, TN. Because of its extensive documentation, the 1930s meander bend cut-off program along the lower Mississippi provides excellent data to examine the hydraulic, hydrologic, and geomorphic adjustment of a large river to intensive engineering.

Implementation of the controversial cut-off program was all but certain following the Great Mississippi Flood of 1927, and thus it was a key part of the 1928 Mississippi River & Tributaries Project. The meander bend cut-off program of the 1930s formally extended upstream (Table 5.5) from Glasscock cut-off in 1933, 48 km upstream of Red River Landing, and finishing in 1942 with Hardin cut-off, 85 km downstream of Memphis. The cut-off program was a grand experiment, as a river engineering project of this scale and intensity had never previously been attempted (Hudson et al., 2008).

Unfortunately the USACE had not thoroughly studied the process of actual channel cut-offs prior to implementation of their plan (Ferguson, 1940). The sole exception occurred along a single meander bend located along the west side of the river between Natchez and Vicksburg. Prior to its cut-off, which curiously occurred during low-flow stage rather than a flood, Yucatan Bend averaged 75.2 m/yr of lateral migration (Table 5.5). Following decades in which the USACE strictly enforced a "no cut-off" policy, Yucatan Bend, an elongated meander neck, was allowed to "naturally" cut-off. The USACE collected copious amounts of hydrographic (survey) and hydrologic (stage) data before, during, and after the cut-off. By 1929 the cut-off was effectively complete, and deemed manageable. Thus, with only two years of study the USACE embarked upon the largest channelization program on Earth.

The channel reach for Mississippi meander bend cut-offs was selected for a couple of reasons. While reducing flooding was the driving motivation, the reach between about Vicksburg to Red River Landing also had the highest rates of natural cutbank erosion and lateral migration along the Mississippi (Hudson and Kesel, 2000). The rapid shifting and shoaling of the river was a constant concern to navigation and steamboat pilots (Humphreys and Abbot, 1863; MRC, 1880; Twain, 1883; Elliott, 1932). When the USACE began its cut-off program most of the meander neck cut-offs were in coarse sandy deposits that rapidly incised and cut-off within a few years. But the pilot channel location along some meander bends occasionally encountered sedimentary resistance. This occurred in the form of stiff Holocene backswamp deposits or Tertiary bedrock. In such cases, the pilot channel would not sufficiently incise, arresting cut-off processes of the meander bend. Such was the case with Giles Cut-off (in 1933), located

Table 5.5 *Channel adjustment of the lower Mississippi River to meander bend cut-offs in association with Quaternary floodplain deposits along channel cutbanks*

Name of meander cut-off[a]	KM u/s HoP[b]	Cut-off year[a]	Bend dist.[a] (km)	Cut-off dist.[a] (km)	Reduced dist.[a] (km)	Migration rate (m/yr): 1880 to cut-off year	Ch. Width (m): 1911	Ch. Width (m): 2016	Quaternary deposits[c] Dominance[d]
Glasscock	552	1933	25.1	7.7	17.4	52.2	690	605	HB
Giles	589	1933	22.5	4.7	17.9	59.0	950	643	HB, Hpm3, Hcom
Rodney	624	1936	16.1	6.6	9.5	76.9	510	2,834	HB, Hpm1
Yucatan	657	1929	19.6	4.2	15.5	75.2	646	1,205	Hpm1, Hchm
Diamond	682	1933	23.5	4.2	19.3	105.2	787	1,051	Hpm1, Hpm2, Hchm
Marshall	721	1934	11.7	5.0	6.8	97.2	829	1,803	Hpm1, Hchm
Willow	745	1934	20.0	7.6	12.4	27.9	644	1,528	Hchm, Hpm1
Sarah	811	1936	13.7	5.2	8.5	70.6	517	1,619	Hpm1, Hchm
Worthington	827	1933	13.0	6.1	6.9	14.6	843	1,865	HB, Hchm
Leland	867	1933	18.0	2.3	15.8	37.1	1,031	3,256	Hpm1, Hchm
Tarpley	871	1935	19.6	5.8	13.8	6.5	700	1,691	Hpm1, Hchm
Ashbrook	884	1935	21.4	3.1	18.3	26.1	533	1,223	Hpm1, Hchm
Caulk	925	1937	27.7	3.2	24.5	12.9	612	645	Hchm, Pvl2, Hpm1
Sunflower	1,006	1942	20.8	4.0	16.7	19.3	536	1,227	Pve2, Hpm1
Jackson	1,011	1941	17.9	3.9	14.0	29.1	799	882	Hchm, Hpm1
Hardin	1,091	1942	30.3	3.1	27.2	53.7	576	1,255	Hpm1, Hchm

Note the large increase in channel width between 1911 and 2016.
[a] From Winkley (1977).
[b] Kilometers upstream of Head of Passes, Louisiana, at mouth of delta.
[c] Quaternary depositional environment being reworked by meander bend at year of cut-off. Depositional environments based on Saucier (1994): HB= Holocene backswamp, Hpm1 = Holocene meander belt 1, Hpm2 = Holocene meander belt 2, Hpm3 = Holocene meander belt 3, Hchm = Holocene channel cut-offs (includes meander neck and chute cut-offs as "clay plugs" or lakes and wetlands), Hcom = Holocene abandoned courses, Pvl2 = late Pleistocene valley train, Pve2 = early Pleistocene valley train.
[d] In order of proportion (~length) along cutbank of meander bend.
Modified from Hudson et al. (2008).

immediately upstream of Natchez. The pilot channel for Giles Cut-off was immediately adjacent to the valley wall that is rimmed with stiff backswamp clay (Saucier, 1994) (e.g., Figure 5.7). The USACE had great difficulty in completing Giles Cut-off. While the pilot channel captured low and average flows within several years, the hard clay prevented further enlargement of the pilot channel and by 1959 the cut-off only carried about half of the high flow discharge. Thus, after the cut-off was opened the USACE was required to dredge an additional ~15,000,000 m^3 of sediment (Table 5.6).

From the perspective of impacts to the hydraulic geometry, the velocity of the discharge pulse increased in the post cut-off era. Velocity measurements prior, during, and after the cut-off period reveals a distinctive pattern of hydraulic adjustment (Figure 5.26). The increase in velocity with discharge occurs at a higher rate during the cut-off period. Channelization of the Mississippi resulted in a permanent change to its velocity regime, as the average velocity in the post-cut-off phase is higher than the pre-cut-off era. Such changes occur because of the increased hydraulic efficiency of the channel, associated with an overall reduction in channel distance and proportional increase in slope (Winkley, 1977; Biedenharn et al., 2017).

5.4.2 Impact of Cut-offs on Hydrology and Flooding

The goal of meander cut-offs is to reduce flooding, and here the hydrologic influence of channel knickpoint incision can be observed with sufficient data collected during the pre- and post-cut-off eras (Winkley, 1977; Toth et al., 1993; Wasklewicz et al., 2004; Biedenharn et al., 2017; Echevarria-Doyle et al., 2020). A comparison of stage levels for a range of discharge magnitudes collected before and after meander neck

Table 5.6 *Development of Giles Cut-off (1933) along the lower Mississippi River, including the progressive increase in flow and sediment dredged*

Date	Natchez Gauge (m)	Flow through cut-off (%)			Dredged (m^3)	Dredge accumulation (m^3)
		Low water	Average	High water		
Initial cut opened 5/25/1933					1,421,213	
6/11/1933	15.3			3	4,083,889	5,505,102
10/8/1934	1.9	10			4,180,681	9,685,783
12/14/1934	6.6		12			
Open for navigation January 1935						
4/23/1935	15.3			20		
8/17/1935	5.4		44			
10/12/1935	1.8	50				
2/25/1936	6.1		40			
5/2/1936	14.1			33		
11/19/1936	6.2		46			
12/5/1936	1.7	60			775,679	10,461,500
1/31/1937	4.2		50			
2/25/1937	17.7			37		
3/13/1937	15.7			42		
7/17/1937	6.6		57			
10/1/1937	1.3	81				
10/18/1937	1.8	76			3,585,392	14,046,968
1938		100	94		1,187,895	15,234,863
1959				58	889,256	16,124,119

From Winkley (1977).

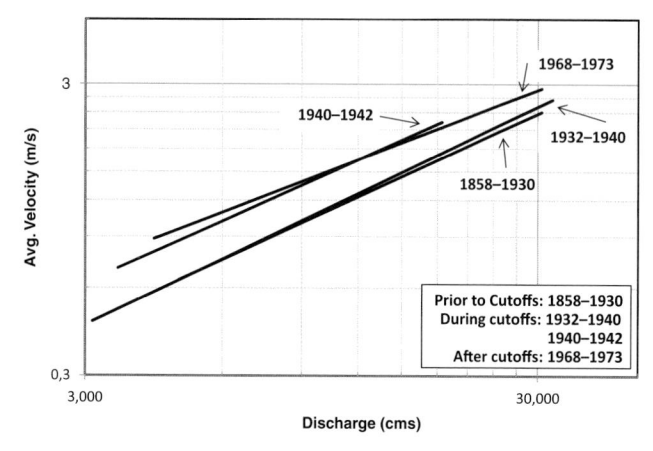

Figure 5.26. Historic changes in average velocity (m/s) with discharge at Vicksburg, MS following meander neck cut-off on the lower Mississippi River. (Source: Winkley, 1977.)

cut-offs shows the abrupt reduction in stage levels (Figure 5.27). At Arkansas City, some ~180 km upstream of Vicksburg the low flow stage data reveals cut-offs resulted in an abrupt ~2–3 m reduction in low flow levels. Prior to the cut-offs, stage levels were increasing, likely because of channel bed aggradation (Hudson and Kesel, 2006; Hudson et al., 2008). But large discharge pulses provide the most insight into the influence of cut-offs on flood hydrology. Indeed, for discharge magnitudes of 39,640 m^3/s and 28,320 m^3/s stage levels abruptly reduced by some ~5 m between 1933 and 1941, and have remained at about the same level up through 2015 (Biedenharn et al., 2017).

The importance of the reduction in stage levels is that the floodplain is less frequently flooded, and inundated for a shorter duration. A comparison of stage levels for the annual hydrograph before (1900–1925) and after (1940–2001) the cut-off program at Vicksburg reveals that average stage levels reduced by ~2 m after the cut-off program (Figure 5.28). Moreover, whereas before the cut-off program flooding began in mid-February and continued to early June, after the cut-offs flooding does not begin until late March and ends by late May (Schramm et al., 2009). Thus, the period of floodplain inundation has decreased by two months, beginning later and ending earlier. Such profound hydrologic changes adversely impact a range of abiotic and ecological processes, such as the connectivity for aquatic organisms that migrate between channels and floodplains with the seasonal flood pulse. Additionally, the shortened inundation period has reduced riparian nutrient sequestration, which is associated with increased downstream phosphorus and nitrogen flux to the Gulf of Mexico and

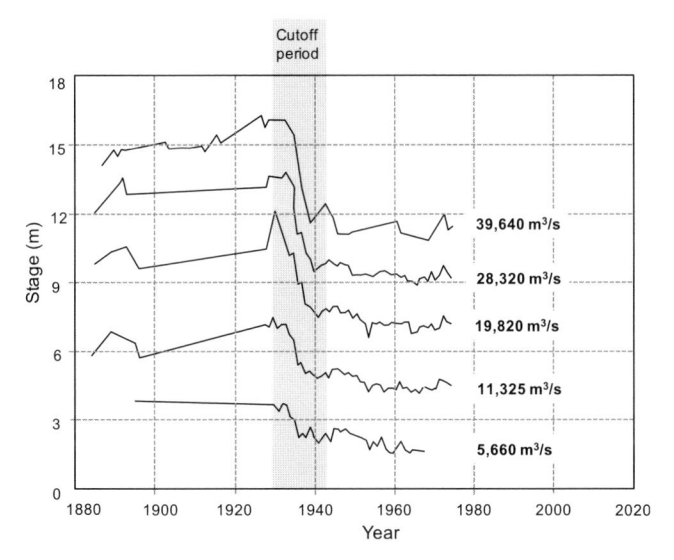

Figure 5.27. Historic adjustment of river stage levels for a range of specific discharge magnitudes, lower Mississippi at Arkansas City, AR (880 KM upstream of Head of Passes, ~20 km downstream of Arkansas River confluence). Flood stage is at 10.2 m. (After Winkley, 1977.)

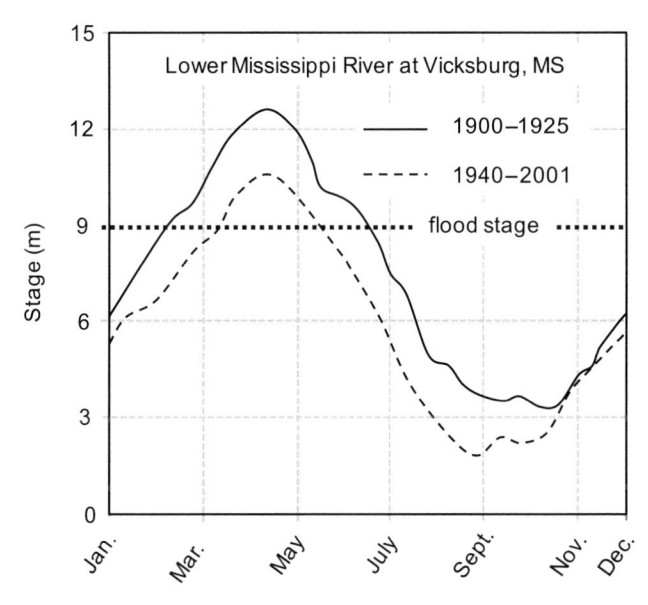

Figure 5.28. Average stage in the lower Mississippi River, Vicksburg, Mississippi. 1900–1925 is before cut-off construction; 1940–2001 is (mainly) after cut-off construction. Horizontal dashed line is bank full stage, the stage at which floodplain inundation begins. (Source: Schramm et al., 2009.)

seasonal growth of the (hypoxic) dead zone (Mitsch et al., 2001; Xu, 2006; Schramm et al., 2009).

Spatially, changes to flooding along the lower Mississippi River varied in response to the cut-offs and other human impacts (Figure 5.29). The reduction in stage levels and flooding was most profound in the middle reaches of the valley, between about Arkansas City to Vicksburg, location for the most intensive channelization efforts (Ferguson, 1940; Winkley, 1977, 1994). The knickpoint zone never reached the northern portions of the valley, upstream of Memphis, TN. Indeed, at Columbus, KY river stage levels actually increased after the cut-offs because of continued aggradation, by more than 1 m, a trend that began after about 1900 (e.g., Hudson and Kesel, 2006; Hudson et al., 2008). And downstream of the cut-off zone, at Red River Landing, following a slight initial reduction in stage, flood levels have actually increased and are now higher than before the cut-offs (Wasklewicz et al., 2004; Muñoz et al., 2018; Remo et al., 2018; Echevarria-Doyle et al., 2020). Additionally, annual low and high stage levels at Natchez, MS (~80 km upstream of RRL) are higher than before cut-offs (Figure 5.30). Higher stage levels between Vicksburg, MS and Red River Landing, LA are possibly caused by the influence of groynes on flow resistance, channel bed aggradation because of the backwater effect from increased rates of sea level rise, and climate change (Chatanantavet et al., 2012; Muñoz et al., 2018; Remo et al., 2018).

5.4.3 Unintended Geomorphic Consequences

While the hydrologic impact of the cut-off program was abrupt and clearly discernible, the knickpoints triggered a suite of unintended geomorphic responses, which required decades to observe, and the channel continues to adjust (Hudson et al., 2008, 2019). That is, the river adjusted in ways that were not intended in the original engineering design. Following the meander bend cut-offs different forms of channel degradation started to occur, which then required other types of management operations. Succinctly put, the management requires managing. It is important to note, however, that the geomorphic adjustments to the lower Mississippi occurred within a distinct sequence, somewhat in agreement with other models of alluvial river adjustment to disturbance (e.g., Schumm, 1991; Simon and Rinaldi, 2006). Thus, the sequence of adjustments following the cut-offs included (1) abrupt (local) channel bed incision, (2) localized downstream channel bed shoaling, (3) upstream knickpoint migration, (4) bank caving and channel bed shoaling, and (5) channel bed widening. Additionally, after some decades sedimentary bars have become vegetated, resulting in the formation of numerous channel islands. Of course, the style and intensity in which these adjustments played out varied reach-by-reach, in part because of the antecedent geomorphology and older sedimentary framework.

In the case of a meander neck cut-off, the sediment produced from the channel bed and bank erosion is transported immediately downstream. This resulted in shoaling and the formation of channel bars, which was problematic for navigation. Thus, the cut-offs necessitated a large-scale increase in channel dredging to ensure that the added coarse sediment would be efficiently transported downstream. Channel knickpoints occur at a local

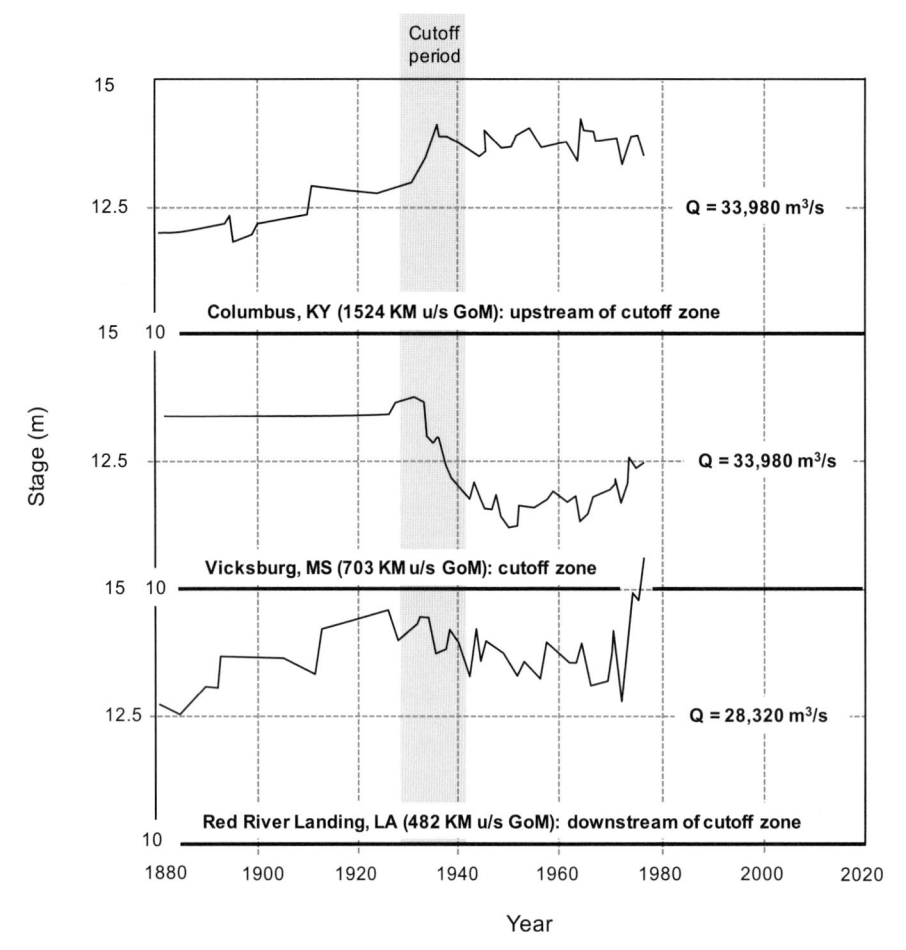

Figure 5.29. Historic changes to flood stage levels in association with the cut-off period along the lower Mississippi River in the 1930s and early 1940s. The gauging stations are for Columbus, KY (upstream of the cut-off zone), Vicksburg, MS (within the cut-off zone), and Red River Landing, LA (downstream of the cut-off zone). (Source: Winkley, 1977.)

channel reach where the slope is over-steepened. In alluvial rivers knickpoints migrate upstream to reduce channel slope. Within about a decade or so, the channel (slope) profile flattens out and stabilizes. And while a single cut-off may be dampened within a decade or so because of other types of mutual adjustments, the occurrence of numerous cut-offs within a relatively short distance and over such a brief period of time was more than the river could accommodate.

Channel bed incision abruptly destabilized the channel banks, as erosion by mass failure (sloughing) became common within the knickpoint zone. It should also be noted that the cut-offs were engineered prior to construction of channel training structures or bank stabilization, such as groynes and concrete revetment. The limited bank stabilization works in place – primarily willow revetment – was largely ineffective. Subsequent to the cut-offs, therefore, the channel was mainly free to adjust to the imposed hydraulic changes. Following channel bed incision channel bank erosion rapidly occurred, especially where dominated by sandy sedimentary materials.

Bank erosion resulted in a much wider river channel. The channel width at Leland bend cut-off, for example, increased in width from 1,031 m before the cut-offs to 3,256 m in 2016 (Table 5.5). The last bend to cut-off, Hardin in 1942, increased in channel width from 576 m to 1,255 m. Giles Cut-off (1933) actually narrowed, from 950 m to 643 m, because the new channel is incised into clayey backswamp deposits resistant to bank erosion and widening.

The observation of the rapid bank caving problem along the lower Mississippi by the USACE was the stimulus for greatly extending channel bank protection by articulated concrete revetment (described above). While concrete effectively is emplaced along most of the lower Mississippi River (Hudson and Kesel, 2006; Echevarria-Doyle et al., 2020), the development of revetment involved decades of trial and error. The design of concrete revetments had already begun in the early 1900s (Markham, 1916), upon the USACE noting a bank erosion problem within the Vicksburg District. The initial installation of concrete revetment along the channel banks, however,

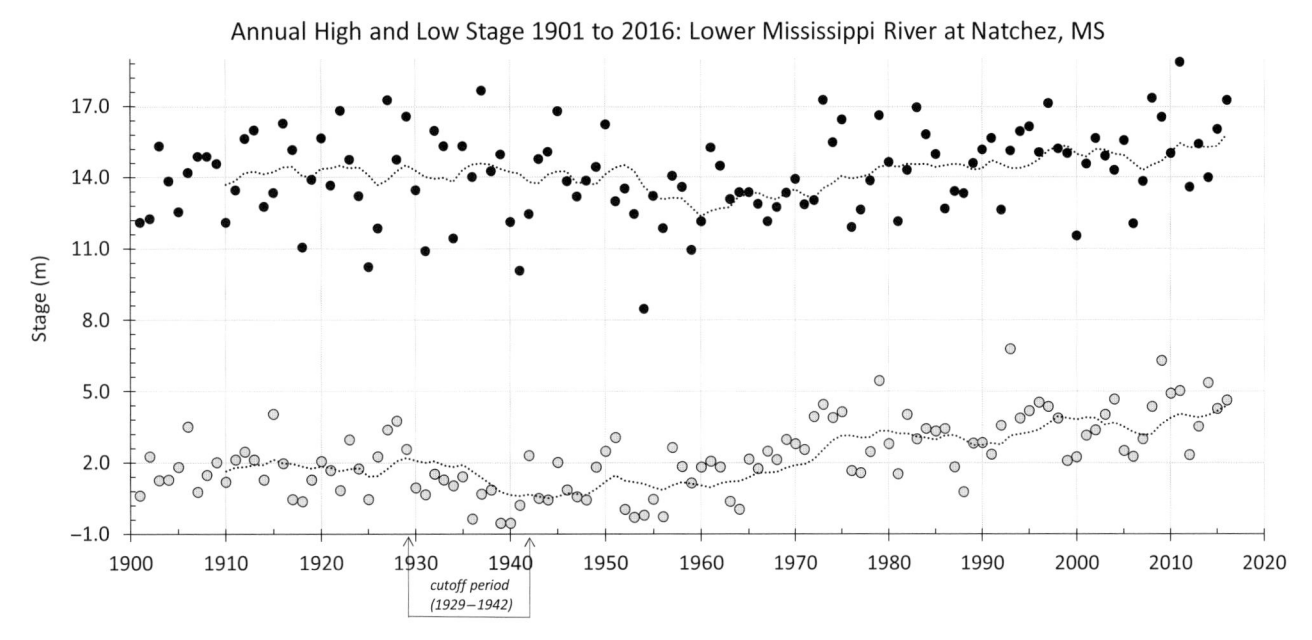

Figure 5.30. Annual high and low stage levels for the lower Mississippi River at Natchez, MS from 1901 to 2016. A ten-year moving average trend line reveals stage level decline associated with the cut-off period and subsequent increasing stage levels, particularly after about 1960. (Author figure. Data source: USACE, Rivergages.com.)

failed to take into consideration key hydraulic processes and sedimentary controls that influence meandering river dynamics. A noted problem was that the revetment did not initially extend to the channel bed thalweg (e.g., Figure 5.16). Thus, high discharge events were able to deeply scour and undermine the cutbanks, at the base of the revetment. Helical flow was accentuated by the smooth concrete, resulting in higher channel velocity at the toe of the cutbank (outside of meander bend). This was especially a problem along sandy channel banks (point bar deposits) because the lack of cohesion resulted in saturated banks sloughing into the channel, regardless of the concrete. By the 1950s the USACE had redesigned the concrete revetment to take into consideration important local-scale hydraulic and sedimentary controls. A massive installation program occurred in the 1950s and 1960s, which included stripping floodplain vegetation and grading channel banks to reduce the slope angle.

In large river basins heavily impacted by humans, multiple types of disturbances are superimposed across different parts of the fluvial system, including basin headwaters and the lower main-stem deposition zone (Schumm and Winkley, 1994; Thorne, 2002; Gregory, 2006; Pinter et al., 2006b). While the revetment program has been effective at reducing bank erosion, it also disrupted the sediment budget of the lower Mississippi River. Natural rates of bank erosion were previously an important source of suspended sediment, as a third of the sediment discharged into the Gulf of Mexico was supplied by the process of bank erosion (i.e., Kesel et al., 1992). Thus, while concrete revetments solved one problem – bank erosion – it greatly reduced sedimentary inputs into the channel and contributed to the channel bed incision problem (e.g., Mosselman et al., 2000; Kiss et al., 2019).

Decadal changes in the channel thalweg in the uppermost section of the lower Mississippi alluvial valley reveals the influence of two major contrasting human impacts: channel bed aggradation caused by upstream sediment supplied from historic land degradation (i.e., Happ et al., 1940, Figure 2.7) and downstream channel incision caused by cut-offs along the main-stem channel (Winkley, 1977; Hudson and Kesel, 2006). A comparison of the thalweg trend between 1930s and 1948 hydrographic survey, for example, reveals the channel without the influence of incision created by the cut-off program, as there were no cut-offs along this segment of the alluvial valley and the downstream knickpoints had not migrated upstream (Hudson et al., 2008). In the lowermost reaches of the Mississippi River, the initial incision that occurred by 1948 (hydrographic survey) was just the beginning (Harmar et al., 2005). The thalweg and associated trend line for 1999 reveals that from ~3 m to ~2 m of incision occurred over about a five-decade period (Figure 5.31). It should be noted that the timing of revetment construction (Hudson and Kesel, 2006) occurred about a decade or so after the large dams were constructed in the upper reaches of the Mississippi basin, along the Missouri River. Since the 1950s, trapping of upstream sediment loads by dams (reviewed in Chapter 4) reduced downstream sediment loads into the lower Mississippi by over 50% (Meade and Moody, 2010).

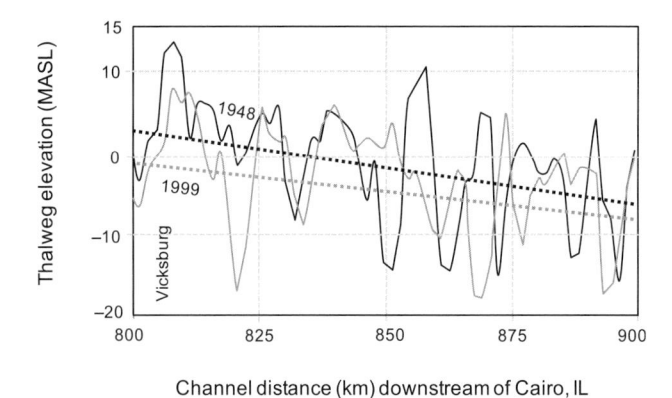

Figure 5.31. Incision of channel thalweg along lower Mississippi River between 1948 and 1999 within the cut-off zone, from upstream of Vicksburg to upstream of Natchez, MS. The 1948 thalweg profile is about ten years after cut-offs, during the early period of incision. The 1999 thalweg data reveals channel incision by an average of 2.5 m. (Source: Hudson et al., 2008.)

Figure 5.32. Historic aerial photo (1:20,000, 1967) of newly developed channel bar along the lower Mississippi River downstream of Leland meander bend cut-off (in 1933). (Photo source: US Geological Survey.)

Thus, the lower Mississippi is responding to multiple types of human impacts.

Adjustment of the lower Mississippi River to hydraulic engineering is not limited to the channel, as other forms of engineering began to have profound influences on other modes of adjustment, with possible consequences to flooding.

The final major type of channel engineering along the lower Mississippi was construction of groyne fields (wing dikes) primarily in the 1970s and 1980s (Smith and Winkley, 1996; Echevarria-Doyle et al., 2020). In response to the navigation problem of having a wide shallow channel in the Vicksburg and Memphis Districts, the USACE sought to "train" the channel with groynes so that the current would be focused in the channel center, creating a narrower self-scouring channel for navigation. Groynes did not need to be installed in the delta, which is much more stable. Groynes are spaced in clusters of 3–10 some ~300–1,000 m apart along sections of the channel that are actively adjusting, or in specific sections to control hydro-sedimentary processes within side channels. In comparison to large engineered European Rivers (e.g., Danube, Rhine, Elbe, Rhone), the placement and spacing of groynes is much less regular along the lower Mississippi, which is likely because it continues to actively adjust its channel.

Increasing flood stages along portions of the Mississippi basin, particularly the lower Missouri and middle Mississippi, stimulated discussion regarding the hydrologic impact of groynes (Watson and Biedenharn, 2009; GAO, 2011; Pinter, 2015a, 2015b; Watson et al., 2013a, 2015a). As physical structures, groynes increase resistance to flow and thus result in higher river stages, especially flood stages. The influence of groynes, therefore, represents an interesting example of an unintended consequence associated with a large flood control project. In the case of engineering along the lower Mississippi, groynes constructed to regulate the channel that was degraded because of flood control engineering (by cut-offs), but potentially have the effect of actually increasing flooding along some channel reaches.

Channel adjustments to the Mississippi continue to unfold even some eight to nine decades after the cut-off period of the 1930s (Hudson et al., 2019). While the US government has not acknowledged the influence of groynes on changes to flood levels (GAO, 2011), other nations are keenly aware of a link. Groynes along the Rhine River in the Netherlands have recently been lowered for the purpose of accomplishing modest (~10 cm) reductions in flood levels (see see Sections 8.2.2, 8.6.3.1).

5.4.3.1 RAPID GROWTH IN VEGETATED ISLANDS

Geomorphically, groynes are contributing to the continued evolution of the lower Mississippi River. A management issue along the lower Mississippi is that groynes have been too effective at trapping sediment. A key impact of the groynes is that, over the past five decades, numerous channel bars and vegetated islands have developed within the main-stem channel of the lower Mississippi, which in some cases is associated with infilling of side channels (LMRCC, 2015; Guntren et al., 2016). The trapping of bed material by the groynes frequently stabilized channel bars, which subsequently became vegetated and trapped further sedimentary deposits. In some instances, islands were initiated as bars formed after meander bend cut-offs (Figure 5.32) by reworking of sediment supplied by knickpoint incision or caving banks (as reviewed above). Sediment stored in channel bars,

Figure 5.33. Islands within the lower Mississippi River (downstream of Natchez, MS), having formed after meander bend cut-offs of the 1930s. The large island is 308 ha, whereas the new (small) island is 6.4 ha within the side channel. Line of sight (arrow) from surface photo view (in Figure 5.34) to active floodplain. The floodplain floods about every year and recently breached a local dike (levee). (Figure: from Hudson et al., 2019. Image: October 11, 2015, Google Earth Pro.)

Figure 5.34. Vegetated island atop recently formed channel bar in the lower Mississippi. The arrow corresponds to that in Figure 5.33, a distance of 210 m to floodplain edge at line of trees in background. (Source: author photo, October 2017.)

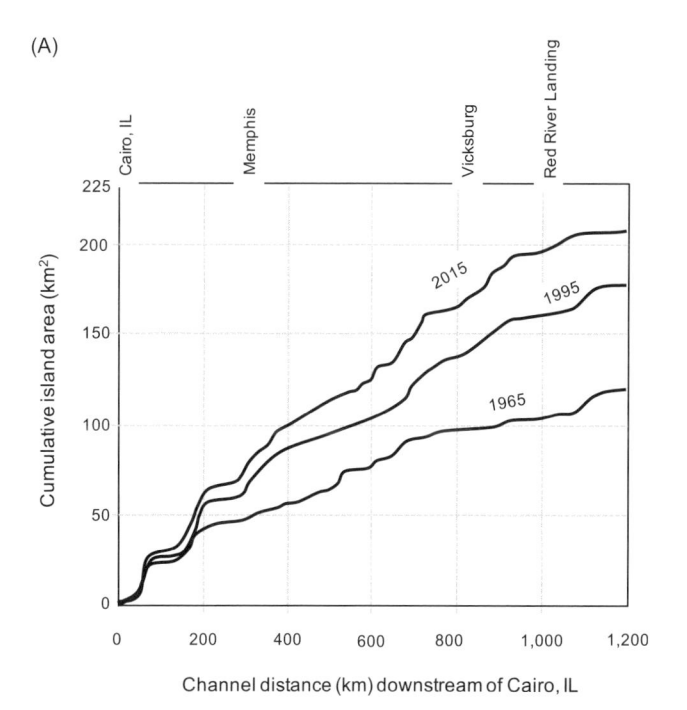

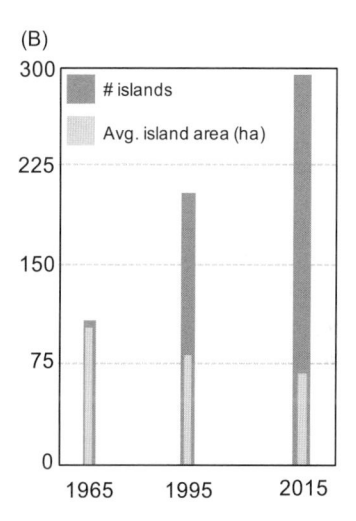

Figure 5.35. Chronology of increase in vegetated islands in the lower Mississippi. (A) Cumulative downstream increase in vegetated islands between 1965 and 2015 (average year). (B) Increase in island number and area. (Source: after Hudson et al., 2019.)

however, was not reworked and assimilated back into the river's active channel bed. Rather, groynes caused bars to stabilize and enlarge, with additional coarse sedimentary inputs as well as overbank (silt–fine sand) deposition (Figures 5.33 and 5.34). Over about the past five decades the number of vegetated islands within the lower Mississippi channel has substantially increased (Hudson et al., 2019), from 112 islands in 1965, 170 islands in 1995, to 204 islands in 2016 (Figure 5.35).

Figure 5.36. Lake, on an island, in the lower Mississippi River, representing important riparian ecosystem values. Bayou Goula towhead island (~35 km downstream of Baton Rouge, LA) is about one hundred years old (Hudson et al., 2019). With time, islands along the lower Mississippi have become more geomorphically and environmentally complex, including a range of water bodies and wetlands that provide crucial habitat for riparian organisms. (Photo source: N. Olsen, by permission.)

The lower Mississippi is currently in a remarkable phase of adjustment, and evolution. As vegetated islands become further established and "welded" to the embanked floodplain with fine-grained overbank sedimentary deposits, woody vegetation encroaches over new alluvial deposits. What was once a dynamic channel becomes a new floodplain surface. Older islands also develop distinctive environmental habitat, including wetlands and small lakes (Figure 5.36), enhancing riparian ecosystem services (LMRCC, 2015). The stabilization and terrestrialization of islands are well established along large heavily engineered lowland rivers in Europe, such as the Rhine, Loire, Elbe, Danube, and Volga. Along somewhat stable reaches of large natural lowland rivers, such as the Amazon, channel islands with extensive lakes and wetlands fringed by natural levees are a staple component of the floodplain geomorphology (Sternberg, 1960; Dunne et al., 1998; WinklerPrins, 2002; Park and Latrubesse, 2017).

Because all large rivers are unique the Lower Mississippi islands will undoubtedly develop differently than islands along other rivers. But government authorities will need to decide how to manage these new riparian lands and address whether they should be allowed to remain as vegetated islands in view of increasing flood levels (Wasklewicz et al., 2004; Muñoz et al., 2018). The USACE has thus far not devoted considerable management to the increased vegetation density associated with the growth of the channel islands (i.e., Hudson et al., 2019; Hudson, 2020), although it does recognize problems with side channel sedimentation and infilling, and has opened up hundreds of notches in groynes (wing dikes) to increase flow and reduce side channel infilling (LMRCC, 2015; Guntren et al., 2016). Floodplain vegetation along the Rhine River in the Netherlands, however, is closely monitored and controlled because of recognition that dense riparian forests increase flow resistance and increase flood levels (Straatsma and Kleinhans, 2018) (see Section 8.3.6).

5.5 CLOSURE: LINKAGES BETWEEN CHANNEL ENGINEERING AND FLOODPLAIN EMBANKMENT

This review has illustrated that channel engineering for flood control influences hydraulic and sedimentary processes that have unintended geomorphic consequences, and which require decades to unfold. For the lower Mississippi this results in a sequence of engineering (management) activities, each designed to address unforeseen management issues that were not part of the original flood control strategy in 1928. Meander bend cut-offs along the lower Mississippi River, for example, resulted in abrupt flood stage reduction as intended. But the increase in channel velocity and flood wave acceleration drove substantial channel bed incision and bank erosion, an unintended geomorphic consequence. The remedy for the latter problem was to use concrete revetment to seal off the river from the floodplain, and to control the channel bed with groynes, which were also to control the channel for navigation. The effects of groyne construction were to increase flow resistance, which initiated the growth of numerous channel bars and vegetated islands, which could be associated with an increase in flood stage (i.e., Wasklewicz et al., 2004; Remo et al., 2018). Thus, along highly engineered lowland rivers the outcome of extensive channel engineering for flood control has been considerable channel degradation, and in some cases increased flood stage.

Interestingly, rapid growth of vegetated islands along the lower Mississippi occurred during a period of substantial sediment load reduction. Thus, the recent phenomenon of vegetated island development along the lower Mississippi illustrates that, as regards fluvial landform construction, one type of hydraulic engineering can offset the other: groynes have countered the influence of upstream dams reducing downstream sediment loads by trapping coarse sediment and building islands (Hudson et al., 2019).

For over a hundred years the vision of flood control for large lowland rivers has included a heavy emphasis on "hard" channel engineering. Much less attention, however, was given to the floodplain deposits in which the channel is formed, or the hydrologic and sedimentary linkages between rivers and floodplains (i.e., flood pulse concept). Indeed, the goal of many flood control systems was to "disconnect" the channel from the floodplain. Such motivations occurred prior to modern scientific understanding of the importance of hydrologic connectivity between rivers and floodplains, even if constrained to a narrower embanked floodplain. New forms of integrated management recognize the importance of the seasonal flood pulse to healthy riparian environments (Chapter 8), its role in floodplain biogeochemical cycles, and that it reduces flood risk.

In Chapter 6 we continue reviewing the impacts of hydraulic engineering to lowland rivers, by transitioning from channels to floodplains, and specifically by examining embanked floodplains that, because of their altered connectivity, have unique morphologic, hydrologic, and sedimentary environments.

6 Dikes and Floodplains: Impacts of Flood Control on Lowland Rivers

[T]he big rise brought a new world under my vision. By the time the river was over its banks we had forsaken our old paths and were hourly climbing over bars that had stood ten feet out of water before ... the dense, untouched forest overhung both banks ... and one could believe that human creatures had never intruded there before.

—Mark Twain (1883)

During the Ohio River flood of 1937 ... a pasture covered with sod gained one-fourth to three-eighths of an inch of silt loam while a clean – tilled cornfield lost the same amount.

—Gilbert White (1945, p. 185)

6.1 DIKES AND FLOODPLAINS

If river flooding is a vital natural process, flood control is a significant environmental impact. In Chapter 5 we reviewed the myriad ways that channel engineering for flood control and navigation impacts rivers, with unintended geomorphic consequences that can take decades to manifest. Internationally, there is a range of hydraulic infrastructure along many lowland rivers designed to "combat" flooding. Traditional rigid engineering structures, unfortunately, were not constructed to interact or sustain the complexity and dynamics of natural biophysical processes common to large lowland floodplains. The sectioning of floodplains into engineered parcels by dike construction, therefore, has adverse consequences to linked fluvial environments and increases flood risk beyond the embanked floodplain (Figure 6.1).

In this chapter, we singularly examine the various influences of dikes to floodplain processes. Specifically, the chapter examines dike impacts to embanked floodplain geomorphology, hydrology, and sedimentology. In combination with the sequence of channel engineering (Chapter 5), flood control embankment drives a continuum of riparian adjustments to lowland floodplains and deltas that requires further management, and in some instances increases the potential for river avulsion.

Flooding wreaks enormous financial and personal cost on societies and is annually the most expensive "natural" disaster (CIRIA, 2013; NOAA, 2019). Because human activities on floodplains adapt to the perception of safety, human flood risk increases after floodplain embankment (White, 1945; WMO, 2004; NRC, 2013; Vergouwe, 2016; Rijkswaterstaat, 2018). After embankment, human settlement and economic activities on floodplains are dependent upon adequate flood control, and dikes are the primary means to assure safety from flooding. Over long timescales the influence of dikes on hydrologic and sedimentary processes, however, can increase human risk to flooding by dike failure. Indeed, the largest flood disasters occur due to dike breach events (e.g., Figure 6.2).

Dikes are passive earthen structures the majority of time, extending for many kilometers along river valleys and deltas. It is only during large infrequently occurring floods that dikes actively influence fluvial processes. Embankment results in unique anthropogenic floodplain processes during flood events (Figure 6.3). Distinctive hydrologic and sedimentary processes thereby influence riparian environments between dikes, outside of dikes, and alongside dikes that results in features unique from natural floodplains (Kolb, 1975; Hesselink et al., 2003). Floodplain hydrology is changed by embankment in several ways, including alterations to floodplain water bodies (hydrography), flood regime (stage, frequency, and duration), groundwater flow, and dike breach events (Jacobson and Oberg, 1997).

The onset of flood control embankment by dike construction sets forward a time-dependent sequence of evolutionary floodplain changes between the dikes, and beyond (Middelkoop, 1997; Hesselink et al., 2003; Hudson et al., 2008). Trapping flood sediment within a much narrower embanked corridor disrupts fundamental patterns of sedimentation and, over long periods, results in overbank aggradation being the dominant mode of floodplain adjustment (Middelkoop et al., 2010). Beyond the dikes, land drainage and settlement drive ground subsidence in flood basins, resulting in flood basin lowering, and requiring further drainage (van de Ven, 2007). The product of many decades and centuries of channel engineering and floodplain embankment, therefore, is a narrow active floodplain characterized by changes to "natural" processes, as well as new fluvial processes that drive embanked floodplain

DIKES AND FLOODPLAINS: IMPACTS OF FLOOD CONTROL ON LOWLAND RIVERS

Figure 6.1. Dike breach and flood event by ice jam mechanisms along the Waal River (Dutch Rhine), February 21, 1799 at Bemmel, ~15 km downstream from NL-DE border. Originally published by C. Josi, Amsterdam 1802. (Source: Rijksmuseum, Amsterdam.)

Figure 6.2. Dike breach and devastating flooding in 1931 along lower Yangtze River, near Nanjing, China. In association with flooding of adjacent Huanghe and Huai Rivers ~181,000 km² of lowland coastal China was flooded for three to six months. The flood was responsible for a combined 3.7 million fatalities and is considered the worst flood disaster in history, among other flood disasters within Chinese coastal lowlands (e.g., 1887, 1911, 1935, 1937, 1938). (Photo by Charles Lindbergh, 1931.)

evolution (Figure 6.4). Because of the association of floodplain processes with specific types of biogeochemical processes, floodplain ecosystem services are also degraded (Nienhuis, 2008, p. 389; Schramm et al., 2009). Over time, the accreted embanked floodplain is "perched" above lower flood basins, increasing the potential for channel avulsion during dike breach events (e.g., Slingerland and Smith, 2004; Zheng et al., 2017).

6.2 FLOOD CONTROL SYSTEMS AND THE CYCLE OF DIKE MANAGEMENT

6.2.1 Dikes and Flood Control

While there are multiple approaches to manage flooding, the oldest and most effective form of flood control technology remains the millennial old earthen dike.

The use of dikes for flood control dates to perhaps the first riparian communities, and extends more than five millennia for complex riparian societies (Hewson, 1860; van Veen, 1948; Butzer, 1976; Kidder and Liu, 2014; Li et al., 2020). Some of Earth's larger rivers, such as the Indus, Nile, Huanghe, and Tigris–Euphrates have been embanked for millennia. Dike systems along lowland rivers enable population expansion, and precede human settlement along large river courses.

It is important to note that the primary motivation to construct dikes is not to protect floodplains from flooding, but rather to develop floodplain lands for agriculture, industry, and human settlement. Dikes provide humans with access to fertile lands and stable water resources for food production, as well as

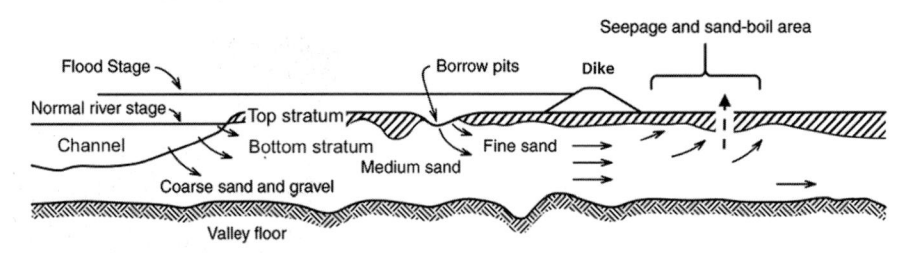

Figure 6.3. Hydrologic pathways and characteristics of a large embanked alluvial river related to flood control and sedimentology, emphasizing hydrologic variation between fine-grained top stratum and coarse-grained bottom stratum. Groundwater flow depicted at flood stage with high hydraulic head. Borrow pit formation is associated with dike (levee) construction and maintenance.

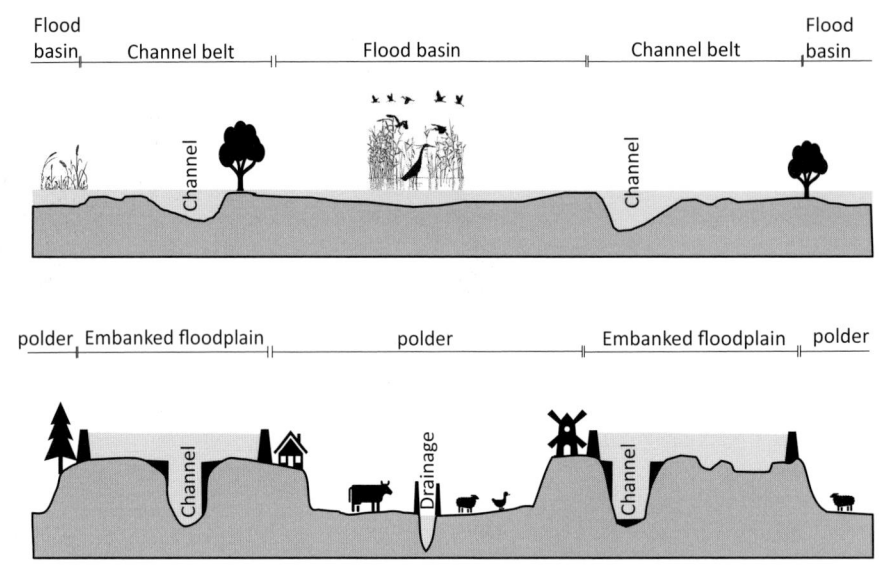

Figure 6.4. Lowland embanked floodplain evolution after dike construction and related drainage for human land use activities over long periods. (After Hesselink, 2002.)

facilitating transportation, economic activities, and residential settlement. Indeed, the benchmark study by White (1945) can succinctly be summarized as "development follows the levee." Nevertheless, competing stakeholders along lowland rivers pressure management organizations to seek a balance between environmental conservation and ensuring the safety of humans residing behind dikes. This is increasingly important in an era of global environmental change where large lowland rivers are populated by increasingly large urban populations, especially in the Asian mega-deltas (Chan et al., 2013).

River dikes are Earth's largest anthropogenic landforms, with some extending for hundreds of kilometers. The longest continuous dike in the world extends 615 km along the west bank of the lower Mississippi River, from the Arkansas River to the Atchafalaya basin. Only the Great Wall of China is a longer continuous anthropogenic landform (USACE, 2000a). As regards the lower Mississippi dikes, collectively the main-stem dike system extends more than 5,600 km along the alluvial valley and delta, and represents about 20% of the total length of registered dikes in the United States.

6.2.1.1 DIKE SYSTEMS AND LARGE FLOOD EVENTS

Dike systems are intricate and require careful planning, analysis, construction, and continuous monitoring and maintenance after completion. In addition to the actual dike structure, other elements of dike systems (Table 6.1) include drainage ditches, relief wells, and pumps to manage underseepage and flood basin inundation landward of dikes (e.g., Van de Ven, 2004; DEFRA, 2005; ILPRC, 2006; Morris et al., 2007; van Heezik, 2008; Rijkswaterstaat, 2009; California DWR, 2010; CIRIA, 2013; DELWP, 2015). Along some rivers, intensive main-stem channel engineering is necessary to protect the dike structure from bank erosion. These options commonly include river stabilization structures such as groynes and revetment (Singer et al., 2008; Kiss et al., 2019), which indeed also serve navigation interests.

Government organizations decide on a specific combination of options to attain the desired flood management goal based on the available technologies, the size and complexity of the area of interest, economics, and stakeholder interests (riparian ecology, cultural heritage, etc.).

Table 6.1 *Common riparian land-based structural measures associated with flood control*

Engineering measures*	Justification	Potential adverse impacts	Environment of impact (channel, embanked floodplain, flood basin)
Flow diversion structures	Flood risk	Channel bed aggradation, overbank burial of wetlands, pollutants	Channel, embanked floodplain, flood basins
Sluice gates	Land drainage	Reduce flood pulse	Flood basins
Ditches	Dike seepage, land drainage	Reduce flood pulse, wetland oxidation, and subsidence	Flood basins
Relief wells	Dike seepage	Reduce flood pulse	Flood basins
Pumps	Land drainage	Reduce flood pulse, wetland oxidation, and subsidence	Flood basins
Dikes (levees)	Flood risk	Reduce flood pulse, embanked floodplain accretion, increase flood risk due to increased settlement, and economic activities	Embanked floodplain

*Channel-based flood control engineering options reviewed in Chapter 5, dams reviewed in Chapter 4. (Author table.)

Table 6.2 *Common types of flood control dikes along rivers*

Type	Purpose and design
Main-stem and tributary dikes	Along main channel and upstream into tributary lowlands; includes flood wall in dense urban settings
Ring dikes	Dikes encircle a small specific area
Setback dikes	Extend landward from a damaged or threatened dike, or a redesigned dike
Local (summer) dikes	Lower, riverward of main-stem dike, mainly for agricultural purposes, designed to variable specifications
Sub-dikes	Small, constructed to control underseepage, enclose a local seepage point (sand boil) on the landward side

Modified from USACE (2000a); Rijkswaterstaat (2018).

As hydraulic infrastructure, dike systems commonly consist of a main-stem dike and a variety of secondary dike types. The complexity of the dike system is related to its design purposes as well as the specific type of dike (Table 6.2). Dikes in rural areas are constructed of earthen (sedimentary) materials obtained from local sources. This represent about 97% of dike types in the United States (U.S. National Levee Database). In denser urban areas where space is limited, flood walls are common and are constructed from either concrete or thick interlocking steel plates driven into the subsurface in either a single or double "I" design. These may be supported by a narrow fringe of earthen material or concrete on either side to prevent scour. The catastrophic 2005 flooding of New Orleans from storm surge caused by Hurricanes Katrina and Rita mainly occurred because of flood walls being pushed over (Section 7.6), as they were improperly designed for the soft peat and organic rich soils comprising the underlying sedimentary foundation (U.S. GAO, 2006a, 2006b; ASCE, 2007; IPET, 2007, 2009; Duncan et al., 2008). Main-stem dikes also extend upstream into tributary basins, and along large tributaries dikes can extend for tens of kilometers because of the backwater effect at confluences. In addition to the main-stem dikes constructed for a specific hydrologic probabilistic frequency (i.e., flood recurrence interval), smaller types of dikes include summer dikes, setback dikes, and dike rings (Table 6.2).

Dikes, as with dams, are critical hydraulic infrastructure that require maintenance and monitoring to assure safety during brief periods in which they are formally (and finally) tested by hydrologic events. The great importance of dikes to safety, economics, and tremendous construction cost suggest dikes would be thoroughly documented, inventoried, and characterized as to their risk of failure. But as is common with aging infrastructure (i.e., Doyle et al., 2008), many dikes are inadequately maintained or neglected (ASCE, 2007). Indeed, national inventories of dikes with information regarding their status and safety are difficult to obtain.

The Netherlands has some ~3,700 km of active flood defenses, which includes dikes for rivers and canals, sea dikes, storm surge barriers, dams, and coastal dunes (Rijkswaterstaat, 2018). The high population density and economic activities in the Netherlands result in dikes being an intricate architectural component of the Dutch landscape, extending along canals into flood basins (Figure 6.5). Successive Rhine River high-water (flood)

Figure 6.5. Canal dike and flood basin subsidence in the Rhine delta (Rijnland), the Netherlands. Higher canal water surface relative to adjacent peat basin, which is bordered by a peat dike. Note abrupt height reduction of ~2 m with a distance of about 15 m from water's edge. (Author photo, 2006.)

events in 1993 and 1995 (recurrence intervals of 50–70 years) revealed weaknesses in the Dutch flood control infrastructure. Thus, dikes were upgraded to higher standards (Figure 6.6), and a period of intensive data gathering occurred to develop comprehensive information regarding the condition and status of each dike segment. This information is utilized to calculate precise flood risk probabilities for a range of forecast climate and land cover change scenarios. This includes specific combinations of factors, such as the dike structure, environmental conditions, land use, economic cost, and human fatalities, among others (Vergouwe, 2016). Nevertheless, considering the many centuries over which the Netherlands has constructed dikes to support a high population, some estimates are that there are thousands of kilometers of undocumented and relict dikes embedded in the Dutch landscape (Lola, 2014).

Large flood disasters can drive fundamental changes to the strategy of flood management (Hudson, 2018). The 1993 and 1995 high water events in the Rhine delta were one of several stimuli to change the vision of flood management in the Netherlands and to adapt the Room for the River strategy (reviewed in Chapter 8). Room for the River is a philosophical shift somewhat away from core tenets of the Delta Plan, a "hard engineering" approach developed in response to the 1953 coastal storm surge flooding in the Netherlands (Hudson et al., 2008). In the United States, large flood events in the Mississippi basin have also had broader – national – implications to US flood management (White, 1945). Several key flood disasters in the Mississippi basin that have substantially influenced policy and practice include the Great Mississippi Flood of 1927 along the lower valley, the Great Flood of 1993 in the upper Mississippi, Missouri, and middle Mississippi Rivers, the 1973 lower Mississippi flood, and the 2005 flooding of south Louisiana and New Orleans by Hurricanes Katrina and Rita (see Section 7.6).

The disastrous failure of the New Orleans flood protection system stimulated intense scrutiny of US flood management,

Figure 6.6. Upgrading Rhine delta main-stem dikes along the Lek River as part of the Room for the River program. These specific operations involved placing a protective clay cap along the dike and extending a clay apron in front and behind (where possible) of the dike to reduce underseepage during flood events. (Author photo, 2007.)

its dike system in particular (ASCE, 2007; NRC, 2013; Hudson, 2018). This resulted in passage of new federal legislation, including the Water Resources Development Act of 2007 (also known as the National Levee Safety Act [NLSA], 2007). The NLSA approved formation of a National Committee on Levee Safety (NCLS) to deliver expert advice to US Congress regarding management of US flood control dikes. An important charge of the Committee was to develop a strategic plan for how to increase the safety of the system of dikes across the United States. Subsequently, the NLSC delivered a report to Congress in 2009 that sought to reshape the policy of flood management for a large nation (NCLS, 2009; Hudson, 2018). Key recommendations of the sweeping 2009 NCLS Report included (1) increase dike flood safety from the 100-year to a 200-year safety level; (2) design dikes around the concept of flood risk rather than static hydrologic levels; (3) establish a national committee to oversee dike safety; (4) provide funding for dike repairs, maintenance, and upgrades; and (5) require inspections with standardized criteria. The flood risk concept, whereby human safety and economic activity are formally considered, has been evolving within the USACE (NRC, 2013) although it is the standard practice in other nations, including the Netherlands (Hudson, 2018).

The intense scrutiny of dike systems in the wake of the 2005 New Orleans flood disaster revealed that the US federal government has woefully inadequate knowledge about the safety status of its dikes (NCLS, 2009; Hudson, 2018). Many US dikes are in deplorable condition and numerous sections are close to failure (ASCE, 2007). This is daunting because there are over 160,000 km of dikes in the United States, although it is not possible to obtain a precise tally. Only 4,500 km of dikes were both constructed and are being managed by the USACE. The USACE maintains an additional 3,500 km of dikes constructed by local interests. And some 15,000 km of dikes constructed by the USACE are maintained and managed by local interests. The majority of US dikes, some 77,000 km, receive no federal oversight in construction or management. About half of US dikes were constructed and are managed by state or local interests rather than to a consistent national standard (NCLS, 2009). Indeed, while the USACE has specific dike construction and management standards (USACE, 1998a, 2000a), otherwise there are no national standards for dike construction and maintenance (USACE, 2018a). Thus, dikes in the United States are managed across a range of governmental scales and standards.

The lack of fundamental information about the status of US dike systems resulted in Congress charging the USACE to develop a master database about the status and characteristics of US flood control dikes. The National Levee Database (http://nld.usace.army.mil/) is intended as a gateway for information regarding the status, monitoring, and maintenance of dikes under the USACE portfolio (Hudson, 2018). The database also facilitates information exchange with relevant stakeholders, such as the National Flood Insurance Program, and researchers. The National Levee Database (as of May 2020) includes 8,611 different levee systems, 45,108 km of levees, and 45,551 different levee structures (Table 6.3). The period of dike construction in the United States peaked in the mid-twentieth century (Figure 6.7). This timing of US levee construction mirrors the timing of US dam construction. A serendipitous observation, with broader relevance, is that the average age of US federal levees is fifty-seven years old, precisely the same average age as US large dams (i.e., Figure 4.8, Section 4.4.1.2). Thus, US riparian environments in the mid-twentieth century were indeed being intensively subjugated by hard engineering structures designed by government organizations.

Table 6.3 *Summary characteristics of U.S. National Levee Database**

Levee systems (#)	8,611 (6,492 are non-USACE)
Total distance of levees (km)	45,108 (22,898 km are USACE Program levees; 22,210 km are non-USACE)
Levee structures (#)	45,551
Average levee age (yr)	57

* Database includes levees constructed and/or managed by the USACE, although it is not comprehensive and continues to be updated.
Author table. Data source: USACE U.S. National Levee Database (https://levees.sec.usace.army.mil/#/, accessed May 2020).

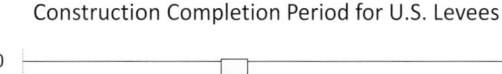

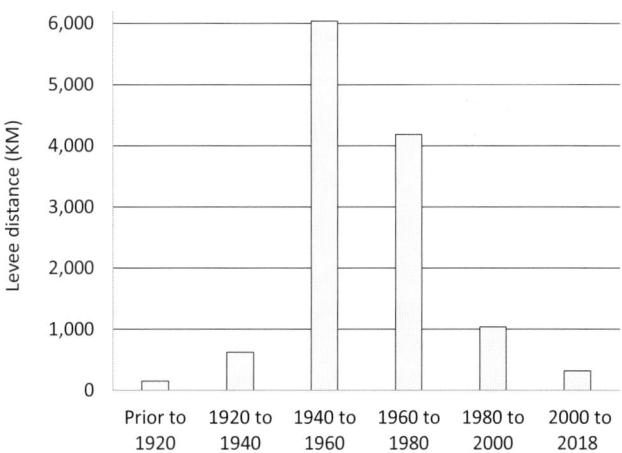

Figure 6.7. Period of US levee (dike) construction in the USACE portfolio, from before 1920 to 2018. Note: construction periods represent original levee construction date or date of significant modification or upgrade. Many levees have a long history spanning many decades. (Author figure. Data source: USACE, 2018a.)

6.2.2 Cycle of Dike Management

Effective management of a flood control system is a large and sophisticated operation (Sayers et al., 2015). The activities of flood control are embedded across different scales of governmental institutions, including local, state/provincial, federal, and international, where rivers serve as national borders. The phases of dike management can be conceptualized along a "cycle" of activities (Figure 6.8) with five distinct phases. Following policy, these include (1) planning and design, (2) construction, (3) monitoring and maintenance, (4) performance assessment, and (5) post-flood operations (e.g., Morris et al., 2007; Pescaroli and Nones, 2016). While the cycle of dike management can be seen as rather generic and applicable to most flood control systems, the specific roles and

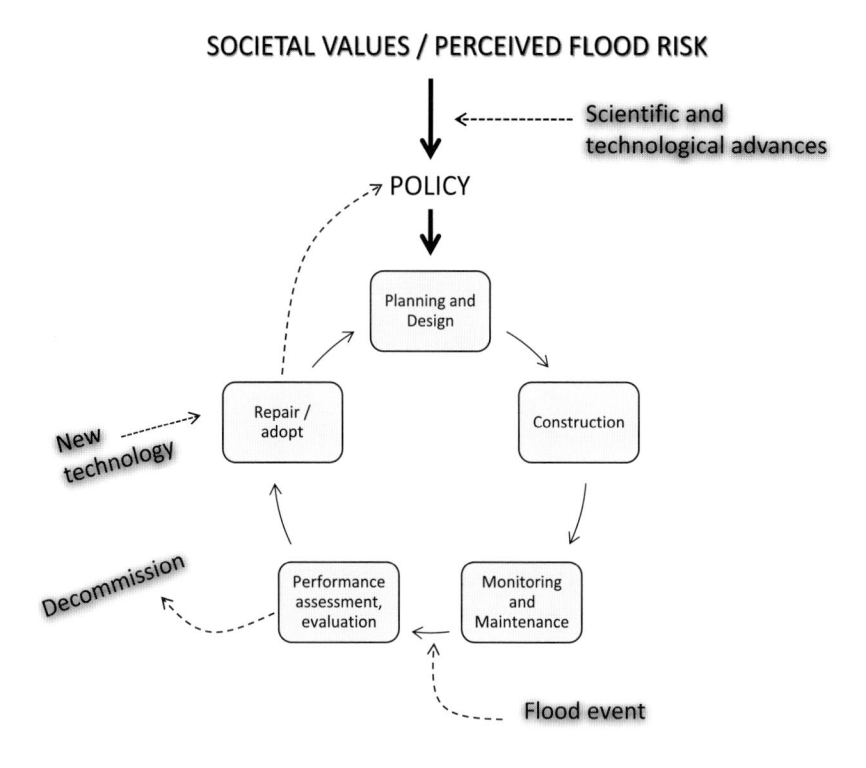

Figure 6.8. Five stages in the cycle of dike management. (Modified from Hudson, 2018.)

responsibilities for associated tasks are often assigned to different governmental institutions and private entities (Morris et al., 2007). Thus, the manner in which the dike cycle unfolds is specific and unique to each river (e.g., James and Singer, 2008).

At the onset of the dike cycle is policy, which develops in response to societal values and national priorities (NRC, 2001). Paradigm changes in policy are common, such as whether federal or local interests should have more influence on the cost and design of dike systems, and often follow major flood disasters (Hudson, 2018). Formally, policies regarding flood protection determine the height (and size) to which dikes are constructed. Dikes constructed to withstand a 250-year flood event, for example, are necessarily higher and larger than dikes constructed to withstand a 100-year flood event. Additionally, the method of flood risk calculation, and whether it is based primarily on hydrologic probability or includes economic and societal dynamics criteria, influences the design of dike systems (NRC, 2001; Vergouwe, 2016; Aerts et al., 2018). Dike design in some nations places a high value on the potential disturbance of endangered species habitat, but less so in other nations. Additionally, stakeholder perception and their level of influence as regards the importance of environmental, agricultural, industry, development, and cultural heritage, among others, greatly influence the spatial configuration of the dike system.

Dike planning and the redesign of existing dike structures is ideally open to advances in science and technology. The design of dikes increasingly incorporates flood risk attributed to global and local environmental change projections, including increased discharge, global sea level rise, and relative sea level rise (Rijkswaterstaat, 2009; USACE, 2009b, 2014a; CIRIA, 2013; Hudson, 2018). New scientific developments in the monitoring of sea level rise, including eustatic (absolute) and relative sea level rise due to ground subsidence, are resulting in an overhaul of some lowland flood control systems, including in New Orleans (USACE, 2019). The rapid development of geospatial technologies, particularly lidar and data derived from frequent UAV (drone) monitoring is increasingly incorporated into all phases of the cycle of dike management (Bakkenist and Flos, 2015; Brauneck et al., 2016).

1. Planning and Design Phase

Aside from small dikes created by local government entities or subsistence communities, the planning and design of main-stem dike systems along larger rivers is carried out by the federal government. This level of involvement is effectively mandated because flooding does not respect provincial and state borders (White, 1945; Barry, 1997). And federal institutions are especially necessary for international rivers, which directly require bilateral cooperation to assure that political borders do not represent the point of dike breach events. Along rivers that serve as national borders, international commissions are often established to coordinate activities over long periods.

Table 6.4 *Important factors to consider in dike design*

Factor	Importance
Hydrology and flood regime	Range in flood stage, duration, seasonality, frequency (recurrence), local-scale sources (yazoo rivers, small tributaries)
Geotechnical	Dike geometry (width, crown, height, slope), dike material (pervious or cohesive)
Valley configuration	Valley width, orientation, gradient, neotectonics, tributary confluences
Floodplain geomorphology and sedimentology	Topography, thickness of top-stratum, secondary channels, paleochannels, channel erosion (process, rates), size of backwater flood basins, sediment type (particle size), sedimentation rate (mm/yr), subsidence (mm/yr)
Human settlement and cultural	Density of settlement, socioeconomic characteristics, native people, cultural heritage (archaeology)
Land use	Property value, types of economic activities, existing infrastructure, utilities, vegetation type and density (i.e., surface roughness)
Environmental	Rare and endangered species, soil type, relation to flood regime, historic pollution (soil contamination)
Groundwater	Hydraulic conductivity, infiltration rate, sources of groundwater (local, main-stem)

After Hudson (2018).

The planning phase takes into account multiple physical and human factors in the design of a dike system that requires consultation with a diverse range of materials and expertise (Table 6.4). A key step in the planning phase involves modeling flood regime changes in response to specified flood safety levels (recurrence intervals), local floodplain hydraulics, and physical characteristics of the proposed dike system, among others. The importance of groundwater flow beneath dikes requires detailed sedimentologic data to estimate hydraulic conductivity to estimate dike underseepage and sand boil potential (USACE, 1998a, b, c; Glynn and Kuszmaul, 2010). At the provincial or state level, cultural heritage surveys are usually mandated prior to landscape clearance, which includes field-based site assessment for archaeological materials (Brown, 2008). Impacts to riparian ecosystems must be specified, and in the case of endangered species a plan to offset the habitat impacts is required to obtain the requisite permits. Key economic factors include property values and the distance to which suitable dike material (soil/sediment) will need to be transported. Combined with land use and floodplain topography, these factors ultimately determine the spatial configuration of the dike system.

Finally, the sedimentary materials, physical dimensions of the dike (height, slope, width of base and crest), as well as features such as drainage ditches, relief wells, and freeboard are considered with regards to the five main causes of dike failure. These include (1) overtopping, (2) surface erosion, (3) internal erosion (piping), (4) underseepage, and (5) mass failure (sliding, slumping).

2. Construction Phase

The actual construction of an earthen dike system along a large river is a massive operation that leaves an extensive footprint upon the riparian landscape. These activities are largely oriented toward preventing erosion by underseepage below the dike and piping through the dike (Table 6.5).

The construction phase often relies upon state, local, and private operators for earthworks (borings, backhoe operations, etc.), mandated to follow federal guidelines to assure the dike is constructed in accordance to specific geophysical criteria (Table 6.5). Here there can be a disconnect between the precision of dike design and that of the activities associated with the actual surface and subsurface environment (Auerbach et al., 2015). Some of the construction activities that can reduce the dike integrity and lead to failure include leaving tree roots at base of dike, penetrating the cohesive top-stratum, and utilizing inappropriate sedimentary fill materials (Table 6.6). Additionally, it is important that the creation of borrow pits, the source of most of the earthen material for dikes, be designed to reduce underseepage or scour of the dike structure (USACE, 2000a). Along large rivers, borrow pit formation requires extensive machinery to scrape off the top-stratum, which in itself can be damaging to the floodplain (Figure 6.9).

3. Monitoring and Maintenance Phase

The monitoring and maintenance phase requires that strict federal guidelines are met so that the dike is not compromised prior to an actual event. Several problems are associated with this phase, including inappropriate land use adjacent to dikes, physical damage to dikes by burrowing animals, the growth of vegetation (especially tree roots) along and atop the dike, settlement of earthen dike materials, cracks in the dike during dry conditions, erosion of dike surface by runoff from rainfall, and subsidence of underlying soils. Critical problems (i.e., slumping) have

Table 6.5 *Geophysical indices of floodplain deposits in relation to dike design to prevent problems with dike failure (underseepage, piping, sliding, erosion, river erosion)*

Index	Purpose	Data source and/or test (setting)
Sedimentary		
Thickness (m) of sedimentary units (top-stratum, bottom-stratum)	Seepage management, availability of suitable material for earthen dike construction	Borings (field), ground penetrating radar (in situ)
Shear strength (τ)	Resistance of floodplain soil and/or dike material	Borings (field), direct or triaxial shear test (in situ or lab)
Atterberg limits (Pl)	Susceptibility of earthen dike material to failure	Plasticity index (lab)
Particle size (μm)	Hydraulic permeability, estimate groundwater flow, cohesive or noncohesive	Borings (field), particle size analysis (lab)
Geomorphic/topographic		
River distance (m) from dike	Local hydraulic pressure, time for seepage/piping to occur	Cartographic/GIS (PC lab)
Borrow pit distance (m) from river and dike	Vulnerability of dikes during flood event	Cartographic/GIS (PC lab)

Based on several sources, including USACE, 2000a; Rijkswaterstaat, 2009; Morris et al., 2007; CIRIA, 2013.
After Hudson (2018).

Table 6.6 *Deficiencies in dikes caused by construction activities*

Deficiency	Possible consequence
Organic matter (tree roots) not stripped from foundation	Differential settlements; shear failure; internal erosion caused by seepage
Highly organic or excessively wet or dry fill	Excessive settlements; inadequate strength
Placement of pervious layers extending completely through embankment	Allows unimpeded through seepage, which may lead to internal erosion and failure
Inadequate compaction of embankment (lifts too thick, haphazard coverage by compacting equipment, etc.)	Excessive settlements; inadequate strength; through seepage
Inadequate compaction of backfill around structures in embankment	Excessive settlements; inadequate strength; provides seepage path between structure and material, which may lead to internal erosion and failure by piping

Source: Table 7.2 from USACE, 2000a.

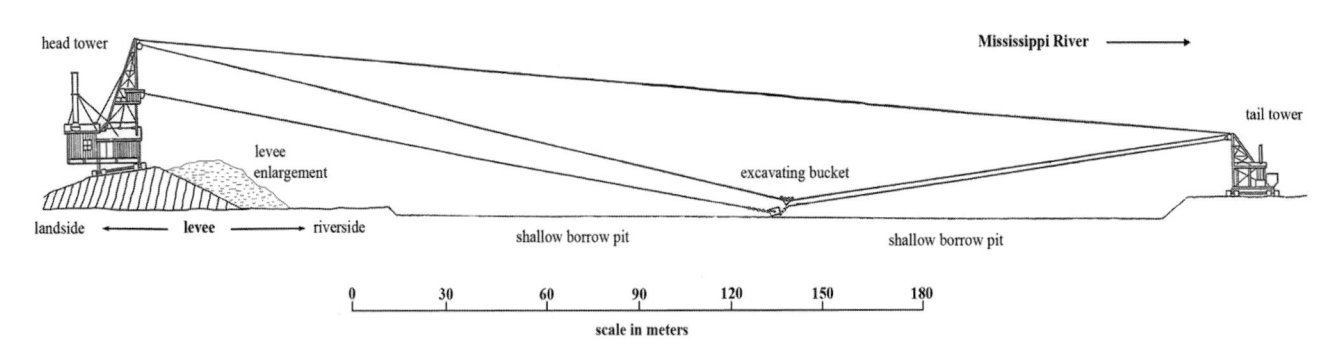

Figure 6.9. The Tower Excavator method of levee construction scrapes cohesive top-stratum floodplain deposits on the riverside, and as a consequence leaves shallow elongated borrow pits. (Source: USACE.)

to be remedied prior to the upcoming flood season. This is an essential phase of the dike cycle and can be challenging because of different governmental organizations being responsible for varying tasks.

Prolonged dry and low periods are not necessarily safe dike conditions, and river managers should be especially vigilant during drought because it also represents an extreme hydrologic event that threatens dike stability. Shrinking and cracking of earthen dike materials during drought increases dike vulnerability to piping and failure during high-water conditions. Dike failure tends to occur during extreme hydrologic conditions, including drought (van Baars, 2005).

This is particularly the case with peat dikes in the Rhine delta in the Netherlands (Figure 6.10), requiring expedient repairs and fortification prior to the onset of seasonal high-water conditions. In 2003 drought triggered failure of a peat dike at Wilnis, NL (Groot-Mijdrecht polder) (Figure 6.10). The dike failure resulted in local flooding of 0.5 m and required evacuation of 1,500 residents, in addition to damaging electrical and gas utilities. While the local water board is formally responsible for dike maintenance, a series of law suits and appeals concluded in 2014 that they were not liable for financial damages.

4. Performance Phase

The performance assessment phase of the dike cycle requires personnel to monitor the integrity of the dike during a flood event using geospatial technologies and field-based operations. The coordination of personnel may be the responsibility of either federal or local-scale government organizations. Federal institutions are usually responsible for approval of a flood fight plan (DeHaan et al., 2012), and also for coordinating specific geotechnical operations during the actual flood. This includes monitoring real-time equipment (well levels), dike sections susceptible to slumping and seepage, coordinating sand bag operations with local contractors and volunteers (Figures 6.11 and 6.12), and public relations.

Figure 6.10. Flood risk by dike failure during dry conditions. (A) Failure of a peat dike in the ring dike of Groot-Mijdrecht polder, Wilnis (NL) during a severe drought, August 26, 2003. Failure occurred after prolonged dry conditions resulted in cracking and weakening of organic dike material, followed by ~10 m of lateral heave (van Baars, 2005). The failure plane occurred at the contact between a ~7 m layer of peat and organic clay atop a sand layer (Photo source: Rijkswaterstaat).
(B) Wetting peat dikes by local water board to reduce cracking and failure risk during severe summer drought of 2018 (i.e., "lessons learned" from the Wilnis peat dike failure). Water boards in the Netherlands are preparing for projections of increased summer dry conditions by increasing dike inspections and implementing measures to mitigate summer drought (Source: Hoogheemraadschap Delfland.)

Figure 6.11. Organization of flood fighters and sand bag operations during the 1995 flood along the Dutch Rhine (Waal River), an event that helped to shape a new vision for flood management in the Netherlands. (Source: Rijkswaterstaat.)

Figure 6.12. Flood wall failure along the London Avenue Canal, New Orleans, pushed over from storm surge associated with Hurricane Katrina. Lateral heave occurred along a failure plane with soft organic clay and peat. To plug the breach required lowering massive (+2,700 kg) sand bags by helicopter. Photo date: September 16, 2005. (Source: USACE New Orleans District.)

5. *Repair and Adapt*

Post-flood operations are initially organized around identification of damaged dike sections (Figure 6.13) with a timeline to complete repairs before the next flood season. Compromised dike sections may be decommissioned and a setback dike (e.g., Table 6.2) constructed on the landward side of the main-stem dike (USACE, 2000b; DeHaan et al., 2012; STOWA, 2012; CIRIA, 2013). After the flood event, the channel bed of larger rivers with considerable hydraulic infrastructure that support dike systems are surveyed to identify problem sections (scour zones), which increases susceptibility to channel bank failure. Management authorities of larger rivers often produce an annual post-flood report that details dike performance and relevant river adjustments. Following larger floods such reports contain considerable hydrologic and geotechnical information crucial to modifying dike management, and of considerable scholarly relevance to characterize embanked floodplain processes (e.g., DeHaan et al., 2012).

Figure 6.13. Dike failure by channel bank erosion, resulting in a ~200 m dike section to slide into east side of lower Mississippi, 125 km upstream of New Orleans River (Ascension Parish). Failure occurred August 23, 1983 after recession of the fourth largest flood of record (at Red River Landing) caused deep channel bed scour. Scale: distance across levee in background from the road to trees is ~60 m. (Source: USACE.)

6.2.2.1 THE PALIMPSEST OF FLOOD MANAGEMENT

Upon being established along a river, dikes evolve such that a cross section represents a palimpsest of natural and anthropogenic floodplain influences (Hudson and Middelkoop, 2015; Hudson, 2018). Changes to dike structures occur for a variety of reasons, including changes in flood regime, intensity of human utilization of embanked lands, changes to dike construction materials, and policy (Rijnland, 2009). These changes often represent new phases in the development of the dike structure, and in some cases literally appear as different sedimentary strata within the dike. The material initially utilized to construct dikes along a river often included local debris and rubble, which resulted in heterogeneous hydraulic conductivity and piping. Dikes along the Sacramento River, for example, were originally constructed by dredging coarse channel bed material that included hydraulic mining debris, and were infamous for piping and breaching (Mount, 1995, p. 209; James and Singer, 2008). Recognition of the problem has resulted in major upgrades that utilize an assortment of materials, including fine-grained sedimentary deposits, concrete and riprap, and riparian vegetation (USACE Sacramento District, 2020).

Changes in the vision and strategy of flood control are often driven by flood disasters that represent hinge-points in the history of flood management, which is ultimately manifest in dike systems. Increased population and economic activities landward of dikes result in increased flood risk and require higher levels of protection, and higher dikes (White, 1945; Pinter et al., 2005). Policy changes stemming from society valuing greater flood protection are a common driver of increased dike fortification. As larger rivers are often more heavily relied upon for commerce, industry, and settlement the changes are manifest on a grand scale with enormous potential environmental consequences. Increased understanding of the process of dike failure by breach and piping in relation to dike sedimentary materials results in changes to dike structures. Higher dikes require compensatory increases in the width of the dike base, with coarser-grained levee material requiring a wider base and more land acquisition, which increases cost. Additionally, substantial cost is associated with fortifying the landward slope with a thick clay drape to reduce problems associated with underseepage and sand boil development.

A New Orleans levee illustrates the palimpsest of flood management continuing to unfold. Since their initial construction in 1717 by the French and with subsequent developments in response to major floods and policy changes, New Orleans levees have grown much higher and wider (Figure 6.14). After the Hurricanes Katrina-Rita flood disaster of 2005, increased scrutiny of New Orleans' levees resulted in some modest improvements and upgrades to the 100-year level of protection, which cost some $14.6 billion. Much of the cost was to reengineer flood walls that had failed by lateral shear (pushed over) because of being emplaced too shallow for the soft organic rich clay and peat (Select Bipartisan Committee, 2006; U.S. GAO, 2006a, 2006b; Seed et al., 2008).

Alarmingly, portions of New Orleans levees will again fall below the federal 100-year recurrence interval level of safety by 2023. By 2073, a combination of levee settlement,[1] ground subsidence, and relative sea level rise will result in New Orleans levees becoming lower by 1.31–1.96 m (Figure 6.15), tremendously increasing flood risk unless some $3.2 billion is spent to upgrade levees to the 100-year recurrence interval level of safety (USACE, 2019). It is worth noting that Rotterdam, NL, which is about the same size as New Orleans, is protected by a flood control system that provides protection to 1:10,000 annual probability. The fundamental difference in flood safety has more to do with societal values than engineering competence.

6.3 SCALE AND PATTERN OF FLOODPLAIN EMBANKMENT

6.3.1 Disconnecting River Floodplains

Floodplain embankment disconnects considerable areas of fluvial lowlands from active flood pulse dynamics (Figure 6.16). Embankments are constructed for specific flood

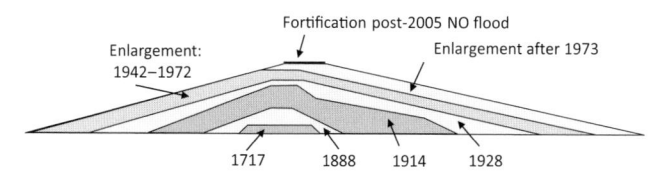

Figure 6.14. Evolution of main-stem Mississippi River levees for New Orleans. Original levees in 1717 were about ~1 m high, 1888 levees were heightened because of several large floods, and in poor condition because of deferred maintenance and repairs following U.S. Civil War. The 1914 levees represent "levees only" policy championed by a strong federal role after formation of Mississippi River Commission in 1880. The levees in 1928 include upgrades following the disastrous 1927 flood, whereas 1942–1972 levee profile represents a large expansion because of major federal flood control "acts" to fortify levees, including the 1965 Lake Pontchartrain & Vicinity, Louisiana Hurricane Protection Project. The levees since 1973 reveal added free-board and backslope stabilization following the enormous flood of 1973. Repairs and modest upgrade occurred after 2005 Hurricanes Katrina-Rita New Orleans flood disaster to work toward completion of 1965 Lake Pontchartrain & Vicinity, Louisiana Hurricane Protection Project. (Source: USACE New Orleans District, with author modifications.)

heights, which is usually expressed by flood recurrence intervals or historic "project floods." The amount of floodplain surface removed from the active flood pulse is dependent upon the height of dikes and indexed to specific flood frequencies such as the 100-year recurrence interval (USACE, 2000a; CIRIA, 2013).

Along lowland river valleys, the area of land inundated by flooding is not linearly related to flood height (Benito and Hudson, 2010). This is because embankment of large lowland rivers is primarily along negative-relief floodplains (e.g., Section 3.5.1). The embanked floodplain area along the river is primarily the active channel belt, with dikes constructed atop natural levees and point bar deposits. This means that the adjacent floodplain surface is at a lower elevation, and commonly includes flood basins separated by older channel belts. Should dike failure occur, substantial floodplain areas are at risk of being inundated by relatively moderate flood events.

The consequences of embankment to floodplain hydrology can be contextualized in several ways, including changes to the width and length of embankments. Additionally, the height of embankment in relation to hydrologic indices, such as flood recurrence intervals, directly relates to environmental change outside of the dike system (Gergel et al., 2002; Pongruktham and Ochs, 2015). That is, the extent of floodplain bottomlands that become disconnected (isolated) from overbank flooding (e.g., Figure 6.16). This is important to consider because it is these lands that are then subject to subsidence because of drainage and agriculture (van Asselen et al., 2009), increasing flood risk when dike breach

---

[1] Levee settlement occurs by consolidation of sedimentary fill materials and fundamentally differs from compaction or ground subsidence. Settlement results in levee height reduction, and mainly occurs soon after construction. For this reason levee height is initially increased according to projected settlement of fill materials, including 0–5% for compacted fill, 5–10 % for semicompacted fill, and 15% for uncompacted fill (USACE, 2000a, Section III). See USACE Engineer Manual No. 1110-1-1904 (Settlement Analysis) for detailed analyses of levee settlement for varying earthen materials and dimensions (grade, height, width).

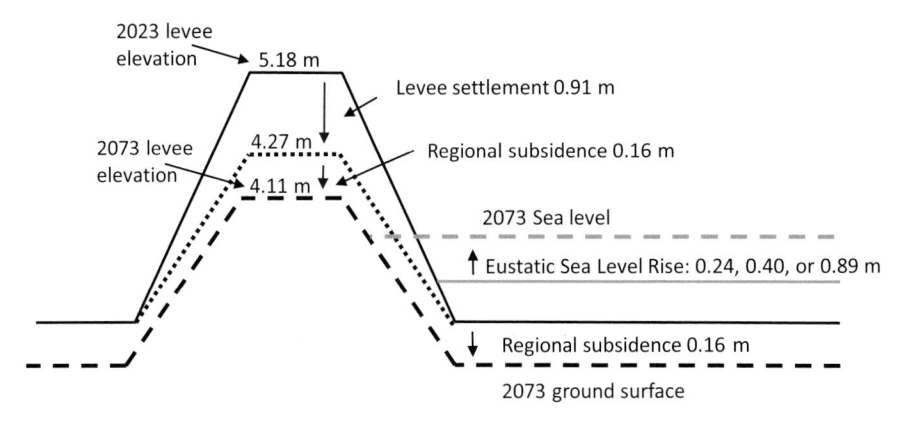

Figure 6.15. Projected changes in levee (dike) elevation in relation to sea level rise between 2023 and 2073 for New Orleans Lake Pontchartrain and Vincity Project (New Orleans dike system). A combination of levee material settlement, ground subsidence, and eustatic sea level rise will result in dikes (levees) becoming from 1.31 m to 1.96 m lower by 2073. (Source: USACE, 2019.)

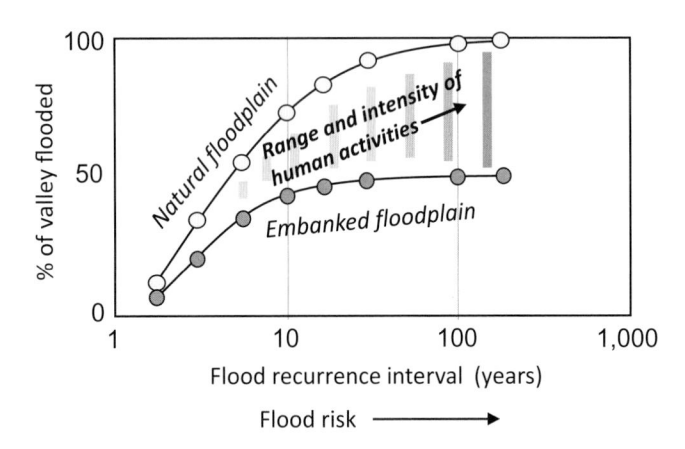

Figure 6.16. Comparison of inundated floodplain area for natural and embanked floodplains at flood recurrence intervals of 2, 5, 10, 25, 50, 100, 200, and 500-year floods. The range and intensity of human and economic activities (esp. agriculture, industry, and settlement) increases with the increasing gap between the two curves. (After Gergel et al., 2002; Benito and Hudson, 2010, with author modifications.)

events do occur. The landward side of the dikes is seldom left in a natural state. Rather, bottomlands outside of the embanked floodplain become available for human use, especially urbanization and agriculture, a direct form of environmental degradation that increases flood risk (White, 1945; Pinter et al., 2016).

Many of Earth's large lowland river valleys in densely populated regions utilized for floodplain agriculture have been degraded by flood control embankment (Higgins et al., 2013). Thus, the proportion of bottomland floodplain subject to flood pulse dynamics and lateral connectivity has greatly reduced. Internationally, the most intensive regions for floodplain embankment by government institutions occur in Europe, Australia, United States, and Asia. To be clear, some form of embankment occurs along most rivers utilized for floodplain agriculture, although in many developing nations minor or local agricultural dikes are common and often not constructed to consistent specifications. Such local dikes are often overtopped and breached by moderate magnitude flood events (WMO, 2004).

6.3.2 The Pattern and Intensity of Floodplain Embankment

The primary control on the pattern of floodplain embankment is the scale of the Holocene (natural) floodplain, particularly valley width. The pattern of embankment varies widely along most larger rivers, with some embanked reaches being very narrow while other reaches are very wide (Figure 6.17). For rivers flowing adjacent to the valley wall, embankment only requires a dike on the opposite side of the river. Along a 245-km-long segment of the lower Mississippi valley between Vicksburg, Mississippi and Baton Rouge, Louisiana, for example, the river flows adjacent to the eastern valley wall that only requires a main-stem dike along the western side of the channel. Factors that result in wide embanked floodplains (i.e., distance between dikes) include zones of active geomorphic processes (e.g., lateral migration, crevasse breaching), confluence of large tributaries, floodplain reaches with frequent sand boils and dike underseepage (requiring setback dikes), and conservation areas associated with critical riparian habitat. Wide embanked floodplains allow floodwaters and sediment to be distributed across a larger area and can somewhat preserve natural riparian ecosystems.

An examination of the pattern of floodplain embankment along most rivers reveals the presence of many narrow reaches where floodwaters and sediment are not distributed across a large area (Figure 6.17). Factors that result in narrow embanked floodplains include geologic controls (resistant lithology), preservation of cultural heritage, urban encroachment onto floodplain lands, transportation corridors (e.g., bridges, roads, railways), and laterally stable channel reaches, especially along deltaic distributaries. The Rhine gorge is a 64-km valley segment between Koblenz and Bingen, DE deeply incised into Rhenish

SPATIAL VARIATION IN WIDTH OF FLOODPLAIN EMBANKMENT

Wide Embanked Floodplain:
Larger tributary valleys, zone of high lateral migration, zone of seepage and sand boil formation, critical riparian habitat (conservation area).

Narrow Embanked Floodplain:
narrow valley (resistant lithology), cultural heritage zone, transportation corridor (bridge), urban encroachment, laterally stable delta distributaries

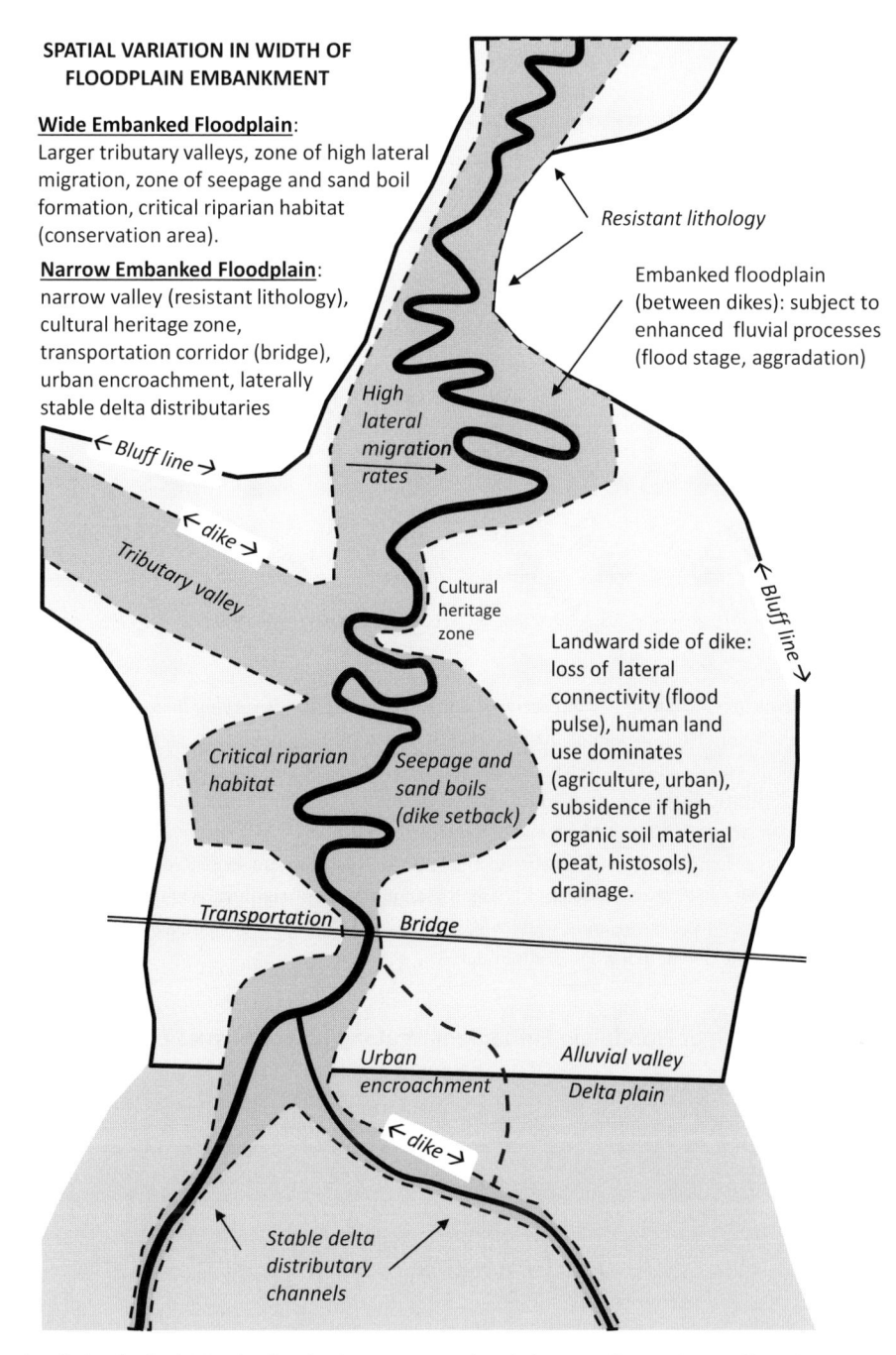

Figure 6.17. Concept of variation in fluvial lowland embankment pattern in relation to valley geology, tributaries, urbanization, bridges, cultural heritage, geomorphology, and sedimentology. (Author figure.)

slate, for example, and requires minimal flood control dikes because of its very narrow floodplain (< ~500 m). Embanked floodplains are particularly narrow at bridge crossings, usually within urban areas. Because of the convenience of access this is often the location of gauging stations with long-term streamflow records. The bottleneck pattern of dikes in these locations does not enable floodwater and sediment to disperse across the floodplain, which can result in higher stage levels during flood events (Heine and Pinter, 2012).

The intensity of embankment can be contextualized by directly comparing the width of the embanked floodplain in relation to the natural (Holocene) valley width. The lower reaches of some of Earth's larger engineered rivers have overall embankment ratios ranging from ~75% to ~90%. That is, most of the valley floor is removed from active flood processes, which leaves only a narrow ribbon of floodplain subject to active fluvial processes (i.e., the embanked floodplain). The lower reaches of the Huanghe and Yangtze Rivers in China, for example, have

Embankment Indices of some larger U.S. Rivers

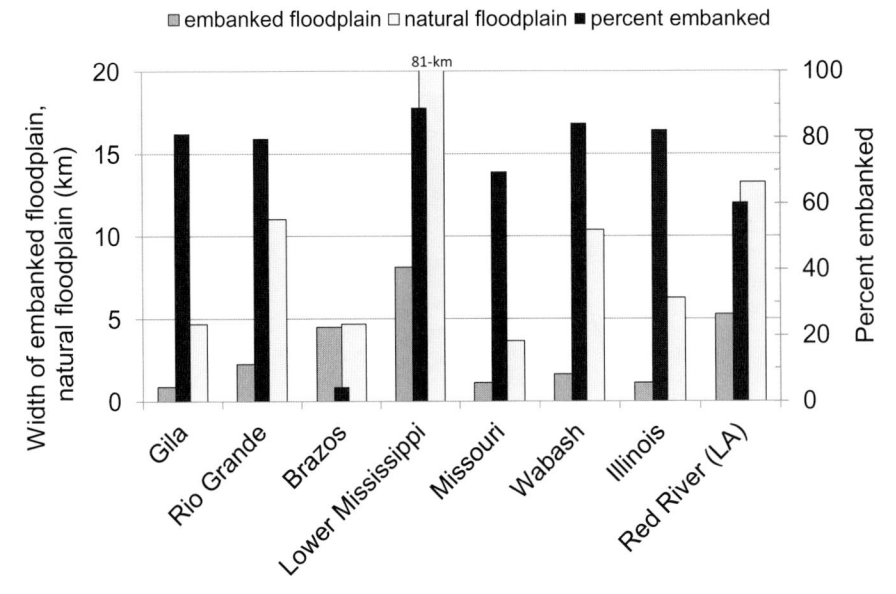

Figure 6.18. Variation in floodplain embankment along lower reaches of several larger US rivers varying in size and physical setting. Data is average width (km) of embanked floodplain (between dikes, including channel), width of natural floodplain (e.g., terrace or valley wall), and percent of embankment (ratio of width of floodplain outside of dikes to natural). Reaches generally extend from river mouth to upstream major tributary along transects within embanked floodplain. Specific distances upstream for each river and # of measurements: Gila: 83.8 km upstream of Colorado River at 10 cross sections; Rio Grande: 190 km upstream of Gulf of Mexico at seven locations; Brazos: 53 km upstream of delta apex at 6 cross sections: lower Mississippi: 757 km upstream of Red River Landing, LA at 109 cross sections; Red: 314 km upstream of Red River Landing, LA at 14 cross sections; Missouri: 195 km upstream of St. Louis, MO at 10 cross sections; Wabash: 83 km upstream of confluence with Ohio River at 10 cross sections (lower 60 km of Wabash has few dikes); Illinois: 244 km upstream of middle Mississippi at 11 cross sections. (Data sources: U.S. Geological Survey topographic maps [1:24,000, 1:100,000] and Google Earth Pro.; author data.)

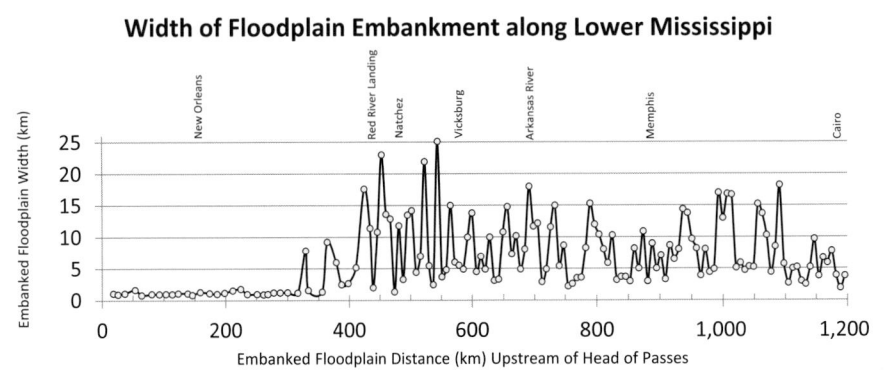

Figure 6.19. Variation in floodplain embankment width along the lower Mississippi alluvial valley and delta. Embanked floodplain width measured tangent to the axis of the embanked floodplain in a GIS. (Author data and figure.)

embankment ratios over 80%, similar to the main-stem Rhine River in Germany and the Netherlands (author data).

Larger rivers in the United States often have an embankment ratio of 60% or higher, and the lower Mississippi is the highest (Figure 6.18). The lower Illinois and lower Missouri have embankment ratios of 82% and 70%, respectively. And some large US rivers have low embankment indices. The Brazos River, the largest river between the Rio Grande and lower Mississippi River, has an embankment index of only 4% (Figure 6.18). The Brazos is mainly embanked within the delta and few dikes exist within the alluvial valley, as the floodplain is largely used as cattle rangeland and is sparsely inhabited. The enormous lower Ohio River has an embankment index of only 29% because of the valley being incised and relatively narrow, with long reaches that do not have dikes.

The lower Mississippi River has an embankment index of 89% within the alluvial valley, which is primarily for the purpose of supporting an enormous agricultural system. The pattern of floodplain embankment for the lower Mississippi is somewhat similar to the pattern of lateral channel migration (i.e., Hudson and Kesel, 2000), and averages 7.3 km in width from Cairo, IL to Vicksburg, MS and 12.7 km from Vicksburg to below Red River Landing. In the delta, embanked floodplain width averages less than 2 km for the last ~300 km until Head of Passes, LA (Figure 6.19). Reach-scale variability in embankment width occurs at large tributaries, such as the Arkansas River, and also narrows where the river is impinged against the valley wall. In the delta, channel adjustment is restricted by cohesive clayey deposits that results in low rates of lateral migration (Kolb, 1963; Hudson and Kesel, 2000), enabling dike emplacement adjacent to channel banks. In addition to geologic inflections (hard points) that constrict the natural alluvial valley, narrow points frequently occur at the location of bridges and urban areas, with potential consequences to flood hydrology. The embanked floodplain at Natchez is only 1.3 km in width, the narrowest embanked floodplain width along the lower Mississippi alluvial valley. The average annual range between annual minimum and annual maximum flow levels (stage) at Natchez is 14.3 m,[2] the greatest range along the lower Mississippi.

6.4 HYDROGRAPHY OF EMBANKED FLOODPLAINS

6.4.1 Changes to Flood Processes

Flood stage is exacerbated by embankment because of floodwaters being constrained to narrower floodplain. The influence of dikes on flood stage was a long-standing debate within the river engineering community. The argument originally put forward by a number of prominent scientists, including Humphreys and Abbot (1863), was that floodwaters contained within a narrower embanked floodplain would cause the river channel bed to incise, thereby reducing flood stage. While there is some theoretical rationale for this position – increased flow depth directly increases channel boundary shear stress – the empirical evidence is not supportive. In contrast, it is increasingly accepted that embankment modestly increases flood height (Belt, 1975; Baldassarre et al., 2009; Pinter et al., 2016; Remo et al., 2018; Tran et al., 2018), although the increase in flood height can be compensated (buffered) by upstream dams and other forms of hydraulic engineering (Pinter et al., 2006b; Remo et al., 2018).

A key role of large river floodplains is floodwater conveyance. Thus, after floodplain embankment floodwaters must go somewhere, and they often go up, as stage increases to compensate for reduced flow width. Flood stage for the lower Po River, Italy (at

[2] Between 1935 and 2016 based on annual maximum and minimum stage levels. Data source: Rivergages.com

(A)

Adjustment of flood depth with length of dike system of Po River, Italy

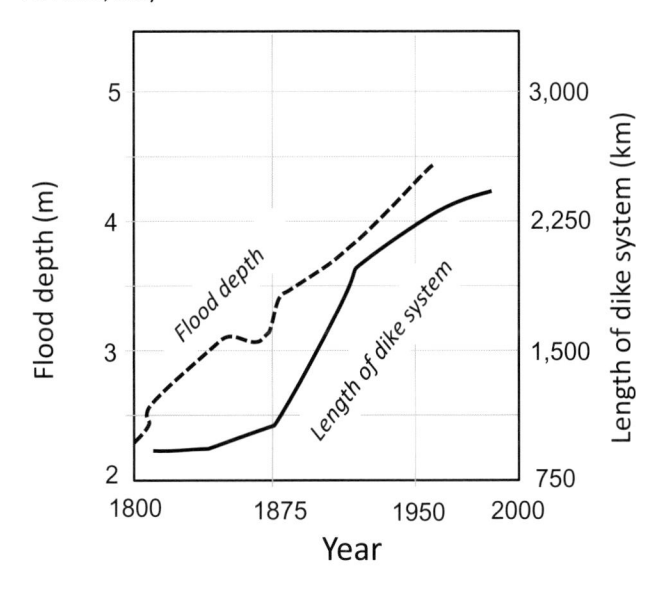

(B)

Comparison of channel cross-sections of Po River at Pontelagoscuro Italy: 1878 and 2005

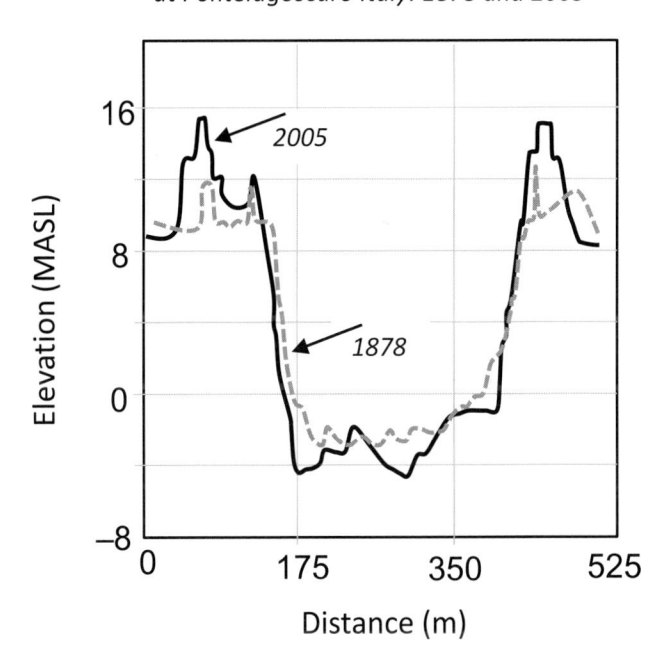

Figure 6.20. Changes in flood depth for lower Po River, Italy with (A) extension of dike system and (B) changes to channel cross sections at Pontelagoscuro, Italy (upper delta plain) showing higher dikes. (After Baldassarre et al., 2009.)

Pontelagoscuro) showed a progressive increase of ~2 m in flood height over a couple centuries with increasing extent of embankment and increased height of dikes (Figure 6.20). The degree that

Wabash River, IL (USA) at Mt. Carmel (USGS 03377500)

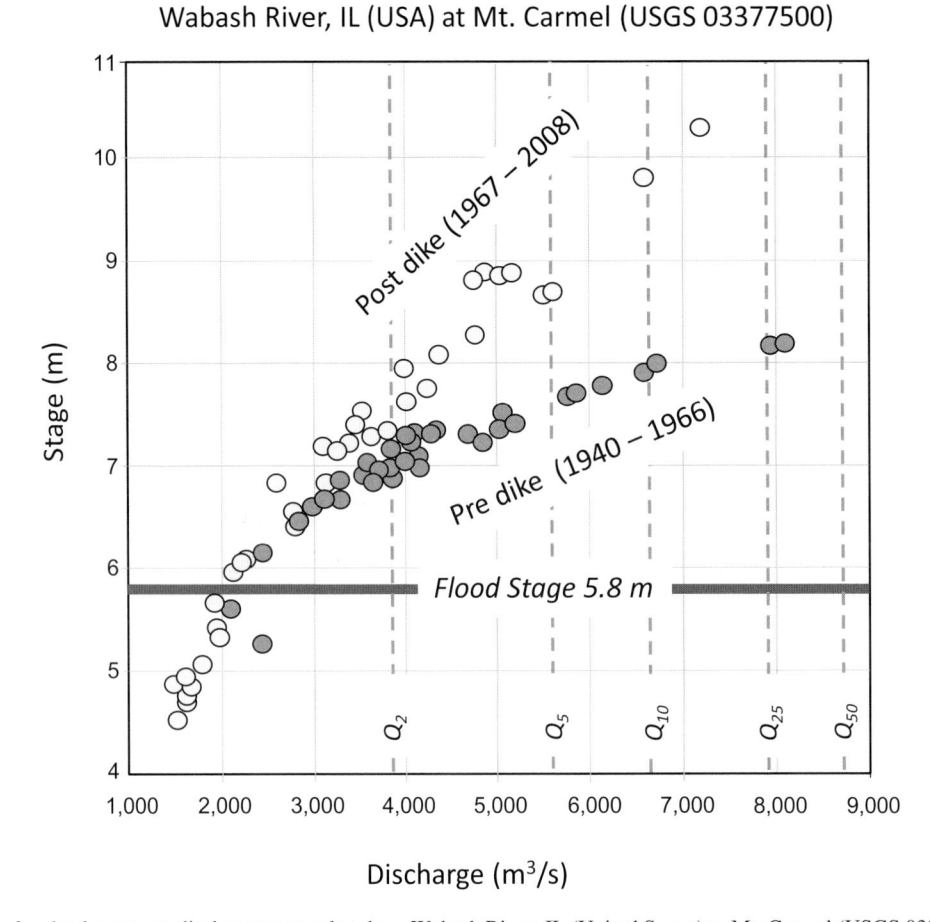

Figure 6.21. Impact of embankments on discharge-stage plot along Wabash River, IL (United States) at Mt. Carmel (USGS 03377500), drainage area of 74,164 km^2, showing divergence and higher stages after embankment, especially for larger floods. The modest ten-year flood event is almost +2 m higher after embankment. (Source: Heine and Pinter, 2012.)

flood stage increases, however, depends upon a range of factors related to the flood hydraulics, such as the change in the width of the embanked floodplain, channel and floodplain boundary roughness (vegetation, topography), slope, and the intensity of channel engineering (Matthai, 1967; Heine and Pinter, 2012; Nagy et al., 2018). Thus, flood stage does not always increase after embankment because other types of channel engineering (e.g., Chapter 5) activities influence the hydraulic geometry, which possibly results in channel bed incision and flood stage lowering (Winkley, 1977; Pinter et al., 2006a; Biedenharn et al., 2017; Echevarria-Doyle et al., 2020).

Examining the influence of embankment on flood height is made challenging because adequate streamflow records often do not precede embankment. In more recently embanked rivers, however, the influence of dikes on flood stage levels can be discerned if stream gauges were installed prior to embankment. And ideally streamflow data is available at a sufficient spatial resolution, from upstream to downstream of the embanked section of interest (Figure 6.21). For a range of lowland rivers in the Midwestern United States, dike emplacement resulted in a clear increase in flood stage after embankment (Gergel et al., 2002; Heine and Pinter, 2012). The influence of dikes (narrow embankment) on stage is more important with larger floods, which is also associated with a slight increase in shear stress and velocity. The influence of dikes on flood stage is limited to the reach scale, occurring immediately within the embanked reach and also somewhat upstream. An important finding is that flood stages remain at pre-disturbance levels downstream of embanked floodplains where floodwaters are able to flow across the natural floodplain (Heine and Pinter, 2012).

These findings are consistent with recent intensive dike construction in the Mekong delta. Between 2000 and 2011, dike construction in the upper Mekong delta resulted in a significant increase in flood levels by 0.68 m (Tran et al., 2018). And, as similar to the Wabash River (United States), downstream reaches in the lower delta that were not embanked did not undergo a significant increase in flood levels (Tran et al., 2018). Combined, these findings from two very different lowland rivers are important because of having direct implications for integrative approaches to floodplain management – specifically, to more

Table 6.7 *Natural and anthropogenic water bodies within embanked floodplains*

Lake type	Characteristics	Geomorphic setting: channel belt, flood basin
NATURAL		
Oxbow (meander and chute)	Natural cut-off with hydrologic connectivity varying with sequence of infilling that is accelerated because of embankment	Channel belt
Topographic lake	Depression: between higher natural levees and alluvial ridges. Valley margin: between bluff line of valley wall and other floodplain topography; includes spring sapping from valley wall	Channel belt, flood basin
Abandoned channel	Channel course or reach created by avulsion	Old channel belt within flood basin
Swale	Point bar accretion and lateral migration	Channel belt
Crevasse / scour	Overbank flood incision into floodplain and accretionary channels (with natural levees) within flood basin	Channel belt, flood basin
ANTHROPOGENIC		
Oxbow (meander and chute)	Engineered cut-offs, function similar to natural cut-offs (described above)	Channel belt
Dike breach	Dike failure and floodplain scour (varies by cohesive or sandy floodplain deposits)	Channel belt, flood basin
Seepage ponds	Dike underseepage and scour on landward side	Channel belt, flood basin
Borrow pit ponds	Adjacent to dike, excavated for dike construction and maintenance (berms possibly reduce connectivity)	Channel belt, flood basin
Canals and ditches	For irrigation, drainage	Channel belt, flood basin

Source: Author table.

precisely design dike construction and dike setback or removal for flood control and restoration activities.

6.4.2 Hydrologic Connectivity of Embanked Floodplain Water Bodies

Hydrologic connectivity of embanked floodplains occurs between the main-stem channel and floodplain water bodies that range from open lakes with continuous connectivity (lotic conditions) to enclosed lakes and wetlands that are only seasonally inundated or connected with large discharge pulses. For engineered rivers that have incised and undergone a reduction in flood frequency and duration, seepage through channel–floodplain conduits becomes a critically important floodplain process. And upon embankment with dikes, floodplain lakes outside of the dikes effectively change from lotic to lentic water bodies, which makes groundwater connectivity with the main-stem channel via coarse-grained channel belt deposits an especially important hydrologic mechanism (Tockner et al., 2000; Thoms, 2003).

6.4.3 Types of Embanked Floodplain Water Bodies

The hydrography of embanked floodplains is comprised of natural and anthropogenic water bodies (Table 6.7). Within the context of an intensive humanly impacted landscape, floodplain water bodies are especially important to the environmental integrity of embanked floodplains (Shields and Knight, 2013; Pongruktham and Ochs, 2015; van der Most and Hudson, 2018). This is particularly the case where channelization (cut-offs) has reduced flooding such that true overbank (natural levee) floods seldom occur. Because of their importance as aquatic habitat for riparian flora and fauna, riparian ecologists have prioritized conservation of lakes within embanked floodplains (Ward and Stanford, 1995; Shields and Knight, 2013). Government agencies and environmental organizations, such as The Nature Conservancy, are directly involved in projects to manage hydrologic connectivity of water bodies in embanked floodplains to conserve and restore riparian ecosystems.

Embanked floodplain water bodies evolve along an accelerated continuum, beginning with their initial formation as an open water environment (e.g., Gagliano and Howard, 1984; Shields and Abt, 1989; Middelkoop, 1997; Dépret et al., 2017). The sequence of change – from open water to infilled wetlands – occurs more rapidly within embanked floodplains because of higher rates of sedimentation. Flood deposits trapped within the narrow embanked riparian corridor subsequently infill floodplain lakes, ultimately resulting in the setting being dominated by hygrophytic wetland and even terrestrial vegetation, depending on changes to the hydrologic regime (Dépret et al., 2017). Between these two endpoints,

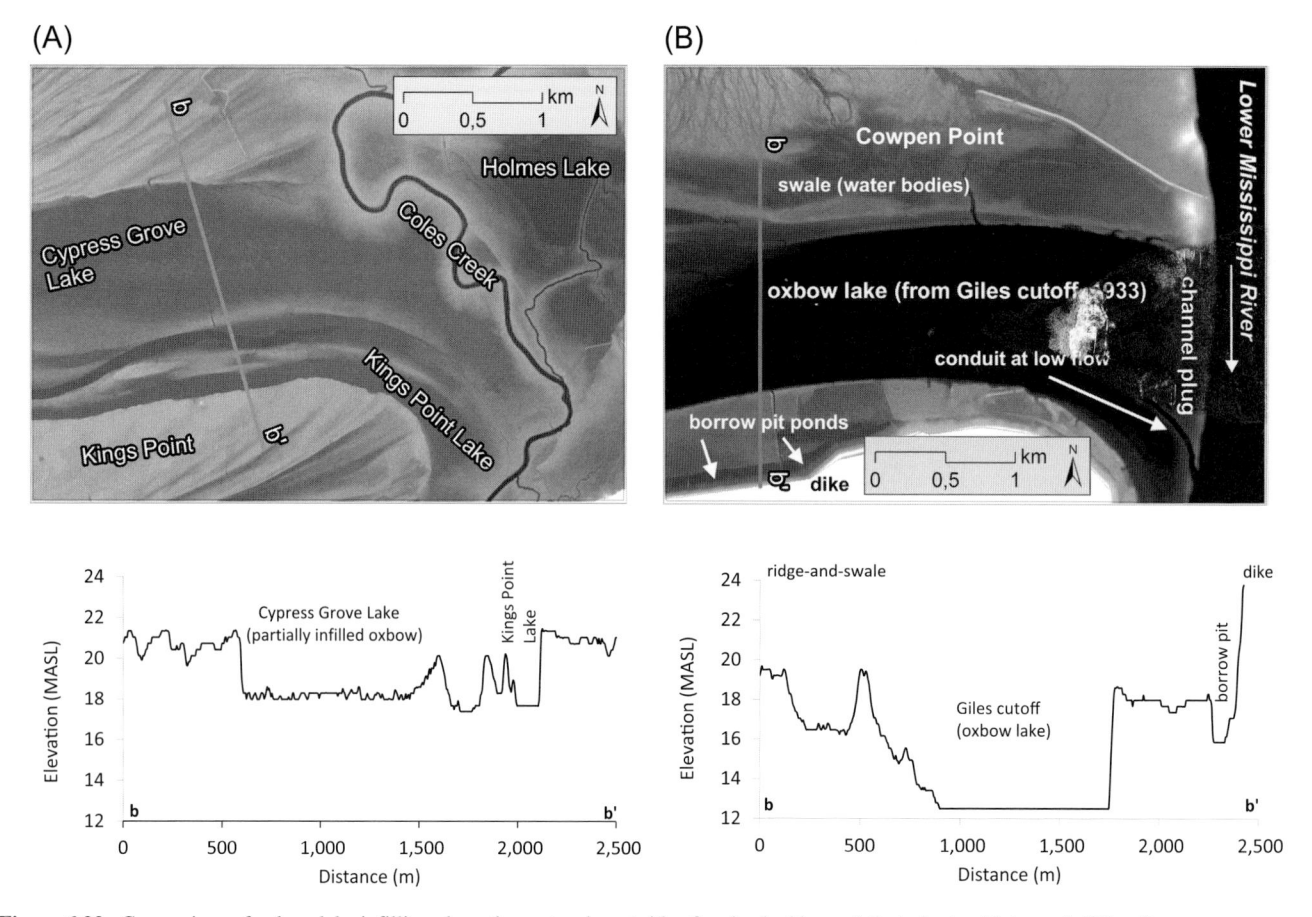

Figure 6.22. Comparison of oxbow lake infilling along the east and west side of embanked lower Mississippi, with lower infilling for more recent cut-off. A: (upper) Waterproof cut-off (in 1884) on east side of valley is being infilled by main-stem as well as local sediment from Coles Creek, which drains adjacent bluff with eroding loess (20 km upstream of Natchez). Waterproof cut-off is considered a natural cut-off, although local residents unsuccessfully attempted to hasten the inevitable cut-off by digging a ditch across the narrow meander neck. (lower) Elevation profile (transect) across cutoff lake and relict meander, revealing variable topography in association with lake infilling. B: (upper) Artificial oxbow lake showing different topographic and environmental features associated with Giles meander neck cut-off (in 1933) and dike embankment (4 km upstream of Natchez). Note small connecting channel (at moderate flow levels) within the channel plug fill. (lower) Elevation profile (transect) across lake and floodplain, revealing variable topography, including borrow pit ponds immediately adjacent to the dike. (Author figure. Data source: Lidar data of State of Louisiana https://atlas.ga.lsu.edu/.)

however, floodplain water bodies range across a spectrum of hydrologic and ecological functioning with varying proportions of open water and wetlands, and have different signatures of hydrologic connectivity (e.g., Shields and Abt, 1989; Hudson et al., 2012; Biedenharn, 2017).

6.4.3.1 NATURAL WATER BODIES

Embanked floodplain water bodies formed by natural processes include (a) oxbow lakes (meander neck and chute cut-off), (b) meander scroll sloughs (swale), (c) abandoned channel (avulsion), (d) depression (neotectonics, deposition), (e) valley margin, and (f) crevasse (overbank scour).

Along lowland rivers, processes of lateral migration form oxbow lakes and arcuate sloughs, respectively, two important mechanisms for the formation of floodplain water bodies (see Chapter 3). Along anabranching river floodplains, the process of avulsion and overbank flooding is the dominant mechanism for lake formation, resulting in elongated arcuate lakes in old channels and smaller lakes associated with scour and crevasse formation. Other water bodies less common, but environmentally important, along large lowland floodplains are depression (sag) lakes and valley margin lakes (Section 3.4.2). Such lakes are topographically formed in flood basins, between high natural levees of the channel belt and the valley wall. The formation of such lakes can be influenced or created by neotectonic subsidence and valley floor warping (Guccione et al., 2000). Water bodies located against the valley wall are commonly supplied water and sediment by small tributaries draining the bounding terrace, as well as spring sapping at the base of the bluff line (Figure 6.22A).

DIKES AND FLOODPLAINS: IMPACTS OF FLOOD CONTROL ON LOWLAND RIVERS

Figure 6.23. Borrow pit pond adjacent to main-stem dike (levee) along lower Mississippi River at Smithfield Landing (483 km upstream of Head of Passes, LA). Photo taken during recession of spring flood wave (note flood line), with floodwaters immediately against dike. (Author photo, March 2010.)

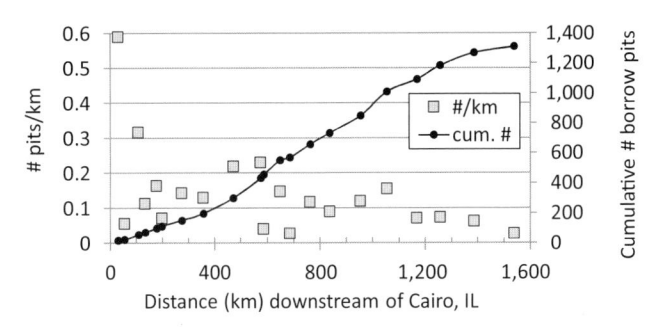

Figure 6.24. Tally of borrow pits larger than 1 acre (0.4 ha) along the main-stem dike line of the lower Mississippi River. (Author data.)

6.4.3.2 ANTHROPOGENIC WATER BODIES

Anthropogenic processes that form water bodies within embanked floodplains include (a) oxbow (meander neck and chute cut-off), (b) wielen (dike breach), (c) borrow pit ponds, (d) groundwater seepage ponds, and (e) various types of drainage canals and ditches. Because along many lowland rivers artificial meander bend cut-offs are part of a larger strategy of flood control, anthropogenic oxbow lakes result in large proportion of embanked floodplain water bodies (e.g., Table 6.7).

A range of flood control engineering ultimately results in formation of several types of unique anthropogenic water bodies. Borrow pits, dike breach ponds, and seepage ponds are directly associated with the actual dike structure. While borrow pit ponds are created by dike construction, dike breach ponds are created by dike destruction. Along some lowland rivers high rates of groundwater seepage can form a seepage pond on the landward side of the dike. Additionally, canals and drainage ditches are commonly associated with agricultural activities or settlement, which are factors that initiated dike construction for floodplain development.

In the context of an intensively humanly altered and constrained (narrower) floodplain, anthropogenic water bodies provide additional hydrologic and environmental diversity to the embanked floodplain (Nienhuis, 2008; Reckendorfer et al., 2013; Biedenharn et al., 2018). Thus, while flood control engineering overall has adverse ecological impacts to lowland rivers, there are also positive benefits from the standpoint of artificial lakes (Figure 6.22B) contributing to the hydrologic and ecological diversity of embanked floodplain environments (Pongruktham and Ochs, 2015).

6.4.3.3 BORROW PIT PONDS AND DIKES

A prime example of an unintended environmental benefit of dike construction is the formation of borrow pit ponds, the most common anthropogenic water body along embanked floodplains. The excavation of sediment for dike construction is directly responsible for borrow pit pond formation and is associated with the history of dike construction. In addition to rates of floodplain sedimentation, which results in infilling (shallowing) of borrow pits, subsequent change is related to the frequency and volume of sediment removal for dike repair and fortification (Yuill et al., 2016a). Important attributes of borrow pit ponds includes serving as valuable recreation, flood water retention, and riparian aquatic habitat (Figure 6.23).

Borrow pit ponds form a chain of shallow, elongated water bodies adjacent to dikes. The overall dimensions of borrow pits should be somewhat scale dependent, based on the size of the dike. The length of individual borrow pit ponds along the lower Mississippi, for example, ranges from about several hundred meters to 2 km and are usually wider than a 100 m (Hudson et al., 2008). Mississippi borrow pit ponds average 7.8 ha in area and 2.1 m in depth, ranging from 4.1 m to 1.2 m (Buglewicz et al., 1988). Some borrow pit ponds are left dormant after dike construction and begin to infill with flood sediments. Other borrow pit ponds, however, are frequently reactivated and excavated by local contractors and government agencies for dike maintenance, with both coarse (sandy) and fine-grained cohesive deposits being in demand. And it should be noted that borrow pits are also created for other purposes, including general aggregate extraction and also to provide clay for brick manufacturing (Berendsen, 1993).

Along the lower Mississippi main-stem dike line there are over 1,300 borrow pits >1 acre (0.4 ha). The density of borrow pits is 0.14 (>1 acre) per km of channel distance, and their spacing effectively results in borrow pits being a constant hydrographic presence along the embanked floodplain (Figure 6.24). Many of the borrow pits were created during an intense period of dike construction and fortification in the late 1930s and 1940s, and then

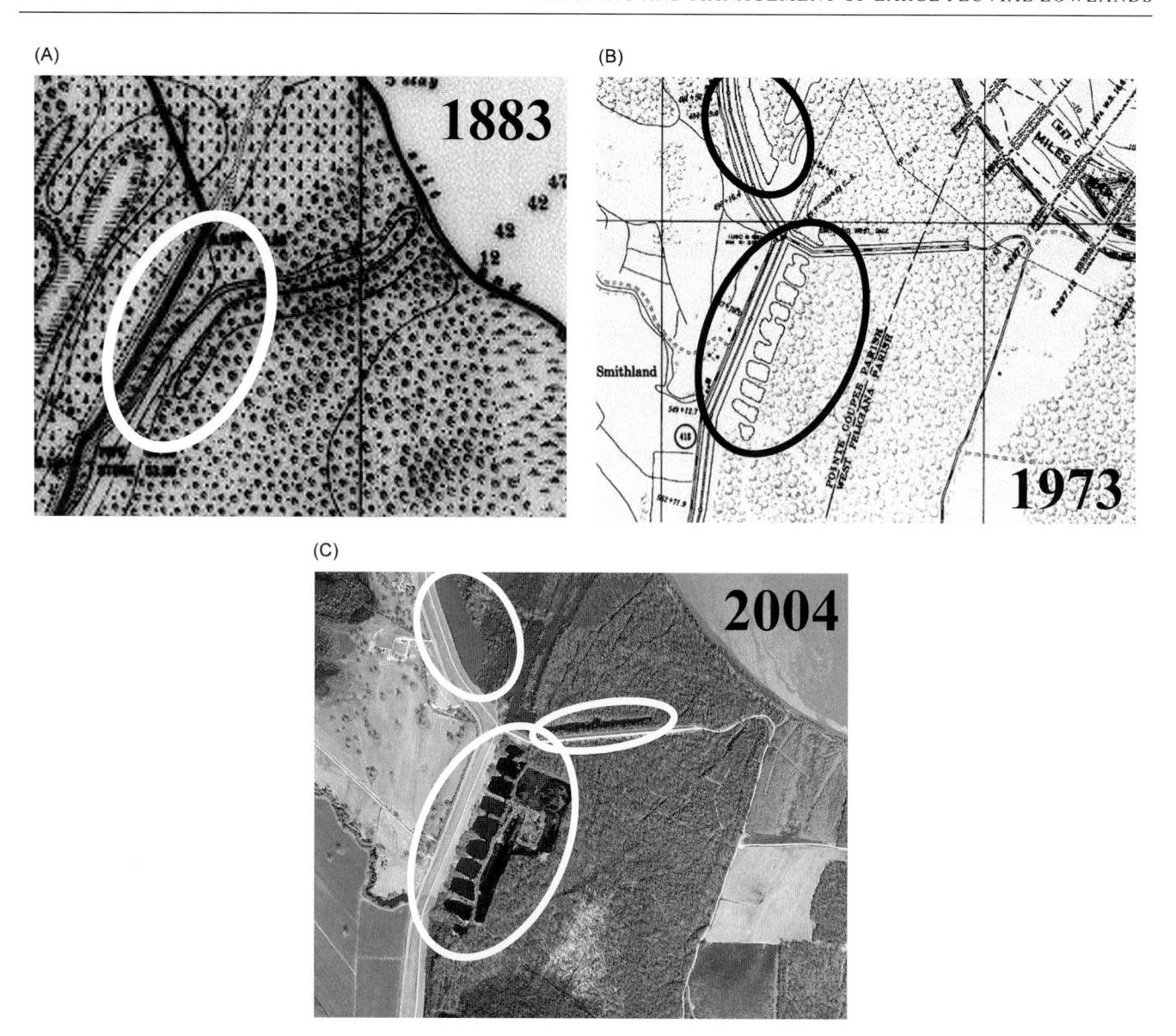

Figure 6.25. Evolution of embanked floodplain landscape with borrow pit expansion along lower Mississippi River embanked floodplain at Smithfield Landing, ~2 km upstream of Raccourci Old River (483 km upstream of Head of Passes, LA). (A) 1883: Hydrographic survey at Smithland, revealing swales but no borrow pits. (Source: Mississippi River Commission, scale 1:20,000.) (B) 1973: Borrow pits expanding over prior surveys after major dike and floodplain engineering efforts of the 1930s and 1940s. (Source: MRC survey, scale: 1:20,000.) (C) 2004: Hydrologic connectivity of borrow pits at sub-flood stage (November) (Source: USDA aerial photograph.) (Modified from Hudson et al., 2013.)

again in the 1970s (Figure 6.25). Following Hurricanes Katrina and Rita's devastation of the flood control system in Southeast Louisiana in 2005, a new phase of borrow pit excavation was initiated in southeast Louisiana because of the demand for suitable clay for dike fortification (Hudson et al., 2008; Hudson, 2018).

Embanked floodplains are increasingly seen less as a type of "no-man's land," but rather one that has valuable environmental benefits and is worth preserving (Biedenharn et al., 2017). In this regard, borrow pit ponds pose an interesting conundrum as concerns the value of incidental anthropogenic landscapes: their formation represents a type of environmental degradation while their preservation provides clear ecological benefits to the floodplain (Shields and Knight, 2013).

The hydrologic and sedimentologic dynamics of borrow pits represent a unique type of geodiversity within the humanly altered embanked floodplain. Thus, there is increasing attention to the environmental values of borrow pits. A key role of borrow pits is that they provide habitat for a range of floodplain aquatic organisms, including fish and water fowl. Along the lower Mississippi, the spawning habitat of the iconic alligator gar (*Atractosteus spatula*), for example, is ideally represented by the shallow and warm aquatic environments of borrow pits

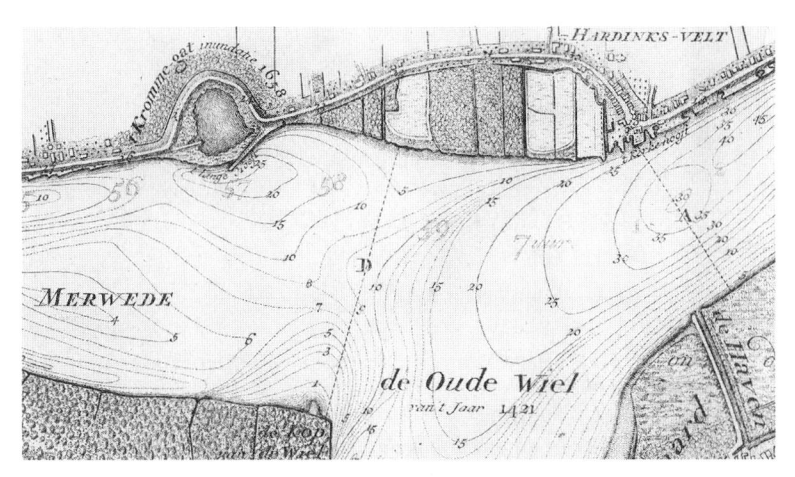

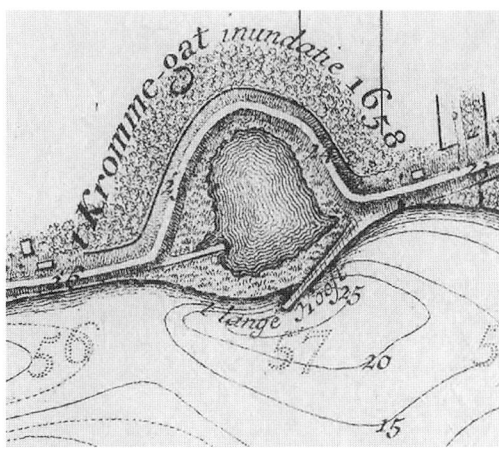

Figure 6.26. Dike breach pond (wiel) along the Merwede River (lower Rhine delta distributary) mapped in 1729, as portrayed in historic cartography (published in 1745). The map depicts a dike breach pond created in 1658 with wetlands (and/or sedimentary deposits) around the inside perimeter of the wiel. The dike is relocated around the dike, allowing open connectivity between the wiel and river. Note also the bifurcation of de Merwede at de Oude Wiel, which was a massive dike breach event during the St. Elizabeth Day flood of 1421. Map: De Rivier de Merwede, van ontrent de Steenen-hoek, Oostwaardsop tot verby het dorp van Sleeuwyk: met den Ouden-Wiel, en de Killen, die uit deselve na den Bies-Bos afloopen (The River Merwede, from about the Steenenhoek, Eastwards up to the village of Sleeuwyk: with the Ouden-Wiel, and the Killen, which run down from the same after the Bies-Bosch) (for 1729 depiction). Scale: 1:10,000, Cartographer: Nicolaas Cruquius. Staten van Holland, 's-Gravenhage. (Utrecht University historic collection.)

(Miranda, 2005; Allen et al., 2014; van der Most and Hudson, 2018). Management of borrow pits for environmental purposes has gained attention and is associated with formal plans in dike construction (USACE, 1998a,b,c). Nevertheless, while borrow pits form distinct ecological niches within the embanked floodplain, the creation of new borrow pits is often controversial.

An excellent example of the controversy associated with borrow pit pond formation is related to flood control of New Orleans and South Louisiana after the 2005 Hurricanes Katrina–Rita disaster. The amount of suitable clayey top-stratum deposits with low organic content required for dike construction and fortification is limited, and in high demand. To complete the levee (dike) repair and fortification project by 2011, the USACE was in urgent need of some 60 million m^3 of suitable clay to fortify the dikes, an amount of clay that would fill up the famous New Orleans Superdome (at 3.5 million m^3) some seventeen times. The USACE identified lands to purchase that had suitable clay and effectively pleaded with property owners to sell their soil to the Corps (USACE – New Orleans District, 2017), in some instances invoking eminent domain.[3] Not surprising, legal cases ensued and are ongoing, with particular dispute over the volume of suitable clay and the distinction between property values in comparison to the market value of bulk clay. In most instances the clay, at $4 per yrd^3 (= 0.765 m^3), is worth much more than the market value of the land (e.g., *National Food & Beverage Co., Inc.* v. *United States*, 2010; *Lake Borgne Basin Levee District and St. Bernard Parish Government* v. *United States*, 2016; *White Oak Realty, LLC and Citrus Realty, LLC* v. *United States*, 2018).

6.4.3.4 DIKE BREACH PONDS: WIELEN

Historic human impacts to landscapes may be relict, while also having important contemporary environmental relevance. Dike breach ponds are an additional type of anthropogenic floodplain water body associated with embanked floodplains, and specifically dikes. Dike breach ponds are formed along the margins of dikes by flood scour (erosion) during dike failure. Sedimentology is important to the formation of dike breach ponds, as dike breach within sandy noncohesive deposits result in a greater amount of scour and larger and deeper ponds than breach in cohesive fine-grained floodplain deposits (Berendsen, 1982, 1993; Jacobson and Oberg, 1997).

While dike breach ponds are abruptly formed during a single flood event, it likely requires centuries before there are sufficient numbers of dike breach ponds that they become a regular component of the embanked floodplain landscape. By definition they do not exist until after embankment and floodplain settlement. Historic cartography and modern geospatial data reveal the presence of old dike breach ponds along lowland European rivers long intensively diked, resulting in the incidence of severe floods being preserved upon the landscape. Dike breach ponds in the Netherlands are referred to as wiel, a common hydrologic feature along the embanked floodplain (Figure 6.26).

Upon formation, the evolutionary trajectory of dike breach ponds is primarily dependent upon the position of the repaired dike structure (Middelkoop, 1997). Dike breach ponds within the embanked floodplain infill with fluvial deposits, which over time

---

[3] Eminent domain refers to the power of the US government to take private property and convert it into public use. The Fifth Amendment provides that the government may only exercise this power if they provide just compensation to the property owners.

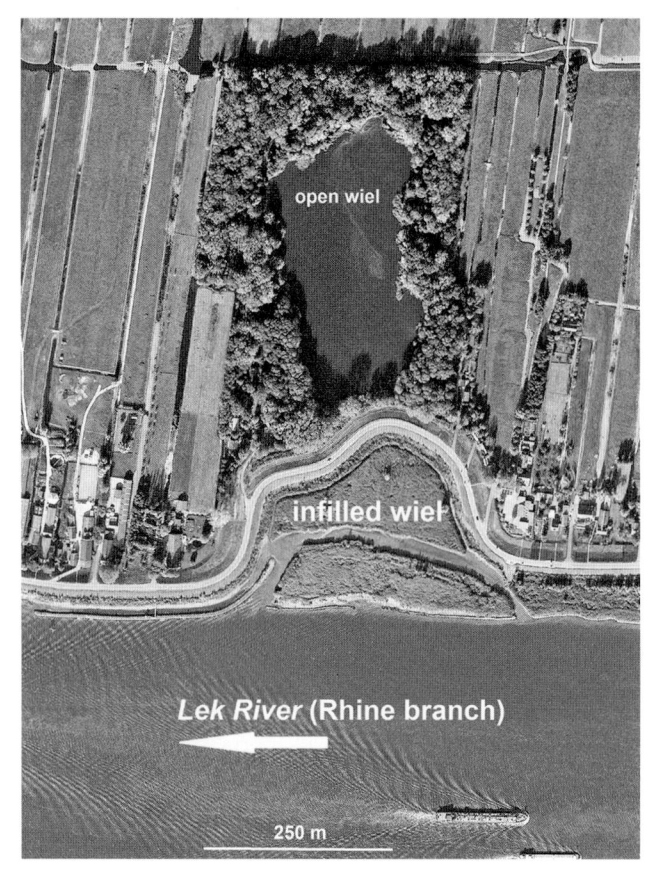

Figure 6.27. Dike breach ponds (wielen) along Lek River, secondary Rhine branch distributary. The wiel (de Bakkerswaal) formed during the St. Elizabeth flood of 1421. The dike was relocated around the wiel resulting in fluvial infilling and wetland whereas landward portions remain an open pond with a depth of over 10 m, providing important aquatic habitat for water fowl within a landscape intensively transformed for agriculture. (Image date, 2005, source: Google Earth Pro.)

becomes a wetland (Figure 6.27). If the dike is relocated between the pond and the river (Figure 6.27), the pond essentially becomes separated from fluvial inputs and becomes a permanent open water environment.

Creation of dike breach ponds results in management issues regarding restructuring the dike line, which involves local stakeholders. The decision of whether to reroute the dike in front or behind the dike scour ponds depends on local land use, pond depth, infrastructure, flood risk, and financial cost of rerouting the dike. With these concerns satisfied, from a contemporary perspective the repaired dike should be routed around the landward side of the pond, resulting in a new water body within the embanked floodplain and thereby enhancing fluvial geodiversity and biodiversity.

6.4.3.5 DIKE SEEPAGE PONDS

Rather than dike overtopping, dike seepage ponds are formed by groundwater flow beneath the dike and associated scour. High water levels increase hydraulic pressure that, in sandy deposits, increases groundwater flow until it is erosive. The depth of dike seepage ponds may be ~10 m. Such features begin as sand boils and subsequently enlarge.

6.5 BETWEEN THE DIKES: SEDIMENTOLOGY OF EMBANKED FLOODPLAINS

6.5.1 Disruption to Overbank Sedimentation

Floodplain embankment over extended periods – centuries – results in an anthropogenic style of embanked floodplain sedimentary deposits (Middelkoop, 1997). In this context, backswamp sedimentation is inconsistent with embanked sedimentation because of the higher energy events. Higher sedimentation rates infill floodplain topographic depressions, such as old channels and swales from river migration, that serve as vital riparian habitat. The magnitude of embanked sedimentation is well illustrated by examining floodplain deposits for rivers with a long history of flood control.

Sedimentary data for intensively humanized Rhine delta distributaries reveals the influence of different management options (i.e., Chapter 5). In addition to being embanked for flood control, the Rhine River is also heavily trained with channel engineering. The result is that the channel is not able to rework recent floodplain deposits with lateral channel migration, resulting in flood deposits effectively being permanently trapped within the embanked floodplain corridor (Middelkoop et al., 2010). Over about the past eight centuries the embanked floodplain of the Rhine delta aggraded by 2–3 m because of sediment trapping between the dikes. This is significant because, in addition to degrading (i.e., burying) floodplain aquatic habitat it reduces the flood water storage capacity of the floodplain. The result is that flood stage increases, as there is less room available to store flood waters during large events (Silva et al., 2001; Hesselink et al., 2003; Glynn and Kuszmaul, 2010), which was one of the drivers for the Room for the River program (see Section 8.6).

Where embanked floodplains are narrow the flood stage is higher and has greater velocity, which results in some fine sediment (wash load) being transported further down the floodplain corridor, where it is eventually deposited in wider segments or flushed out to shallow coastal environments (Middelkoop et al., 2010). Because boundary shear stress increases linearly with flood stage (i.e., flow depth), coarser flood sediment is deposited within distant reaches of the embanked floodplain, resulting in a departure to the classical lateral fining trend (e.g., Pizzuto, 1987). Sedimentation within backswamp environments from the epic 2011 flood along the lower Mississippi River, for example, consisted of 18% fine sand, considerably coarser than the usual clayey flood deposits (Heitmuller et al., 2017).

The dual industrial and agricultural utilization of the Rhine embanked floodplain is illustrated in the pattern of surficial particle size data along the Nederrijn-Lek River (Wolfswaard polder) (Figure 6.28A). The elevation profile reveals considerably higher rates of aggradation within the summer dikes, which were mainly constructed after the mid-eighteenth century as the river channel stabilized. Between the summer and winter dikes the fine-grained top-stratum deposits were scraped off for brick manufacturing, revealing floodplain "patches" at different elevations (Figure 6.28 B). Additionally, overbank sedimentation does not exhibit a classic pattern of lateral fining of particle size (e.g., Pizzuto, 1987; Magilligan, 1992; Hudson and Heitmuller, 2003), which is attributed to removal of the fine-grained surface deposits. In this case, the coarsest particle size is actually located closest to the distant high winter dike, because here the excavation (stripping) has penetrated the top-stratum (e.g., Figure 6.3) and extended into the coarser channel bed deposits (bottom stratum). Sedimentary disruption to top-stratum floodplain deposits, particularly adjacent to dikes, can increase flood risk by increasing dike underseepage.

The influence of floodplain engineering on the broader pattern of embanked floodplain deposition over extended periods can be appreciated by considering flood sedimentation at the single-event scale (Middelkoop and Asselman, 1998; Thonon et al., 2007). Data for the Waal and IJssel Rhine delta distributaries (Figure 3.41) for moderate magnitude floods over 2001–2004 provides a means to directly compare the intensity of floodplain embankment on sedimentation (Thonon et al., 2007). The average sedimentation for the Waal was 1.44 kg/m^2 and the IJssel was 1.50 kg/m^2. Whereas sedimentation is relatively the same, the pattern of sedimentation is significantly different. While the Waal section does not exhibit a lateral trend in sedimentation, the amount of sedimentation along the IJssel clearly decreases across the floodplain with distance from the channel. This difference in sedimentation is likely because of the intensity of floodplain embankment. The Waal has an F/C ratio (floodplain width to channel width ratio) of 3.0 whereas the smaller IJssel River section has an F/C ratio of 9.0. Thus, in comparison with the IJssel, the Waal is much more intensively engineered and proportionately floodwaters are trapped within a much smaller area, disrupting classic spatial flood sedimentation patterns.

While it is important to consider the role of flood control engineering on floodplain sedimentation it is also important to consider, especially for large rivers, that simultaneous human activities in upstream portions of the drainage basin can influence downstream floodplain aggradation. The most severe period of land degradation and soil erosion in the Rhine basin occurred from the late Bronze Age through the Roman era, prior to floodplain embankment (Kooijmans, 1974; Hoffmann et al., 2009; Middelkoop et al., 2010). The sediment produced during this period is distributed across the larger surface of the Rhine delta rather than being concentrated between dikes.

A prime example of extreme soil erosion and downstream embanked floodplain aggradation is illustrated along the lower Huanghe valley and delta in China, with origins of the problem already in the mid-Holocene (Kidder and Zhuang, 2015). Upland human activities and accelerated soil erosion in the Chinese Loess Plateau was responsible for the Huanghe globally having the highest sediment load of any river. The high silt load results in large amounts of downstream channel bed aggradation, with rates as high as 0.1 m/yr. Constriction of the channel cross section results in more frequent overbank events, and increases embanked floodplain aggradation. The result of this long-term period of high floodplain sedimentation has resulted in the Huanghe channel belt being perched some 3–15 m above the adjacent flood basins (Slingerland and Smith, 2004; Kidder and Liu, 2014; Zhang et al., 2016). Verification that this geomorphic condition is unstable can be seen in flood deposits that bury abandoned dike structures from several millennia BP (Kidder and Liu, 2014; Kidder and Zhuang, 2015).

The high rate of embanked floodplain aggradation along the Huanghe River is not just historic – it continues to be a significant management problem (UNESCO-IHP, 2011). A comparison of survey profiles across the floodplain reveals aggradation over about the past five decades (Figure 6.29). The channel bed has aggraded by 2.8 m whereas the floodplain has aggraded by 1 m. Much of the aggradation occurs within an especially narrow reach of the embanked floodplain, between the secondary low (summer) dikes. Although downstream sedimentation rates have decreased because of the reduction in upstream sediment loads (Wang et al., 2005), until several decades ago the delta was prograding at an average annual rate of 42 km^2 year (Zhang et al., 1990).

While sediment loads have declined (Section 4.3.2.3), the perched Huanghe embanked floodplain is a considerable management challenge. In addition to the problem of accelerated aggradation burying riparian environments, it also reduces dike safety and increases flood risk, especially as related to the process of dike breach events and channel avulsion (Section 7.4.2).

6.5.2 Infilling and Terrestrialization of Floodplain Lakes

The process of embanked floodplain sedimentation varies for different floodplain components (Section 3.5.2). As open water environments oxbow lakes are essential elements of the embanked floodplain that are sequentially degraded by higher rates of embanked sedimentation. While artificial meander cut-offs initially resemble natural oxbow lakes in function and shape, there is an important distinction. As with dike breach ponds within the embanked floodplain, cut-offs (oxbow or chute) that are not managed undergo higher rates of infilling and environmental change than cut-offs outside of the embanked floodplain. This is because they remain actively connected to the main-stem channel at flow levels below flood stage, and because sediment is constrained to a

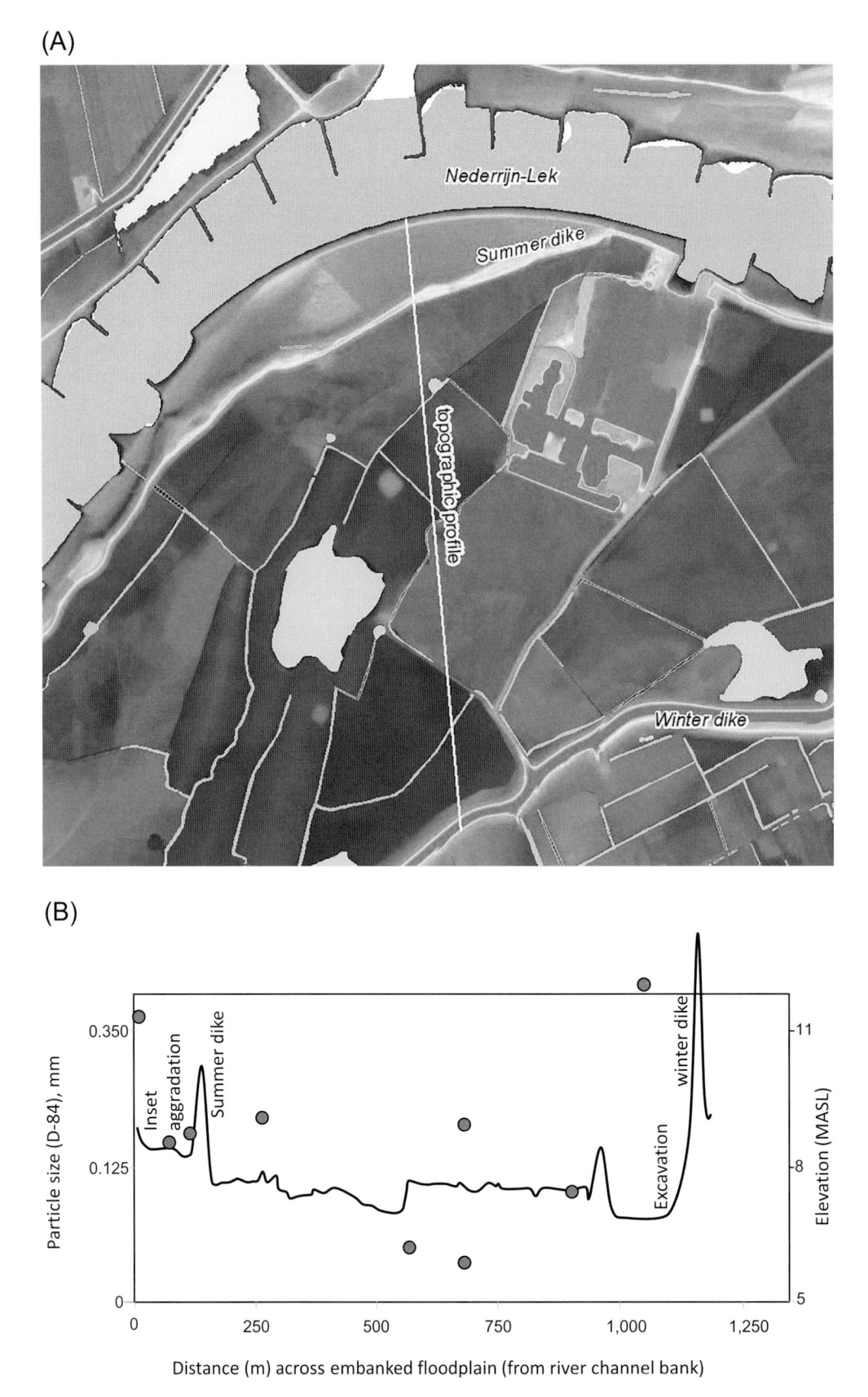

Figure 6.28. Common floodplain landscape along the Rhine delta specifically the Nederrijn-Lek distributary. (A) The "patchy" floodplain topography illustrates the dual activities of agriculture and industry, scraping top-stratum deposits for brick manufacturing (factory surrounded by ring dike) and summer dike embankment for agriculture. (B) Lateral grain-size trend of d_{84} (mm) along the topographic profile depicted with lidar elevation data, revealing both disrupted particle size trend and floodplain topography. (Author figure. Source: elevation data, lidar DEM AHN-1, Author field data.)

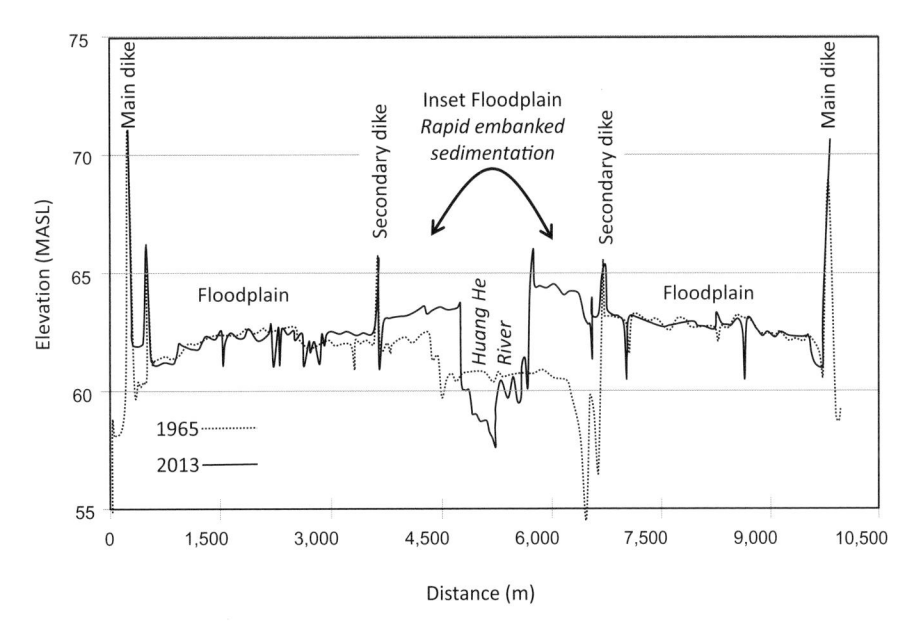

DIKES AND FLOODPLAINS: IMPACTS OF FLOOD CONTROL ON LOWLAND RIVERS 191

Figure 6.29. Cross section of embanked floodplain of the Huanghe (Yellow) River, which has rapidly aggraded by high sedimentation rates driven by accelerated soil erosion and land degradation of the Chinese Loess Plateau. Channel narrowing attributed to engineering. (Source: Zhang et al., 2016.)

smaller area. An important control on oxbow infilling is the angle of the upstream and downstream cut-off channel relative to the main-stem channel, with rates of sedimentation likely being related to a zone of flow separation at the cut-off channel entrance (Fisk, 1944; Gagliano and Howard, 1984; Shields and Abt, 1989; Rowland et al.., 2010; Dépret et al., 2017; Ondruch et al., 2018). Other factors that influence artificial lake infilling are related to engineering activities associated with the cut-off as well as post-cut-off development of the main-stem channel. Related to engineering activities to ensure the cut-off is not reactivated, embankments and groynes may be installed at the entrance to the cut-off channel, which reduces coarse-grained channel plug sedimentation. Additionally, main-stem channel incision reduces cut-off connectivity and fine-grained sedimentation. This scenario can result in cut-off terrestrialization by woody vegetation encroaching over lake bed deposits (Dépret et al., 2017).

As an example of the high rates of oxbow lake terrestrialization within embanked floodplains, the channel cut-off program of the 1928 Mississippi River & Tributaries Project created sixteen meander neck cut-offs in the 1930s (Section 5.4.1). Not only did this substantially change the river pattern and geomorphic processes, it also resulted in an abrupt change to the embanked floodplain hydrology. The new artificial oxbow lakes are contained within a relatively narrow floodplain, ranging from 1.3 km to 25.3 km. This concentrates sedimentation within the embanked floodplain. The amount of sedimentary infilling of oxbow lakes since the year of cut-off is indicated by the presence of active batture channels (e.g., Figure 6.30). Rapid extension of

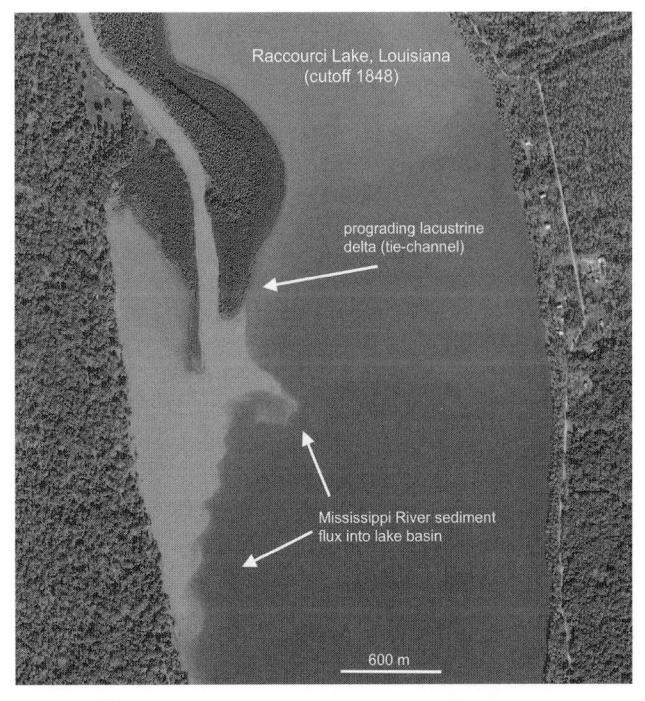

Figure 6.30. Sediment flux into Raccourci Lake (Old River) below flood stage through batture channel (tie-channel) connection. Raccourci Lake, Louisiana is within the lower Mississippi embanked floodplain (~25 km downstream of Red River Landing). The state of Louisiana engineered the cut-off in 1848 to reduce downstream flood levels, the only artificial Mississippi River cut-off not engineered by the federal government. (Author figure, after Rowland et al., 2010. Image: November 23, 2015, Google Earth Pro.)

Table 6.8 *Comparison of embanked oxbow lake terrestrialization along the lower Mississippi River*

Name	1963 KM/A/HOP	Year of cut-off	Area (km^2) of cut-off	Oxbow Lake Area (km^2)	% Wetland	Embanked (yes/no)
False River	417	1699–1722[1]	29.1	12.9	55.7	No
Raccourci Lake (Old River)	481	1848[1]	32.1	20.6	35.9	Yes
Lake Mary	518	1776[1]	16.4	11.3	31.1	No
Glasscock Cut-off	552	1933[2]	17.3	7.7	55.8	Yes
Giles Cut-off	589	1933[2]	21.4	5.4	74.6	Yes
Waterproof cut-off	608	1884	24.3	3.28	86.5	yes
Lake Saint John	610	Prehistoric[1]	17.2	9.2	46.6	No
Rodney Cut-off	624	1936[2]	8.2	2.8	65.3	No
Lake Bruin	642	Prehistoric[1]	15.3	12.2	20.4	No
Yucatan Cut-off	657	1929[2]	12.7	8.5	32.5	Yes
Diamond Cut-off	682	1933[2]	18.5	4.3	76.7	Yes
Marshall Cut-off	721	1934[2]	9.7	0.5	94.8	Yes
Eagle Lake	744	1866[1]	26.9	17.9	33.5	No
Willow Cut-off	745	1934[2]	12.9	3.7	71.4	Yes
Lake Providence	784	Prehistoric[1]	12.4	6.0	51.5	No
Sarah Cut-off	811	1936[2]	7.1	1.3	81.5	Yes
Worthington Cut-off	827	1933[2]	11.0	1.6	85.4	Yes
Lake Lee	847	1858[1]	13.5	7.3	46.3	Yes
Lake Chicot	856	Prehistoric[1]	22.1	17.4	21.3	No
Leland Cut-off	867	1933[2]	18.6	8.5	54.0	Yes
Tarpley Cut-off	871	1935[2]	13.7	2.3	83.5	Yes
Ashbrook Cut-off	884	1935[2]	11.4	5.1	55.0	Yes
Caulk Cut-off	925	1937[2]	17.0	6.0	64.5	Yes
Sunflower Cut-off	1,006	1942[2]	17.1	6.5	62.0	Yes
Jackson Cut-off	1,011	1941[2]	14.3	3.7	73.9	Yes
Hardin Cut-off	1,091	1942[2]	32.4	15.5	52.0	Yes

[1] Source: Gagliano and Howard (1984).
[2] 1928 Mississippi River & Tributaries Project, Winkley (1977).
Area of terrestrialization (lake to wetland conversion) expressed as a percentage of the original surface area. Oxbow lake area (km^2) estimated from 1911 channel dimensions and compared with USGS National Hydrographic Data in ArcGIS.
Source: Hudson et al. (2008).

batture channels (tie-channels) and associated infilling within the cut-offs after the 1930s resulted in much new lake area being converted to marsh and forested wetlands. Batture channels prograde into oxbow lakes, forming small lacustrine deltas. As batture channels further prograde into the lake basin a sequence of wetland vegetation develops, ranging from initial marsh to willows, and then a cypress tupelo swamp (Gagliano and Howard, 1984; Saucier, 1994; Rowland et al., 2012; Ondruch et al., 2018).

The artificial oxbow lakes created by the lower Mississippi meander neck cut-off program of the 1930s and early 1940s have rapidly evolved from open water environments to predominantly being wetland environments (Hudson et al., 2008). On average, 66% of lake area has converted to wetlands. Some of the artificial cut-off oxbow lakes have become terrestrialized by 94% (Table 6.8). While oxbow lake processes are usually dynamic, in comparison to rates of change of oxbow lakes within natural floodplains this is a remarkably high rate of evolution. For comparison, oxbow lakes located outside of the lower Mississippi embanked floodplain, although being considerably older with most predating dike construction, retain much more of their open water surface (Figure 6.31). On average, the much older prehistoric oxbow lakes have terrestrialized by 40% (Table 6.8).

Perhaps due to a longer history of embankment differences exist between infilling and terrestrialization of lower Mississippi oxbow lakes with artificial meander cut-offs of European rivers. It is not surprising that within long embanked rivers in Europe, few cut-off lakes exist with extensive open water environments. Indeed, oxbow lakes created from the famous Tulla cut-offs (1817–1876) along the upper Rhine (e.g., Chapter 5), for example, have largely evolved into wetlands (Figure 6.32).

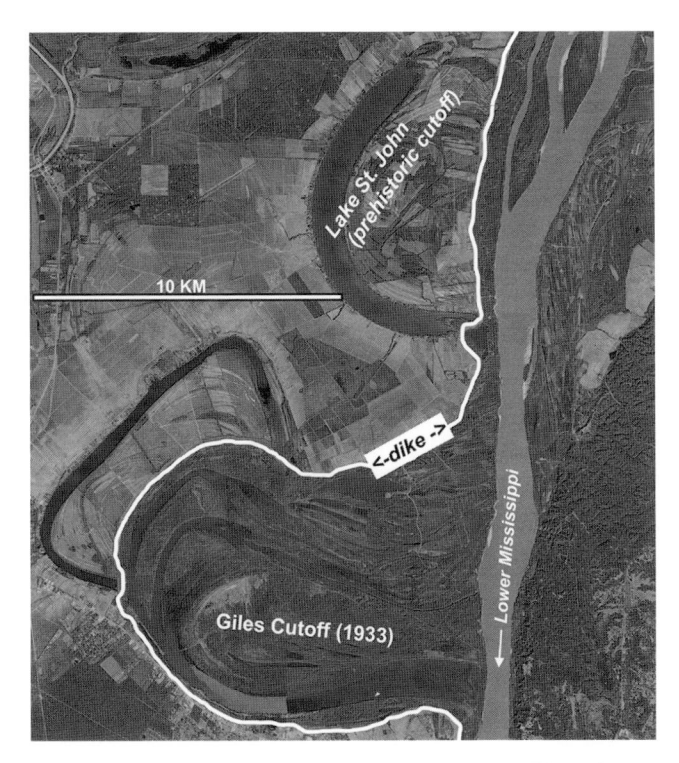

Figure 6.31. Two meander neck cut-offs (dashed lines) illustrating embankment influence on oxbow lake infilling near Natchez, MS. Lake St. John is a prehistoric meander bend cut-off outside of dikes (solid line), whereas Giles cut-off (in 1933) is within the embanked floodplain. (Author figure, image October 2015, Google Earth Pro.)

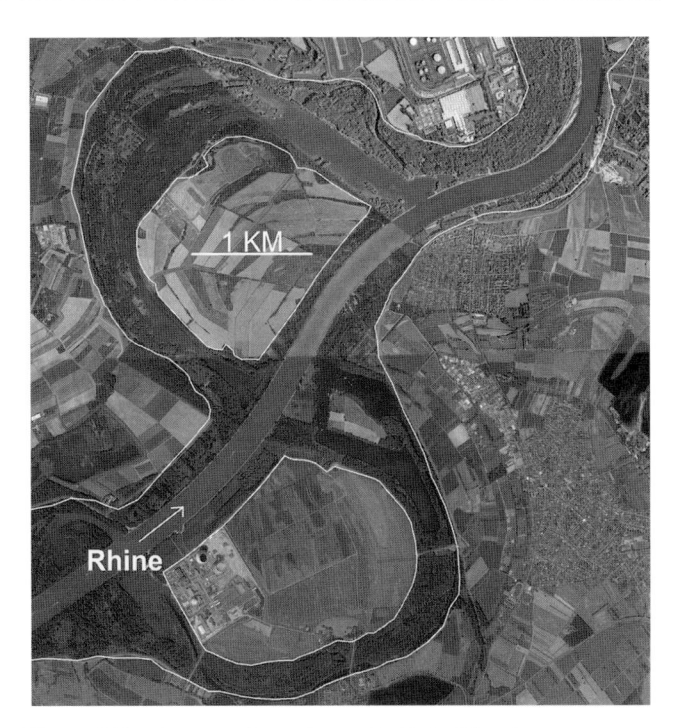

Figure 6.32. Two meander neck cut-offs (dashed lines) near Oberhausen-Rheinhausen, Germany within upper Rhine embanked floodplain (~30 km upstream of Karlsruhe). The sinuous channel segment was straightened as part of the Tulla river training program in the mid-nineteenth century. Most of the lake surface has been converted to wetland, with some reaches remaining open to support industrial activities. (Author figure, image June 2016, Google Earth Pro.)

Exceptions occur where decisions were made to leave certain bends open, such as at urban areas and industrial activities to retain a connection to the main-stem Rhine. Along the Rhône River in France, some of the artificial oxbow lakes created in the mid-nineteenth century were managed such that infilling rates were slower. This occurred by emplacing underwater dikes (embankments) to reduce bed load inputs and formation of a coarse sedimentary channel plug (Dépret et al., 2017).

6.6 BEYOND THE DIKES: SAND BOILS, DIKE BREACH, AND SEDIMENTATION

6.6.1 Dike Breach Processes

In addition to dikes being responsible for unique embanked floodplain deposits, their interaction with the dike and floodwaters results in unique anthropogenic fluvial sedimentary features along – and beyond – the margins of the dikes. Of key importance is management of active flood hydrologic processes in relation to the floodplain sedimentology. This is particularly important as to whether the dike is located atop cohesive or coarse-grained deposits (Figure 6.33).

Floodplain hydrology is directly influenced by dike structures, which has implications to dike stability and inundation beyond the dike margin (Middelkoop, 1997; Remo et al., 2018). Dikes are destabilized by flood waters in four primary ways, including (1) water piping through the dike, (2) water overtopping the dike, (3) underseepage below the dike, and (4) wave processes atop the dike (Figure 6.34). Saturation of dike materials for extended periods is always a concern, especially with dikes comprised of coarse-grained materials (Alexander et al., 2013). Such conditions can result in subsequent enlargement of macropores, resulting in water piping and scour through the sedimentary matrix of the dike, destabilizing the structure (USACE, 2000a; Glynn and Kuszmaul, 2010; Vergouwe, 2016). Prolonged saturated conditions can also result in dike slumping, tremendously weakening the dike structure. Additionally, underseepage beneath dikes in permeable sands and gravels associated with buried channel belts (e.g., point bar deposits) can ultimately result in sand boils beyond the dike (e.g., Figure 6.33). This is especially prevalent with high flood stage events because of the greater hydraulic head. High flood stage is also a problem to dike stability because of wave attack on the riverside, and especially in the case of water overtopping the dike crest.

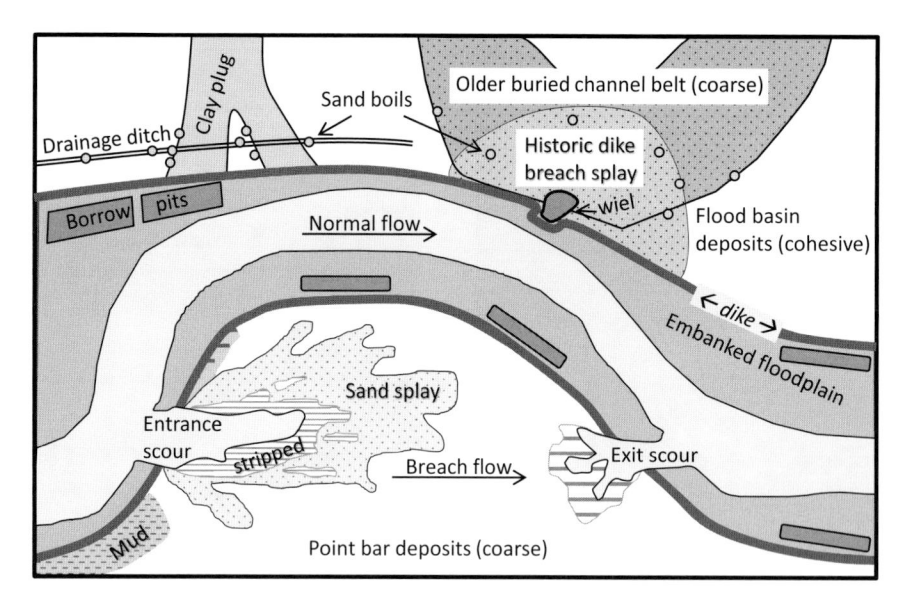

Figure 6.33. Embanked floodplain sedimentology and hydrography associated with dike breach events combining concepts developed from the lower Missouri, Mississippi, and Rhine Rivers. (Author figure based on Kolb, 1975, Berendsen, 1982, and Alexander et al., 2013.)

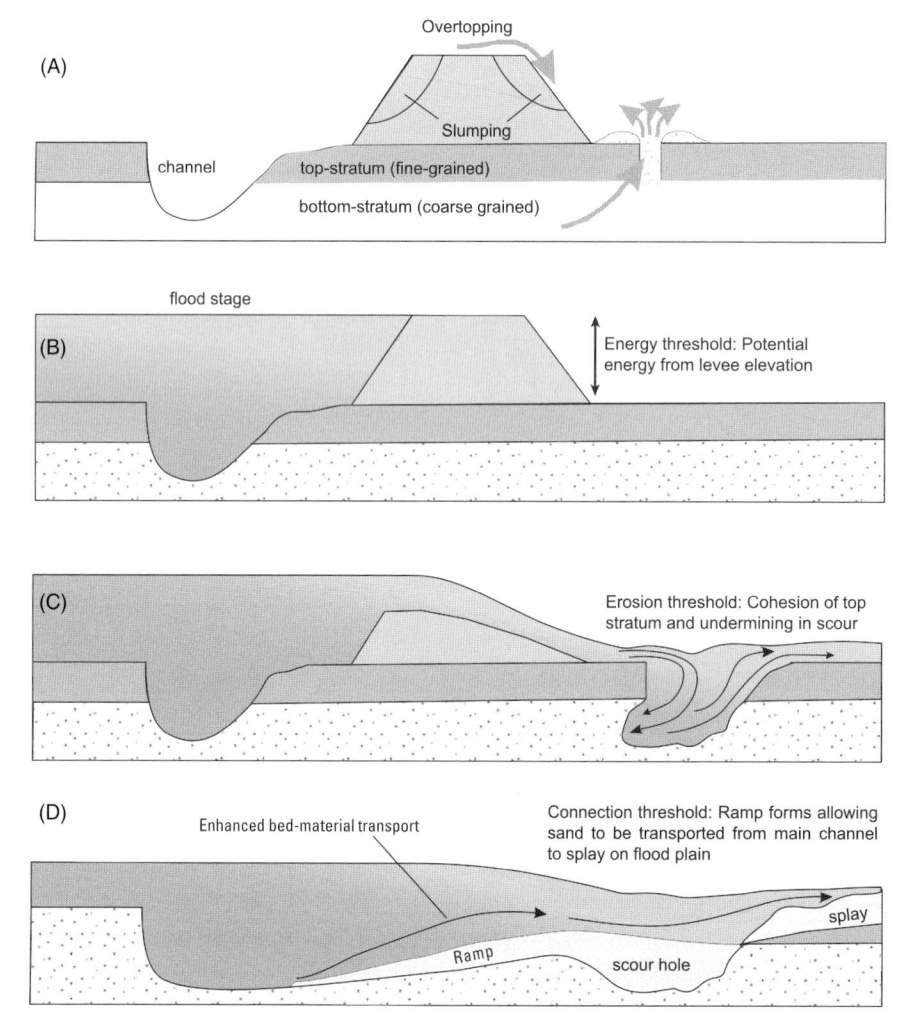

Figure 6.34. Stages of dike (levee) breach progression: (A) Failure mechanisms, including overtopping, slumping, dike underseepage, and sand boil; (B) dike height creates local potential energy grade and local threshold that, when exceeded, releases local energy; (C) overtopped or failed dike concentrates streamflow, resulting in high velocity and floodplain scour through cohesive top-stratum into underlying noncohesive sandy deposits, representing a second erosional threshold that increases scour; (D) connection threshold occurs when floodplain scour extends upward and intersects the main channel, resulting in direct ramp-like access of suspended and traction sediment onto floodplain. (Source: modified from Alexander et al., 2013.)

As flood waters overtop the dike there are two key influences on the overall magnitude of floodplain scour (Jacobson and Oberg, 1997). The first is the hydraulic head or potential energy, set up by the height of the dike above the adjacent floodplain (Figure 6.34). In this case, a stable dike retains the high energy gradient (slope), which assures that scour will be considerable on the landward side (Alexander et al., 2013). Additionally, the resistance of the floodplain deposits is important. A thin cohesive top-stratum provides less resistance to erosion, resulting in deep scour of underlying sandy deposits. In such cases, channel bar deposits (bottom-stratum) are eroded and assimilated with flood sediments, and deposited beyond the dike. With dike failure the energy slope decreases, although the lack of a barrier between the river and floodplain results in sandy channel bed material being assimilated into splay deposits.

6.6.2 Sand Boils and Seepage

Dikes are directly responsible for sand boil formation and underseepage, unique hydrologic processes that only occur along embanked floodplains (Vergouwe, 2016). The two processes occur for several reasons, including the higher hydraulic head (pressure) during flood events set up by the dike height, in combination with heterogeneous floodplain deposits with high hydraulic conductivity (Mansur et al., 1956; Krinitzsky and Wire, 1964; Kolb, 1975; Martin et al., 2017; García Martínez et al., 2020). Additionally, land use activities that rupture cohesive fine-grained top-stratum, such as borrow pit formation on the riverside of the dike, increases groundwater flow into the alluvial aquifer, thereby leading to seepage and possibly sand boils on the landward side of dikes. Large floods along the Mississippi River over the past century, especially the 1937, 1957, 1973, 1993, 2008, and 2011 floods, resulted in considerable local flooding by underseepage and sand boils. During the 2008 flood along the lower Mississippi River, for example, forty sand boils were identified along a 150 km long river reach, which inundated over 40,000 ha. This is despite the USACE having installed a number of relief wells and drainage ditches ahead of the flood.

During a flood, the trapping of flood waters behind dikes results in a sharp increase in hydraulic head between the embanked floodplain and landward side of dikes. Underseepage occurs where the landward side of the dike lacks a sufficiently thick cohesive top-stratum, or lacks an adequete artificial clay apron constructed by floodplain engineering. In the case of groundwater flow through alluvial pipes (macropores), the concentrated flow can generate sufficient force to cause erosion and transport sediment to the surface (Mansur et al., 1956; Kolb, 1975; Li et al., 1996; Glynn and Kuszmaul, 2010; García Martínez et al., 2020), where the water emerges as a central orifice (Figure 6.35). The transported sediment is deposited within a conical mound of well-sorted fine to medium sized sand deposited in steeply (35°–45°) dipping layers (Figure 6.36). The size of sand boils is controlled by the flood duration and hydraulic head, and ranges between 0.5 m and 10 m in diameter and is about 0.5 m in height (Kolb, 1975; Li et al., 1996).

Sand boils tend to be located within about 100 m of the dike. A steeper hydraulic head caused by a high stage can result in more sand boils, which often exfiltrate along drainage ditches on the landward side (Figure 6.35). In addition to the hydraulic infrastructure, sedimentologic controls include the presence of underlying pervious sediments as well as the thickness of the fine-grained top-stratum deposits on either side of the dike (García Martínez et al., 2020). Sand boils are more likely to develop where the top-stratum is thin or where holes (pits) exist on the riverside of the dike. Additionally, sand boils are often concentrated along the contact line between cohesive clayey channel infill deposits and coarse-grained deposits (i.e., Figure 6.33). In exceptional cases persistent sand boils further erode and becomes enlarged, eventually forming a dike seepage pond. During the 1993 flood (high water) event in the Rhine delta ~120 sand boils were identified along the three main

Figure 6.35. Photo of sand boil along Rhine dike during the 1995 flood. (Source: Rijkswaterstaat archive.)

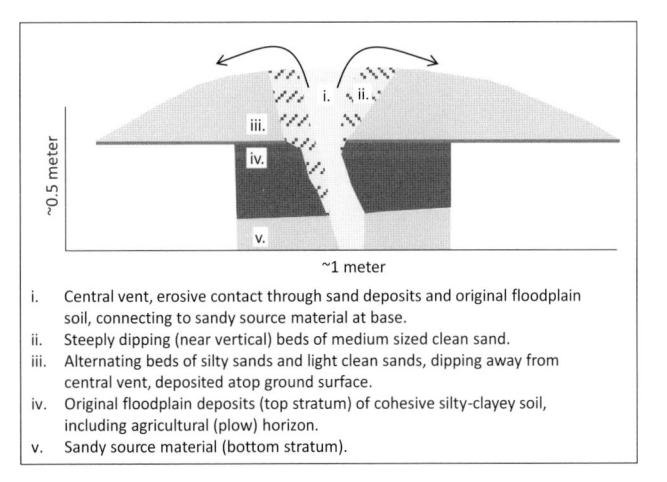

Figure 6.36. Sand boil sedimentology along lower Mississippi River after the 1993 flood. (After Li et al., 1996.)

i. Central vent, erosive contact through sand deposits and original floodplain soil, connecting to sandy source material at base.
ii. Steeply dipping (near vertical) beds of medium sized clean sand.
iii. Alternating beds of silty sands and light clean sands, dipping away from central vent, deposited atop ground surface.
iv. Original floodplain deposits (top stratum) of cohesive silty-clayey soil, including agricultural (plow) horizon.
v. Sandy source material (bottom stratum).

distributaries, and during the 1995 event about 180 sand boils were observed. The sand boils tended to occur along drainage ditches and at the boundary between sandy and clayey sedimentary deposits (Vergouwe, 2016).

The epic 2011 flood along the lower Mississippi River was the highest flood of record, and produced numerous sand boils (DeHaan et al., 2012). Within the coarse channel belt deposits at Cairo, IL, the head of the lower Mississippi Valley, several sand boils were located adjacent to the dike structures, including three "mega" sand boils that were up to 4 m in diameter (Morton and Olson, 2015). Sand boils, and especially mega boils, are a serious threat to dike stability. Managing sand boils is a staple of the "flood fight" during high-water conditions, and usually involves a team of trained professionals and volunteers to look for seepage below and through the dike, and landward. Modern methods continue to employ teams of flood fighters, but also with the assistance of UAV (drones). When identified, the severity of sand boils is assessed by its size (diameter) and flow of water, and especially as to whether there is a sediment plume (DeHaan et al., 2012). Increasing sediment transport through the pipe indicates increasing underground cavity enlargement, which can result in dike failure by collapse. The universal approach to managing sand boils are sand bags, stacked in rings around the sand boil to a height that decreases the hydraulic gradient, reducing erosive underseepage. Interestingly, during the 2011 flood within the upper delta, several sand boils and seepage zones developed ~1,000 m from the dike, flooding agricultural lands. This indicated considerable macropores of pervious sediments within the Pleistocene stratigraphy. While the importance of such distant hydrologic connectivity is usually seen as a local flood problem, it should also be noted that seepage enables distant (outside the embanked floodplain) floodplain water bodies to receive fresh pulses of river water.

6.6.3 Dike Breach Events

In addition to influencing the pattern of sedimentation within the embanked floodplain, dikes also influence sedimentation outside of the embanked floodplain (e.g., Figure 6.33). In this case, sandy splay deposits are formed from dike breach events along the dike margins.

For rivers that have been diked for hundreds of years, the result of numerous dike breaches produces a coarse edge of flood deposits on the landward side of the dike. Much of the dike margin along the Rhine delta has coarse splay-like deposits because of the many historic dike breach events. Between 1700 and 1927, for example, there were almost 500 dike breach events, with an average of four events per year occurring between 1750 and 1800 (van Veen, 1948). The accumulation of coarser flood deposits atop the fine-grained cohesive sediment has consequences to floodplain land use for subsequent generations.

The Betuwe region of the Netherlands is a narrow (~5–15 km) fringe of fluvial land situated between the two main distributaries of the Rhine River, the Waal and Nederrijn-Lek Rivers. Historically the Betuwe was frequently flooded, resulting in coarser sediment accumulating within the flood prone lands (Egberts, 1950; Kooijmans, 1974; Van Dinter and Van Zijverden, 2010). The dike breach events occurred along the main Rhine River distributary channels as well as smaller distributary channels, such as the Linge (Figure 6.37). Because the Waal and Nederrijn-Lek are separated by a narrow fringe of land, the cumulative effect of flooding over hundreds of years was the blanketing of the soil in coarser flood deposits (silt and fine sand rather than primarily clay) and sandy ridges (Egberts, 1950; Berendsen, 2008, pg. 99). The subsequent soil development can be considered somewhat anthropogenic. Interestingly, the coarser soil texture is ideal for fruit production for which the Betuwe is known, an important sector of the highly profitable Dutch horticulture industry.

Development of the distinctive anthropogenic dike breach flood mechanism can be examined by considering historic and recent events along rivers that have not been embanked for as long as many of the rivers in Europe and Asia. Along the Mississippi River, for example, the historic system of mainline dikes was incredibly ineffective from the time of its construction until the 1930s, when it was rebuilt following the Great Flood of 1927. But prior to the 1927 flood the dikes were frequently breached. The cause of dike breaches was not necessarily the magnitude of discharge events, but often the dire condition of the dike system because of the lack of coordinated management.

For many decades mainline dikes along the lower Mississippi were poorly maintained, and constructed to unequal heights and standards (Elliot, 1932, vol. II). Thus, it is not surprising that the dikes were continually breached by the frequent flood events that occurred between the late 1800s and early 1900s. The flood of 1893, for example, caused fifteen documented crevasses in Louisiana. The reasons for the dike breaches included crawfish leaks, rice flumes, and waves (Table 6.9). The Rescue Crevasse, upstream of New Orleans, had a width of 380 m and a maximum discharge of 2,990 m^3/s, larger than the average discharge of the Rhine River. While the 1927 flood discharge was the largest in North American history, as with the 1893 flood breach causes, the lower standards and general poor condition of dikes in relation to different types of human activities were also to blame for numerous dike breaches.

The 1927 dike breach events have had lasting impacts to the embanked floodplain geomorphology and hydrology of the lower Mississippi that are directly related to the flood control infrastructure (Figures 6.38 and 6.39).

In terms of the influence of large floods on policy and management, the epic 2011 flood along the lower Mississippi River – the highest flood of record – is destined to have but a marginal

Table 6.9 *Crevasse (dike breach) events associated with the 1893 flood along the lower Mississippi in Louisiana*

Name	KM/B/C/L-R	Date open (1893)	Date close (1893)	Max. Width of breach (m)	Max Q (m³/s)	Cause of breach
Mc Collum	1,343-L	June 13, 21:00	June 14, 20:00	21	95	Crawfish leak
Rescue	1,406-L	June 23, 10:30	Not closed	380	2,990	Wave wash
Uncle Sam	4,450-L	June 18, 21:30	June 20, 04:00	12	53	unknown
Reserve	1,482-L	June 18, 19:00	June 21, 18:00	39	415	Crawfish leak
Magnolia	1,576-R	June 15, 09:00	June 25, 18:00	5		Crawfish leak
St. Clair	1,582-L	June 14, 12:00	June 15, 11:00	3		Crawfish leak
Harlem	1,619-L	June 20, 09:00	June 30, 02:00	57		Rice flume
Pierre Casse	1,630-L	June 20, 09:00	June 20, 17:00	4		Unknown
Barthelemy	1,633	June 14, 12:00	June 15, 06:00	5		Rice flume
Adam's crevasse	799-R	May 11				
Kiger's	813-R	May 14				
Mathew's Bend	818-R	May 15				
Wyly's	877-R	May 23				

Source: Mississippi River Commission, Fourth District, Map Showing Crevasses and Area Overflowed, Flood of 1893, Plate XVIII.

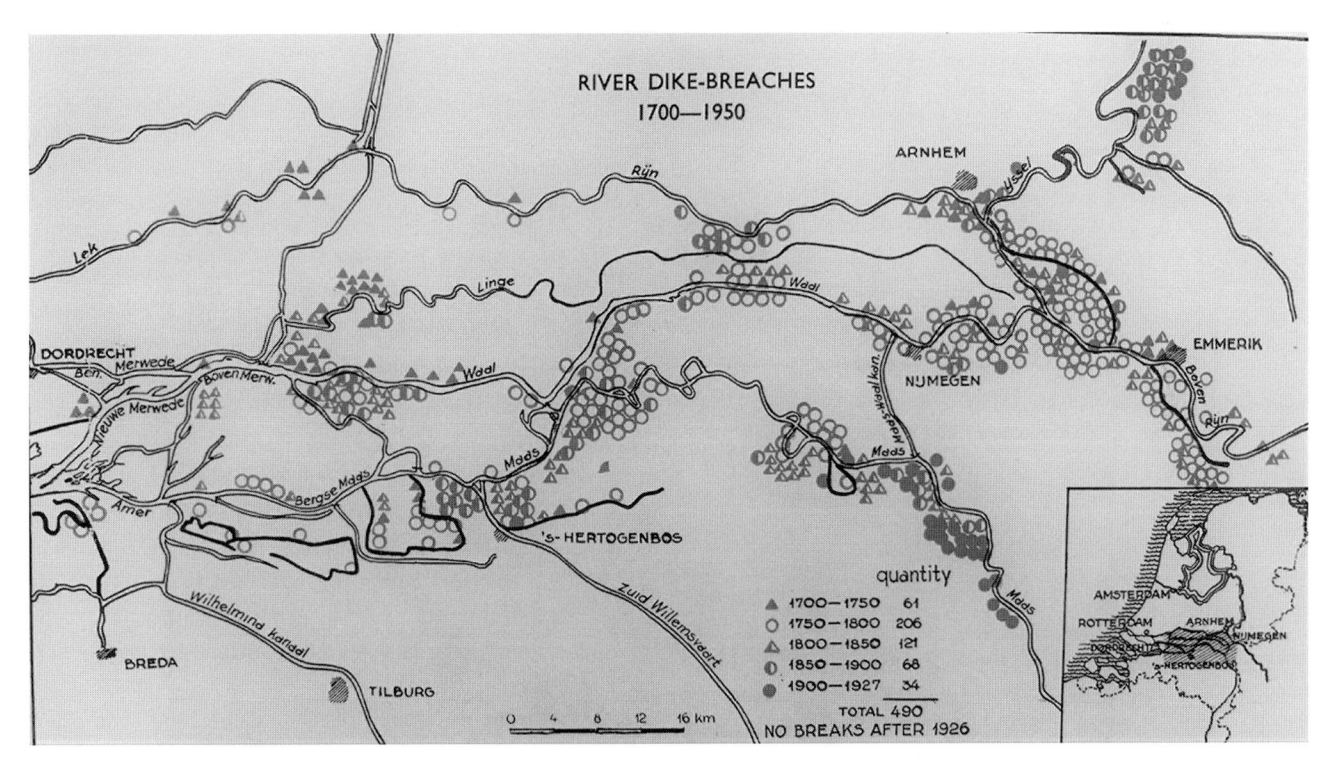

Figure 6.37. Historic dike breach events within the Rhine delta between 1700 and 1950, located in the Betuwe region between the Lek-Nederrijn (to north) and Waal River (to south). (Source: From van Veen, 1962.)

influence on flood management policy. This is because the flood control system effectively worked as intended (Camillo, 2012), providing a measure of vindication to the 1928 Mississippi River & Tributaries project and flood management strategy by the U.S. Army Corps of Engineers.

6.6.3.1 ICE JAM FLOODS

In high-latitude rivers ice jams are an important mechanism for dike breach flooding (Lindenschmidt et al., 2019). Because of the threat to vital flood control and transportation infrastructure (e.g., bridges), much is known about the mechanisms of ice dam (ice jam) flooding

Figure 6.38. Beginning of crevasse breaching dike at Mounds Landing, Mississippi. Flow convergence and overtopping corresponds to stage C in Figure 6.34. (Image from: "The Floods of 1927 in the Mississippi Basin," Frankenfeld, H. C., 1927 Monthly Weather Review Supplement # 29. Licensed by CC.)

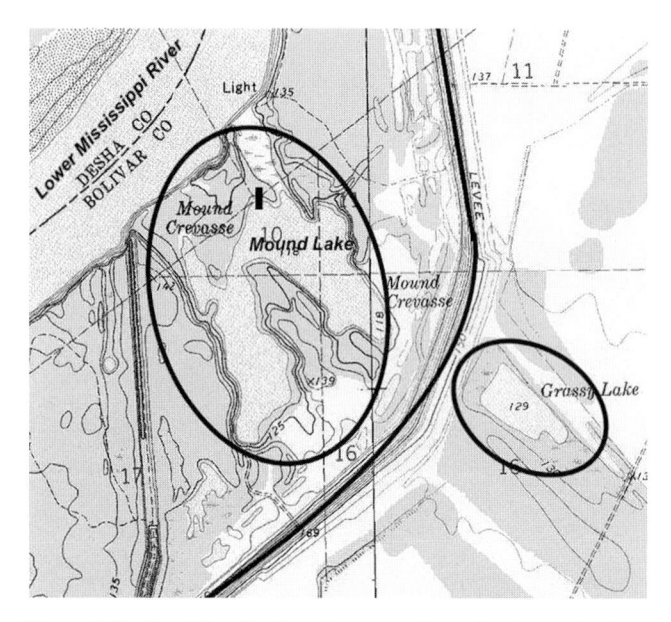

Figure 6.39. Resulting dike breach topography and hydrography from Mound Crevasse (1927), inside (Mound Lake) and outside (Grassy Lake) of the main-stem dike. The contemporary size of Mound Lake is 4.78 km^2. Realignment of the dike outside (around) the main scour hole. (Source: USGS topographic map, Mound Crevasse [scale, 1:24,000].)

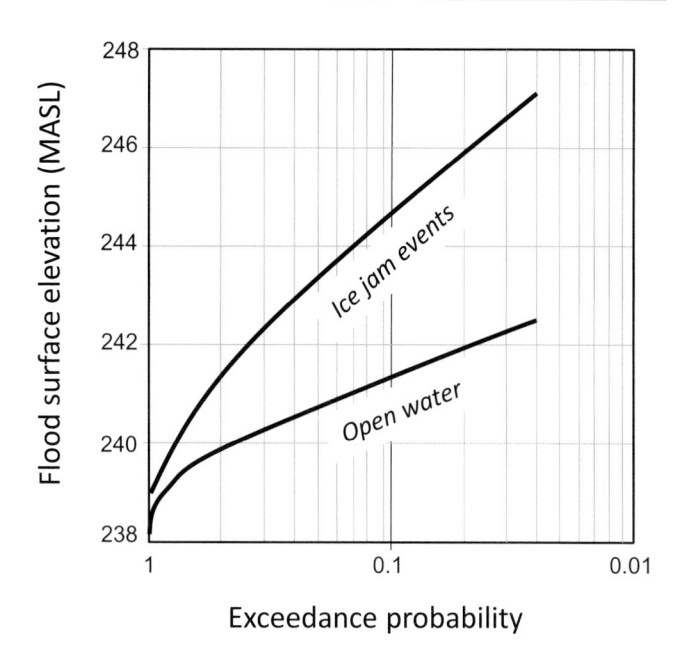

Figure 6.40. River stage response to ice jam events for specific exceedance probability, Athabasca River, Canada. (Source: Lindenschmidt, 2018.)

(Rokaya et al., 2018). Ice dam flooding is driven by upstream sources and in situ sources.

In situ ice forms as the river water surface freezes over and forms frazil ice, at slightly below freezing temperatures. Frazil ice is the formation of numerous low-density ice crystals within the (liquid) water column, resulting in the river having a cloudy white appearance. Sustained freezing temperatures result in formation of small ice slabs, which subsequently coalesce into a larger ice sheet at the water surface. An additional important step in formation of river ice is "anchor ice," which forms at the channel bed and banks. Anchor ice forms because of supercooling of coarse channel bed substrate (e.g., larger cobbles, boulders), tree limbs, infrastructure, or other promontories along channel banks. Movement of anchor ice then results in direct erosion of the channel bed and/or banks, with possible consequences to associated infrastructure.

The second mechanism by which ice jams form is receipt of upstream ice blocks becoming "jumbled up" at a specific river reach (Garver and Cockburn, 2009). As downstream flow continues, it drives upstream increases in river stage (Figure 6.40). In addition to floodwaters overtopping the dike, a considerable danger with ice jams is that pressure from streamflow pushes ice blocks into river banks and associated infrastructure, such as bridges and dikes, and damages their structural integrity by scour or lateral stress (pushing).

Of course, the worst scenario is a combination of in situ ice formation and upstream sources. This usually occurs where there is a constriction, reduction in slope, or where the channel is constrained by infrastructure – such as a bridge cross section. For a given flow-exceedance probability this results in higher flood stages in comparison to open water stage events (i.e., Figure 6.40).

In addition to hydrologic processes, hydraulic infrastructure along the Rhine was historically associated with formation of many ice jam events. Prior to the early 1800s some riverside farmers attempted to gain new agricultural lands by extending makeshift "local" groynes (kribben) into the river channel. The idea, as reviewed in Chapter 5, is that the groynes would trap sediment and result in formation of new lands. In many cases, however, such structures resulted in shallows that led to rapid ice formation during cold winters. The ice thickening caused streamflow to back up and flow at a higher stage. Ultimately, particularly in combination with upstream-derived discharge pulses, the ice thickening caused damage by piling up high against the dike, occasionally triggering dike breach floods (e.g., Figure 6.1).

6.6.3.2 INFILLING OF DIKE BREACH PONDS

Dike breach ponds (wiel) in different stages of evolution are common along the Dutch Rhine and other rivers that have long been embanked. Hydrologic connectivity and sedimentation influences dike breach pond evolution, and are markedly different on either side of the dike (Middelkoop, 1997). A comparison of two wielen formed over the last few centuries at Wamel, downstream of Nijmegen, NL, provides a good characterization of the distinct evolution of these unique anthropogenic hydrographic features, as previously examined by Middelkoop (1997). A wiel formed by a dike breach event in 1726 subsequently had the dike located to the outside of the wiel, such that it is located on the river-side of the dike and remains hydrologically connected to the Waal River. The original scour depth was 8.5 m, with the pond bottom identified as a sharp contact between the channel belt deposits and +5 m of fine-grained humic-rich lake-fill sediment. The pond is hydrologically connected to the Waal when discharge exceeds a threshold of 7,900 m^3/s, much higher than the average annual discharge (at Lobith, ~50 km upstream at the German border). Distinct sedimentary laminations reveal individual flood events, and vary from 1-cm to 5-cm thick dark gray to black, and 0.2-cm to 3.5-cm thick brown-gray laminae. Overall, clay comprises from 30% to 60% of the deposits. In contrast, a wiel located on the land-side of the dike, which formed on the night of January 26, 1781, has less than half of the sedimentary infill. The lack of hydrologic connectivity results in an absence of clastic deposits, as lake-fill deposits have, by contrast, much higher amounts of black organic-rich clay.

6.6.3.3 DIKE BREACH SEDIMENTOLOGY

The sedimentary signature of an individual dike breach event resembles a circular crevasse splay, with eroded sediment from the dike and the breach scour hole deposited as coarse-grained sediment, with an overall lateral fining trend ranging from coarse sand to silts (e.g., Gomez et al., 1997; Jacobson and Oberg, 1997; Middelkoop, 1997) (Figure 6.33). The particle size of such deposits is coarser than the underlying "natural" floodplain deposits. The circular (horseshoe) pattern of dike breach deposits differs somewhat from "natural" crevasse deposits, the latter having an elongated incisional dendritic pattern extending further beyond the channel breach.

Floodplain sedimentology is an important control on the severity and extent of dike breach sedimentation (Nelson and Leclair, 1996; Gomez et al., 1997; Middelkoop, 1997; Magilligan et al., 1998; Alexander et al., 2013). In particular, cohesive fine-grained floodplain deposits contrast with coarser channel belt deposits. Thus, dike breach splay deposits are larger where sandy channel belt deposits underlay dikes. The sand within the old channel belts can be easily scoured and reworked – ultimately redistributed within the splay. Thick cohesive clayey top-stratum deposits, however, are inherently resistant to scour. Dike breach events occurring over such deposits, therefore, do not contribute as much sediment to the splay deposits, and also result in less overall damage to the dike structure.

The Great Flood of 1993

A contemporary example of interrelations between dike breach events with floodplain sedimentology and geomorphology is the Great Flood of 1993 in the American Midwest (Figure 6.41), along the Lower Missouri and the Middle Mississippi Rivers (see Section 3.2.1.1 for hydroclimatology of 1993 flood). The event was historic and exceeded flood stage heights for nearly all stations, with peak flood magnitudes ranging from >100-year to 500-year recurrence intervals (Jacobson and Oberg, 1997; Schalk and Jacobson, 1997; Pitlick, 1997). The severity of the event caused dikes to be breached in numerous locations, and most dikes failed. Overall, 93,000 km^2 within the Missouri and upper and Middle

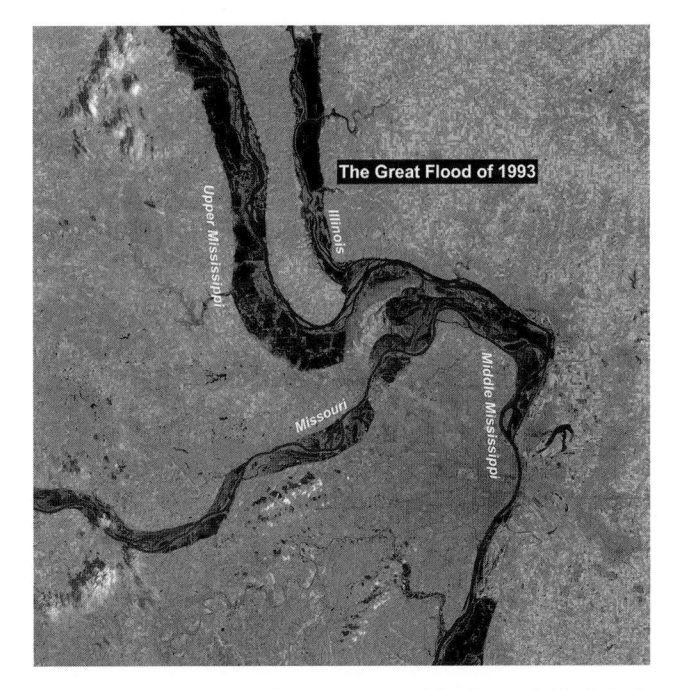

Figure 6.41. Great Flood of 1993 along the Missouri and Mississippi River basins. The lower Missouri, Mississippi, and Illinois Rivers just after the flood crest on August 19, 1993. St. Louis, MO is below Missouri and Mississippi Rivers' confluence. (Source: NASA Earth Observatory, Landsat 5.)

Table 6.10 *Floodplain erosion and sedimentation from multiple dike breach events from the Great Flood of 1993 along the middle Mississippi River, Miller City, Illinois*

Geomorphic mapping unit	Area (km^2)	Depth (cm)	Volume (m$^3 \times 10^6$)
Thick sand deposits	7.1	60	4.2
Medium-thick sand deposits	15.2	30	4.6
Thin sand deposits	52.5	10	5.2
Stripped surfaces	1.8	30	0.5
Scours	5.8	Variable	5.2
Net deposition			8.2

Source: Jacobson and Oberg (1997).

Mississippi basins were flooded, caused by the failure of 1,082 of the 1,576 federal and nonfederal dikes (Interagency Floodplain Management Review Committee, 1994; Gomez et al., 1997).

Within the lower Missouri, the dike breaches considerably altered the floodplain sedimentology and geomorphology along the dike margins. Sand was deposited as a continuous blanket atop the embanked floodplain and at the focus of dike breach events (Schalk and Jacobson, 1997). Some 300 km^2 of agricultural land was covered in sand at thicknesses >0.6 m. Dike breach deposits as thick as 2 m were common, with some as thick as 6 m. The maximum depth of the numerous scour holes ranged from 6.5 m to 15 m.

A factor in the catastrophic response of the lower Missouri concerns the sensitivity of the floodplain. The sedimentology of the Holocene Missouri valley is dominated by sandy channel belt deposits, which were locally scoured and reworked at the focus of the dike breach. Indeed, in addition to deposition, floodplain erosion was common in front of splay deposits, and resulted in numerous scour holes (Galat et al., 1998). Scour holes were created by the high hydraulic head produced from the dike (levee) height in relation to floodplain bottoms, with flow intensified through the dike breach opening (Jacobson and Oberg, 1997). Much of the sediment deposited from dike breach events was locally derived from erosion of the floodplain and dike.

The longer-term geomorphic response of the dike breach events along the lower Missouri also has positive environmental effects, which is key because of the overall degradation and loss of riparian bottomlands associated with embankment. Galat et al. (1998) identified 466 erosional scour holes created from the 1993 flood event along the Lower Missouri, which ranged in size from 0.07 ha to 32 ha. The scour holes enhance the floodplain geodiversity, ultimately providing aquatic habitat that supports greater biodiversity (Galat et al., 1998).

In contrast to the lower Missouri, the downstream geomorphic and sedimentary response to the flood along the middle Mississippi was more variable because of the heterogeneous floodplain sedimentology (Jacobson and Oberg, 1997; Magilligan et al., 1998). Along floodplain sections with coarser-grained channel belt deposits dike breach events were more extensive, resulting in substantial amounts of erosion and sedimentation. At Miller City, IL, for example, just upstream of the head of the lower Mississippi valley, extensive dike breach scour holes, floodplain stripping, and sedimentation were common (Table 6.10). Indeed, the size of the dike breach accommodated as much as 28% of the total volume of Mississippi discharge and had a velocity as high as 300 cm/s. This amount of energy resulted in a scour zone 2.2 km long with a depth of 2 m. Overall, some 8.2 million m^3 of sand was deposited across 75 km^2 of agricultural fields.

But impacts of the 1993 flood were much different along other reaches of the middle Mississippi floodplain, illustrating that it is not possible to generalize about the geomorphic impacts of large flood events (Knox and Daniels, 2002; Heitmuller et al., 2017). Floodplain reaches with thick cohesive floodplain deposits were found to be inherently resistant to floodplain stripping and scour hole formation (Gomez et al., 1997; Magilligan et al., 1998). Indeed, despite the severity of the flood event little sediment was actually deposited at dike breach splays along much of the middle Mississippi. The breach deposits were characterized as horseshoe-shaped splays with most of the sand deposited along the rim, a common depositional expression of dike breach events (e.g., Berendsen, 1982). With the exception of areas immediately adjacent to the dike breach splay, the average thickness of flood deposits was <5 mm (Gomez et al., 1997; Magilligan et al., 1998). The source of the deposits was suspended sediment, which partially derived from erosion and transport of eroded dike material. Additionally, there was little variability between splay deposits and dike material, which had a median particle size (d_{50}) of ~0.4 mm and ~0.5 mm, respectively. Adjacent sections of floodplain that were outside of the dikes had floodplain deposits with a median diameter (d_{50}) of ~0.15 mm, marginally finer than flood deposits inside of the embanked floodplain with a d_{50} of ~0.18 mm (Figure 6.42). Thus, in comparison to the above reported findings for the Miller City breach event (Table 6.10), little of the sediment deposited along extensive reaches of the middle Mississippi derived from erosion and reworking of the existing floodplain deposits.

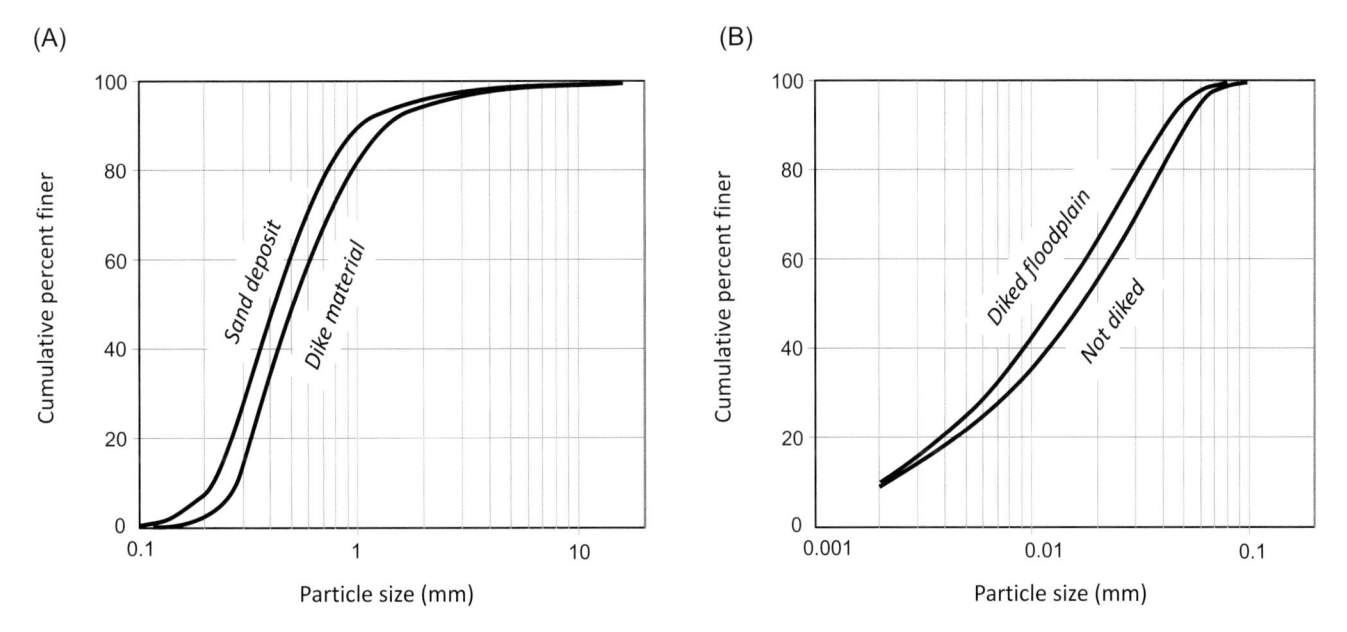

Figure 6.42. Composite floodplain deposits associated with the Great Flood of 1993: (A) comparison of sand deposits with somewhat coarser dike breach material and (B) comparison of diked to non-diked deposits. (Data source: Gomez et al., 1997.)

Without flood control infrastructure most large-scale mechanized floodplain agriculture would not exist. Thus, erosion of floodplain lands and deposition of coarse sediments across agricultural fields by a dike breach event is yet another unintended consequence of flood control, and an additional management challenge. Along the Missouri River, more than 1,800 km^2 of agricultural floodplain land was impacted by the erosion and deposition from the 1993 flood. In the short term after a flood event the question is what to do with the sand. The coarse flood sediments deposited atop agricultural lands must be removed if the land is to recover and become agriculturally viable. Along the Missouri River, the removal of sand and efforts required to return the floodplain to agricultural productivity after the 1993 event cost the government $1.2 million/km^2 (Jacobson and Oberg, 1997). In many developed nations, these costs are ultimately borne by taxpayers via government programs to assist farmers after such disasters, but in regions with less robust government institutions the cost can be borne by the farmer and local community. Or the agricultural floodplain is not restored, and the depositional signature of the dike breach flood event remains as a legacy to the anthropogenic sedimentary record (i.e., Cohen et al., 2009; James, 2013).

6.7 EMBANKED FLOODPLAIN EVOLUTION AND END-MEMBER REGULATED RIVERS

Over extended periods of time the construction of flood control systems drives the evolution of lowland floodplains, converting them into a coupled human–natural system, largely controlled by societal demands (Hesselink, 2002; Hudson and Middelkoop, 2015). Such rivers are hydrologically, sedimentologically, and morphologically distinct from natural rivers, although they may include a large range of environmental attributes in support of stakeholder interests and societal values. Some engineering modifications enhance certain biological attributes, such as borrow pits providing aquatic habitat, but the enormous reduction in the lateral extent of hydrologic connectivity is a substantial loss of riparian habitat, and a great reduction to riparian nutrient sequestration (Nienhuis, 2008; Schramm et al., 2009). Additionally, sedimentary infilling of floodplain wetlands and shallow water bodies further degrades habitat within the embanked floodplain. The loss of these water bodies not only represents a considerable decline in the ecological health of the embanked floodplain, it also represents a reduction in the capacity of floodplains to store flood waters and reduce downstream flood risk.

The preceding reviews (i.e., Chapters 5 and 6) of channel engineering and flood control impacts to rivers elucidate general conclusions regarding the continuum of fluvial adjustment in response to a somewhat typical sequence of hard engineering for flood control and channel stabilization.

River channels and floodplains adjust along a geomorphic continuum in response to a sequence of engineering (e.g., Chapter 5), in most cases beginning with dike construction (Figure 6.43 and Table 6.11), which often requires hard engineering for channel stabilization (Kiss et al., 2019). The timescale for changes to initially manifest within the lower reaches of large alluvial rivers depends upon a number of influences, and as a whole is rather abrupt (several years to a couple of decades, not many decades). Endogenic controls, particularly the older

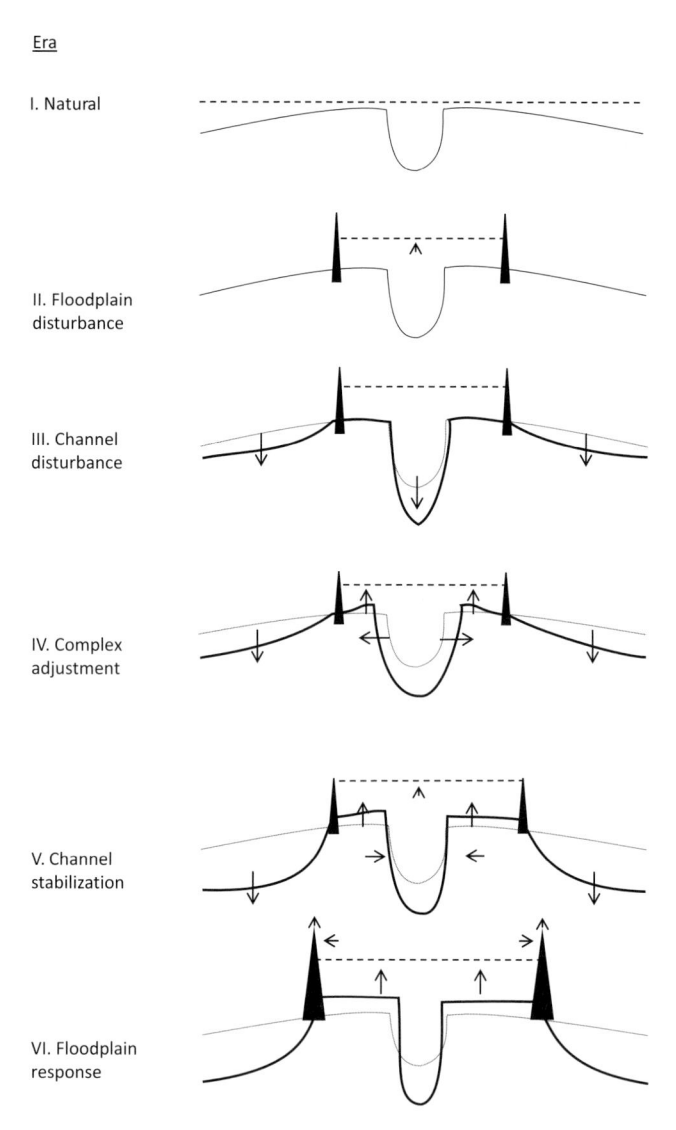

Figure 6.43. Continuum of lowland floodplain adjustment in response to the sequence of flood control engineering (see Table 6.11). (Modified from Hudson et al., 2008.)

Era:
I. Natural
II. Floodplain disturbance
III. Channel disturbance
IV. Complex adjustment
V. Channel stabilization
VI. Floodplain response

sedimentology, influence the sensitivity of a particular river reach to change. The diversity and combination of drivers, therefore, result in varying timescales for rivers to respond to flood control engineering. In general terms, the varied timescales of response can be appreciated by reviewing European and North American rivers, such as the lower Mississippi and Rhine. While the response of large lowland North American rivers to flood control may be in its infancy in comparison to many European rivers, the sequence has been initiated. Many larger European lowland rivers have been modified by hydraulic engineering for centuries, if not longer. Some have effectively reached the status of end-member regulated rivers with their morphodynamics entirely controlled by engineering. And the physical and environmental integrity of some larger rivers is being restored to undo historic mismanagement (e.g., Tulla cut-offs).

The fundamental requirement to reduce human flood risk means that dikes are the first significant form of engineering along lowland rivers. This abruptly segments the floodplain into two distinctive zones (Figure 6.43-I): embanked floodplains and detached flood basins. The embanked floodplain undergoes an enhanced flood pulse, which alters natural floodplain hydrologic and sedimentologic processes. In general, embanked floodplains undergo higher flood stages and increased sedimentation, with patterns that vary from natural floodplains. Embanked floodplain sedimentation can be characterized by coarser sediments and a less pronounced pattern of lateral fining. The detached flood basins are removed from the flood pulse, except for dike underseepage and groundwater flow that provide some hydrologic connectivity to oxbow lakes and some other floodplain water bodies. Unique anthropogenic floodplain features that are only associated with dikes form along the border (dike line) of embanked floodplains and flood basins, specifically borrow pits, dike breach ponds, and sand boils (Mansur et al., 1956). Main-stem sedimentation into the flood basin only occurs during extreme floods overtopping dikes, including dike breach splays (Figure 6.43-II).

Drainage channels and pumps are commonly associated with construction of a dike system (USACE, 1998a, 1998b, 1998c, 2000a), supporting human land use (e.g., settlement, agriculture). The drainage of flood basins often results in land subsidence by oxidation and dewatering (compaction), with subsidence rates being highly variable and related to the intensity of land use as well as flood basin sedimentology (see Section 7.3), particularly organic matter content and clay (Stouthamer and van Asselen, 2015).

In addition to dikes, channel straightening by meander neck cut-offs is also a common approach to flood control (Figure 6.43-III). The reduced sinuosity and slope increase create a channel knickpoint that directly increases boundary shear stress. In large alluvial rivers knickpoints are inherently unstable, and drive upstream channel bed incision as a form of negative feedback to reduce channel slope. Flood stage is lowered within the knickpoint zone, which is also associated with an adjustment to the hydraulic geometry and notably an increase in velocity (Section 5.4.1). Of course, the overall magnitude and rate of knickpoint incision is related to the channel substrate (particle size) as well as the presence of older sedimentary deposits.

Although "local" channel knickpoints are anticipated at a meander neck cut-off, in many cases the knickpoints initiate several unintended geomorphic consequences with management implications. Specifically, the channel bed incision results in alluvial banks becoming over-steepened. This is especially a problem along reaches with sandy bank material (e.g., point bar deposits). In such cases bank erosion ensues. The bank erosion

Table 6.11 *Continuum of lowland floodplain adjustment in response to the sequence of flood control engineering (see Figure 6.43)*

Era	Management / Engineering	Condition / Response
I. Natural	N/A	Channel: stable, includes sediment exchange (erosion/aggradation) with floodplain
		Floodplain: complex topography and dynamic lateral connectivity (surface and hyporheic)
II. Floodplain Disturbance	Dikes and land drainage	Channel: no strong response to dike construction
		Floodplain: increased flood stage and sedimentation of embanked floodplain during large discharge events, dikes result in loss of surface lateral connectivity with channel and associated floodplain processes, hyporheic connectivity with the channel (perhaps increased at high river stage)
III. Channel Disturbance	Channel straightening (cut-offs)	Channel: knickpoint formation (increase in slope and stream power) associated with meander bend cut-offs, rapid channel bed incision
		Floodplain: creation of artificial oxbow lakes within embanked floodplain, embanked sedimentation results in infilling of low swales and sloughs and diminished floodplain topography, land drainage and loss of sedimentation results in subsidence of distant floodplain
IV. Complex Adjustment	Monitoring	Channel: bank erosion (lateral) is dominant due to over-steepened banks, particularly along noncohesive floodplain deposits, rapid reworking of floodplain deposits.
		Floodplain: continued aggradation of low depressions and rapid infilling of artificial oxbow lakes within the embanked floodplain, continued subsidence and hyporheic connectivity of distant floodplain require further land drainage
V. Channel Stabilization	Bank protection, groynes and revetments	Channel: narrowing and greater stability (locked in place)
		Floodplain: absence of floodplain reworking accelerates aggradation of coarser flood deposits and loss of surface connectivity paths (sloughs, swales), accelerated infilling of new oxbow lakes, continued subsidence of floodplain on landside of dikes, drainage problems caused by sand boils and seepage during flood events
VI. Floodplain Response	Dike stability and fortification, setback	Channel: moderate incision due to loss of sediment from floodplain reworking, but lateral stability maintained with groynes and revetments
		Floodplain: accelerated aggradation of coarser flood deposits within embanked floodplain reduces flood water storage, resulting in higher flood (overbank) stage. Increased threat of channel avulsion and vulnerability of distant floodplain prompts new management solutions

Source: Modified from Hudson et al. (2008).

results in an increase in channel width, and as bank erosion continues it can threaten dikes. The eroded bank material results in channel bed shoaling, and is a problem to navigation. Dredging is therefore required to deepen the channel bed. To stabilize channel banks concrete revetment and riprap is installed along channel banks. To ensure the thalweg is "trained" in a specific location and to engineer a "self-scouring" channel bed, groynes (wing dikes) are installed along the river. Groynes are especially effective at narrowing the channel and reducing lateral activity, but they also ultimately result in accretion of coarse sediment and formation of lateral channel bars (Figure 6.43-IV). With continued sedimentary accretion by overbank processes and as vegetation becomes established atop channel bars, the bars merge with the adjacent floodplain (Figure 6.43-V). Along some rivers low floodplain surfaces are utilized for human uses, such as agriculture.

At this stage, the river and floodplain are highly trained and have a distinctive anthropogenic sedimentary and geomorphic signature. Overbank floodplain aggradation is dominant (Figure 6.43-VI) because the river has very little lateral mobility and floodplain deposits are not able to be reworked and assimilated into the active sediment load. A consequence of embanked floodplain aggradation is reduction to floodplain topographic complexity (e.g., geodiversity) as shallow water bodies have infilled. The reduction in water bodies and higher floodplain surface means that there is less room for floodwater storage, driving flood stages higher. Additionally, the combination of the higher embanked floodplain surface by aggradation and the lower flood basin surface by subsidence increases differences between lateral and down-valley slope values. Thus, from the perspective of flood management this stage of the continuum is associated with increased risk of a dike breach event, which can eventually result in an avulsion.

6.8 CLOSURE: TOWARD FLOOD BASIN MANAGEMENT

Along many lowland rivers dikes are the border between dynamic embanked floodplains and disconnected flood basins, many of which are ecologically, hydrologically, and geomorphically degraded by agriculture and settlement. Effective management of large flood basins has taken on added importance in the context of global environmental change, which includes sea level rise, hydroclimatic change, increased human population density, and land use change across fluvial lowlands (Reuss, 1998; Twilley et al., 2016; Best and Darby, 2020; Minderhoud et al., 2020). In combination with a strong motivation to conserve and restore riparian and fluvial deltaic ecosystems, these factors increase the pressure on the overall system, and the consequences of dike failure. Large flood basins are complex environments in which to manage and require engineering attentive to balancing the needs of stakeholder interests with knowledge of fluvial-deltaic sedimentology.

In Chapter 7 we examine approaches and challenges in managing flood basins within densely populated areas, particularly impacted by sea level rise, subsidence, and coastal storm surge, including New Orleans.

7 Flood Basins and Deltas: Subsidence, Sediment, and Storm Surge

> [L]ife outside of New Orleans took me through the vortex to the whirlpool of despair ... [Ignatius J. Reilly]
> —John Kennedy Toole (1980)

7.1 A CHALLENGING ENVIRONMENT

Beyond the margins of the active channel belt, expansive lowland flood basins are the most challenging fluvial environments to manage. In the context of most human uses, the local sedimentology, soils, topography, and connectivity with adjacent water bodies make for a complex hydrology. The variety of mechanisms that contribute to land inundation in these settings makes it curious that they are so intensively utilized. While hydrologic connectivity across large fluvial lowlands is a fundamental natural process, in the context of human welfare flooding is hazardous, and largely unacceptable. To effectively manage flood basins requires knowledge of hydrologic and sedimentary processes influenced by upstream and downstream controls. Importantly, flood prevention along large fluvial–deltaic settings requires coastal storm surge management. And flood basin inundation is exacerbated by the most quiescent of geomorphic processes – subsidence.

Effective hydrologic management of flood basins requires a sophisticated understanding of the spatial and temporal dynamics of different inundation mechanisms, sediment transport, and flood basin geomorphology and sedimentology. And flood basin management cannot be separated from competing environmental and societal interests, both inside and outside of the basin (Figure 7.1). Hydrologically, flood basins are inundated by seasonal discharge pulses fed by upstream sources for extended durations, but also by small "local" tributaries, groundwater, and intense pluvial events occurring in the lowermost reaches of the drainage basin. Geomorphically, flood basins are more hybrid than are usually characterized and include a mosaic of topographic and sedimentary features, including thick cohesive deposits, peat domes, and moribund distributaries associated with channel avulsion processes (e.g., Saucier, 1994; Stouthamer and Berendsen, 2001). As reviewed in Chapter 3, flood basins within larger river valleys usually have considerable backswamp environments, which are topographically situated between raised channel belts and the bluff line of the river valley. Along large deltas, flood basins are located within former interdistributary bay environments, which transition from freshwater swamp forest to brackish coastal marsh. The coast, therefore, is not a sharp boundary but instead can be considered along a gradient of change in hydrology, vegetation, and salinity. The delta plain includes a range of coastal morphologies and processes that influence its vulnerability to flooding by storm surge events (Coleman and Wright, 1971; Penland et al., 1988). The morphology of tidal- and fluvial-dominated deltas, for example, which have wide river mouths and interdistributary basins, enables marine waters to surge far inland.

In this chapter we focus on the management of flood basins within fluvial lowlands and the coastal zone. For many decades floodplain sciences lagged behind the capabilities of engineers to alter flood basin hydrology and geomorphology, contributing to substantial biophysical degradation of lowland environments. Floodplain engineering by pumping, drainage, and embankment caused indirect (i.e., unintended) changes to floodplain environments that, because of the scale of large fluvial lowlands, required decades to be observed. In some instances, flood risk increased, such as with the onset of subsidence. Increasingly, however, integrated floodplain sciences involving some combination of geomorphic, hydrologic, and ecosystem processes, are utilized to examine mismanagement by traditional "hard engineering" approaches. The "flood pulse paradigm" (i.e., Chapter 3) can be seen as a key conceptual tool in the movement toward integrated management and recognizes the importance of maintaining natural-like hydrologic and sedimentary dynamics (i.e., geodiversity) to also reduce flood disasters. Even in dense urban settings, for example, maintaining a high groundwater table and providing water with (urban) space, rather than continuous pumping of groundwater, is seen as a valuable approach to reduce subsidence. Such a perspective is important when considering the high concentration of populations and range of human activities subject to flood risk within lowland rivers (e.g., Chan et al., 2013).

Figure 7.1. Operation of Bonnet Carré Spillway for floodwater diversion during the 2016 flood of lower Mississippi River (to the right), upstream of New Orleans (looking south, downstream). The spillway was only opened eight times between 1937 and 1997, but seven times since 2008 (2008, 2011, 2016, 2018, 2019, 2019, 2020). The spillway structure is 2.36 km long. (Photo: January 10, 2016, U.S. Army Corps of Engineers, New Orleans Division.)

7.2 FLOOD DIVERSION STRUCTURES: ENVIRONMENTAL IMPACTS AND STAKEHOLDER CONTROVERSY

7.2.1 Types of Flood Diversion Structures

The hydrologic management of flood basin environments includes engineering and planning to control inundation processes. Controlling water in flood basins usually occurs by a combination of drainage, pumping, and flood diversion structures. Ditches redirect floodwaters to where it can be removed by gravity drainage or pumping, and also ensures that groundwater seepage does not represent a hazard to the integrity of dike structures or agriculture and settlement. Pumping is usually reserved for larger events and sustained wet periods. In contrast to pumps and ditches, which move water out of flood basins, flood diversion structures move floodwaters into flood basins. Although they have existed for centuries and have a legacy associated with a "hard" engineering approach, flood diversion structures remain an important mode of modern flood control.

Flood diversion structures take advantage of expansive lowland flood basins to attenuate the downstream flood wave, reducing the flood crest height. Flood diversion structures are designed to divert a specific amount of water while the flood wave travels downstream, reducing the height of the flood crest. Within the flood basin, the water control infrastructure often integrates "natural" floodplain geomorphic features. Former river courses and oxbow lakes, for example, can be utilized as storage for water pumped from flood basin surfaces, and then transported to larger storage basins or pumped back into the main river channel following passage of the flood wave.

Diversion structures function by siphoning floodwaters into the basin at a predefined discharge threshold. This involves floodwaters flowing over a spillway weir, which is usually a concrete structure built at a specific elevation precisely calibrated to the stage-discharge curve at the channel cross section (Singer and Aalto, 2009). The reduction in the downstream flood wave can then be controlled by the size (area) of the spillway opening and the duration in which it is open. The actual opening of the spillway may be automated or require manual operation.

The diversion of so much water represents an abrupt change in the hydraulics of both the flood basin and river channel. Subsequently, this results in a temporary change in shear stress and the competence of floodwaters to transport sediment (Snedden et al., 2007), which may be geomorphically manifest as either erosion or deposition within the river channel and flood basin, requiring further management options. The diversion and subsequent deposition of coarse sediment, for example, results in aggradation, hinders the functioning of engineered structures and reduces floodwater storage capacity. The coarse sedimentary deposits require removal by dredge operators or local contractors. The diversion of fine sediment (wash-load) can influence the flood basin biogeochemistry because of organic and inorganic pollutants being adsorbed to colloidal sediment (~clay), potentially degrading associated ecosystems (Springborn et al., 2011). The operation of flood diversion structures, therefore, should take into consideration the discharge–sediment dynamics of the main-stem river (Mossa, 1996; DeLaune et al., 2003; Day et al., 2012).

There are two main types of flood diversion structures (Table 7.1), backwater areas and bypass structures. Backwater areas are a parcel of lowland adjacent to the main-stem channel, which is reserved for storage of floodwaters exceeding a specific magnitude (stage elevation). Following recession of the flood crest, a sluice or floodgate is opened and the waters drain (by gravity) back into the channel. Pumping may be required, depending upon topography and local conditions.

The second type of diversion structure is a flood bypass structure. As the term suggests, floodwaters are allowed to bypass a specific reach along the main-stem river and flow through an engineered channel within the flood basin. The diverted floodwaters may reconnect with the main-stem channel downstream, and usually they are designed to bypass a specific urban area. The time associated with floodwaters disconnecting and then reconnecting with the main-stem channel results in downstream flood wave attenuation. In the case of deltas, the floodwaters are often funneled to a distributary channel course, which then drains into an interdistributary basin and coastal marsh before eventually flowing into the ocean.

A unique approach to flood basin diversions occurs in the Rhine delta, a labyrinth of active and abandoned distributary channels and flood basins. The approach takes advantage of distributaries and numerous flood subbasins formed between natural levees and channel belts, as well as different drainage districts (van de Ven, 2004). The advantage is that when

Table 7.1 *Types of flood diversion structures along large lowland rivers**

Location	Type	Operations	Examples
Alluvial valley	Flood storage (retention)	Temporary storage in large floodplain (backswamp, lake/wetland) basins, pumping may be required to move water back into main-stem channel. Passive and active weir operation, with trigger mechanisms including threshold (elevation) stage or discharge overspill, removal of pins/bays	Yazoo Basin (Mississippi); Poyang Lake (Yangtze)
	Flood bypass; can include engineered trough, simple lowland basins, semi-engineered structure to guide direction of floodwaters, dike walls to retain in corridor, or old channel	Floodwater diverted via lowland paths in flood basin, partial engineering to direct water flow, reconnect downstream of inlet Passive and active spillway weir, with trigger mechanisms including threshold (elevation) stage or discharge overspill, removal of pins/bays, fuse/explosives or machinery to remove earthen dike material	Red River Bypass (Red, ND), New Danube (Vienna), New Madrid (Mississippi), Sutter Bypass (Sacramento)
Delta	Flood bypass	Floodwater does not reenter the main-stem river, flushed to coast, utilizes natural distributary or engineered flood trough. Passive or active weir operation.	Atchafalaya (Mississippi), Bonnet Carré, (Mississippi), IJssel River** (Rhine delta), Morganza (Mississippi), Yolo Bypass (Sacramento)

*Does not include bedrock troughs or tunnels engineered for smaller rivers, **includes Hondsbroeksche Pleij (Nederrijn-IJssel in Rhine delta)
Source: Author table.

floodwater is diverted into a flood basin by a diversion structure, or dike breach, it fills up the flood basin – much like a bathtub – but does not threaten to flood adjacent subbasins. In the case of the Rhine delta, the slope – oriented east to west – along the delta plain assures that the floodwaters slowly flow westerly (coastward) and then pile up against the bounding dike at the western side of the subbasin, where it may be routed into adjacent polder basins, or pumped into larger canals for drainage into rivers.

The above scenario is exemplified by the historic Alblasserwaard polder system (Nillesen and Kok, 2015), which straddles several embanked distributaries upstream of Rotterdam (Figure 7.2). The polder developed through trial-and-error over six centuries, but ultimately an approach of impoldering by dikes and ditches has resulted in a sustainable water management system that has preserved a cultural wet landscape underlain by thick peat, and which is being adapted to an integrated system of water and flood management to accommodate projected environmental change (Niellsen and Kok, 2015; Rivierland Waterschap, 2017). Polders in the Netherlands drain by gravity into ditches and then are pumped into canals, historically by windmills (Figure 7.3), where water is temporarily stored before being discharged by gravity and sluice gates into distributary channels. Some windmills utilized an Archimedes screw pump, an approach that continues to be utilized albeit much larger motorized pumps (Figure 7.4). Historically, this approach developed rather haphazardly because of the various periods that embankment and drainage of flood basins occurred, and also because of numerous water management districts that developed over the late medieval period (van Veen, 1948; Lambert, 1985). The later centralization of the water boards assured their coordination and consolidation, although they remain independent and democratic. This approach to water managements – subbasins and water boards – is uniquely Dutch and has remained remarkably effective. While historically they predate the Netherlands, water boards continue to be a core local-scale governmental component of the modern Dutch strategy to manage floods within the Rhine delta.

7.2.2 Flood Bypass on a Massive Scale: The Lower Mississippi

As much as hydrologic management of flood basins involves the redistribution of water, it is also environmental management. As part of its 1928 Mississippi River & Tributaries Project (1928 MR&T) the U.S. Army Corps of Engineers developed a system of flood diversion structures (MRC, 2008). The system along the

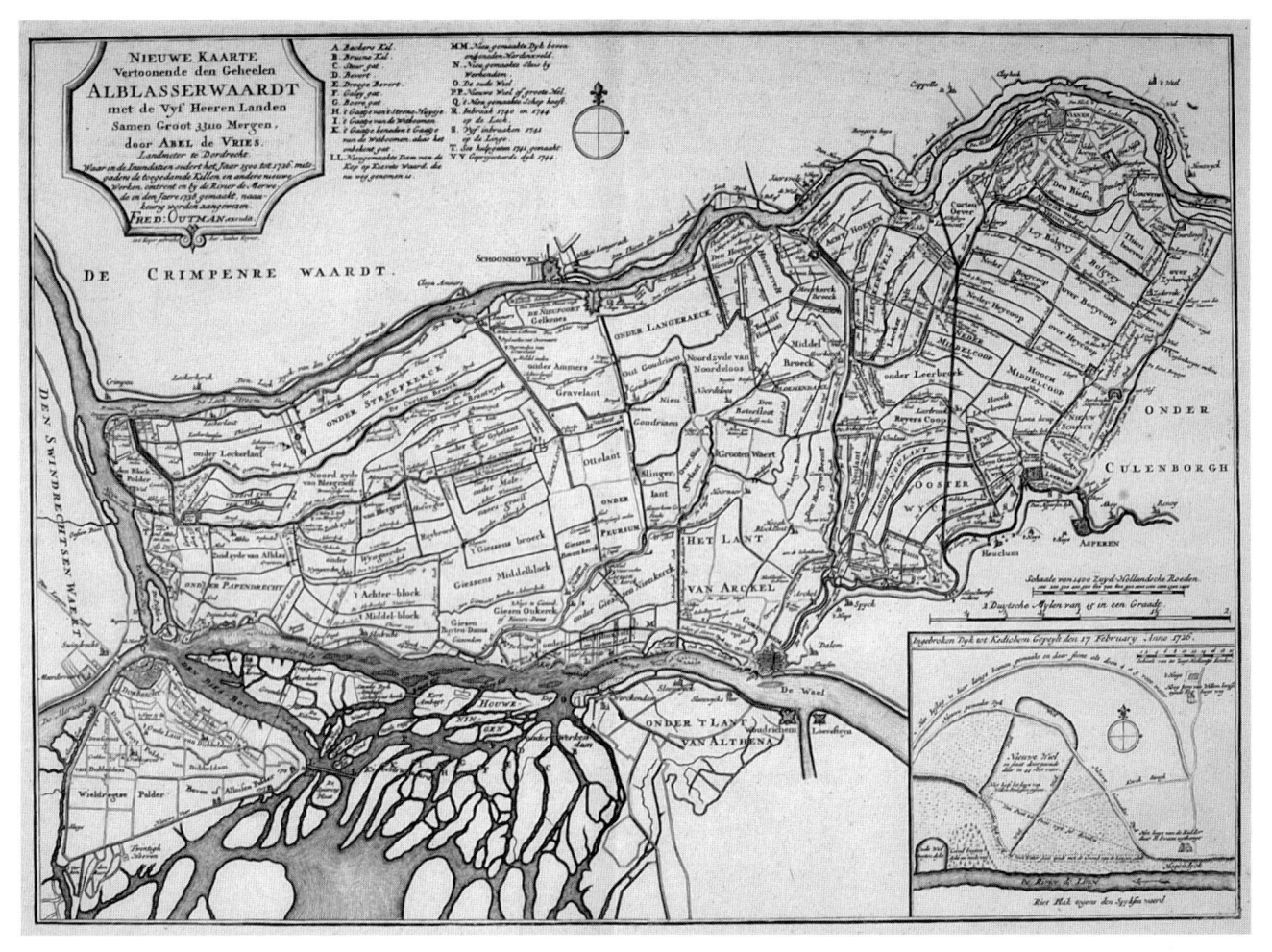

Figure 7.2. Historic map of Alblasserwaard and Vijfheerenlanden polders. The map shows individual polders within the structure of ditches and canals, between embanked Rhine River distributaries. To the south, west, and north are the Waal and Merwede, Noord, and Lek Rivers, respectively. The future Biesbosch nature reserve, initiated by dike breach events during the St. Elizabeth Day floods (1421–1423), is shown as a prograding inland delta (bottom left of map). Bottom right an inset of the dike breach at Kedichem. North is to the top. Surveyor: Abel de Vries, Cartographer: Jacobus Keyser; published in 1726. (Source: Archives of the Gemeente Dordrecht, Zuid Holland. Collection Dordracum Illustratum [# 551_15011].)

lower Mississippi utilizes subbasins in the alluvial valley and delta as flood backwater areas and flood bypass structures, and involves a combination of engineered structures and natural geomorphic features. Formal components of the main-stem Mississippi River flood bypass system include the New Madrid Floodway, Morganza Floodway, and Bonnet Carré Spillway. The 1928 MR&T also includes four backwater areas for temporary storage of floodwaters at tributary river mouths, including the St. Francis, White, Yazoo and Red rivers. Additionally, the Old River Control Structure allocates main-stem Mississippi River discharge into the Atchafalaya basin, which includes the West Atchafalaya Floodway and the Lower Atchafalaya Floodway. The initial development of the 1928 MR&T project was in response to the epic 1927 flood. While the 1928 MR&T project has not yet been completed, since 1927 it has successfully prevented subsequent main-stem flooding derived from upstream sources (DeHaan et al., 2012).

7.2.2.1 YAZOO FLOOD BASIN CONTROVERSY: ENVIRONMENT VERSUS AGRICULTURE

There is often great tension between stakeholders concerning the management of flood basins for different land uses, especially agriculture and nature conservation (White, 1945; Harrison, 1950; Nienhuis and Leuven, 2001; Lorimer and Driessen, 2014; Berkowitz et al., 2020). An excellent example of this issue is represented by flood management of the Yazoo basin, located within the eastern margins of the lower Mississippi alluvial valley (Figure 7.5), where it is playing out on a massive scale, hydrologically, ecologically, and governmentally.

Prior to human modifications the Yazoo was one of Earth's great flood basins, a pulsing, respiring biophysical hydrosystem. The Yazoo is the largest (34,590 km^2) subbasin of the lower Mississippi alluvial valley (Saucier, 1994), seasonally inundating extensive tracts of bottomland forest. Although 76% of the Yazoo's once extensive bottomland hardwood forests has been

Figure 7.3. Photo of historic windmills and drainage canal at Kinderdijk in lower (northwest) Alblasserwaard polder system, the Netherlands (upstream of Rotterdam). Kinderdijk dates to the early medieval period; the windmills were constructed in the mid-eighteenth century. (Photo: Rijkswaterstaat.)

Figure 7.4. Modern Archimedes screw to drain Alblasserwaard polder system into Lek River, upstream of Rotterdam. (Photo: Rijkswaterstaat.)

replaced with agriculture, the remaining 24% includes some of the largest remaining tracts in North America (Klimas et al., 2011). Ironically, conservation of the remaining unique hydroecosystem required direct intervention from two US Presidents – both reluctant conservationists – separated by a century of floodplain engineering and environmental change.

Like most large lowland geomorphic settings, the Yazoo basin is hardly a homogenous environment and its flood hydrology is complex. The drainage boundaries of the Yazoo basin are both natural and anthropogenic, which creates unique issues related to flood control and environmental management. The easterly and northeasterly drainage border are the bounding Pleistocene and Tertiary terraces along the valley margins whereas the main-stem Mississippi dike system delineates its westerly and southerly drainage border, and it is here where the Yazoo River finally joins the lower Mississippi River at Vicksburg, Mississippi.

Within the basin there are several reaches of partially buried main-stem Mississippi River paleo-channels, as well as yazoo-style rivers that drain adjacent terraces. Indeed, "yazoo" is the textbook term referring to parallel flowing channels within flood basins, usually several meters below the level of the active main-stem channel belt. The impeded drainage of these yazoo rivers historically contributed to extensive flooding across the basin. Inundation of the basin also occurs locally from exfiltration of groundwater from the Mississippi alluvial aquifer (Krinitzsky and Wire, 1964). Intense pluvial events characterize the precipitation regime, as the proximity of the warm Gulf of Mexico contributes to the basin's high annual rainfall (140 cm). Soils largely derive from clayey overbank deposits up to ~20 m in thickness (Saucier, 1994). Cohesive soil results in low rates of infiltration (Krinitzsky and Wire, 1964), and during cooler winter and spring months 50–70% of rainfall flows as runoff rather than percolating into the soil.

But the major control on flooding and the drainage "problem" in the Yazoo basin are lower Mississippi flood events. The Yazoo basin always served as a natural backwater area for Mississippi River floodwaters, reducing downstream flood crests. Because the main-stem dikes along the Mississippi essentially eliminated much of the natural flood pulse, the only inlet for Mississippi floodwaters is the Yazoo basin outlet (at Vicksburg). At high stage levels (>21.3 MASL) the Mississippi River backs up into the tortuous Yazoo drainage network, resulting in inundation of extensive areas of natural and agricultural landscapes (U.S. EPA, 2008).

Historically, the annual flood pulse dynamics supported an ecosystem with high levels of biodiversity closely related to the basin's floodplain hydrology and geomorphology (Klimas et al., 2005). The basin contains extensive swamp forests with characteristic bald cypress (*Taxodium distichum*) and water tupelo (*Nyssa aquatica*). These forests support valuable *keystone* wildlife, species deemed essential to the maintenance and integrity of an ecosystem (Dahl et al., 2009). This includes the Louisiana Black Bear (*Ursus americanus luteolus*), recently (in 2016) delisted from the federal endangered species list. While the Louisiana Black Bear is a keystone species, as concerns conservation it is also the ultimate – and original – example of "charismatic megafauna." In 1902 the Louisiana Black Bear became internationally characterized as "the" teddy bear following a botched hunting expedition to the Yazoo basin by US President Theodore Roosevelt. Since that moment the basin underwent much deforestation for timber and agriculture, and is persistently under heavy pressure – despite its hydrologic limitations – for agriculture. Roosevelt's swamps are preserved as a national forest in his name. And that so much of the Yazoo basin forests survived the great timber extraction that swept through the fluvial lowlands of the Southern United States in the early- to mid-1900s (Wohl, 2004), however, is largely because of the immortalization of Roosevelt's bear.

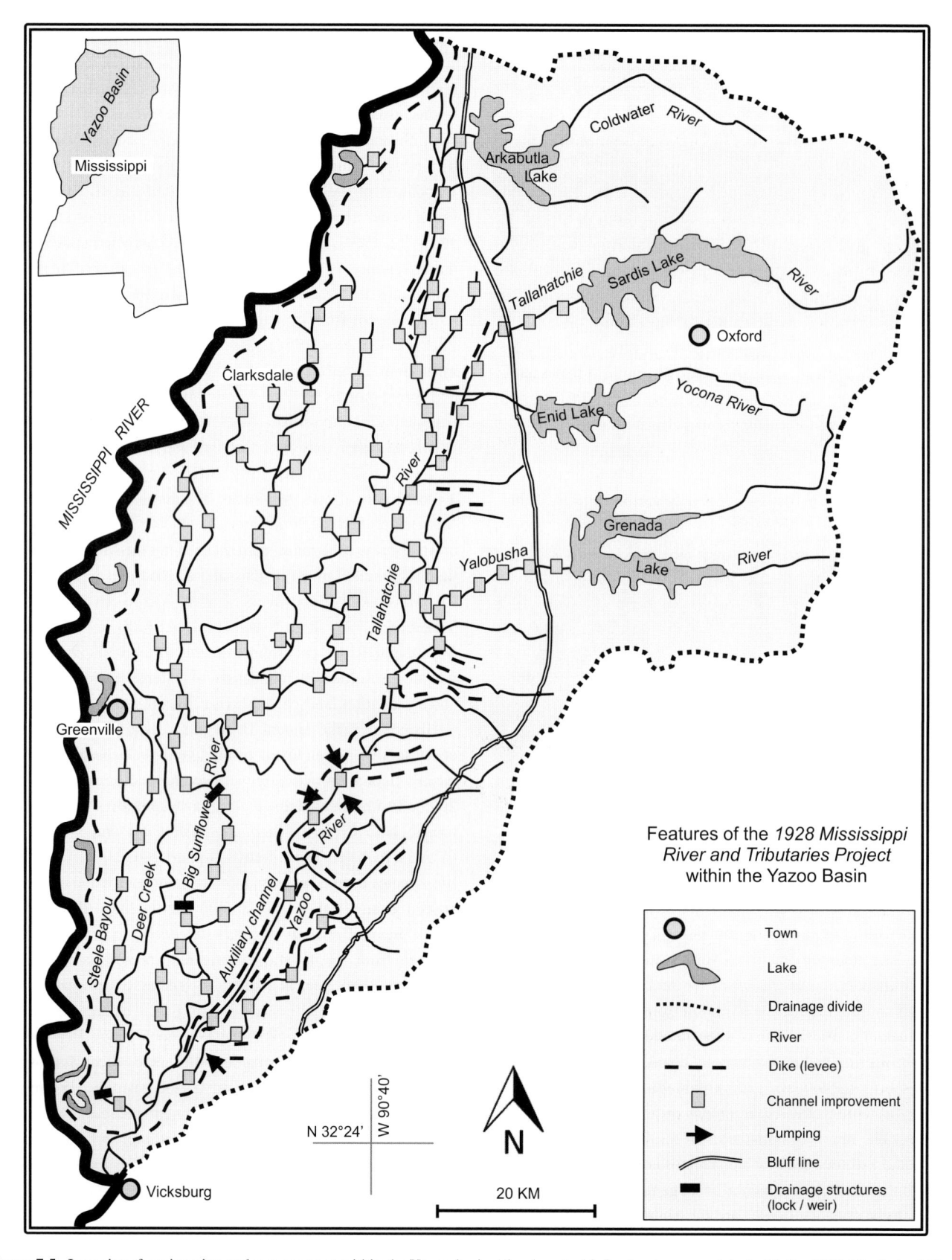

Figure 7.5. Intensity of engineering and management within the Yazoo basin. Flood control infrastructure associated with the 1928 Mississippi River & Tributaries Project indicated, including pumps, drainage structures, channel engineering ("improvement"), and dikes. Dams and storage reservoirs constructed atop adjacent bluffs. (Author figure based on U.S. Army Corps of Engineers.)

The Yazoo basin has an extensive floodplain ecosystem intricately linked to a network of wetlands and water bodies created from cut-offs, avulsions, secondary channels, arcuate sloughs from meander scroll, and sunken depressions (Klimas et al., 2005). The wetlands serve as prime bird habitat along a biannual hemispheric migratory route between the Arctic Circle and South America. The U.S. Fish and Wildlife Service has identified 258 bird species within the Yazoo backwater area, including forty-one rare bird species (U.S. FWS, 2001). Spring flooding coincides with key moments in the migration by providing habitat for breeding and foraging. The wetlands support a rich assemblage of invertebrates that serve as prime food source, such as water insects, which explode in population with the onset of inundation.

While floodplain water bodies within the Yazoo basin have been degraded, they continue to support a diverse range of fishes having a life cycle intricately linked to the (diminished) annual flood pulse. Fish require variable depth, currents, and temperatures for different components of their life history. By reducing the frequency, depth and duration of inundation (i.e., reducing flow variability) floodplain engineering has directly reduced suitable aquatic habitat. Larval fish hatched in the backwaters are able to survive for about a week from nutrients supplied by the attached yolk sack. Beyond about a week, however, the survival of fry depends upon obtaining adequate food sources within the shallow floodplain water bodies. Invertebrates on the flooded forest floor live off detrital food sources (e.g., fungi, bacteria), and thereby represent prime food sources for juvenile fish. The longer that juvenile fish are able to spend in floodplain water bodies, the healthier and more diverse are the fish stock in the main-stem channel.

The signature fish for the lower Mississippi floodplain is the alligator gar (*Atractosteus spatula*), which annually requires a couple of months in shallow flooded backwater areas for spawning and nourishment (Allen et al., 2014). Alligator gar (*Atractosteus spatula*) are large fish, growing longer than 2 m in length at maturity, and can live more than ~50 years. But their large size and slow rate of reproduction – they do not spawn until about ten years of age – means that they are also a delicate species, sensitive to the hydrologic alteration of their physical habitat (Allen et al., 2014).

As is common to lowland floodplain settings, the annual flood pulse dynamics of the Yazoo basin is essential to nutrient and biogeochemical cycling of the lower Mississippi, and thus North America. And the impact of flood control engineering, especially by the 1928 MR&T Project, is directly responsible for a reduction in floodplain sequestration of carbon, nitrogen, and phosphorus. Denitrification, for example, requires microbial processes in an anaerobic (saturated) environment. Reduction in the flood pulse and seasonal inundation has reduced denitrification and proportionately increased the downstream transport of nitrogen to the Gulf of Mexico (Schramm et al., 2009), thereby contributing to expansion of the dead zone (Mitsch et al., 2001).

Flood management in the United States is like a pendulum, shifting back and forth with changes in the priorities of major federal agencies (Hudson, 2018)

For decades, flood problems in the Yazoo basin stymied governmental and nongovernmental stakeholders (USACE, 1982). In recent decades the basin has been a battlefield between major federal agencies, the U.S. Corps of Engineers versus the U.S. Fish and Wildlife and the U.S. Environmental Protection Agency. Indeed, the proposed drainage of the Yazoo basin had its roots in the 1941 Flood Control Act, which amended the 1928 MR&T to allow for an extension of dikes and other types of flood control infrastructure into the Yazoo basin (Figure 7.5). This resulted in construction of several upland reservoirs, as well as engineering and straightening of a number of channels, requiring much concrete and riprap for bank protection because of erosion problems. The internal drainage of 305,000 ha of the basin is largely controlled by a floodgate at the lower (southern) end of the basin, the Steele Bayou Gravity Control Structure, initially completed in 1969 (USACE, 1988). Subsequently, by 1978 a much larger drainage complex had been created with completion of additional channel engineering, retention areas and gate structures, including Little Sunflower Drainage, Muddy Bayou, Eagle Lake, and Yazoo Backwater Levees. These followed two major floods in the early 1970s and after decades of complaints from local interests and their congressional representatives. Of course, the major impetuses for these works were agricultural. Within the Yazoo basin farming is mainly limited to cotton and soy cultivation, which can withstand alternating periods of inundation and drying. And, soy farming can be profitable during a short growing season, which often occurs during the late spring and summer, but dependent upon when the Army Corps of Engineers opens the floodgate to drain the Yazoo basin.

Steele Bayou drainage structure allows water from the Yazoo basin to drain into the lower Mississippi River, but only when the stage of the Mississippi is below 21.3 MASL. During Mississippi River high discharge events the structure is closed and Mississippi floodwaters are blocked from entering the basin. As long as the Yazoo basin waters are low this mainly works, but many times when the structure is closed the interior of the basin is simultaneously flooded because of local/regional hydrologic sources and groundwater seepage (Figure 7.6). This creates a perilous situation that increases flood risk, both within the basin and for the lower Mississippi because of reduced storage of its floodwaters. And although in its natural state the Yazoo slowly drained into the lower Mississippi, installation of the floodgate abruptly increases the hydraulic head, such that spring drainage season is now a high-energy event requiring additional management and maintenance of the engineered channel. Soon after construction was completed, a scour problem in the downstream channel was observed. Flood waters flowing over the floodgate

Figure 7.6. Backwater flooding of Yazoo basin from the epic 2011 flood along lower Mississippi. During high discharge events Mississippi floodwaters back into Yazoo basin, flooding bottomland forest and agricultural lands, reducing downstream flood crests. (Source: Delta Regional Authority.)

Figure 7.7. Managing the management. Erosion repair to Steele Bayou drainage channel that manages floodwaters in the Yazoo basin. The channel required 204 tons of riprap for bank stabilization, which erodes with floodwater release. The floodgate creates a hydraulic head and shear stress that initiates erosion of the outlet channel, requiring additional maintenance and monitoring. (Photo: 2014, U.S. Army Corps of Engineers.)

generated considerable channel shear stress, creating a major scour hole some 200 m wide and 13 m deep, compromising the integrity of the floodgate structure (USACE, 1988) and requiring constant maintenance (Figure 7.7).

Completion of the Steele Bayou Gravity Drainage Structure was not enough to satisfy agricultural interests, and it was long the intention the structure would be supported by hydraulic pumps (USACE, 1982; U.S. EPA, 2008). As part of its 1928 MR&T project the USACE very much intended to drain the Yazoo basin. A plan to construct enormous pumps was put forward in 1941, but was never realized. Again in 1960 a plan was put forward, but was tabled because of complications in subsequent federal Water Acts. And the 1986 Water Resources Development Act required federal projects to receive a minimum of 25% contributing financial support from "local" partners. The finances never materialized, as cost–benefit analyses did not demonstrate the (agricultural) economics were viable. The 1986 requirement of federal and local cost sharing was a pivotal moment in U.S. flood management because of giving local constituencies more control over which types of projects were implemented in their area of interest, a movement against the strong (top down) federal role in flood control. But assuming sufficient funds were available, local and federal stakeholder interests are often not aligned. The obvious problem here is that local stakeholder interests do not always have a broader – spatial or temporal – view on hydrologic and environmental issues, and may lack the expertise to evaluate the scientific basis of such projects.

Two decades after the 1986 Water Resources Development Act, the U.S. Army Corps of Engineers put forward a new plan to drain the Yazoo basin; the Yazoo Backwater Area Reformulation Main Report and Final Supplemental Environmental Impact Statement. Its 2007 plan, a revision of their massive plan put forward in 2000 in response to comments from nongovernmental and governmental stakeholders, would have seen construction of the world's largest pumps (at the time) with a capacity of 481 m^3/s to finally control the hydrology of the Yazoo basin and protect its agricultural interests (U.S. EPA, 2008). The project area spanned 257,040 ha in the lower (southern) reaches of the Yazoo basin. About half the project area, 133,400 ha, is subject to inundation every two years, representing a strong flood pulse for such an enormous area. Within this area the project included 61,200–93,430 ha of floodplain wetlands and bottomland forests. These unique hydro-ecosystems, as noted above, represent vital aquatic habitat for flora and fauna, and provide essential ecosystem services by improving the quality of groundwater and surface waters (Klimas et al., 2011).

Significant opposition to the USACE's 2007 plan occurred from nongovernmental environmental stakeholders and other federal agencies, including the U.S. Fish and Wildlife Service and the U.S. EPA. Section 404c of the 1972 Clean Water Act grants the U.S. EPA authority to deny the permit of projects that would degrade surface waters or associated ecosystems. The U.S. EPA concluded that 27,000–37,000 ha of wetlands would be *unacceptably* adversely impacted by the drainage project, and recommended against its implementation. The plan was finally outright rejected, somewhat ironically, by President George W. Bush – with much fanfare and media coverage – whose father, George H. W. Bush in 1988 established a federal policy of "no net loss of wetlands."

In contrast to the approach adapted by the USACE, the U.S. EPA evaluated the USACE 2007 plan based on their broader – national – perspective on the overall degradation of flood basin wetlands across the entire lower Mississippi alluvial valley, which has already been reduced by 80%. And they did so by employing a modern conceptual scientific framework, the flood pulse paradigm (e.g., Junk et al., 1989; Hupp et al., 2008), in combination with a range of geospatial datasets and statistical models to characterize the potential impacts to the basin's interrelated hydrology and ecology. Additional justification for canceling the Yazoo drainage project was realization that removal of such a large area of backwater inundation reduced the capacity of the floodplain to store floodwaters. This would have effectively diminished natural flood diversion structures, and thereby increased downstream flood risk. That such a justification had become paramount enough to influence cancelation of a major federal project championed by an enormous federal agency speaks to the change in mind-set occurring in flood management along the lower Mississippi and the United States. In part, this can be seen as the influence of new integrated – "room for the river" – management approaches, such as have been implemented along the lower Rhine in the Netherlands.

Finally, floodplain sciences had caught up with the ability of engineers to degrade floodplain environments and processes. Or so it seemed. Following a change in stewardship of the U.S. EPA after the 2016 presidential election, the priorities of the responsible federal agencies changed. An updated version of the 2007 plan to install pumps and drain the Yazoo basin was put forward by the U.S. Army Corps of Engineers and approved by the U.S. EPA January 11, 2021, just days before a new U.S. President – with different environmental priorities – was installed in office. Although implementation of the plan is pending the outcome of a lawsuit by environmental organizations, the "pendulum" has again shifted because of changes in federal environmental priorities.

7.2.2.2 MORGANZA SPILLWAY AND THE ATCHAFALAYA

Further downstream, in the Mississippi delta, flow diversion and flood control occur through the Atchafalaya and Pontchartrain interdistributary basins located along the western and eastern portions of the delta, respectively (Klimas et al., 2011). Topographically, the basins are formed between a major meander belt and the bounding Pleistocene bluffs. The Atchafalaya River, which forms at the apex of the delta, is geo-engineered to utilize an old meander bend cut-off to siphon ~30% of the Mississippi's discharge and sediment into its basin via the Old River Control Structure. Flow within the Atchafalaya basin occurs through various channels and eventually empties into the Gulf of Mexico via a natural and human made outlet. The Morganza Spillway is located just 20 km downstream (river distance) of the Old River Control Structure, along the west bank of the Mississippi. The Morganza Spillway is among the largest flood diversion structures in the world, consisting of a concrete weir with several main control gates. At a length of 1,300 m, the spillway can accommodate a maximum of 17,000 m^3/s of discharge. For scale, this is about seven times the average annual discharge of the Rhine River. Unlike Mississippi waters and sediment transported via the Old River Control Structure, however, the floodwaters and sediment siphoned off by the Morganza Spillway are not morphologically connected to the main-stem Atchafalaya River. While the floodwaters eventually make their way through the basin and are flushed into the Gulf of Mexico within a couple of weeks, most of the sediment is effectively trapped within the Atchafalaya basin. Rates of sedimentation are overall high, but vary greatly. Over 2000–2003 hydrologic years, for example, sedimentation rates ranged from 2 mm/yr to 42 mm/yr, with the variability controlled by the duration of inundation and the hydrologic connectivity of interior subbasins (Hupp et al., 2008). For comparison, the Atchafalaya sedimentation is much higher than the often reported <2 mm/yr of many freshwater flood basins. Indeed, the Atchafalaya basin is rapidly infilling and evolving from an open water swamp to a closed wetland and marsh ecosystem. This has direct consequences to the aquatic habitat of the basin's unique assemblage of flora and fauna. The livelihoods of small local Atchafalaya swamp communities are somewhat dependent upon such natural resources, especially its iconic crawfish (*Procambarus clarkii*).

7.2.2.3 BONNET CARRÉ SPILLWAY: FLOODWATER PULSES AND LAKE PONTCHARTRAIN ENVIRONMENT

Just 75 km south (river distance) of the Morganza Spillway is Bonnet Carré Spillway (Figure 7.8). Bonnet Carré Spillway is a floodwater diversion structure located along the eastern side of the Mississippi that flows into Lake Pontchartrain, 20 km upstream of New Orleans. The Bonnet Carré was constructed in 1931 as a vital component of the 1928 MR&T to save New Orleans from flooding. The spillway was intended to be constructed on the east side of the river at about the location of the Bonnet Carré Crevasses, which functioned from 1848 to 1874 (Kesel and McGraw, 2015). It was not uncommon for the crevasses and dike breaches to function for several months, inundating wetlands and flooding surrounding communities, which occasionally included New Orleans. To protect the structure from the threat of cutbank erosion the USACE located the spillway just downstream of the historic crevasse sites, which hydraulically is equivalent to locating it on the opposite side of the river (Kesel and McGraw, 2015). The location of the spillway at a bend rather than a cutbank results in opposite hydraulics, as helical flow is oriented toward the spillway. For this reason, a large amount of coarse sediment is deposited into the spillway trough. A comparison of sediment samples shows similar particle size for crevasse and spillway deposits. The median particle size of Bonnet Carré Spillway deposits is very fine sand (0.08 mm) while the median particle size of adjacent crevasses deposits is fine sand (0.1 mm), with spillway deposits being less well sorted than crevasse deposits (Kesel and McGraw, 2015). Some 7 million m^3 of sediment, mainly sand and silt, is deposited within the

Figure 7.8. (A) Diversion of record 2011 floodwater into Bonnet Carré spillway by removal of large wooden pins to open bays. (B) Coarse sedimentary splay deposits within the Bonnet Carré Spillway along the lower Mississippi River, upstream of New Orleans. The structure is 2.36 km in width. Date of photo: January 10, 2016. (Photo source: USACE, New Orleans Division.)

floodway during an average spillway event (USACE New Orleans Division). Most of the coarse-grained sediment is rapidly deposited within the first two km of the spillway opening (Figure 7.8), prior to entering Lake Pontchartrain, and is a valuable natural resource rapidly excavated by local operators for various construction activities.

The impact of flood diversion through the Bonne Carré Spillway not limited to Lake Pontchartrain, however, as it also impacts the main-stem channel. At the point of diversion, the removal of a large percentage of discharge results in a reduction in stream competence, as the velocity and shear stress abruptly decline. This results in deposition of coarse bed load immediately downstream of the structure, which can result in the formation of a large bar deposit. Such was the case during the 2011 flood along the lower Mississippi, an epic event that set new stage and discharge records along most gauging stations, although resulting in low amounts of floodplain sedimentation (Heitmuller et al., 2017). The diversion of 16% of the discharge through the Bonnet Carré spillway for a period of six weeks resulted in the deposition of 9.1 million tons of sand along a 13 km reach of channel immediately downstream of the structure. While much of the sediment was removed by natural sediment transport processes during the subsequent high-water period, the initial low-water period required dredging because of the threat to navigation during this period (Allison et al., 2013). Channel bed aggradation caused by spillway openings represents another unintended consequence associated with flood control that requires further management operations. Flood basin management by large-scale flood diversion structures, therefore, must be coordinated with river channel management.

In terms of fine-grained sediment, the timing of the siphoning of floodwaters into the flood basin is a fundamental decision to be made by flood control organizations, which can have considerable consequences to humans and flood basin (and lake) environments. The Bonnet Carré Spillway is activated when discharge exceeds a specific threshold stage height – at about 35,000 m^3/s – a stage with an elevation of 5.2 MASL. The ultimate decision, however, is made by the Chief of the USACE at the lower Mississippi Valley Division in Vicksburg, Mississippi (USACE, 2012). The operation of the structure occurs by the removal of some 7,000 large wooden creosote coated pins (beams) within 350 bays (Figure 7.8). The pins are vertically placed along the top of the concrete spillway, forming a wall of pins. The amount of water diverted is controlled by the number of pins removed from the structure by a crane, which in total requires about 36 h. The maximum discharge that the Bonnet Carré can siphon off of the lower Mississippi is 7,079 m^3/s, more than three times the average discharge of the Rhine River and a volume of water that would fill the New Orleans Super Dome (indoor football stadium) in about 8 min. The actual amount of water diverted through the spillway depends on the upstream discharge and the forecasted flood crest, shape of the flood wave, and the magnitude and duration of the flood event. Since the Bonnet Carré flood floodwater diversion structure was completed in 1931 the Spillway has been opened eleven times, on average once every eight

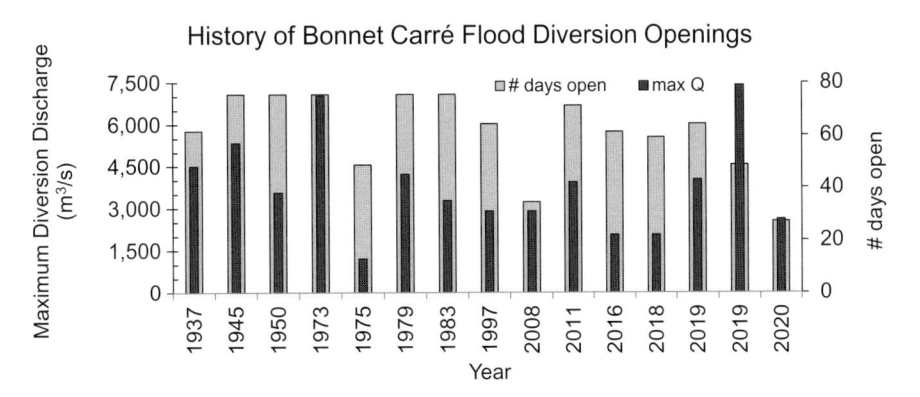

Figure 7.9. Discharge (m³/s) and duration (days) of floodwater diversion events for the Bonnet Carré Flood Diversion Structure since construction in 1937. Maximum diversion discharge m³/s is calculated by the ideal flow rate for each bay (at 20.2 m³/s per bay) multiplied by number of open bays. Maximum spillway design capacity is 7,079 m³/s with all 350 bays open. (Author figure. Data source: U.S. Army Corps of Engineers, New Orleans Division.)

years, including 2020 (Figure 7.9). The importance of the Bonnet Carré diversion structure is appreciated when considering its downstream flood protection: The last main-stem river flood of New Orleans occurred in 1927.

Floodwater diversion through Bonnet Carré spillway is a major hydrologic and sedimentary event, and an ecological disturbance. This is because the floodwaters rush abruptly into the adjacent Lake Pontchartrain, an enormous tidal basin (1,630 km²) formed between an old main-stem Mississippi delta lobe (St. Bernard), the New Orleans metropolitan area, and the southern extent of a northern fringing Pleistocene terrace (Saucier, 1963). Lake Pontchartrain forms a low salinity estuarine basin with local freshwater inputs and a couple of narrow passages to the Gulf of Mexico. Salinity ranges from 0.2 ppt to 0.8 ppt, being more saline toward the east where it flows through the rigolets and eventually the Gulf of Mexico. Freshwater pulses derived from the mid-continent of North America represent an ecological disturbance to the estuarine ecology, especially benthic organisms (Brammer et al., 2007). The hardest hit industry associated with freshwater flood diversions is Louisiana's legendary oyster industry. While fish swim to saltier waters, oysters (*Ostreidae*) native to Lake Pontchartrain are unable to withstand extended freshwater periods, which results in high mortality. In addition to the salinity change, diversion of Mississippi floodwater also significantly increases turbidity, oxygen, pH, and inputs of large amounts of phosphorus, nitrogen, and organic carbon derived from the agricultural core of the United States. Thus, the decision to open the spillway, while relieving downstream flood risk, has consequences that require further management and planning.

An important consideration of when to operate a spillway should take into account suspended sediment dynamics, as well as sediment quality (Mossa, 1996; DeLaune et al., 2003; Allison and Meselhe, 2010). Assuming flood safety is assured, the inherent relationships between discharge and suspended sediment presents management organizations with different sediment diversion options. Coarse-grained suspended sediment is mainly sand supplied from bed material and its entrainment is discharge (energy) dependent. Therefore, the peak volume of suspended sand simultaneously occurs during the maximum discharge (i.e., flood crest). The opening of the diversion structure at the flood crest, therefore, results in copious amounts of sand being diverted into the flood basin. The large (relative) size of sand results in abrupt deposition upon being transported into the spillway and the creation of splay-like sand deposits (i.e., Figure 7.8). It is well established, however, that suspended sediment and discharge exhibit nonlinear (hysteresis) relationships. For most rivers this means that the wash-load (silt/clay) component of the suspended sediment load peaks before discharge, because it is supply dependent rather than energy dependent (Mossa, 1996). That is, the source of fine-grained suspended sediment does not derive from entrainment of the channel bed, but rather from upstream runoff flushing silt and clay (i.e., soil erosion) into rivers. The lag-time between peak sediment and peak discharge is scale dependent, and increases as drainage area increases (Mossa, 1996; Hudson, 2003). The lower Mississippi, for example, has a lag of a couple of months during large flood events (Figure 7.10), whereas the lower reaches of less formidable lowland rivers, such as the Sacramento River, have a lag of perhaps a week during large flood events. Along large rivers, therefore, the opening of diversion structures prior to the peak of the flood crest results in disproportionate amounts of wash-load (fine silt/clay) being diverted into flood basins before the flood wave arrives (Mossa, 1996). With minimal flow velocity wash load can be flushed through the flood basin, which can be integrated into diversion structure operations (Caernarvon Delta and Diversion Study, 2014).

Depending on the quality and quantity of the wash-load, the operation style of the diversion structure can have important consequences to flood basin environments. Suspended clay has adsorbed nutrients and metals, and other pollutants, and thus its diversion into flood basins can diminish water quality and

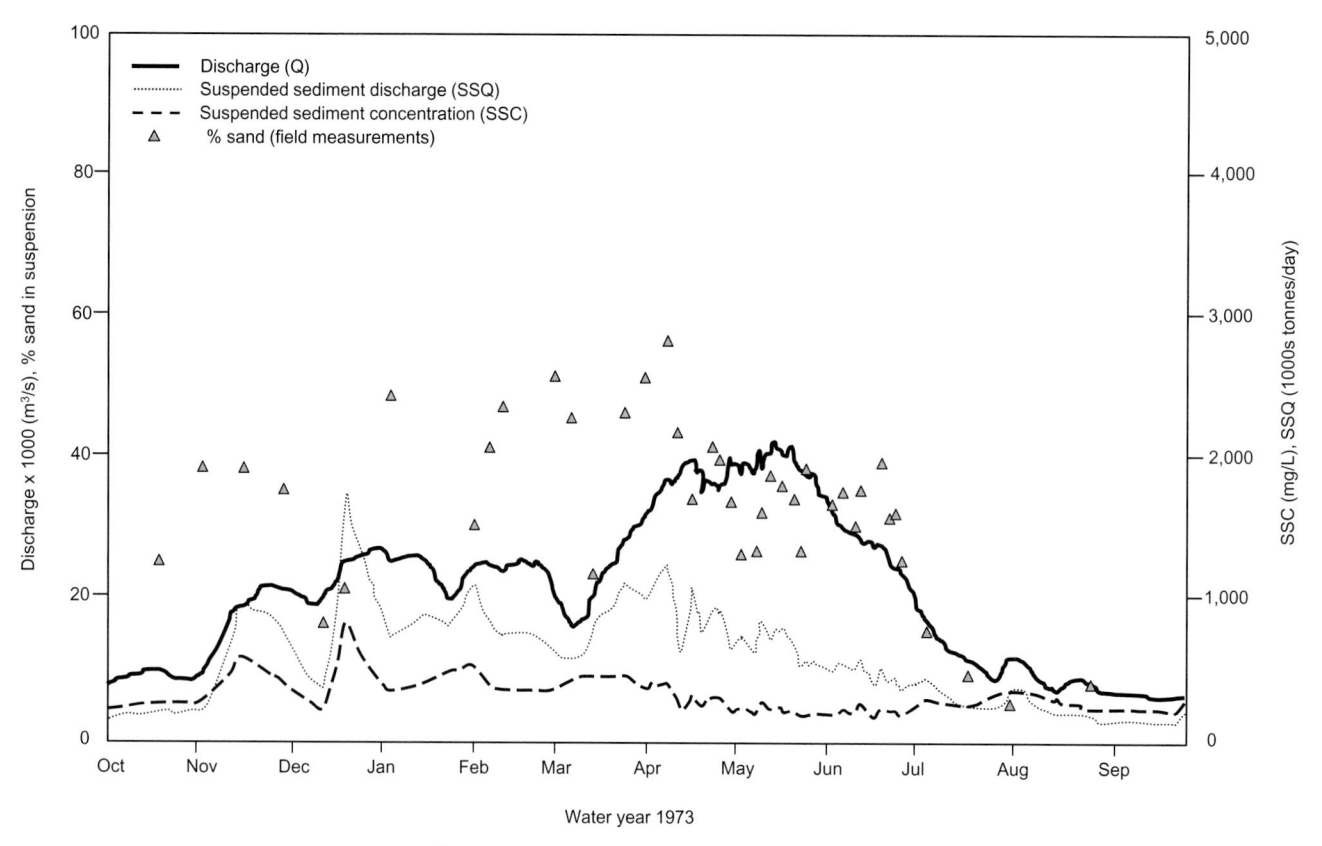

Figure 7.10. Relationships between discharge (m³/s) and suspended sediment concentration (mg/L) for the lower Mississippi at Baton Rouge during flood year 1973, displaying pronounced negative hysteresis with suspended sediment leading discharge by nearly two months. (Source: Mossa, 1996.)

benthic sedimentary layers if deposited rather than flushed through the basin. This can be seen as degrading associated ecosystems, and in the case of flood basins utilized for agriculture, a potential human health hazard.

7.2.3 Sacramento Flood Bypass: Mercury, Salmon, and Legacy of Hydraulic Gold Mining

The issue of polluted sediment and flood basin management is well illustrated along the lower Sacramento River basin in California (i.e., Section 2.2.2.2). The Yolo Bypass diverts floodwaters from the heavily engineered Sacramento River (Singer et al., 2008), 22 km upstream of Sacramento and Davis, California, into a westerly flood basin (Figure 7.11). Originally conceived in 1870 because of substantial flooding that curtailed development in the lower reaches of the basin, the Yolo Bypass structure was completed by the U.S. Army Corps of Engineers in the 1920s. The structure receives the upstreammost input into Yolo Basin is from Fremont Weir, a passive (gravity) overflow structure that results in floodwaters spilling into the Yolo basin when Sacramento River flood crest exceeds 10.1 MASL (California DWR, 2010). Floodwaters flowing over Fremont Weir transport a large amount of coarse suspended sediment, sand and coarse silt, which is abruptly deposited as proto-levees (Figure 7.12). The formation of such discreet features can be considered as a typical anthropogenic floodplain deposits associated with alluvial valleys engineered for flood control (i.e., Hudson et al., 2008; Singer and Aalto, 2009). Such deposits can eventually threaten the integrity of the diversion structure if not removed by local operators.

Figure 7.11. The flooded Yolo Bypass along the lower Sacramento River (February 23, 2017). The view is southwest from west Sacramento. (Photo source: USFWS Photo/Steve Martarano. Licensed under CC.)

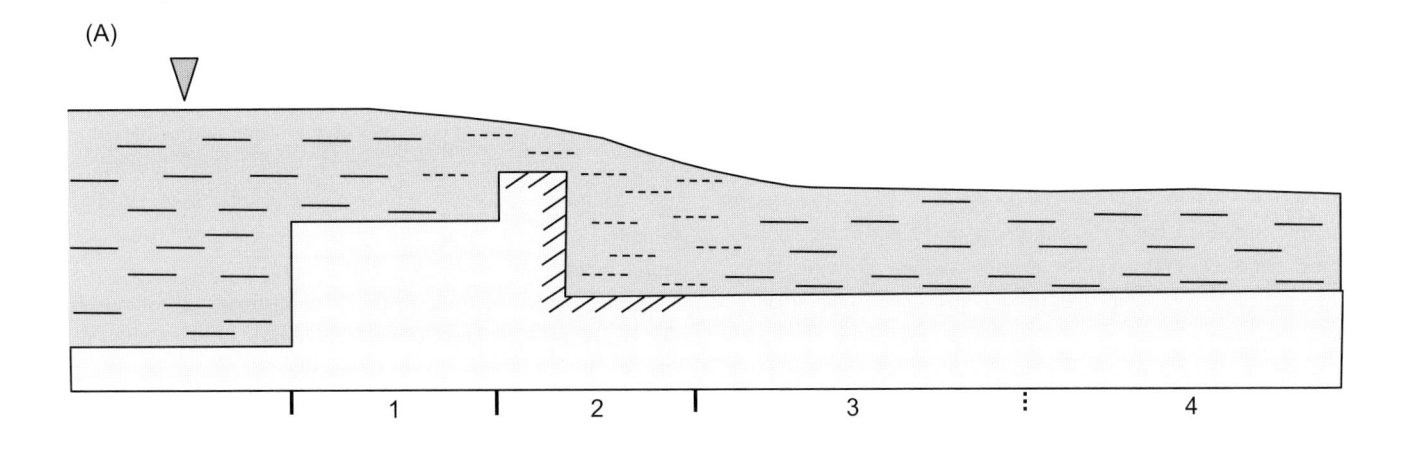

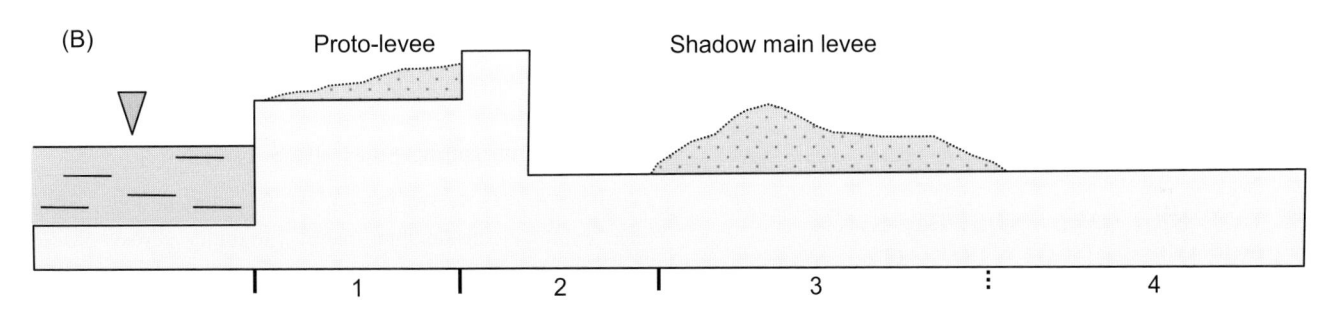

Figure 7.12. Schematic of passive spillway structure similar to Fremont Weir, the upstream spillway entrance into Yolo River flood bypass at flood stage (A) and resulting sedimentary deposits (B). (Source: Singer and Aalto, 2009.)

Upon overflowing Freemont Weir, floodwaters flow within a trough-like structure parallel to the Sacramento River for 66 km before eventually draining into the Sacramento-San Joaquin Delta at Rio Vista. While it has not come close to being realized, the structure ultimately has a design capacity of 9,700 m^3/s at its inlet, which is 80% of the project design floodwater. The transport capacity of the flood basin increases downstream of Sacramento, and attains a maximum discharge of 15,500 m^3/s at its outlet. The increased downstream discharge capacity is the result of contributions of two additional flood diversion structures and because of several tributaries draining into the basin along the eastern flanks of the adjacent Coast Range, which also supply additional sediment directly into the floodway basin. Although the Yolo Bypass is generally regarded as a biologically diverse natural area important to both aquatic ecosystems and floodplain farming, especially rice, it's become threatened because of an influx of polluted sediments.

Transport of fine-grained sediment into the Yolo Bypass has resulted in the deposition of a high amount of mercury (Hg), degrading the ecosystem and threatening agricultural lands. The major source of mercury is from historic upstream gold mining in the Sierra Nevada foothills, and mercury mines in the westerly Coast Range (noted in Section 2.2.2.2). Collectively within the Sacramento basin there are hundreds of abandoned gold mines and fifty-two abandoned mercury mines, which include a U.S. EPA Superfund site. Mining was extensive across the foothills and was effectively unregulated from the late 1840s to the early 1900s, with some limited mining continuing until the 1950s (Domagalski et al., 2004). The abandoned mines and slag waste produce a continuous supply of mercury into the larger tributaries of the Sacramento River by hydrologic and geomorphic processes. While the annual input of ~1,000 kilotons of suspended sediment is effectively balanced with outputs to San Francisco Bay, large amounts of mercury are concentrated within Yolo Bypass wetlands. Over a ten-year period (1993–2003), for example, an accounting of upstream inputs at Fremont Weir, storage within the bypass basin, and outputs to San Francisco Bay revealed that a net of ~500 kg of mercury has accumulated within the basin, which is mainly attributed to two large flood events (Springborn et al., 2011). This is detrimental to associated ecosystems with clear biological magnification occurring in larger predatory fish species. The U.S. EPA identifies the safe mercury level in tissue to be 0.3 ppm (U.S. EPA, 2001), a level exceeded by some larger fish in the middle and lower reaches of tributary rivers.

Management options to reduce mercury levels in the Yolo Bypass include settling basins and ponds in the lower reaches of tributaries having especially high Hg levels, such as Cache Creek (2,900 km^2), as well as dredging and removal of contaminated sediments (California DWR, 2010). Mercury levels are "diluted" in the Yolo Bypass when mixed with the substantial flux of Sacramento River floodwater and sediments. Here, federally endangered Chinook salmon (*Oncorhynchus tshawytscha*), for example, thrive in the basin's nutrient-rich floodwater habitats during overflow events, with juvenile salmon growing larger (0.7% per day, by weight) than those constrained to the adjacent heavily engineered Sacramento River. The flood pulse, therefore, represents an important management tool for conservation of a valuable aquatic species. Unfortunately, Yolo Bypass salmon also have higher levels of mercury (3.2% per day higher) in tissue samples than similar aged fishes in the adjacent main-stem river. (Henry et al., 2010). Thus, management organizations intending to utilize the floodwater habitat of the Yolo Bypass as a nursery for juvenile salmon are confronted with a dilemma: Geoengineering the basin's flood pulse dynamics as a conservation tool both nourishes and harms that which is being conserved.

7.3 LOWER LOWLANDS: HUMAN ACTIVITIES, SEDIMENTOLOGY, AND SUBSIDENCE

7.3.1 Flood Basins and Drivers of Subsidence

Flood basins are characterized as benign geomorphic environments. But they're moving, slowly, because of subsidence. Within lowland fluvial systems, flooding is exacerbated – and increasingly caused – by subsidence (Yuill et al., 2009), among the most quiescent of geomorphic processes. Subsidence refers to the lowering of a surface relative to a fixed vertical datum, a process that has obvious relevance to surficial hydrology and land inundation driven by fluvial, atmospheric, and marine mechanisms. Over the past decade or so the importance of subsidence as a control on flooding has gained greater public recognition, especially because of its role in the 2005 New Orleans flood disaster by Hurricanes Katrina-Rita.

Subsidence is driven by a variety of processes, which along large lowland fluvial settings is often exacerbated by human impacts (Yuill et al., 2009). Three causes of subsidence include (1) compaction of soft sedimentary layers by sediment loading, (2) oxidation of organic surficial layers, and (3) neotectonics. The first two processes are greatly exacerbated by a range of human activities, such as wetland drainage, pumping, and a reduction in sediment delivery. As these factors are caused by humans, they can also be (somewhat) managed and mitigated.

The third driver of subsidence, neotectonics, is associated with the downward slippage and sinking of older and deeper stratigraphic units (Dokka, 2006; Törnqvist et al., 2008), and represents a natural "background" form of subsidence inherent to fluvial–deltaic systems (i.e., Section 3.6). Neotectonics and isostatic adjustment to continental glaciation and sea level rise, however, contribute to "background" subsidence rates that must be recognized before a human cause can be associated with subsidence. Most large fluvial–deltaic systems cross active faults, which, in association with sediment loading, results in continued subsidence along the downthrown side of the fault line (Cohen, 2003; Dokka, 2006; Törnqvist et al., 2008; Panin, 2015). The upstream limits of both the Rhine and the Mississippi deltas, for example, are located at the boundary of large fault systems. Sediment loading downstream of the fault results in constant subsidence of Tertiary and Quaternary deposits. Although the sinking of sedimentary deposits downstream of the fault zone is considered a natural process, it is important to modern-day hydrologic processes and management of respective fluvial–deltaic environments.

7.3.2 Human Impacts on Delta Subsidence

From the perspective of fluvial–deltaic sedimentology, subsidence exhibits considerable spatial variability across different depositional environments, which are also associated with different forms of human activities (Allison et al., 2016; Jones et al., 2016). The highest rates of subsidence are usually within organic-rich interdistributary flood basin environments underlain by clay and peat layers of variable thicknesses. Flood basins within alluvial valleys have lower rates of subsidence than in deltas, which is largely a result of being less thick and having had more time to consolidate. The lowest rates of subsidence are usually immediately above channel belts, which include dense sandy–silty deposits associated with point bars and natural levees. Indeed, low rates of subsidence along natural levees are yet another reason why they remain the most viable settings for human settlement within fluvial lowlands.

7.3.2.1 INTERNATIONAL DELTA SUBSIDENCE RATES

A review of subsidence rates across a large range of different deltas reveals the importance of anthropogenic drivers, including urbanization, infrastructure construction, pumping and drainage for agriculture, and pumping and fluid withdrawal for groundwater and hydrocarbon extraction (Table 7.2). These factors are well documented and considered significant enough to alter local surficial hydrology and flood processes, and in many instances have resulted in landscape-scale adjustments to large river deltas (Galloway et al., 1999; Allison et al., 2016).

Compaction of soft unconsolidated clay and organic material is among the largest driver of subsidence along large floodplains and deltas. Although it varies by minerology, shape, and size, the

Table 7.2 *Annual subsidence rates along major lowland rivers and deltas impacted by human activities*

River Delta	Rates (mm/yr)	Notes	Source
Rhine (Netherlands)			
	0.1–0.3	Natural (Holocene)	Cohen, 2003
	1.7–70	Drainage and peat oxidation	Cuenca et al., 2007
	6.7–11.7	Drainage and peat oxidation	Querner et al., 2012
Mississippi (USA)			
	0.1	Natural (Holocene)	Törnqvist et al., 2008
	>10	Drainage and compaction	Day and Giosan, 2008
	10–15	Drainage and compaction	Shinkle and Dokka, 2004
Ebro (Spain)			
	4–5	Natural (Holocene)	Stanley, 1997
	3		Sanchez-Arcilla et al., 1998
	1–3.2	Sediment reduction (storage in upstream reservoirs)	Ibàñez et al., 1997
Yangtze (China)			
	10	Groundwater pumping, urban construction (Shanghai)	Xiquing, 1998
	10–65	Groundwater pumping, urban construction (Shanghai)	Gong et al., 2008
Nile (Egypt)			
	0.0–5.0	Natural (Holocene)	Stanley and Warne, 1998
	4–5	Natural (Holocene)	Stanley, 1997
	0.0–8.0		Becker and Sultan, 2009
	5.0		Fugate, 2014
Po (Italy)			
	1.0	Natural (Holocene)	Carminati et al., 2005
	1–3	Natural (Holocene)	Stanley, 1997
	2–60	Groundwater pumping, fossil fuel extraction	Overeem and Syvitski, 2009
	37–250	Groundwater pumping, fossil fuel extraction	Fabris et al., 2014
	6–18	Management of subsidence	Fabris et al., 2014
Huanghe (China)			
	25–252	Aquaculture, groundwater pumping	Higgins et al., 2013
	3–5	Groundwater pumping, fossil fuel extraction	Fan et al., 2006
	12–30	Fossil fuel extraction	Liu et al., 2015
	10	Urbanization and groundwater pumping	Xiquing, 1998
Pearl (China)			
	2.5–6	Urbanization (building weight and groundwater pumping)	Wang et al., 2012
	15.1–31.0	Urbanization (building weight and groundwater pumping)	Chen et al., 2012
Irrawaddy (Myanmar)			
	3.6–6	Groundwater pumping	Overeem and Syvitski, 2009
Trinity (USA)			
	4.5–42.2	Groundwater pumping, fossil fuel extraction	Galloway et al., 1999; Bawden et al., 2012
Chao Phraya (Thailand)			
	13–150	Groundwater pumping, aquaculture	Syvitski et al., 2009
	80–120	Groundwater pumping	Phien-weg et al., 2006
Sacramento (USA)			
	25–75	Groundwater pumping	U.S. Geological Survey, 2010
	15–51	Groundwater pumping	Rojstaczer et al., 1991
Fraser (Canada)			
	1–2	Natural (Holocene)	Mazzotti et al., 2009
	3–8	Dike construction (sediment starvation)	Mazzotti et al., 2009

Table 7.2 (cont.)

River Delta	Rates (mm/yr)	Notes	Source
Danube (Romania)	1.5	Natural (Holocene)	Panin, 1999
Rhône (France)			
	2–7	Natural (Holocene)	Stanley, 1997
	0.5–4.5	N/A	Hensel et al., 1999
	1	Natural (Holocene)	Suanez and Provensal, 1996
Mekong (Vietnam)			
	5–30	Groundwater pumping, shrimp farms, canals	Anthony et al., 2015
	10–40	Groundwater pumping, farming	Erban et al., 2014

Source: Author table.

dry bulk density of sand is around 1.6 g/cm^3 with a porosity of 30%. Clay deposits are more porous and less dense, commonly having a dry bulk density as low as 1.1 g/cm^3 and a porosity of 60%. In addition to the lower porosity and higher density of sand, the round shape of sand particles (and coarse silt) results in very low rates of compaction (Karlsson and Hansbro, 1981). Thus, subsidence is usually not a problem atop thick sandy channel belts and natural levee crests. Because of the process of slack-water deposition and the physical properties of clay, it is much more prone to compaction. Foremost is that the small size (<0.004 mm) and platy shape of clay particles results in a high surface area. Additionally, the porosity of clay is especially high where deposited within brackish and marine settings because of flocculation of individual clay particles. Because of the very small size of clay, it is colloidal and thus inherently has a negative charge. Upon entering positively charged brackish water, individual clay particles "clump" together. Because of the platy shape of individual clay particles, the clumping creates a larger aggregate particle size with a higher porosity. Unfortunately, as the particles are only loosely organized it creates an inherently unstable foundation for subsequent sedimentary loading, particularly denser silts and sands prograding atop the clay base (see Section 3.7.2.3). The result is that flocculated clay dewaters with compaction, resulting in high rates of subsidence (e.g., Coleman and Wright, 1971; Törnqvist et al., 2008; Bianchi et al., 2009; Stanley and Clemente, 2014; Panin, 2015).

7.3.2.2 PEAT: GEOPHYSICAL PROPERTIES AND SUBSIDENCE PROCESSES

In addition to clay, subsidence of peat is also vitally important to understand with regard to flooding across large floodplains and deltas. Peat formally requires more than 75% organic matter, considerably higher than organic soil (>25%). Peat varies considerably according to age and composition, the latter determined by the vegetation type comprising the wetlands as well as the environmental controls (temperature, hydrology, chemistry, etc.). Peat is classified as fibric, hemic, or sapric, the latter form representing the densest and most decomposed form of peat (Boelter, 1969; Silc and Sanek, 1977; Huat et al., 2011, 2014). Much peat within active – healthy – flood basins is fibric, which is not dense and contains well-preserved plant structures with individual vegetation types being clearly identifiable. The most common plant types comprising peatlands are *phragmites austrilus* and *spartina altiniflora*, which are dominant in freshwater and salt-water environments, respectively. Tropical peat in coastal swamp forests is often comprised of mangroves (Wüst et al., 2003), including red (*Rhizophora mangle)*, black (*Avicennia germinans)*, and white (*Laguncularia racemose*).

If undisturbed, peat environments are mainly stable and have low rates of subsidence, in part because of continued growth of new vegetation at the surface. But with sediment loading, especially following a crevasse or an avulsion, peat compresses and results in high rates of subsidence (van Asselen et al., 2009). Detailed field measurements and mathematical modeling reveals that after sediment loading there is initially a lag in compaction rates (Figure 7.13). But thereafter compaction (as a percentage of porosity) occurs very quickly such that the highest rates of peat compaction occur within about one hundred years. Depending on the density of the peat, its compaction ultimately ranges from 15% to 34% after sediment loading, with peat containing clay having lower compaction rates. Compaction then exponentially declines and then levels off for the next several thousand years. From the standpoint of fluvial dynamics peat compaction creates new accommodation space for clastic sedimentary deposits, and facilitates channel avulsion (subsequently discussed). Because fluvial and deltaic flood basins are often characterized by a stacking of clastic and organic layers associated with fluvial and coastal dynamics, cohesive clay and peat deposits often overlay such that subsidence rates exhibit considerable vertical variability (Stanley and Clemente, 2014). While consideration of such processes over extended timescales may seem less applicable to modern-day processes, it is nonetheless absolutely

Table 7.3 *Geophysical properties associated with peat classification*

Type	Fiber content (%)	Color	Structure and decomposition	Porosity (%)	Bulk density (g/cm^3)	Hydraulic conductivity (m/day)
Fibric	>66	Yellow-tan to light brown	Low decomposition with distinct fibrous structure. Individual plant fibers are retained, plant identification is easy.	>90	<0.09	>1.3
Hemic	33–66	Brown	Intermediate degree of decomposition and moderate humification. Plant fibers range from relatively distinct to not identifiable.	85–90	0.09–0.2	1.3–0.01
Sapric	<33	Brown to brown-black	High degree of decomposition and humification. No visible plant fibers. Mushy consistency.	<85	>0.2	<0.01

Compiled from Boelter, 1969; Silc and Sanek, 1977; Wüst et al., 2003; Huat et al., 2011, 2014.

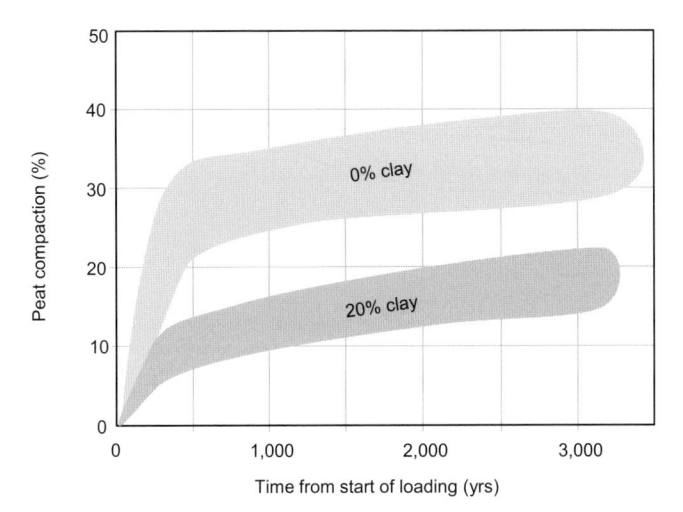

Figure 7.13. Model results showing range in compaction (%) for two 6-m thick peat sequences over several thousand years following sediment loading. The two scenarios reveal the influence of clay, as pure peat undergoes higher rates of compaction than peat with clay (at 20% by weight). For either case, rapid hydrologic and environmental change can be expected soon after the onset of compaction. For both peat sequences loss on ignition (LOI) was 0.8. (Modified from van Asselen et al., 2011.)

critical. This is because the age of peat determines background subsidence rates that predominate within a particular site (van Asselen et al., 2011), thereby explaining many subsidence and inundation problems associated with human settlement across different types of flood basin environments (Erkens et al., 2015; Allison et al., 2016).

From the standpoint of engineering, the geophysical qualities of peat are critically important to the design of flood control infrastructure. The bulk density of peat is considerably lower than clastic soils, and ranges from 0.09 g/cm^3 for very young and undecomposed peat to 0.2 g/cm^3 for strongly humified and decomposed peat (Table 7.3). The higher bulk density of sapric peat results in lower rates of compaction. Additionally, the hydraulic conductivity of sapric peat is very low (<0.01 m/day), whereas the hydraulic conductivity of fibric peat is two orders of magnitude higher (>1.3 m/day). Thus, engineers must be careful when making generalizations about peat as it pertains to groundwater flow, and dike seepage. Fibric peat is among the least stable of environments in which to establish flood control infrastructure, as the high rates of hydraulic conductivity lead to dike instability associated with underseepage, whereas the low bulk density and high fiber content result in high rates of compaction and oxidation, and high rates of subsidence (FAO, 1988; Huat et al., 2014).

Oxidation of organic material is especially a problem along floodplains and deltas with extensive wetlands, where thick peat and organic soil layers have accumulated for thousands of years. As long as the water table remains close to the surface, subsidence is minimal. But upon diking and drainage, especially by mechanisms such as ditching and pumping, the surface and subsurface flow paths are altered and the water table lowers. This results in organic soil that primarily developed under anaerobic conditions to become dominated by aerobic conditions. The outcome is rapid degradation of peat and soil organic matter, as organic carbon is oxidized and converted to carbon dioxide (CO_2). Additionally, further decomposition occurs by biological activity, including microflora, fungi, and bacteria. In most cases, CO_2 rapidly degasses. From the standpoint of soil composition, the decomposition of organic matter results in a loss of soil mass and an increase in soil porosity (Yuill et al., 2009). Additionally, dewatering of the pore space eliminates positive pore pressure (Stouthamer and van Asselen, 2015). Thus, the combined effect is that the soil undergoes an increase in stress from the overlying layers, such that subsidence is promoted and accelerated because of a reduction in the strength of the soil matrix. Such impacts can extend for hundreds to thousands of square meters and ultimately be significant at the landscape scale (Figure 7.14). The most abrupt

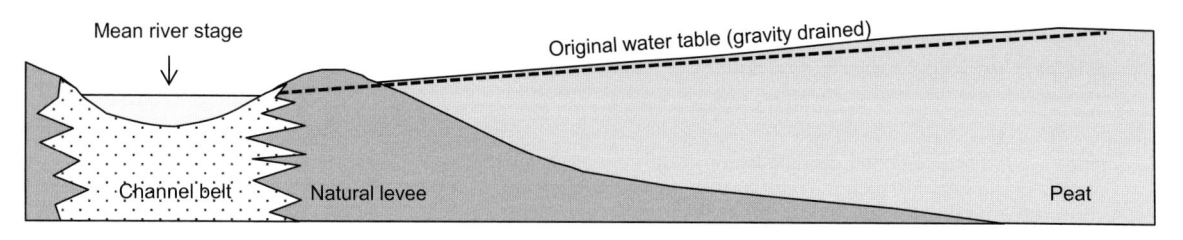

Stage I: The original undisturbed peat lands have a stilted water table, which is gravity drained.

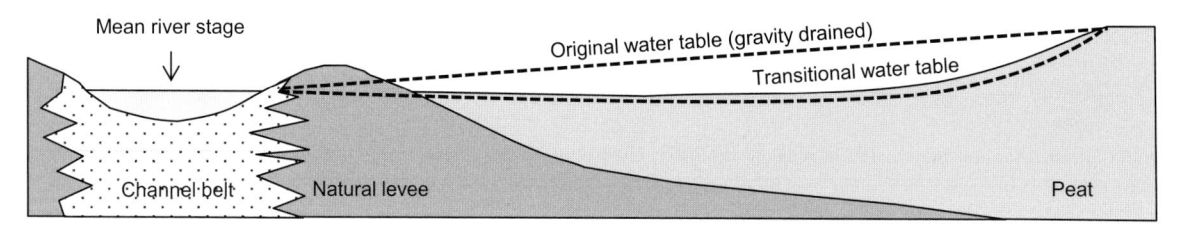

Stage II: Disturbance by drainage ditches causes rapid subsidence of peat surface and loss of gravity drainage, requiring empoldering and pumping to drain land.

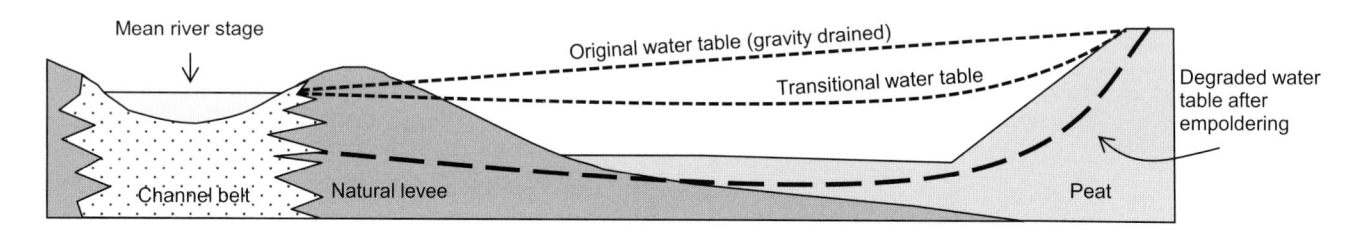

Stage III: Empoldering and pumping has caused deep subsidence of peat surface below mean river stage and a geometric land surface profile.

Figure 7.14. Model of flood basin evolution by peat oxidation and land subsidence caused by agricultural drainage. (After multiple sources, including FAO, 1988; Galloway et al., 1999; van de Ven, 2004.)

period of response (i.e., highest rates of subsidence) occurs with decades to perhaps a hundred years of drainage, although subsidence rates eventually decline as the pore space is reduced by compaction (FAO, 1988; Huat et al., 2011).

7.3.2.3 INFLUENCE OF WETLAND DRAINAGE AND FOSSIL FUEL ACTIVITIES ON SUBSIDENCE

Petroleum activities are nearly ubiquitous along many large fluvial–deltaic settings (e.g., Figure 7.15). The withdrawal of fossil fuels up to several thousand meters below the surface commonly results in a subsidence bowl ranging from several hundred to a couple of thousand meters wide and several meters deep (Figure 7.16). The subsidence is driven from deep below the subsurface and fundamentally occurs because of the pore-pressure change associated with withdrawal of fluid and mass of the overburden (Whittaker and Reddish, 1989; Olea and Coleman, 2014; Liu et al., 2015). And it is not only fluid withdrawal that is damaging, as much of the infrastructure and activities associated with maintenance and operation of fossil fuel activities also drives subsidence. Conventionally, creation of canals and associated spoil banks fundamentally changes local wetland hydrology such that marsh becomes disconnected from active fluvial–deltaic processes (Figure 7.17). In contrast to fluid withdrawal, this activity drives subsidence from above, primarily by oxidation of peat and organic soil (Whittaker and Reddish, 1989; Ko and Day, 2004; Yuill et al., 2009; Jones et al., 2016). The near ubiquitous presence of fossil fuel activities along large fluvial lowlands suggests its larger impacts to ground subsidence should be systematically appraised.

Oxidation of organic material is particularly problematic where drainage has long been a component of land use, such as for farming and expansion of agricultural lands in peat basins (FAO, 1988). In the Netherlands, for example, for many centuries the drainage of peat domes has simultaneously occurred with creation of polders, which are discreet hydrologic land parcels defined by drainage, ditches, pumping, and/or dikes. For many parts of the Netherlands the high rate of subsidence is also associated with dikes disconnecting the flood basins from active fluvial processes. Along the branches of the Rhine delta in the Netherlands, considerable subsidence, ranging from 2 m to 4 m, has occurred since about 1200 AD, associated with land reclamation efforts by poldering and dike construction (van Veen, 1948; van de Ven, 2004). Such activities have resulted in dramatic change across many deltas that have sustained human activities for long periods. Juxtaposition of extensive ground subsidence immediately adjacent to aggraded embanked floodplains (Section 6.5) has resulted in pronounced changes to flood basin landscapes (Figure 7.18).

Figure 7.15. Petroleum infrastructure in lower Mississippi River valley near Natchez, MS. (Author photo, 2015.)

The experience with subsidence in the Netherlands, unfortunately, did not result in a learned lesson for many places, until it was too late. Along the lowermost reaches of the Mississippi alluvial valley and delta, annual subsidence rates range from 5 mm/yr to 50 mm/yr, which is mainly anthropogenic (Shinkle and Dokka, 2004; Jones et al., 2016). In the Sacramento Delta land subsidence caused by surface drainage has resulted in an average of ~3 m of subsidence and a maximum of ~11 m since the early 1900s (Galloway et al., 1999).

7.3.2.4 ASIAN MEGA-DELTAS AND SUBSIDENCE

Subsidence of the Asian mega-deltas (Table 7.2) was especially rapid over the twentieth century, and continues at alarming rates (Syvitski et al., 2009; Minderhoud et al., 2020; Best and Darby, 2020). Here subsidence is related to land drainage and groundwater pumping associated with agricultural expansion, especially including rice production, fish farming, and increasingly cattle production (Gong et al., 2008; Chen et al., 2012; Wang et al., 2012; Higgins et al., 2013; Erban et al., 2014; Liu et al., 2015; Allison et al., 2016; Best and Darby, 2020). In the Huanghe delta, as with several other Asian mega-deltas, the drive to increase protein production has resulted in extensive fish farming further exacerbating the subsidence problem. In some instances, subsidence was as high as 110 mm/yr (Chan et al., 2013).

In other locations, rapid urbanization associated with fundamental demographic shifts has resulted in massive urban sprawl,

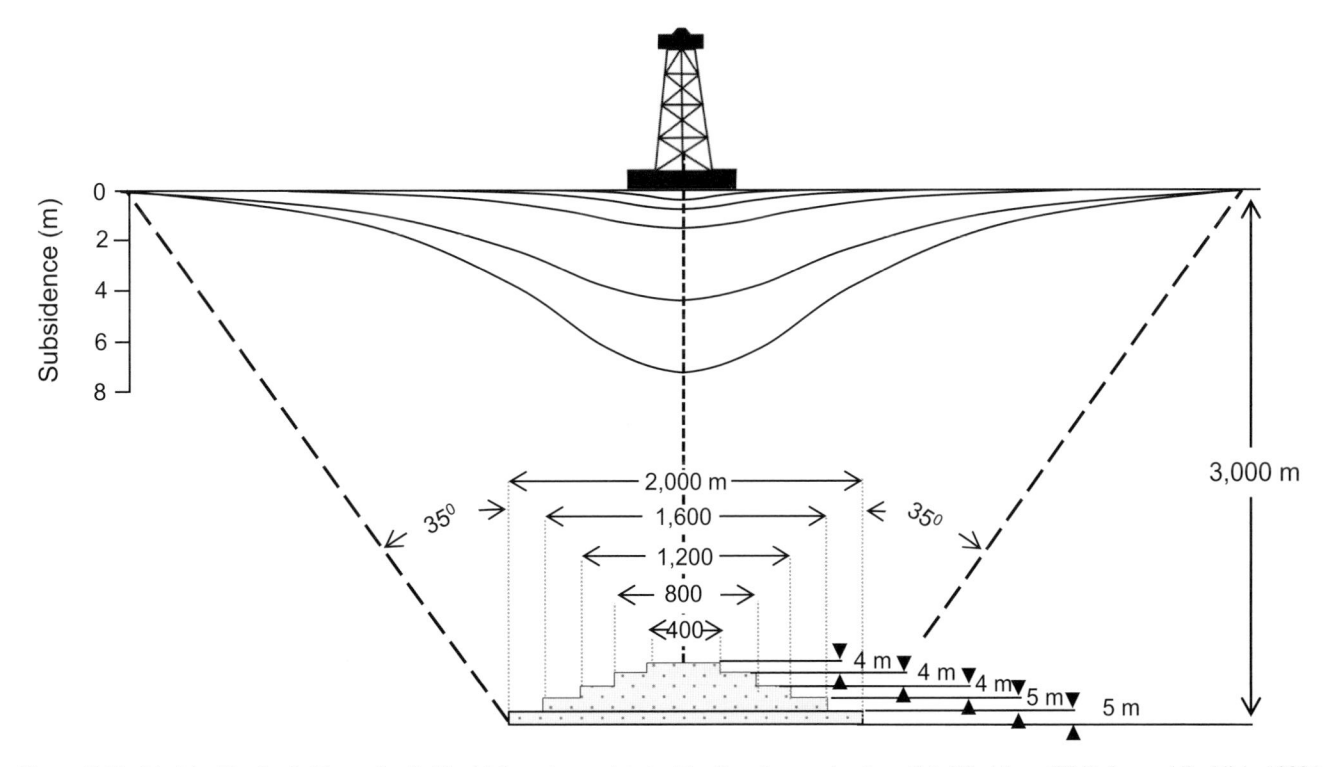

Figure 7.16. Model of land subsidence for fluid withdrawal associated with oil and gas extraction. (Modified from Whittaker and Reddish, 1989.)

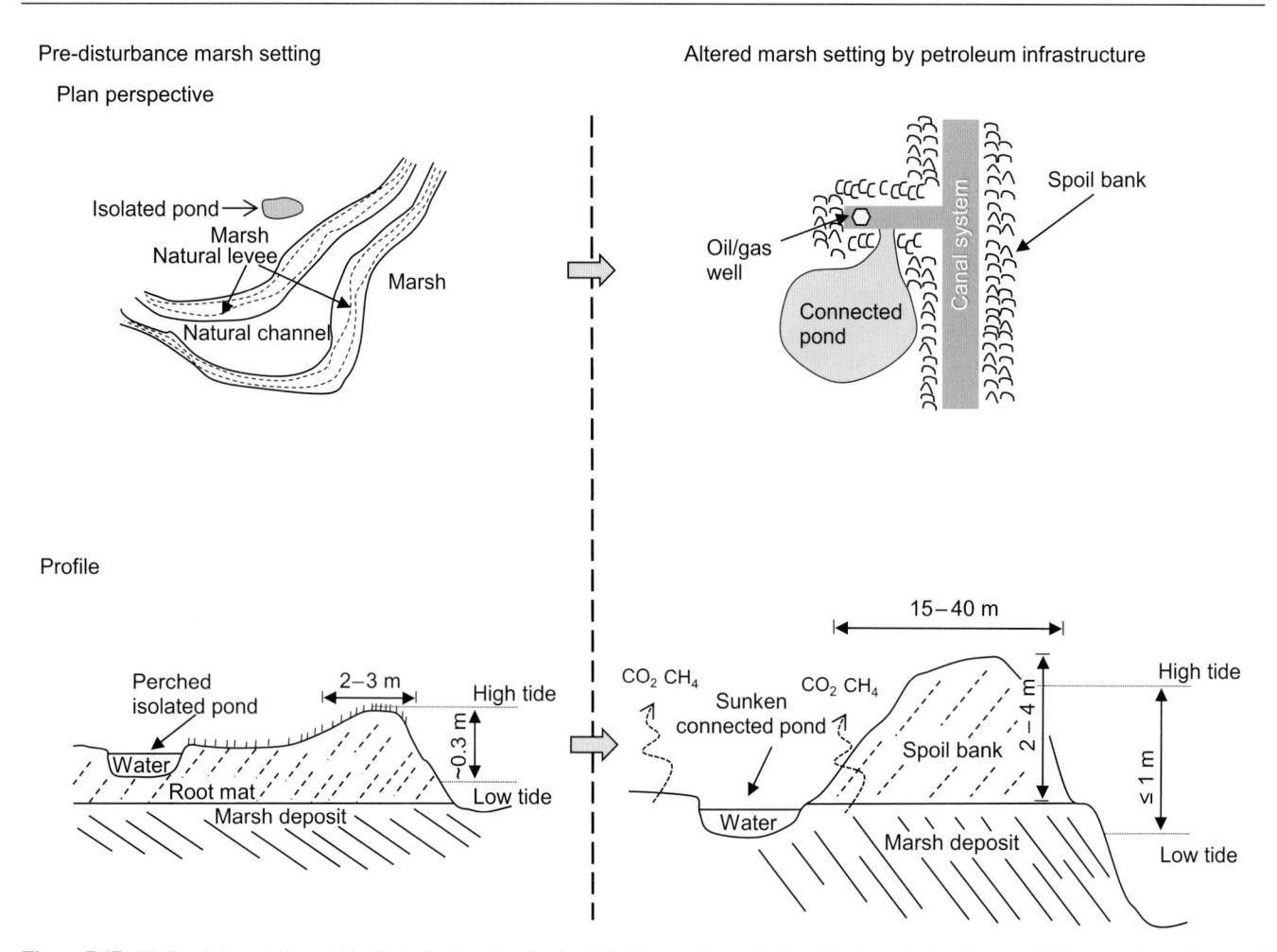

Figure 7.17. Wetland degradation and hydrologic alteration for fossil fuel extraction activities. Marsh oxidation drives subsidence at the surface, and is associated with degassing of carbon dioxide and methane. (Modified from Ko and Day, 2004.)

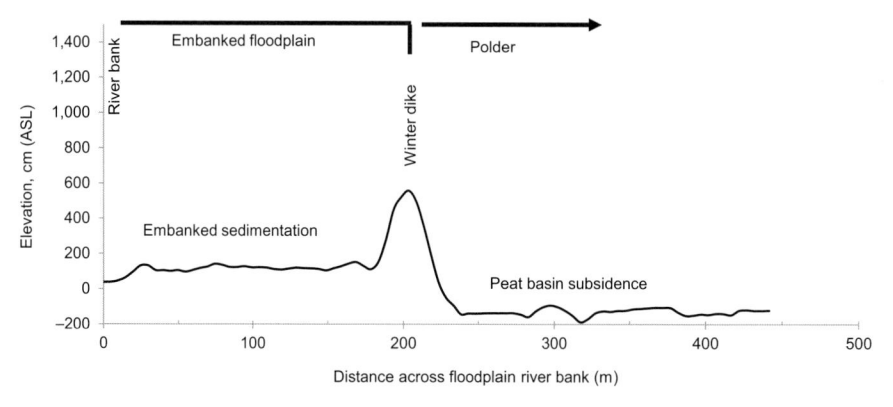

Figure 7.18. Topographic profile across the Lek River, Rhine distributary, NL illustrating substantial difference in land elevation (~3 m) between embanked floodplain in contrast to subsidence of adjacent peat basin. (Author figure, data source: lidar AHN-1, NL.)

which increases the weight of the overburden but is also associated with an especially high rate of groundwater withdrawal. Substantial subsidence associated with the development of Shanghai within the Yangtze delta (Figure 7.19), for instance, is associated with a change in the pore space of clayey sediment, which varies between older and more recent deposits. The pore space of older and deeper (~40 m) deposits decreased by more than one-third after consolidation (Figure 7.20). Of course, here the problem is associated with ~2.0 m of subsidence since 1920, tremendously increasing flood risk to a rapidly growing dense urban population (Gong et al., 2008). In Shanghai, much of the compaction-caused subsidence is attributed to direct

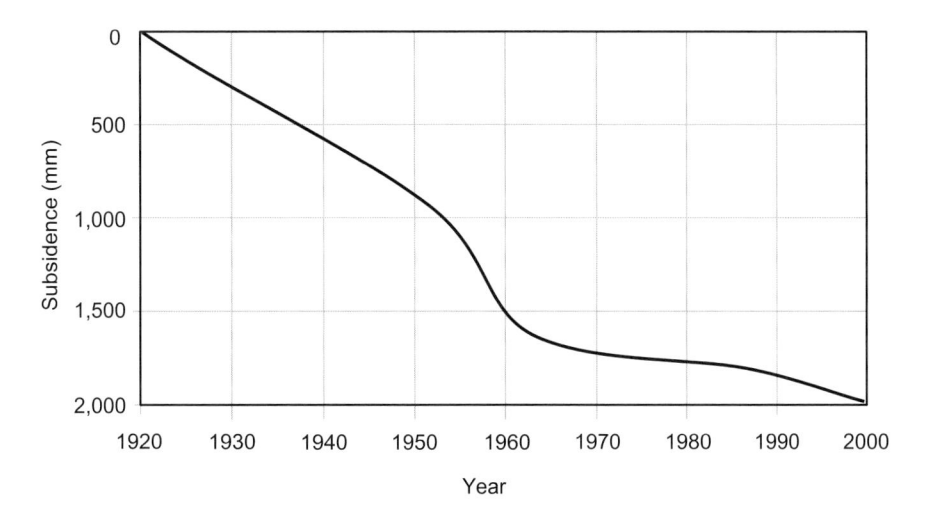

Figure 7.19. Subsidence of Shanghai associated with groundwater withdrawal and urbanization during the twentieth century. (Modified from Gong et al., 2008.)

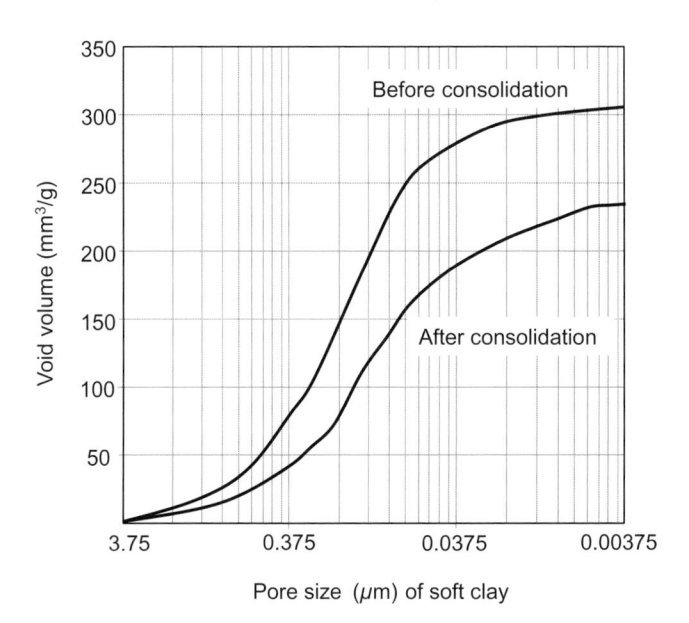

Figure 7.20. Change in pore space with sediment loading in Shanghai, Yangtze River delta. Upper line represents soft clay at depth of 4 m below land surface. Lower line is 40 m at a loading of 200 kPa. (Modified from Gong et al., 2008.)

urbanization and groundwater extraction (fluid withdrawal), causing the sedimentary pore space to collapse. For seventy-five years the annual subsidence of urban Shanghai increased, reaching a maximum of 120 mm/yr in 1965. This was attributed to high population growth after the 1950s, and especially an emphasis on construction of tall buildings and heavy industry. The increase in industry and urban growth both required large amounts of groundwater, while the increasingly dense cities with tall buildings directly contributed to (sediment) loading and compaction of the soft deltaic clay (Gong et al., 2008). Since the 1970s, however, measures to manage subsidence, including reducing groundwater withdrawal and also direct injection of water into key subsurface units, have reduced the annual rate of subsidence and it has now leveled off to about 10 mm/yr.

Other large cities within the Asian mega-deltas have similar problems. Bangkok has subsided by 2.0 m since 1970. In some locations the annual subsidence rate is 100 mm/yr, which dramatically increases vulnerability of urban populations to watershed driven flooding, as well as coastal storm surge events and sea level rise (Phien-wej et al., 2006).

7.4 AVULSION AND RIVER MANAGEMENT

7.4.1 From Dike Breach to New Channel Belt

The problem with accelerated land subsidence is not limited to the area in which it occurs, as it can influence more substantial fluvial adjustments, including avulsion. This is particularly the case where subsiding flood basins are adjacent to high channel belts that have aggraded because of embankment, increasing concern with dike breach risk. While dike breach events are singularly important geomorphic mechanisms and challenges to management, they should be viewed as part of a continuum of fluvial adjustment.

An avulsion is usually triggered by a flood event, with the actual driving conditions having been set up over longer periods. A crevasse initially incises a new channel atop the channel bank and flanking natural levee, and develops multiple active channels in a dendritic pattern. As overbank flooding continues minor crevasse channels are abandoned within the overall splay complex while a dominant crevasse channel extends into the flood basin. Further adjustments to the crevasse morphology influences avulsion potential (Jones and Schumm, 2009). From the perspective of energy, as its draws

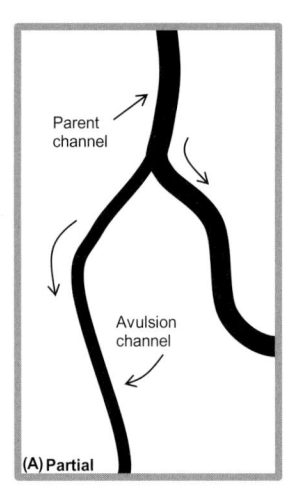

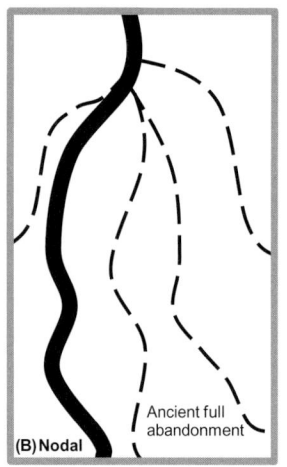

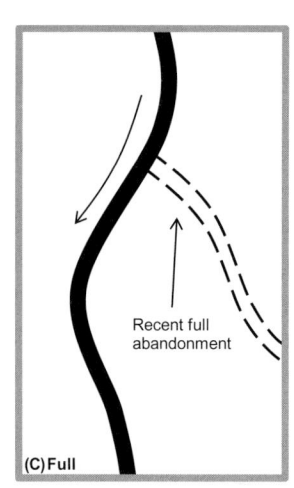

Figure 7.21. Different avulsion styles, including (A) partial, (B) nodal, and (C) full. Partial avulsion results in discharge being shared between multiple channels and formation of an anabranching channel pattern, which may be anastomosing if it rejoins. Nodal avulsion refers to repeated avulsion occurrence along the same segment, such as the delta apex or valley slope inflection. (Author figure, modified from Slingerland and Smith, 2004.)

additional discharge from the main-stem channel the crevasse channel bed incises until an equilibrium is reached between sediment transport and stream power (i.e., slope x discharge). As the crevasse channel enlarges and incises it accommodates an increasing amount of discharge, although its slope decreases. Thus, a large flood may create an initial crevasse channel, but if the slope between the flood basin and the main-stem channel bank is not sufficiently high the crevasse will be unable to enlarge to the point of capturing the entire discharge of the main-stem channel.

Several key factors are important to consider in development of an avulsion, including (1) hydraulic factors controlled by slope differences between the natural levee and flood basin, (2) flood basin sedimentology and topography, and (3) the location of old channel belts that can be reoccupied during a flood event. These factors are influenced by neotectonics, sea level rise, hydro-climatology, as well as human activities and local disturbances that influence the style of avulsion (Figure 7.21).

An important control on the location of avulsions along a channel belt is the ratio of the cross-valley gradient to the down-valley gradient. The cross-valley gradient is effectively the slope between the crevasse channel bed (or embanked floodplain surface) to the flood basin. The down-valley gradient is the slope of the valley floor, the space the new channel would accommodate. A higher slope ratio suggests that an avulsion is more probable, and the threshold for avulsion may occur at slope ratios ranging from about 3 to 5 (Törnqvist and Bridge, 2003). A higher slope ratio is created by aggradation within the embanked floodplain as well as flood basin subsidence. Additionally, it is important to consider whether there is sufficient flood basin space to accommodate a new channel belt.

7.4.2 Impacts of Avulsion

While dikes and other forms of hydraulic engineering certainly reduces the incidences of crevasse formation, historically they're not always effective at preventing avulsions (Berendsen and Stouthamer, 2001; Slingerland and Smith, 2004). Such has been the case with the Huanghe (Yellow) River in China where numerous avulsions have occurred, in some cases even resulting in partial connection with the Yangtze River located to the south (Syvitski and Kettner, 2011; Kidder and Zhuang, 2015).

Extreme flooding in 1855 caused the Huanghe to breach its dikes, and within a couple of days the entire discharge was diverted into the adjacent northern flood basin (Figure 7.22). At the moment of avulsion, the channel bed of the Huanghe was from 7 m to 10 m higher than the adjacent flood basin, resulting in a high slope ratio between the down-valley to cross-valley slope profiles (e.g., Törnqvist and Bridge, 2003; Slingerland and Smith, 2004). The flood was responsible for tens of thousands of fatalities within a densely populated area of 31,000 km^2. The flood basin was abruptly transformed, with rapid sedimentation and incision and development of new channels. The river mouth ultimately relocated an astonishingly 495 km to the north (Figure 7.22).

While the Huanghe avulsion in 1855 was triggered by a large flood, the conditions had been set up for centuries (Kidder and Liu, 2014, 2017). Moreover, because of the dikes having to continually be constructed higher, upon diversion of the floodwaters outside of the embankment, the Huanghe was unable to reenter its channel belt. Avulsion was inevitable.

While the 1855 avulsion was the last major avulsion of the Huanghe, smaller avulsions and subdelta formation have continued. In 1976, for example, the main channel of the Huanghe avulsed along its lowermost ~50 km, upstream of the coast (Figure 7.23). The avulsion resulted in a shorter distance to the

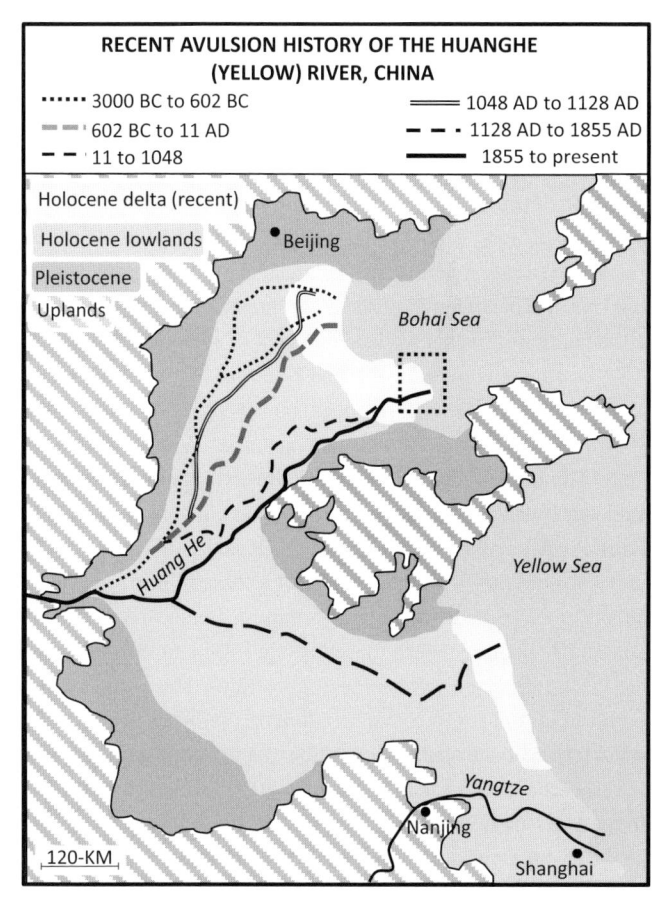

FLOOD BASINS AND DELTAS: SUBSIDENCE, SEDIMENT, AND STORM SURGE 227

Figure 7.22. Recent avulsion history of the Huanghe (Yellow) River delta, China showing rapid avulsions over historic periods. The box at the mouth of the delta references Figure 7.23. Lowermost Yangtze River shown to the south. (Author figure, based on Xue, 1993 and Zheng et al., 2017.)

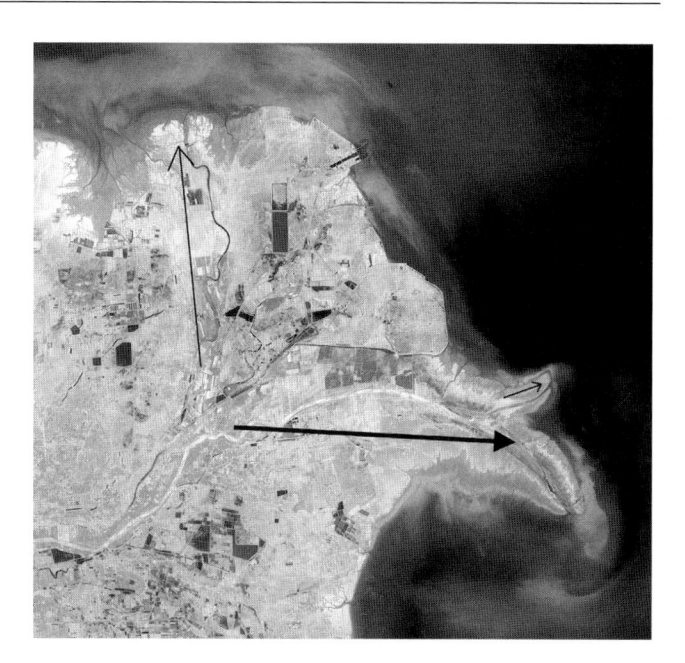

Figure 7.23. Satellite imagery of the Huanghe River delta showing rapid fluvial deltaic progradation caused by channel avulsion and sedimentation, increasing land area by several hundred square kilometers between 1976 and 2017. (Image: ETM Landsat 7+, October 7, 1999, NASA Earth Observatory.)

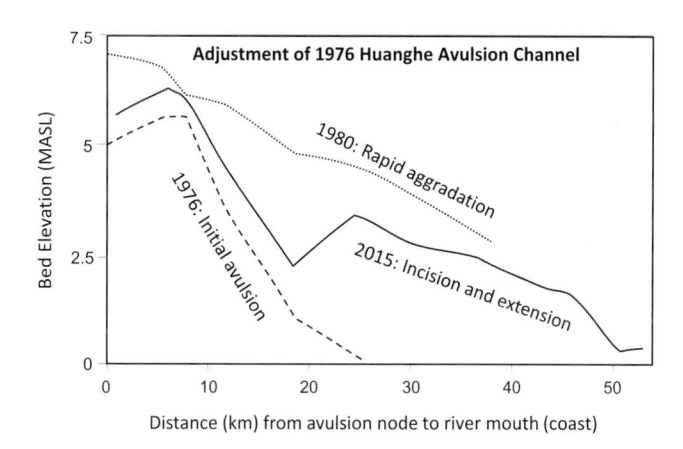

Figure 7.24. Longitudinal channel thalweg profiles of Huanghe River delta, from 1976 to 2015. (After Zheng et al., 2017.)

sea by some ~25 km. Within only several years high amounts of sedimentation resulted in thalweg aggradation and rapid river mouth progradation, as revealed in the long-profiles of the channel thalweg (Figure 7.24). By 2015, the new channel of the Huanghe delta had extended over 25 km, with an overall reduction in the slope of the long-profile. The most rapid adjustment of the Huanghe channel in response to avulsion occurred in less than a decade, and thereafter exponentially decreased (Zheng et al., 2017). While sediment loads in the Huanghe have tremendously declined (Chapter 2, Chapter 4) over the past fifty years because of improved upstream land management and dams (Wang et al., 2015), the Huanghe River will continue to adjust its profile and build new deltaic lands because of planned artificial dike breaches as part of a larger plan to manage coastal deltaic sedimentation (Giosan et al., 2014; Zheng et al., 2017).

River channel avulsion is a fundamental geomorphic process with considerable geologic, environmental, and societal implications. Over the past several thousand years the Huanghe has experienced seven main-stem avulsions and numerous smaller avulsions (Syvitski and Kettner, 2011). The loss of many lives associated with flooding and channel avulsion in densely populated fluvial lowlands, such as along the lower Yangtze and Huanghe Rivers in China, highlights the importance of understanding fundamental processes that drive river avulsion to develop effective management strategies.

7.4.2.1 ST. ELIZABETH DAY FLOODS OF 1421–1424: IMPACTS TO FLOOD BASIN ENVIRONMENTS

In Medieval Holland, a well-documented example of a partial channel avulsion occurred along the Rhine delta during the St. Elizabeth Day Floods of 1421–1424 (Figure 7.25). Although the

Figure 7.25. The St. Elizabeth Day Flood of 1421 in Brabant, Netherlands, with dike breach indicated. Note flood height in relation to lower surfaces. Portion of original painting by Meester van de Heilige c. 1490. (Image: photo of painting in Rijksmuseum, Amsterdam.)

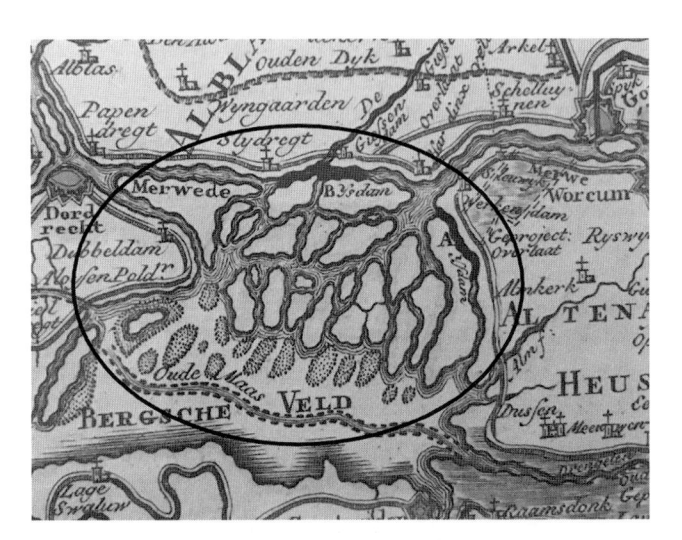

Figure 7.26. Biesbosch area in about 1755 showing avulsion development from 1421 dike breach event, illustrating absence of dominant channel and incomplete basin infilling. (Source: Kaart va den Loop der Rivieren de Rhyn, de Maas, de Waal, de Merwe, en de Lek, door de Provinctien van Gelderland, Holland en Utrecht. In, Atlas of the XVII Nederlandsche Provintien, 1755. Cartographer: Cornelis Velsen. Published by Isaak Tirion, Amsterdam. Author photo of original map.)

Rhine delta had been embanked for flood control for a couple of centuries, the combination of upstream flood pulse and a large coastal storm surge was too much for the dikes. The flooding was accentuated by tidal storm surge that resulted in abrupt dike breaches (Gottschalk, 1975; Lambert, 1985; Kleinhans et al., 2010). The ensuing dike breaches were responsible for the loss of thousands of lives. The flooding resulted in a partial avulsion that created an impressive crevasse splay subdelta within a subsiding flood basin. The topography and environments created by the rapid sedimentation formed a wetland complex of some 60 km^2.

To be sure, the dike breach and flooding of the St. Elizabeth Day events were also influenced by negligence of the flood control system and poor land management. Prior to the event, the reclaimed peat polder lands had already experienced considerable subsidence of a couple of meters (i.e., increasing the slope ratio). Additionally, the dikes were neglected and in poor condition. The combination set the stage for a catastrophic response as the storm surge and flood event occurred. For nearly 200 years after the breach events the river was allowed to partially flow into the crevasse splay channels (Figure 7.26). The sedimentation produced a thick (~0.5–5.0 m) unit of sedimentary deposits characterized by a sandy prograding river mouth with dynamic crevasse channels (Kleinhans et al., 2010). Historic maps and surveys of the topography reveal that for hundreds of years thereafter the river did not have a single dominant main channel, as a dynamic anabranching pattern ensued. The coarser channel and sedimentary splay deposits are capped by 0.5–2.0 m of flood basin clay and tidal overbank sediments deposited in a lagoonal environment. The dike breach that formed the Biesbosch also had upstream consequences. The sediment deposition and shoaling resulted in increased flow resistance and higher upstream stage levels for a couple of centuries, thereby requiring further management actions (Kleinhans et al., 2010).

In an ironic twist that requires centuries to occur, what was a catastrophic flood disaster of great human and natural proportions is now an important environmental preserve for biodiversity and recreation, and has an important role in flood management (Figure 7.27). The Biesbosch is a national park that consists of a large assemblage of freshwater wetlands and polders including numerous channels and lakes in different stages of development, somewhat preserving the splay-like pattern. While modern flood control and agricultural land use efforts in the twentieth century greatly reduced the geodiversity of the Biesbosch, the fluvial bifurcations and islands retain some of the character associated with the dike breach event of 1421.

7.4.3 Avulsions and Flood Basin Sedimentology

In addition to hydraulic factors and management issues, flood basin sedimentology is also an important control on avulsions. And not all initial avulsions are successful, as some of them fail.

Figure 7.27. Oblique air photo of National Park Biesbosch, NL showing typical channel bifurcation with "natural" area (center) that stores floodwaters and lower polder lands protected by dikes (right of channel). (Image source: Rijkswaterstaat.)

Resistant cohesive clayey and peaty flood basins can have sufficient strength such that full development of an avulsion is prevented (Stouthamer and Berendsen, 2001; Makaske et al., 2007). That is, the underlying sedimentology of the flood basin influences whether the crevasse will be able to ultimately scour the flood basin and create a channel large enough to capture the entire discharge of the main-stem river.

The importance of floodplain sedimentology in influencing avulsion success can be appreciated by considering natural avulsions. The Schoonrewoerd River in the Rhine delta, a minor distributary of the Waal, is a prime example of flood basin sedimentology preventing the development of a full avulsion (Kooijmans, 1974; Stouthamer and Berendsen, 2001; Makaske et al., 2007). The Schoonrewoerd channel belt meanders for some 35 km within the flood basin between the Nederrijn-Lek River and the Waal River. The channel formed rapidly, but then functioned for only about 100 years, from 3,900 BP to 3,800 BP. The channel was never able to sufficiently enlarge within the resistant cohesive flood basin deposits to capture the Waal River. The Schoonrewoerd ultimately had a stream power of only 0.5–1.8 W/m². Nevertheless, the surface morphology of the relict channel encased within cohesive flood basin muds evidently provided important flood protection, as the higher natural levees provided modest relief in an otherwise wet landscape, as testified by continued human occupation from the Middle Bronze Age to the Iron Age (Kooijmans, 1974).

7.4.3.1 ATCHAFALAYA: MISSISSIPPI DELTA AVULSION CASE STUDY

Research that has formally evaluated the slope ratio threshold approach to understand the formation of avulsions identifies a wide range of values, pointing to a combination of factors that control channel avulsion (Törnqvist and Bridge, 2003; Aslan et al., 2005). Along the lower Mississippi, for example, avulsions are associated with slope ratio indices spanning from 16 to 110, much higher than the reported 3–5 range. The slope ratio where the active channel avulsed from Bayou Lafourche – the youngest paleo distributary channel belt – is 39, whereas the slope ratio at the nodal point of the Atchafalaya River avulsion – the most recent avulsion – is 35. The Atchafalaya avulsion is interesting because it is active, illustrating a variety of factors associated with avulsions and floodplain sedimentology in the context of a complex management scheme (Figure 7.28).

While the slope advantage of the Atchafalaya is certainly important to its avulsion, other factors have also been important, particularly the hydrology and historic geomorphology. In the centuries prior to the artificial cut-off of the bend in 1831 by Captain Shreve (at Turnbull Bend/Island), the sandy deposits of ancient meander belts facilitated the westerly migration of the main-stem Mississippi toward the Red River channel belt. The flood hydrology at the Mississippi River and Red River confluence has a large seasonal range, with a difference between the average low and official flood stage of 10.5 m,[1] and a duration ranging from about a month to three months overbank. Frequent crevasses are expected to occur. Additionally, the floodplain sedimentology is easily erodable sand from former channel belts and crevasse deposits (Figure 7.29). Thus, a crevasse at the right location (having sufficient gradient) was likely to erode the coarse sediments and establish a channel sufficient to capture discharge from the main-stem channel.

Since its formation in the early to mid-1700s, the Atchafalaya River has rapidly evolved into a major river, and is developing typical channel belt sedimentologic units (Fisk, 1952; Coleman, 1988). The upper reaches of the floodplain have developed formidable natural levees, with a height and width sufficient to support ample floodplain agriculture above the lower lying flood basins (Figure 7.30).

The importance of contrasting cohesive and noncohesive sedimentology on avulsions within the vicinity of Old River and the Atchafalaya basin can be appreciated by considering that other large crevasses and several avulsions ultimately failed. Several crevasse channels extend from the main-stem Mississippi into the cohesive deposits of the Atchafalaya basin, but were unable to capture the entire discharge of the Mississippi (Aslan et al., 2005; Nienhuis et al., 2018). Bayou Fordache (Figure 7.28), for example, is located at Morganza, a meander bend that has experienced repeated large crevasses and the diversion of floodwaters into the Atchafalaya basin. To further highlight the importance of sedimentology in limiting the scour and development of a large channel, it was already noted in Chapter 5 that the success of

---

[1] Mississippi River at Red River Landing, USGS 01120.

Floodplain Geomorphology and Hydrography: Lower Mississippi, vicinity of Old River, Louisiana

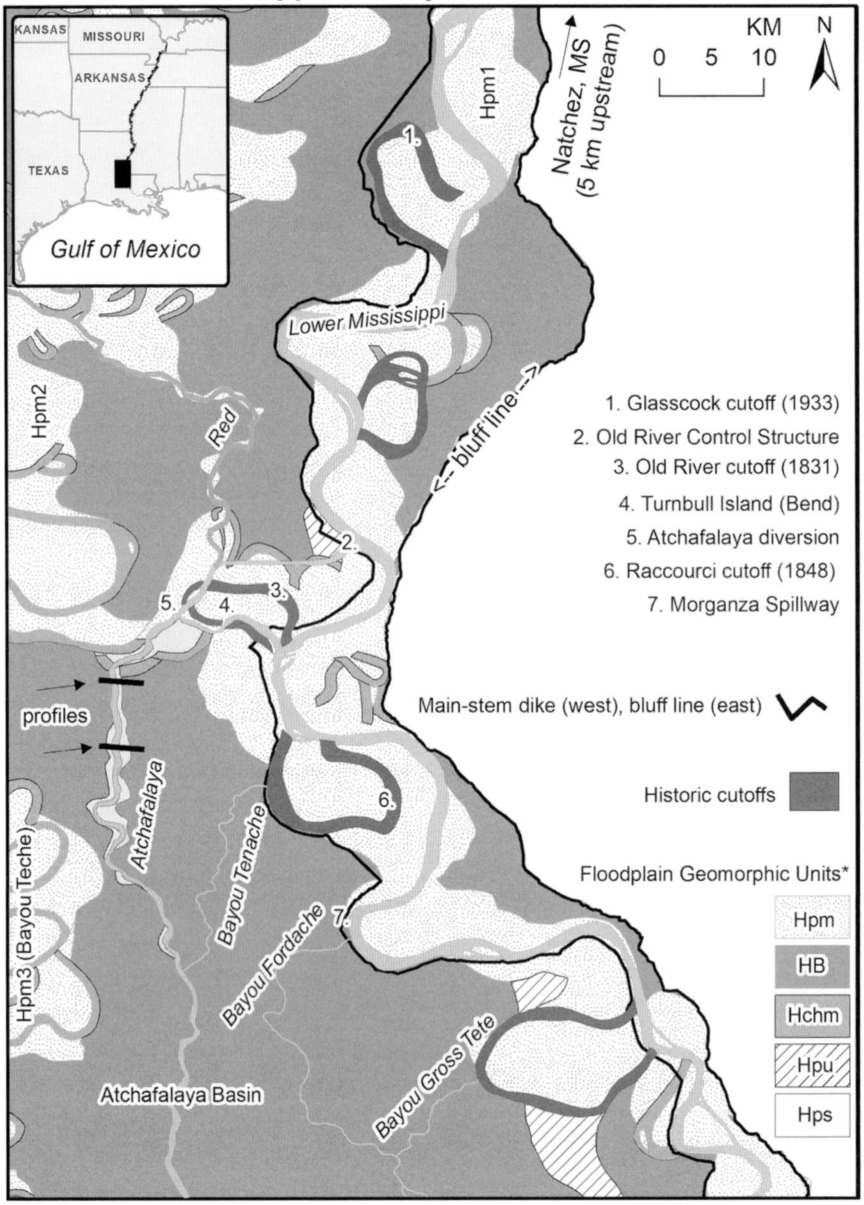

1. Glasscock cutoff (1933)
2. Old River Control Structure
3. Old River cutoff (1831)
4. Turnbull Island (Bend)
5. Atchafalaya diversion
6. Raccourci cutoff (1848)
7. Morganza Spillway

*Floodplain geomorphic units from Saucier, 1994: Hpm1, Hpm2, Hpm3, Red River = Holocene meander belts; Hpu = pointbar deposits buried by backswamp deposits, HB = flood basin and backswamp deposits, Hps = Atchafalaya meander scroll deposits; Hchm = infilled channel deposits

Figure 7.28. Upper portions of Atchafalaya River basin and vicinity, including Mississippi River meander belts and failed avulsion channels. Upper and lower profiles are indexed to Figures 7.29 and 7.30, respectively. (Author figure, compiled from various sources, including Quaternary geomorphology from Saucier, 1994.)

meander cut-offs in the 1930s was somewhat dependent upon sedimentology. In particular, the pilot channels across the meander necks of Giles and Glasscock Cut-offs (Figure 7.28), upstream of Old River, were excavated within thick cohesive backswamp deposits (Ferguson, 1940; Winkley, 1977). The thick clay deposits resisted fluvial incision of the pilot channel, requiring much additional dredging and several years to capture the full discharge and complete the cut-off of the meander bends (e.g., Table 5.5).

The diversion of Mississippi River floodwaters into the Atchafalaya basin at the location of the Morganza Spillway is no longer a natural phenomenon, but an essential component of the U.S. Army Corps of Engineers flood control infrastructure.

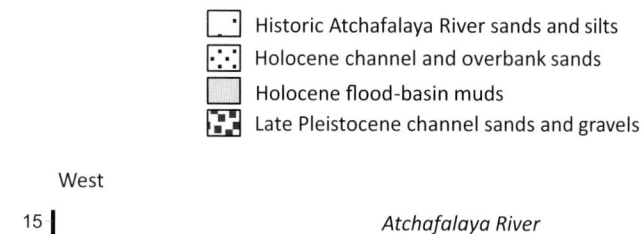

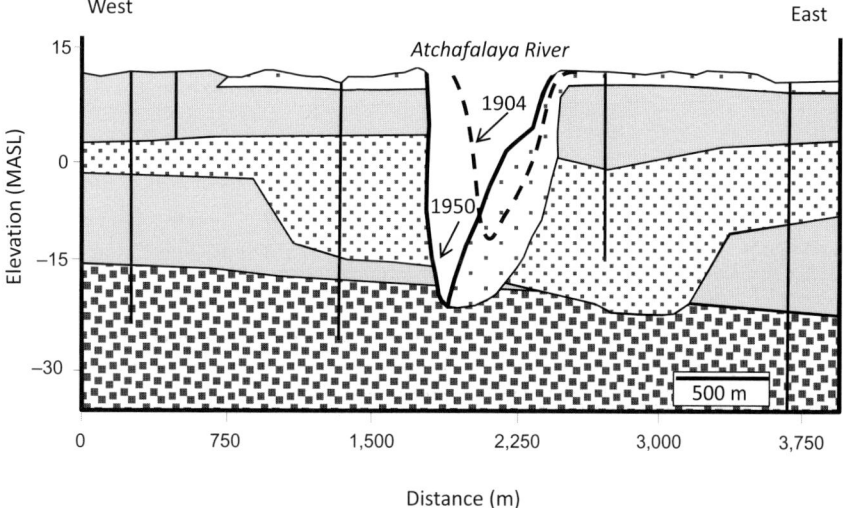

Figure 7.29. Profile of sedimentology associated with Atchafalaya avulsion, 10 km downstream of Red River Landing. Vertical lines demarcate coring depth. (From Aslan et al., 2005.)

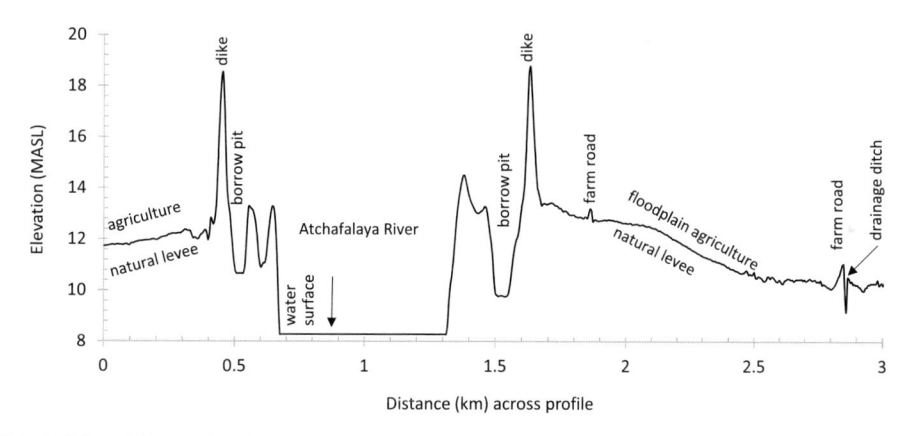

Figure 7.30. Profile of Atchafalaya River and adjacent floodplain topography (20 km downstream of Old River), showing development of natural levees. (Author figure, source: Lidar DEM from ATLAS, Department of Geography & Anthropology, Louisiana State University.)

And it is vital to the diverse ecosystems associated with the Atchafalaya basin (i.e., Figure 1.5). While the infrequent diversion of floodwaters through the Morganza control structure somewhat replicates episodic flood pulses, the larger environmental and flood management problems in the Atchafalaya basin are associated with the continual diversion of Mississippi River sediment into the basin via the Old River Control Structure. Such constant – daily – sediment inputs over the past seventy-five years have resulted in profound changes to the basin's geomorphology and ecology (Hupp et al., 2008). This occurs because of high sedimentation rates and subsequent basin infilling, which has resulted in a net accretion of ~2.5 billion tons of sediment since 1932. Excessive sedimentation is converting open freshwater swamp forests to bottomland forest (Mossa, 2015b). The conversion of water to land by diversion of large pulses of water and sediment into flood basins has consequences to ecosystems, stakeholder livelihoods, and flood management.

7.4.4 Managing Channel Bifurcations and Avulsions

Identifying a critical threshold of the ratio between cross-valley and down-valley slopes to trigger avulsions is attractive to geomorphologists. And it would also be useful to management

Table 7.4 *Factors to consider in the management of channel avulsions and bifurcations*

Location	Issue (concern/problem)	Approach
Nodal point (bifurcation)	Discharge/sediment conveyance, channel bank erosion (expansion)	Groynes, weirs, bank protection, channel bed armor
Upstream	Knickpoint incision and migration, stage level variability (+/−)	Sediment dumping and channel bed fortification, bank protection, dike heightening
Downstream (old channel)	Aggradation, ecosystem degradation, loss of navigability	Pumps to maintain minimal "flushing" discharge, dredging
Downstream (new channel)	Rapid change, including channel adjustment/bank erosion (esp. widening), smaller channels aggrade	Groynes, bank protection, dikes, adaptive management goals in response to rapid fluvial adjustment
Flood basin	Increased flooding (esp. duration), sedimentation, creation of new wetlands	Flood control structures (see Table 7.1), sluice gates, pumps, dikes

Source: Author table.

organizations as a tool to help predict when such fundamental changes to flood regime and ecosystems occur, enabling effective measures to be implemented. There are a range of factors to consider in regard to management of channel avulsions (Table 7.4). Managing channel avulsions occurs in several locations, including (1) the point of bifurcation, (2) within the flood basin, (3) upstream of the avulsion node, and (4) downstream of the avulsion node.

At the location of imminent avulsion, the first management decision is whether the crevasse will be allowed to further enlarge and draw water. Assuming that the crevasse flow can be managed, crevasses can usually be infilled after the floodwaters recede, and most natural crevasses "heal" after about a decade or so. If the avulsion is imminent, a decision needs to be made as to which type of avulsion will be permitted. In some cases it may be preferable to completely abandon a channel, but in many instances it is beneficial to allow both channels to function. After the proportion of discharge is determined, suitable "hard" structures, such as weirs and bank protection revetment and groynes, need to be installed to "train" the new river so that it does not become too large. Further, discharge in the old channel should at least be minimally maintained such that there remain adequate sediment flushing events, which is also dependent upon the size of the channel bed material. In practice this is not an easy concept to manage, especially when a larger percentage of the sediment load consists of gravel and coarse sand. Wash-load (silt/clay) can easily be flushed downstream with pumps and occasional larger flows. In many instances, however, channel dredging will become a mainstay of channel maintenance along the old channel, especially if local interests such as towns and port activities depend on access to navigable waters.

A channel avulsion results in a channel bifurcation, with water shared between two courses. Prior to fluvial sedimentologists having adequate radiocarbon dating to determine the frequency and timing of channel avulsions, it was believed that channel avulsions could effectively occur "in the cover of night" (geologically speaking), in response to a single large flood event. The rapid increase in Mississippi River discharge being diverted into the Atchafalaya River was one of the motivations for constructing the Old River Control Structure (Figure 7.28), particularly since it followed the infamous 1927 flood. Between 1850 and 1950 the amount of Mississippi River discharge flowing into the Atchafalaya River increased from <10% to about 30%. While not actually based on science, the 30% flow volume was ultimately accepted as the "latitude flow" to maintain and prevent an abrupt Mississippi avulsion when the Old River Control Structure was completed in 1963 (McPhee, 1989; Barry, 1997; Mossa, 2015b; Barnett, 2016). Since then, fluvial sedimentologists have established that even abruptly formed avulsions take many decades to finalize. And in many instances, such as anabranching rivers and fluvial deltas, the discharge is shared between a couple of river courses for several centuries or more (Törnqvist et al., 1996; Stouthamer and Berendsen, 2001; Phillips, 2011, 2012). Upstream of the nodal point of avulsion, the largest danger is development of a channel knickpoint, which can migrate upstream and generate channel degradation (Section 5.3.1). Such knickpoints can often be managed by the dumping of coarse channel bed material, such as gravel and cobbles, immediately upstream of the point of avulsion.

In addition to managing an avulsion at the delta apex, bifurcations at the mouth of the Mississippi River have been managed for nearly 150 years. The first jetties along the lower Mississippi (US) were constructed of willow fascine mattress in 1879 by Captain Eads at Head of Passes, Louisiana, where the Mississippi splits into several distributaries and discharges into the Gulf of Mexico (Figure 7.31). This was among the most important hydraulic engineering modifications to a US river. And it was vital to the future of New Orleans, as the narrower channel created a somewhat self-scouring channel and reduced

FLOOD BASINS AND DELTAS: SUBSIDENCE, SEDIMENT, AND STORM SURGE 233

Figure 7.31. Head of Passes, Louisiana, bifurcation point of the lower Mississippi at the river mouth. To the right is Southwest Pass and to the left is South Pass. For scale, "bar" is ~10 m in length. (Photo: John Ruskey, with permission.)

Figure 7.32. Bifurcation point (splitsingspunt) of Rhine River near delta apex, 5 km downstream of NL-DE border. The point of bifurcation marks the beginning of the River Waal (left) and the Pannerdens Canal (right), which then further bifurcates into the Nederrijn-Lek and IJssel Rivers. For scale, "bar" is 10 m in length. (Photo source: Rijkswaterstaat.)

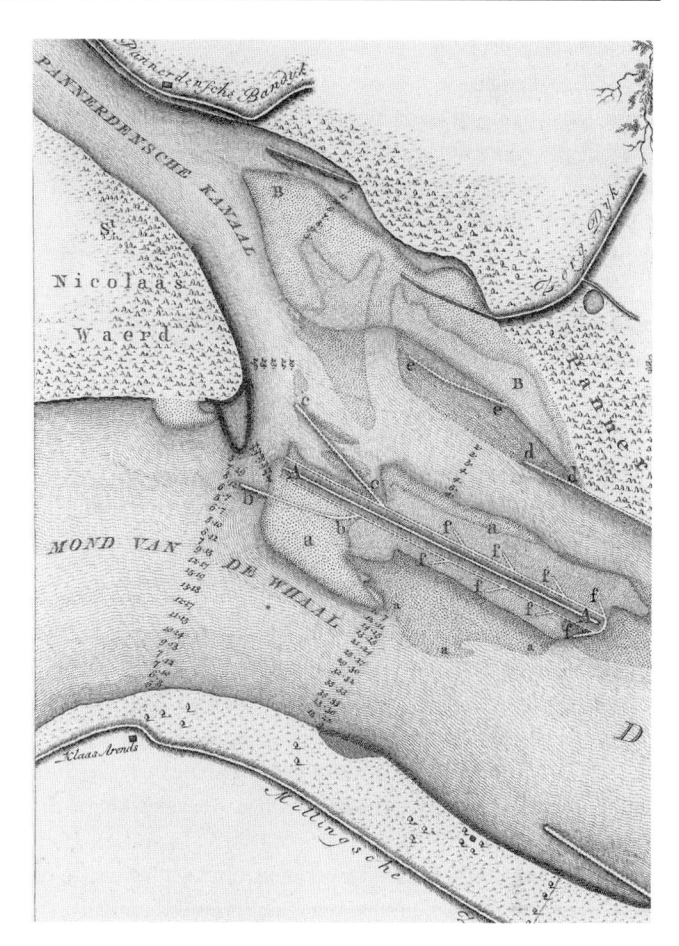

Figure 7.33. Historic map (1749) of Pannerdensch Kanaal (Pannerdens Canal) revealing dynamic channel at time of canal engineering, with wider channel and active sedimentary bars. Immediately upstream is the Bijlandsch Kanaal (not shown), the meander neck cut-off (1773–1776) of the main-stem Waal. (Map source, National Archives, NL.)

dredging operations and tremendously improved navigability to the port of New Orleans (Barry, 1997). Nevertheless, channel bed aggradation continues to be a problem to navigation, requiring continued dredging for reliable navigation.

Management of the passes also includes monitoring and closing upstream crevasses so that reduced stream power does not result in channel bed aggradation, enabling the continued flushing of bed material through the passes (Kemp et al., 2014). While certainly necessary to facilitate navigation, this latter management approach is somewhat at odds with "integrated" approaches to floodplain management that result in intentional levee breaches to increase hydrologic connectivity between rivers and flood basins. In the lower Mississippi delta, the engineering of sediment diversion channels to construct artificial subdeltas is the major management tool to build land that will buffer storm surge events and replenish coastal ecosystems (Kemp et al., 2014; Nienhuis et al., 2018) (Section 8.5).

Immediately downstream of the Netherlands–German border the Rhine River bifurcates at the splitsingspunt into the Waal River and Pannerdens Canal (Figure 7.32), which subsequently bifurcates 11 km downstream into the Nederrijn-Lek and IJssel Rivers. The IJssel then flows northwesterly to the IJsselmeer. Discharge allocation of Rhine River distributaries was established upon completion of the Pannerdens Canal in 1709. The discharge was fixed at 2/3, 2/9, and 1/9 for the Waal, Nederrijn, and IJssel Rivers, respectively, and has remained remarkably stable over the past centuries (van de Ven, 2007, p. 104). The engineering of the bifurcation was required because, historically, at the location at the hinge point between the alluvial valley and delta apex, the channel was subject to considerable shoaling and lateral adjustment (Figure 7.33). Attaining channel stability at the bifurcation (splitsingspunt) required innovative groyne structures, in addition to subsequent channel alignment with a large meander bend cut-off (1773–1776) of the main-stem Waal (Bijlandsch Kanaal).

Since the fixing of the Rhine discharge within the three main distributaries, downstream flood control infrastructure and economic activities have become tightly linked to a precise discharge (flood level). Therefore, changes to the channel bed levels (incision), recent climate change projections, and hydrologic forecasts have resulted in the construction of innovative, adaptive "hard engineering" measures to assure that the discharge allocation remains constant with future projected hydrologic variability.

7.5 URBAN FLOODING AND MANAGEMENT CHALLENGES

7.5.1 Urban Flood Basin Environments

Considering the inherent challenges associated with hydrologic management of flood basins it is surprising that anyone would choose to live in such a setting. But this is what roughly half a billion do. These complexities become especially challenging in the context of large urban populations. In comparison to rural settings, managing flooding in dense urban settings requires different infrastructure and conceptual approaches (Table 7.5). These factors influence various flood mechanisms, which include riverine, coastal storm surge, groundwater, and overhead rainfall (pluvial) events (Jha et al., 2012; Chan et al., 2013; Rotterdam Climate Proof Initiative, 2013). The latter, especially in view of the high amount of impervious cover inherent to urban areas, can generate flash flooding within a few hours. The worst-case scenario is when flooding occurs by a combination of these factors.

Floodwaters do not drain upslope. This truism is always important to keep in mind because it hints at the most fundamental pillar of urban–fluvial settings; they exist in depressions below sea level. Thus, flood control infrastructure for cities located in lowland flood basins is designed for one of two roles. First, infrastructure must be designed to keep water from coming into the city. Second, infrastructure must be designed to move water out of the city. And to do its job flood control infrastructure must simultaneously function. Infrastructure to prevent water from coming into the city includes floodwalls, dikes, and bypass structures. Infrastructure to move water out of the city includes ditches, drains, sluice gates, canals, retention basins, and pumps. Each of these features can only be utilized in one of two roles, and when they are not well maintained and coordinated it can have dire hydrologic consequences.

The functioning of flood control infrastructure in dense urban settings is complex, with narrow margins for error. In urban settings, small areas include many thousands of residents and substantial economic activities, all subject to factors that alter hydrologic processes (Table 7.5). Specifically, this includes transportation infrastructure, including bridges, roads, and rail, which represent a narrow belt of higher ground that alters the natural pathways of land drainage (van Riel, 2011; van den Brink et al., 2014). The impact of urban infrastructure is that it frequently obstructs surface flow and causes ponding, but also serves as a berm to reduce inundation of properties located on the other side. Additionally, large urban areas substantially alter the groundwater hydrology of aquifers. The penetration of the urban footprint into the subsurface immediately alters shallow groundwater flow paths whereas many urban settings also heavily rely upon groundwater for industry and consumption. The imprint of tall office building foundations, subterranean parking garages, and especially tunnels for viaducts, trains, and automobiles restrict natural groundwater flow paths such that surface drainage is impeded, which can cause inundation of structures designed for a lower water table.

From the standpoint of deeper groundwater sources, changes in rates of groundwater withdraw for industry can have considerable impact. The topic is important enough that it was highlighted in the EU's Flood Directive (EC, 2007), and has received an increasing amount of attention following large riverine flood events in the 2000s, which were especially damaging and costly because of surface connectivity to groundwater. Old industrial cities in northern Europe, for example, previously withdrew large amounts of groundwater for many decades and centuries. This resulted in lowering of the water table and subsidence of older basement strata. And for some cities the phreatic zone dropped by tens of meters. Subsequent urban expansion adjusted to the lower water table. But now that many of these cities are far less industrial, and because they more carefully regulate their groundwater resources, the water table below many old industrial cities is rebounding.

An analysis by the British Geological Survey, for example, revealed that up to 900,000 properties within southern England are subject to groundwater flooding from alluvial aquifers. Indeed, across the low-lying London Basin the phenomenon of groundwater rebound has resulted in persistent urban flood problems, which threatens building foundations and urban infrastructure. In some instances, the lag time associated with groundwater flow results in flooding by exfiltration days after the rainfall event and discharge event have crested, and persist for weeks (Macdonald et al., 2012). In the case of urban London, by 1965 the water table, which under natural conditions is at the level of the Thames River (~0 MASL), had dropped to as low as −75 MASL, well into the deeper chalk (Late Cretaceous) aquifer. Changes to the urban economy and better regulation of groundwater pumping, however, resulted in a water table rebound, ~50 m by 2007. While restoration of the water table helps to reduce subsidence of the London Clay (Eocene), the renewed groundwater flow is also a threat to communication and transportation infrastructure, including the great London

Table 7.5 *Issues associated with flooding in urban areas within flood basins*

Category	Impact/problem	Consequences
Groundwater withdrawal (pumping)	Subsidence	Lower ground surface, increased flood vulnerability
Pump failure		Capacity and possible inundation (inoperable)
Impervious cover	Increased runoff	Surface flooding due to drainage capacity overwhelmed
Drainage (ditches/canals)	Infilling by sedimentation or rubbish (due to poor maintenance, improper use)	Reduce drainage capacity, local flooding
Infrastructure (bridges, roads, sewage)	Alters runoff paths	Ponding of water (no drainage), lack of water
Subsurface construction (metro/subway, tunnels, pipelines, deep foundations)	Blocks and alters groundwater flow paths	Exfiltration, water table fluctuation
Reduction in groundwater withdrawals (especially in old industrial areas)	Rebound of water table surface (increased height)	Exfiltration, restrict drainage (infiltration), change in strength of aquifer, destabilization of engineering and urban structures
Urban construction	Subsidence, groundwater flow paths disrupted	Weight of large buildings compacts surface, water table fluctuation and impacts to structures

Compiled from various sources, including van Riel, 2011; Goudie, 2013, figures 5.20 and 5.21; Jha et al., 2012; Rotterdam Climate Proof Initiative, 2013.

Underground. Additionally, the change in the aquifer geochemistry can result in corrosion of buried structures.

Across the North Sea in the Netherlands, at the terminus of the Rhine delta, large cities such as Rotterdam and The Hague have diverse management strategies concerned with groundwater flooding. Here the threat of groundwater flooding is not driven by water table rebound, but by the urban footprint and sea level rise. Altered groundwater flow paths caused by large buildings and transportation infrastructure block and redirect natural flow paths. This is especially significant in the western Netherlands, where the majority of the population and economic activities are located, because for most areas the natural water table should be above the ground surface. Only because of an extensive and intricate network of drainage canals and pumps is lowland Netherlands able to maintain "dry feet." But urbanization continues to present challenges. Groundwater inundation commonly manifests as low surfaces remaining wet or ponded, which can be noticed by streets not draining or hydrophytic vegetation establishing itself in gardens and parks. More troubling to residents is the flooding of parking garages, and especially their cellars. The persistence of the problem has resulted in a range of very specialized management approaches at the neighborhood scale, where water boards increase pumping to offset the upward creeping water table. An entire home repair industry has developed to provide residential-scale waterproofing and pumps for cellars and gardens, a costly requirement imposed on property owners when liability cannot be directed at the city council or local water board.

7.5.2 Asian Mega-Delta Case Study: 2011 Chao Phraya Flooding of Bangkok

With regard to large urban populations and flooding, Asian mega-deltas are the most vulnerable and sensitive to flooding and flood risk. As noted, the Asian mega-deltas represent the proverbial "perfect storm" regarding natural setting and human-imposed environmental change (Cruz et al., 2007; Auerbach et al., 2015), which includes a rapidly growing urban population in megacities that collectively have a population of almost half a billion. The upstream portions of most of the drainage basins have undergone considerable deforestation, and dams have also trapped substantial amounts of sediment and reduced downstream sediment supply (Chapter 4). Combined with enormous freshwater withdrawal to support intensive delta agriculture, increasingly converting to protein production, subsidence rates are especially high. But a further compounding problem confronting the Asian megacities is that the urban infrastructure has substantially contributed to flood risk. Some of the drainage infrastructure was constructed during former colonial rule and is degraded, and has not been considerably upgraded to protect against sea level rise projections and high local subsidence rates.

The above-mentioned issues are illustrated by the 2011 Chao Phraya River flooding of Bangkok in southern Thailand (Figure 7.34). The large vulnerable urban population residing within a changing and fragile environment had all of the ingredients for a catastrophic disaster, and even seemed to have been foretold several years earlier by the IPCC report on Asian

Figure 7.34. Bangkok flooding in 2011 (October 18). Workers repair a collapsed canal embankment in the flooded outskirts of Bangkok. (Source: Pathum Thani province.)

mega-deltas (Cruz et al., 2007). The devastating flood resulted in 813 fatalities. The financial cost was US$42 billion (Gale and Saunders, 2013), among the world's most expensive freshwater flood events.

The Chao Phraya has a complex drainage network that contributes to its style and pattern of flooding. In part, this is because several large tributaries join the main-stem Chao Phraya in the lower reaches of the basin, and then flow as yazoo-style rivers within the large alluvial valley. A major confluence is located just 75 km upstream of Bangkok, at Ayutthaya, where the northeasterly draining Pa Sak River joins the Chao Phraya River. The Pa Sak River includes a large dam, the Pa Sak Dam, which has a key role in protecting Bangkok from flooding. Additionally, the delta is geomorphically segmented into an older upstream delta and, toward the coast, a lower and younger delta with a number of active distributaries. The Chao Phraya finally drains into the Gulf of Thailand through its main-stem channel and two prominent distributaries. The largest distributary is the westerly flowing Tha Chin River, which splits from the Chao Phraya 150 km upstream of Bangkok.

The Chao Phraya delta is enormous (11,430 km^2) in view of the modest size of its drainage basin (159,650 km^2), which in itself is roughly one-third the entire area of Thailand. The delta has a population of 16.2 million, with 12 million located in Bangkok (1,569 km^2). In addition to upstream flooding being influenced by deforestation and engineering modifications to the flow regime, the location of Bangkok only ~30 km from the Gulf of Thailand means that it is also threatened by coastal erosion and storm surge events. The latter is increasingly a threat because of the high rates of land subsidence caused by groundwater pumping for rural agriculture and urban consumption, which has resulted in the land lowering by 2–3 m over the twentieth century (Syvitski et al., 2009; Davivongs et al., 2012).

The flood of 2011 was initiated by an especially wet monsoon made more severe because of having been preceded by several minor tropical cyclones. Whereas 80% of the annual rainfall is provided during the monsoon months (May to October), the 2011 monsoon season had 145% of the annual rainfall. The occurrence of a minor El Niño event likely further accentuated the monsoon (Gale and Saunders, 2013). But the distribution of annual rainfall is what is most significant to the 2011 flood, and the southern portions of the Chao Phraya basin were associated with the highest positive rainfall anomalies. In the case of Bangkok, 2,073 mm of rainfall was received, more than 500 mm above its annual average of 1,543 mm (Gale and Saunders, 2013). The discharge of the Chao Phraya for the 2011 flood was 4,945 (m^3/s), which was only exceeded in 1995 by a peak discharge of 5,461 (m^3/s). Nevertheless, the flood stage, 2.53 MASL, was the highest ever recorded in Bangkok and more than 2 m higher than the 1995 event (Figure 7.35A and B). The large disparity between stage and discharge between 1995 and 2011 points to surficial and structural influences on flooding.

All of the rainfall and runoff had to go somewhere, and it ended up flooding some 30,000 km^2 of the lower alluvial valley and delta plain. During the 2011 flood, the main zone of inundation started some ~300 km upstream of Bangkok, at the confluence of the Yam and Non Rivers. Bangkok experienced flooding from the main-stem river, but also because the flood wave was flowing within the alluvial valley and delta plain. Thus, while the Chao Phraya River created flooding as it flowed through the city, the larger flood wave "attacked" Bangkok from the outside.

While a large flood of the Chao Phraya in 2011 was inevitable, considering the volume and distribution of rainfall received across the drainage basin, the flooding of Bangkok was made much worse because of the dysfunctional urban flood control system. In broad terms the flood control system failed in three ways: (1) lack of maintenance and capacity of existing structures, (2) degradation of the natural system, especially ground subsidence within Bangkok, and (3) poor communication and coordination across different water management sectors, as well as between the government and key stakeholders.

To move floodwaters Bangkok has 1,682 canals that total 2,604 km, 6,180 km of drainage pipes, 158 pumping stations with a maximum capacity of 1,683 (m^3/s), 7 large underground pipes that have a capacity of 155 (m^3/s), and 25 water retention areas to store some 12,800,000 (m^3). Thus, initially it seems that Bangkok has an impressive flood defense infrastructure (Figure 7.35A and B). But the reality is that the urban water control infrastructure was highly dysfunctional during the 2011 flood event, which was in part a function of its evolution over different historical periods spanning dynastic, colonial, and postcolonial legacies (Changxing et al., 2007; Davivongs et al., 2012; Komori et al., 2012; McGrath et al., 2013).

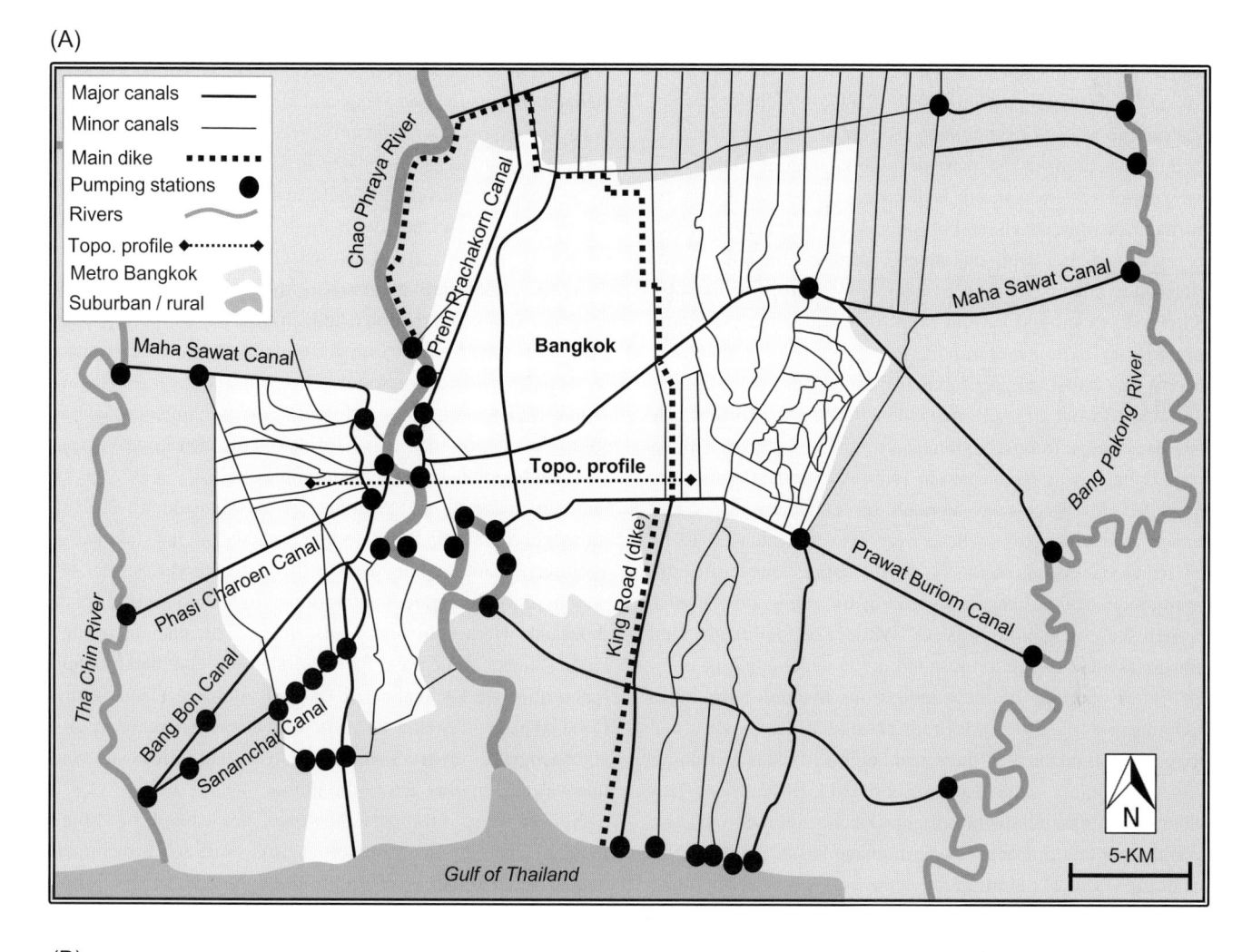

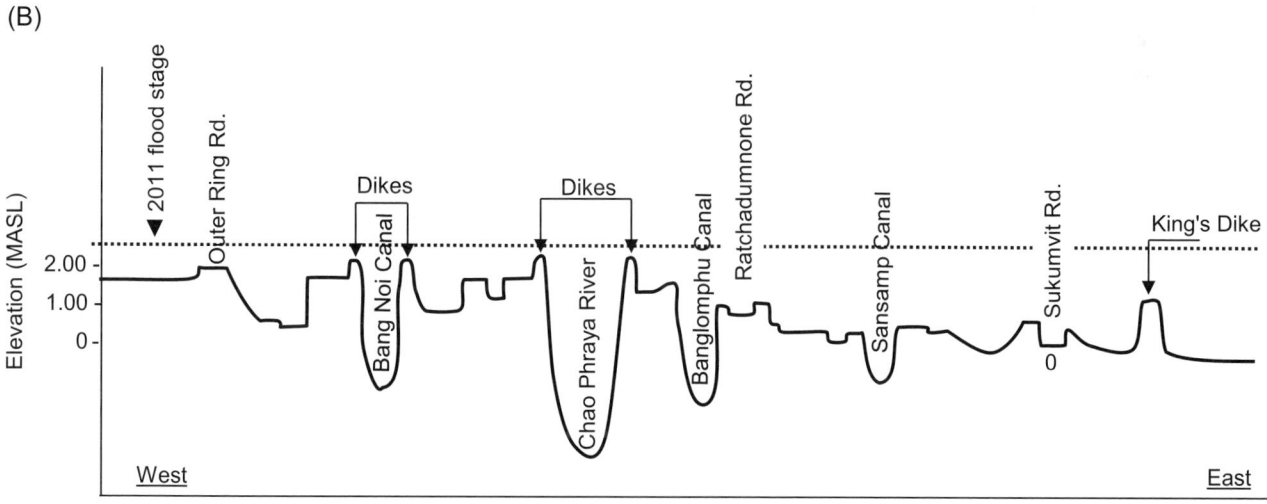

FLOOD BASINS AND DELTAS: SUBSIDENCE, SEDIMENT, AND STORM SURGE 237

Figure 7.35. (A) Canal and drainage system to manage flooding for metropolitan Bangkok, along lower Chao Phraya River. Canals were constructed over a 500-year period associated with dynastic, colonial, and national initiatives for drainage, military, and irrigation and water management. (Source: Bangkok Metropolitan Agency.) (B) Topographic profile of Bangkok in relation to 2011 flood stage and geometric profile associated with dense urban deltaic environment along a large river (see Figure 7.35A for location of profile). The 2011 flood stage peaked at 2.53 MASL, the highest recorded for the Chao Phraya in Bangkok. Transportation infrastructure represents barriers and conduits for floodwaters. (Source: Bangkok Metropolitan Agency.)

In old cities, flood control is a palimpsest of management strategies and hard engineering structures developed over hundreds of years. The problems with Bangkok's flood control infrastructure are as contemporary as they are deeply rooted in historic issues common to many developing nations. Bangkok's flood control system was not well coordinated in terms of the movement of floodwaters and utilities. Most of the engineering structures were not well maintained, resulting in either poor performance during flooding or complete failure. Canals were infilled with muck and various types of urban debris, reducing their capacity to transmit runoff. Perhaps just as problematic concerns the actual origin of many of the canals. The majority of the east-to-west aligned canals were constructed for military purposes, to move troops, beginning after the first monarch of the current dynasty was established in 1780. Additional canals were constructed for large-scale agricultural irrigation, especially intensive rice production. These were also not intended to be used for flood management. To make matters more difficult, numerous ditches and canals were built by indigenous peoples for small-scale irrigation purposes. While also facilitating land drainage, rapid urbanization resulted in the degradation or complete burial of many of these indigenous hydraulic structures (Davivongs et al., 2012; McGrath et al., 2013). Finally, and perhaps most fundamental, the actual heights of dikes and floodwalls were simply inadequate for the 2011 flood crest. This problem is attributed to the absence of consideration of two key anthropogenic impacts on the drainage basin: (1) the influence of high rates of deforestation in the headwaters accelerating the flood wave and (2) the increase in flood stage (i.e., stage–discharge adjustment) occurring with subsidence of the delta plain. This latter problem directly influenced flood defense because the height of dikes and floodwalls are designed for a specific elevation (flood stage). Subsidence, therefore, lowers the height of dikes and results in them being overtopped by floods that originally would have been below the height of the dike. The tremendous, and ongoing, subsidence of the delta plain in the vicinity of Bangkok is caused by a culprit familiar to large delta cities – rapid urbanization and industrial growth requiring substantial pumping to supply groundwater (Galloway et al., 1999; Phien-wej et al., 2006; Changxing et al., 2007; Chan et al., 2013; Allison et al., 2016).

Large floods can have global financial consequences. The rapid urbanization of Bangkok is also to blame for the extremely high cost of the flood. In Bangkok's northeastern sector the federal government and municipality invested several billion (USD) to expand Thailand's industrial output. It worked, and a number of multinational companies located production facilities in Bangkok, making it a major Southeast Asian industrial hub and an important strand in the global web of industrial supply chains (Haraguchi and Lall, 2015). Some 25% of all computer hard drives, for example, are produced by one company, Western Digital, in the Bangkok industrial complex. Bangkok also manufactures automobiles for a number of Japanese companies (Honda, Toyota, Nissan, etc.), and also manufactures automobile parts for export. Unfortunately, most of the land in Bangkok dedicated for industrial estates is situated within backswamp, away from the main channel belt, in a low-lying interdistributary flood basin associated with high subsidence, tremendously increasing the city's flood risk (McGrath et al., 2013).

As the floodwaters approached, modern shiny factories comprising a major international industrial complex were protected by humble earthen dikes fortified with sandbags, a feeble attempt to stop a +2 m wall of water. The result is that factories were swamped, knocking a key industrial supply chain out of commission. The cost of computer hard drives increased on average by 10%, internationally. The automobile industry was especially hard hit, reducing 2011 production in Thailand by 20%. But more substantially, the flooding resulted in an international shortage of automobile parts utilized for production in other nations, curtailing global automobile production several hundred thousand (Haraguchi and Lall, 2015). The manufacturing of Honda trucks and SUVs, for example, was completely stopped for several weeks in the US state of Alabama. Thus, a major flood disaster in Southeast Asia, exacerbated because of global anthropogenic environmental change, had financial consequences that rippled across the globe.

Not all flood problems, however, are caused by environmental change. Coordination across various governmental bodies is essential to effective flood control. In the case of the 2011 Bangkok flood, communication between different governmental scales, such as municipal and federal, was sorely lacking (Cohen, 2012; Nair et al., 2014). An important aspect of this is that the upstream reservoirs initially held back floodwaters for downstream agricultural uses. But this only resulted in a much larger release of water during the heaviest period of inundation, later in the monsoon, which further exacerbated the flooding of Bangkok. Within Bangkok, municipal representatives did a poor job of communicating its "flood fight" strategy to local residents. This fostered local distrust of the intentions of municipal engineers, resulting in an "every man for himself" outcome. A number of dikes were sabotaged and thereby breached, for example, because local interests believed they could reduce the height of the flood crest. Additionally, wealthier residents arranged to have their automobiles and valuable possessions moved to higher ground to avoid flooding, which often involved having their car/s parked atop highway overpasses. This act, in particular, resulted in a tremendous clogging and shutdown of the transportation system. Not only did this hamper access of government flood defense workers but it also made it nearly impossible for many residents to move their personal items outside of the core urban flood zone, thereby increasing the cost of flood damages.

7.6 HURRICANE KATRINA FLOODING OF NEW ORLEANS: CUMULATIVE EFFECTS OF RIVER BASIN MISMANAGEMENT

If the chaotic and shoddy state of flood control systems is of concern to large populations living in the developing world (Tellman et al., 2021), it is interesting to then consider large urban delta populations in developed nations, especially the richest nation on Earth. And here the lessons of New Orleans flooding along the lower Mississippi delta are of particular interest. New Orleans has a metropolitan area of 465 km^2 and a population of ~400,000, although it has been declining since the 1960s. While Bangkok is an order of magnitude larger in size and population than New Orleans, fundamentally both cities are similar in that the core business district and oldest sections are located along the banks of the river, atop natural levees. Recent urban settlement and industry occurs in suburbs, which have expanded into flood basins below the elevation of the natural levee. Of course, an important distinction between the flooding of Bangkok and New Orleans is that the latter occurred by a coastal storm surge generated from an enormous tropical cyclone, not by floodwaters supplied by the upper drainage basin.

New Orleans is effectively an urban island surrounded by a mosaic of wetlands, lakes, bays, and riverine waterways (Figure 7.36). The original layout of New Orleans in the early 1700s was aligned along the arcuate Mississippi River natural levees. The natural levees are massive, stable, and several meters above sea level. In the mid-1800s urban expansion had started to migrate down the flank of the natural levee into the surrounding flood basins. This required development of a flood control infrastructure to drain the land, but also pumping of groundwater. Additionally, natural bayous that flowed into Lake Pontchartrain, such as Bayou St. John, were engineered to drain more distant portions of the New Orleans flood basin. While originally metropolitan New Orleans was entirely at or above sea level, accelerated subsidence has resulted in about half of New Orleans now being located below sea level, with portions of the city being >3 m below sea level (Campanella, 2007). Subsidence of the flood basins has literally moved humans and economic activities from higher to lower elevations with greater risk, thereby tremendously increasing the vulnerability of New Orleans to flooding (Campanella, 2008, pg. 327).

7.6.1 Subsidence in New Orleans

Spatial and temporal variability in New Orleans' subsidence rates are directly related to sedimentology and human activities (Kolb and Saucier, 1982; Campanella, 2007; Jones et al., 2016). In New Orleans, the 0-MASL demarcation roughly corresponds to the gradational boundary between natural levees and flood basins (Figure 7.37). Natural levee deposits associated with the active lower Mississippi River and the Metairie Ridge have undergone little subsidence. The flood basin hinterlands, in Kenner and Metairie, however, have undergone some ~2 m of subsidence (Figure 7.38). The combination of groundwater pumping and ditch and canal drainage caused subsidence by lowering of the water table. The drop in the water table resulted in flood basin deposits being exposed to subaerial conditions, causing oxidation of organic material. Also, fluid withdrawal reduced the particle pore pressure, resulting in sedimentary compaction and a loss of pore space. Subsidence started with wetland drainage and canal

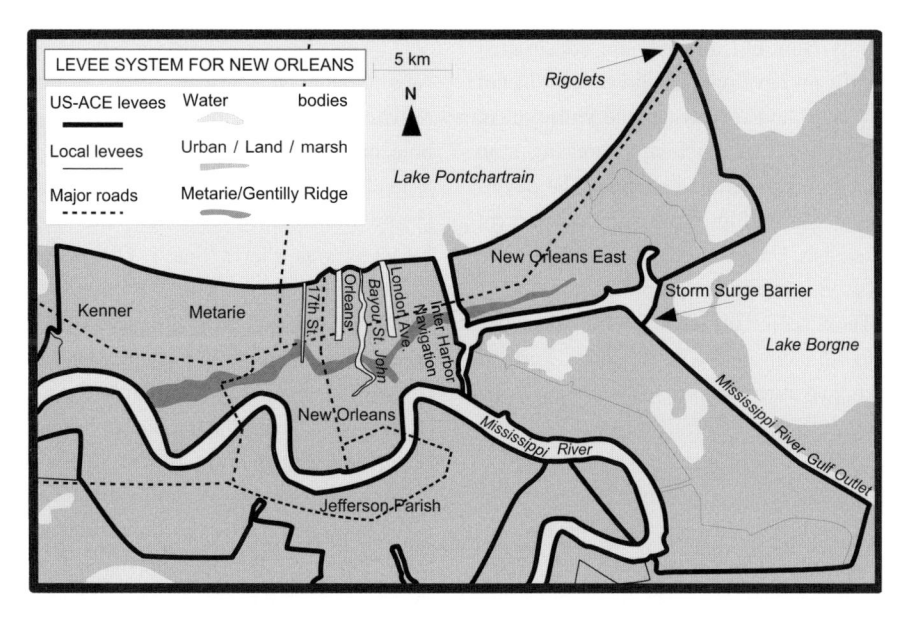

Figure 7.36. Flood control system of New Orleans metropolitan area. (Author figure, compiled from various sources.)

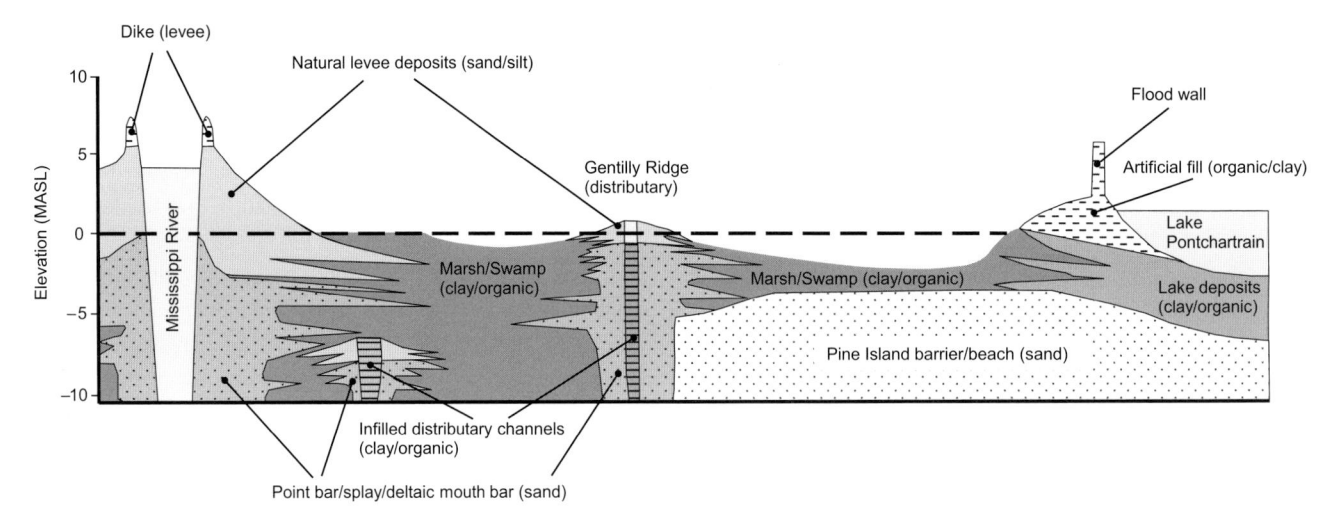

Figure 7.37. Geomorphic profile of New Orleans, from the lower Mississippi natural levee to Lake Pontchartrain. (Based on several sources, including Saucier, 1994; Seed et al., 2008.)

construction in the mid-1800s. But most subsidence occurred after the 1950s and was related to groundwater pumping to support suburban housing and industry (Figure 7.38).

The presence of a large dense layer of coarse sand associated with a paleo beach ridge (Pine Island Barrier, mid-Holocene) in central and eastern New Orleans serves as a control to reduce further subsidence. But where this unit is absent, in Kenner and Metairie, high subsidence rates are expected to continue (Kolb and Saucier, 1982; Whittaker and Reddish, 1989; Jones et al., 2016).

Perhaps the most problematic area for subsidence concerns the Lake Shore district, the northern edge of the city along Lake Pontchartrain, which has annual average subsidence rates ranging from 15 mm to 25 mm. Historically, the flood basins graded into Lake Pontchartrain as an extensive wetland – mainly a freshwater cypress forest. By the mid-1800s high-end waterfront properties for wealthy New Orleanians were constructed along the lakeshore. Significant development, however, did not occur until after World War II, and was associated with typical American subdivisions comprised of ranch-style housing atop concrete slabs laying directly on the ground surface. The development was only possible because of an extensive embankment constructed along the lake edge. Unfortunately, the embankment was entirely of dredge spoil material comprised mainly of organic marsh deposits. These sediments have low bulk densities ranging from 0.5 g/cm^3 to 1.5 g/cm^3 (Seed et al., 2008). Further, the spoil deposits were dumped atop thick organic-rich lacustrine deposits ranging from 3.1 m to 9.3 m in thickness, which suggests that high subsidence rates will persist.

Subsidence of New Orleans flood basins had direct consequences to its surficial hydrology, and effectively means that floodwaters can no longer naturally drain into Lake Pontchartrain or the Gulf of Mexico. The hydrologic and topographic context of New Orleans, therefore, means that the city can be flooded by several mechanisms, including the Mississippi River, Gulf of Mexico storm surge, Lake Pontchartrain, groundwater, and pluvial events associated with midlatitude fronts, local intense convectional rainfall, and precipitation deriving from tropical cyclones. As a result, the city is surrounded and embedded within a flood control infrastructure designed to both keep water from flowing into the city, but also to move water out of the city. To accomplish this the city employs a system of pumps, canals, floodgates, levees, and flood walls constructed, maintained, and operated by a combination of the U.S. Army Corps of Engineers and local interests, including the local levee boards (Independent Levee Investigation Team, 2006).

7.6.2 A Historic Precedent

New Orleans has always been aware of its vulnerability to flooding. Since being founded in 1718, Mississippi levee breaches have been a persistent problem – and especially curtailed development of northwestern sectors of the city – in what is now the suburb of Kenner. One of the more infamous levee breaches was Pierre Sauvé Crevasse in 1849 (Figure 7.39). Located just upstream of the city center, the crevasse flooded the core sector of New Orleans. And interestingly, the spatial extent of flooding was influenced by a winding moribund distributary channel belt, locally referred to as Gentilly (or Metairie) Ridge, which continues to be important to New Orleans flood hydrology. Depending upon what side of the ridge you live on relative to the source of inundation, Gentilly Ridge either exacerbates flooding or provides flood protection. In the case of Pierre Sauvé Crevasse, the ridge greatly exacerbated the flood problem in central New Orleans because of trapping floodwaters in the core sector of the city, between the natural levee and the Metairie

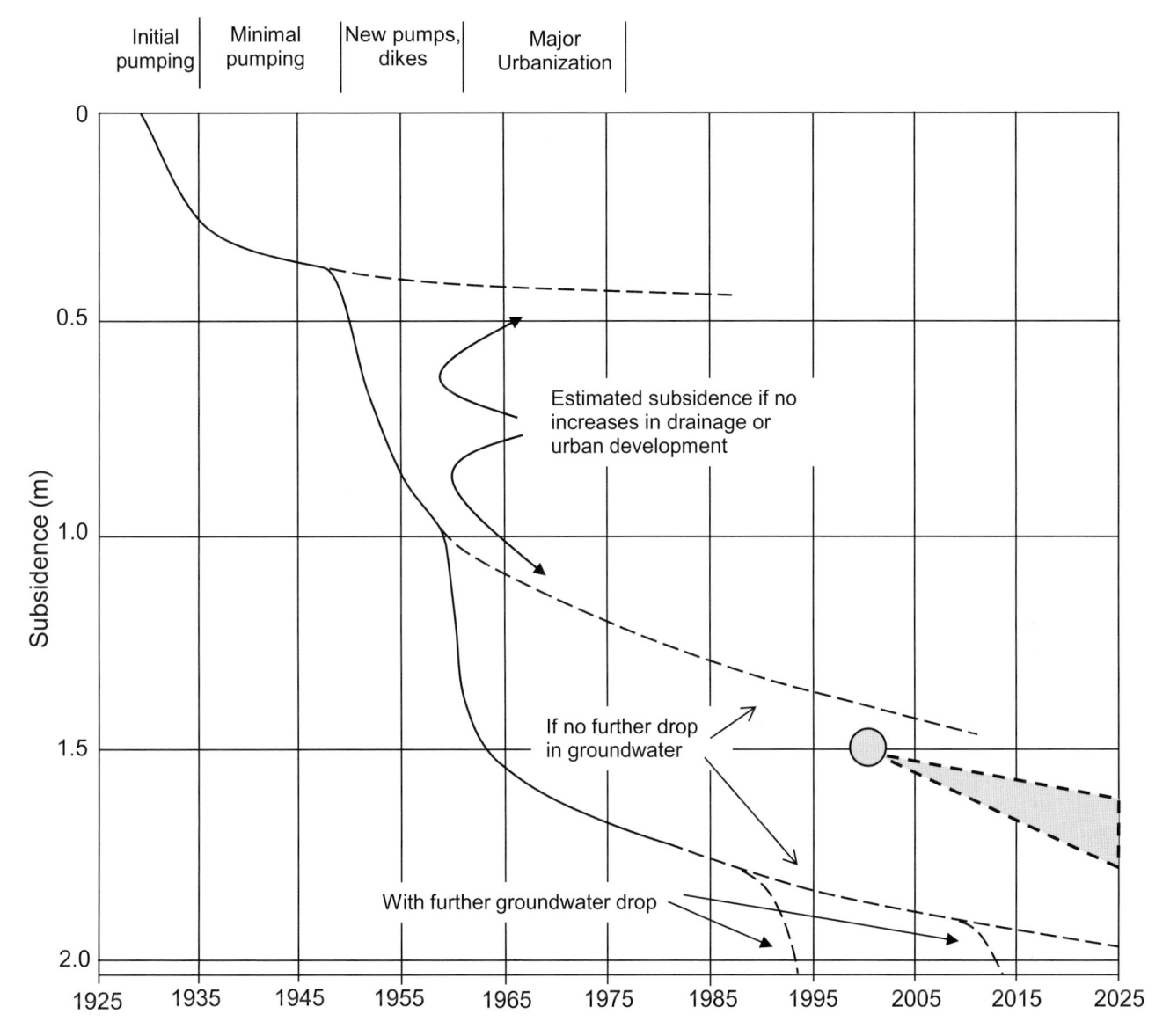

Figure 7.38. Subsidence in suburban New Orleans associated with drainage and urban development. (Sources: subsidence in Kenner from Kolb and Saucier, 1982; subsidence for Metairie from comparison of 1897 topographic data with 1999 LiDAR topographic data; and projected low (0.5 mm/yr) and high (10.0 mm/yr) subsidence range from geophysical data from Shinkle and Dokka, 2004.)

Ridge. The flooding persisted for nearly two months (longer than the 2005 flooding), which was a result of the difficulty in shutting down the crevasse but also because the water could only escape to Lake Pontchartrain via a small canal that fed into Bayou St. John. While much of the area flooded by the 1849 Pierre Sauvé Crevasse was swampland (now suburban Kenner and Metairie), it also resulted in extensive flooding of central New Orleans and partially extended up the Mississippi natural levee. Indeed, flood heights in central New Orleans were actually higher in 1849 than the 2005 flood from Hurricanes Katrina-Rita. During the 2005 flood, Gentilly Ridge was not completely inundated and provided a sliver of dry land for an otherwise drowned city.

7.6.3 New Orleans Flood Control System

Mechanical pumps for flood control were initially installed in the early 1800s, but it was not until the late 1800s and early 1900s that New Orleans had a viable system of drainage pumps. It should be noted, however, that the pumps pulled water from the land (ditches) and discharged into the canal system, which flowed north and drained, by gravity, into Lake Pontchartrain. This approach is inherently flawed because the stage (elevation) of Lake Pontchartrain is frequently above the elevation of canals, especially for storm surge events occurring during spring tides, when drainage is most crucial. Alarmingly, floodgates were not installed between the canals and Lake Pontchartrain (IPET, 2009). Incredibly, this

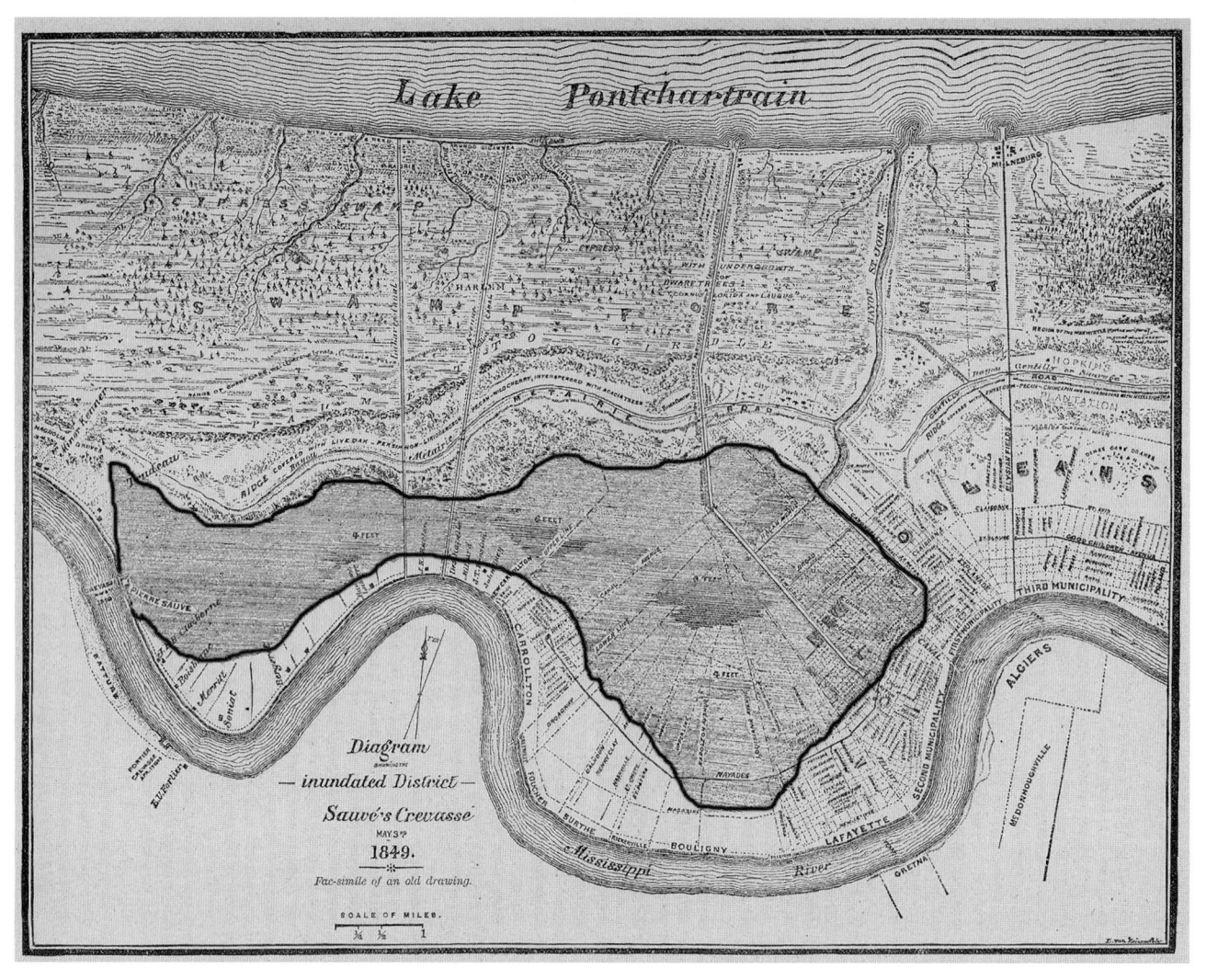

Figure 7.39. Extent of historic New Orleans flooding from Pierre Sauvé Crevasse in 1849. Flooding occurred between lower Mississippi natural levee and natural levee of Metairie Ridge, a moribund distributary that extends through midtown. Note: natural drainage to Lake Pontchartrain. (Source: Historic New Orleans Collection.)

drainage strategy continued until after the 2005 Hurricanes Katrina-Rita flood disaster, when for the first time floodgates and pumps were installed at the entrance of the canals into Lake Pontchartrain. In addition to pumps installed at the canal inlet to Lake Pontchartrain, additional pumps have been installed to move urban floodwaters from canals to wetlands in the southeast (the Gulf Intracoastal Waterway West Closure Complex). The pumps have a maximum discharge capacity of 541 m^3/s, the largest pumps in the world (even larger than proposed for the Yazoo basin). The pump drains floodwaters from industrial canals to the Bayou aux Carpes marsh wetland, which is of significant environmental importance. Unlike the proposed pumps for the Yazoo basin, however, the New Orleans pumps are in compliance with section 404c of the Clean Water Act. Thus, they do not significantly degrade the surrounding wetlands, which is crucial because of their role in protecting New Orleans from storm surge events.

1965 Lake Pontchartrain and Vicinity Project relative to hurricane parameters.

Prior to directly reviewing various processes and failures associated with the 2005 flooding of New Orleans, we should first review an additional key development in the evolution of the New Orleans flood control system. Four decades prior to Hurricane Katrina, the Lake Pontchartrain and Vicinity Hurricane Protection Project (aka the "1965 Pontchartrain Project") was passed by Congress as part of the federal Flood Control Act of 1965 (IPET, 2007). The 1965 Pontchartrain Project was to fund a major expansion of flood control infrastructure for New Orleans, especially construction of levees and flood walls along the south shore of Lake Pontchartrain, New Orleans east, and southeast New Orleans. Unfortunately, the 1965 Pontchartrain Project was only designed to prevent flooding from storm surge events and precipitation associated with a "project hurricane" (Table 7.6), a "hypothetical" moderate magnitude

Table 7.6 *Geophysical characteristics of Standard Project Hurricane for 1965 Lake Pontchartrain and Vicinity Project in comparison with Category*-II, -III, -IV, and -V hurricane parameters*

Parameter	Standard project hurricane for coastal Louisiana**	Category-II	Category-III	Category-IV	Category-V
Central pressure (Mb)	935	965–979	945–964	921–944	<921
Wind speed (km/h)	161	154–177	178–208	209–251	>251
Radius of maximum winds (km)	48	N/A	N/A	N/A	N/A
Average forward speed (km/h)	11	N/A	N/A	N/A	N/A
Storm surge height (m)	3.4–4.0	1.8–2.6	2.7–3.8	3.9–5.6	>5.6

* Saffir–Simpson hurricane wind scale.
** Modeled for the 1965 Lake Pontchartrain and Vicinity Project. Compiled from various sources, including U.S. GAO (2006a, 2006b).

tropical cyclone by comparison to very large hurricanes which had – in "reality" – already impacted New Orleans (U.S. GAO, 2005, 2006a, 2006b; IPET, 2009). The project hurricane was, by comparison, comparable to a fast-moving category-III hurricane (Saffir–Simpson scale). The storm surge from a project hurricane ranges from 3.4 m to 4.0 m, not considering the influence of flood control infrastructure on storm surge heights (i.e., funneling effect). A slow-moving category-III hurricane which could stall over the city would generate too much precipitation and local runoff for flood protection afforded by the 1965 Pontchartrain Project. Thus, it was never the intention that the 1965 Pontchartrain Project would provide protection from hurricanes at the scale of Katrina and Rita, much less a category-V event.

The 1965 Pontchartrain Project was supposed to have been completed in thirteen years, by 1978. The threat of hurricane-generated flooding associated with significant amounts of urban rainfall and/or storm surge was well known to New Orleanians, as experienced in 1915, 1940, 1947, 1965, and 1969, such that the events are part of their identity. Indeed, the signature New Orleans cocktail is a *Hurricane*. Prior to 2005, the last significant flooding of New Orleans occurred in 1969 by a storm surge generated by Hurricane Camille, a category-V hurricane. When Hurricane Katrina made landfall in August 2005 completion of different segments of the project ranged from 60% to 90% (U.S. GAO, 2005), with an estimated final project completion of 2015. It was unfortunate that not enough of the 1965 Pontchartrain Project was completed before Hurricane Camille in 1969, but it was tragic and inexcusable that the 1965 Pontchartrain Project was incomplete by landfall of Hurricanes Katrina and Rita in 2005 (U.S. GAO, 2006a,b).

7.6.4 Hurricane Katrina: Landfall and Storm Surge

Hurricane Katrina made landfall August 29, 2005 upstream of the mouth of the Mississippi River as a category-III hurricane with winds at 200 kmph, having rapidly downgraded from a category-V event on August 28. The storm tracked back into the Gulf of Mexico and made landfall again as a category-III event with winds at 190 kmph along the border of Louisiana and Mississippi, east of New Orleans. But directly it was the storm surge and rainfall that were most significant to flooding. Katrina produced rainfall totals over New Orleans of 330 mm. The storm surge from Hurricane Katrina ranged from 3.0 m to 5.8 m along coastal southeastern Louisiana, which was somewhat higher than the storm surge range for a category-III hurricane and also somewhat higher than what was supposed to be provided by the New Orleans flood control system (3.4–4.0 m). And less than a month after Katrina, while the city was still flooded, Hurricane Rita made landfall to the west, near the Texas–Louisiana border. This effectively resulted in a one-two punch that compounded and extended New Orleans' tragedy, resulting in renewed inundation that prevented engineers from completing emergency repairs. Hurricane Rita's storm surge in the vicinity of New Orleans was 2.1 m, although the storm – mercifully – produced little precipitation.

In addition to New Orleans, Hurricane Katrina's storm surge impacted a much broader swath of the Gulf Coast, which was especially hard hit east of the storm track. The storm surge caused flooding and devastation to many communities along the bays and rivers of this extensive lowland swampy region, extending some ~300 km from central Louisiana to western Florida, where Pensacola experienced a storm surge of 1.6 m. The storm surge was especially devastating east of the hurricane eye wall, and established new records across coastal Mississippi, including Pascagoula (5.5 m), Biloxi (6.7 m), Gulfport (7.5 m), Long Beach (7.8 m), and Bay St. Louis (7.6 m). And at Pass Christian, 25 km east of the eye of the hurricane, the surge was 8.5 m, a new record storm surge for the United States. Incidentally, the storm surge from Hurricane Katrina broke the previous US storm surge record, which had also been set at Pass Christian (in 1969, from Hurricane Camille). Thus, even for a region accustomed to flooding and devastation from storm surge, the Hurricane Katrina storm surge was massive.

Part of the reason for the enormous storm surge is that prior to landfall Hurricane Katrina was a category-V hurricane, having rapidly intensified as the eye of the storm crossed over a warm loop current ring, a large (~300 km diameter) clockwise rotating swath of warm water that pinches off from the loop current in the southeastern Gulf of Mexico. Upon separating from the main loop current, the warm rings drift northwesterly toward Louisiana and Texas at 3 km/day to 4 km/day, eventually circulating around the Gulf of Mexico basin in four to six months. The depth of loop current rings – to about 1,000 m – and their large size provides tropical cyclones with additional energy (i.e., warm water) and increases its wind speed, the main driver of storm surge height. Thus, while Hurricane Katrina was formally a category-III storm when it made landfall, the storm surge that advanced into the bays and bayous along the Gulf Coast was a residual of a category-V hurricane (i.e., Table 7.5).

While New Orleans was not flooded by the Mississippi River it should be noted that the strength of the storm surge was great enough that it resulted in an abrupt rise in upstream river stage (Figure 7.40). Hydrologic data for several gauging stations reveals the height of the storm surge and its influence on the stage of a major river. The eye of the storm made landfall on the west bank of the lower Mississippi River at Buras-Triumph at 7:10 AM on August 29. The storm surge moved upstream and caused a 3.6 m rise in stage at New Orleans, 162 km upstream of Head of Passes. At New Orleans the abrupt rise in river stage caused by the coastal storm surge was 0.6 m below flood stage and 1.3 m below the height of the levee. And the surge continued to move upstream, and incredibly caused a 3.3 m rise in stage at the Bonne Carré Spillway (before gauge stopped recording), the main structure designed to prevent New Orleans from flooding by main-stem river floods. In fact, the influence of the storm surge can be detected as far upstream as Baton Rouge, 376 km upstream of Head of Passes, where it caused a 1.4 m rise in stage (before gauge stopped recording). It was simply good fortune that the Mississippi was at low stage and not at high water when Hurricane Katrina made landfall. But that the storm surge could directly impact the hydrology of a major fluvial system, such as the lower Mississippi, hundreds of kilometers upstream, is a testament to the strength of the storm surge produced by Hurricane Katrina, and an ominous indication of the severity in which smaller rivers and associated communities were impacted by the event along the Gulf Coast.

The storm surge primarily attacked the flood control system of New Orleans from the southeast and from Lake Pontchartrain. Of course, the deltaic landscape that the storm surge moved across was quite different than that for which the 1965 Pontchartrain Project was designed, or when Camille made landfall in 1969. Notably it was lower, because of ground subsidence and the reduction in topography. And for this reason, in comparison to even five decades ago there was far less wetland vegetation to provide resistance to the storm surge, as some of the highest rates of coastal erosion in the world (e.g., Barras et al., 2003; Couvillion et al., 2017) had transformed expansive marshlands to open water (Figure 7.41).

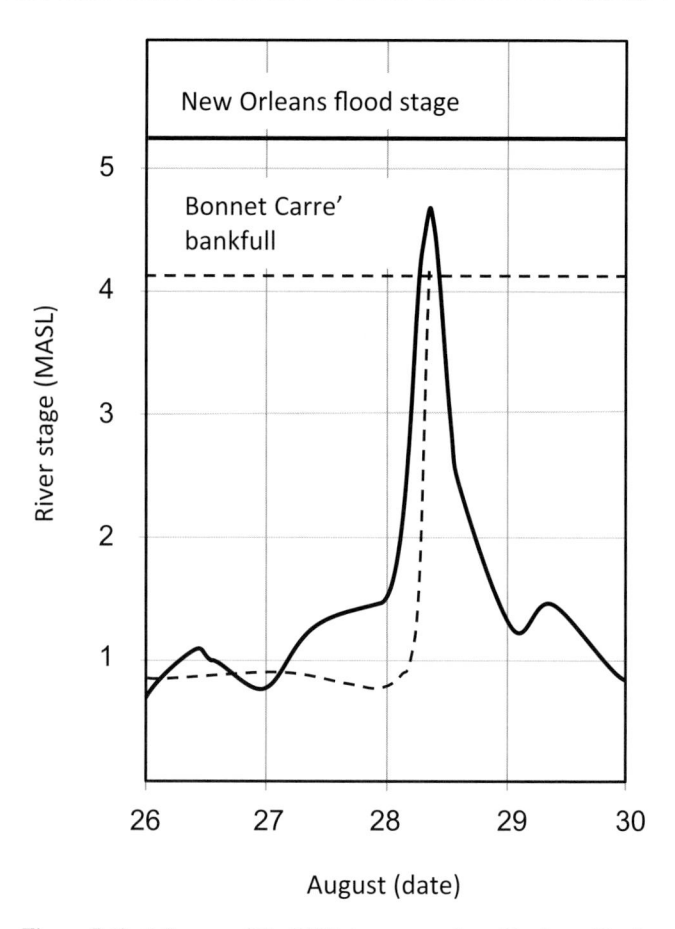

Figure 7.40. Influence of the 2005 storm surge from Hurricane Katrina on Mississippi River stage (elevation) for New Orleans (Carrolton gauge, COE #: 0130) and at Bonnet Carré Spillway (COE #: 01280). Solid line refers to New Orleans data whereas the dashed line refers to data for Bonnet Carré station, which was made inoperable by the storm surge. (Source: U.S. Army Corps of Engineers.)

The surge flowed through Rigolets Pass, the narrow entrance into Lake Pontchartrain, and moved to New Orleans lakeshore where it set new stage records. The height of the surge decreased as it moved west, from 3.9 m along the eastern portions of the lakefront to 2.4 m along the Metairie and Kenner lakefront. The lake rose and surged into the heart of the city via the 17th St., London, Orleans, and Inter Harbor Navigation Canals, in addition to Bayou St. Jean. Additionally, as Hurricane Katrina moved north of the city it produced a second storm surge associated with wind reversal piling up lake water against the southern edge of Lake Pontchartrain. From the southeast the surge was funneled from the Gulf of Mexico into Lake Borgne and the Mississippi River Gulf Outlet (MR-GO), whereby it then moved into the city's core via the Inter Harbor Navigation and Industrial Canals. Large sections of the levee (dike) along Lake Borgne

Figure 7.41. Subsidence in the Mississippi delta showing geodetic survey bench mark and dead live oak sinking into the ocean. (Photo: 2003, courtesy of Lane Lefort, New Orleans by permission.)

Figure 7.42. View along 17th St. canal, looking south to city center. The floodwall along the east (left) side of the canal failed, resulting in extensive flooding along the New Orleans lakeshore. Floodwalls along the west (right) side held, resulting in Metairie having less extensive flooding. (Source: U.S. Army Corps of Engineers.)

were below the design stage because of the 1965 Pontchartrain Project being incomplete (ILIT, 2006; U.S. GAO, 2006a, 2006b), thereby resulting in an incomplete ring of floodwalls and levees to protect the Ninth Ward and Chalmette.

7.6.5 Failure of Levees and Flood Walls

The levees along MR-GO failed in multiple locations. They were either overtopped because of being too low due to subsidence or they were breached because of being of insufficient strength. The reason for this is because the US-ACE constructed the levees from "local" dredge spoil obtained from the actual excavation of MR-GO, which was highly organic material. The levees along MR-GO had some of the highest rates of subsidence within the Mississippi delta (>30 mm/yr). And some of the material was comprised of erodable sand and lightweight shell materials, the latter which especially has insufficient strength to withstand the shear produced from the high storm surge events (ILIT, 2006; Rogers, 2008; Seed et al., 2008; IPET, 2009).

Overall, there were fifty-three failures of levees and flood walls in the greater New Orleans flood control system, with twenty-three occurring directly within New Orleans, resulting in flooding of 83% of the city (ILIT, 2006; U.S. GAO, 2006a; Seed et al., 2008). Spatially, flooding was most severe along the lakeshore and in southeastern portions of the city where the floodwaters exceeded 3.3 m in depth. The lakeshore area was flooded from Lake Pontchartrain, caused by failures in the flood walls and levees that aligned the drainage canals. Three major breaches (with deep scour), one major breach on the 17th St. canal and two large breaches on the London St. canal were responsible for 80% of the floodwaters. The Orleans Canal levee, for example, did not breach, but instead a 100 m section was not completed because of dysfunctional relations between federal and local entities. This reduced the height of flood protection from 3.8 m to 2.1 m, a reduction of 1.7 m. In southeastern New Orleans, floodwaters initially surged in from Lake Borgne and MR-GO and flooded the Ninth Ward and Chalmette by breaching low earthen levees. And then the surge entered from Lake Pontchartrain via the Inner Harbor Navigation Canal, breaching the flood walls along the Industrial Canal and further inundating the Ninth Ward.

Other sectors of the city received lower levels of inundation because of the spatial configuration of interior New Orleans subbasins. Although having undergone significant subsidence, Metairie and Kenner were actually not heavily flooded because the flood walls held along the west side of the 17th St. canal (Figure 7.42), and thus they received 0 m to 3.3 m of flooding, mainly from rain water and because of the failure of the drainage pumps. And, in comparison to 1849, central New Orleans benefitted greatly from Metairie Ridge, as the old moribund distributary blocked the advance of floodwaters from the lakeshore canals. Central New Orleans was flooded mainly by rain water.

7.6.6 Causes and Consequences of Flood System Failure

While the storm surge and wave loading exceeded some of the engineering design levels of the levees and flood walls, it was not much greater and for much of the system it was less than what the infrastructure should have withstood. The failure, however, was disproportionate. In part this is because the 1965 Pontchartrain Project was never completed. But also, in the context of an urban setting within a subsiding flood basin, the engineering structures that had been completed were inappropriately designed.

A glaring mistake by the U.S. Army Corps of Engineers is that they did not build the levees and flood walls according to their own official published safety standards (Rogers, 2008; Seed et al., 2008). The proverbial smoking gun, which faults the U.S. Army Corps of Engineers and their design of New Orleans flood control infrastructure, rather than the magnitude of the hydrologic event, can be appreciated by examining sedimentary data pertaining to construction of the 17th St. canal levee and flood wall, which singularly was responsible for much of the flooding across New Orleans (ASCE, 2007; Duncan et al., 2008; Rogers, 2008; Seed et al., 2008). First, for engineering flood walls in soft organic-rich deposits the sheet metal (I-wall) should have extended to a depth of 9.5–14.0 m. Instead, the Corps extended sheet metal to only 5.2 m. This greatly reduced project costs by roughly $100 million (U.S. GAO, 2005), but it also compromised the integrity of the flood control system and the safety of many thousands. Secondly, the shear strength (kPa) of underlying sediment was substantially less than advised by USACE guidelines (Figure 7.43). Most troubling was that at the depth between −3.0 MASL and −4.8 MASL, the average sample shear strength (kPa) of the levee center line was less than or equal to the recommended guidelines, which ranged from 18 KPa to 15 KPa. This section is considered to be the source of the underseepage and failure, as it consists of fibric peat (former marsh) with high rates of hydraulic conductivity (i.e., Table 7.3), rather than dense sand or stiff clay (clastic). An additional problem is the low shear strength of sediment underlying the toe of the levee, which is the point at which levee underseepage and piping initially occurs (e.g., Chapter 6, dike design). At the toe of the levee, the shear strength of the

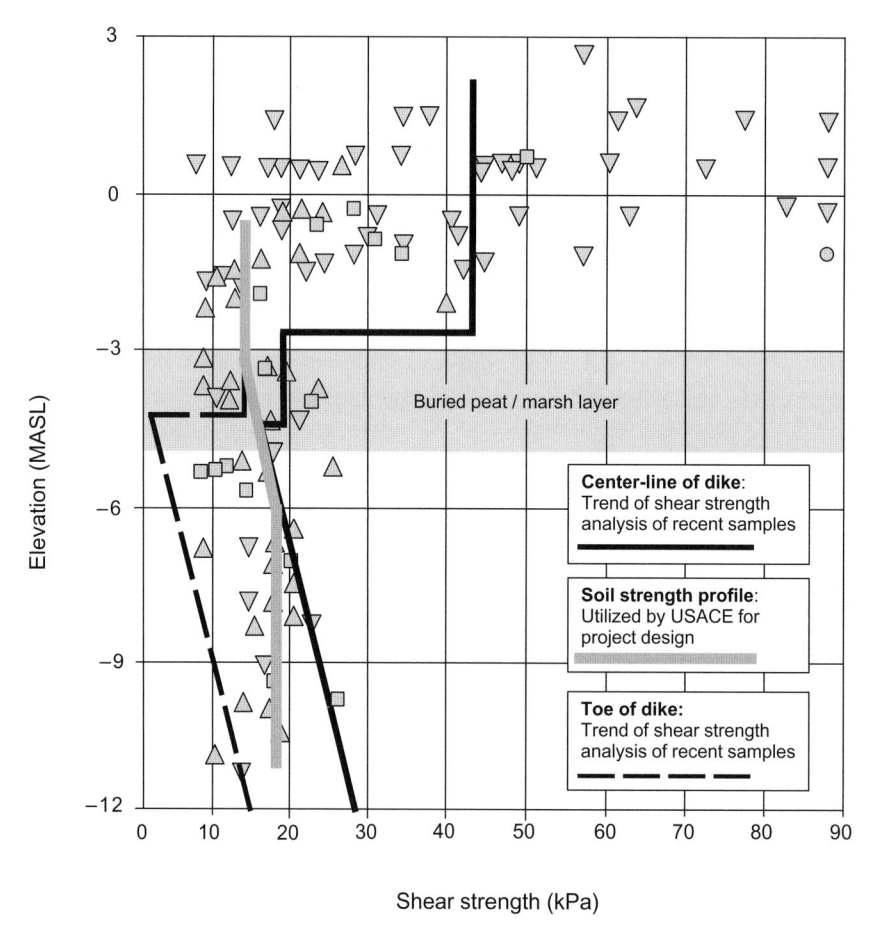

Figure 7.43. Strength of sediment in relation to safety margins for samples underlying the failure location of the 17th St. canal in New Orleans. Sedimentary material underlying the flood wall was often not as strong as the design recommended by U.S. Army Corps of Engineers. (Based upon Duncan et al., 2008; Rogers et al., 2008; Seed et al., 2008.)

Table 7.7 *Causes and consequences of flood management failures associated with the New Orleans Hurricane Katrina flood disaster*

Category	Specific cause	Consequence
Structural (design)	Absence of floodgates at entrance to drainage canals	Storm surge funneled into the network of city canals
	Numerous km of canals	Multiple pathways and opportunities for failure
	MR-GO	Funneled storm surge into city
	Dikes too low	Overtopped by high surge
	Flood control system not completed	Increased vulnerability because of low dikes and flood walls, or flood protection absent
Engineering	Dikes and floodwalls installed in improper soil type	Overtopping of dikes and flood walls because of lower surface, shearing, and displacement of weak soil stratum
	Dikes constructed by inappropriate soil/sediment type	Subsidence (lowering) of dike, breach/failure of dike
	Pumps too low	Pumps were flooded and failed, resulting in no drainage
	Flood wall poorly designed	Bent by high shear stress, leak between adjoining sheets/slabs
Organizational	Poor coordination between local, state/regional, and federal actors	Maintenance and monitoring substandard because of uncertainty over responsibility
	Conceptual mismatches	Organizations not suited for some construction and maintenance operations, conflicts of interest
Natural	Category-V hurricane, track/path of storm in relation to urban pattern	High storm surge entered city from several directions, Lake Pontchartrain and Gulf of Mexico
	High subsidence	Low lying surface, with much below sea level
	Topography (inverted)	Bowl-shaped topography, distributary channel blocked flow, inherent urban drainage problems
Historical	Shoddy maintenance	Infrastructure in poor condition
	Accelerated wetland loss/coastal erosion	Increased vulnerability because of less of a coastal buffer
	Water table lowering because of excessive pumping	Increased subsidence by compaction and oxidation of organic soils leading to increased flood vulnerability

Table compiled from various sources, including Dixon et al., 2006; ILIT, 2006; U.S. GAO, 2006a, 2006b; Rogers, 2008; Seed et al., 2008; IPET, 2009.

majority of sediment samples at the 17th St. canal levee ranges from 15 kPa to 1 kPa, considerably less than the design shear strength of 16 kPa (Dixon et al., 2006; Rogers, 2008).

Post-flood "forensic" fieldwork revealed the nature of the failure of the 17th St. canal (and many others), and that it occurred along the thin 1.8 m (−3.0 to −4.8 MASL) peat layer. The mechanism of failure for the flood wall (built into the levee) at this location of the 17th St. canal was that it was pushed (laterally) by the force of the canal water along the 1.8 m thick weak failure plane. It wasn't breached by overtopping, as the actual canal water was 1.5 m below the top of the flood wall. Thus, the designed levees and flood walls should have never have been constructed within such weak sedimentary layers. They never stood a chance.

Failure of the New Orleans flood control system associated with the storm surge of Hurricane Katrina and Hurricane Rita ranks among Earth's most disastrous failure of an engineered system (Seed et al., 2008). And it was also one of Earth's most expensive natural disasters, with an estimated cost of US$150 billion. More significant were the lives lost – 1,464 in New Orleans. Additionally, 450,000 people were displaced, and many evacuees never returned. The post-Katrina demographic character of New Orleans is considerably different, with nearly ~200,000 fewer residents.

The flood inundated 83% of New Orleans, in addition to hundreds of square kilometers beyond the city limits. The cause of the failure has been reviewed by engineers and geoscientists from the government and private sector (Dixon et al., 2006; U.S. GAO, 2006b; Seed et al., 2008). The broader causes of failure are the magnitude of the natural disaster itself, the organizational structure of agencies charged with construction, maintenance, and monitoring of the flood control infrastructure (Table 7.7). And especially, the engineering design failed to effectively take into account the sedimentological variability of New Orleans' geomorphology, as well as the potential for human impacts to the flood basin environments.

Since Katrina, much of New Orleans' flood control infrastructure has been rebuilt to higher standards (i.e., Figure 7.44 and Figure 7.45). But although New Orleans has encountered several major hurricanes within the past five decades, debate continues about the actual hurricane category (i.e., Table 7.6) and flood

Figure 7.44. Aerial photo of newly constructed pumps and floodgate at London St. Canal as it flows into Lake Pontchartrain. The system simultaneously prevents storm surge waters from flowing into the city while urban floodwaters (i.e., from rainfall) are pumped out of the city. The added protection has stimulated new housing behind fortified floodwalls. (Photo source: U.S. Army Corps of Engineers.)

Figure 7.45. Aerial view of new Inter-Harbor Lake Borgne Surge Barrier (August 2010) intersection with Mississippi River Gulf Outlet (MR-GO), which has subsequently been sealed off from the Gulf of Mexico. MR-GO was a major culprit in New Orleans flooding due to storm surge funneling from the Gulf of Mexico into city canals. (Source: U.S. Army Corps of Engineers.)

recurrence interval to which New Orleans should be protected. To prevent another Katrina-like flood event from occurring, New Orleans requires a new way of thinking, one that is explicitly aligned with scientific projections of global environmental change and sea level rise, precipitation regime, as well as a much more detailed understanding of how to manage local natural resources, such as groundwater – and especially subsidence.

7.7 CLOSURE: REFLECTION ON URBAN FLOOD DISASTERS

Despite being located in opposite sides of the globe and being very different types of hydro-climatic events, the 2005 flooding of New Orleans and the 2011 flooding of Bangkok share much in common. Generally, these include (1) a woefully inadequate governmental organization exacerbated because of inherent tensions between federal and regional/local stakeholders, (2) flood control infrastructure that was inappropriately designed and was in poor operating condition, (3) increased flood vulnerability because of the spatial configuration of urban and suburban development, and (4) a high-energy hydrologic event occurring over a sinking delta.

Several of the events described above are unfortunately somewhat similar to Hurricane Harvey flooding of Houston, Texas in 2017. Hurricane Harvey made landfall in August 2017 as a category-IV storm over southwest Louisiana and Southeast Texas. The wettest hurricane in US history (National Hurricane Center, 2018) generated over 1,000 mm of rainfall during a four-day period, with maximum rainfall totals of over 1,539 mm along the Texas–Louisiana border at Nederland, Texas. The ensuing flooding set numerous records across southeastern Texas (Watson et al., 2018), with the US$125 billion in financial cost about equal to Hurricane Katrina flooding of New Orleans. Some 30,000 homes were flooded, a problem made worse because of the high area of impervious cover associated with extensive suburbanization, in addition to mismanagement of hydraulic infrastructure including dams and drainage canals (Zhang et al., 2018).

The prevention of more flood disasters to large urban–fluvial settings requires a new approach to flood management. Specifically, this calls attention to integrated flood management, which embodies a user-specific combination of hard and soft approaches to flood control as well as balancing the needs of human safety with environmental concerns and economic interests. To examine this, in Chapter 8 we return to the densely inhabited Rhine delta in the Netherlands. Managing water and flooding is in the Dutch consciousness. The practice of flood control in the Netherlands is something both historic as much as it is modern and state-of-the-art. And the Dutch are at the forefront of a new approach to flood management that especially focuses on subsidence, which is diffusing to an international range of settings.

8 Toward Integrated Management of Lowland Rivers

> Undue attention to stressors risks underestimating the intricate interplay of environmental, political, and sociocultural resilience in limiting the damages of collapse or in facilitating reconstruction.
>
> —K. W. Butzer (2012)

8.1 NEED FOR A NEW APPROACH

The vision and approach to river and flood management began to change before the end of the twentieth century, by about the 1990s. After many decades, cracks had formed in the "hard" engineering approach to flood control, which disconnected rivers from floodplains and separated flood control from environmental management. Indeed, scientific evidence regarding the consequences of hard (structural) engineering to driving increased flood risk and riparian degradation had been gathering weight for over a decade prior to the new millennium (e.g., Williams and Wolman, 1984; Downs et al., 1991; Toth et al., 1993; Smith and Winkley, 1996; Graf, 2001; Nienhuis and Leuven, 2001; NRC, 2002). In combination with emergence of a robust climate change science the stage was set for a new approach. It took time, but government agencies followed, especially following the stimulus of large floods, such as the 1993 events on the Mississippi and Rhine Rivers.

The prior four chapters detailed a variety of problems concerning how lowland rivers have been adversely impacted by flood control and hydraulic engineering, ranging from dams, channel engineering, embanked floodplains, and flood basins and deltas. Utilizing this logical sequence, this chapter outlines a range of management measures for heavily modified lowland rivers that have been shown to work, which combine to reduce flood risk and enhance biodiversity. Thus, while the previous chapters examined a range of largely negative impacts to rivers, in this chapter we look at positive outcomes.

Despite the somewhat uneven pace of advancement, much is known about the options that collectively comprise integrated river basin management (IRBM) or integrated floodplain management (IFM) (Table 7.1). Many case studies provide worthy experiences to guide future management efforts.

Most integrated river basin management is occurring within Europe, North America, and Australia. A number of large river basins in Asia, Africa, and South America are increasingly being impacted by construction of large hydroelectric dams (Chapter 4), especially large rivers flowing into the Asian mega-deltas and the Amazon (Chan et al., 2013; Latrubesse et al., 2017). Even for nations with sophisticated governmental organizations and scholarly networks, implementation of scientifically grounded river management options is a cumbersome process, particularly because of economic and stakeholder interests. The development of plans is time consuming, often requiring expensive new methods of engineering and management tailor designed for specific channel segments. Implementation of integrated river basin management (IRBM) is occurring, however, basin-by-basin, and at different rates. While international organizations are essential in coordinating the management of large river basins that drain multiple nations, lowland fluvial management is ultimately a national enterprise (Hudson, 2018). Thus, there is much variability in effectiveness on a country-by-country basis. The progressive European Union's Water Framework (2000/60/EC) and Floods Directives (2007/60/EC), for example, has missed a number of targets and remains an ongoing project because of considerable variability in implementation by member nations (i.e., Voulvoulis et al., 2017).

Integrated river basin management is primarily driven by three goals: (i) flood risk reduction, (ii) improved functioning of environmental systems, and (iii) sustaining and enhancing economic activities. A key challenge is getting diverse stakeholders to agree to a common vision (Downs et al., 1991; Havakes et al., 2015; Hudson, 2018). As much as the approach views the three as being interrelated, flood risk reduction is always the highest priority. Subsequently, a host of individual approaches to environmentally oriented fluvial management is possible, according to stakeholder demands (Table 8.1). The premier example of IRBM occurs within the Rhine basin, broadly coordinated by the International Commission for the Protection of the Rhine. The gold standard for IRBM is the "Room for the River" program along the Dutch Rhine.

Table 8.1 *Strategies and approaches common to integrated river basin management (IRBM)**

Strategy	Approaches
Governance and organization	Stakeholder engagement, international coordination, determination of specific roles and responsibilities across varying governmental scales, communication and information flow, establishing land use policy and practice (including non-structural management measures)
River channel engineering	Groyne lowering and removal, sediment replenishment/nourishment, channel deepening, channel widening, large woody debris, old channel and/or meander bend reconnection, revetment removal and/or vegetation planting
Embanked floodplain	Side channel creation, lake reconnection, floodplain lowering (sediment removal), minor dike lowering or removal, main dike heightening or setback, seepage control, removal of obstructions (bridge alteration), ring-dikes and berms for industry, roughness management (esp. vegetation lowering)
Flood basins and deltas	Floodwater storage, lake reconnections, freshwater and sediment diversions, dike breach, subdelta creation, subsidence and groundwater management, coastal fortification
Urban	Flood water routing, urban water storage, groundwater management, pumps and canals, nonstructural means (e.g., zoning, property buyout, financial incentives, evacuation protocols and emergency preparedness)

*IRBM is preferred over "integrated flood management" (IFM) because of its explicit watershed-scale perspective. IFM often pertains to a more local (floodplain) application, although in principle it should also include a basin-scale perspective. Based upon actual practice and implementation this writer views IRBM as appropriate and sees no need to further distinguish.
Source: Author table.

Figure 8.1. Construction phase of Room for the River program along Waal River at Nijmegen, NL, near delta apex. The primary focus along Nijmegen is to increase flood conveyance by removal of hydraulic "bottlenecks" where bridges were too narrow, construct side channels (as shown) to increase space for water and enhance biodiversity, and urban landscape redesign. (Photo: Rijkswaterstaat, June 11, 2014.)

The foremost example of Room for the River occurs along the Waal River at Nijmegen (Figure 8.1). The overarching goal of the Nijmegen segment was to attain symbiotic linkages between flood control, urbanization, and nature enhancement. For centuries Nijmegen segment was a problem area for flooding and, historically, was the site of numerous major dike breach events, including by ice jam (e.g., Section 6.6.3). The adapted measures included implementing several large-scale options, such as new side channel creation for floodwater retention and nature enhancement, floodplain lowering (lower flood stage), dike heightening, dike setback, seepage control, bridge and infrastructure modification to remove flow obstructions, and urban redesign, among others (Figure 8.2). The Urban River Park, which includes the Zalige Bridge, was internationally heralded and has received awards related to new urbanism and landscape architecture.[1]

[1] 2018 Landezine International Landscape Award.

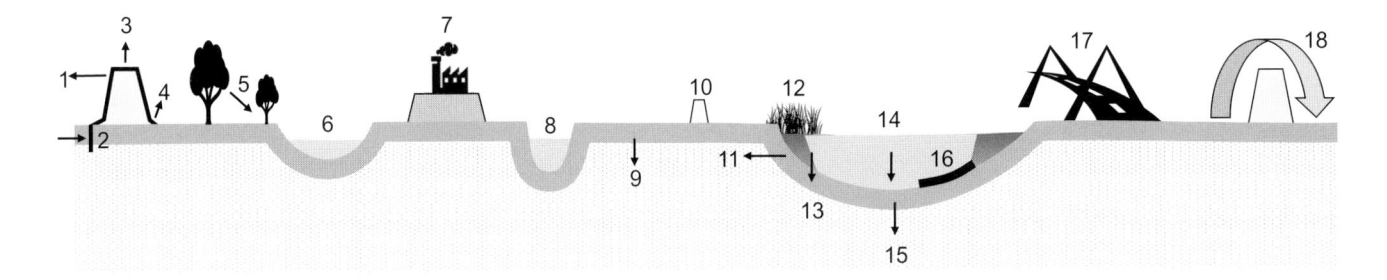

1. Dike setback
2. Obstruct lateral seepage
3. Dike raising
4. Dike strengthening
5. Vegetation (~roughness) maintenance
6. Reconnection of lakes (floodwater retention)
7. Ring dike / berm construction (industry)
8. Side channel creation
9. Floodplain lowering (top stratum excavation)
10. Removal of minor (summer) dikes
11. Channel widening
12. Bank revegetation (protection/nature enhancement)
13. Groyne lowering
*14. Sediment (bed load) replenishment
15. Channel deepening
16. Channel armoring / bed protection
17. Bridge design alteration (flow obstruction removal)
18. Floodwater retention outside of embanked floodplain

Figure 8.2. Primary structural measures for Room for the River flood risk reduction and nature enhancement. *Sediment replenishment (#14) is applied to upstream sections of the Rhine River for channel bed management, and not in the delta for flood risk reduction. (Modified from Middelkoop and van Aselen, 1999 and Silva et al., 2001.)

8.2 REAL CHANNEL IMPROVEMENT

8.2.1 Preliminary Considerations

Conventional hydraulic engineering frequently used the term channel "improvement" related to channelization bank protection, and snag removal to increase navigability and flood conveyance, while it actually resulted in massive degradation of fluvial environments.

Measures associated with integrated management of river channels includes removal of bank protection (revetment), sediment replenishment (dumping), reintroduction of large woody debris, meander bend reconnection, and the lowering or removal of groynes. The decision to go with a specific integrated approach is dependent upon the geomorphic, hydrologic, and ecological characteristics of the channel reach being managed, as well as human utilization (e.g., navigation, urban).

Utilizing an integrated approach to managing river channels is a sophisticated operation because of the need to reduce flood risk and enhance riparian biodiversity while ensuring a navigable channel for shipping (Bravard and Gaydou, 2015). A common objective with regard to integrated approaches to managing river channels is to enhance the geodiversity or physical integrity (i.e., Graf, 2001) of the channel bed and banks. This directly relates to improved riparian habitat. The guiding principle is that rivers should not be straight and smooth, but instead should be rough and dynamic. These latter two qualities create micro- and meso-scale habitat for aquatic and terrestrial riparian flora and fauna. As many lowland rivers engineered for flood control are also engineered for navigation, they are effectively "locked" in place by hard structures such as concrete revetment, channel armoring, and groynes. The requirement that navigable rivers have stable channel beds and banks effectively means that the river will not be allowed to freely meander across the floodplain. The alteration of a heavily engineered river, therefore, has narrow margins with which to restore natural-like channel morphodynamics.

8.2.2 Groyne Lowering

Channel groynes are ubiquitous along lowland rivers engineered for navigation. Groyne construction results in channel narrowing, and the goal is that the river becomes somewhat "self-scouring," reducing the need for dredging (Section 5.3.3). But groynes are also associated with considerable deposition of coarse bed load along the shallow margins of alluvial channels. In many instances groynes result in the infilling of side channels, which is often the motivation for their construction. The infilling of side channels, however, results in less space for water and degrades and removes aquatic habitat, tremendously reducing the biodiversity of the riverine environment (Meyer et al., 2013; Guntren et al., 2016; van Denderen et al., 2019b). Additionally, an increasing amount of evidence shows that flood stage increases following construction of groynes (Pinter et al., 2006a). Thus, the removal or lowering of groynes is among the most common approaches to integrated river management (Figure 8.3), as it lowers flood stages and increases riparian geodiversity.

Figure 8.3. Groyne lowering along Waal River in the Rhine delta, Room for the River. (Source: Rijkswaterstaat.)

Groyne lowering directly influences hydraulics within the main channel and within the groyne field. At high-stage levels groyne lowering reduces flow resistance, which then reduces the backwater effect of groynes and results in flood stage reduction (Straatsma and Kleinhans, 2018). Additionally, during moderate flood stages flow velocity increases within the groyne field, resulting in transport of trapped coarser bed material and erosion of lateral bars within the groyne field. This then creates additional space for water, and enhances the benthic habitat. An opposite pattern occurs within the narrower main (navigation) channel, as the reduction in flow velocity results in some channel bed aggradation. This may be an unintended consequence that requires additional management operations. The lowering of groynes along the main branch (Waal River) of the Rhine River in the Netherlands, for example, is expected to reduce flow velocity and sediment transport within the navigation channel. The outcome is an expected increase in channel dredging by ~10% (van Vuren et al., 2015), which needs to be incorporated into the larger management plan and annual operating budget.

Groyne lowering has become a key measure to reduce flood levels, including for the Netherlands' Room for the River approach to integrated management (Silva et al., 2001). Lowering groynes is relatively inexpensive and directly reduces flood levels (Silva et al., 2004). Groyne lowering along the Rhine River in the Netherlands was designed to reduce flood levels by 5–15 cm along the Waal and Ijssel Rivers, whereas along the Lek-Nederrijn River groyne lowering is expected to reduce flood levels by 10 cm. While this amount of lowering could seem inconsequential, when combined with other measures the result is a lowering of flood stage by some 35–50 cm, which enables the channel to accommodate larger discharge events. Other channel modification procedures include continued management to control floodplain roughness (e.g., vegetation height), which also influences flood levels. Sophisticated hydraulic models illustrate that a combination of approaches can result in ~0.5–1.5 m of flood reduction, varying with flood magnitude and spatially along the river (Straatsma and Kleinhans, 2018).

8.2.3 Sediment Replenishment/Nourishment

In intensively engineered river reaches channel bed incision can threaten the integrity of hydraulic infrastructure, navigation, and flood control systems by undermining structures and channel banks (Dunbar et al., 1999; Habersack et al., 2013; Quick et al., 2020). Some reaches of the Rhine River in Germany and the Netherlands, for example, incised by as much as 7 m after the cut-off period, and several segments continue to incise at rates ranging from 10 mm/yr to 15 mm/yr (Quick et al., 2020). Channel bed incision contributes to river degradation in several ways. This includes an upstream migrating knickpoint that can threaten the stability of bridges, locks and other hydraulic infrastructure. The disturbed channel bed also degrades the benthic habitat of macro-invertebrates as well as fish spawning habitat. Additionally, channel bed incision lowers the average river stage, which lowers the local water table of the alluvial aquifer. This reduces the wetland water levels and creates opportunities for woody vegetation encroachment and habitat terrestrialization, creating additional management challenges concerning riparian wetland and aquatic environments (Uehlinger et al., 2009).

Therefore, for sediment-starved rivers it is often important to artificially replenish coarse bed material to arrest channel bed incision and to restore gravels to enhance the benthic zone ecosystem (Götz, 2008; Kondolf et al., 2014b; Habersack et al., 2016; Frings et al., 2019). This commonly involves dumping specific sized sediments along channel reaches undergoing channel bed incision, and is often coordinated with sediment dredging operations at problem areas, such as downstream

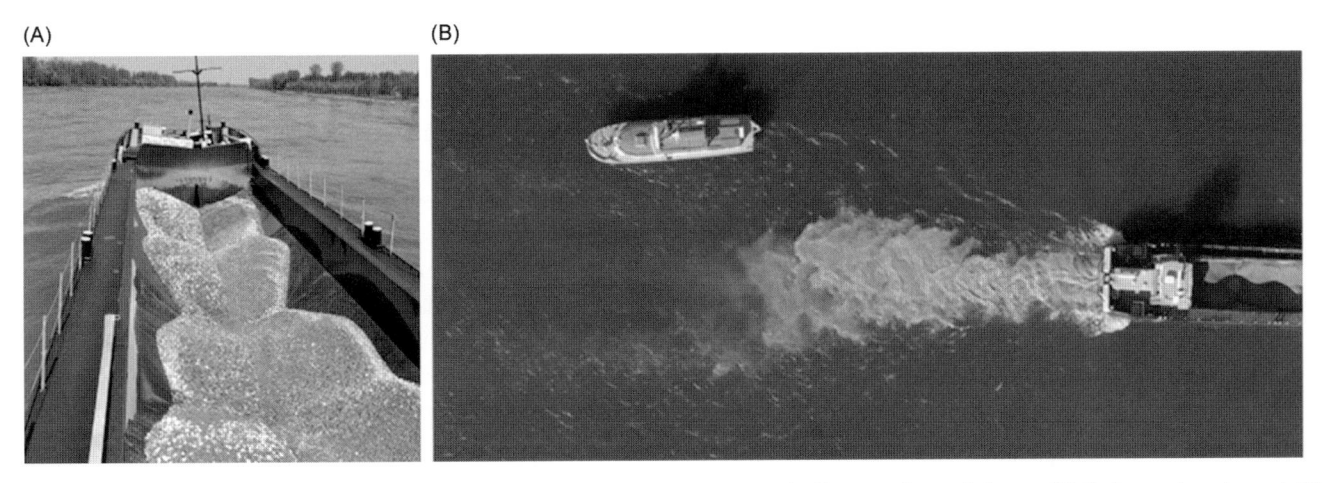

Figure 8.4. Sediment replenishment along the Rhine River, near Freiburg, Germany. (A) Coarse sediment in barge. (B) Sediment dumping to infill reaches undergoing scour because of coarse sediment reduction. (Source: German Federal Waterways and Shipping Administration.)

of reservoir impoundments that trap fluvial sediment (Kondolf et al., 2014b). Sediment replenishment is a staple management approach along intensively engineered rivers in Europe, including the Elbe, Danube, and Rhine (Götz, 2008; Koenig et al., 2012).

Because all sediment derived from the Alpine Rhine headwaters and High Rhine tributaries (Aare R.) are trapped in Lake Constance and a cascade of twenty-one main-stem dams (see Figure 4.20), downstream sediment replenishment is vital to manage the incision problem to assure navigation and fortify aquatic habitat. Coarse sediment is "surgically" dumped (Figure 8.4) into incised channel reaches along the Rhine River (Götz, 2008). A twofold strategy of sediment replenishment is utilized on the Rhine, including sediment replenishment for ecological channel bed restoration within the impounded section and sediment replenishment for channel bed erosion control in the free-flowing section, downstream of Iffezheim Dam (Götz, 2008; Frings et al., 2014, 2019). The erosion control program involves the largest amount of sediment, and includes sand and gravel inputs downstream of Iffezheim Dam to be assimilated into the active bed load (Frings et al., 2019). Additionally, a "patch and repair" strategy is used that directly dumps gravel and cobbles over local scour zones, with the idea it will create a nonerodable armor layer. This strategy was used for a channel section downstream of Iffezheim (km 338–352) and a section along the Lower Rhine (km 665–857) (Frings et al., 2019). The sediment is piled into barges and released into the river daily (Götz, 2008; Kondolf et al., 2014b; Frings et al., 2019). The operation immediately downstream of Iffezheim Dam accounts for the largest flux of gravel and cobble into the Rhine main-stem, at 0.348 Mt/yr between 1991 and 2010 (Frings et al., 2019). Frequent hydrographic monitoring by river authorities provides high-resolution channel bathymetric data to quantitatively assess the effectiveness of the sediment management plan and its overall importance to the sediment budget of the Rhine River (Frings et al., 2019). Sediment replenishment is a valuable component of the Rhine River sediment budget (Figure 8.5) and is equal to 1.681 Mt/yr, a substantial sediment source, although a large proportion of the sediment is indeed recycled between dredging and dumping (Frings et al., 2019).

The sediment nourishment approach in the Rhine River is mainly deemed effective, although river managers have to be vigilant and constantly intake new measurements to assess changes to the depth and location of scour zones. Channel bed erosion along the Lower Rhine has declined since the mid-1980s, and some locations are aggrading. The approach is now a regular component of the Rhine River basin management regime, which is necessary to balance economics, flood control, and sustainability for intensively engineered rivers (Quick et al., 2020).

8.2.4 Removal of Bank Protection Works

Bank erosion is a key component of fluvial sediment budgets along large alluvial rivers (Kesel et al., 1992; Dunne et al., 1998), and detrimental to channel stability. Because large lowland rivers are often utilized for navigation, bank erosion is among the first modes of fluvial channel adjustment that is managed. River banks are "protected" by articulated concrete revetment and riprap (Section 5.3.2), which then reduces sediment loads and causes channel bed degradation (Mosselman et al., 2004; Kiss et al., 2019). The "sealing up" of river banks (i.e., removal of natural roughness) greatly reduces riverine geodiversity, degrading aquatic and terrestrial biodiversity. Thus, a goal of some integrated river management approaches is to allow the river to again freely migrate, to increase natural biodiversity (Lambeets

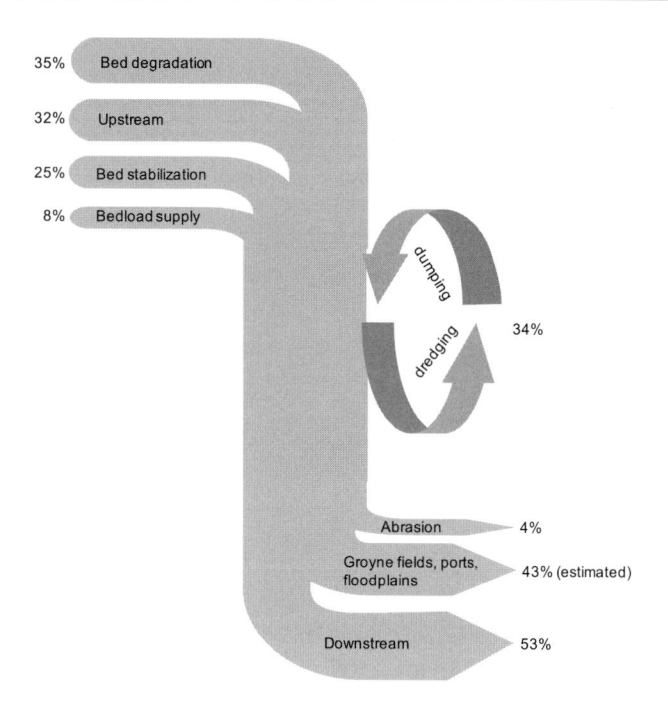

Figure 8.5. Channel bed sediment budget along Rhine River in northern Germany (between 640 km and 865 km), emphasizing the role of sediment dumping and reworking within groynes from 1991 to 2010. Total sediment flux (i.e., 100%) = 1.26 Mt/yr. (From Frings et al., 2014.)

et al., 2008). Additionally, allowing rivers to migrate laterally results in a diversity of depositional and erosional channel bed and bank features important to aquatic and terrestrial flora and fauna (Figure 8.6), and is therefore a key component of river "rewilding" strategies (Brown et al., 2018).

While the environmental value of removing concrete revetment to restore natural channel banks is well understood, freely migrating rivers do not necessarily reduce flood risk (Sparks, 1995). Additionally, a laterally migrating channel is problematic for navigation because reworking bank material results in channel bed shoaling and shifting bars. In general, channel bank revetment results in deeper pools at the apex of bends by about 5–20% (Thorne, 1989). Thus, removal of revetment can be expected to result in an overall shallowing of the river channel. Nevertheless, for rivers not heavily utilized for navigation the removal of bank protection measures is a common approach to enhancing fluvial geodiversity and biodiversity. For scenarios that require bank stability with an aim to increase channel roughness, "softer" measures include replacing concrete and stone riprap with geotextile materials, tree root wads, and logs (Table 8.2). These measures are often combined with bank revegetation and dike setback, with the latter often requiring the purchase or long-term lease of floodplain lands from riparian land owners.

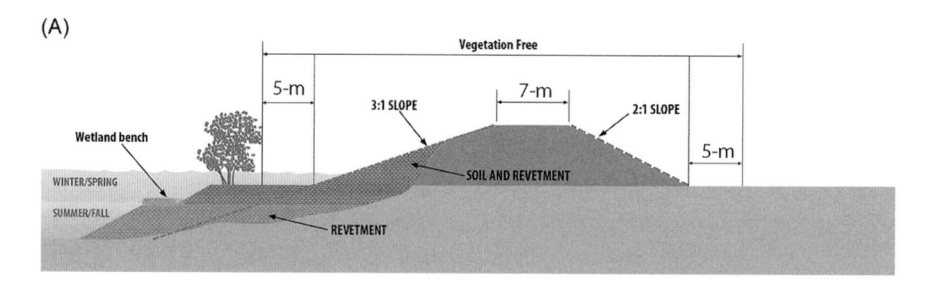

Figure 8.6. Approaches to environmental river bank management. (A) Design modification to channel banks using geotextile and vegetation to replace concrete revetment and riprap, part of a new program for bank protection to protect flood control infrastructure and enhance environmental attributes of river banks along Sacramento River, California. (B) Implanting large woody debris in shallow reaches along channel banks for protection and creation of aquatic habitat in the Netherlands. (Figure sources: (A) California Department of Water Protection [California DWR], 2010; (B) Rijkswaterstaat, 2016.)

Table 8.2 *Techniques and benefits of utilizing environmentally friendly "soft" bank protection rather than "hard" measures (riprap, concrete revetment)*

Techniques: a range of techniques can be used, depending upon the type and characteristics of the river:
– Removal of hard bank protection and reprofiling of banks to avoid the need for additional protection
– Removal of hard bank protection and replacement with soft engineering technique
– Planting suitable native plants to enhance existing habitats, and create new habitats
– Creation of aquatic ledges instead of hard bank protection
– Create backwaters and wetlands to increase habitat diversity
– Create buffer strips to improve or preserve riparian and bankside habitats, and increase bank stability
Benefits: Improvements to river's edge and bank habitats can deliver a wide range of direct and indirect benefits:
– Direct benefits to plants, invertebrates, birds and animals which live on the banks and riparian zone
– Improvements to in-channel habitats for aquatic plants, fish, and invertebrates
– Improvements to the physical habitat (geodiversity) conditions of the watercourse, including the creation of a more natural bank profile and the creation of more varied habitat niches, including varied sedimentary deposits
– Restoration of natural processes, including erosion and deposition
– Improvements to the aesthetic value of the watercourse and improvements to its recreational value
– Reduction in maintenance costs of hard defences

After European Center for River Restoration (ECRR, 2018).

The removal of concrete revetment and riprap to promote a freely migrating river obviously requires consideration of other types of floodplain management. Two fundamental considerations concern rates of floodplain reworking (lateral bank erosion), and the proximity of floodplain infrastructure. The location of dikes in proximity of eroding channel banks is especially a concern, as they can be undermined by cutbank erosion. Prior to the removal of revetment, a thorough geomorphic assessment of erosion rates should be carried out at a reach by reach scale. Such work commonly includes a combination of historic maps and modern hydrographic surveys to directly measure erosion rates prior to construction of bank protection measures, in association with hydraulic modeling over a range of flow magnitudes that include future (projected) hydroclimatic scenarios (Larsen, 2007). Because the process of bank erosion along large lowland rivers is driven more by mass failure (bank sloughing) than hydraulic entrainment, such analysis should incorporate sedimentologic data of channel bank material (e.g., Dunbar et al., 1999).

Additionally, land use and land cover within the embanked floodplain should be factored into the project design and budget. In many cases floodplain lands will need to be purchased, or property owners compensated for a loss of productivity because of floodplain erosion or land use change, and legal cases will ensue. An important quality of embanked lands is that they harbor a great deal of riparian biodiversity within an otherwise humanized landscape. Thus, in some instances specific patches of floodplains will need to be protected from erosion. Embanked floodplains may also include important cultural heritage, including prehistoric and historic material culture deemed worthy of preservation (Brown, 2008; NSW, 2014). That many provinces and states mandate "cultural resource surveys" prior to floodplain modification helps to ensure that such lands are prioritized for conservation, and has been a stimulus for floodplain geomorphic research.

While removal of revetment and riprap occurs most often in smaller rivers, some larger rivers are also seeing the limited removal of hard bank protection measures, especially for environmental motivations (BSTAC, 2013). Channel bank material and dynamics of the Sacramento River provides ideal habitat for the bank swallow (*Riparia riparia*). Bank swallows burrow into coarser cubank material, especially within the lenses of silty–sandy flood deposits that are exposed along natural levee cutbanks (Figure 8.7).

Prior to the 1970s the Sacramento River was noted for its lateral migration and classic meandering pattern (Larsen, 2007). But extensive revetment construction, especially after federal passage of the Sacramento River Bank Protection Project (SRBPP) Phase-I in Section 203 of the Flood Control Act of 1960, and subsequent passage of Phase-II in 1974, effectively locked the river in place. The consequences of revetment construction were to "seal" over bank swallow burrows (Figure 8.8), precipitously reducing bank swallow populations (BSTAC, 2013).

Limited removal of concrete revetment is occurring along selected reaches of the Sacramento River, in line with California's new vision for flood management (CSFMP, 2013). The result of limited revetment removal in the mid-1990s was an increase in bank swallow habitat along river reaches allowed to freely migrate (Givetz, 2010). The decision of where to allow freely migrating bank sections was guided by examining over five decades of historic channel positions analyzed with hydraulic models (Larson, 2007; USACE, 2014b). Much revetment along the Sacramento River will remain, however, because of the necessity in protecting flood control infrastructure (dikes).

8.2.5 Channel Reconnection

The reconnection of former channel segments is among the most high-profile approaches to river restoration (Figure 8.9). Channel reconnection occurs at the scale of former channel courses or at the scale of individual channel reaches, particularly meander

Figure 8.7. (A) Bank swallow (*Riparia riparia*) habitat within an eroding alluvial cutbank along the Sacramento River. Note numerous burrows (small holes) within the upper coarse sandy/silty layers. (B) Bank swallow and close-up of habitat within sandy deposits. (Source: BSTAC, 2013.)

Figure 8.9. Channel reconnection along Kissimmee River in south Florida. Recently reconnected meandering segment in foreground, and C-38 canal (background) was the Kissimmee River after it was straightened in the 1960s. (Source: U.S. Army Corps of Engineers.)

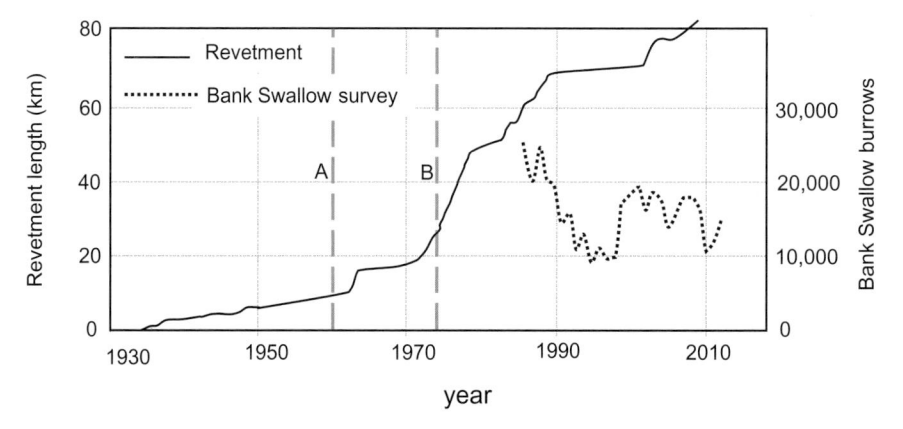

Figure 8.8. Cumulative length of rock revetment installed along middle Sacramento River between Red Bluff and Colusa over about 160 km of river length, between 1935 and 2008. Bank Swallow (*Riparia riparia*) counts from 1986 to 2012. Vertical line A refers to the initial authorization of SRBPP Phase 1 in 1960, while vertical line B refers to authorization of SRBPP Phase 2 in 1974. (Source: BSTAC, 2013.)

bends. That most channel reconnection efforts include reaches that had previously been cut-off for flood control represents an admission of failure of one of the most environmentally destructive types of flood control measures (see Sections 5.3.1 and 5.4.1).

The main purpose of channel reconnection is to enhance riparian biodiversity and to redistribute streamflow over a larger area, which slows down the discharge pulse and lowers stage levels. It should be noted that many artificial cut-offs and channel segments become vital components of the floodplain ecosystem, serving as water bodies and wetlands with diverse hydrologic functioning (Lorenz et al., 2016; Dépret et al., 2017). The decision, therefore, to reconnect cut-offs or old channel segments with the active channel system should be formally evaluated from the standpoint of the overall changes to the riparian ecosystem, as some artificial cut-offs provide important biological functions such as critical habitat for fish spawning and other ecosystem services (Miranda, 2005; Shields and Knight, 2013; Mossa, 2015a).

Channel reconnection is conceptually straight forward but in practice is complex and intricate, often requiring many years to complete. The actual hydrologic and geomorphic reconnection, once initiated, may only require several years to restore physical functionality, which includes restoration of the bed load regime. But ecological recovery may take much longer. While fish and some benthic communities may recover within several years, some benthic organisms may not return if basin boundary conditions have changed. Additionally, reestablishing diverse assemblages of vegetation communities associated with variable hydrologic flow levels can take decades. As with many environmental restoration projects the largest amount of time with channel reconnection is likely associated with coordination between government agencies and riparian stakeholders (Warne et al., 2000; Downs and Gregory, 2004; Lorenz et al., 2016). Sequentially, channel reconnection measures include (1) initial baseline geomorphic and biological studies of the prior system as well as the altered floodplain and channel, (2) removal of artificial and natural sedimentary fill deposits, (3) reconnection with the active channel, and (4) monitoring the geomorphic and biological adjustment of the channel reach for years thereafter (Table 8.3).

Table 8.3 *Management steps for design of channel reconnection project*

Step	Goals	Activities
Baseline study	Establish geomorphic and hydrologic dynamics, link between geodiversity and biodiversity	Background characterization and analysis of historic streamflow regime, sediment dynamics, and geomorphic activity; hydraulic geometry and planform dynamics; aquatic biology in relation to geomorphic reaches (bed material, channel bed, banks), biologic sampling
Property/land cover assessment, acquire permissions, zoning and policy	Acquire suitable lands to accommodate new fluvial regime (flooding, sedimentation, channel activity)	Stakeholder meetings, purchase and/or lease property, compensation to floodplain land owners
Removal of sedimentary fill and organic materials	Prepare for variable flow conditions	Excavate channel to target geometry
Reconnection of old channel	Sustainable channel (does not infill or scour)	Old channel: Weir removal, breach of channel fill plug Pilot channel: construct blockage (weir, sediment) to force flow into old channel Upstream/downstream: possible training groynes and bank protection Biologic sampling, benthic macroinvertebrates, fish, vegetation
Geomorphic adjustment	Reactivation of bed load, channel bars, bank erosion	Monitor channel bathymetry (ADCP) in reconnected channel, upstream/downstream of reconnection, establish reference cross sections for repeat surveying, monitor changes to flow stage (pressure transducers)

Source: Author table.

8.3 EMBANKED FLOODPLAIN MEASURES

8.3.1 Preliminary Considerations

Integrated river basin management requires hydrologic management across varying types of riparian environments. From the perspective of flood risk reduction the hydrologic goal is to attenuate the flood pulse by redistributing floodwaters across a larger surface, thereby reducing the flood peak and increasing the lag time (Figure 8.10). Thus, enhancement of hydrologic connectivity is essential, which involves reconnecting floodplain water bodies and wetlands at varying flow thresholds (Tockner et al., 2000; Thoms, 2003). The driving motivation for many restoration projects, after reducing flood risk, is to conserve or restore floodplain ecosystem processes (Pedroli et al., 2002; van Looy et al., 2003). This should ideally embody an environmental flows approach, and include managing riparian habitat types at discharge magnitudes ranging from base flow to larger flood events.

At the local scale there are many successful cases of embanked floodplain management, and because of its effects being compartmentalized to specific floodplain "patches" it is probably the most straightforward type of integrated management (e.g., Nienhuis, 2008).

Four key measures utilized in embanked floodplain management include (1) side channel creation, (2) lake reconnection, (3) floodplain lowering (sediment removal), and (4) dike setback (Table 8.1). These approaches are seldom created in isolation, but rather as a multi-tiered approach to integrated floodplain management. Other measures could also include dike seepage control, ring dike and/or berm creation for specific uses (industry), removal of low (summer) dikes, and removal of blockages. The latter category includes measures such as redesigning bridges and other lateral blockages to floodwater flow. While a costly measure, the redesign of bridges is especially effective where floodwaters are blocked during large events. This was an important measure carried out in the Room for the River project at Nijmegen, NL (Figures 8.1).

8.3.2 Side Channel Creation

The creation of artificial side channels has proven to be a successful approach at flood height reduction while enhancing the environmental functions of embanked floodplains (Mosselman, 2001; Pander et al., 2015; Schober et al., 2015). While engineered (excavated) side channels are herein emphasized, in practice side channels utilized in floodplain restoration projects may exist along a continuum, from pure "natural" channels to completely engineered channels. Along the Rhône River, France, for example, both excavated channels and former cut-off channels were utilized for floodwater storage and ecological rehabilitation (Amoros et al., 2005).

Engineered side channels are narrower than main-stem channels and along larger rivers may extend up to a kilometer or so in length. Side channels are not designed to transport large discharge volumes. In the Netherlands engineered side channels convey at most 3% of flood discharge, resulting in minor flood level reduction (Mosselman, 2001). Thus, as with other approaches to flood level reduction (i.e., Table 8.1) it is the combination of measures at a specific site that results in significant flood level reduction overall (Figure 8.11). Side channels are excavated within the existing embanked floodplain, which often represents thick embanked deposits (Section 6.5.1). Side channels may be located in floodplain reaches that have also undergone lowering by scraping (removal) of embanked top-stratum flood deposits. Because of concern with erosion of floodplain dike structures side channels along the Waal River in the Netherlands must be located >100 m from dikes, or located >50 m from dikes with thick cohesive top-stratum deposits (Mosselman, 2001). For some side channel projects it is also necessary to set back the adjacent dike to create sufficient space for floodwaters and to allow for some adjustment (erosion) of the side channel.

Hydrologically, side channels are usually connected with the main-stem flow by a weir at the upstream and downstream inlet and outlet, respectively. The height of the water needed to overtop the weir, either to infill the water body at the upstream site or to drain the water body at the downstream site, is set at a height related to a specific flow threshold to accomplish flood risk goals (floodwater retention) and ecosystem functions. At higher flows, side channels temporarily remove flood waters from the main-stem channel, and then result in a slow release of floodwaters after crest passage, attenuating the flood peak and

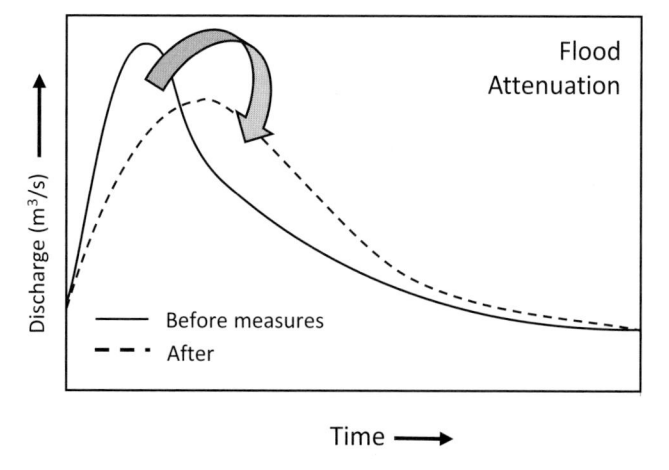

Figure 8.10. Comparison of single-event flood hydrograph before and after flood attenuation measures. (Source: U.S. Geological Survey.)

Figure 8.11. Side channel construction along the upper Rhine, downstream of Heidelberg, DE. Narrow valley imposes constraints on side channel design for floodwater retention. (Source: Kuhn et al., 2016. Licensed by CC.)

Table 8.4 *Guidelines for design of floodplain side channels*

Length	250–50,000 m, dependent upon channel dimensions of main river
Width	10–100 m, with variation
Shape	Elongated, slightly curved, horseshoe shaped
Side slopes	1:10 inner bend, 1:3 outer bend
Depth	Deeper outer bends (to 8 m) and shallower inner bends (3 m), depth should increase downstream (slope 1:100 to 1:1,000)
Location	Utilize existing historic channels if possible, excavate channel plug deposits
Distance to river	Isolated lakes, further from river; single connection channels close to river
Management	Bank stabilization for high-discharge events, possible dredging/removal of sediment after flooding, grazing management important where embanked floodplain utilized by cattle

After Buisje et al. (2001).

reducing downstream flood risk. At low stage, side channels are effectively lacustrine water bodies or wetlands that serve important ecological functions to riparian and aquatic organisms that vary from main-stem channel functions. In some cases, side channels may completely dry up during prolonged low flow periods.

The design of side channels can be complicated by the dual goals of flood risk reduction and nature enhancement (Pander et al., 2015), and in view of their geomorphic evolution (Amoros et al., 2005; Table 8.4). The cross-sectional morphology should be trapezoidal, with a high width-to-depth ratio (shallow and wide) to reduce bank erosion. A key factor in the function and life-span of side channels is the entrance and exit angle of the side channel relative to side channel length (length ratio) (Mosselman, 2001; van Denderen et al., 2019b). Uncertainty exists in the eventual evolution of side channels because of variable and changing hydrogeomorphic conditions (e.g., Silva et al., 2004; van Denderen et al., 2019a). For engineered side channels to be sustainable they usually require hard structures, including bank protection (riprap) and groynes to prevent erosion and to train the flow during large events. Side channel typology is related to morphologic indices (cross-section vs. planform), specifically whether side channels are dominated by bed load (gravel), suspended sand, or wash load (silt/clay) deposition (van Denderen et al., 2019a)

Morphologic changes to floodplain side channels is influenced by vegetation, hydrologic, and sediment regime that results in adjustments to environmental habitat over varying time-scales (Buisje et al., 2001; Mosselman, 2001). Over longer (decadal) timescales sedimentation reduces floodwater storage capacity of side channels, which should be taken into account during the design phase. Additionally, sedimentary infilling changes the physical habitat for aquatic organisms, and ultimately result in woody vegetation displacing aquatic macrophytes or open water conditions (Buisje et al., 2002), with terrestrialisation being enhanced along reaches where groundwater levels have lowered (Amoros et al., 2005). Thus, maintenance is an important component of floodplain management with engineered side channels, and in particular should include the removal of recently deposited coarse sediments before they are anchored by encroaching woody vegetation. Without continued maintenance and sediment excavation engineered side channels in the Rhine delta are estimated to be viable for only about fifteen years (Mosselman, 2001).

8.3.3 Floodplain Lowering

A less common approach to integrated floodplain management is floodplain lowering, which increases the capacity of the floodplain surface to temporarily store floodwaters. The scraping and removal of cohesive top-stratum deposits lowers the floodplain elevation and directly increases floodwater storage. Hydraulic modeling illustrates that flood levels along the Rhine are lowered by about 0.35 m for low and moderate-sized events when some 15% of a floodplain reach was lowered. The scraping of floodplain deposits creates new riparian wetlands and ponds, albeit with an unnatural rectilinear pattern. Here it is important to note that it is "embanked floodplain deposits" that are removed (Buisje et al., 2002; Baptist et al., 2004; Arroyave and Crosato, 2010). Such deposits artificially accumulated over long time periods because of the trapping of flood waters between dikes. Thus, while creation of extensive pits by scraping may appear to be degradation of the floodplain environment, it can also be seen as a movement toward restoration of the floodplain to its baseline riparian status.

In the Netherlands, embanked floodplain deposits were removed to a depth of 0.5–1.0 m as part of the Room for the River program. The suitability of a floodplain reach for sediment scraping is highly dependent upon riparian stakeholders, including environmental and socioeconomic (Straatsma et al., 2018). Considerable embanked sediment scraping in the Rhine delta was coordinated with brick factories interested in clayey deposits as raw material (Van der Meulen et al., 2009). Historically, cohesive floodplain deposits along the Dutch Rhine had been scraped away for brick manufacturing. The tall chimneys of old abandoned brick kilns are common along the Dutch Rhine and are considered part of the "cultural landscape," worthy of preservation within the embanked floodplain.

As with side channel creation, floodplain lowering should not be viewed as a permanent fix because it also increases floodplain sedimentation. After a period of about 15 years, higher sedimentation along floodplain sections lowered along the Waal River resulted in new natural levee formation and associated vegetation succession, and reduction in floodwater storage. Thus, over decadal time-scales artificial scraping of floodplain deposits followed by higher rates of embanked floodplain deposition can be considered a type of "cyclic floodplain rejuvenation" (Baptist et al., 2004; Geerling et al., 2008).

8.3.4 Lake Reconnection

The reconnection of rivers to floodplain lakes, including oxbow lakes and valley side lake basins, is an important method of restoring hydrologic connectivity and has been widely utilized across many lowland rivers (Ollero et al., 2015; Peng et al., 2019). The approach commonly involves engineering passive weir structures that enable hydrologic exchange between floodplain lakes and rivers at specific stage levels. This approach differs from channel reconnection, in that the old channels remain as lakes, but now with a semi-natural managed flood pulse. The threshold discharge of connectivity can be engineered at a flow level that supports a specific volume of floodwater retention and ecosystem functions.

The Danube River floodplain has been extensively impacted by engineering (Hohensinner et al., 2007; Constantinescu et al., 2015; ICPDR, 2015; Habersack et al., 2016). Along the Austrian Danube, for example, reconnected floodplains resulted in a reduction in flood peak that is related to the overall area of reconnection, and was equal to 110 m^3/s reduction per kilometer of channel distance (Figure 8.12). In some cases the peak discharge was reduced by 1,000 m^3/s for a flood event with a 100-year recurrence interval, a volume equal to about 10% of the peak discharge (Schober et al., 2015; Hein et al., 2018). The physical characteristics of the floodplain are important to the volume of floodwater retention, including floodplain roughness and topographic patterns within the floodplain, such as obstructions caused by old channel belts and natural levees.

8.3.4.1 RECONNECTION OF FLOODPLAIN LAKES AND ECOSYSTEM SERVICES

Reconnecting lakes and old channels to active flow during high-stage events has considerable benefit to riparian ecosystem services (Table 8.5). Indeed, riparian ecosystem services often rapidly recover upon hydrologic reconnection, including aquatic habitat restoration, biogeochemical cycling, water purification, and flood crest reduction, among others (Nienhuis, 2008; Zehetner et al., 2008; Richards et al., 2017; Peng et al., 2019). Along the upper Danube River, for example, 46% of the total types of fish species recolonized the reconnected floodplain water body within only two months. Within four years, the species diversity within the reconnected lake was effectively as diverse as those lakes that had remained connected (Pander et al., 2015). Such efforts also have a larger influence on riparian ecosystems because of supporting longitudinal connectivity during the seasonal flood pulse (Ward and Stanford, 1995; Tockner et al., 2000; Reckendorfer et al., 2013).

8.3.5 Dike Setback

The practice of removing riparian lands protected from flooding to again flood seems paradoxical to flood control. But the problem with floodplain embankment and channel engineering is that for some rivers it increases flood levels, increasing flood risk (Belt, 1975; Baldassarre and Castellarin, 2009; Remo et al., 2018). Indeed, perhaps the signature type of integrated management along large lowland rivers is the setback of dikes, giving "more room for the river." Where dike setback has occurred, it

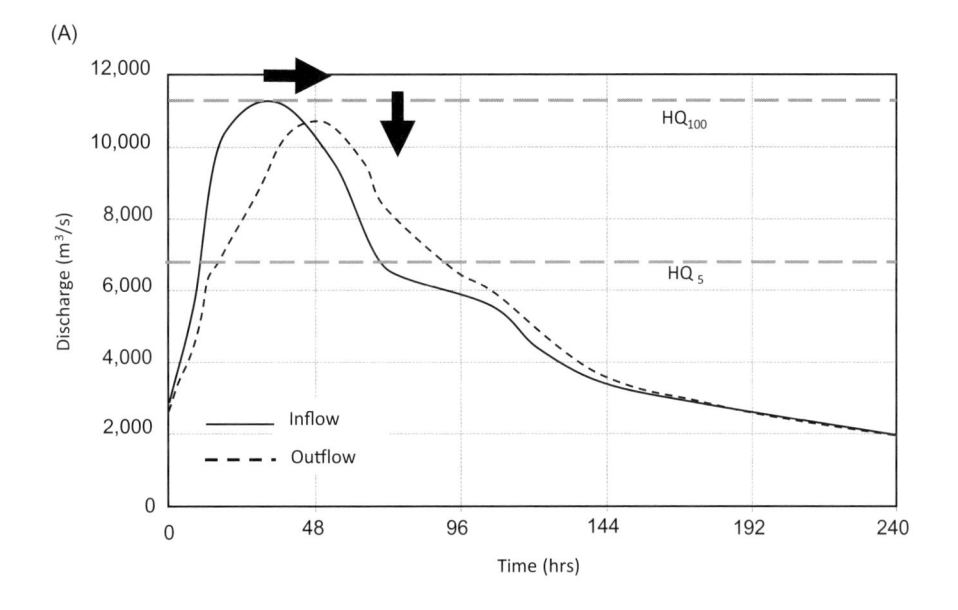

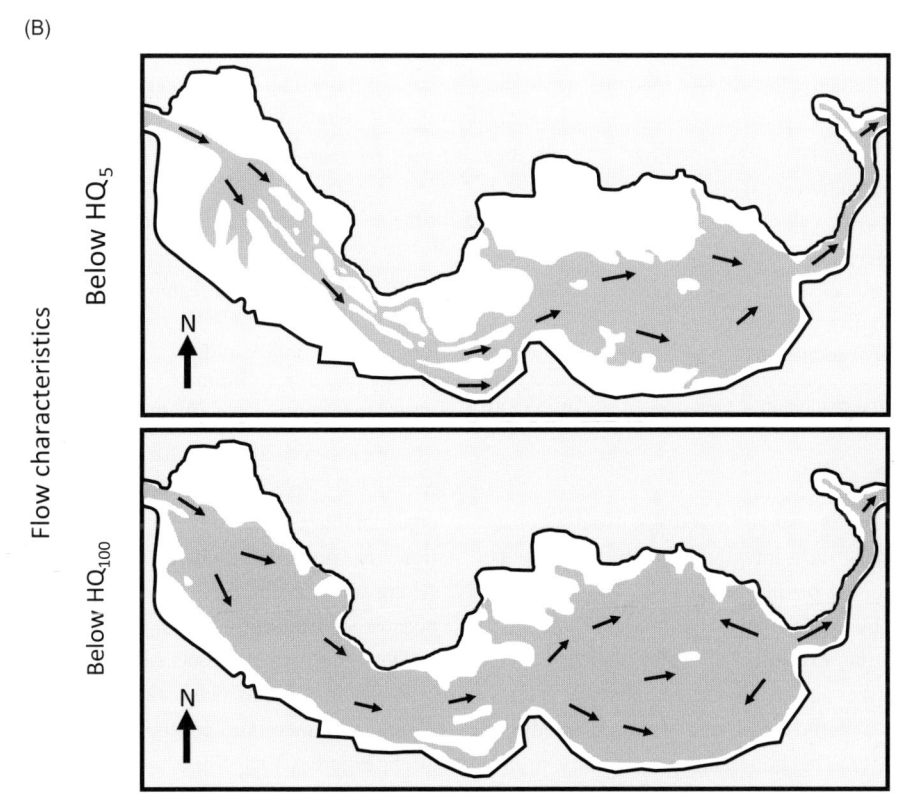

Figure 8.12. Downstream change in discharge (A) due to enhanced floodwater retention (B) along the Danube River at Machland, Austria, downstream of Vienna. (After Schober et al., 2015; Hein et al., 2018.)

has often received considerable media attention. The flood risk goal of dike setback is to reduce flood levels by increasing the volume of floodwater retention within the floodplain (Jacobson et al., 2015; Schober et al., 2015; Smith et al., 2017). The setback of dikes also increases dike safety by reducing flow velocity and hydraulic pressure on dikes during flood events (Schober et al., 2015), as well as reducing dike underseepage and sand boil formation that cause local-scale inundation. Dike setback measures are usually combined with other integrated flood management approaches, such as side channel creation and groyne

Table 8.5 *Ecosystem service benefits associated with channel and lake reconnection*

Ecosystem services		Rating	Evidence
Provisioning	Water storage	Medium	Does not directly create new storage volumes, but connects these volumes with the river to restore the water balance of the system.
	Fish stocks and recruiting	Medium	Diversity and coverage of aquatic habitats created by the measure may result in increased fish stocks.
	Natural biomass production	Low	Potential for utilization of reed biomass in reconnected oxbows.
Regulatory and maintenance	Biodiversity preservation	High	Improvement of both aquatic and riparian habitats can result in improved populations of water birds, amphibian, reptilian and mammal species.
	Climate change adaptation and mitigation	Medium	Increased retention volumes for peak flows and improved water storage in the river system in droughts can participate in climate change adaptation and mitigation.
	Groundwater / aquifer recharge	Medium	Increase groundwater level and surface for natural groundwater recharge.
	Flood risk reduction	Medium	Although the retention capacity of the single application is usually not large, the cumulative result could be high if systematically applied.
	Erosion / sediment control	High	By increasing the length of the river, therefore decreasing flow velocity, prevents and mitigates bank and bed erosion, as well as favoring sediment deposition.
	Filtration of pollutants	Medium	Can increase path of the polluted water and contact filtrating surface. Riparian vegetation plays a filtration role of pollutants.
Cultural	Recreational opportunities	Low	Creation of places for sport fishing.
	Aesthetic / cultural value	Medium	Restoration of the natural riparian landscape.
Abiotic	Navigation	Low	Some navigation for large connected features.
	Geological resources	None	
	Energy production	None	

Modified from Natural Water Retention Management (NWRM, 2018), European Commission – Environment.

lowering (i.e., Figure 8.3). As such, dike setback usually constitutes a minor component of most integrated floodplain projects.

Environmentally, the goal of dike setback is to increase the floodplain geodiversity by restoration of the flood pulse across a larger area, enhancing the exchange of water, nutrients, and organisms between the river and the floodplain. Nutrient sequestration is also enhanced with the larger area of pulsed floodplain (e.g., Mitsch et al., 2001). The floodplain geodiversity is also enhanced by scour and deposition during flood events, increasing floodplain habitat types that support different types of riparian flora and fauna. Additionally, in the case of dike setback projects that also include removal of hard bank protection measures (e.g., revetment, riprap, groynes), the channel is able to rework floodplain deposits and the subsequent erosion and deposition increase channel bank, bed, and floodplain geodiversity.

Prior to any dike setback extensive analysis is required that weighs the benefits and possible disadvantages associated with flood risk reduction, ecological benefits, geomorphic activity, and economics. The decision of where to set back dikes requires considerable input and engagement with local stakeholders, particularly property owners and environmental organizations. Along some reaches the land use no longer permits dike setback, because of cultural heritage, urbanization, or industrial uses.

The reduction in flood height (stage) associated with dike setback projects has been shown to vary, but often ranges from several centimeters to perhaps greater than 1 m (Jacobson et al., 2015; Figure 8.13). While such a reduction in total flood height may seem modest considering the cost associated with dike setback projects, even several centimeters can be a significant reduction in flood risk when considering the magnitude and frequency of flood events. Studies that have modeled varying setback distance scenarios (Jacobson et al., 2015; Schober et al., 2015; Straatsma and Kleinhans, 2018) commonly find that flood stage levels decrease more as the setback distance increases, with the greatest reduction in flood stage associated with dike removal (Dierauer et al., 2012; Guida et al., 2013; Jacobson et al., 2015). Additional variation in flood level response occurs because of changes in the floodplain topography with distance, as various

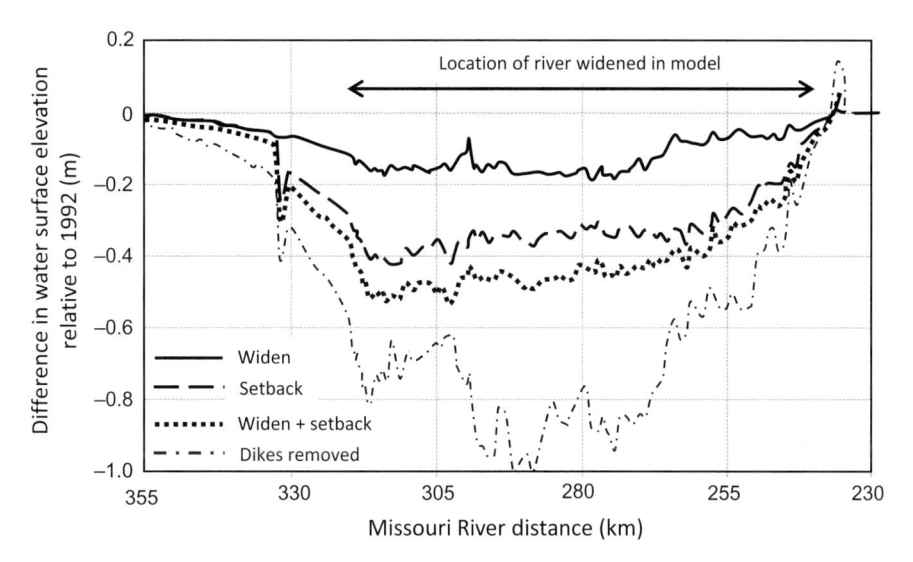

Figure 8.13. Modeled changes to water surface elevations for moderate flood event (ten-year RI) along lower Missouri River in relation to dike setback, dike removal, and channel widening. (From Jacobson et al., 2015.)

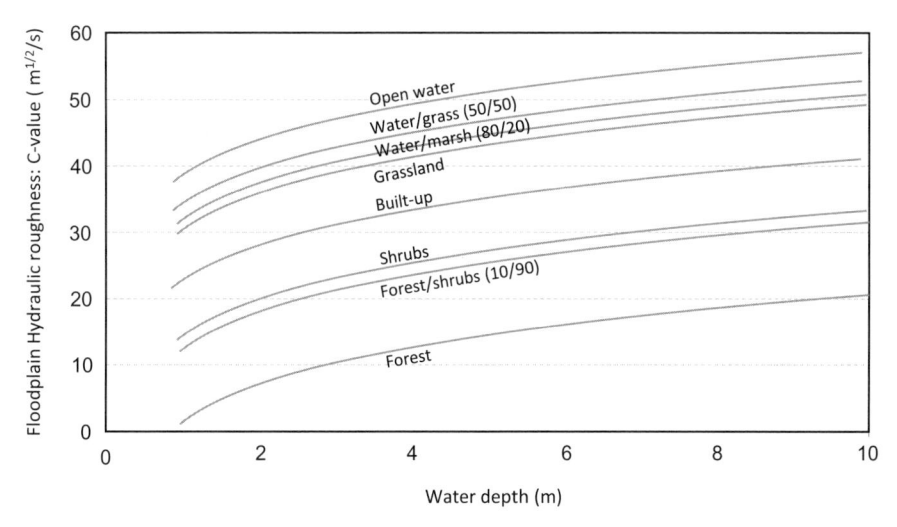

Figure 8.14. Changes to hydraulic roughness (C-value in $m^{1/2}/s$) in relation to flood water depth for different types of floodplain surfaces, including vegetation and builtup areas. Note that rougher surfaces have lower C-values, while higher C-values are those of "smoother" surfaces. (After Wolters et al., 2001.)

topographic features, such as old meander belts and natural levees, can result in pockets of backwater that actually result in slight increases in flood stage, particularly during moderate magnitude flood events (Jacobson et al., 2015).

8.3.6 Role of Floodplain Vegetation

An important consideration in integrated floodplain management projects is how to manage the change in land cover, especially riparian vegetation. Since many restoration projects are initiated in part to improve floodplain environments, the end result is the growth of riparian vegetation to enhance the natural functioning of the embanked floodplain (Hanberry et al., 2015). The decision to allow the growth of "natural" vegetation, however, needs to be carefully considered because of increasing resistance (i.e., floodplain roughness) to floodwater conveyance (Figure 8.14). Hydraulic roughness reduces flood wave velocity, and in combination with vegetation growth over several decades can actually increase flood stage (Makaske et al., 2011). As vegetation becomes denser and higher, sedimentation rates increase, which infills shallow water bodies and reduces floodwater retention (Buisje et al., 2002).

Floodplain forests are especially important to improving riparian environments, including ecosystem functions and carbon storage (Hanberry et al., 2015). Woody vegetation, however, has the largest floodplain roughness (Figure 8.14). Hydraulic modeling study of the Maas River (NL) found that flood levels increased by about 1 m after a twenty-year period of forest growth (Figure 8.15), which is consistent with findings along major distributaries of the Rhine River (Baptist et al., 2004; Straatsma et al., 2009; Makaske et al., 2011; Straatsma and Kleinhans, 2018).

Findings in the Netherlands that account for vegetation growth as a change in floodplain roughness are generally consistent with studies in other regions. Hydraulic simulation of changes in moderate flood levels along the lower Missouri River shows the clear influence of vegetation growth. An increase in embanked floodplain roughness, from agricultural (Manning's "n" = 0.05) to very dense young forest of willow and cottonwood ("n" = 0.2), resulted in flood stage levels increasing by 0.14 m despite the dike being setback by about 1.5 km (Jacobson et al., 2015).

Floodplain management that fosters vegetation growth should be carefully implemented so as to not increase flood risk. Vegetation management should always be incorporated into strategies to reduce flood levels of embanked floodplains. Such measures should permit a range of land cover types and practices, including forest management by timber harvesting, as well as cultivation and grazing practices attuned to the new flood regime of the embanked floodplain (Figure 8.16).

8.4 FLOOD BASINS AND DELTAS: MANAGEMENT CHALLENGES

In comparison to management of embanked floodplains, successful management of flood basins and deltas is more challenging because of the immense size and diversity of the landscapes and ecosystems impacted, and because many are at or below sea level. Many flood basins are undergoing accelerated rates of human exacerbated subsidence, which is especially urgent because of declining sediment loads (Syvitski et al., 2009; Chan et al., 2013; Allison et al., 2016). Additionally, the lowermost reaches of fluvial–deltaic systems require consideration of coastal management because of the influence of storm surge events and increased salinity.

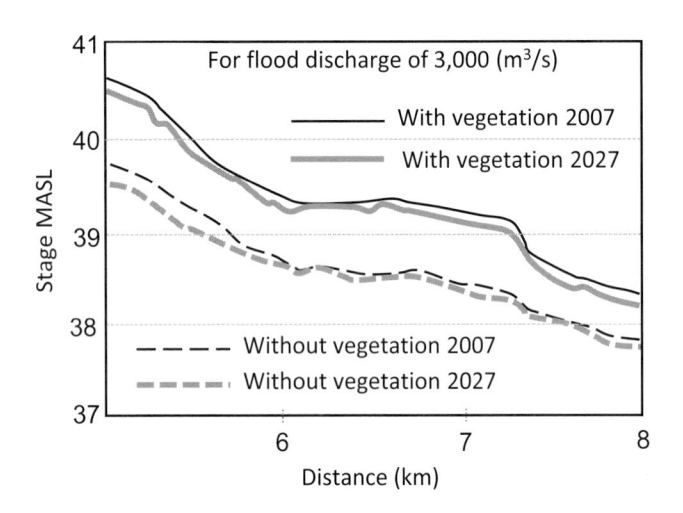

Figure 8.15. Modeled change to flood stage in relation to vegetation growth and floodplain roughness along Maas River, the Netherlands. (After Arroyave and Crosato, 2010.)

Figure 8.16. Floodplain and riverine landscape along IJssel River, the Netherlands. Different surfaces result in varying floodplain hydraulic roughness values, including open water, grassland, and forest/shrubs. The influence of embanked floodplain surface roughness on flood stage levels means that landscape and vegetation management is also flood management. Compare surface types to Figure 8.14. (Source: Rijkswaterstaat.)

Recognition of a sediment reduction problem, a subsidence problem, and a sea level rise problem means that, globally, deltas are highly endangered (Syvitski et al., 2009; Giosan et al., 2014; Allison et al., 2016; Best and Darby, 2020), and indeed in a state of crisis (Bravard, 2019). In this regard, the Rhine and the Mississippi represent two very different examples of deltas under stress (Table 1.1). While some specific processes and applications can be compared between the Rhine and Mississippi regarding channel and embanked floodplain processes, the coastal–deltaic environments are fundamentally different (Hudson et al., 2008; Hudson, 2018). The Mississippi continues to have a great deal of natural features worthy of conserving, and necessary to reduce flood risk. The Rhine, alternatively, has very little that is natural. Areas within the Rhine delta that have "natural environmental" attributes, such as the Biesbosch wetlands, have an anthropogenic origin (Section 7.4.2.1).

8.5 THE LOUISIANA EXPERIENCE: SUBDELTAS AND SEDIMENT DIVERSION

8.5.1 Context: Coastal Wetland Loss

The rate of Mississippi deltaic and coastal land loss in Louisiana is alarming, and is among the world's highest. Approximately 4,900 km^2 deltaic wetlands have been lost in Louisiana since the early twentieth century (Figure 8.18). Rates of land loss increased between the early 1900s to the 1940s, from 17.4 km^2/yr to 40 km^2/yr. During the 1970s land loss was the highest, reaching ~100 km^2/yr, beore declining to about 43 km^2/yr. The main causes of wetland loss in Louisiana are attributed to sediment decline and subsidence, which result in vegetation succession and dieback and ultimately the conversion of wetlands to open water environments. Dams and channel engineering are associated with a precipitous reduction in sediment supply to the coastal zone (Kesel, 2003), from 400 million tonnes/yr to 145 million tonnes/yr (Meade and Moody, 2010). Other causes include flood control dikes that prevent flood sediments from replenishing wetlands. Additionally, wetland drainage for human uses, such as petroleum industry and groundwater withdrawal, and storm surge events generated by tropical cyclones have further contributed to subsidence and land loss (Barras et al., 2008; Alexander et al., 2012; Allison et al., 2016; Day et al., 2016; Couvillion et al., 2017).

8.5.2 Legacy as Lesson: Historic Models of Deltaic Wetland Construction

Freshwater and sediment diversions have a crucial role in the maintenance and restoration of deltaic wetlands within Louisiana, and internationally (Giosan et al., 2014; Kemp et al., 2014; Amer

Figure 8.17. Flood disaster or new floodplain restoration measure? Aerial view of dike breach along lower Feather River near Nicolaus, California (Sacramento River basin). Dikes breach events revitalize floodplain geodiversity and ecohydrology and can be viable floodplain restoration measures. (Source: USACE, January 1997.)

Integrated management of flood basins and deltas often includes diversion of water and sediment with the intention of simulating natural physical processes. While flood stage reduction is the highest priority, diversion of water and sediment also occurs to construct new fluvial land to offset subsidence and coastal erosion, and to conserve wetlands. Common approaches to managing flood basins and deltas include flow and sediment diversion structures, controlled dike breach events (Figure 8.17), subdelta creation, and groundwater management. Drought conditions require specific management approaches in flood basins (noted in Section 6.2.2, Figure 6.10), particularly wetting dikes constructed of organic-rich materials to prevent cracking and failure. And for large urban populations residing within fluvial deltas, flood management options are increasingly integrated into urban design in a strategy commonly known as "living with water," which may include a combination of flood water routing, pumping, precision groundwater management, and urban water storage.

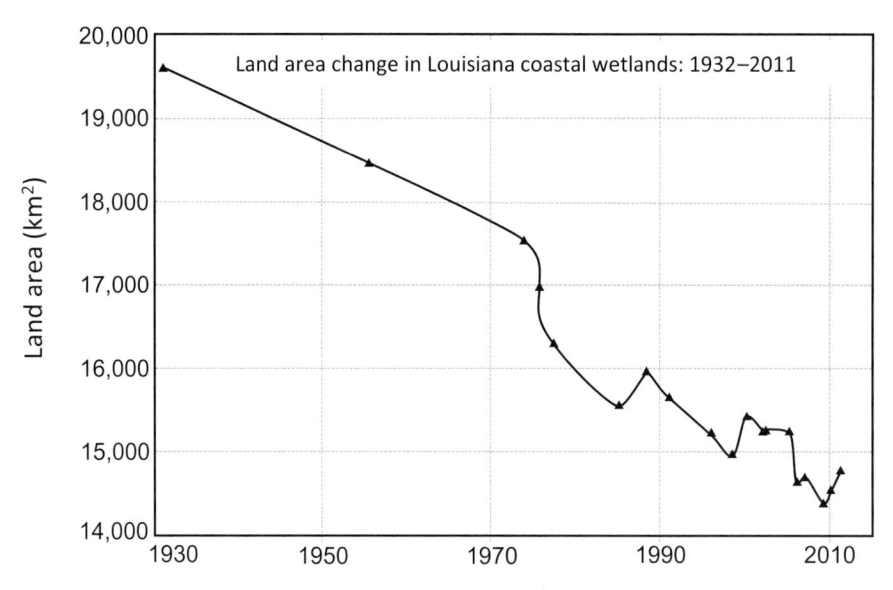

Figure 8.18. Change in the land area of coastal Louisiana between 1932 and 2010. (Figure source: Alexander et al., 2012 and Couvillion et al., 2017.)

et al., 2017). The design of wetland rehabilitation projects can be fine-tuned by considering historic examples of freshwater and sediment diversions that were unintentionally triggered by human activities. In the Mississippi delta, several historic case studies have been informative in shaping the conceptual approach, each of which was triggered by a different motivation. These include (1) the accidental formation of Cubits Gap subdelta by channel breach, (2) construction of the Wax Lake delta associated with flood risk reduction, and (3) the creation of new wetlands associated with freshwater diversion for salinity management.

While the overall trajectory of land loss within the Mississippi delta is daunting (Blum and Roberts, 2012), it is important to note that several serendipitous events historically occurred that inadvertently resulted in new wetland construction. In particular, diversion of floodwaters into bays and interdistributary basins by crevasse-like events and through artificial dredge channels has inadvertently resulted in new wetland construction. Such wetlands can be investigated to guide sediment diversion structures to construct new wetlands in the Mississippi delta, and other subsiding delta locations around the world. While the benefits of restoring and creating new wetlands in Louisiana would seem obvious, for a variety of reasons the issue has long been complex and controversial. As much as any single cause, the issue that looms largest is the seemingly benign oyster (*Crassostrea virginica*), which annually accounts for thousands of jobs and several hundred million dollars in revenue.

8.5.2.1 WAX LAKE DELTA

The highest rates of new land gain in the Mississippi delta are located within Wax Lake Delta, perhaps the future Mississippi delta lobe and part of the larger Atchafalaya Delta complex (Figure 8.19). The Atchafalaya River extends 160 km through a dense network of anastomosing channels and shallow lakes (Hupp et al., 2008; Mossa, 2015b). Wax Lake Outlet is an artificial channel created as part of a flood control project in 1941. The outlet was created by dredging a channel across an old Mississippi River meander belt (Teche ridge) that served as the southern topographic border between marine and freshwater environments. The cut through the delta lobe provided an efficient outlet for freshwater to flow directly into Atchafalaya Bay. The outlet was originally designed to divert about one-third of the Atchafalaya River flow, which would then reduce flood stages within the Atchafalaya Basin, particularly for Morgan City. Recent estimates, however, reveal that the Wax Lake Outlet has an average discharge of 2,667 m^3/s – 46% of the total flow of the Atchafalaya River.

The formation of new deltaic lands is related to sedimentation during large flood events. The massive 1973 Mississippi River flood resulted in considerable sediment deposition, such that Wax Lake Delta literally emerged from the bay (van Heerden and Roberts, 1980; Roberts, 1998). The sedimentation resulted in the development of a classic bayhead-style delta. The delta is characterized by sandy islands deposited between bifurcating distributary channels, which are rapidly creating sandy–silty flanking natural levees, the primary deltaic land form. Recent flood events document the continued evolution of Wax Lake Delta (Carle et al., 2015). The distributary channels continue to prograde into the bay, increasing in size as they accommodate larger flows (Shaw et al., 2013). The size of the delta land (above −0.6 m elevation) was already 51.1 km^2 in 1997. And as with the 1973 flood, the extreme 2011 flood also resulted in considerable new land construction across Wax

Table 8.6 *Characteristics of four subdeltas within the modern Mississippi (Balize) delta lobe*

Subdelta	Year of initiation	Mississippi discharge (%)	Max. area (km^2)	Growth (km^2) per year	Depth of accommodation space (m)
Baptiste Collette	1874	2.6	57.2	0.79	2–6
Cubits Gap	1862	13.2	193.1	2.33	2–10
Garden Island Bay	1891	4.0	124.0	4.00	2–8
West Bay	1839	4.4	300.05	3.23	4–10

From Wells et al. (1983).

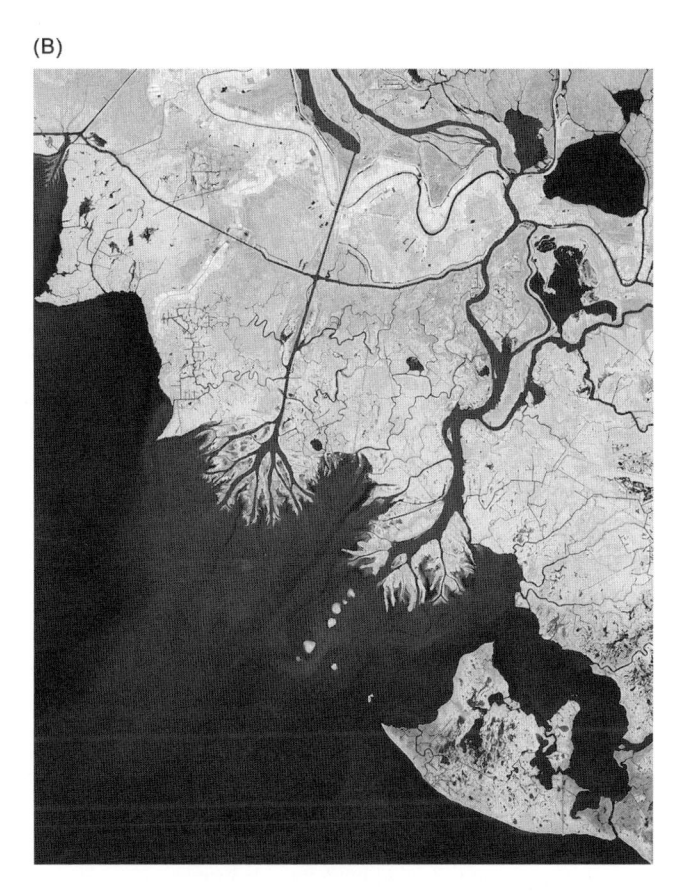

Figure 8.19. Satellite imagery depicting land growth in the Atchafalaya delta complex between 1983 (A) and 2010 (B). The engineered Wax Lake Delta is to left and the Atchafalaya Delta is to right. For scale, the width of the 2010 subaerial Wax Lake Delta is about 13 km. (Source: NASA.)

Lake delta, which is rapidly being colonized by deltaic plant communities. Specifically, the 2011 flood sediment pulse was responsible for the creation of 4.9 km^2 of new delta wetlands (Carle et al., 2015).

8.5.2.2 CUBITS GAP SUBDELTA

The Mississippi delta includes a variety of well-documented cases of artificial freshwater and sediment diversions that influenced wetlands, some of them unintentional (Yuill et al., 2016b). The active delta lobe includes a number of smaller subdeltas created by levee breach that provide useful historic analogues for deltaic wetland construction. Present Mississippi subdeltas vary in their evolutionary stage of development, with several prominent subdeltas having formed over about the past hundred and fifty years (Table 8.6). A well-known example of an artificial diversion occurred with the formation of Cubits Gap subdelta in the mid-nineteenth century (Welder, 1955; Wells et al., 1983; Roberts, 1998). Cubits Gap subdelta is located on the eastern side of the Mississippi River, 6 km upstream of Head of Passes (Figure 8.20). The subdelta was initiated in 1862 when oyster fishermen cut a "gap" in the river bank to reduce travel distance to access oyster reefs (Ockerson, 1883; Welder, 1955). The

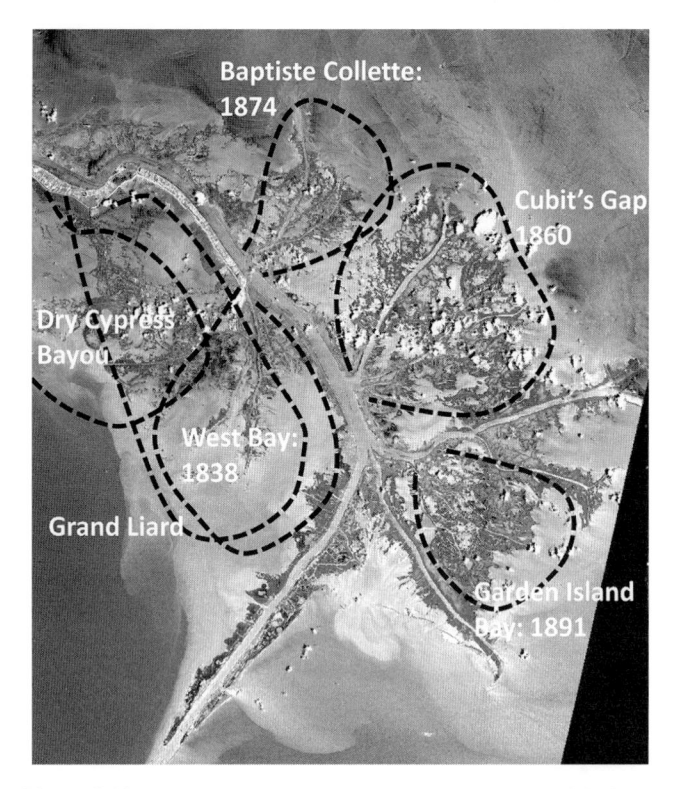

Figure 8.20. Modern subdeltas along the active Mississippi delta lobe showing year subdelta was initiated (if known). A large proportion of the active Mississippi delta lobe is formed by fairly recent (<200 years) subdelta, providing a historical analogy for design of sediment diversion structures. (Source: Imagery: NASA, subdelta data: Wells et al., 1983.)

decision to cut the notch in the river bank substantially reduced Cubits travel distance to his oyster leases, initially. But unfortunately for the fishermen, and fortunately for the sake of deltaic wetland creation, the notch unleashed abrupt crevasse-like deltaic sedimentation processes, with classic prograding distributary mouth bar, bifurcating channels, and formation of additional crevasse subdeltas (e.g., Wright and Coleman, 1974). The oyster reefs were ultimately buried by deltaic deposits.

After the initial cut, Cubits Gap subdelta rapidly evolved (Figure 8.21). Within twenty years a distinctive subdelta had formed, a microcosm of a major delta lobe with bifurcating and prograding channels, natural levees, and subaqueous deposits resulting in bay infilling. By 1877, the channel width and maximum channel depth at the inlet was already 950 m and 46 m, respectively. Large floods in the late 1800s drove further progradation, and by 1905 the subdelta was about 125 km^2. In terms of discharge, by 1915 Cubits Gap transported 12.5% (1,790 m^3/s) of the total Mississippi River discharge. Cubits Gap subdelta reached its zenith at about 1950, and was associated with a land area of about 200 km^2. But soon after 1950 Cubits Gap subdelta started to rapidly deteriorate, as shoaling of coarse sediment reduced the maximum channel depth to about 11 m (Welder, 1955). After the 1950s, the distributary network was

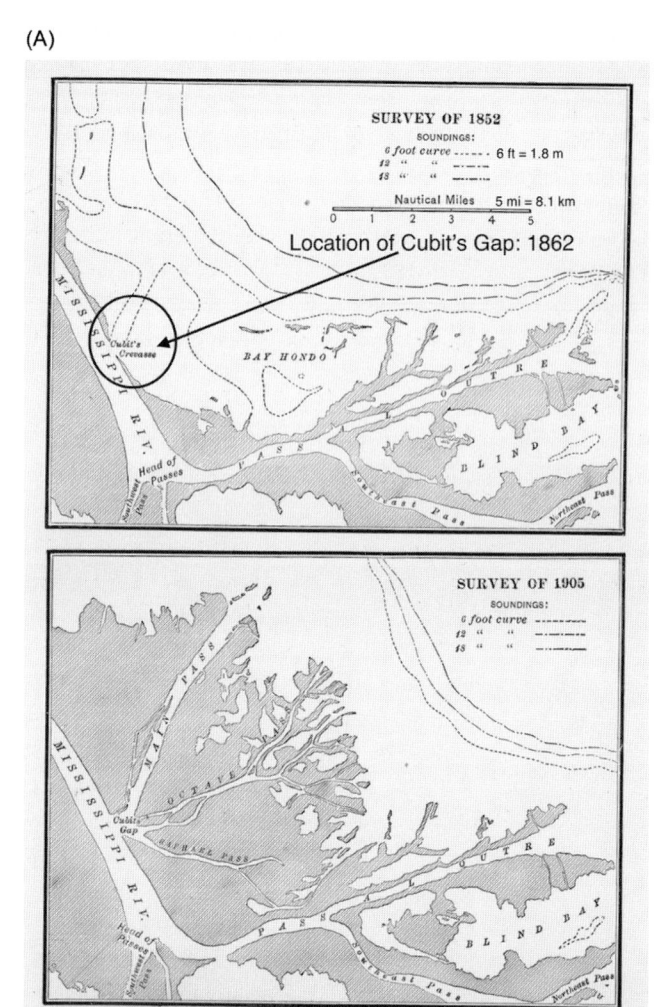

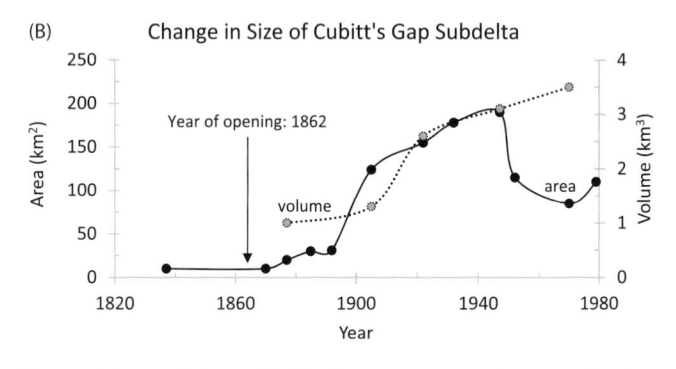

Figure 8.21. Evolution of Cubits Gap subdelta. (A) Growth of land at Cubits Gap, Mississippi River Delta, between 1852 and 1905. Note change in bathymetry, including soundings, (B) Area (km^2) and volume (km^3) of subdelta bay-fill. (Source: (A) In "Nautical Charts" by George Rockwell Putnam, 1908. 162 pp. NOAA's Historic Coast & Geodetic Survey, Louisiana, Mississippi Delta, Cubits Gap, 1908. (B) Wells et al., 1983.)

overextended and became hydraulically inefficient. Combined with subsidence of soft clay and marsh deposits this resulted in rapid wetland deterioration and open water expansion within the interdistributary networks. By the 1970s, the size of Cubits Gap subdelta had reduced to less than half of its maximum size. As interdistributary marshlands eroded, the remaining wetlands are located along the margins of the skeletal framework of the subdelta, which are comprised of denser and coarser channel and natural levee sediments. Since the 1970s, discharge and channel dimensions at the point of Cubits Gap breach have slightly increased, with a width of 1,150 m and a discharge ranging from 2,711 (m^3/s) to 627 (m^3/s) during high (April) and low (November) flow periods, respectively (Esposito et al., 2013). The current subdelta supports about 125 km^2 of deltaic wetlands.

Human-initiated freshwater and sediment diversions effectively harness fluvial–deltaic processes to create large amounts of new wetlands that are virtually undistinguishable from their natural analogue. Whether initiated by natural flood crevasse processes or by humans, Mississippi subdeltas generally have a life span ranging from 100 to 200 years (Roberts, 1998). Cubits Gap was an uncontrolled accident, created for one reason that led to positive (mainly) gains, although not appreciated for many decades.

8.5.2.3 CAERNARVON FRESHWATER DIVERSION STRUCTURE

Recognition and concerns over fluvial–deltaic wetland deterioration caused by increased salinity and vegetation dieback and their impacts to the associated marsh ecology has stimulated a variety of coastal protection and restoration plans. Within the Mississippi delta there are a variety of diversion structures, some are passive while others require the operation of control gates. An important case study in the Mississippi delta is the Caernarvon Freshwater Diversion structure (Figure 8.22), located on the east side of the Mississippi River ~25 km downstream from New Orleans (Caernarvon Delta and Diversion Study, 2014). Interestingly, the diversion structure is located where the dike (levee) was dynamited during the Great Flood of 1927 to relieve New Orleans from flooding (Day et al., 2016). Authorization for Caernarvon is provided in the U.S Flood Control Act of 1965 and Water Resource Development Acts of 1974, 1986, and 1996. The structure is fully operational by opening five gates to divert Mississippi River discharge into a passive conveyance canal that flows about a kilometer into a labyrinth of lakes, channels, and shallow bays (Figure 8.22). Before emptying into the Gulf of Mexico the diverted water and sediment flow into Breton Sound, an enormous (311 km^2) estuary undergoing some of the highest rates of coastal erosion in the world (Morton et al., 2016; Couvillion et al., 2017).

The Caernarvon Freshwater Diversion structure started operation in 1991 with the objective of increasing commercial and recreational fishing and wildlife by reducing marsh salinity, thereby enhancing vegetation growth and reducing marsh loss. While a variety of species would benefit, to be clear, the main target is to increase the production of oysters. The main design strategy was to divert freshwater so that saline waters would be pushed coastward, providing a larger area of productive brackish water, essential for oyster cultivation. The structure would accomplish this by siphoning off up to 225 m^3/s of surface water directly from the Mississippi River, in addition to sediments and nutrients. It is important to note that the structure was designed to divert freshwater to reduce high salinity levels caused by salt water intrusion. It was understood that the structure would also divert some Mississippi River sediment, but sedimentary processes were not factored into the design of the diversion structure. And, unlike the Bonnet Carré diversion structure upstream of New Orleans (see Section 7.2.2.3), Caernarvon was not designed to reduce flood levels.

Figure 8.22. Caernarvon Freshwater Diversion structure, located at the head of Breton Sound Estuary. The outlet canal from the diversion structure flows into Big Mar, an open water lake environment being converted into wetlands. Big Mar drains into channels and bays that eventually empty into Breton Sound and the Gulf of Mexico. (Source: U.S. Geological Survey, February 7, 2017.)

After more than a decade of monitoring and analysis of salinity data the structure had mainly accomplished its goals. Salinity in the bay had reduced, as the 5 ppt and 15 ppt isohalines migrated coastward. And oyster production increased, as did other shellfish and finfish. The Breton Sound fishery clearly benefitted from the diversion structure and reduced salinities (LDNR, 2006; Day et al., 2009).

An unintended (and welcome) outcome was that the Caernarvon diversion project also increased the area of wetlands, with growth and species type following the salinity line, such that freshwater, brackish, and saltwater marsh vegetation also migrated coastward (LDNR, 2006). But while the wetland vegetation increased in area, it was not to last long.

In 2005, Hurricane Katrina's winds and storm surge made a direct impact to the restoration area, causing considerable erosion of the newly created wetlands. Post-impact studies revealed that, although wetlands had expanded there was only minimal sedimentary substrate to anchor plant roots (Kearney et al., 2011; Lopez et al., 2014). The suspended sediment transported through the diversion structure was not distributed far beyond the outlet, across the marsh. In effect, the diversion structure operation did not take into account suspended sediment dynamics (Section 7.2.2.3) as regards the timing of discharge relative to the transport of suspended sands and silt/clay (wash load). With regard to operation of the diversion structure for land gain, it was mismanaged. Additionally, the natural hydrologic regime of the Mississippi River was not utilized to create "flood pulses" across the marsh, which are considered essential to biological productivity (Tockner et al., 2000; Day et al., 2009). Thus, while the diversion structure was effective at initially enhancing the deltaic ecosystem, because the sedimentology was neglected it was a short-term gain.

Realization that a lack of marsh sedimentary deposits increased vulnerability of the wetland ecosystem to storm events resulted in a new management strategy for the Caernarvon Freshwater Diversion Structure. The Louisiana Coastal Protection and Restoration Authority was created in December 2005, just months after the Hurricanes Katrina-Rita flood disaster. Following the Water Resources Development Act of 2007, the LCP&RA was charged with developing and implementing the new diversion strategy that takes into account the suspended sediment dynamics.

The outcome of over a decade of operation, including the loss of wetlands by Hurricane Katrina, resulted in tangible "lessons learned" as regards the operation and management of Caernarvon Freshwater Diversion structure, and ultimately a much more "integrated" plan. The new strategy includes planting specific types of marsh vegetation, and barrier island restoration. But the key change to the diversion strategy is to explicitly design its operation for sediment transport to build land mass (USACE, 2013). The new strategy also seeks to replicate natural flow variability by including a range of small and larger discharge pulses (Wang et al., 2017). To maximize sediment transport to Breton Sound the new strategy factors in the suspended sediment transport dynamics in relation to Mississippi River discharge pulses (Day et al., 2009). As with Cubits Gap, the larger discharge pulses and building of sedimentary land mass means that oyster habitat will be lost (Wang et al., 2017). Thus, to address stakeholder concerns the project includes funding to purchase and acquire easements from key stakeholders, including oyster fishermen.

8.5.3 Applying Historic Lessons: Sediment, Strategy, and Stakeholders

The range of historic and contemporary experiences with artificial and natural sediment diversions are the driving rationale for new wetland restoration plans for coastal Louisiana (CPRA, 2017). Although controversial among some stakeholders, especially commercial fisheries and the channel dredge industry, the state and federal government has agreed that sediment diversion structures within the Mississippi delta are the key management tool to protect and restore wetlands. Of note is that the Hurricane Katrina flooding of New Orleans increased the urgency to address the topic of wetland loss, but mainly in the interest of flood protection, as wetlands are considered critical to buffering storm surge generated by tropical cyclones. But the conservation of existing wetlands and creation of new wetlands are also considered important to protect the overall deltaic ecosystem and commercial fishery, over longer timescales.

The issue is more complex than the natural processes alone, as coastal environments and people connected to the landscape will be impacted by construction and operation of sediment diversion structures (Day et al., 2012; Twilley et al., 2016; Science, 2021, p. 334.). The commercial fishing industry is adamantly opposed to sediment diversions, and contend fishery harvests will decline, and adversely impact the livelihood of people strongly linked to the industry (Ko et al., 2017). Large sediment diversions are projected to reduce salinity levels, displacing flora and fauna associated with a specific salinity further offshore (Soniat et al., 2013; Rose et al., 2014; Wang et al., 2017). The scope of the controversy can be appreciated by considering that Louisiana has the largest seafood industry in the United States, which is annually a multibillion dollar industry.

While the science indeed supports a decline in oyster harvests over a short term (Soniat et al., 2013; Wang et al., 2017), implementation of the diversion strategy ensures that fishery resources are available into the next century if coupled with sensible ecosystem and aquatic habitat management measures (Soniat et al., 2013; CPRA, 2017) (CPRA, 2017). Additionally, the option of not conserving wetlands increases rates of relative sea level rise and results in higher wetland salinities, and therefore higher rates of fluvial–deltaic wetland loss. While the land gain within the Mississippi delta caused by breaches provides some reason for optimism, it needs to be considered against future sea level rise scenarios (Figure 8.23). Models that examine specific diversion scenarios in relation to land gain complement empirical (historical) studies for the initial decades, and then level off.

8.5.3.1 SEDIMENT DIVERSION STRUCTURES: DESIGN AND OPERATION

The Louisiana Coastal Protection and Restoration Authority will sponsor two main sediment diversion structures, in coordination with the U.S. Army Corps of Engineers, to potentially begin operation as soon as 2024. The structures will be located in Barataria Bay and Breton Sound, on the west and east of the main Mississippi River, respectively (Figure 8.24). Since the 1930s, Breton and Barataria basins have lost ~1,800 km^2 of wetlands, some of the highest rates of land loss in the world. The Breton diversion project will have a capacity to transport as much as 1,000 (m³/s), whereas the Barataria basin diversion project will have a capacity to divert as much as as 2,100 (m³/s), roughly the average annual discharge of the Rhine.

Operation of the diversions will take advantage of state-of-the-science knowledge of suspended sediment dynamics and utilize an "environmental flows" approach to manage specific habitat types. Rather than opening the diversion structure during the rising and falling limb of the hydrograph, the structure should be opened during the rising limb, and closed after the peak, which is ideal for managing oyster habitat (Figure 8.25). This management approach results in 56% of the water being diverted

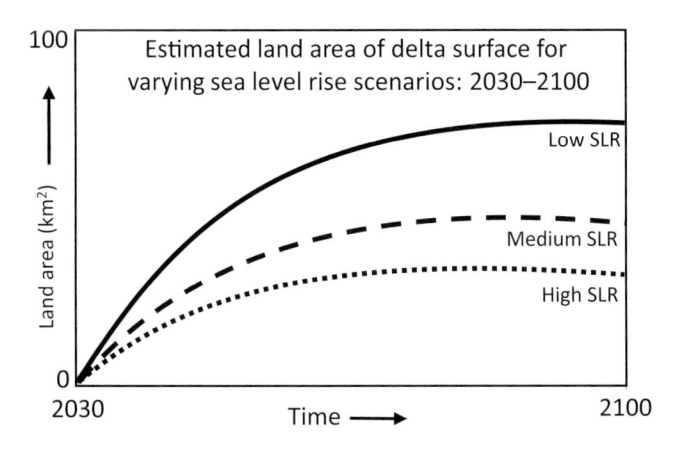

Figure 8.23. Land gain from 2030 to 2100 for modeled sediment diversion structure with Q = 7079 m³/s, illustrating diminishing trajectory and stabilization of land area gain for varying sea level rise (SLR) scenarios. The X-axis extends from 2030 to 2100 and the Y-axis extends from 0 to 100 km². (After Rutherford et al., 2018.)

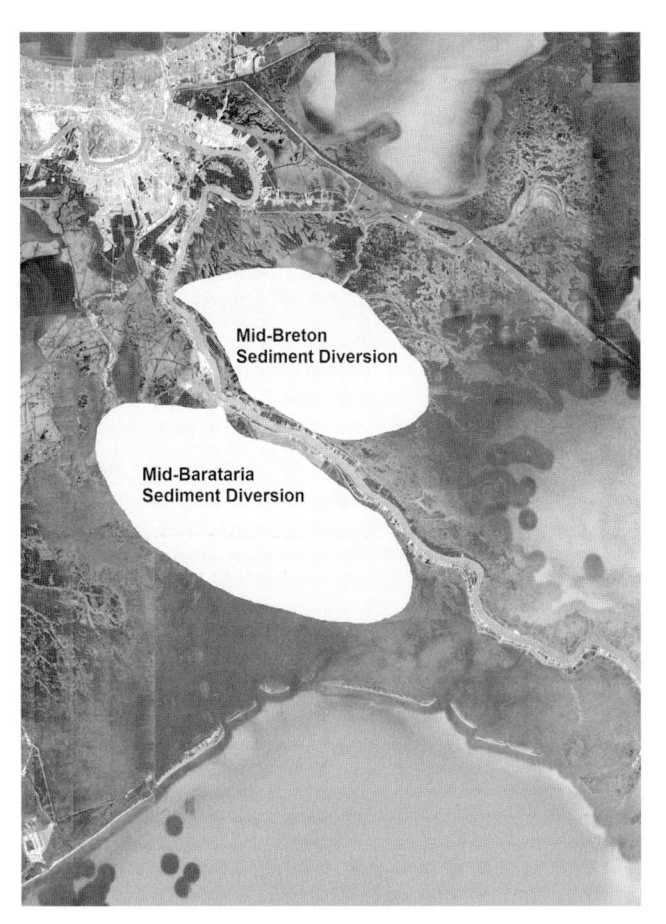

Figure 8.24. Location of proposed Mississippi delta sediment diversion structures for mid-Baratarria and Breton Sound, with modeled area of influence. (Author figure, based on Louisiana Coastal Protection and Restoration Authority projections.)

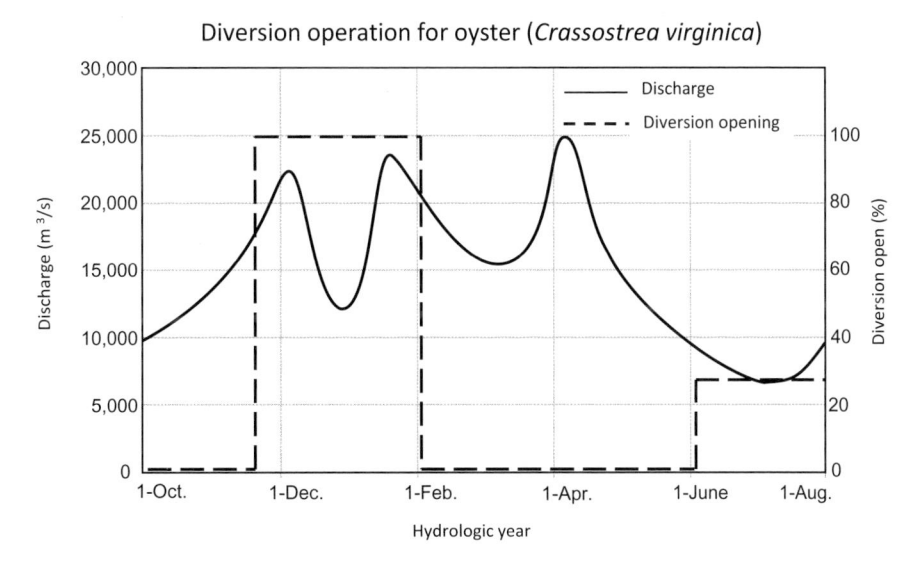

Figure 8.25. Hydrograph typology for Mississippi River in context of environmental flows and ecological maintenance of oyster habitat within the delta. (Source: Peyronnin et al., 2017.)

Table 8.7 *Recommendations for operation strategy of large sediment diversion structures: years 0–10*

Entity	Management considerations
Distributary channel network	Requires minimum of five years to establish distributary network, including progradational sedimentary facies.
Basin hydrology	To prevent unintentional flooding, detailed hydrologic monitoring and water budget is necessary to understand synergy between basin water level fluctuations because of inputs (groundwater, precipitation), outputs (drainage), storage, and discharge inputs.
Erosion	Some initial local erosion is expected of friable weak marsh deposits upon diversion of jet plume pulses.
Climatology	Dominant "local" precipitation mechanism needs to be assessed, as well as synergy with discharge inputs. Influence of cold fronts, tropical cyclones, and convectional storms on water setup and wave energy.
Vegetation	Loss of vegetation could initially occur because of stress due to adjusting to new pulsed hydrologic regime, especially during growing season. Diversion operations ideally active during dormant season.
Fish & wildlife	As with vegetation, fish and wildlife may suffer stress with changing hydrologic (water level, velocity), salinity, nutrients, and higher turbidity. Diversion operations should occur gradually to allow different species to adapt.

Sources: Peyronnin et al. (2017), with author modifications.

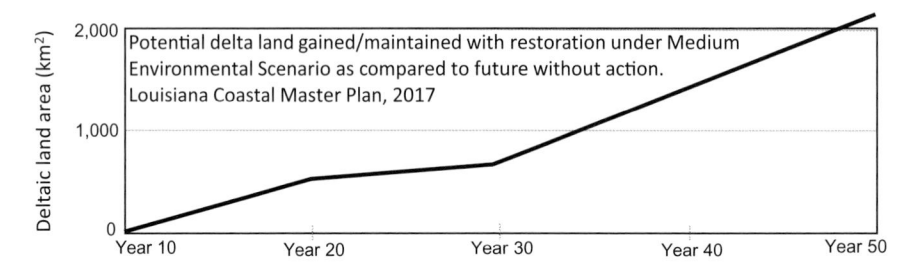

Figure 8.26. Potential new delta lands retained and gained over a fifty-year period with implementation of Louisiana's Coastal Master Plan. Actual land building benefits are expected to continue far beyond the fifty-year period of implementation. The projection is for a medium environmental scenario (IPCC sea level rise projection). (Source: CPRA, 2017.)

but 72% of available sediment being diverted into the marsh (Peyronnin et al., 2017). The reason for the higher proportion of sediment diverted is because of the phenomenon of clockwise hysteresis, which refers to most of the fine-grained sediment (wash load) being transported ahead of the flood wave. Additionally, some maintenance flow may be required during Mississippi low-flow periods (summer in Louisiana) to prevent saltwater intrusion.

In accord with an integrated approach to management, the design and operation of the diversion structures will need to be flexible, such that scientific advances and policy changes can be effectively incorporated into the operation (e.g., WMO, 2004). The diversion strategy should expect to change as the wetland basin adjusts to sedimentary accretion and ecotopes evolve, especially during the first five to ten years (Table 8.7). Detailed temporal and spatial monitoring during the early phase is critical so that unexpected changes can be detected, and modification to management efficiently implemented. During the initial phases of operations, the distributary channel network will need to develop, and can be expected to be a dynamic process, with shifting channels, shoaling, and some erosion (Shaw et al., 2013). Additionally, some vegetation dieback may occur, and fisheries production may initially decline because of abrupt changes in salinity and nutrients (e.g., Soniat et al., 2013; Wang et al., 2017). A detailed water budget is required so that the inputs, outputs, and storage within the basin are accounted for prior to discharge pulses to avoid unintended flooding. Additionally, the synergism between sediment and discharge inputs with the hydro-climatic regime should be established. The large size of the basins means that storms and cold fronts, for example, can generate winds and waves that alter the intended pattern of sediment dispersal.

The Louisiana Coastal Protection and Restoration Plan was approved by the state legislature in 2012 and is being implemented, with the sediment diversion structures slated for construction after final approval. The plan would ultimately conserve and restore about 2,000 km² of coastal wetlands over a fifty-year period (Figure 8.26). The actual amount of wetland conservation and creation will of course be highly dependent upon climate and sea level change scenarios, in addition to future land use, planning, and demographic changes. In Louisiana, relative sea level rise is highly variable because of spatial

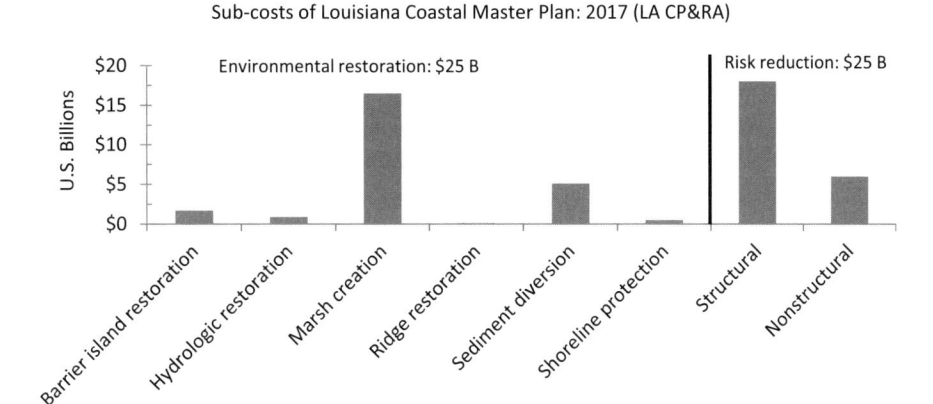

Figure 8.27. Sub-cost of different approaches within the Louisiana Coastal Master Plan, to be implemented by the Louisiana Coastal Protection & Restoration Authority. (Source: Louisiana Coastal Master Plan.)

differences in rates of ground subsidence, which vary from 4 to >20 mm/yr (Morton et al., 2016). While the sediment diversion structures are the main approach within Barataria Bay and Breton Sound, the larger Louisiana Coastal Master Plan also utilizes a variety of other approaches, such as barrier island restoration, marsh vegetation planting, and dredging. Additionally, half of the budget (US$25 billion) is dedicated to reducing flood risk, which include traditional structural (e.g., dike fortification, groynes, sea walls) and nonstructural approaches (Figure 8.27). The latter category includes measures such as property buyout, zoning, spatial planning, and emergency preparedness.

Coastal restoration activities for the Mississippi delta are organized by the Louisiana Coastal Protection and Restoration Authority, a state entity created in the wake of Hurricanes Katrina and Rita flood disasters. The Louisiana CPRA serves as the coordinator for numerous projects associated with preserving and restoring the Mississippi delta. A key role of the Louisiana CPRA is to serve as a link between many local and state entities with federal organizations, particularly the U.S. Army Corps of Engineers. But as the LCPRA is a state entity, its activities are solely confined to the Mississippi delta and have no direct jurisdiction upstream of the delta, or outside of Louisiana. The Louisiana CPRA, therefore, lacks an institutional framework to facilitate a truly integrated approach to fluvial and coastal management for the Mississippi delta.

Freshwater and sediment diversion structures represent a type of geoengineering, hydraulic structures designed to stimulate natural-like hydrologic and sedimentologic processes to restore wetlands. Diversion structures vary from classical hard engineering infrastructure (e.g., dikes, groynes, sea walls) in that they are utilized to harness natural processes for the construction of new wetlands (DeLaune et al., 2003; Day et al., 2012; Giosan et al., 2014). Thus, rather than the conventional engineering approaches that "disconnected" water from landscapes, sediment diversion structures foster "connectivity" of rivers with flood basins.

Freshwater and sediment diversion structures are increasingly seen as the key instrument for the conservation and restoration of fluvial–deltaic wetlands for a range of deltas across the globe and are especially necessary because of the problem of global sediment decline, climate change, and sea level rise that threatens to drown many of the world's deltas (Cruz et al., 2007; Syvitski et al., 2009; Kearney et al., 2011; Cohen, 2012; Shaw et al., 2013; Giosan et al., 2014; Allison et al., 2016; Best and Darby, 2020).

Because of fundamental differences in deltaic processes and the status of different deltaic lands it is not recommended to utilize a single type of structure from one delta to another. Four such approaches include (Figure 8.28) (i) channelization within the Danube delta, (ii) crevasse splays in the Mississippi delta, (iii) subdeltas within the Atchafalaya delta, and (iv) the new delta lobe construction that mimics the avulsion process in the Huanghe (Yellow) River delta (Giosan et al., 2014). Each type of structure should be employed within a management framework that considers the detailed discharge and suspended sediment dynamics (i.e., Mossa, 1996), deltaic geometry (receiving basin), ecotope characteristics, marine energy (wave and tidal) regime, as well as rates of subsidence and sea level rise (Table 8.8). Additionally, because human-influenced deltas include a range of associated land use types and economic activities, management organizations should be prepared to engage with a range of stakeholders to identify conservation and restoration priorities.

Table 8.8 *Considerations in the design of freshwater and sediment diversions to offset deltaic land loss*

Riverine considerations	Sediment size	Sand, silt, clay distribution and requirements for land gain and vegetation stabilization
	Discharge variability	Importance of large discharge pulses for sediment volume and distribution
	Main channel	Sediment aggradation near point of withdrawal, possible dredging required
Design and operation	Discharge of diversion structure	Influences distance of transport across wetlands
	Hard structures	Spatial pattern of sediment distribution
	Timing of sediment diversion (opening/pumping)	In relation to flood wave and sediment – discharge dynamics, including wash load vs. suspended sand
	Design of diversion structure	Passive gravity (e.g., weir) or active (pumping), length of trough
	Seasonality of release	Related to vegetation colonization, wind patterns, wave regime
Environmental	Ecologic	Turbidity and salinity change disrupt aquatic habitat, possible endangered species
	Natural resource	Synchronized to "natural" ecosystem processes for flora (growth stages) and fauna (spawning, feeding, migration)
	Riverine water quality	Nutrient loading and other pollutants, change in biogeochemistry of receiving basin
	Wetland type	Marsh, mangrove, forest, open water in relation to salinity levels
Receiving basin	Geometry of receiving basin	Accommodation space: depth and area of basin, distance required to transport sediment
	Sea level rise projections	Utilizing a low, medium, or high SLR projection to influence operation of diversion structure
	Energy of receiving basin	Tidal regime, wave regime
Stakeholders	Economic	Human uses (e.g., fishing, oyster production), local industry, shipping and navigation
	Cultural	Disruption to livelihoods, historic and archaeological materials, hunting/fishing season

Author table from multiple sources.

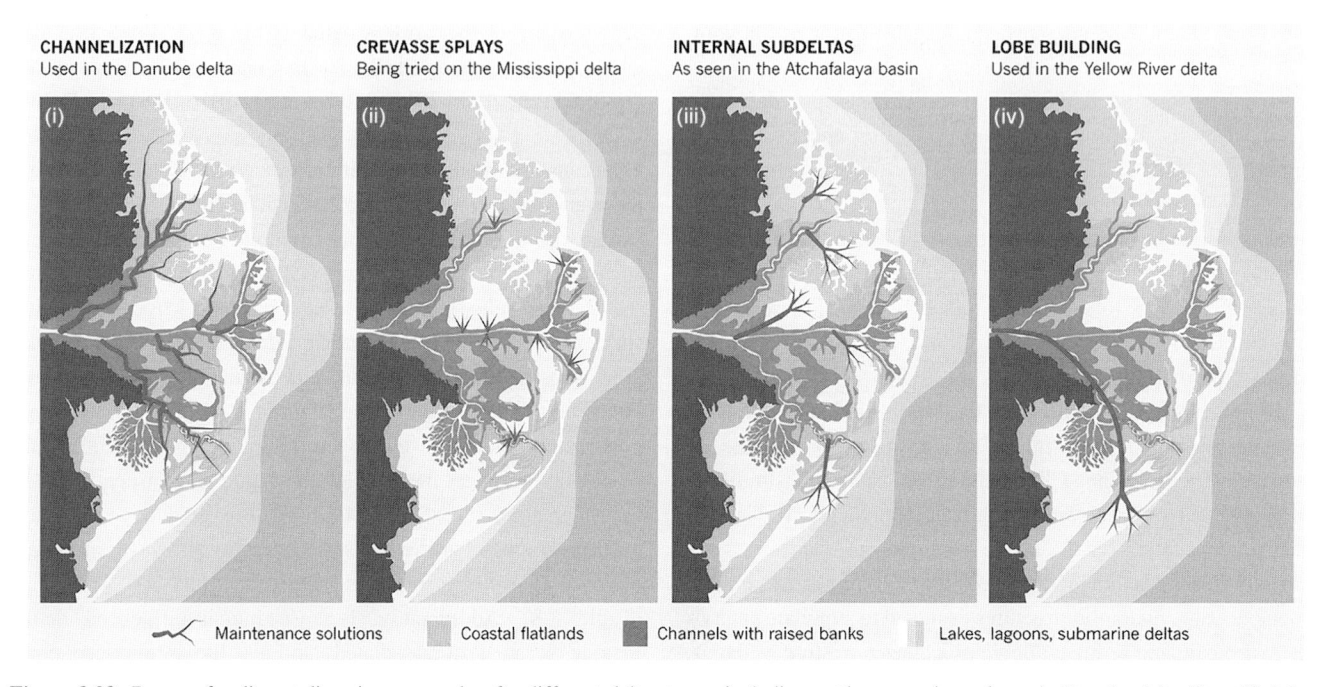

Figure 8.28. Range of sediment diversion approaches for different deltas types, including cutting new channels, as in Danube delta (i), artificial breaches to simulate crevasse processes, as with Mississippi delta (ii), channeled diversion within interior lakes and lagoons, as in Atchafalaya basin (iii), and larger channels that create new progradational delta lobes, as in Huanghe (Yellow) River delta (iv). Morphodynamics of each respective delta varies, which influences suitable type of sediment diversion. (From Giosan et al., 2014.)

8.6 INTEGRATED MANAGEMENT: EU WATER FRAMEWORK DIRECTIVE AND ROOM FOR THE RIVER

8.6.1 Increasing Connectivity and Floodwater Retention

A prime example of an organization focused toward a basin-scale integrated watershed framework is the International Commission for the Protection of the Rhine. The ICPR provides a crucial institutional link between the nine nations within the Rhine basin, and helps to coordinate implementation of the European Union's Water Framework Directive (EC/60/200) and Floods Directive (EC/60/2007), which collectively work to reduce flood risk and improve the basin's environmental conditions.

A key measure to reduce flood levels is to increase lateral connectivity between rivers and riparian lands, as floodplains include a range of suitable environments for floodwater retention (ICPR, 2016). By 1977, the highly regulated and idustrialized Rhine retained only 55 million m^3 of floodwaters along its entire main stem (Figure 8.29). By 1995 floodwater retention had already increased to 160 million m^3, which serves as the reference year for restoration targets. Over a ten-year period, between 1995 and 2005, the floodwater retention increased from 160 to 212 million m^3, and by another 17 million m^3 by 2010. Thus, floodwater retention along the Rhine has substantially increased in the past decades, and now at 340 million m^3 (ICPR, 2020a) is close to the EU Water Framework Directive target of 361 million m^3 by 2020.

Reconnection of floodplains is a vital step toward the goal of reduced flood risk for future climate change and also to improve riparian ecosystem health (ICPR, 2015a, 2015b). Between 2000 and 2012 122 km^2 of floodplain had been reactivated along the Rhine (Figure 8.30), which was on schedule for reaching an overall target of 160 km^2 by 2020 (ICPR, 2013a, 2020a). Most of the floodplain that has been reactivated is located within the Upper Rhine and the delta, as these segments also have the largest overall areas of floodplain available for reconnection.

The surface area that is reactivated includes a variety of natural and anthropogenic water bodies, including polders and oxbow lakes, some of which had been artificially cut off for flood control. Between 2000 and 2012, a total of eighty water bodies along the Rhine were reconnected (Figure 8.30). Ultimately, the ICPR has set a goal for a total of 100 floodplain water bodies to be reconnected in order to meet the EU WFD floodwater and ecosystem health targets by 2020 (Figure 8.30). The ICPR target of 100 reconnected water bodies was indeed exceeded by 2020 with some 150 reconnected water bodies, with the largest increase between 2012 and 2020 occurring in the Rhine delta (ICPR, 2020a).

While the measures to enhance lateral connectivity along the Rhine and tributaries are largely on target, measures to increase the geodiversity of river banks (Figure 8.30) is far behind its stated 2020 target of 800 km of linear channel bank (ICPR, 2013a, 2020a). Increasing the structural diversity of channel banks is important to a variety of aquatic and terrestrial organisms that utilize alluvial bank habitat. The edge of alluvial rivers is shaped by a combination of active erosional and depositional processes that result in banks with varying roughness and habitat conditions created by recently deposited sediments as well as older heterogeneous alluvium exposed along eroding banks. Unfortunately, the near-constant concrete revetment and riprap along the Rhine River and larger tributaries has effectively sealed off the channel banks, which represents a great degradation in the overall status of the riparian habitat. While there are a range of "soft" options available to increase the geodiversity of channel banks while maintaining bank stability (e.g., geotextiles, vegetation strategy), little channel bank along the Rhine River has been restored. In 2018, the total length of restored channel banks along the Rhine was a mere 166 km, and further progress is slow because of stakeholder interests, including municipalities and industry who insist on specific "hard" property lines along riparian lands (ICPR, 2020a).

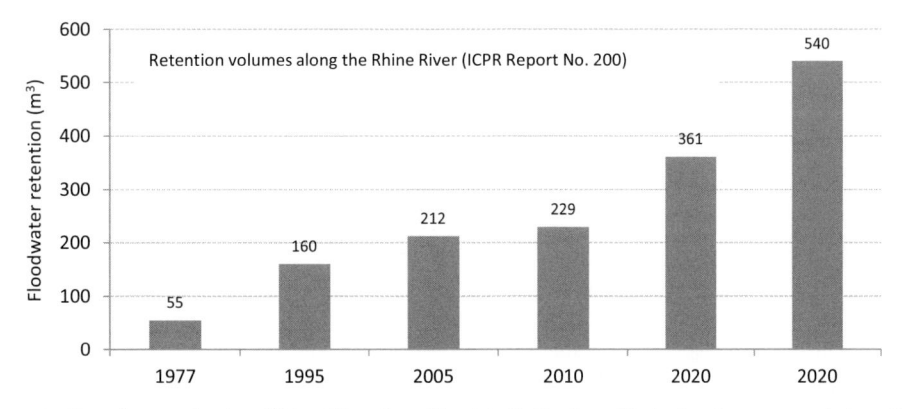

Figure 8.29. Floodwater retention along main-stem Rhine River from Basel to Rotterdam, illustrating improvement since the late 1970s. The two 2020 columns represent minimal and optimal numbers to meet EU WFD targets (i.e., goals for 2020). (Source: ICPR, 2013a.)

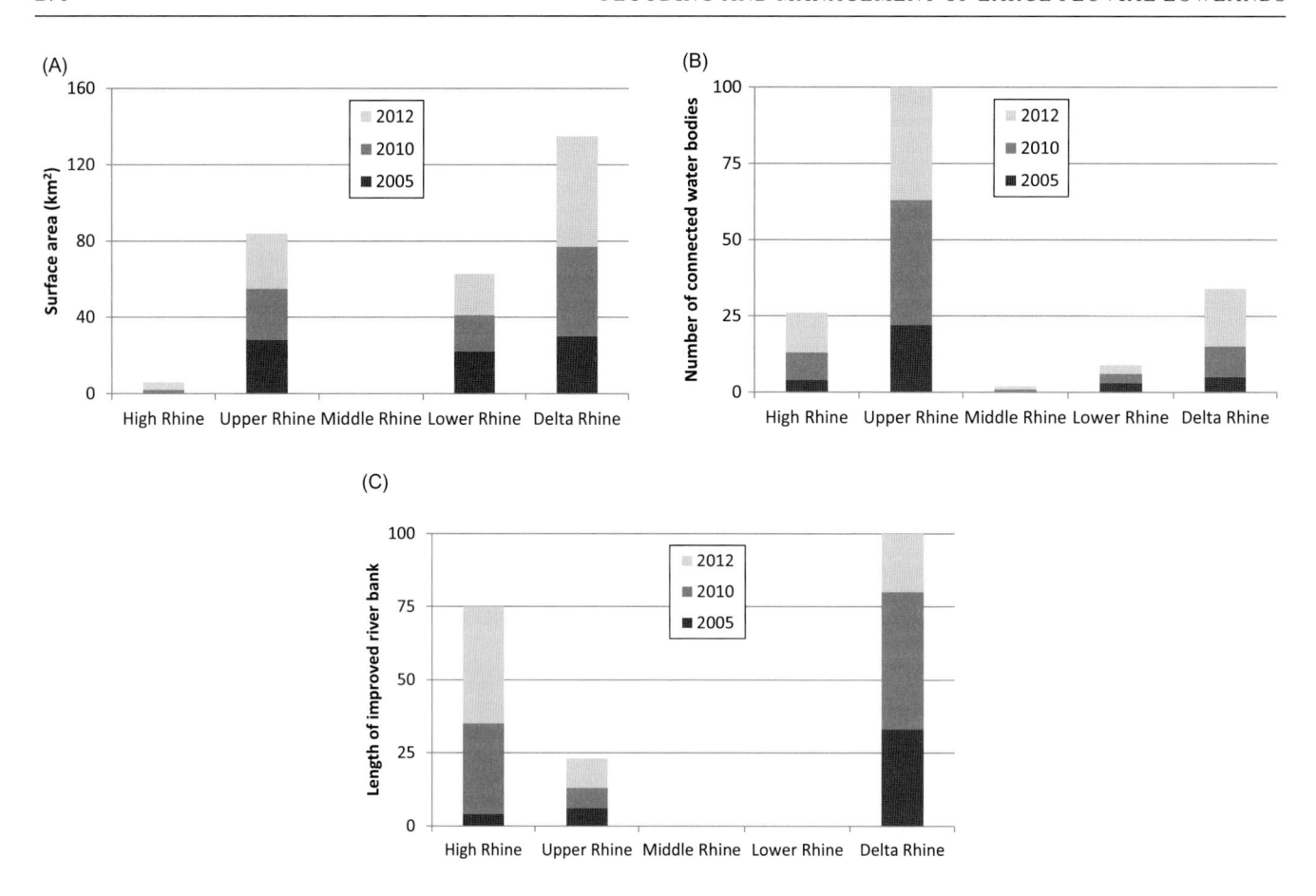

Figure 8.30. Progress in riparian environmental restoration efforts for different segments of the main-stem Rhine River. (A) Increasing size of reactivated floodplain areas between 2005 and 2012. The 2020 EU WFD target for 2020 is 160 km^2 of floodplain waters reconnected throughout the Rhine basin. (B) Number of reconnected water bodies between 2005 and 2012 for different segments of the Rhine basin. The target for the EU WFD 2020 is 100 reconnected water bodies throughout the Rhine basin. (C) The length of improved and restored river banks between 2005 and 2012 for different segments of the Rhine basin. The target for the EU WFD 2020 is 800 km of improved river bank. (Source: ICPR, 2013a)

8.6.1.1 RESTORING SALMON TO THE RHINE

Channel and floodplain restoration is vital to improving the biodiversity of the Rhine basin. A keystone species is the Atlantic salmon (*Salmo salar*), which were historically abundant within the Rhine basin, spawning in headwater tributaries as distant as the Swiss Alps (ICPR, 2015c, 2018). But hyperindustrialization of the Rhine basin and associated habitat decline wiped out the Atlantic salmon by about 1950. In addition to pollution and overfishing, the salmon decline in the Rhine basin was largely caused by obstructions associated with channel engineering of weirs and locks for navigation, which supports one of the world's busiest inland waterways. The Rhine upstream of its confluence with the Main River has a total of twelve locks, with two upstream of Basel. The Main River has a total of thirty-four locks and the Mossell has twenty-eight locks. Additionally, the construction of hydroelectric power plants along Rhine tributaries was especially detrimental to fish migration. In addition to dams and locks within the main-stem river and larger tributaries, there are also numerous weirs and minor impoundment structures within the smaller tributaries that have greatly reduced habitat for salmon migration and spawning.

As of the early 1990s salmon have returned to the Rhine basin (Figure 8.31), which is largely due to restoration efforts driven by the ICPR. In 1991, ICPR launched its Salmon 2000 program to reintroduce Atlantic salmon to the Rhine. To be clear, because of being a keystone species Salmon 2000 is also another way to enhance the overall environmental attributes of the Rhine basin, while also being in accord with the EU Watershed Framework Directive. With the targets for salmon restoration in the Rhine well off the ICPR's Salmon 2000 goal, the program was continued with but a slight change in name; Salmon 2020. While the number of salmon in the Rhine are orders of magnitude below historic populations, since 1990 formal counts of adult salmon have shown a large increase, reaching a high of 800 in 2007. A reason for the return of salmon is due to massive investment, including the installation of fish ladders around locks and dams along the

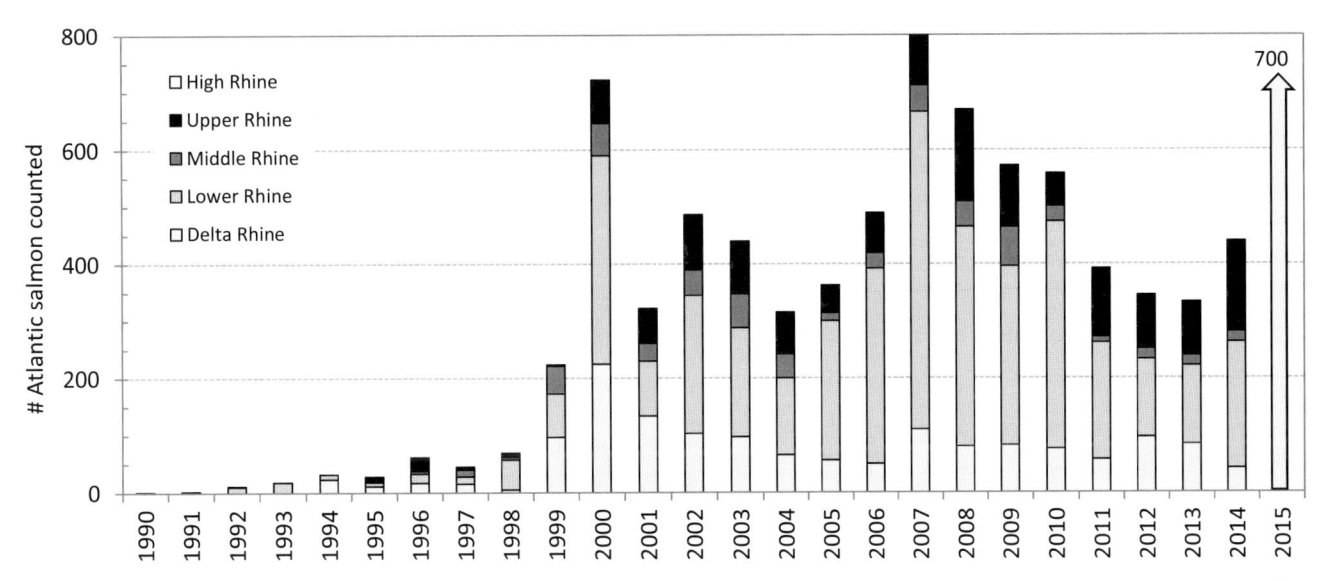

Figure 8.31. Changes in frequency of adult Atlantic salmon (*Salmo salar*) between 1990 and 2015 within different segments of the Rhine basin. Salmon not identified in the High Rhine (count = 0), except for years 2012 and 2013 with a count of 2 and 2 (not in figure). Total salmon count for 2015 was over 700 for entire Rhine system. (Data source: ICPR, 2015c, 2018.)

Figure 8.32. Fish ladder along the Nederrijn en Lek (Rhine) River in the Netherlands at the Hagestein lock. This diminutive structure is responsible for the entire stock of Atlantic salmon making it upstream, and is the first blockage fish encounter along their migration to upstream spawning grounds in the Swiss Alps. The fish ladder (passage) is 400 m long and 10 m wide and winds around the Hagestein lock (background). The "ladder" consists of a series of 10 m long basins with 16 cm high steps (as viewed in photo). (Source: Rijkswaterstaat, October 13, 2004.)

main-stem Rhine and its larger tributaries (Figure 8.32). The largest fish ladder in Europe was completed in 2006 around the hydroelectric dam at Gambsheim in the Upper Rhine.

Variability in salmon counts is attributed to different reporting methods, illegal fishing, and construction projects at hydroelectric power plants. While salmon counts declined after 2007, the completion of riverine construction projects and a new fish passage in 2013 is attributed to an increase in counts. In 2015, 700 salmon were counted in the Rhine, with 175 reported in the Upper and High Rhine that utilized the fish passage around the Iffezheim hydroelectric powerplant. Other fish stocks are also improving, including sea trout, sea lamprey, river lamprey, and shad (ICPR, 2018). By 2017 the number of salmon had again declined, to about 550 (ICPR, 2020a), well below the Salmon 2020 goal. But the vision for a thriving and sustainable wild salmon population in the Rhine is strong, and the program is now enveloped within the new "Rhine 2040" program in accord with recent EU initiatives to remove structural riparian blockages and to increase the number of free flowing river sections (ICPR, 2020b).

8.6.1.2 INFLUENCE ON RHINE FLOOD LEVELS

As a whole, in comparison with the reference year of 1995 flood stages will reduce by 2020 along the entire main-stem Rhine River (see Rhine basin map Figure 3.40), from Basel to Rotterdam (Figure 8.33). The flood stage reductions targeted for 2020 vary by hydrogeomorphic segment, including the Upper Rhine (Basel to Mainz), Middle Rhine (to Bonn), Lower Rhine (to Lobith), and the delta (ICPR, 2013a, 2015a). The minimum average flood stage of the 100-year event will reduce from as little as 5 cm in the Lower Rhine to 17 cm in the Middle Rhine segment. The maximum average flood stage reduction of the 100-year event will be 15 cm in the Lower Rhine to 44 cm in the delta. Some individual reaches of the Rhine delta are expected to have as much as 56 cm of flood stage reduction for the 100-year event. For the high average reduction, flood stage will reduce between 5 cm and 44 cm, with the largest stage reductions again being in the delta.

Flood stage reductions and restoration efforts vary by hydrogeomorphic segment of the Rhine basin. Even along the main-stem Rhine, downstream of Basel, there is much variability in channel and floodplain characteristics before entering the delta. Thus, different approaches need to be applied for varying reaches that respect hydrogeomorphic and environmental differences, as well as the extensive legacy of human impacts.

8.6.2 Upper Rhine Case Study

Flood stage reduction in the Upper Rhine, between Basil (CH) and Mannheim (DE), is predominantly by floodwater retention in the polders and lakes within the embanked floodplain (Herget et al., 2005; Uehlinger et al., 2009; Vorogushyn and Merz, 2013; Kuhn et al., 2016). The storage of floodwaters is designed to not only reduce flood stage, but also to slow down the flood wave (Figure 8.34).

The geomorphic characteristics of the Upper Rhine are heavily influenced by the legacy of meander bend cut-offs, mainly in the nineteenth century. The total reduction in channel length from the infamous Tulla cut-offs was 81 km between Basel and Worms (DE). This resulted in an abrupt reduction in sinuosity and a compensatory increase in channel slope that, as well as with further sediment reductions, initiated a century or so of channel bed incision. After the cut-offs, intensive engineering along the channel and floodplain followed. These activities included bank protection works such as revetment and riprap, groynes to train the channel, as well as the construction of dikes and dams. Subsequent to controlling the channel of the Rhine River for navigation, floodplain development ensued for agriculture, industry, and settlement. The movement of dikes along the channel bank resulted in flooding only being able to inundate a narrow riparian corridor ranging from 1 km to 2 km in width, a great reduction from the natural width of 10–12 km. Following the Treaty of Versailles (1919), France constructed ten dams between 1928 and 1977, further reducing sediment loads and flow variability. The channel bed incision resulted in a lowering of the alluvial aquifer, which when combined with the loss of flood pulse contributed to degradation of shallow riparian water bodies, and subsequent change in terrestrialization of riparian lands with woody vegetation. Thus, by the 1980s the upper Rhine had been severely degraded, resulting in not only a reduction in biodiversity but also an increase in downstream flood risk. The loss of some 130 km^2 of natural floodplain constrained floodwaters to a narrow corridor, resulting in an

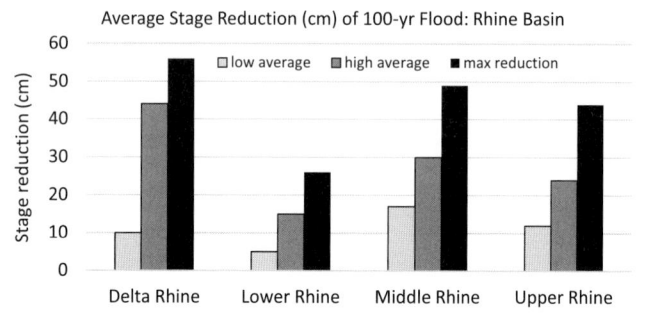

Figure 8.33. Average projected flood stage reduction by hydrogeomorphic province along the main-stem Rhine for the 100-year flood event for 2020 targets after all flood control options have been implemented. (Source: ICPR, 2013a.)

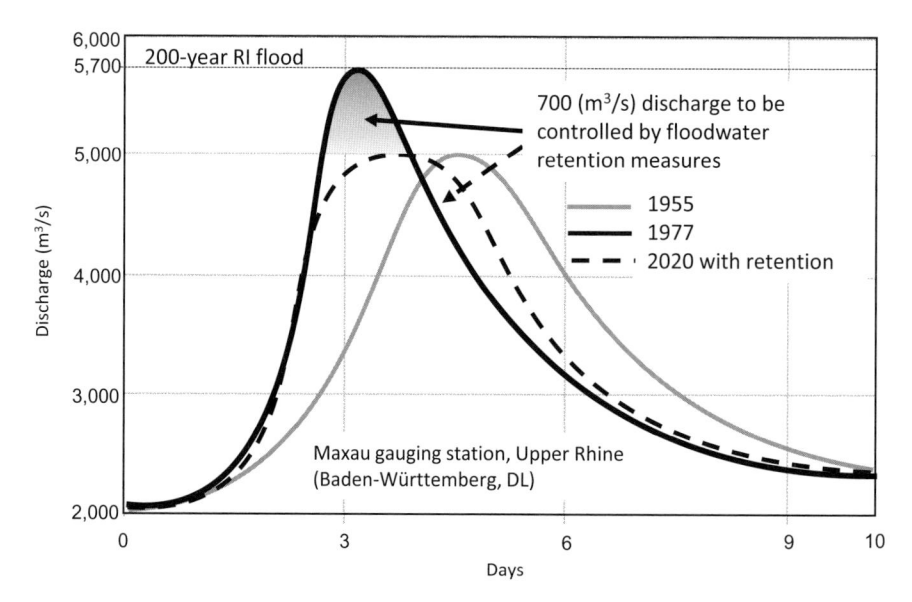

Figure 8.34. Projected changes in flood hydrograph after floodwater retention measures are implemented along the Upper Rhine. (Source: Kuhn et al., 2016.)

increased height of flood peaks. Moreover, because of the shortening of the Rhine channel, the higher main-stem flood peaks coincided with timing of flood peaks from two major downstream tributaries, the Neckar River and the Main River (See Figure 3.40, Rhine basin).

The Integrated Rhine Program is focused along the Upper Rhine River, primarily within the German state of Baden-Württemberg (Kuhn et al., 2016). The program was approved in 1996 to provide flood protection and aims to reduce the flood discharge of the 200-year RI by 700 m^3/s, resulting in a flood peak similar to what existed prior to development (Figure 8.34). The program includes thirteen floodplain restoration measures to reduce flood risk and increase riparian biodiversity. Flood risk is to be reduced by retaining floodwater within polders, dike relocations, weirs, and by emergency operations of power plants (French side). In total, 273 million m^3 of floodwater will be retained between Worms and Basel within the Upper Rhine, combined with the retention areas in France (left bank), Rhineland-Palatinate, and Baden-Württemberg. While not yet completed, some 45% of the basin area is already in place.

A test of the strategy already occurred in early summer 2013, when a devastating flood disaster was prevented with a specific flood stage reduction of 24 cm along the Upper Rhine (Kuhn et al., 2016). While the 2013 event along the Rhine River was largely contained, the cost of a major flood within the dense conurbations of the upper Rhine is estimated to exceed €7 billion, in addition to the loss of life. Thus, the cost of the Integrated Rhine Program, at €1.47 billion (in 2015 within Baden-Württemberg), should be considered a bargain.

8.6.3 Rhine Delta: Room for the River

At the basin terminus, the Netherlands completed its Room for the River program along the Rhine delta in 2015. The program is a major integrated style of flood management comprised of thirty-four individual projects (Rijkswaterstaat, 2018), specific to an intensively industrialized large river delta. Most projects consisted of a combination of options tailor-made for site-specific conditions, flood reduction targets, and environmental restoration goals. The measures are designed to accommodate a peak discharge event of 16,000 (m^3/s).

The Netherlands Room for the River program is designed around several broad approaches for a range of distributaries within the delta (Figure 8.35), including (1) the increased retention of floodwater in upstream locations (mainly within Germany), (2) floodwater retention in floodplain areas to reduce discharge magnitudes, and (3) increased flood discharge capacity of the channel (Wolters et al., 2001). Of the three, most of the thirty-four measures carried out for the Room for the River project administered by the Rijkswaterstaat, the Dutch state water authority, were designed to increase flood discharge capacity. These include measures such as lowering groynes, removing obstacles (e.g., "bottle necks," especially in urban zones), and river channel deepening (Figure 8.2). Because of the expense and uncertainty of benefits, large retention reservoirs were not constructed. Flood retention reservoirs are primarily located upstream in the Lower and Upper Rhine.

8.6.3.1 MEASURES UTILIZED, STAGE REDUCTION LEVELS

While the idea of moving a dike back (dike setback) captured the attention of the international flood science community, dike setback is actually a minor overall component of the Room for the River program. Only four of the individual projects primarily utilize dike setback to reduce flood levels (Table 8.9). This is because the removal and construction of dikes and associated land acquisition is expensive, and the process is often complicated because of local stakeholders. Additionally, hydraulic modeling illustrates that increasing the channel and floodplain discharge capacity was often the most effective way to accommodate larger flood events (Straatsma and Kleinhans, 2018). Several measures were undertaken to reduce flood stage by floodplain storage in oxbow lakes and polders, including side channel creation (Silva et al., 2004). Seventeen of the individual projects utilize floodplain lowering as the primary measure to low flood levels, which with vegetation management and groyne modification are the most effective measures. Lowering the height of channel groynes was found to be both cost and hydrologically effective. The program lowered 462 groynes by 1 m between Nijmegen and Gorinchem. Additionally, forty groynes have been completely removed or replaced by other more environmentally friendly measures, such as "longitudinal dams" that do not impede water flow and thus do not increase flood stage (see Figure 5.20, upper).

8.7 MANAGING EXPECTATIONS: INTEGRATED RIVER BASIN MANAGEMENT IN CONTEXT

Integrated river basin management is a change in the vision and mechanics in the governmental stewardship of essential natural resources. Because integrated flood management explicitly combines flood reduction, environmental enhancement, and stakeholder involvement, the approach has never been about restoration of watershed processes to a "pristine" baseline. It is first – and foremost – a flood risk reduction strategy, and thereafter it is a compromise between environmental enhancement and economic activities guided by societal priorities. Although

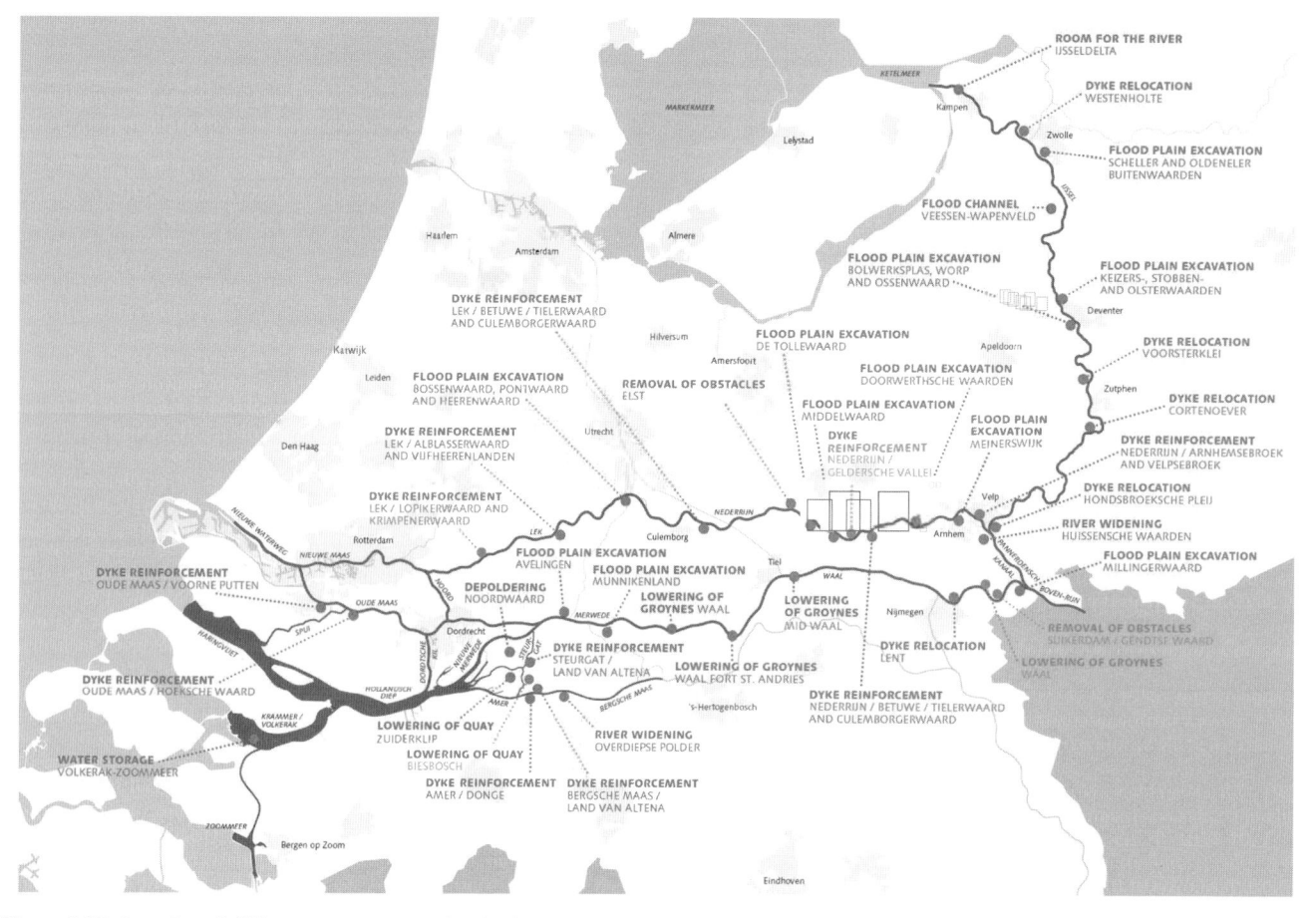

Figure 8.35. Location of different measures associated with the Dutch Room for the River program. (Source: Rijkswaterstaat.)

there remains "room" for improvement (Jonkman et al., 2018), the approach is gaining momentum. The last decade has seen the adaption of many tenets of IRBM approaches applied to a diverse array of watersheds and lowland rivers (Adamson, 2018; Zevenbergen et al., 2018).

The "purest" form of integrated river basin management requires a watershed-scale perspective so that inherent spatial linkages (i.e., longitudinal connectivity) throughout the drainage basin are integrated, including headwaters, transfer zones, and the deposition zones (e.g., Figure 2.1). This implies, for example, that land cover types across basin headwaters are considered from the perspective of how they influence downstream hydrologic, sedimentary, and riparian ecological processes. Within the transfer zone, along major tributary axes and main-stem river valley, channel and floodplain engineering are modified to accommodate larger flow volumes that enhance lateral connectivity and floodwater retention. Within the deposition zone, in addition to creating more space for water storage within the channel and floodplain, engineering for higher sea levels and coastal storm surge is integrated into wetland restoration and urban planning within flood basin environments.

While challenges remain, the Rhine basin is a premier example of an *attempt* at integrated river basin management for a large international river intensively utilized by a range of stakeholders. That is, it is embedded in reality. The context of the Rhine's location is important to consider as regards the specific strategy, and its effectiveness. It should be noted that the Rhine benefits from having robust institutions across all governmental scales (local to international), and is located within an affluent region with extensive scientific expertise. But across many river basins, attempts at "integrated" styles of flood management, while well intended, fall short. There are many examples of one-off environmentally friendly floodplain management measures, such as oxbow lake reconnection or remeandering. While such measures provide limited ecosystem and floodwater retention benefits, they often lack a true basin-scale perspective. This means a rigorous evaluation of the impact of such measures on flood risk reduction has not been factored into its design and broader regional environmental goals. Often such measures are not designed in the context of upstream and downstream measures carried out by other organizations and states. Nevertheless, while such local-scale

Table 8.9 *Room for the River program in the Netherlands: Primary measures to lower flood stage levels and expected minimum reduction*[1]

River KM	Rivers	Location of measure	Type of measure	Minimal required flood stage reduction (cm) by 2020[2]
865	Bovenrijn/Waal/	Rijnwaarden	Floodplain lowering	11
871	Merwwedes	Millingerwaard	Removing obstacles	9
871		Millingerwaard	Floodplain lowering	9
871		Suikerdam	Removing obstacles	8
878		Bemmel	Floodplain lowering	5
882		Lent	Dike relocation	27
897		Afferdensche and Deetsche Waard	Floodplain lowering	6
867		Waalbochten	Groyne lowering	8
887		Midden-Waal	Groyne lowering	12
916		Waal Fort St. Andries	Groyne lowering	8
934		Beneden Waal	Groyne lowering	6
948		Munnikenland	Floodplain lowering	11
955		Avelingen	Floodplain lowering	5
964		Noordwaard	Depoldering (dike relocation)	30
968		Noordwaard	Floodplain lowering	17
871	Pannerdensch Kanaal,	Huissen	Floodplain lowering	8
883	Nederrijn/Lek	Meinerswijk	Floodplain lowering	7
893		Doorwerthsche Waarden	Floodplain lowering	2
898		Renkumse Benedenwaard	Floodplain lowering	18
898		Veerstoep Lexkesveer	Removing obstacles	18
908		Middelwaard	Floodplain lowering	3
911		De Tollewaard	Floodplain lowering	6
917		Machinistenschool Elst	Removing obstacles	5
946		Vianen	Floodplain lowering	6
878	IJssel	Hondsbroekse Pleij	Dike relocation	46
918		Cortenoever	Dike relocation	35
930		Voorster Klei	Dike relocation	29
943		Bolwerksplas	Floodplain lowering	17
947		Keizerswaard	Floodplain lowering	10
957		Fortmonder- and Welsumerwaarden	Floodplain lowering	6–8
961		Veessen-Wapenveld	High water channel	63
977		Scheller and Oldenelerwaarden	Floodplain lowering	8
978		Spoorbrug Zwolle	Removing obstacles	6
980		Westenholte	Dike relocation	15
980		Westenholte	Deepening the summer bed	29

[1] Based on Room for the River (Rijkswaterstaat) from ICPR, 2015b Part A (table: Annex 11-2).
[2] Overall aim is to increase discharge capacity of the Rhine delta to 16,000 (m^3/s). Thus, only targeted water level reduction per measure is included. Measures do not take into account overall floodwater retention volume.

measures may be somewhat patchwork they can be seen as "working with nature" and can provide tangible benefits that are (mainly) a step in the right direction. Incrementally, the combination of many local-scale efforts can have important effects on flood risk reduction and enhancing environmental attributes of lowland riparian floodplains.

In addition to extensive scientific research the Room for the River program has attracted considerable international media attention, including visits of political delegations and civil engineers from large metropolitan areas following major flood disasters. The Netherlands approach that combines increased safety and environmental restoration with spatial planning is attractive to planners and landscape architects. The juxtaposition of the issues of climate change and sea level rise with increased urban delta populations has the Netherlands well positioned to financially benefit from their expertise. Nations with high populations residing within delta lowlands regularly consult with the Netherlands, particularly the Rijkswaterstaat, Deltares, local water boards, and private consultants. The flood-vulnerable nations of Bangladesh, Indonesia, and Vietnam, for example, are adapting a number of measures utilized by the Room for the River program, albeit locally tailored to their unique riparian and coastal environments.

The management of the lower Mississippi continues to rely primarily upon a traditional hard engineering approach for flood control mandated nearly a century ago with the 1928 Mississippi River & Tributaries Project, and the 1965 Lake Pontchartrain and Vicinity, Louisiana Hurricane Protection Project. As of writing they both remain unfinished. Nevertheless, progress is being made, incrementally. Within its great alluvial valley, a number of floodplain connectivity projects have been implemented, including reconnecting oxbow lakes and converting patches of floodplain agriculture into wetlands (LMRCC, 2015; Pongruktham and Ochs, 2015). Many of these activities occur under a guise to improve habitat for fish and water fowl, often for recreational purposes (i.e., fishing and hunting). Governmental organizations such as the USACE and USFWS as well as nongovernmental organizations, especially The Nature Conservancy, stimulate floodplain conservation by purchasing and leasing misused riparian lands, and by funding scientific research. These efforts improve flood pulse dynamics (i.e., lateral connectivity) and associated biogeomorphic processes, crucial for nutrient retention within the riparian zone, reducing nutrient transfer to the coastal zone (Mitsch et al., 2001; Schramm et al., 2009; Twilley et al., 2016). Additionally, over the past twenty years some ~200 notches have been engineered within existing groynes (wing dikes) within the Memphis and Vicksburg Districts, and several in the New Orleans District. The lower surface of the notches is engineered to the low water reference plane, which is defined at a flow duration of 97%. The key reason for creating the notches is to provide flow through secondary channels to reduce sedimentation and infilling (e.g., Guntren et al., 2016), so that vegetated islands do not become part of the floodplain. This approach increases velocity variability (i.e., geodiversity) of side channels and fosters increased biodiversity, such as a higher number of fish species and macroinvertebrates. Maintaining functioning side channels also results in water storage, helpful for flood management (i.e., Buisje et al., 2001; Silva et al., 2001).

Within the Mississippi delta, the new plan to restore coastal Louisiana is at least integrated from its perspective to reduce coastal storm surge flood risk by restoring coastal wetlands. And it considers basin-scale suspended sediment-discharge dynamics for operation of sediment diversion structures and constructing new wetlands. This alone is a big step forward, and lessons learned as a result of the 2005 Hurricanes Katrina-Rita flood disaster are appropriately being applied. Nevertheless, the Mississippi delta program is stymied because of difficulty to get key stakeholders – including industry, navigation, fisheries, and conservation – to agree to a common strategy of implementation (Twilley et al., 2016). For example, the installation of sediment diversion structures, the key measure to conserve and restore coastal wetlands in Louisiana, continues to remain a topic of great debate among fishing,

Figure 8.36. End of the river, where the Rhine meets the North Sea. Closing of the mouth of the Rhine (Maas) River ~20 km downstream of Rotterdam (looking upstream). Closure of Maeslantkering (the great arms) is fully automated and only occurs when a coastal storm surge event is projected to exceed +2.6 m (NAP) in Rotterdam. Operation of the entire structure takes about 6.5 h, from flood warning to full closure. For scale, each arm weighs 6,800 tons and is 237 m in length (longer than the Eiffel Tower). The structure was completed in 1997 and was the final component of the Delta Works flood control project. The Maeslantkering has been activated two times for actual storm surge events, in 2007 and 2018 (Source: Rijkswaterstaat.)

shipping and dredging, and environmental organizations. Thus, the economic and governmental dimensions of integrated management within the Mississippi delta remain woefully incomplete (Hudson, 2018).

The Room for the River program in the Netherlands combines both soft and hard approaches to flood management. Interestingly, while considered progressive the strategy is "integrated" within the existing Delta Works, an internationally acclaimed example of a hard engineering approach to flood management (Figure 8.36). At a cost of €2.1 billion, the Room for the River project was completed on time and under budget. The real value of the program will be appreciated as the environmental targets established in the EU Water Framework are realized and, considering the high population density and economic activities within the Rhine delta, immediately after the next major high-water (flood) event.

9 Lessons Learned: Future Management of Fluvial Lowlands

9.1 LOWLAND RIVER MANAGEMENT IN REVIEW

The science of river management has never been more important than today. Society is intricately connected to lowland river environments long influenced by a range of human impacts, and which are increasingly threatened by different forms of local and global environmental change. Thus the urgent need for robust forms of integrated management to restore and conserve riparian ecosystems while minimizing flood risk.

In previous chapters we examined specific approaches to river and floodplain management that addressed different types of human impacts to fluvial systems. Chapter 9, therefore, synthesizes and summarizes key ideas regarding future environmental management of large lowland rivers (Figure 9.1). While ultimately the purpose of this treatise is forward looking – providing insights into improved river management – an overarching theme is that "you can't move forward without knowing where you've been."

The preceding five chapters examined many instances of river management for flood control and channel engineering for an international range of lowland rivers, especially the Mississippi and the Rhine. An overarching point is that hydraulic infrastructure designed for flood control and navigation has unintended consequences to geomorphology and riparian environments, which requires decades to unfold. Because of an abundance of data and technologies such case studies provide valuable lessons regarding sustainable lowland river management (Table 9.1).

9.2 KEY LESSONS LEARNED

9.2.1 The Continuum of Geomorphic Adjustment and Sequence of Management Activities

Effective flood management is a long-term endeavor that requires multiple management approaches in the context of a changing environment. Management infrastructure utilized in flood control and channel engineering (e.g., Table 5.1) requires time before its secondary impacts on hydraulic and sedimentary processes are ultimately manifest upon the fluvial landscape (Schumm and Winkley, 1994; Gregory, 2006; Hudson et al., 2008). The response of the system to flood control activities occurs along a continuum of geomorphic adjustment, which also degrades channel and riparian environments. The continuum of geomorphic adjustment then drives a sequence of management activities to conserve and restore riparian environments (Figures 5.1 and 6.43).

The single greatest disturbance caused by flood control engineering is river straightening by meander bend cut-offs, which trigger hydraulic, geomorphic, hydrologic, and ecological responses that unfold over varying timescales. A meander bend cut-off results in an immediate formation of a channel knickpoint. Subsequently, channel bed incision and reduction to the slope profile (energy gradient) requires perhaps a decade or two. While the reduction in flood stage occurs soon after a cut-off (at a local scale), the lowering of flood stage is ultimately caused by channel bed incision and increased velocity or shear stress (Winkley, 1977; Harmar et al., 2005; Hudson and Kesel, 2006; Biedenharn et al., 2017). Within two to three decades channel bed incision destabilizes channel banks, thereby resulting in rapid bank erosion and channel widening.

The lesson here is that rivers should never be straightened, as meander bend cut-offs should no longer be used for flood control. In the case of recently straightened rivers, agencies should efficiently plan monitoring and management activities accordingly to prevent excessive degradation. This should include enhanced bed roughness and bank structures to dampen the impact of the channel knickpoint such that it does not migrate upstream. Additionally, bank protection structures should be temporally installed until channel bed incision subsides and the slope profile has stabilized.

In addition to channel geomorphic change, meander bend cut-offs also result in the formation of floodplain water bodies, as each cut-off creates a new oxbow lake (Section 6.4.3). In the context of heavily engineered rivers such water bodies are valuable riparian environments. Within about five decades artificial

Figure 9.1. "Vision" of Room for the River program in the Netherlands and ideal combination of "hard" and "soft" flood control. Herein presented is the case study of Nijmegen, portraying linkages between flood control, urbanization, and nature enhancement. (Source: Rijkswaterstaat.)

oxbow lakes show signs of sedimentary infilling and terrestrialisation, as open water environments transition to wetlands.

The continuum of geomorphic adjustment necessitates a specific sequence of engineering and management operations, which in some instances drives further fluvial adjustment. It is well established that groyne (wing dike) construction results in lateral channel bed aggradation and overall channel narrowing. In the Rhine delta, for example, groyne construction resulted in the (initial) formation of lateral sedimentary bars. Subsequent vertical accretion of fine-grained sediments transformed the lateral bars into new floodplain units. Thus, groynes stabilized a once dynamic channel that previously had numerous active channel bars and islands, resulting in formation of a single thread channel that was narrower, although with increased riparian lands.

While the Rhine provides an excellent historical analogue and an "end-member" engineered river, the Mississippi represents a system in an active state of adjustment. Since construction of groynes along the lower Mississippi in the (primarily) 1970s and 1980s the river has formed hundreds of vegetated islands, representing the trapping of coarse sediment within the riparian corridor (Hudson et al., 2019). While representing channel degradation the creation of new vegetated lands positively enhances the riparian ecosystem. A major concern, however, is that the formation of the islands has reduced space for water, causing increased flood stages (i.e., Wasklewicz et al., 2004; Remo et al., 2018). Decades after groynes were constructed fluvial adjustment of the lower Mississippi continues, requiring further management and engineering operations with potential geomorphic consequences to riparian environments.

Longer timescales – many decades to centuries – are required before floodplain sedimentology and geomorphology is significantly impacted by embankment and associated land reclamation activities (Hesselink et al., 2003). Such impacts are commonly observed in Old World rivers. The trapping of flood sediments within the embanked floodplain results in higher rates of aggradation, and eventually the burial of riparian water bodies. Additionally, over long periods the cumulative impact of numerous dike failure events results in unique anthropogenic features along the dike line, such as water bodies created by breach events (wielen), borrow pit ponds, coarse splay deposits, and sand boils.

Although dikes are effective at reducing flooding most of the time, channel avulsions do still occur, and represent a substantial societal risk. Because dike construction is ultimately responsible for sediment being deposited within a narrower corridor, over long periods it results in formation of a perched channel belt. Such conditions directly increase the risk of avulsion (i.e., increased slope ratio), although dependent upon the flood basin sedimentology. The Huanghe River is an excellent example of sedimentary aggradation within the embanked floodplain, especially juxtaposed against subsiding flood basins that are intensively cultivated and have high population densities. Indeed, the Huanghe is the world's most avulsion-prone large engineered. While sediment loads have greatly reduced over past decades, the avulsion problem along the Huanghe is largely driven by excessive embanked floodplain sedimentation, caused by a legacy of upstream land mismanagement and soil erosion (Slingerland and Smith, 2004; Giosan et al., 2014; Kidder and Zhuang, 2015).

Beyond the dikes, within flood basins, ground subsidence occurs because of floodplain drainage necessitated to support various human land use activities (i.e., settlement and agriculture) and because dikes effectively eliminate sedimentation.

Table 9.1 *Overview of "lessons learned" in relation to management options, with reference to some corresponding sections*

Key lessons learned	Management issues and applications (supporting section #)
Continuum of geomorphic adjustment and the sequence of engineering	The variety of engineering options (Sections 5.1–5.3), riparian vegetation as an unintended geomorphic and environmental consequence (Sections 4.34, 5.4.3, 8.3.6), influence of dikes on sand boils, seepage, and sedimentation (Sections 6.4–6.6), embanked floodplain evolution (Section 6.7), subsidence causes and problems (Section 7.3), the design of flow and sediment diversion structures to restore coastal wetlands (Section 8.5)
Role of sedimentology	Cohesive vs. noncohesive bank material (Section 5.2), coarse sediment replenishment (Section 8.2.3), scale and pattern of embanked floodplains (Section 6.3.2), sedimentology of embanked floodplains (Section 6.5), influence of sedimentology on avulsions (Section 7.4.3), opening of flood diversion structures (Section 7.2), subsidence and sedimentology of flood basins and deltas (Section 7.3), flow diversion structures and Caernarvon marsh loss and gain (Section 8.5)
Urbanization and flood control	Legacy of New Orleans 2005 flood disaster (Section 7.5), subsidence in urban flood basins (Section 7.3), Asian megacities and 2011 flooding of Bangkok (Section 7.5), Rotterdam and floodwater retention (Section 8.6)
Linkages between geodiversity and biodiversity	Channel revetments (Section 5.3.2), impacts of cut-offs on hydrology and flooding (Section 5.4), impact of groynes and development of islands (Section 5.4.3), hydrography of embanked floodplains (Section 6.4), dike breach events and sedimentation (Section 6.5), flow diversion into Lake Pontchartrain (Section 7.2.2), removal of bank revetment (Section 8.2.4), embanked floodplain reconnection (Section 8.3)
Dam removal and reservoir sediment management	Science of dam removal (Section 4.4.3); streamflow, sediment, channel, and ecosystem impacts (Sections 4.3.3–4.3.5), river recovery to dam removal (Section 4.4.4), emerging science of dam removal (Section 4.4.3), changes to reservoir sediment management (Section 4.5)
The palimpsest of lowland floodplains: The past is not history and continues to influence contemporary fluvial environmental processes	Hydraulic gold mining in California (Sections 2.2, 2.4, 7.2.3), Palimpsest of flood management (Section 6.2), creation of dike breach ponds (Section 6.4.3), trouble with avulsions (Section 6.6), dike breach ponds and sedimentation (Section 6.5), Bangkok flooding and historic flood control infrastructure (Section 7.5.2), Cubit's Gap and wetland construction (Section 8.5.2)

In such cases, ground subsidence becomes the dominant geomorphic process within flood basins and coastal wetlands. While perhaps ranging from 2 mm to >100 mm annually (Section 7.3 and Table 7.2), over many decades such rates of ground lowering increases flood risk. This occurs because dikes designed for a specific flood height are overtopped by smaller floods with a higher frequency of occurrence. Additionally, within deltaic wetlands ground subsidence results in salt water inundation and vegetation dieback, directly exacerbating processes of coastal erosion. In such cases, the key management approach developed over the past couple of decades are sediment diversion structures (Section 8.5.3), which harness natural fluvial processes to create new wetlands.

9.2.2 The Role of Sedimentology

While understanding the hydrology of a river system is essential, the success or failure of lowland hydraulic engineering is also dependent upon formal consideration of its sedimentology (Schumm and Spitz, 1996). Importantly, the foundation for massive hard structures, such as dikes, groynes, and revetment, utilizes recent (Holocene) floodplain deposits, and in some cases are anchored to the Pleistocene strata. The stability and proper functioning of hydraulic infrastructure, therefore, is dependent upon the sedimentary matrix in which they are embedded. The key distinction related to floodplain sedimentology is whether the deposits are coarse-grained noncohesive (sandy/gravelly) or fine-grained cohesive (clay/slit) deposits (Section 5.2).

The extent and orientation of embanked floodplains in relation to the Holocene valley is dependent upon sedimentary and geomorphic variability, especially the pattern of coarse meander belts in relation to fine-grained flood basins (Section 6.3). The stability of dikes and flood walls is highly related to the hydraulic permeability and cohesion of underlying deposits (Section 6.6). Dikes constructed within thin, fine-grained top-stratum sedimentary deposits are subject to considerable underseepage during flood events, and flooding from the formation of sand boils (Section 6.6). The pattern of cohesive deposits such as backswamp and clay plugs (channel fill deposits) in relation to dike alignment influences the spatial distribution of sand boil formation, which often occurs along the contacts of clay plugs with coarser deposits. Buried channel belts also provide a conduit for high rates of groundwater flow, which can inundate land areas kilometers from the dike line. Dike integrity is compromised with prolonged underseepage and sand boil activity. Over long time periods the floodplain embankment results in a distinctive sedimentology associated with higher rates of embanked sedimentation (Section 6.5) as well as scour and sedimentation beyond the margins of embanked floodplain associated with dike failure events (Section 6.6). Additionally, along rivers with thick embanked floodplain deposits, the space for floodwater is reduced, which increases flood stage and flood risk.

The importance of sedimentology to dike safety is not limited to underseepage, however, but also to stability of the dike structure (Section 7.5.3). In the case of the New Orleans Hurricanes Katrina and Rita flooding, many of the flood wall failures were not caused by breaching (over the top); they were caused by excessive shear resulting in lateral displacement (pushing) of flood walls. The specific failure mechanism occurred because flood wall foundations were not embedded deeply enough into the subsurface, especially where embedded in organic peat rather than clastic sedimentary materials.

Construction of massive concrete revetment, draped along channel banks to protect sloughing (erosion), also requires consideration of sedimentary deposits. Early efforts by the U.S. Army Corps of Engineers to reduce bank erosion were hampered by the high erodibility of sandy point bar deposits. The lack of sedimentary cohesion of coarser-grained channel belt deposits results in higher rates of bank sloughing upon pore saturation. This was also related to the scour and deepening of the channel bed at the toe of the revetment, ultimately requiring the design to extend from channel bank to channel thalweg. Cohesive floodplain deposits, especially in the delta with low unit stream power, are inherently more stable and provide sufficient resistance to sloughing, requiring much less maintenance.

Changes in channel positions, including channel cut-offs and avulsions, are also dependent upon sedimentary controls (Section 7.4.3). The efficiency of engineering meander bend cut-offs is dependent upon coarse point bar deposits, which are rapidly eroded and allow an efficient capture of the main-stem channel into the pilot channel (Section 5.2). In the case of cut-offs constructed within cohesive backswamp deposits, the thick cohesive deposits are resistant to erosion, requiring much additional (and expensive) dredging and engineering. Historic geomorphic antecedents point to this issue, particularly the number of minor (incomplete) avulsion channels in the Dutch Rhine and Mississippi delta that were unable to capture the full discharge of the main-stem channel because of the resistant peat and clayey flood basin deposits (Section 7.4.3). Indeed, the Atchafalaya River represents a relatively new avulsion of the Mississippi River, which initially developed rapidly because of erodible sandy deposits from older channel belts. In contrast, other large crevasses and failed avulsions within cohesive flood basin deposits of the Atchafalaya were unable to sufficiently incise to capture a large proportion of Mississippi River discharge (Aslan et al., 2005).

For some intensively engineered sediment-starved rivers, such as the Danube and the Rhine, the lack of sufficient quantities of coarse bed material results in significant channel bed incision, which undermines hydraulic engineering structures (Section 8.2.3). In such cases channel bed incision is managed, in part, by replenishing coarse bed material in problem river reaches with sediment dredged from upstream reservoir deposits (e.g., Götz, 2008; Kondolf et al., 2014a; Habersack et al., 2016; Bravard, 2019).

The active sediment load is also a consideration with regard to the suspended sediment load and the design of flow diversion structures, in part to mitigate subsidence (Section 8.5.2). Key to the proper functioning of flow diversion structures is consideration of the suspended sediment–discharge dynamics (Section 7.2.2). The consideration of wash load (silt/clay) in comparison to suspended sand transport is particularly important to consider with regard to the opening of freshwater and sediment diversion structures (i.e., Mossa, 1996). To actually build new land, the opening of diversion structures needs to be timed to optimize the dispersal of suspended sediment across wetlands. This was an important lesson learned in the Mississippi delta with the Caernarvon diversion structure. Hurricane Katrina caused considerable erosion of newly developed wetlands because the marsh vegetation was not anchored into sedimentary deposits. Since becoming aware of this problem, the operation and management of the Caernarvon diversion structure has been redesigned and synchronized with the timing of flood and sediment pulses to enhance land building to offset subsidence and sea level rise (Section 8.5.2).

9.2.3 Institutions of Management, Instruments of Policy

To effectively implement policy the management of large lowland rivers requires coordination of activities across a range of governmental and spatial scales, including large urban areas (Ward et al., 2013). Large rivers require robust federal organizations because of the scale of operations and because of the need to have a "helicopter view" of various activities distributed across different regions and states within the basin. Additionally, because large rivers are likely to intersect multiple nations, robust transnational organizations are needed to formally negotiate competing national interests. In this context, institutions such as the International Commission for the Protection of the Rhine (ICPR) are especially important to facilitate knowledge exchange and coordinate management activities across different states (and languages) within the basin. The ICPR gained its charter in 1815 and is the oldest international river management organization. The strength of the ICPR is illustrated in its way of facilitating coordination of flood management, and also activities such as environmental restoration, emergency response to toxic spills, navigation, and spatial planning. The activities of the ICPR are now largely harnessed to facilitate implementation of the EU Water Framework Directive (2000/60/EC) and Floods Directive (2006/60/EC), an integrated flood management strategy for the whole of the European Union.

In general terms, governmental organizations that oversee flood management activities are implementing a policy vision that stems from societal values. The EU Water Framework Directive and Floods Directive mandates all European river basins attain specific levels of flood risk reduction and riverine environmental health defined by specific targets and thresholds unique to each basin. To meet the WFD objectives each nation is required to create three sequential river basin management plans. The first two river basin management plans were due in 2011 and 2015, and the third is due in 2021. After feedback from the European Commission, the final deadline for nations to comply with WFD objectives is 2027.

While the EU WFD has yet to be completed, the sweeping scope of the policy has greatly increased awareness regarding the linkages between environmental change and flooding. Additionally, the policy has stimulated research in allied interdisciplinary fields, such as improved flood risk analysis, riparian ecosystem health, and public policy analysis with reagard to the role of stakeholders in integrated river basin management. Unfortunately, a similar vision for river management has not developed within the United States, where the practice and policies of river basin management are much more fragmented and vary considerably basin-by-basin, and can fluctuate like a pendulum according to changing priorities of major federal agencies (Hudson, 2018). Such is the case with the proposed drainage of the Yazoo basin wetlands, which would damage riparian hydrology and ecosystem processes to benefit agriculture. The on-again/off-again saga has continued for eight decades (Section 7.2.2.1).

Government management organizations have only recently grasped the implications of the long-term impacts to lowland rivers caused by land cover change and the construction of large-scale hydraulic infrastructure projects. Upon recognition of geomorphic and environmental problems, the design and implementation of new styles of flood management and riparian restoration can require decades. The lower Mississippi and the lower Rhine provide interesting comparison. The 1928 Mississippi River & Tributaries Project, the US government's response to the devastating 1927 flood on the lower Mississippi, has not yet been completed. The Delta Works project, the Netherlands government's response to the devastating 1953 flood on the Rhine delta, was completed by 1997. The Room for the River project, the Netherlands environmentally oriented flood risk reduction plan (partially in response to the high-water events of 1993 and 1995), was initiated in 2007 and completed in 2015 (under budget). And upon completion of such large projects it can require additional decades before the hydraulic infrastructure is tested with actual large flood events. The Maeslantkering, the massive metal arms that close off the Rhine River to North Sea storm surge (Figure 8.36), has only been tested twice by actual events (2007 and 2018).

Importance differences between the Rhine and the lower Mississippi (Table 1.1) as pertains management options should also be noted. The Rhine is considerably smaller than the lower Mississippi, which likely influences the amount of time to develop structural engineering. But the much higher population density and intensity of economic activities along the Rhine, in comparison to the lower Mississippi, results in much tighter margins for management to work with, and actually presents fewer management options than are possible for the lower Mississippi system.

A vital lesson learned with regard to effective implementation of integrated floodplain management is the inclusion of local-scale perspectives that include riparian and urban stakeholders. The Mississippi and the Dutch Rhine provide an interesting comparison of how local-scale perspectives are included in a broader – national scale – plan. The Dutch approach includes robust independent water boards at the local scale. Importantly, the Netherlands constitution mandates that water boards include representation from different local stakeholders (e.g., commercial, agricultural, environmental, residential). This enables the development of a finely tuned flood management strategy within a format that facilitates approaches that prioritize flood risk as well as important societal values.

9.2.4 Urbanization and Flood Control

Because large fluvial lowlands are increasingly supporting large urban populations, of key importance to future flood risk reduction is the rethinking of flood control and urban planning.

While adaptable and integrated styles of governance are required for urban areas (e.g., Ward et al., 2013), the actual physical structure of cities also needs to be explicitly considered in regard to flood management. This includes modifying structural bottlenecks – bridges – to increase flood conveyance, such as with Nijmegen, NL for the Room for the River project. The rapid growth of large cities within some of the most vulnerable deltas requires new styles of management if such populations are going to sustainably reside within such fragile environments. The flood disasters in Bangkok (2011) and New Orleans (2005) provide several lessons regarding future directions in urban planning. Expansion of urban areas should proceed only with knowledge of flood basin sedimentology and subsidence processes, which cannot be assumed of property developers and city planners. Additionally, several approaches adapted in the Netherlands, particularly in Rotterdam, represent promising directions toward the design of sustainable urban flood management.

An efficient system of urban canals to move floodwaters out of the city is as important as fortifying dikes to prevent floodwaters from moving into the city. This includes removal of urban runoff derived from rainfall. This lesson was very clear in the review of the devastating Bangkok flood of 2011 (Section 7.5.2). Ditches need to be viewed as vital hydraulic infrastructure, and maintained such that rubbish and vegetation do not obstruct floodwater drainage. Additionally, urban canals in flood basins should never be gravity drained (i.e., free flowing), but instead should be impounded with a flood wall and discharged into flood basins by pumps during large events. While this became clear as regards the 2011 Bangkok flood, it was excruciatingly obvious in the case of New Orleans. Four major canals intended to discharge floodwaters into Lake Pontchartrain were merely gravity drained, thereby becoming conduits for floodwaters to move into the city because of the record height of the storm surge (Section 7.6).

Precision groundwater management is required in urban delta areas to minimize and manage land subsidence (Table 8.2), a delicate operation because of the concern with exacerbating flood risk (Minderhoud et al., 2020). The key strategy to manage subsidence in urban areas is to maintain higher groundwater levels to both reduce oxidation of organic-rich soil and peat layers, and to maintain positive pore pressure to reduce compaction. Such measures require a network of drains and pumps, as well as meticulous monitoring. One strategy to reduce land subsidence is to directly route runoff (if clean) into underground aquifers, such as practiced in Shanghai to manage the "subsidence crisis."

A key lesson learned over the past decades is that the philosophy of urban flood control has been misguided. Instead of "fighting" water, cities should "live with water." This means that the urban design and city structure should develop ways in which water can be stored and integrated within the city structure. Along the Dutch Rhine, Rotterdam (Figure 9.2) has implemented innovative strategies designed to reduce its flood risk (City of Rotterdam, 2015). The measures include fortifying dikes and storm surge barriers, but also modifying urban space to act as a "sponge," which includes surface storage, underground car parks, infiltration zones, and green spaces. The idea is that the urban sponge temporarily stores water within the city during the height of a flood or large rainfall event, and then slowly releases the water into the Rhine River after passage of the flood crest. Key among these measures are water plazas – large sunken areas normally utilized for recreation but which can store vast amounts of urban runoff. Additionally, large underground parking garages are designed to store massive volumes of floodwaters. Lastly, green roofs also help to manage local urban flooding from direct pluvial events. Rotterdam has an ambitious goal of being "climate proof" by 2025, which includes being resilient to urban flooding caused by riverine or pluvial events. Among a range of measures, the city has added 185,000 m^2 of new green roof space and constructed 10,000 m^3 of water storage structures. Other large delta cities should carefully review the example of Rotterdam to discern which measures could also be employed to their situation.

9.2.5 Linkages between Geodiversity and Biodiversity

A key reason why integrated flood management represents such an advance over older traditional forms of flood control is that it explicitly acknowledges linkages between geodiversity and biodiversity. Riparian environmental management has not always included geodiversity within its management goals and targets. Facets of the concept, however, were included in earlier water resource management policy (1971 U.S. Clean Water Act) that requires maintenance and restoration of "physical integrity" as much as biological integrity (Graf, 2001).

Dimensions of fluvial geodiversity important to biodiversity include hydrologic, sedimentologic, and floodplain topographic variability (Sections 3.4–3.7). While these forms of geodiversity share some interdependence, each has a distinct role with regard to the maintenance of ecosystem processes (NRC, 2002). Because some flood control engineering fundamentally degrades the natural flow regime by reducing the extent and duration of inundation (Sections 5.4 and 6.3), vital riparian ecological and biogeochemical processes are diminished as related to hydrologic connectivity. This was observed for the lower Mississippi with regard to the impacts of cut-offs and dike embankment, and

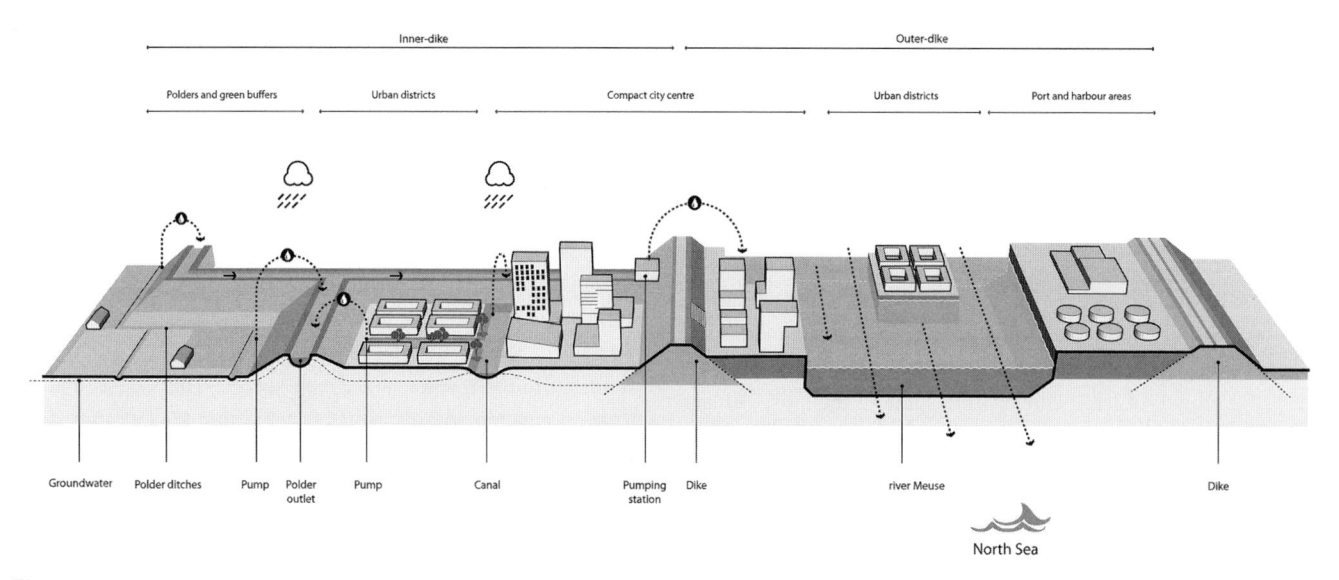

Figure 9.2. Integrated urban approach to flood management within a sinking delta city, the Rotterdam approach to urban flood management along the Rhine delta, Netherlands, Netherlands, which includes space for storage of surface waters and groundwater management to minimize subsidence. (Source: City of Rotterdam Climate Initiative, City of Rotterdam, 2015.)

applies more broadly to other lowland rivers engineered for flood control. The restoration of flood pulse dynamics to enhance riparian biodiversity, therefore, is an acknowledgment of the importance of this key facet of fluvial geodiversity. Indeed, the setback and removal of dikes to increase the area of inundation is a fundamental approach within integrated flood management and is effective from the standpoint of increasing riparian biodiversity and ecosystem services (Section 8.3, Table 8.5).

While the impacts of some channel engineering measures to river environments, such as cut-offs, have long been realized, the impact of channel groyne to flood hydrology and riparian biodiversity has more recently been appreciated. Groynes directly increase flow resistance, reducing channel velocity that results in deposition of coarse bed material, therein trapped within groyne fields. Ultimately, this results in the formation of lateral bars and, eventually, vegetated islands. Groynes are very effective at channel narrowing and the development of hydraulic conditions that improves channel navigability. But groynes also directly result in the terrestrialization of aquatic environments (Section 5.4.3). While the complete removal of groynes is problematic (and unlikely) because of the importance of navigable rivers for commerce, there is now a realization that groynes were overbuilt along many rivers, being too densely spaced and constructed too high. Thus, groyne lowering is a key measure of integrated flood management strategies to reduce flood levels (Section 8.2.2). The measure is also important because coarse sediment trapped within groyne fields is re-assimilated into the active bed load, which is essential to rehabilitation of the lower reaches of sediment-starved rivers.

In the context of flood basins, an important approach to flood risk reduction and nature enhancement are flow diversion structures (Table 7.1). The opening of a spillway or diversion structure simulates a flood pulse, a vital ecologic disturbance. Aside from sedimentary deposits, such operations also abruptly alter water quality, which can impact aquatic fauna and flora. In coastal flood basins the freshwater pulse results in an abrupt reduction to salinity levels that can temporarily stress benthic organisms, such as shellfish, while other organisms benefit from the supply of nutrients (Sections 7.2.2 and 8.5).

Effective management of flow and sediment diversion structures to create new wetlands requires consideration of a number of specific criteria that relate to basin-scale and local-scale controls. Sediment diversion should always take into account the inherent evolutionary character of deltaic wetlands as represented within the "delta cycle," cognizant that each delta is unique.

Of key importance to managing diversion structures is consideration of suspended sediment dynamics, especially because of the hysteresis phenomenon between discharge and suspended sediment load. Hysteresis is especially important to consider along large rivers because the timing of sand transport varies in comparison to the timing of wash load (silt/clay), with the latter peaking perhaps weeks before the flood crest (Section 7.22). This lesson was blatantly obvious after the 2005 Hurricanes Katrina and Rita impact in the Mississippi delta, where it was revealed that the failure to prioritize sediment deposition resulted in new wetlands forming without effective sedimentary substrate to withstand coastal storm surge events (Section 8.5.2). The high storm surge and wind energy, therefore, resulted in substantial erosion of the newly created wetlands

Lessons learned from examination of subdeltas created by historic artificial water and sediment diversion in coastal

Louisiana documents that new deltaic wetlands can be created rapidly, with substantial land gain in two to three decades. The formation of Cubits Gap (in 1862) and Wax Lake Outlet (in 1955), for example, revealed a dramatic increase in new land creation (Section 8.5.2). Information gleaned from sedimentary analysis of historic and modern breach events is being utilized to improve the design of new artificial flood diversion structures with the purpose of creating new wetlands to reduce coastal flood risk. Such structures represent hydraulic infrastructure vital to flood and environmental management across large fluvial lowlands that can reduce future urban flood risk. In particular, they take into consideration the suspended sediment dynamics within an evolving deltaic landscape, which means that as land area grows the operation of the structure needs to be modified to accommodate a changing environment. This crucial base of knowledge can be (prudently) applied to other sinking deltaic landscapes (e.g., Syvitski et al., 2009; Giosan et al., 2014; Minderhoud et al., 2020), especially the Asian mega-deltas that are rapidly being degraded due to upstream sediment reduction and downstream subsidence within coastal deltaic flood basins (Table 7.2, Sections 7.3.2 and 8.5).

9.2.6 The Palimpsest of Lowland Floodplains: The Past Is Not History

Individual fluvial events and anthropogenic activities are "draped" atop older floodplain surfaces, which results in each large fluvial lowland being a unique palimpsest (Section 6.6.2), a "stratigraphic column" of geomorphic events and management ideas that evolves through time (Gregory, 2010; Hudson and Middelkoop, 2015). While many new ideas and techniques have been employed, historic management approaches embedded within the floodplain become the active surface in which new strategies are implemented.

The impacts of historic geomorphic changes and management activities require (lag) time to develop, and often unfold in unanticipated ways that require further management operations. Many such changes take on new importance and roles from the perspective of the overall suite of processes and environments associated with lowland fluvial settings.

The impacts of the anthropogenic levee breach at Cubits Gap (Section 8.5.2) to the creation of new Mississippi delta wetlands is a prime example of a historic – legacy – event that has functioned and evolved over many decades, and which is important to contemporary and future deltaic environments (Wells et al., 1983). Historic – anthropogenically influenced – subdeltas exist along many large deltas heavily impacted by humans, such as the Huanghe delta and the Rhine delta (Section 6.6.3).

Historic dike breach events result in a geomorphic, sedimentary, and hydrologic imprint upon the fluvial landscape that has ecological importance (Sections 6.4 and 6.5). The dike system along the Dutch Rhine, for example, includes numerous wetlands and ponds (wielen) formed centuries ago by dike breach events. While wielen are relict hydro-geomorphic features that would not exist without embankment (dikes), they provide essential aquatic habitat to contemporary (and future) riparian ecosystems. Additionally, such anthropogenic water bodies are important to recreation and enable humans to "experience" nature, providing a societal link with hydraulic infrastructure and historic environmental events.

Along rivers that have been under human influence for long time-scales, relict infrastructure associated with former settlement, transportation, or river engineering activities continue to influence modern floodplain processes. Such features are often abandoned, and commonly become partially buried by channel bed aggradation or floodplain accretion. These fixtures effectively become permanent components of the floodplain. Old channel groynes, for example, trap bed material and become "welded" into the floodplain after subsequent fine-grained overbank deposition and vegetation succession. Old flood control dikes, especially low dikes formerly constructed by farmers or other local entities, often remain on the floodplain and influence sedimentation and floodplain hydrology during smaller flood events. Along some rivers, considerable embanked floodplain aggradation occurs between low (summer) dikes, burying wetlands, altering floodplain soil and sedimentology (Section 6.6), and resulting in appreciable "legacy" sedimentary deposits (Middelkoop, 1997; James, 2013; Kidder and Zhuang, 2015).

New flood and environmental management strategies designed for specific rivers to address different forms of global environmental change (e.g., climate, sea level, land cover) must directly contend with such legacy deposits. In the case of the Room for the River program in the Netherlands, embanked floodplain (legacy) deposits were scraped away to increase floodwater retention (Section 8.3.3). Thus, although the legacy deposits may have been removed, the removal of such deposits is a modern management approach that was conceptually (and temporally) "draped" upon the older management approach represented by historic loodplain embankment.

9.3 LOOKING FORWARD: CONCERN AND OPTIMISM IN INTEGRATED FLOODPLAIN MANAGEMENT

As the concept of integrated river basin management has become firmly established in the twenty-first century, lessons learned from decades – and centuries – of trial-and-error experiments with environmental management and flood control provides causes for concern, and reason for optimism.

9.3.1 Causes for Concern

Despite a sophisticated knowledge of lowland fluvial processes in general, several specific concerns exist with regard to the conservation and restoration of environmental functions, while reducing human flood risk. These include (1) global environmental change, (2) data gaps, and (3) the fortitude of governmental institutions that implement management.

9.3.1.1 GLOBAL ENVIRONMENTAL CHANGE

Global environmental change is both an overarching and specific concern regarding future management of lowland rivers. The three key facets of global environmental change include climate, sea level, and land cover change. All three occur in the context of a rapidly growing population, a population increasingly located within urban settings along coastal margins, including half a billion within delta megacities (Best and Darby, 2020). A major concern is that such populations are increasingly vulnerable because of being concentrated within areas having inadequate flood control infrastructure, in addition to subsidence increasing flood risk.

Climate change presents new challenges to river and floodplain management and the steady march toward a 2°C global rise in temperature by 2050 is fraught with concerns. Climate change is manifest in several ways, including changes in annual precipitation totals, seasonal variability in precipitation (and mechanism), winds and temperature and evapotranspiration. Such changes can ultimately enhance or buffer other forms of environmental change across fluvial lowlands. While the large decline in Mekong River basin sediment flux is mainly attributed to upstream dams, changes to tropical cyclone activity over the past decades have also significantly reduced coastal sediment flux – amplifying the problem of upstream sediment trapping in reservoirs (Darby et al., 2016). Because large river basins extend across varying zones of Earth's general circulation, climate change needs to be considered from a watershed perspective. The pronounced latitudinal zonation to climate change especially has implications to north–south aligned watersheds. Climate change occurring over basin headwaters, for example, impacts fluvial lowlands differently than climate change occurring directly over flood basins and deltas. This issue represents a massive management challenge to the Indus River basin. With forecast wetter conditions in upper headwaters and drier conditions in the lower valley and delta, effective reservoir management will be essential to continued agricultural productivity (Laghari et al., 2012).

Most of the attention to climate change and flood risk concerns potential problems with too much water. But drier conditions are also a problem to lowland river management, especially during summer months. Subsidence rates in drained wetlands increase during prolonged dry conditions because of oxidation of organic material (Brouns et al., 2014). Dikes constructed of sedimentary deposits with high organic content and peat are especially vulnerable to reduced rainfall because of differential shrinkage of the very material that comprises the dikes (i.e., peat, organic soil). This can result in the development of cracks in the dike structure, perhaps leading to failure even during low flow conditions (Section 6.2.2). Such was the case in 2003 in the Netherlands when an organic (peat) dike failed during a summer drought (van Baars and van Kempen, 2009). Fortunately, the failure provided useful management lessons for the Rijkswaterstaat and local water boards, which have since been actively applied, including during the record extreme summer drought of 2018.

Lowland subsidence is also a major concern because it increases relative sea level rise, which thereby increases human vulnerability to large storm surge events, as well as a myriad range of other hydrologic problems. Across many humanized lowland river deltas relative sea level rise greatly exceeds absolute sea level rise, a fact that is usually not well integrated into coastal–deltaic management plans. In 2009, the US Congress mandated that federal coastal flood defenses be designed to account for IPCC sea level rise projections, calculated at 3.2 mm/yr in 2018. But subsidence rates within the Mississippi delta is often five times that of global sea level rise, ranging from 5 mm/yr to up to 28 mm/yr. Frequent updates to (absolute) sea level rise projections by the IPCC continue to raise the probability that a full 1.0 m rise in sea level by 2100 is a very real possibility (Oppenheimer et al., 2019). This will have substantial consequences across fluvial–deltaic landscapes (Church et al., 2013), especially those with high populations and utilized for intensive food production (Cruz et al., 2007; Redfern et al., 2012; Best and Darby, 2020: Minderhoud et al., 2020).

In addition to sea level rise and coastal storm surge by large events, increased salinization of shallow coastal groundwater aquifers is also of concern. Much of the Netherlands' highly profitable and sophisticated greenhouse food production system is located within a subsiding peatland, which is already below sea level and immediately adjacent to the coast, representing an imminent management problem. The salinity of groundwater is rising, especially in months when pluvial inputs into the freshwater aquifer are below average (i.e., Mazi et al., 2013). An additional facet of coastal subsidence and sea level rise is that saline water is able to flow many kilometers into the delta hinterland via a network of canals and irrigation ditches, damaging crops and reducing yields. This is a pressing concern in the Asian mega-deltas where the production of rice and aquaculture relies upon increasing amounts of quality fresh groundwater (Redfern et al., 2012; Erban et al., 2014; Best and Darby, 2020).

Conversion of natural vegetation to agricultural and urban land uses is also a form of global environmental change

(Foley et al., 2005), and the encroachment of agriculture into lowland river valleys is particularly worrisome. While European and North American rivers have been herein emphasized because of the history and intensity of hydraulic engineering and especially floodplain embankment, within a couple of decades a much larger proportion of material on this topic will surely come from lowland rivers in Southeast Asia, South America, and Africa because of worrisome land cover change projections (e.g., Scanlon et al., 2007; FAO, 2017; UN WWDR, 2020). An increasing amount of agricultural lands in these regions will require irrigation that will rely upon hydraulic infrastructure linked to dams and groundwater pumping.

Expansion of large-scale mechanized agriculture across these regions will likely not ignore the tremendous fertility provided by lowland floodplain soils, which therefore necessitates construction of dike systems and creation of embanked floodplains. Hopefully, responsible organizations will consider the potential for riparian environmental degradation and increased flood risk herein reviewed in this treatise. History, unfortunately, does not suggest they will. It is worth noting that even after the Great Flood of 1993 on the Missouri and Mississippi Rivers (Sections 3.2.1.1 and 6.6.3.3), within a decade new settlements (suburbs) and floodplain agriculture had already encroached upon the flood-ravaged riparian lands (Pinter, 2005), illustrating that development *always* follows the dike (White, 1945).

The main cause of the global reduction in sediment yield is the proliferation of dam construction that occurred over the twentieth century (Section 4.2) (Biermans et al., 2011). Some 1.2 trillion tons of sediment are stored in reservoirs (Syvitski et al., 2009). The tremendous reduction in sediment transport to the coastal zone has consequences to lowland fluvial systems. Many large deltas simply do not have enough sediment to be maintained in view of (even moderate) sea level rise projections (i.e., Figure 1.4) (Blum and Roberts, 2012; Giosan et al., 2014; Best and Darby, 2020). While some deltaic land construction approaches can be locally effective, there is not enough sediment to be distributed across the entire delta plain. Thus, government organizations charged with managing large deltas, such as the Mississippi, Mekong, and Danube, should understand that the consequences of reduced sediment supply and rising sea levels will be delta drowning, resulting in wetland degradation and a loss of ecosystems, and a substantial disruption to associated human and economic activities (Best and Darby, 2020).

Despite what is already known about the adverse impacts of dams to sediment loads and fluvial–deltaic environments, dam construction is occurring at an alarming pace. New dam construction is especially occurring in Southeast Asia and South America, and more limited in sub-Saharan Africa, in settings where much less is known about the fluvial systems (Latrubesse et al., 2017; Zarfl et al., 2019; Huđek et al., 2020).

An overarching position of this treatise is that the extent and scope of hydraulic infrastructure embedded across Earth's watersheds *is* global environmental change. For some large fluvial lowlands the problem of reduced sediment flux caused by upstream dams must be offset by other forms of hydraulic infrastructure, including dikes and drainage to reduce flooding caused by ground subsidence and sediment diversion structures to build new wetlands (Giosan et al., 2014). To effectively manage large fluvial lowlands requires attention to surficial forms of global environmental change as much as atmospheric forms.

9.3.1.2 INFORMATION GAPS

In view of the rapid pace of global environmental change it is crucial that managers have a thorough knowledge of how specific types of hydraulic infrastructure will perform in relation to a dynamic range of fluvial processes.

Globally, most streamflow and sediment is discharged from rivers in developing nations in the tropics. But the paucity of fundamental monitoring to analyze fluvial systems has long been a concern. This is not only a problem for river basins in developing nations, but also with the exploding dam construction in southeastern Europe (Chapter 4). This data gap relates to the full range of physical processes, especially adequate hydrologic (i.e., stage, streamflow) and sediment load data, as well as biological monitoring. The data gap will become a larger problem in future decades as large lowland tropical rivers become increasingly relied upon for navigation and floodplain agriculture, settlement, and industry. Without such crucial information fundamental questions regarding the impact of climate change on fluvial processes cannot be rigorously assessed. Such topics include, for example, whether climate change results in channel bed incision or channel bed aggradation, the height to which new dikes should be constructed for specific levels of flood safety, the time span for rivers downstream of new dams to be impacted, as well as how hydraulic infrastructure impacts specific facets of aquatic and riparian ecosystems (i.e., changes to benthic habitat for macroinvertebrates, nutrient sequestration, fish spawning in floodplain water bodies). While the problem is more acute in developing nations because of their greater vulnerability to global environmental change (Best and Darby, 2020), it also looms large in developed nations. The dam construction boom in the Balkans is occurring with very scant pre-dam, dam construction, or post-dam environmental monitoring (Huđek et al., 2020). Thus, it will be challenging – and in some cases impossible – to understand the impacts of the dams on the environment or to assess the potential for the structures to be adversely impacted by future environmental change. It will also be very difficult to assess whether such rivers in southeastern Europe are able to comply with the EU Water Framework Directive. And with the acknowledged importance of long-term data sets for analyzing the impacts of global environmental change, it raises concern about removal of streamflow and sediment gauging stations with long-term records in the United States (USGS, 2018).

Of key importance to the design of hydraulic infrastructure is floodplain sedimentology, which represents an enormous data gap. Few large lowland rivers have adequate subsurface sedimentary data. Such data is obtained from floodplain corings, and commonly includes indices of grain size, organic content, mapping and dating of old channel belts, as well as other geophysical proxies. While ground-penetrating radar and satellite imagery analysis increasingly expand in capability, the primary way that sedimentologic data is obtained is through field-based floodplain coring. Such fieldwork is labor intensive, expensive, and requires expertise and training.

Government organizations involved in the design of flood control systems should strive to obtain as much sedimentary data as possible. For large river floodplains and deltas this requires sustained initiatives to collect (and sufficiently analyze) a sufficient number of corings in key geomorphic environments. This point was recently emphasized during the design of the massive new flood control project on the lower Sacramento River, where it was revealed that the old system of dikes had been constructed with little knowledge of the underlying sedimentology (USACE, 2015). The sedimentary database for the Rhine delta in the Netherlands includes detailed descriptions of over several hundred thousand corings, the most extensive of any large delta, facilitating a high level of research that enables different theories about human and environmental impacts on lowland floodplains to be rigorously tested (Berendsen and Stouthamer, 2001; Middelkoop et al., 2010).

9.3.1.3 INSTITUTIONAL FORTITUDE

Not only is the issue of data gaps a problem, data must be intelligently harnessed within a framework of effective policy and management, an issue of urgent concern given global environmental change projections. Flood control infrastructure is ultimately designed and implemented across a range of governmental scales, from local to federal to international (in some cases), which means that specific tasks are assigned to different entities. The exchange of information and assigned tasks must be appropriately suited for the governmental entity. Different models of governmental structure for flood control exist and vary from top-down federal controlled to a more distributed structure, whereby local (and regional) stakeholders have important input into the design and operation of floodplain management and associated hydraulic infrastructure.

In developing nations flood management may initially be designed and financed by federal sponsors (often with international development aid), but thereafter government institutions often lack adequate resources to *sustain* monitoring and maintenance programs that are ultimately essential to the success or failure of a flood control project (CIRIA, 2013; World Bank, 2017). Additionally, the transfer of projects to different governmental entities can result in a number of issues that makes the system less effective. This can result in periods in which no entity assumes responsibility for project finalization, which results in increased vulnerability to flood events. Additionally, different standards in monitoring can result in poor information transfer and communication about the status and operation of infrastructure. A failure to plan and finance long-term maintenance can result in flood control systems degrading over time, resulting in their ineffectiveness for flood events at specific design frequencies. These issues were paramount in the devastating Chao Phraya River flooding of Bangkok in 2011 (Section 7.5.2). However, they were also apparent in the 2005 flooding of New Orleans by Hurricanes Katrina-Rita (Section 7.6). In this case the City's flood control infrastructure had not met the standards of the Lake Pontchartrain and Vicinity Hurricane Protection Project, and much was mismanaged (GAO, 2005, 2006b; Rogers, 2008).

To ensure flood management systems are effective over longer timescales, specific maintenance plans should be included in the initial project design. Further, it is key that specific maintenance and monitoring operations are assigned to specific entities prior to project construction, and, if possible, such roles should be embedded into the actual regulations and legal framework concerning the flood control project (World Bank, 2017). The nexus of these issues (reviewed above) are magnified in developing nations, especially within large lowland rivers and deltas in the tropics and subtropics, particularly rapidly growing Asian megacities (Section 7.5) (Cruz et al., 2007; Best and Darby, 2020).

Highly developed nations can also lack the institutional fortitude to carry out effective flood management in view of global environmental change. The United States has extensive scientific expertise and overall strong governmental institutions associated with designing and maintaining flood control projects, but changes in the vision of both environmental and flood risk policy is causing the United States to fall behind other regions, such as the European Union. A key example was the recent (2017) overturning of Executive Order 13690 (2015), which would have moved the United States toward an "integrated" EU-style of flood management by adapting higher flood safety standards and enhancing riparian environments inline with modern climate change science (Hudson, 2018). The failure to implement EO 13690 is especially shortsighted given the relatively recent flood disasters generated by tropical cyclones, including flooding of the Houston region (Hurricane Harvey in 2017), New York and New Jersey (Hurricane Sandy in 2010), and New Orleans and the Gulf Coast (Hurricanes Katrina-Rita, 2005) (Section 7.6), among others. Currently there is uncertainty about the future direction of flood management policy in the United States A clear position on flood management and global environmental change is needed, in part because flood control approaches in the United States have historically been adapted by other nations. Organizations such as the U.S. Army Corps of Engineers generously accommodate and actively facilitate international knowledge exchanges through expert workshops and visits, in addition to formal memorandums of understanding with other nations (Hudson, 2018).

9.3.2 Reasons for Optimism

It is always preferable to finish on a positive note. While there are reasons to be pessimistic as regards the future of sustainable management of large fluvial lowlands, there are also many "silver linings," and even some reason for optimism.

9.3.2.1 INTERDISCIPLINARITY

An important development since the new millennium is the increasing interdisciplinarity of floodplain science practice and research, especially at the interface of ecology, hydrology, geomorphology, sedimentology, and hydraulic engineering. The crux of many problems associated with flood control along lowland rivers is that prior approaches were very disciplinary, and especially driven by engineering. The complexity and diversity of flood and ecosystem processes along large lowland rivers require an interdisciplinary perspective to implement effective management. Following the epic work of Junk et al. (1989), and others, on the "flood pulse concept," aquatic biologists began to more explicitly consider hydrologic and geomorphic dynamics associated with active floodplain processes along large lowland rivers. In addition to an interest in understanding how physical floodplain processes impact aquatic ecosystems, recognition of the impact of flood control activities on biogeochemical cycling, especially carbon, nitrogen, and phosphorus, is a key point of intersection between floodplain ecologists and geomorphologists that look to restore such vital riparian processes (e.g., Mitsch et al., 2001; Nienhuis, 2008; Schramm et al., 2009).

Geomorphology has been at the forefront of interdisciplinary approaches to environmental floodplain restoration. As a discipline, its temporal scope, which views the past while looking to the future, makes it well poised to contribute to interdisciplinary approaches to integrated flood management. The contribution of geomorphology spans a range of floodplain processes and environments, ranging from flood risk and avulsion processes to contemporary hydro-ecological processes. Ecosystem restoration of floodplain water bodies, a common approach to integrated flood management, requires an understanding of sedimentary dynamics, hydrology, and evolutionary sequence of riparian water bodies and wetlands. Flood sedimentary processes ultimately drive physical changes to floodplain water bodies that impact the type of biota that can be supported. Thus, geomorphologists provide information essential to ecologists examining the suitability of a lake environment to support specific aquatic fauna as well as how to engineer lake restoration, which may require weirs or pumps to provide adequate exchange of water and nutrients. Engineering flow and sediment diversion structures requires understanding the evolution of geomorphic environments in which wetland flora and fauna inhabit. Historic models developed from analysis of subdelta geomorphology provide a rich sedimentary and ecological archive and natural baseline of environmental information relevant to future restoration strategies.

Social scientists have also been "integrated" into interdisciplinary approaches to environmental floodplain management, and the rapid growth of the field of "ecosystem services" (Section 8.3.4) is an especially important development. The quantification of economic and societal values for specific floodplain biophysical processes is essential to developing broader public support for environmental floodplain management. The framing of ecosystem values as concerns river restoration has been cited as vital to gaining financial support (e.g., bond passage) to pay for restoration activities (Richards et al., 2017; Straatsma et al., 2019). Additionally, designing flood control infrastructure from the perspective of flood "risk" rather than a strict flood frequency (probabilistic) approach requires the explicit integration of social sciences, especially in urban settings where property values and population densities are much higher (WMO, 2004). The inclusion of a social science perspectives is especially important to dam removal. For dam removal to be successful requires deep and sustained engagement with community stakeholders across a range of societal issues (Heinz Center, 2002; Lejon et al., 2009), representing a key area for collaboration between natural and social scientists. The modern approach to fluvial management, therefore, is much more "open" than in former decades. As the interrelations and linkages between different modes of floodplain sciences have been (fairly well) established, the focus today is more toward solutions rather than specific disciplinary outcomes.

9.3.2.2 DAM REMOVAL AND RESERVOIR MANAGEMENT

An important trend in Europe and North America is the increasing frequency with which dams and obstructions (weirs) are being removed (Sections 4.4 and 4.5), reducing hydrologic fragmentation and increasing longitudinal connectivity. For dams located in the headwaters of large river basins this is ultimately a positive sign as regards lowland rivers, as sediment release and transport downstream may help to (marginally) offset decades of sediment decline. Nongovernmental organizations – such as

Dam Removal Europe and Rivers International – play an important role because they shine the spotlight on the issue and provide up to date tallies on the status of dam removal. Relatedly, dam removal is increasingly driven because of requirement to be in compliance with specific water resource policies, such as the EU Water Framework and Floods Directives.

Dam removal is an emerging science and effectively mandates an interdisciplinary approach. The main motivation for dam removal is ecologic, especially fish (esp. salmon) habitat restoration. The interdisciplinary dimensions of restoration of aquatic habitat can be appreciated when considering the fact that it requires understanding (among others) bed load transport, biological criteria for spawning habitat, and channel bed morphology and dynamics (Doyle et al., 2003b; Foley et al., 2017; Ritchie et al., 2018). Additionally, the procedure for removing the dam structure requires engineers and geomorphologists to consider reservoir sediment reworking in relation to the style of removal. While there are valid concerns related to the impacts of dam removal, river channels quickly recover after dam removal (years not decades) and riparian ecosystems also improve (Section 4.4.4).

While the increased frequency of dam removal is important, it is mainly for smaller rivers and many dams are effectively permanent fixtures upon the riparian landscape. Therefore, a promising development is reservoir sediment management. This can require existing dam structures to be altered so that reservoir release schedules can ideally "sluice" sediment during large discharge events. For dams that have already lost considerable storage capacity because of sedimentation, reservoir drawdown and "flushing" can be effective, although there are downstream environmental concerns associated with the quantity and quality of reactivated reservoir deposits (Graf et al., 2011; Kondolf et al., 2014a).

9.3.2.3 TECHNOLOGY

A true reason for optimism is the fast pace of improved technologies for monitoring and data analysis. Over the past couple of decades, the "tools" for monitoring, data collection, and modeling have substantially improved while becoming less costly (e.g., Kondolf and Piégay, 2016; Boothroyd et al., 2020). This occurs in the realm of remote sensing and geospatial technologies, in addition to field-based instrumentation. In some instances such advances facilitate analyses that could not have been conceived a couple of decades ago.

Remote sensing technologies provide the opportunity to obtain increasingly high-resolution environmental information over large spatial scales. Landsat imagery extends back to the 1970s, providing an opportunity at temporal comparisons for over four decades. The Quickbird (DigitalGlobe) satellite system, for example, obtained high-resolution (0.65 m) multispectral imagery with frequent acquisition (~2 days). Widely available internet-based geospatial analysis (e.g., Google Earth Engine) can be utilized to automatically "digitize" channel banks over different time periods at much higher spatial resolutions than available even a decade ago (Boothroyd et al., 2020). But the use of lidar DEMs, which is becoming increasingly available, provides the highest spatial topographic resolution (~0.2 m vertical). Lidar is essential to floodplain mapping and has revolutionized hydraulic modeling and flood analysis, as well as quantification of geomorphic change over short time-scales. The combination of field-based sediment samples to calibrate remote sensing turbidity signals provides accurate estimates of suspended sediment flux. Unmanned aerial vehicles (i.e., drones) have become standard to organizations responsible for mapping and monitoring hydraulic infrastructure and fluvial environments. In addition to drones providing detailed user-specified imagery to map river and floodplain environments, they are also increasingly utilized in "flood fights" to identify problem areas for dike reinforcement.

Ground-based instrumentation increasingly extends the capabilities of small organizations and teams to collect high-resolution environmental information at high spatial and temporal frequencies. In part, this is because of improvements in data loggers and battery life, enabling monitoring stations to be set up in remote locations without the problem of frequent assessment and retrieval. Pressure transducers that provide highly detailed recording of flood stage variability can be installed in distant settings and retrieved every year or so. Turbidity sensors obtain reliable estimates of wash load (silt/clay) sediment flux. The calculation of velocity, discharge, and sediment load increasingly utilizes acoustic Doppler current profiler (ADCP) technology (Kostaschuk et al., 2005), enabling efficient measurements of large rivers in remote locations.

9.3.2.4 SCIENCE-BASED POLICY AND MANAGEMENT

The reasons for optimism noted are only helpful if appropriate policy and effective management are implemented. The institutional fortitude of many organizations charged with carrying out tasks related to managing fluvial lowlands is a concern, but there are strong strands of improvement. Primarily this is concerned with an increasing use of fluvial science to create good policy, which supports an integrated approach to floodplain management.

Integrated flood management is clearly the way forward, and always has the goal of first reducing human flood risk, with economic and environmental interests determined by societal and stakeholder priorities. Such an approach recognizes the importance of coordinating different scales of governments, and the inclusion of local stakeholders. While it takes time (a decade or so) for science to become infused into policy and management, that an increasing number of government agencies are adapting an integrated approach to flood management is in itself an important statement about the recognition of utilizing the latest scientific concepts to address societal flood risk and restore riparian ecosystems.

The European Union's Water Framework Directive and Floods Directive is a key example of an integrated approach to river management (Section 8.6). Being continental in scope, the WFD is ambitious. It is also comprehensive in that it takes a drainage basin perspective that considers how management and environmental change in headwaters impact lowland floodplains. In the European Union, such an approach is championed by the International Commission for the Protection for the Rhine, which facilitated cooperation between eight member nations. With revisions and updates (e.g., Jager et al., 2016; Helmholtz Centre for Environmental Research, 2017), the EU WFD is increasingly held up as the new international standard for integrated flood management (Jager et al., 2016; EC, 2019), and has the potential to have a much broader impact on the future of river management.

9.4 IT'S NOT TOO LATE

Optimism and concern are sensible sentiments when considering the future prospects of large lowland riverine health. Large lowland rivers that have been impacted for decades and centuries by flood control and related engineering activities cannot be restored to a pristine condition. In particular, many coastal deltaic wetlands will never have enough sediment to rebuild the land that is being lost because of (relative) sea level rise. But with investment, their ecological and geomorphic conditions can improve, and selected reaches of lowland rivers and deltas can even become quite good. While it is too late for complete restoration to a natural status, we should consider that by implementing effective management now, large lowland rivers can be substantially improved even in the face of impending global environmental change. Some human-dominated rivers can also support natural environmental functions and attain a new stable condition with regards to geodiversity and biodiversity, while continuing to support societal values. Additionally, it's not too late for many large lowland rivers that are not yet dominated by intensive hydraulic infrastructure, especially some large tropical rivers. The imposed impacts and changes that many large lowland rivers have already undergone, and herein reviewed, provide a range of lessons to guide organizations in their stewardship of these invaluable natural resources.

Bibliography

Aalto, R., Maurice-Bourgoin, L., Dunne, T., Montgomery, D. R., Nittrouer, C. A., and Guyot, J.-L. 2003. Episodic sediment accumulation on Amazonian flood plains influenced by El Niño/Southern Oscillation. *Nature* 425, 493–497.

Aalto, R., Lauer, J. W., and Dietrich, W. E. 2008. Spatial and temporal dynamics of sediment accumulation and exchange along Strickland River floodplains (Papua New Guinea) over decadal to centennial timescales. *Journal of Geophysical Research: Earth Surface* 113, F01S04. doi: 10.1029/2006JF0006277.

Abernethy, B., and Rutherfurd, I. D. 2001. The distribution and strength of riparian tree roots in relation to riverbank reinforcement. *Hydrological Processes* 15, 63–79.

Abizaid, C. 2005. An anthropogenic meander cutoff along the Ucayali River, Peruvian Amazon. *Geographical Review* 95, 122–135.

Adams, J. 1980. Active tilting of the United States mid-continent: geodetic and geomorphic evidence. *Geology* 8, 442–446.

Adams, P. N., Slingerland, R. L., and Smith, N. D. 2004. Variations in natural levee morphology in anastomosed channel floodplain complexes. *Geomorphology* 61, 127–142. doi: 10.1016/j.geomorph.2003.10.005.

Adamson, M. 2018. Flood risk management in Europe: the EU "Floods" directive and a case study of Ireland. *International Journal of River Basin Management* 16, 261–272.

Aerts, J. C. J. H., Botzen, W. J., Clarke, K. C., et al. 2018. Integrating human behaviour dynamics into flood disaster risk assessment. *Nature Climate Change* 8, 193–199. doi: 10.1038/s41558–018–0085-1.

Ahmed, A. A., and Ismail, U. H. A. E. 2008. Sediment in the Nile River System. UNESCO International Hydrological Programme, International Sediment Initiative. Khartoum, Sudan.

Ahn, J., Yang, C. T., Boyd, P. M., Pridal, D. B., and Remus, J. I. 2013. Numerical modeling of sediment flushing from Lewis and Clark Lake. *International Journal of Sediment Research* 28, 182–193.

Alcamo, J. D., Van Vuuren, D. P., Cramer, W. et al. 2005. Changes in ecosystem services and their drivers across the scenarios. In Sinh, B. T., Hammond, A., Field, C. et al. (eds.), *Ecosystems and Human Well-Being: Scenarios*. Island Press, Washington, DC, Chapter 9, pp. 297–373.

Alexander, J., Fielding, C. R., and Pocock, G. D. 1999. Flood deposits of the Burdekin River, tropical north Queensland, Australia. In Marriott, S. G., and Alexander, J. (eds.), *Floodplains: Interdisciplinary Approaches*, vol. 163. Geological Society London Special Publication, London, pp. 27–40.

Alexander, J. S., Wilson, R. C., and Green, W. R. 2012. A brief history and summary of the effects of river engineering and dams on the Mississippi River system and delta. U.S. Geological Survey Circular, 1375.

Alexander, J. S., Jacobson, R. B., and Rus, D. L. 2013. Sediment transport and deposition in the lower Missouri River during the 2011 flood. U.S. Geological Survey Professional Paper 1798-F. doi: 10.3133/PP1798F.

Alexander, M. A., Kilbourne, K. H., and Nye, J. A. 2014. Climate variability during warm and cold phases of the Atlantic Multidecadal Oscillation (AMO) 1871–2008. *Journal of Marine Systems* 143, 14–26.

Allen, Y., Kimmel, K. M., and Constant, G. C. 2014. Alligator gar movement and water quality patterns on the St. Catherine Creek National Wildlife Refuge floodplain. U.S. Fish and Wildlife Service, Conservation Office Report, Baton Rouge, LA.

Allison, M. A., and Meselhe, E. A. 2010. The use of large water and sediment diversions in the lower Mississippi River (Louisiana) for coastal restoration. *Journal of Hydrology* 387, 346–360.

Allison, M. A., Kuehl, S. A., Martin, T. C., and Hassan, A. 1998. The importance of floodplain sedimentation for river sediment budgets and terrigenous inputs to the ocean: insights from the Brahmaputra–Jamuna rivers. *Geology* 26, 175–178.

Allison, M. A., Vosburg, B. M., Ramirez, M. T., and Meselhe, E. A. 2013. Mississippi River channel response to the Bonnet Carré Spillway opening in the 2011 flood and its implications for the design and operation of river diversions. *Journal of Hydrology* 477, 104–118.

Allison, M. A., Törnqvist, T., Amelung, F., et al. 2016. Global risks and research priorities for coastal subsidence. *Eos* 97. doi: 10.1029/2016EO055013.

Amer, R., Kolker, A. S., and Muscietta, A. 2017. Propensity for erosion and deposition in a deltaic wetland complex: implications for river management and coastal restoration. *Remote Sensing of Environment* 199, 33–50.

American Rivers. 2020. American Rivers Dam Removal Database. https://figshare.com/articles/dataset/American_Rivers_Dam_Removal_Database/5234068 (accessed May 5, 2020).

American Society of Civil Engineers (ASCE). 2007. The New Orleans hurricane protection system: what went wrong and why. American Society of Civil Engineers, Hurricane Katrina External Review Panel.

Amissah, G. J., Kiss, T., and Fiala, K. 2018. Morphological evolution of the lower Tisza River (Hungary) in the 20th century in response to human interventions. *Water* 10, 884.

Amoros, C., Elger, A., Dufour, S. et al. 2005. Flood scouring and groundwater supply in rehabilitated side-channels of the Rhône River, France: Sedimentation and aquatic vegetation responses. *Large Rivers, Archives für Hydrobiologie* Suppl., 15, 1–4.

ANCOLD. 2020. Registrar of large dams in Australia. Australian National Committee on Large Dams. www.ancold.org.au/ (accessed May 4, 2020).

Andrews, J. 1938. Goyder's line: a vanished frontier. *Australian Geographer* 3, 32–36. doi: 10.1080/00049183808702187.

Annandale, G. W., Morris, G. L., and Karki, P. 2016. *Extending the Life of Reservoirs: Sustainable Sediment Management for Dams and Run-of-River Hydropower*. Directions in Development – Energy and Mining. World Bank, Washington, DC.

Annual Report of the Chief of Engineers. 1890. Improvement of the navigation of Red River, Louisiana, and of certain rivers in Louisiana, Texas, Mississippi, Arkansas, and Tennessee; and water-gauges on the Mississippi and its principal tributaries. Report of J. H. Willard, Appendix W.

Annual Report of the Chief of Engineers. 1904. Western rivers: improvement of certain rivers and waterways in Louisiana, Texas, Arkansas, Indian Territory, and Mississippi Tributary to Mississippi River. Annual Reports for the War Department for Fiscal Year ending in June 30, 1904, part 1, volume 5, Washington, DC, pp. 382–415.

Anonymous. 1828. Letter to members of the Board of Internal Improvements on the importance of removing the raft on the Atchafalaya River. United States. Works Progress Administration of Louisiana, March 4, 1928, p. 3, c. 2–4.

Anthony, E., Brunier, G., Besset, M., et al. 2015. Linking rapid erosion of the Mekong River delta to human activities. *Scientific Reports* 5, 14745. doi: 10.1038/srep14745.

Archer, M. W., Pracheil, B. M., Otto, A. E., and Pegg, M. E. 2019. Fish community response to in-channel woody debris in a channelized river system. *Journal of Freshwater Ecology* 34, 351–362.

Areneo, D., and Villalba, R. 2014. Variability in the annual cycle of the Río Atuel streamflows and its relationship with tropospheric circulation. *International Journal of Climatology* 35, 2948–2967.

Armstrong, C., Mohrig, D., Hess, T., George, T., and Straub, T. M. 2013. Influence of growth faults on coastal fluvial systems: examples from the late Miocene to Recent Mississippi River Delta. *Sedimentary Geology* 301, 120–132.

Arnaud, F., Piégay, H., Schmitt, L., Rollet, A. J., Ferrier, V., and Béal, D. 2015. Historical geomorphic analysis (1932–2011) of a by-passed river reach in process-based restoration perspectives: the Old Rhine downstream of the Kembs diversion dam (France, Germany). *Geomorphology* 236, 163–177.

Arroyave, V. J. A., and Crosato, A. 2010. Effects of river floodplain lowering and vegetation cover. *Water Management* 163, 457–467. doi: 10.1680/wama.900023.

ASDSO. 2016. The cost of rehabilitating our nation's dams: a methodology, estimate and proposed funding mechanisms. Prepared by a Task Committee of the Association of State Dam Safety Officials. https://damsafety.org/state-performance (accessed 5 April 2020).

Ashmore, P. J. 1991. How do gravel-bed rivers braid? *Canadian Journal of Earth Sciences* 28, 326–341.

Ashmore, P. J. 1996. Mid-Channel bar growth and its relationship to local flow strength and direction. *Earth Surface Processes and Landforms* 21, 103–123.

Aslan, A., and Autin, W. J. 1998. Holocene flood-plain soil formation in the southern lower Mississippi Valley: implications for interpreting alluvial paleosols. *Geological Society of America Bulletin* 110, 433–449.

Aslan, A., and Autin, W. J. 1999. Evolution of the Holocene Mississippi River floodplain, Ferriday, Louisiana: insights on the origin of fine-grained floodplains. *Journal of Sedimentary Research* 69, 800–815.

Aslan, A., and Blum, M. D. 1999. Contrasting styles of Holocene avulsion, Texas Gulf Coastal Plain, U.S.A. In Smith, N. D., and Rogers, J. (eds.), *Fluvial Sedimentology VI*. International Association of Sedimentologists, Special Publication 28. Blackwell, Oxford, pp. 293–308.

Aslan, A., White, W. A., Warne, A. G., and Andguevara, E. J. 2003. Holocene evolution of the western Orinoco Delta, Venezuela. *Geological Society of America Bulletin* 115, 479–498.

Aslan, A., Autin, W. J., and Blum, M. D. 2005. Late Holocene avulsion history of the Mississippi River, south Louisiana, U.S.A. *Journal of Sedimentary Research* 75, 648–662.

Aspen Institute. 2002. *Dam Removal: A New Option for a New Century*. Program on Energy, the Environment, and the Economy.

Asselman, N. E. M. 1999. Suspended sediment dynamics in a large drainage basin: the River Rhine. *Hydrological Processes* 13, 1437–1450.

Asselman, N. E. M., and Middelkoop, H. 1998. Temporal variability of contemporary floodplain sedimentation in the Rhine–Meuse delta, The Netherlands. *Earth Surface Processes and Landforms* 23, 595–609.

Asselman, N. E. M., Middelkoop, H., and Van Dijk, P. M. 2003. The impact of changes in climate and land use on soil erosion, transport and deposition of suspended sediment in the River Rhine. *Hydrological Processes* 17, 3225–3244.

Auerbach, D. A., Deisenroth, D. B., McShane, R. R., McCluney, K. E., and Poff, N. L. 2014. Beyond the concrete: accounting for ecosystem services from free-flowing rivers. *Ecosystem Services* 10, 1–5.

Auerbach, L. W., Goodbred, S. L., Jr., Mondal, D. R., *et al.* 2015. Flood risk of natural and embanked landscapes on the Ganges–Brahmaputra tidal delta plain. *Nature Climate Change* 5, 153–157. doi: 10.1038/nclimate2472.

Autin, W. J., Burns, S. F., Miller, B. J., Saucier, R. T., and Snead, J. I. 1991. Quaternary geology of the lower Mississippi Valley. In Morrison, R. B. (ed.), *Quaternary Nonglacial Geology*, vol. K-2. Geological Society of America, The Geology of North America, Boulder, CO, pp. 547–581.

Avakyan, A. B. 1998. Volga-Kama cascade reservoirs and their optimal use. *Lakes and Reservoirs Research and Management* 3, 113–121. doi: 10.1111/j.1440-1770.1998.tb00038.x.

AWA. 2020. Large dams fact sheet. Australian Water Association. www.awa.asn.au/ (accessed May 7, 2020).

Baker, R. G., Schwert, D. P., Bettis, E. A., and Chumbly, C. A. 1993. Impact of Euro-American settlement on a riparian landscape in northeast Iowa, Midwestern USA: an integrated approach based on historical evidence, floodplain sediments, fossil pollen, plant macrofossils and insects. *The Holocene* 3, 314–323.

Bakkenist, S., and Flos, S. 2015. Dijkinspectie met drones: Verkenning van de Mogeikheden voor het gebruik van drones bij inspectie en het beheer van waterkeringen, STOWA Report 2015-09, Amersfoort, NL.

Baldassarre, G. Di and Castellarin, A. 2009. Analysis of the effects of levee heightening on flood propagation: example of the River Po, Italy. *Hydrological Science Journal* 54, 1007–1017.

Bank Swallow Technical Advisory Committee (BSTAC). 2013. Bank swallow (Riparia riparia) conservation strategy for the Sacramento River Watershed, California. Version 1.0. www.sacramentoriver.org/bans/ (accessed September 15, 2018).

Baptist, M. J., Penning, W. E., Duel, H., *et al.* 2004. Assessment of the effects of cyclic floodplain rejuvenation on flood levels and biodiversity along the Rhine River. *River Research and Applications* 20, 285–297.

Barnett, J. F., Jr. 2017. *Beyond Control: The Mississippi River's New Channel to the Gulf of Mexico*. University Press of Mississippi, Oxford.

Barras, J. A., Bernier, J. C., and Morton, R. A. 2008. Land area change in coastal Louisiana – a multidecadal perspective (from 1956 to 2006). U.S. Geological Survey Scientific Investigations Map 3019, scale 1:250,000, 14 p. pamphlet.

Barras, J. A., Beville, S., Britsch, D. *et al.* 2003. Historical and projected coastal Louisiana land changes: 1978–2050. U.S. Geological Survey, Open-File Report 2003–334.

Barry, J. A. 1997. *Rising Tide: The Great Mississippi Flood of 1927 and How It Changed America*. Simon and Schuster, New York.

Batalla, R. J. 2003. Sediment deficit in rivers caused by dams and instream gravel mining. A review with examples from NE Spain. *Review Cuaternario y Geomorfologia* 17, 79–91.

Batalla, R. J., Kondolf, G. M., and Gomez, C. M. .2004. Reservoir-induced hydrological changes in the Ebro River basin, NE Spain. *Journal of Hydrology* 290, 117–136.

Batalla, R. J., Vericat, D., and Martínez, T. I. 2006. River-channel changes downstream from dams in the lower Ebro River. *Zeitschrift für Geomorphologie/Supplement* 143, 1–15.

Bawden, G. W., Johnson, M. R., Kasmarek, M. C., Brandt, J., and Middleton, C. S. 2012. Investigation of Land Subsidence in the Houston-Galveston Region of Texas by using the Global Positioning System and Interferometric Synthetic Aperture Radar, 1993–2000. U.S. Geological Survey, Scientific Investigations Report, 2012–5211, 88pp.

Baxter, R. M. 1977. Environmental effects of dams and impoundments. *Annual Review of Ecology and Systematics* 8, 255–283. doi: 10.1146/annurev.es.08.110177.001351.

Beach, T. 1994. The fate of eroded soil: sediment sinks and sediment budgets of agrarian landscapes in southern Minnesota, 1851–1988. *Annals of the Association of American Geographers* 84, 5–28.

Beach, T., Luzzadder-Beach, S., Dunning, N., and Cook, D. 2008. Human and natural impacts on fluvial and karst depressions of the Maya Lowlands. *Geomorphology* 101, 301–331.

Becker, R. H., and Sultan, M. 2009. Land subsidence in the Nile Delta: inferences from radar interferometry. *The Holocene* 19, 949–954.

Bednarek, A. T., and Hart, D. D. 2005. Modifying dam operations to restore rivers: ecological responses to Tennessee River dam mitigation. *Ecological Applications* 15, 997–1008.

Beilfuss, R. D. 2018. The Zambezi Delta (Mozambique). In Finlayson, C. M., Everard, M., Irvine, K. *et al.* (eds.), *The Wetland Book II: Distribution, Description, and Conservation*. Springer, Dordrecht, pp. 1233–1242, https://doi.org/10.1007/978-94-007-4001-3_195.

Bellin, J. N. 1764. *Suite du cours du fleuve St. Louis depuis la rivière d'Iberville jusq'à celle des Yasous, et les parties connues de la Rivière Rouge et la Rivière Noire*. [Paris] [Map] Retrieved from the Library of Congress Geography and Map Division, Washington, DC. www.loc.gov/item/74693009/.

Belt, C. B., Jr. 1975. The 1973 flood and man's constriction of the Mississippi River. *Science* 189, 681–684.

Benedetti, M. M. 2003. Controls on overbank deposition in the Upper Mississippi River. *Geomorphology* 56, 271–290.

Benito, G., and Hudson, P. F. 2010. Flood hazards: the context of fluvial geomorphology. In Alcántara-Ayala, I., and Goudie, A. (eds.), *Geomorphological Hazards and Disaster Prevention*. Cambridge University Press, Cambridge, pp. 111–128.

Benito, G., and Machado, M. J. 2012. Floods in the Iberian Peninsula. In Kundzewicz, Z. W. (ed.), *Changes of Flood Risk in Europe*. IAHS Special Publication 10. IAHS Press and CRC Press, Balkema.

Benke, A. C., and Wallace, J. B. 2003. Influence of wood on invertebrate communities in streams and rivers. In Gregory, S. V., Boyer, K. L., and Gurnell, A. M. (eds.), *The Ecology and Management of Wood in World Rivers*. American Fisheries Society Symposium 37. American Fisheries Society, Bethesda, MD, pp. 149–177.

Benke, A. C., Henry, R. L., Gillespie, D. M., and Hunter, R. J. 1985. Importance of snag habitat for animal production in southeastern streams. *Fisheries* 10, 8–13.

Benson, M. A., and Thomas, D. M. 1966. A definition of dominant discharge. *Hydrological Sciences Journal* 11, 76–80. doi: 10.1080/02626666609493460.

Berendsen, H. J. A. 1982. De genese van het landschap in het zuiden van de provincie Utrecht, een fysisch-geografische studie. Dissertatie, Utrechtse Geografische Studies 25.

Berendsen, H. J. A. 1993. De ontwikkeling van het Nederlandse rivierengebied. Stichting Geologische Aktiviteiten 26 (2), 49–76.

Berendsen, H. J. A. 2008. Landschappelijk Nederland: De fysisch-geografische regio's, 4th Ed. Van Gorcum, Assen, NL.

Berendsen, H. J. A., and Stouthamer, E. 2001. *Paleogeographic Development of the Rhine-Meuse Delta, the Netherlands*. Koninklijke Van Gorcum. 3 colored maps, CD-Rom.

Berkowitz, J. F., Johnson, D. R., and Price, J. J. 2020. Forested wetland hydrology in a large Mississippi River tributary system. *Wetlands* 40, 1133–1148. doi: 10.1007/s13157-019-01249-5.

Berri, G. J., Ghietto, M. A., and García, N. O. 2002. The influence of ENSO in the flows of the Upper Paraná River of South America over the past 100 years. *Journal of Hydrometeorology* 2, 57–65.

Best, J. L., and Darby, S. E. 2020. The pace of human-induced change in large rivers: stresses, resilience, and vulnerability to extreme events. *One Earth* 2, 510–514.

Best, J. L., Ashworth, P. J., Sarker, M. H., and Roden, J. E. 2007. The Brahmaputra-Jamuna River, Bangladesh. In Gupta, A. (ed.), *Large Rivers: Geomorphology and Management*. John Wiley & Sons, Chichester, pp. 395–433.

Bianchi, T. S., Allison, M. A., and Karl, D. M. 2009. Large-river delta-front estuaries as natural "recorders" of global environmental change. *Proceedings of the National Academy of Sciences of the United States of America* 106, 8085–8092.

Biedenharn, D. S., Thorne, C. R., and Watson, C. C. 2000. Recent morphological evolution of the lower Mississippi River. *Geomorphology* 34, 227–250.

Biedenharn, D. S., Allison, M. A., Little, C. D., Jr., Thorne, C. R., and Watson, C. C. 2017. Large-Scale Geomorphic Change in the Mississippi River from St. Louis, MO, to Donaldsonville, LA, as revealed by Specific Gage Records. MRGandP Report No. 10. U. S. Army Corps of Engineers, Mississippi Valley Division, Vicksburg, MS. doi: 10.21079/11681/22744.

Biedenharn, D. S., Killgore, K. J., Little, C. D., Murphy, C. E., Jr., and Kleiss, B. A. 2018. Attributes of the lower Mississippi River Batture. MRGandP Technical Note No. 4. U.S. Army Corps of Engineers, Vicksburg, MS.

Biemans, H., Haddeland, I., Kabat, P., *et al.* 2011. Impact of reservoirs on river discharge and irrigation water supply during the 20th century (Fig. 3, Cumulative capacity of global dam storage over the 20th century). *Water Resources Research* 47, W03509. doi: 10.1029/2009WR008929.

Blamey, R. C., Ramos, A. M., Trigo, A. M., Tome, R., and Reason, C. J. C. 2017. The influence of atmospheric rivers over the South Atlantic on winter rainfall in South Africa. *Journal of Hydrometeorology* 11, 127–142. doi: 10.1175/JHM-D-17-0111.1.

Blum, M. D., and Roberts, H. H. 2009. Drowning of the Mississippi Delta due to insufficient sediment supply and global sea-level rise. *Nature Geoscience* 2, 488–491.

Blum, M. D., and Roberts, H. H. 2012. The Mississippi Delta region: past, present, and future. *Annual Review of Earth and Planetary Sciences* 40, 655–683.

Blum, M. D., and Törnqvist, T. E. 2000. Fluvial response to climate and sea level change: a review and look forward. *Sedimentology* 47, 2–48.

Blum, M. D., Toomey, R. S. III, and Valastro, S., Jr. 1994. Fluvial response to Late Quaternary climatic and environmental change, Edwards Plateau, Texas. *Palaeogeography, Palaeoclimatology, Palaeoecology* 108, 1–21.

Boelter, D. H. 1969. Physical properties of peats as related to degree of decomposition. *Soil Science Society of America Proceedings* 33, 606–609.

BoM. 2012. Record-breaking La Niña events. Bureau of Meteorology, Australian Government, August 2012. www.bom.gov.au/climate/enso/history/La-Nina-2010-12.pdf (accessed February 23, 2020).

BoM. 2016. What is La Niña and how does it impact Australia? Bureau of Meteorology, Australian Government, August 2016. www.bom.gov.au/climate/updates/articles/a020.shtml (accessed February 23, 2020).

Bonnema, M., Hossain, F., Nijssen, B., and Holtgrieve, G. 2020. Hydropower's hidden transformation of rivers in the Mekong. *Environmental Research Letters* 15, 044017.

Boothroyd, R. J., Williams, R. D., Hoey, T. B., Barrett, B., and Prasojo, O. A. 2021. Applications of Google Earth Engine in fluvial geomorphology for detecting river channel change. *WIREs Water* 8, e21496. doi: 10.1002/wat2.1496.

Born, S. M., Genskow, K. D., Filbert, T. L., Hernandez-Mora, N., Keefer, M. L., and White, K. A. 1998. Socioeconomic and institutional dimensions of dam removals: the Wisconsin experience. *Environmental Management* 22, 359–370.

Bouse, R. M., Fuller, C. C., Luoma, N., Hornberger, M., Jaffe, B. E., and Smith, R. 2010. Mercury-contaminated hydraulic mining debris in San Francisco Bay. *San Francisco Estuary and Watershed Science* 8. doi: 10.15447/sfews.2010v8iss1art3.

Bowman, M. 2002. Legal perspectives on dam removal. *BioScience* 52, 739–747.

Brammer, A. J., Rodriguez del Rey, Zo., Spalding, E. A., and Poirrier, M. A. 2007. Effects of the 1997 Bonnet Carré Spillway opening on infaunal macroinvertebrates in Lake Pontchartrain, Louisiana. *Journal of Coastal Research* 23, 1292–1303.

Brandao, I., Mannaerts, C. M., and Saraiva, A. C. F. 2017. Seasonal variation of phytoplankton indicates small impacts of anthropic activities in a Brazilian Amazonian reservouir. *Ecohydrology and Ecobiology* 17, 217–226, doi: 10.1016/j.ecohyd.2017.04.001.

Brandt, S. A. 2000. Classification of geomorphological effects downstream of dams. *Catena* 40, 375–401.

Brauneck, J., Pohl, R., and Jupner, R. 2016. Experiences of using UAVs for monitoring levee breaches. In *IOP Conference Series: Earth and Environmental Science*, vol. 46, 6th Digital Earth Summit, July 7–8, 2016, Beijing, China.

Bravard, J.-P. 2019. *Sedimentary Crisis at the Global Scale 2: Deltas, a Major Environmental Crisis*. John Wiley & Sons, Chichester.

Bravard, J.-P., and Gaydou, P. 2015. Historical development and integrated management of the Rhône River Floodplain, from the Alps to the Camargue Delta, France. In Hudson, P. F., and Middelkoop, H. (eds.), *Geomorphic Approaches to Integrated Floodplain Management of Lowland Fluvial Systems in North America and Europe*, Springer-Verlag New York, New York, pp. 289–320.

Bravard, J.-P., Goichot, M., and Gaillot, S. 2013. Geography of sand and gravel mining in the lower Mekong River: first survey and impact assessment. *EchoGéo* 26. doi: 10.4000/echogeo.13659.

Breitburg, D., Levin, L. A., Oschlies, A., *et al.* 2018. Declining oxygen in the global ocean and coastal waters. *Science* 359, 7240. doi: 10.1126/science.aam7240.

Bridge, J. S. 2003. *Rivers and Floodplains: Forms, Processes, and Sedimentary Record*. John Wiley & Sons, Chichester.

Brierley, G. J., Ferguson, R. J., and Woolfe, K. J. 1997. What is a fluvial levee? *Sedimentary Geology* 114, 1–9.

Bristow, C. S., Skelly, R. L., and Ethridge, F. G. 1999. Crevasse splays from the rapidly aggrading sand-bed, braided Niobara River, Nebraska: effect of base level rise. *Sedimentology* 46, 1029–1047.

Brooks, G. R. 2000. Channel changes along the lower reaches of major Mackenzie River tributaries. In Dyke, L. D., and Brooks, G. R. (eds.), *The Physical Environment of the Mackenzie Valley, Northwest Territories: A Base Line for the Assessment of Environmental Change*. Geological Survey of Canada, Bulletin 547, pp. 159–166.

Broothaerts, N., Verstraeten, G., Notebaert, B., *et al.* 2013. Human impact on floodplain geoecology. A Holocene perspective for the Dijle catchment, Central Belgium. In *Abstract Volume 8th International Conference on Geomorphology (IAG)*, August 27–31, 2013, Paris.

Brouns, K., Verhouven, J. T. A., and Hefting, M. M. 2014. Short period of oxygenation releases latch on peat decomposition. *Science of the Total Environment* 481, 61–68. doi: 10.1016/j.scitotenv.2014.02.030.

Brown, A. G., 2008. Geoarchaeology, the four dimensional (4D) fluvial matrix and climatic causality. *Geomorphology* 101, 278–297.

Brown, A. G., Lespez, L., Sear, D. A., *et al.* 2018. Natural vs anthropogenic streams in Europe: history, ecology and implications for restoration, river-rewilding and riverine ecosystem services. *Earth-Science Reviews* 180, 185–205.

Bucala, A. 2014. The impact of human activities on land use and land cover changes and environmental processes in the Gorce Mountains (Western Polish Carpathians) in the past 50 years. *Journal of Environmental Management* 138, 4–14.

Bucala-Hrabia, A. 2017. Long-term impact of socio-economic changes on agricultural land use in the Polish Carpathians. *Land Use Policy* 64, 391–404.

Buglewicz, E. G., Mitchell, W. A., Scott, J. E., Smith, M., and King, W. L. 1988. *A Physical description of main stem levee borrow pits along the lower Mississippi River*. Lower Mississippi River Environmental Program, Report 2, US Army Corps of Engineers, Mississippi River Commission, Vicksburg, MS.

Buisje, T. A. D., Roozen, F. C. J. M., Grift, R. E., and van Geest, G. J. 2001. Stagnant water bodies. In Wolters, H. A., Platteeuw, M., and Schoor, M. M. (eds.), *Guidelines for Rehabilitation and Management of Floodplains: Ecology and Safety Combined*. Rijkswaterstaat, RIZA Report No. 2001.059, pp. 71–88.

Buisje, T. A. D., Coops, H., Staras, M., *et al.* 2002. Restoration strategies for river floodplains along large lowland rivers in Europe. *Freshwater Biology* 47, 889–907.

Bull, L. 1997. Magnitude and variation in the contribution of bank erosion to the suspended sediment load of the River Severn, UK. *Earth Surface Processes and Landforms* 22, 1109–1123.

Burnett, A. W. and Schumm, S. A. 1983. Alluvial-river response to neotectonic deformation in Louisiana and Mississippi. *Science* 222, 49–50.

Burt, T. P., Bates, P. D., Steward, M. D., Claxton, A. J., Anderson, M. G., and Price, D. A. 2002. Water table fluctuations within the floodplain of the River Severn, England. *Journal of Hydrology* 262, 1–20.

Butzer, K. W. 1971. Recent history of an Ethiopian delta: the Omo River and the level of Lake Rudolph. Research Paper 136, Department of Geography, University of Chicago.

Butzer, K. W. 1976. *Early Hydraulic Civilization in Egypt: A Study in Cultural Ecology*. University of Chicago Press, Chicago and London, 134 pp.

Butzer, K. W. 2012. Collapse, environment, and society. *Proceedings of the National Academy of Sciences* 109, 3632–3639. doi: 10.1073/pnas.1114845109.

Butzer, K. W., and Helgren, D. M. 2005. Livestock, land cover, and environmental history: The Tablelands of New South Wales, Australia, 1820–1920. *Annals of the Association of American Geographers* 95, 80–111.

Butzer, K. W., Abbott, J. T., Frederick, C., and Lehman, P. 2008. Soil-geomorphology and identification of "wet" cycles in the Holocene record of north-central Mexico. *Geomorphology* 101, 237–277.

Caernarvon Delta and Diversion Study. 2014. Final report submitted to the Mississippi River Delta coalition. Science Contingency Funds, Louisiana, pp. 7–9, June 5, 2014.

California Department of Water Resources (California DWR). 2010. Fact sheet – Sacramento River flood control project weirs and flood relief structures. California Division of Flood Management, Sacramento, California. www.water.ca.gov/newsroom/docs/WeirsReliefStructures.pdf (accessed February 15, 2016)

California Statewide Flood Management Planning Program (CSFMP). 2013. California's flood future: recommendations for managing the state's flood risk. Final Report, November 2013. State of California, The Natural Resources Agency, Department of Water Resources.

Camillo, C. A. 2012. *Divine Providence: The 2011 Flood in the Mississippi River and Tributaries Project*. Mississippi River Commission, Vicksburg, MS.

Campanella, R. 2007. *Geographies of New Orleans: Urban Fabrics before the Storm*. Center for Louisiana Studies, University of Louisiana at Lafayette.

Campanella, R. 2008. *Bienville's Dilemma: A Historical Geography of New Orleans*. Center for Louisiana Studies, University of Louisiana, Lafayette, Louisiana (U.S.).

Canadian Dam Association. 2019. Dams in Canada. Ottawa, Canada.

Cannatelli, K. M., and Crowe-Curran, J. 2012. Importance of hydrology on channel evolution following dam removal: case study and conceptual model. *ASCE Journal of Hydraulic Engineering* 138, 377–390.

Carle, M. V., Sasser, C. E., and Roberts, H. H. 2015. Accretion and vegetation community change in the Wax Lake Delta following the historic 2011 Mississippi River flood. *Journal of Coastal Research* 31, 569–587.

Carling, P., Jansen, J., and Meshkova, L. 2014. Multichannel rivers: their definition and classification. *Earth Surface Processes and Landforms* 39, 26–37.

Carlson, S. D., and Guccione, M. J. 2010. Short-term uplift rates and surface deformation along the Reelfoot Fault, New Madrid Seismic Zone. *Bulletin of the Seismological Society of America* 100, 1659–1677.

Carminati, E., Doglioni, C., and Scrocca, D. 2005. Magnitude and causes of long-term subsidence of the Po Plain and Venetian Region. In Fletcher, C. A., Spencer, T., Da Mosto, J., and Campostrini, P. (eds.), *Flooding and Environmental Challenges for Venice and Its Lagoon: State of Knowledge*. Cambridge University Press, Cambridge, pp. 21–28.

Caviedes, C. N. 2001. *El Niño in History: Storming through the Ages*. University Press of Florida, Gainesville.

Cazanacli, D., and Smith, N. D. 1997. A study of morphology and texture of natural levees, Cumberland Marshes, Saskatchewan, Canada. *Geomorphology* 25, 43–55.

Cencetti, C., and Tacconi, P. 2005. The fluvial dynamics of the Arno River. *Giornole di Geologia Applicata* 1, 193–202. doi: 10.1474/GGA.2005-01.0-19.0019.

Chamberlain, E. C., Törnqvist, T. E., Shen, Z., Mauz, B., and Wallinga, J. 2018. Anatomy of Mississippi Delta growth and its implications for coastal restoration. *Science Advances* 4. doi: 10.1126/sciadv.aar4740.

Chan, F., Mitchell, G., Adekola, G., and Mcdonald, A. T. 2013. Flood risk in Asia's urban mega-deltas drivers, impacts and response. *Environment and Urbanization ASIA* 3, 41–61.

Chang, H., Chiu, M., Chuang, Y., *et al.* 2017. Community responses to dam removal in a subtropical mountainous stream. *Aquatic Sciences* 79, 967–983.

Changxing, S., Dian, Z., Lianyuan, Y., Bingyuan, L., Zulu, Z., and Ouyang, Z. 2007. Land subsidence as a result of sediment consolidation in the Yellow River Delta. *Journal of Coastal Research* 23, 173–181.

Charlton, R. 2008. *Fundamentals of Fluvial Geomorphology*. Routledge, London, 234pp.

Chatanantavet, P., Lamb, M. P., and Nittouer, J. A. 2012. Backwater controls of avulsion location on deltas. *Geophysical Research Letters* 39, L01402. doi: 10.1029/2011GL050197.

Chen, C., Lin, H., Zhang, Y., and Lu, Z. 2012. Ground subsidence geohazards induced by rapid urbanization: implications from InSAR observation and geological analysis. *Natural Hazards and Earth Systems Science* 12, 935–942. doi: 10.5194/nhess-12-935-2012.

Chen, L., Wei, W., Fu, B.-J., and Lu, Y. 2007. Soil and water conservation on the Loess Plateau in China: review and perspective. *Progress in Physical Geography* 31, 389–403. doi: 10.1177/0309133307081290.

Chen, Y., Syvitski, J. P. M., Gao, S., Overeem, I., and Kettner, A. J. 2012. Socio-economic impacts on flooding: a 4000-year history of the Yellow River, China. *AMBIO* 41, 682–698. doi: 10.1007/s13280–012-0290-5.

Cheng, L., Opperman, J. J., Tickner, D., Speed, R., Guo, Q., and Chen, D. 2018. Managing the three Gorges Dam to implement environmental flows in the Yangtze River. *Frontiers in Environmental Science* 6, 64. doi: 10.3389/fenvs.2018.00064.

Chiew, F. H. S., and McMahon, T. 2002. Global ENSO-streamflow teleconnection, streamflow forecasting and interannual variability. *Hydrological Sciences Journal/Journal des Sciences Hydrologiques* 47, 505–522.

Chin, A., Harris, D. L., Trice, T. H., and Given, J. L. 2002. Adjustment of stream channel capacity following dam closure, Yegua Creek, Texas. *Journal of the American Water Resources Association* 38, 1521–1531. doi: 10.1111/j.1752-1688.2002.tb04362.x.

Church, J. A., Clark, P. U., Cazenave, A., *et al.* 2013. Sea level change. In Stocker, T. F., Qin, D., Plattner, G.-K., *et al.* (eds.), *Climate Change 2013: The Physical Science Basis*. Contribution of Working Group I to the Fifth Assessment Report of the Intergovernmental Panel on Climate Change. Cambridge University Press, Cambridge, 1137–1216.

Church, M. 1983. Pattern of instability in a wandering gravel bed channel. In Collinson, J. D., and Lewin, J. (eds.), *Modern and Ancient Fluvial Systems*. International Association of Sedimentologists, Special Publication, 6, Oxford, pp. 169–180.

Church, M. 2002. Geomorphic thresholds in riverine landscapes. *Freshwater Biology* 47, 541–557.

Church, M. 2015. Channel stability: morphodynamics and morphology of rivers. In Rowinski, P., and Radecki-Pawlik, A. (eds.), *Rivers – Physical, Fluvial and Environmental Processes*. Series: GeoPlanet: Earth and Planetary Sciences. Springer, Switzerland, pp. 282–321.

CIRIA. 2013. *The International Levee Handbook (C731)*. CIRIA, London.

Citterio, A., and Piégay, H. 2009. Overbank sedimentation rates in former channel lakes: characterization and control factors. *Sedimentology* 56, 461–482.

City of Rotterdam. 2015. Rotterdam climate initiative. www.010duurzamestad.nl/ (accessed November 20, 2018).

Clifford, N. J. 1993. Formation of riffle – pool sequences: field evidence for an autogenetic process. *Sedimentary Geology* 85, 39–51.

Coastal Protection and Restoration Authority of Louisiana (CPRA Louisiana). 2017. *Louisiana's Comprehensive Master Plan for a Sustainable Coast.* Coastal Protection and Restoration Authority of Louisiana, Baton Rouge, LA.

Coe, M., Latrubesse, E., Ferreira, M., and Amsler, M. 2011. The effect of deforestation and climate variability on the stream flow of the Araguaia River. *Biogeochemistry* 105, 119–131.

Cohen, E. 2012. Flooded: an auto-ethnography of the 2011 Bangkok flood. *ASEAS – Austrian Journal of South-East Asian Studies* 5, 316–334.

Cohen, K. M. 2003. *Differential Subsidence within a Coastal Prism: Late-Glacial Holocene Tectonics in the Rhine-Meuse Delta, The Netherlands.* Netherlands Geographical Survey 316, Utrecht.

Cohen, K. M., Stouthamer, E., and Berendsen, H. J. A. 2002. Fluvial deposits as a record for Late Quaternary neotectonic activity in the Rhine-Meuse delta, the Netherlands. *Netherlands Journal of Geosciences / Geologie en Mijnbouw* 81, 389–405.

Cohen, K. M., Stouthamer, E., and Berendsen, H. J. A. 2009. Riviersystemen in de polders van het Rivierengebied Zand. In Cohen, K. M., Stouthamer, E., Hoek, W. Z., Berendsen, H. J. A., and Kempen, H. F. J. (eds.), *Banen – Zanddieptekaarten van het Rivierengebied en het IJsseldal in de provincies Gelderland en Overijssel.* Arnhem, Provincie Gelderland, pp. 55–68.

Coker, H. E, Hotchkiss, R. H., and Johnson, D. A. 2009. Conversion of a Missouri River dam and reservoir to a sustainable system: sediment management. *Journal of the American Water Resources Association* 45, 815–827. doi: 10.1111/j.1752-1688.2009.00324.

Coleman, J. M. 1969. Brahmaputra River: channel processes and sedimentation. *Sedimentary Geology* 3, 129–239.

Coleman, J. M. 1988. Dynamic changes and processes in the Mississippi River delta. *Geological Society of America Bulletin* 100, 999–1015.

Coleman, J. M., and Gagliano, S.M. 1964. Cyclic sedimentation in the Mississippi river deltaic plain. *Gulf Coast Association of Geological Societies, Transactions* 14, 67–80.

Coleman, J. M., and Roberts, H. H. 1989. Deltaic coastal wetlands. *Geologie en Mijnbouw (Netherlands Journal of Geosciences)* 68, 1–24.

Coleman, J. M., and Wright, L. D. 1971. Analysis of major river systems and their deltas, procedures and rationale, with two examples. Louisiana State University, Coastal Studies Institute, Technical Report, 95.

Coleman, J. M., and Wright, L. D. 1975. Modern river deltas, variability of processes and sand bodies. In Broussard, M. L. (ed.), *Deltas, Models for Exploration.* Houston Geological Society, Houston, TX, pp. 99–149.

Coleman, J. M., Roberts, H. H., and Huh, O. K. 1986. Deltaic landforms. In Short, N. M., and Blair, R. W., Jr. (eds.), *Geomorphology from Space: A Global Overview of Regional Landforms.* National Aeronatuics and Space Administration, Scientific and Technical Branch, Washington, DC, pp. 317–352.

Coleman, J. M., Roberts, H. H., and Stone, G. W. 1998. Mississippi River delta: an Overview. *Journal of Coastal Research* 14, 698–716.

Coleman, J. S. M., and Budikova, D. 2013. Eastern U.S. summer streamflow during extreme phases of the North Atlantic oscillation. *Journal of Geophysical Research: Atmospheres* 118, 4181–4193.

Collier, M., Webb, R. H., and Schmidt, J. C. 1996. A primer on the downstream effects of dams. U.S. Geological Survey, Circular 1126.

Colloff, M. J. 2014. *Flooded Forest and Desert Creek: Ecology and History of the River Red Gum.* CSIRO Publishing, Collingwood, Australia, 312 p.

Comeaux, M. 1970. The Atchafalaya River raft. *Louisiana Studies* 9, 217–227.

Constantine, J. A., Dunne, T., Ahmed, J., Legleiter, C., and Lazarus, E. D. 2014. Sediment supply as a driver of river meandering and floodplain evolution. *Nature Geoscience* 7, 899–903. doi: 10.1038/NGEO2282.

Constantinescu, Ş., Achim, D., Rus, I., and Giosan, L. 2015. Embanking the lower Danube: from natural to engineered floodplains and back. In Hudson, P. F., and Middelkoop, H. (eds.), *Geomorphic Approaches to Integrated Floodplain Management of Lowland Fluvial Systems in North America and Europe.* Springer, New York, pp. 265–288.

Coomes, O. T., Abizaid, C., and Lapointe, M. 2009. Human modification of a large meandering Amazonian river: genesis, ecological and economic consequences of The Masisea cutoff on the Central Ucayali, Peru. *AMBIO: A Journal of the Human Environment* 38, 130–134.

Couvillion, B. R., Beck, H., Schoolmaster, D., and Fischer, M. 2017. Land area change in coastal Louisiana 1932 to 2016. U.S. Geological Survey Scientific Investigations Map 3381, 16 p. pamphlet.

Crist, E., Mora, C., and Engelman, R. 2017. The interaction of human population, food production, and biodiversity protection. *Science* 356, 260–264.

Cruz, R. V., Harasawa, H., Lal, M., et al. 2007. Asia. Climate Change 2007: impacts, adaptation and vulnerability. In Parry, M. L., Canziani, O. F., Palutikof, J. P., van der Linden, P. J., and Hanson, C. E. (eds.), *Contribution of Working Group II to the Fourth Assessment Report of the Intergovernmental Panel on Climate Change.* Cambridge University Press, Cambridge, pp. 469–506.

Cuenca, M. C., Hannsen, R., and van Leijn, F. 2007. Subsidence due to Peat Decomposition in the Netherlands – Kinematic Observations from Radar Interferometry, TU Delft Research Project, presented in Frascati, Italy.

Dai, P., and Tan, B. 2017. The Nature of the Arctic Oscillation and diversity of the extreme surface weather anomalies it generates. *Journal of Climate* 30, 5563–5584. doi: 10.1175/JCLI-D-16-0467.1.

Dahl, T. E., Swords, J., and Bergeson, M. T. 2009. Wetland inventory of the Yazoo Backwater Area, Mississippi – Wetland status and potential changes based on an updated inventory using remotely sensed imagery. U.S. Fish and Wildlife Service, Division of Habitat and Resource Conservation, Washington, DC.

Dam Removal Europe. 2018. https://damremoval.eu/ (accessed January 9, 2019).

Daniels, J. M. 2002. Drainage network adjustment following channelization, Homochitto River Basin, Mississippi. In Steinberg, M. K., and Hudson, P. F. (eds.), *Cultural and Physical Expositions: Geographic Studies in the Southern United States and Latin America.* Geoscience and Man, 36. Louisiana State University, pp. 291–308.

Daniels, J. M., Leigh, D. S., and Carson, E. C. 2019. Holocene paleohydrology and paleofloods in the Driftless Area. In Carson, E. C., Rawling, J. E., Daniels, J. M., and Attig, J. W. (eds.), *The Physical Geography and Geology of the Driftless Area: The Career and Contributions of James C. Knox.* Geological Society of America Special Paper 543, pp. 75–92.

Darby, S. E., Leyland, J., Kummu, M., Reaseanen, T. A., and Lauri, H. 2013. Decoding the drivers of bank erosion on the Mekong River: the roles of the Asian monsoon, tropical storms, and snowmelt. *Water Resources Research* 49, 2146–2163. doi: 10.1002/wrcr.20205.

Darby, S. E., Hackney, C., Leyland, J. et al. 2016. Fluvial sediment supply to a mega-delta reduced by shifting tropical-cyclone activity. *Nature* 539, 276–279. doi: 10.1038/nature19809.

Davivongs, V., Yokohari, M., and Hara, Y. 2012. Neglected canals: deterioration of indigenous irrigation system by urbanization in the West Peri-Urban area of Bangkok Metropolitan Region. *Water* 4, 12–27. doi: 10.3390/w4010012.

Day, G., Dietrich, W. E., Rowland, J., and Marshall, A. 2009. The depositional web on the floodplain of the Fly River, Papua New Guinea. *Journal of Geophysical Research* 113, F01S02. doi: 10.1029/2006JF000622.

Day, J. H., et al. 2009. The impacts of pulsed reintroduction of river water on a Mississippi delta coastal basin. *Journal of Coastal Research* 54, 225–243.

Day, J. H., Cable, J. E., Lane, R. R., and Kemp, G. P. 2016. Sediment deposition at the Caernarvon crevasse during the Great Mississippi Flood of 1927: implications for coastal restoration. *Water* 8. doi: 10.3390/w802003.

Day, J. W., and Giosan, L. 2008. Geomorphology: survive or subside? *Nature Geoscience* 1, 156–157. doi: 10.1038/ngeo137.

Day, J. W., Hunter, R. F., Keim, R., et al. 2012. Ecological response of forested wetlands with and without Large-Scale Mississippi River input: implications for management. *Ecological Engineering* 46, 57–67.

de Almeida, G. A. M., and Rodríguez, J. F. 2011. Understanding pool-riffle dynamics through continuous morphological simulations. *Water Resources Research* 47, W01502. doi: 10.1029/2010WR009170.

de Bruin, D. 2006. Similarities and differences in the historical development of flood management in the alluvial stretches of the lower Mississippi Basin and the Rhine Basin. *Irrigation and Drainage* 55, 23–54.

D'Haen, K., Dusar, B., Verstraeten, G., Degryse, P., and de Brue, H. 2013. A sediment finger-printing approach to understand the geomorphic coupling in an eastern Mediterranean mountainous river catchment. *Geomorphology* 197, 64–75.

DeHaan, H., Stamper, J., and Walters, B. 2012. Mississippi River and tributaries system 2011 post-flood report. U.S. Army Corps of Engineers, Mississippi Valley Division, Vicksburg, MS, 4 plates.

DeLaune, R. D., Jugsujinda, A., Peterson, G. W., and Patrick, W. H. 2003. Impact of Mississippi River freshwater reintroduction on enhancing marsh accretionary processes in a Louisiana estuary. *Estuarine Coastal and Shelf Science* 58, 653–662.

DEFRA. 2005. *Making Space for Water: Taking Forward a New Government Strategy for Flood and Coastal Erosion Risk Management Forward in England*. DEFRA, London.

DELP Victoria, AU. 2016. *Decommissioning Dams: A Guide for Owners*. Department of Environment, Land, Water and Planning, State of Victoria, Australia.

Denevan, W. M. 2000. *Cultivated Landscapes of Native Amazonia and the Andes*. Oxford University Press.

Dennis, J. V., Jr. 1990. Kampuchea's ecology and resource base: natural limitations on food production. In Ablin, D. A., and Hood, M. (eds.), *Revival: The Cambodian Agony*. Routledge, New York, pp. 208–238.

Department of Environment, Land, Water and Planning (DELWP). 2015. *Levee Management Guidelines*. The State of Victoria, Melbourne.

Dépret, T., Riquier, J., and Piégay, H. 2017. Evolution of abandoned channels: insights on controlling factors in a multi-pressure river system. *Geomorphology* 294, 99–118.

Desloges, J. R., and Church, M. 1989. Wandering gravel bed rivers. *Canadian Geographer* 33, 360–364.

Dettinger, M. 2011. Climate change, atmospheric rivers, and floods in California – a multimodel analysis of storm frequency and magnitude changes. *Journal of the American Water Resources Association* 47, 514–523. doi: 10.1111/j.1752-1688.2011.00546.x.

Diamond, J. 2006. *Collapse: How Societies Choose to Fail or Succeed*. Penguin, New York.

Dieppois, B., Pohl, B., Rouault, M. New, M., Lawler, D., and Keenlyside, N. 2016. Interannual to interdecadal variability of winter and summer southern African rainfall, and their teleconnections, *Journal of Geophysical Research: Atmospheres* 121, 6215–6239. doi: 10.1002/2015JD024576.

Dieras, P. L., Constantine, J. A., Hales, T. C., Piégay, H., and Riquier, J. 2013. The role of oxbow lakes in the off-channel storage of bed material along the Ain River, France. *Geomorphology*. doi: 10.1016/j.geomorph.2012.12.024.

Dierauer, J., Pinter, N., and Remo, J. W. 2012. Evaluation of levee setbacks for flood-loss reduction, Middle Mississippi River, USA. *Journal of Hydrology* 450–451, 1–8. doi: 10.1016/j.jhydrol.2012.05.044.

Ding, L., Chen, L., Ding, C., and Tao, J. 2018. Global trends in dam removal and related research: a systematic review based on associated datasets and bibliometric analysis. *Chinese Geographical Science* 29, 1–12.

Dirmeyer, P. A., and Kinter, J. L. III. 2009. The "Maya Express": floods in the U.S. Midwest. *Eos, Transactions, American Geophysical Union* 90, 101–102. doi: 10.1029/2009EO120001.

Dixon, T. H., Amelung, F., Ferretti, A., *et al.* 2006. Subsidence and flooding in New Orleans. *Nature* 441, 587–588. doi: 10.1038/441587a.

Dokka, R. K. 2006. Modern-day tectonic subsidence in coastal Louisiana. *Geology* 34, 281–284.

Domagalski, J. L., Slotton, D. G., Alpers, C. N., *et al.* 2004. Summary and synthesis of mercury studies in the Cache Creek Watershed, California, 2000–01. U.S. Geological Survey Water-Resources Investigations Report, 03-4335.

Dominquez, J. M. L. 1996. The São Francisco strand plain: a paradigm for wave-dominated deltas? *Geological Society London Special Publications* 117, 217–231.

Doolittle, W. D. 2000. *Cultivated Landscapes of Native North America*. Oxford University Press.

Dorozynski, A. 1975. After the dam the depression?. *Nature* 255, 570. doi: 10.1038/255570a0.

Dos Santos, V., Stevaux, J., and Assine, M. 2017. Fluvial processes in attachment bars in the upper Paraná River, Brazil. *Revista Brasileira de Geomorfologia* 18, 484–499. doi: 10.20502/rbg.v18i3.1135.

Dotterweich. 2008. The history of soil erosion and fluvial deposits in small catchments of central Europe: deciphering the long-term interaction between humans and the environment-A review. *Geomorphology* 101, 192–208.

Downs, P. W., and Gregory, K. J. 2004. *River Channel Management: Towards Sustainable Catchment Hydrosystems*. Arnold, London.

Downs, P. W., Gregory, K. J., and Brookes, A. 1991. How integrated is river basin management? *Environmental Management* 15, 299–309.

Doyle, M. W., Harbor, J., and Stanley, E. H. 2003a. Toward policies and decision-making for dam removal. *Environmental Management* 31, 453–465.

Doyle, M. W., Stanley, E., and Harbor, J. 2003b. Channel adjustments following two dam removals in Wisconsin. *Water Resources Research* 39,

Doyle, M. W., Stanley, E., Havlick, D., *et al.* 2008. Aging infrastructure and ecosystem restoration. *Science* 319(5861), 286–287.

Dreibrodt, S., Lubos, C., Terhorst, B., Dammc, B., and Bork, H.-R. 2009. Historical soil erosion by water in Germany: Scales and archives chronology, research perspectives. *Quaternary International* 222, 80–95.

Duda, J. J., Johnson, R. C., Wieferich, D. J., Wagner, W. J., and Bellmore, J. R. 2020. USGS dam removal science database v3.0 (ver. 3.0, January 2020). U.S. Geological Survey data release.

Dudley, W. T. 2004. Hydraulic geometry relations for rivers in coastal and central Maine. U.S. Geological Survey, Scientific Investigation Report 2004-5042, Augusta, Maine.

Dugan, P. J., Barlow, C., Agostinho, A. A., *et al.* 2010. Fish migration, dams, and loss of ecosystem services in the Mekong basin. *AMBIO* 39, 344-348. doi: 10.1007/s13280-010-0036-1.

Dunbar, J. B., Torrey, V. H. III, and Wakeley, L. D. 1999. A case history of embankment failure: Geological and geotechnical aspects of the Celotex levee failure, New Orleans, Louisiana. Final Report. US Army Engineer Research and Development Center, Technical Report GL-99-11, Vicksburg, MS.

Duncan, J. M., Brandon, T. L., Wright, S. G., and Vroman, N. 2008. Stability of I-walls in New Orleans during Hurricane Katrina. *Journal of Geotechnical and Geoenvironmental Engineering* 134, 681–692.

Dunne, T., and Aalto, R. E. 2013. Large river floodplains. In Shroder, J. F. (ed.), *Treatise on Geomorphology*, vol. 9. Academic Press, San Diego, CA, pp. 645–678.

Dunne, T., Mertes, L. A. K., Meade, R. H., Richey, J., and Forsberg, B. R. 1998. Exchanges of sediment between the floodplain and channel of the Amazon River in Brazil. *Geological Society of America Bulletin* 110, 450–467.

Dusar, B., Verstraeten, G., D'Haen, K., Bakker, J., Kaptijn, E., and Waelkens, M. 2012. Sensitivity of the Eastern Mediterranean geomorphic system towards environmental change during the Late Holocene: a chronological perspective. *Journal of Quaternary Science* 27, 371–382.

East, A. E., Logan, J. B., Mastin, M. C., *et al.* 2018. Geomorphic evolution of a gravel-bed river under sediment-starved versus sediment-rich conditions: river response to the world's largest dam removal. *Journal of Geophysical Research: Earth Surface* 123, 3338–3369. doi: 10.1029/2018JF004703.

Echevarria-Doyle, W., Biedenharn, D. S., and Little, C. D., Jr. 2020. Lake Providence to Old River geomorphology assessment. Final report. U.S. Army Corps of Engineers, Mississippi Valley Division Mississippi River Geomorphology and Potamology Program Report No. 32, Vicksburg, MS.

Egberts, H. 1950. De Bodemgesteldheid van de Betuwe. Stichting voor Bodemkartering Wageningen. Versl. Landbouwk Onderzoek No. 56.1 9, ' 'S Gravenhage, NL.

El Bakry, M. 1994. Net radiation over the Aswan high dam lake. *Theoretical and Applied Climatology* 49, 129–133.

Elliott, C. M., DeLonay, A. J., Chojnacki, K. A., Jacobson, R. B., 2020. Characterization of Pallid Sturgeon (Scaphirhynchus albus) spawning habitat in the lower Missouri River. *Journal of Applied Ichthyology* 36, 25–38.

Elliott, D. O. 1932. The improvement of the lower Mississippi River for Flood Control and Navigation, vol. 1–3. US Army Corps of Engineers Waterways Experiment Station, Vicksburg, MS.

Elliott, T. 1986. Deltas. In Reading, H. G. (ed.), *Sedimentary Environments and Facies*. Blackwell Scientific Publications, Oxford, pp. 113–154.

El-Mahdy, M. E., Abbas, M. S., and Sobhy, H. M. 2019. Development of mass-transfer evaporation model for Lake Nasser, Egypt. *Journal of Water and Climate Change* 12, 223–237. doi: 10.2166/wcc.2019.116.

El-Nashar, W. Y., and Elyamany, A. H. 2018. Managing risks of the Grand Ethiopian Renaissance Dam on Egypt. *Ain Shams Engineering Journal* 9, 2383–2388.

Eltahir, E. A. B. 1996. El Niño and the natural variability in the flow of the Nile River. *Water Resources Research* 32, 131–137.

Eltahir, E. A. B., and Wang, G. 1999. Nilometers, El Niño, and climate variability. *Geophysical Research Letters* 26, 489–492. doi: 10.1029/1999GL900013.

Endris, H. S., Lennard, C., Hewitson, B., Dosio, A., Nikulin, G., and Artan, G. A. 2019. Future changes in rainfall associated with ENSO, IOD and changes in the mean state over Eastern Africa. *Climate Dynamics* 52, 2029–2053. doi: 10.1007/s00382-018-4239-7.

Enfield, D. B., Mestas-Nunez, A. M., and Trimble, P. J. 2001. The Atlantic multidecadal oscillation and its relation to rainfall and river flows in the continental U.S. *Geophysical Research Letters* 28, 2077–2080.

Erban, L. E., Gorelick, S. M., and Zebker, H. A. 2014. Groundwater extraction, land subsidence, and sea-level rise in the Mekong Delta, Vietnam. *Environmental Research Letters* 9(8). doi: 10.1088/1748-9326/9/8/084010.

Erickson, C. L., and Walker, J. H. 2009. Pre-Columbian causeways and canals as landesque capital. In Snead, J., Erickson, C., and Darling, A. (eds.), *Landscapes of Movement: Trails, Paths, and Roads in Anthropological Perspective*. University of Pennsylvania Press, Philadelphia, pp. 232–252.

Erkens, G. Bucx, T., Dam, R., de Lange, G., and Lambert, J. 2015. Sinking cities: an integrated approach to solutions. In *Disaster Risk, edited by Global Facility for Disaster Reduction and Recovery, The Making of a Riskier Future: How Our Decisions Are Shaping Future*. Case Study C 90–99. World Bank, Washington, DC.

Erskine, W. D. 1992. Channel response to large-scale river training works: Hunter River, Australia. *River Research and Applications* 7, 261–278.

Erskine, W. D., and Webb, A. A. 2003. Desnagging to resnagging: new directions in river rehabilitation in southeastern Australia. *River Research and Applications* 19, 233–249.

Escarameia, M. 1998. *River and Channel Revetments: A Design Manual*. Thomas Telford, London.

Espa, P., Batalla, R. J., Brignoli, M. L., Crosa, G., Gentili, G., and Quadroni, S. 2019. Tackling reservoir siltation by controlled sediment flushing: impact on downstream fauna and related management issues. *PLoS ONE* 14, e0218822. doi: 10.1371/journal.pone.0218822.

Esposito, C. R., Georgiou, I. Y., and Kolker, A. S. 2013. Hydrodynamic and geomorphic controls on mouth bar evolution. *Geophysical Research Letters* 40, 1540–1545. doi: 10.1002/grl.50333.

European Centre for River Restoration (ECRR). 2018. www.ecrr.org/RiverRestoration/Whatisriverrestoration/tabid/2614/Default.aspx (accessed October 5, 2018).

European Commission (EC). 2019. Fitness check of the Water Framework Directive, Groundwater Directive, Environmental Quality Standards Directive and Floods Directive. December 10, 2019.

European Council (EC). 2000. Directive 2000/60/EC of the European Parliament and of the Council of 23 October 2000 establishing a framework for Community action in the field of water policy.

European Council (EC). 2007. Directive 2007/60/EC of the European Parliament and of the Council of 23 October 2007 on the assessment and management of flood risks.

European Environment Agency (EEA). 2018. *European Waters: Assessment of Status and Pressures 2018*. European Environmental Agency. EEA Report No 7/2018. Publications Office of the European Union, Luxembourg.

Evans, J. E. 2015. Contaminated sediment and dam removals: problem or opportunity? *Eos* 96. doi: 10.1029/2015EO036385.

Everitt, B. 1993. Channel responses to declining flow on the Rio Grande between Ft. Quitman and Presidio, Texas. *Geomorphology* 6, 225–242.

Executive Order 13690. 2015. Establishing a Federal Flood Risk Management Standard and a process for further soliciting and considering stakeholder input, January 30, 2015.

Fabris, M., Achilli, V., and Menin, A. 2014. Estimation of subsidence in Po Delta area (Northern Italy) by integration of GPS data, high-precision leveling and archival orthometric elevations. *International Journal of Geosciences* 5, 571–585. doi: 10.4236/ijg.2014.56052.

Fan, H., Huang, H., and Zeng, T. 2006. Impacts of anthropogenic activity on the recent evolution of the Huanghe (Yellow) River Delta. *Journal of Coastal Research* 22, 919–929.

Fan, H., Hea, D., and Wang, H. 2015. Environmental consequences of damming the mainstream Lancang-Mekong River: a review. *Earth Science Reviews* 146, 77–91.

FAO. 2011a. *AQUASTAT Country Profile – Cambodia.*. Food and Agriculture Organization of the United Nations (FAO), Rome, Italy.

FAO. 2011b. *AQUASTAT Transboundary River Basins – Indus River Basin*. Food and Agriculture Organization of the United Nations (FAO), Rome, Italy.

FAO. 2011c. *Irrigation in Southern and Eastern Asia in figures – AQUASTAT Survey*. Food and Agriculture Organization of the United Nations (FAO), Rome, Italy.

FAO. 2016a. *AQUASTAT Country Profile – Egypt*. Food and Agriculture Organization of the United Nations (FAO), Rome, Italy.

FAO. 2017. *The Future of Food and Agriculture – Trends and Challenges*. U.N. Food and Agricultural Organization, Rome, Italy.

Farrell, K. M. 1987. Sedimentology and facies architecture of overbank deposits of the Mississippi River, False River Region, Louisiana. In Etheridge, F. G., Flores, R. M., and Harvey, M. D. (eds), *Recent Developments in Fluvial Sedimentology*. Society of Economic Paleontologists and Mineralogists Special Publication 39, pp. 111–120.

Faulkner, D. J. 1998. Spatially variable historical alluviation and channel incision in west-central Wisconsin. *Annals of the Association of American Geographers* 88, 666–685.

Ferguson, H. B. 1940. *History of the Improvement of the Lower Mississippi River for Flood Control and Navigation, 1932–1939*. Mississippi River Commission, Vicksburg, MS.

Ferguson, R. I. 1986. Hydraulics and hydraulic geometry. *Progress in Physical Geography* 10, 1–31.

Ferguson, R. I. 1987. Hydraulic and sedimentary controls on channel pattern. In Richards, K. (ed.), *River Channels: Environment and Process*. Basil Blackwell Ltd, Oxford, pp. 129–158.

Ferguson, R. J., and Brierley, G. J. 1999. Levee morphology and sedimentology along the lower Tuross River, south-eastern Australia. *Sedimentology* 46, 627–648.

Fernández Garrido, P. 2018. Connecting fish, rivers, and people: Swimways around the world from local to global. Regional Conference on river habitat restoration for inland fisheries in the Danube River Basin and adjacent Black Sea áreas. Bucharest, November 13–15, 2018.

Ficchì, A., and Stephens, L. 2019. Climate variability alters flood timing across Africa. *Geophysical Research Letters* 46, 8809–8819. doi: 10.1029/2019GL081988.

Filizola, N., Latrubesse, E. M., Fraizy, P., Souza, R., Guimarães, V., and Guyot, J.-L. 2014. Was the 2009 flood the most hazardous or the largest ever recorded in the Amazon? *Geomorphology* 215, 99–105.

Fischenich, J. C. 2003. Effects of riprap on riverine and riparian ecosystems. ERDC/EL TR-03-4, U.S. Army Corps of Engineer Research and Development Center, Vicksburg, MS.

Fisher, W. L., Brown, L. F., Jr., Scott, A. J., and McGowen, J. H. 1969. *Delta Systems in the Exploration of Oil and Gas*. Bureau of Economic Geology, University of Texas at Austin.

Fisk, H. N. 1944. *Geological Investigation of the Alluvial Valley of the lower Mississippi River*. Mississippi River Commission, Vicksburg, MS.

Fisk, H. N. 1947. *Fine-Grained Alluvial Deposits and Their Effects on Mississippi River Activity*. Mississippi River Commission, Vicksburg, MS.

Fisk, H. N. 1952. *Geological Investigation of the Atchafalaya Basin and the Problem of the Mississippi River Diversion*. U.S. Army Waterways Experiment Station, Mississippi River Commission, Vicksburg, MS.

Florsheim, J. L., and Dettinger, M. D. 2015. Promoting atmospheric-river and snowmelt-fueled biogeomorphic processes by restoring river-floodplain connectivity in California's Central Valley. In Hudson, P. F., and Middelkoop, H. (eds.), *Geomorphic Approaches to Integrated Floodplain Management of Lowland Fluvial Systems in North America and Europe*. Springer-Verlag New York, New York, pp. 119–141.

Florsheim, J. L., and Mount, J. M. 2003. Changes in lowland floodplain sedimentation processes: pre-disturbance to post-rehabilitation, Cosumnes River, CA. *Geomorphology* 56, 305–323.

Florsheim, J. L., Mount, J. F., and Chin, A. 2008. Bank erosion as a desirable attribute of rivers. *BioScience* 58, 519–529.

Foley, J. A., Botta, A.A., Coe, M.T., and Costa, M.H. 2002. El Niño–Southern oscillation and the climate, ecosystems and rivers of Amazonia, *Global Biogeochemical Cycles* 16, 1132. doi: 10.1029/2002GB001872.

Foley, J. A., DeFries, N., Asner, R., *et al.* 2005. Global consequences of land use. *Science* 309, 570–574.

Foley, M. M., Warrick, J.A., Ritchie, A., *et al.* 2017a. Coastal habitat and biological community response to dam removal on the Elwha River. *Ecological Monographs* 87, 552–577. doi: 10.1002/ecm.1268.

Foley, M. M., *et al.* 2017b. Dam removal: listening in. *Water Resources Research* 53, 5229–5246. doi: 10.1002/2017WR020457.

Folk, R. L., and Ward, W. C. 1957. Brazos River bar: a study in the significance of grain size parameters. *Journal of Sedimentary Petrology* 27, 3–26.

Food and Agricultural Organization of the United Nations (FAO). 1988. *Nature and Management of Tropical Peat Soils*. Food and Agricultural Organization of the United Nations, FAO Soils Bulletin, 59. Rome.

Fovet, O., Ndom, N., Crave, A., and Pannard, A. 2020. Influence of dams on river water-quality signatures at event and seasonal scales: The Sélune

River (France) case study. River Research and Applications. doi: 10.1002/rra.3618.

Fox, C. A., Magilligan, F. J., and Snedden, C. S. 2016. "You kill the dam, you are killing a part of me": dam removal and the environmental politics of river restoration. *Geoforum* 70, 93–104.

Franklin, P., Dunbar, M., and Whitehead, P. 2008. Flow control on lowland river macrophytes: a review. *Science of the Total Environment* 400, 369–78.

Frazier, D. E. 1967. Recent deltaic deposits of the Mississippi River: their development and chronology. *Transactions Gulf Coast Association of Geological Society of America* 17, 287–311.

Friedkin, J. F. 1945. *A Laboratory Study of the Meandering of Alluvial Rivers*. U.S. Army Engineer Waterways Experiment Station, Vicksburg, MS.

Frings, R. M., Gehres, N., Promny, M., Middelkoop, H., Schuttrumpf, H., and Vollmer, S. 2014. Today's sediment budget of the Rhine River channel, focusing on the upper Rhine Graben and Rhenish Massif. *Geomorphology* 204, 573–587.

Frings, R. M, Hillebrand, G., Gehres, N., Banhold, K., Schriever, S., and Hoffmann, T. 2019. From source to mouth: basin-scale morphodynamics of the Rhine River. *Earth-Science Reviews* 196, 102830.

Fu, B., Wang, S., Liu, Y., Liu, J., Liang, W., and Miao, C. 2017. Hydrogeomorphic ecosystem responses to natural and anthropogenic changes in the Loess Plateau of China. *Annual Review of Earth and Planetary Sciences* 45, 223–224.

Fu, K. D., He, D. M., and Lu, X. X. 2008. Sedimentation in the Manwan reservoir in the Upper Mekong and its downstream impacts. *Quaternary International* 186, 91–99.

Fu, R., and Li, W. 2004. The influence of the land surface on the transition from dry to wet season in Amazonia. *Theoretical and Applied Climatology* 78, 98–110.

Fugate, J. M. 2014. Measurements of land subsidence rates on the northwestern portion of the Nile Delta using radar interferometry techniques. MSc thesis (supervisors R. H. Becker and M. Sultan), Department of Geology, University of Toledo.

Gagliano, S. M., and Howard, P. C. 1984. The neck cutoff oxbow lake cycle along the lower Mississippi River. In Elliot, C. M. (ed.), *River Meandering: Proceedings of the Conference Rivers '83*. American Society of Civil Engineers, New Orleans, pp. 147–158.

Galat, D. L., Fredrickson, L. H., Humburg, D. D., *et al.* 1998. Flooding to restore connectivity of regulated, large-river wetlands. *BioScience* 48, 721–733.

Gale, E. L., and Saunders, M. A. 2013. The 2011 Thailand flood: climate causes and return periods. *Weather* 68, 233–237. doi: 10.1002/wea.2133.

Galloway, D. C., Jones, D. R., and Ingebritsen, S. E. 1999. Land subsidence in the United States. U.S. Geological Survey, Circular, 1182.

Galloway, W. D. 1975. Process framework for describing the morphologic and stratigraphic evolution of deltaic depositional systems. In Broussard, M. L. (ed.), *Deltas, Models for Exploration*. Houston Geological Society, Houston, TX, pp. 86–98.

Ganti, V., Zhongxin, C., Lamb, M. P., Nittrouer, J. A., and Parker, G. 2016. Testing morphodynamic controls on the location and frequency of river avulsions on fans versus deltas: Huanghe (Yellow River), China. *Geophysical Research Letters* 41, 7882–7890. doi: 10.1002/2014GL061918.

García, N. O., and Mechoso, C. R. 2005. Variability in the discharge of South American rivers and in climate. *Hydrological Sciences Journal* 50. doi: 10.1623/hysj.50.3.459.65030.

García Martínez, M. F., Gottardi, G., Marchi, M., and Tonni, L. 2020. On the reactivation of sand boils near the Po River major embankments. In Calvetti, F., Cotecchia, F., Galli, A., and Jommi, C. (eds.), *Geotechnical Research for Land Protection and Development*. CNRIG 2019. Lecture Notes in Civil Engineering, vol. 40. Springer, Cham, pp. 328–337.

García-Ruiz, J. M. 2010. The effects of land uses on soil erosion in Spain: a review. *Catena* 81, 1–15.

Garver, J. I., and Cockburn, J. M. H. 2009. A historical perspective of ice jams on the lower Mohawk River. In *Proceedings from the 2009 Mohawk Watershed Symposium*, Schenectady, NY, vol. 1, pp. 25–29.

Gasparini, N. M., Fischer, G. C., Adams, J. M., Dawers, N. J., and Janoff, A. M. 2016. Morphological signatures of normal faulting in low-gradient alluvial rivers in south-eastern Louisiana, USA. *Earth Surface Processes and Landforms* 41, 642–657. doi: 10.1002/esp.3852.

Gastaldo, R. A., Allen, G. P., and Huc, A.-Y. 1995. The tidal character of fluvial sediments of the modern Mahakam River delta, Kalimantan, Indonesia. *Special Publications International Association of Sedimentologists* 24, 171–181.

Gebrehiwot, S. G., Ellison, D., Bewket, W., Seleshi, Y., Inogwabini, B.-I., and Bishop, K. 2019. The Nile Basin waters and the West African rainforest: rethinking the boundaries. *WIREs Water* 6, e1317. doi: 10.1002/wat2.1317.

Geerling, G. W., Kater, E., van den Brink, E. K. C., Baptist, M. J., Ragas, A. M. J., and Smits, A. J. M. 2008. Nature rehabilitation by floodplain excavation: the hydraulic effect of 16 years of sedimentation and vegetation succession along the Waal River, NL. *Geomorphology* 99, 317–328

Gergel, S. E., Dixon, M. D., and Turner, M. G. 2002. Consequences of human-altered floods: levees, floods, and floodplain forests along the Wisconsin River. *Ecological Applications* 12, 1755–1770.

Gibson, S., Osorio, A., Creech, C., et al. 2019. Two pool-to-pool spacing periods on large sand-bed rivers: mega-pools on the Madeira and Mississippi. *Geomorphology* 328, 196–210.

Gilbert, G. K. 1917. Hydraulic mining debris in the Sierra Nevada. U.S. Geological Survey Professional Paper 105. Washington, DC.

Gimeno, L., Nieto, R., Vázquez, M., and Lavers, D. A. 2014. Atmospheric rivers: a mini-review. *Frontiers in Earth Science* 2. doi: 10.3389/feart.2014.00002.

Giosan, L., Syvitski, J., Constantinescu, S., and Day, J. 2014. Climate change: protect the world's deltas. *Nature* 516, 31–33.

Givetz, E. H. 2010. Removing erosion control projects increases bank swallow (Riparia riparia) population viability modeled along the Sacramento River, California, USA. *Biological Conservation* 143, 428–438.

Gleason, C. J. 2015. Hydraulic geometry of natural rivers: a review and future directions. *Progress in Physical Geography: Earth and Environment* 39, 337–360.

Glynn, M. E., and Kuszmaul, J. 2010. Prediction of piping erosion along Middle Mississippi River levees – an empirical model. Final Report, U.S. Army Corps of Engineers, Waterways Experiment Station, ERDC/GSL TR-04-12, Vicksburg, MS.

Gomez, B. 2006. The potential rate of bed-load transport. *Proceedings of the National Academy of Sciences USA* 103(46): 17170–17173.

Gomez, B., and Marron, D. C. 1991. Neotectonic effects on sinuosity and channel migration, Belle Fourche River, Western South Dakota. *Earth Surface Processes and Landforms* 16, 227–335. doi: 10.1002/esp.3290160304.

Gomez, B., Phillips, J. D., Magilligan, F. J., and James, L. A. 1997. Floodplain sedimentation and sensitivity: summer 1993 flood, Upper Mississippi River Valley. *Earth Surface Processes and Landforms* 22, 923–936.

Gong, S. L., Li, C., and Lang, S. L. 2008. The microscopic characteristics of Shanghai soft clay and its effect on soil body deformation and land subsidence. *Environmental Geology* 6. doi: 10.1007/s00254-008-1205-4.

Goodbred, S. L. J., and Saito, Y. 2012. Tide-dominated deltas. In Davis, R. A. J., and Dalrymple, R. W. (eds.), *Principles of Tidal Sedimentology*. Springer, Dordrecht, pp. 129–149.

Górski, K., van den Bosch, L. V., van de Wolfshaar, K. E., *et al.* 2012. Post-damming flow regime development in a large lowland river (Volga, Russian Federation): implications for floodplain inundation and fisheries. *River Research and Applications* 28, 1121–1134.

Götz, E. 2008. Improved sediment-management strategies for the sustainable development of German waterways. Sediment Dynamics in Changing Environments (Proceedings of a symposium in Christchurch, New Zealand, December 2008). International Association of Hydrological Sciences, Publ. 325, pp. 540–549.

Gottschalk, M. K. E. 1971–1977. *Stormvloeden en Rivieroverstromingen in Nederland*, vol. 3. Van Gorcum & Comp. N.V., Assen.

Goudie, A. S. 2013. *The Human Impact on the Environment: Past, Present and Future*, 7th ed. Wiley, London.

Gouw, M. 2007. Alluvial architecture of the Holocene Rhine-Meuse delta (The Netherlands) and the lower Mississippi Valley (USA). Netherlands Geographical Studies 364. KNAG/Faculteit Ruimtelijke Wetenschappen Universiteit Utrecht.

Guan, B., and Waliser, D. E. 2015. Detection of atmospheric rivers: evaluation and application of an algorithm for global studies. *Journal of Geophysical Research: Atmospheres* 120, 12514–12535. doi: 10.1002/2015JD024257.

Guan, B., Molotch, N. P., Waliser, D. E., Fetzer, E. J., and Neiman, P. J. 2010. Extreme snowfall events linked to atmospheric rivers and surface air temperature via satellite measurements. *Geophysical Research Letters* 37, L20401. doi: 10.1029/2010GL044696.

Graf, W. L. 1977. The rate law in fluvial geomorphology. *American Journal of Science* 277, 178–191.

Graf, W. L. 1999. Dam nation: a geographic census of large American dams and their hydrologic impacts. *Water Resources Research* 35, 1305–1311.

Graf, W. L. 2001. Damage control: dams and the physical integrity of America's rivers. *Annals of the Association of American Geographers* 91, 1–27.

Graf, W. L. 2002. Rivers, dams, and willow flycatchers: a summary of their science and policy connections. *Geomorphology* 47, 169–188.

Graf, W. L. 2003. The changing role of dams in water resources management. Universities Council on Water Resources, *Water Resources Update* 126, 54–59.

Graf, W. L. 2006. Downstream hydrologic and geomorphic effects of large dams on American rivers. *Geomorphology* 79, 336–360.

Graf, W. L., Wohl, E., Sinha, T., and Sabo, J. L. 2011. Sedimentation and sustainability of western American reservoirs. *Water Resources Research* 46. doi: 10.1029/2009WR008836.

Graham, O. P. 1992. Survey of land degradation in New South Wales, Australia. *Environmental Management* 16, 205–223. doi: 10.1007/BF02393826.

Grall, C., Steckler, M. S., Pickering, J. L., *et al.* 2018. A base-level stratigraphic approach to determining Holocene subsidence of the Ganges–Meghna–Brahmaputra Delta plain. *Earth and Planetary Science Letters* 499, 23–36.

Grams, P. E., and Schmidt, J. C. 2002. Streamflow regulation and multi-level flood plain formation: channel narrowing on the aggrading Green River in the eastern Uinta Mountains, Colorado and Utah. *Geomorphology* 44, 337–360.

Gray, J. R., and Landers, M. N. 2014. Measuring suspended sediment. In Ahuja, S. (ed.), *Comprehensive Water Quality and Purification*, vol. 1. Elsevier, Philadelphia, PA, pp. 157–204.

Gregory, K. J. 2006. The human role in changing river channels. *Geomorphology* 79, 172–191.

Gregory, K. J. 2010. *The Earth's Land Surface: Landforms and Processes in Geomorphology*. Sage, London.

Grenfell, M., Aalto, R., and Nicholas, A. 2012. Chute channel dynamics in large, sand-bed meandering rivers. *Earth Surface Processes and Landforms* 37, 315–331. doi: 10.1002/esp.2257.

Grenfell, M. C., Nicholas, A. P., and Aalto, R. 2014. Meditative adjustment of river dynamics: the role of chute channels in tropical sand-bed meandering rivers. *Sedimentary Geology* 301, 93–106. doi: 10.1016/j.sedgeo.2013.006.007.

Greulich, S., Franklin, S. B., Wasklewicz, T., and Grubaugh, J. 2007. Hydrogeomorphology and forest composition of sunrise towhead island in the lower Mississippi River. *Southeastern Naturalist* 6, 217–234.

Grill, G., Lehner, B., Lumsdon, A. E., MacDonald, G. K., Zarfl, C., and Liermann, C. R. 2015. An index-based framework for assessing patterns and trends in river fragmentation and flow regulation by global dams at multiple scales. *Environmental Research Letters* 10, 1.

Grimardias, D., Guillard, J., and Cattanéo, F. 2017. Drawdown flushing of a hydroelectric reservoir on the Rhône River: impacts on the fish community and implications for the sediment management. *Journal of Environmental Management* 197, 239–249. doi: 10.1016/j.jenvman.2017.03.096.

Guarino, J. 2013. Tribal advocacy and the art of dam removal: the lower Elwha Klallam and the Elwha Dams. *American Indian Law Journal* 11, 114–145.

Guccione, M. J., van Arsdale, R. B., and Lynne, H. H. 2000. Origin and age of the Manilla high and associated Big Lake "sunklands" in the New Madrid seismic zone, northeastern Arkansas. *Geological Society of America Bulletin* 112, 579–590.

Guida, R. J., Swanson, T. L., Remo, J. W. F., and Kiss, T. 2013. Strategic floodplain reconnection for the Lower Tisza River, Hungary: opportunities for flood-height reduction and floodplain-wetland reconnection. *Journal of Hydrology* 521, 274–285.

Guimarães Nobre, G., Jongman, B., Aerts, J., and Ward, P. J. 2017. The role of climate variability in extreme floods in Europe. *Environmental Research Letters* 12, 084012. doi: 10.1088/1748-9326/aa7c22.

Guirguis, K., Gershunov, A., Shulgina, T., Clemesha, R. E. S., and Ralph, F. M. 2019. Atmospheric rivers impacting Northern California and their modulation by a variable climate. *Climate Dynamics* 52, 6569–6583.

Guntren, E. L. M., Oliver, A. J. M., and Keevin, T. M. 2016. Change in lower Mississippi River Secondary Channels: an Atlas of Bathymetric and Photographic Data. Mississippi River Geomorphology and Potamology Program, Report No. 8. St. Louis Division, U.S. Army Corps of Engineers.

Gurnell, A. M., Piegay, H., Swanson, F. J., and Gregory, S. V. 2002. Large wood and fluvial processes. *Freshwater Biology* 47, 601–619.

Gurnell, A. M. Corenblit, D., de Jalón, D. G., *et al.* 2016. A conceptual model of vegetation-hydrogeomorphology interactions within river corridors. *River Research and Applications* 32, 142–163.

Gupta, A. 2007. Introduction. In Gupta, A. (ed.), *Large Rivers: Geomorphology and Management*. John Wiley & Sons, Chichester, pp. 1–5.

Gupta, N., Kleinhans, M. G., Addink, E. A., Atkinson, P. M., and Carling, P. A. 2014. One-dimensional modeling of a recent Ganga avulsion: assessing the potential effect of tectonic subsidence on a large river. *Geomorphology* 213, 24–37.

Ha, D. T., Ouillon, S., and van Vinh, G. 2018. Water and suspended sediment budgets in the lower Mekong from High-Frequency Measurements (2009–2016). *Water* 10, 846. doi: 10.3390/w10070846.

Habersack, H., Jäger, E., and Hauer, C. 2013. The status of the Danube River sediment regime and morphology as a basis for future basin management. *International Journal of River Basin Management* 11, 153–166.

Habersack, H., Hein, T., Stanica, A., *et al.* 2016. Challenges of river basin management: current status of, and prospects for, the River Danube from a river engineering perspective. *Science of the Total Environment* 543, 828–845.

Hackney, C. R., Darby, S. E., Parsons, D. R., *et al.* 2020. River bank instability from unsustainable sand mining in the lower Mekong River. *Nature Sustainability* 3, 217–225. doi: 10.1038/s41893–019–0455-3.

Hanel, M., Rakovec, O., Markonis, Y., *et al.* 2018. Revisiting the recent European droughts from a long-term perspective. *Scientific Reports* 8, 9499. doi: 10.1038/s41598–018-27464-4.

Hanberry, B. B., Kabrick, J. M., and He, H. S. 2015. Potential tree and soil carbon storage in a major historical floodplain forest with disrupted ecological function. *Perspectives in Plant Ecology, Evolution and Systematics* 17, 17–23.

Hannan, A. 1969. Study of Mississippi River bends. Unpublished doctoral dissertation, Colorado State University, Ft. Collins, CO.

Hanson, S., Nicholls, R., Ranger, N., *et al.* 2011. A global ranking of port cities with high exposure to climate extremes. *Climatic Change* 104, 89–111. doi: 10.1007/s10584-010-9977-4.

Happ, S. C., Rittenhouse, G., and Dobson, G. C. 1940. Some principles of accelerated stream and valley sedimentation. US Department of Agriculture, Technical Bulletin 695.

Haraguchi, M., and Lall, U. 2015. Flood risks and impacts: a case study of Thailand's floods in 2011 and research questions for supply chain decision making. *International Journal of Disaster Risk Reduction* 14, 256–272.

Harbor, D. J., Schumm, S. A., and Harvey, M D. 1994. Tectonic control of the Indus River. In Schumm, S. A., and Winkley, B. (eds), *The Variability of Large Alluvial Rivers*. American Society of Civil Engineers, New York, pp. 161–176.

Hardiman, S. C., Dunstone, N. J., Scaife, A. A., *et al.* 2018. The asymmetric response of Yangtze river basin summer rainfall to El Niño/La Niña. *Environmental Research Letters* 13, 024015. doi: 10.1088/1748-9326/aaa172.

Harmar, O. P., Clifford, N. J., Thorne, C. R., and Biedenharn, D. S. 2005. Morphological changes of the lower Mississippi River: geomorphological response to engineering intervention. *River Research and Applications* 21, 1107–1131.

Harrison, L. R., Dunne, T., and Fisher, G. B. 2015. Hydraulic and geomorphic processes in an overbank flood along a meandering, gravel-bed river: implications for chute formation. *Earth Surface Processes and Landforms* 40, 1239–1253. doi: 10.1002/esp.3717.

Harrison, R. W. 1950. Flood control in the Yazoo-Mississippi Delta. *Southern Economic Journal* 17, 1481558.

Havakes, H., Koster, M., Dekking, W., Uijterlinde, R., Wensink, W., and Walker, R. 2015. *Water Governance: The Dutch Water Authority Model*. Dutch Water Authorities, The Hague.

Hayden, B. 1988. Flood hydroclimatology. In Baker, V. R., Kochel, R. C., and Patton, P. C. (eds.), *Flood Geomorphology*. John Wiley & Sons, New York, pp. 13–26.

He, S., Gao, Y., Li, F., Wang, H., and He, Y. 2017. Impact of Arctic Oscillation on the East Asian climate: a review. *Earth-Science Reviews* 164, 48–62.

Heath, R. C. 1983. Basic ground-water hydrology. U.S. Geological Survey Water-Supply Paper 2220.

Heimann, D. C. 2016. Generalized sediment budgets of the lower Missouri River, 1968–2014. U.S. Geological Survey Scientific Investigations Report 2016–5097. doi: 10.3133/sir20165097.

Hein, T., Funk, A., Pletterbauer, F., et al. 2018. Management challenges related to long-term ecological impacts, complex stressor interactions, and different assessment approaches in the Danube River Basin. River Research and Applications. doi: 10.1002/rra.3243.

Heine, R. A., and Pinter, N. 2012. Levee effects upon flood levels: an empirical assessment. *Hydrological Processes* 26, 3225–3240. doi: 10.1002/hyp.8261.

Heinz Center. 2002. Dam removal: science and decision making. Final Report. The H. John Heinz III Center for Science, Economics, and the Environment.

Heitmuller, F. T., and Greene, L. E. 2009. Historical channel adjustment and estimates of selected hydraulic values in the lower Sabine River and lower Brazos River Basins, Texas and Louisiana: U.S. Geological Survey Scientific Investigations Report 2009–5174, 143 p.

Heitmuller, F. T., Hudson, P. F., and Kesel, R. H. 2017. Overbank sedimentation from the historic AD 2011 flood along the lower Mississippi River, USA. *Geology* 45, 107–110.

Helmholtz Centre for Environmental Research – UFZ. 2017. New standards for better water quality in Europe: researchers present recommendations for revision of the EU Water Framework Directive. ScienceDaily, February 27, 2017. www.sciencedaily.com/releases/2017/02/170227100740.htm (accessed May 2020).

Hendon, H. H., Thompson, D. W. J., and Wheeler, M. C. 2007. Australian rainfall and surface temperature variations associated with the Southern Hemisphere Annular Mode. *Journal of Climate* 40, 2452–2467.

Hennessy, K. J., Suppiah, R., and Page, C. M. 1999. Australian rainfall changes, 1910–1995. *Australian Meteorological Magazine* 48, 1–13.

Henry, R. E., Sommer, T. R., and Goldman, C. R. 2010. Growth and methylmercury accumulation in juvenile Chinook salmon in the Sacramento River and its floodplain, the Yolo Bypass. *Transactions of the American Fisheries Society* 139, 550–563.

Hensel, P. E., Day, J. W., and Pont, D. 1999. Wetland vertical accretion and soil elevation change in the Rhône River Delta, France: the importance of riverine flooding. *Journal of Coastal Research* 15, 668–681.

Herget, J., Eckhard, B., Coch, T., Dix, E., Eggenstein, G., and Ewald, K. 2005. Engineering impact on river channels in the River Rhine catchment. *Erdkunde* 59, 294–319.

Hesselink, A. W. 2002. History makes a river. Morphological development of the embanked floodplains of the Rhine and Meuse in the Netherlands in historical time. Netherlands Geographical Studies 292. KNAG/Faculteit Ruimtelijke Wetenschappen Universiteit Utrecht.

Hesselink, A. W., Weerts, H. J. T., and Berendsen, H. J. A. 2003. Alluvial architecture of the human-influenced river Rhine, The Netherlands. *Sedimentary Geology* 161, 229–248.

Hewson, W. 1860. *Principles and Practice of Embanking Lands from River-Floods as Applied to the "Levees" along the Lower Mississippi River.* J. J. Reed, New York.

Hey, R. D., and Tovey, N. K. 1989. Processes of bank failure. In Hemphill, R. W., and Bramley, M. E. (eds.), *Protection of River and Canal Banks*. CIRIA/Butterworths, London, pp. 7–39.

Hickin, E. J. 1977. Hydraulic factors controlling channel migration. In Dacidson-Aenott, R., and Nicking W. (eds.), *Research in Fluvial Systems*. Proceedings of the 5th Guelph Geomorph Symposium, Geobooks, Norwich, pp. 59–72.

Hickin, E. J. 1984. Vegetation and river channel dynamics. *Canadian Geographer* 28, 111–126.

Hickin, E. J., and Nanson, G. C. 1975. The character of channel migration on the Beatton River, north-east British Columbia, Canada. *Bulletin of the Geological Society of America* 86, 487–494.

Hickin, E. J., and Nanson, G. C. 1984. Lateral migration rates of river bends. *Journal of Hydraulic Engineering, ASCE* 110, 1557–1567.

Higgins, J. M., and Brock, W. G. 1999. Overview of reservoir release improvements at 20 TVA dams. *Journal of Energy Engineering* 125, 1–17.

Higgins, S. Overeem, I., Tanaka, A., and Syvitski, J. P. M. 2013. Land subsidence at aquaculture facilities in the Yellow River delta, China. *Geophysical Research Letters* 40, 3898–3902. doi: 10.1002/grl.50758.

Hirschboeck, K. K. 1988. Flood hydroclimatology. In Baker, V. R., Kochel, R. C., and Patton, P. C. (eds.), *Flood Geomorphology*. John Wiley & Sons, New York, pp. 27–49.

Hirschboeck, K. K. 1991. Climate and floods, in National Water Summary 1988–1989, Hydrologic Events and Floods and Droughts. U.S. Geological Survey Water Supply Paper 2375, pp. 67–88.

Hockenos, P. 2018. A Balkan dam boom imperils Europe's wildest rivers. Yale Environment 360. Published at the Yale School of Forestry and Environmental Studies. https://e360.yale.edu/features/a-balkan-dam-boom-imperils-europes-wildest-rivers (accessed October 14, 2019).

Hodgkins, G. A., Whitfield, P. H., Burn, D. H., et al. 2017. Climate-driven variability in the occurrence of major floods across North America and Europe. *Journal of Hydrology* 552, 704–717.

Hoffmann, T., Erkens, G., Gerlach, R., Klostermann, J., and Lang, A. 2009. Trends and controls of Holocene floodplain sedimentation in the Rhine catchment. *Catena* 77, 96–106.

Hoffmann, T., Schlummer, M., Notebaert, B., Verstraeten, G., and Korup, O. 2013. Carbon burial in soil sediments from Holocene agricultural erosion, Central Europe. *Global Biogeochemical Cycles* 27, 828–835.

Hohensinner, S., Egger, G., Haidvogl, G., Jungwirth, M., Muhar, S., and Schmutz, S. 2007. Hydrological connectivity of a Danube river-floodplain system in the Austrian Machland: changes between 1812 and 1991. In, Floodplain Protection, Restoration, Management. Why and how? Editions Lavoisier SAS, Paris, pp. 53–69.

Holbrook, J., and Schumm, S. A. 1999. Geomorphic and sedimentary response of rivers to tectonic deformation. *Tectonophysics* 305, 287–306.

Holmes, R. R., Jr., Koenig, T. A., and Karstensen, K. A. 2010. Flooding in the United States Midwest, 2008. U.S. Geological Survey Professional Paper 1775.

Hooke, J. M. 1979. An analysis of the processes of river bank erosion. *Journal of Hydrology* 42, 39–62.

Hooke, J. M. 1986. The significance of mid-channel bars in an active meandering river. *Sedimentology* 33, 839–850.

Hori, K., and Saito, Y. 2007. Classification, architecture, and evolution of Large-River Deltas. In Gupta, A. (ed.), *Large Rivers: Geomorphology and Management*. John Wiley & Sons, Chichester, pp. 75–96.

Hortle, K. G., and Nam, S. 2017. Mitigation of the impacts of dams on fisheries – a primer. Mekong Development Series No. 7. Mekong River Commission, Vientiane Lao PDR.

Houben, P. 2008. Scale linkage and contingency effects of field-scale and hillslope-scale controls of long-term soil erosion: anthropogeomorphic sediment flux in agricultural loess watersheds of Southern Germany. *Geomorphology* 101, 172–191.

Huang, H. Q., and Nanson, G. C. 1998. The influence of bank strength on channel geometry: an integrated analysis of some observations. *Earth Surface Processes and Landforms* 23, 865–876.

Huang, Q. G., and Nanson, G. A. 2007. Why some alluvial rivers develop an anabranching pattern. *Water Resources Research* 43. doi: 10.1029/2006WR005223.

Huat, B. B. K., Kazemian, S., Prasad, A., and Barghchi, M. 2011. State of an art review of peat: general perspective. *International Journal of the Physical Sciences* 6, 1988–1996. doi: 10.5897/IJPS11.192.

Huat, B. B. K., Prasad, A., Asadi, A., and Kazemian, S. 2014. *Geotechnics of Organic Soils and Peat*. CRC Press.

Huđek, H., Žganecc, K., and Puscha, M. T. 2020. A review of hydropower dams in Southeast Europe – distribution, trends and availability of monitoring data using the example of a multinational Danube catchment sub-area. *Renewable and Sustainable Energy Reviews* 117, 109434. doi: 10.1016/j.rser.2019.109434.

Hudson, P. F. 2002. Pool-riffle morphology in an actively migrating alluvial channel: the lower Mississippi River. *Physical Geography* 23, 154–169.

Hudson, P. F. 2003. Event sequence and sediment exhaustion in the lower Pánuco basin, Mexico. *Catena* 52, 56–76.

Hudson, P. F. 2004a. Geomorphic context of the prehistoric Huastec floodplain environments: lower Pánuco basin, Mexico. *Journal of Archaeological Science* 31, 653–668.

Hudson, P. F. 2004b. The influence of the El Niño Southern Oscillation on suspended sediment load variability in a seasonally humid tropical setting: Pánuco Basin, Mexico. *Geografiska Annaler: Series A, Physical Geography* 85, 263–275.

Hudson, P. F. 2017. Water engineering. In Richardson, D., Castree, N., Kobayashi, A., Liu, W., Goodchild, M., and Marston, R. (eds.), *International Encyclopedia of Geography: People, Earth, Environment and Technology*. American Association of Geographers, Wiley-Blackwell. DOI: 10.1002/9781118786352.wbieg0868.

Hudson, P. F. 2018. Towards integrated flood management along the lower Rhine and Mississippi Rivers and the international legacy of the 2005 New Orleans Hurricanes Katrina–Rita flood disaster. *International Journal of River Basin Management* 16, 273–285.

Hudson, P. F. 2020. Biogeomorphic evolution of lower Mississippi islands: 1965–2015. In Boersema, M. P., Schielen, R. M. J., van Eijsbergen, E., and Rinsema, J. G. (eds.), *Managing Changing Rivers: NCR Days 2020 Proceedings*. Netherlands Centre for River Studies, Publication 44–2020.

Hudson, P. F., and Colditz, R. R. 2003. Flood delineation in a large and complex alluvial valley, lower Pánuco basin, Mexico. *Journal of Hydrology* 280, 229–245.

Hudson, P. F., and Heitmuller, F. T. 2003. Local- and watershed-scale controls on the spatial variability of natural levee deposits in a large fine-grained floodplain: lower Pánuco basin, Mexico. *Geomorphology* 56, 255–269.

Hudson, P. F., and Kesel, R. H. 2000. Channel migration and meander-bend curvature in the lower Mississippi River prior to major modification. *Geology* 28, 531–534.

Hudson, P. F., and Kesel, R. H. 2006. Spatial and temporal adjustment of the lower Mississippi River to major human impacts. *Zeitschrift für Geomorphologie, Supplementband* 143, 17–33.

Hudson, P. F., and Middelkoop, H. 2015. The Palimpsest of river-floodplain management and the role of geomorphology. In Hudson, P. F., and Middelkoop, H. (eds.), *Geomorphic Approaches to Integrated Floodplain Management of Lowland Fluvial Systems in North America and Europe*. Springer-Verlag, New York, pp. 337–350.

Hudson, P. F., and Mossa, J. 1997. Suspended sediment transport effectiveness of three large impounded rivers, U.S. Gulf Coastal Plain. *Environmental Geology* 32, 263–273.

Hudson, P. F., Middelkoop, H., and Stouthamer, E. 2008. Flood management along the Lower Mississippi and Rhine Rivers (The Netherlands) and the continuum of geomorphic adjustment. *Geomorphology* 101, 209–236.

Hudson, P. F., Heitmuller, F. T., Leitch, M. B. 2012. Hydrologic connectivity of oxbow lakes along the lower Guadalupe River, Texas: the influence of geomorphic and climatic controls on the "Flood Pulse Concept." *Journal of Hydrology* 414, 174–183.

Hudson, P. F., Sounny-Slitine, M. A., and LaFevor, M. 2013. A new longitudinal approach to assess hydrologic connectivity: embanked floodplain inundation along the lower Mississippi River. *Hydrological Processes* 27, 2187–2196.

Hudson, P. F., van der Hout, E., and Verdaasdonk, M. 2019. (Re)Development of fluvial islands along the lower Mississippi River over five decades, 1965–2015. *Geomorphology* 331, 78–91.

Humphreys, A. A., and Abbot, H. L. 1863. Report upon the physics and hydraulics of the Mississippi River, upon the protection of the alluvial region against overflow; and upon the deepening of the mouths. In Lippincott, J. B. (ed.), *Professional Papers of the Topographical Engineers, 4*. Philadelphia, PA, 456 p.

Hundey, E. J., and Ashmore, P. E. 2009. Length scale of braided river morphology. *Water Resources Research* 45, W08409. doi: 10.1029/2008WR007521.

Hurrell, J. W. 1995. Decadal trends in the North Atlantic Oscillation: regional temperatures and precipitation. *Science* 269, 676–679.

Hupp, C. P., Demas, C. R., Kroes, D. E., Day, R. H., and Doyle, T. W. 2008. Recent sedimentation patterns within the central Atchafalaya Basin, Louisiana. *Wetlands* 28, 125–140.

Ibàñez, C., Canicio, A., Day, J. W., and Curcó, A. 1997. Morphologic development, relative sea level rise and sustainable management of water and sediment in the Ebre Delta, Spain. *Journal of Coastal Conservation* 97, 191–202. doi: 10.1007/BF02908194.

Ibáñez, C., Caiola, N., Rovira, A., and Real, M. 2012. Monitoring the effects of floods on submerged macrophytes in a large river. *Science of the Total Environment* 440, 132–139.

IKSE. 2005. Die Elbe und ihr einzugsgebiet: Ein geographisch-hydrologischer und wasserwirtschaftlicher überblick (The Elbe and its catchment area: a geographical-hydrological and water management overview). Internationale Kommission zum Schutz der Elbe, Magdeburg. www.ikse-mkol.org/fileadmin/media/user_upload/D/06_Publikationen/07_Verschiedenes/2005_IKSE-Elbe-und-ihr-Einzugsgebiet.pdf (accessed June 5, 2020).

Independent Levee Investigation Team (ILIT). 2006. Investigation of the New Orleans flood protection systems in Hurricane Katrina on August 29, 2005, Final Report, July 31, 2006, www.ce.berkeley.edu/~new_orleans/ (accessed April 18, 2016).

Ingebritsen, S. E., Ikehara, M. E., Galloway, D. L., and Jones, D. R. 2000. Delta subsidence in California; the sinking heart of the state. US Geological Survey Fact Sheet 005–00.

Interagency Floodplain Management Review Committee. 1994. A blueprint for change. Sharing the challenge: floodplain management into the 21st century. Final Report to the Administration Floodplain Management Task Force, June 1994, Washington DC.

Interagency Levee Policy Review Committee (ILPRC). 2006. The national levee challenge: levees and the FEMA map modernization initiative. Stakeholder Committee Report on Levees. 2006. www.fema.gov/library/viewRecord.do?id=2677 (accessed February 5, 2016).

Interagency Performance Evaluation Task Force (IPET). 2007. Performance evaluation of the New Orleans and Southeast Louisiana Hurricane Protection System. Volume III – The Hurricane Protection System. U.S. Army Corps of Engineers. http://biotech.law.lsu.edu/katrina/ipet/.html (accessed May 15, 2016).

Interagency Performance Evaluation Task Force (IPET). 2009. Performance evaluation of the New Orleans and Southeast Louisiana Hurricane Protection System – Final Report, U.S. Army Corps of Engineers. http://permanent.access.gpo.gov/lps71007/ (accessed May 18, 2016)

International Commission for the Protection of the Danube River (ICPDR). 2015. Flood risk management plan for the Danube River Basin District. Document number 4.5, May 28, 2015, Vienna, Austria.

International Commission for the Protection of the Rhine (ICPR). 2013a. *The Rhine and Its Catchment: An Overview*. Koblenz, Germany.

International Commission for the Protection of the Rhine (ICPR). 2013b. Implementation of the sediment management plan. Report on the State of Implementation by End 2013. Report No. 212. Koblenz, Germany.

International Commission for the Protection of the Rhine (ICPR). 2015a. Strategy for the IRBD Rhine for adapting to climate change. Report No. 219. Koblenz, Germany.

International Commission for the Protection of the Rhine (ICPR). 2015b. Internationally coordinated flood risk management plan for the International River Basin District of the Rhine, Part A. December 2015. Koblenz, Germany.

International Commission for the Protection of the Rhine (ICPR). 2015c. Summary report on the Rhine Measurement Programme Biology 2012/2013 and National Assessments according to the WFD. Report No. 232. Koblenz, Germany.

International Commission for the Protection of the Rhine (ICPR). 2016. Assessment of flood risk reduction (Action Plan on Floods, Action Target 1) with due regard to types of measures and receptors of the Directive 2007/60/EC (FD) – synthesis report. Report No. 236. Koblenz, Germany.

International Commission for the Protection of the Rhine (ICPR). 2018. Salmon is progressing. www.iksr.org/en/topics/ecology/plants-and-animals/fish/salmon-is-progressing/ (accessed August 8, 2018)]

International Commission for the Protection of the Rhine (ICPR). 2020a. Assessment Rhine 2020. Koblenz, Germany.

International Commission for the Protection of the Rhine (ICPR). 2020b. Rhine 2040. The Rhine and its catchment: sustainably managed and climate-resilient. 16th Rhine Ministerial Conference, February 13, 2020, Amsterdam.

Ionita, M., Rimbu, N., Lohmann, G. 2011. Decadal variability of the Elbe River streamflow. *International Journal of Climatology* 31, 22–30.

IWPDC. 2020. Moving on up – India now world's 5th largest hydropower producer. International Water Power and Dam Construction, June 1, 2020.

Jacobson, R. B., and Oberg, K. A. 1997. Geomorphic changes on the Mississippi River flood plain at Miller City, Illinois, as a result of the flood of 1993. U.S. Geological Survey Circular 1120-J.

Jacobson, R. B., Blevins, D. W., and Bitner, C. J. 2009. Sediment regime constraints on river restoration – an example from the Lower Missouri River. In James, L. A., Rathburn, S. L., and Whittecar, G. R. (eds.), *Management and Restoration of Fluvial Systems with Broad Historical Changes and Human Impacts*. Geological Society of America Special Paper 451, Boulder, CO, pp. 1–22.

Jacobson, R. B., Linder, G., and Bitner, C. 2015. The role of floodplain restoration in mitigating flood risk, Lower Missouri River, USA. In Hudson, P. F., and Middelkoop, H. (eds.), *Geomorphic Approaches to Integrated Floodplain Management of Lowland Fluvial Systems in North America and Europe*. Springer-Verlag, New York, pp. 203–243.

Jager, N. W., Challies, E., Kochskämper, E. *et al.* 2016. Transforming European water governance? participation and river basin management under the EU Water Framework Directive in 13 member states. *Water* 8, 156.

James, L. A. 1989. Sustained storage and transport of hydraulic mining sediment in the Bear River, California. *Annals of the Association of American Geographers* 79, 570–592.

James, L. A. 2006. Bed waves at the basin scale: implications for river management and restoration. *Earth Surface Processes and Landforms* 31, 1692–1706.

James, L. A. 2011. Contrasting geomorphic impacts of pre-and post-Columbian land-use changes in Anglo America. *Physical Geography* 32, 399–422.

James, L. A. 2013. Legacy sediment: definitions and processes of episodically produced anthropogenic sediment. *Anthropocene* 2, 16–26.

James, L. A., and Lecce, S. A. 2013. Impacts of land-use and land-cover change on river systems. In Shroder, J. (editors in chief) and Wohl, E. (ed.), *Treatise on Geomorphology*, vol. 9. Fluvial Geomorphology. Academic Press, San Diego, CA, pp. 768–793.

James, L. A., and Marcus, W. A. 2006. The human role in changing fluvial systems: retrospect, inventory and prospect. *Geomorphology* 79, 152–171.

James, L. A., and Singer, M. B. 2008. Development of the lower Sacramento valley flood-control system: historical perspective. *Natural Hazards Review* 2008, 125–135.

James, L. A., Singer, M. B., Ghoshal, S., and Megison, M. 2009. Historical channel changes in the lower Yuba and Feather Rivers, California: Long-term effects of contrasting river-management strategies. In James, L. A., Rathburn, S. L., and Whittecar, G. R. (eds.), *Management and Restoration of Fluvial Systems with Broad Historical Changes and Human Impacts*. Geological Society of America Special Paper 451, Boulder, CO, Chapter 5, pp. 57–81. doi: 10.1130/2009.2451(04).

James, L. A., Phillips, J. D., and Lecce, S. A. 2017. A centennial tribute to G.K. Gilbert's Hydraulic Mining Débris in the Sierra Nevada. *Geomorphology* 294, 4–19.

Jha, A. K., Bloch, R., and Lamond, J. 2012. *Cities and Flooding: A Guide to Integrated Urban Flood Risk Management for the 21st Century*. World Bank, Washington, DC.

Jiang, B., Wang, F., and Ni, G. 2018. Heating impact of a tropical reservoir on downstream water temperature: a case study of the Jinghong Dam on the Lancang River. *Water* 2018, 10, 951.

Jiménez-Muñoz, J. C., Mattar, C., Barichivich, J., *et al*. 2016. Record-breaking warming and extreme drought in the Amazon rainforest during the course of El Niño 2015–2016. *Scientific Reports* 6, 33130. doi: 10.1038/srep33130.

Johnson, W. C. 1994. Woodland expansions in the Platte River, Nebraska: patterns and causes. *Ecological Monographs* 64, 45–84. doi: 10.2307/2937055.

Johnson, W. C. 2002. Riparian vegetation diversity along regulated rivers: contribution of novel and relict habitats. *Freshwater Biology* 47, 749–759.

Johnstone, G., and Sithirith, M. 2018. Tonle Sap: fisheries management case study. In Finlayson, C. M., Everard, M., Irvine, K. *et al*. (eds.), *The Wetland Book*. Springer, Dordrecht, pp. 1067–1073.

Jones, C., An, K., Blom, R. G., *et al*. 2016. Anthropogenic and geologic influences on subsidence in the vicinity of New Orleans, Louisiana. *Journal of Geophysical Research: Solid Earth* 121, 3867–3887. doi: 10.1002/2015JB012636.

Jones, J. A. A. 1989. *Global Hydrology: Processes, Resources, and Environmental Management*. Taylor & Francis, London.

Jones, L. S., and Schumm, S. A. 2009. Causes of avulsions: an overview. In Smith, N. D., and Rogers, J. (eds.), *Fluvial Sedimentology VI*. The International Association of Sedimentologists, Special publication 28. Blackwell, Oxford, pp. 171–178.

Jonkman, S. N., Voortman, H. G., Jan Klerk, W., and van Vuren, S. 2018. Developments in the management of flood defences and hydraulic infrastructure in the Netherlands. Developments in the management of flood defences and hydraulic infrastructure in the Netherlands. *Structure and Infrastructure Engineering* 14, 895–910. doi: 10.1080/15732479.2018.1441317.

Jordan, C., Tiede, J., Lojek, O., *et al*. 2019. Sand mining in the Mekong Delta revisited – current scales of local sediment deficits. *Scientific Reports* 9, 17823. doi: 10.1038/s41598-019-53804-z.

Jorgensen, D., and Renofalt, B. M. 2013. Damned if you do, dammed if you don't: debates on dam removal in the Swedish media. *Ecology and Society* 18, 18.

Julien, P. Y. 1995. *Erosion and Sedimentation*. Cambridge University Press, Cambridge.

Junk, W. J., Bayley, P. B., and Sparks, R. E. 1989. The flood pulse concept in river-floodplain systems. In Dodge, D. P. (ed.), *Proceedings of the International Large River Symposium (LARS)*, Honey Harbour, Ontario, Canada. . Special Publication of Canadian Fisheries and Aquatic Sciences, Ottawa, Canada, pp. 110–127.

Kandus, P., and Quintana, R. D. 2018. The Paraná River Delta. In Finlayson, C. M., Everard, M., Irvine, K. *et al*. (eds.), *The Wetland Book*. Springer, Dordrecht, pp. 813–821.

Karlsson, R., and Hansbro, S. 1981. *Soil Classification and Identification*, 1st ed. Swedish Academy for Building Research, Stockholm, Sweden.

Kasse, C., van Balen, R. T., Bohncke, S. J. P., Wallinga, and Vreugdenhil, M. 2016. Climate and base-level controlled fluvial system change and incision during the last glacial–interglacial transition, Roer river, the Netherlands – western Germany. *Netherlands Journal of Geosciences – Geologie en Mijnbouw* 96, 71–92.

Kearney, M. S., Riter, A. J. C., and Turner, E. R. 2011. Freshwater river diversions for marsh restoration in Louisiana: twenty-six years of changing vegetative cover and marsh area. *Geophysical Research Letters* 38, L16405. doi: 10.1029/2011GL047847.

Keller, E. A., and Melhorn, W. N. 1978. Rhythmic spacing and origin of pools and riffles. *Geological Society of America Bulletin* 89, 723–730.

Kemp, G. P., Willson, C. J., Rogers, D., and Binselam, A. 2014. Adapting to change in the Lowermost Mississippi River: implications for navigation, flood control and restoration of the delta ecosystem. In *Perspectives on the Restoration of the Mississippi Delta*. doi: 10.1007/978-94-017-8733-8_5.

Kerr, R. A. 2000. A North Atlantic climate pacemaker for the centuries. *Science* 288(5473), 1984–1986.

Kesel, R. H. 2003. Human modifications to the sediment regime of the lower Mississippi River flood plain. *Geomorphology* 56. doi: 10.1016/S0169-555X(03)00159-4.

Kesel, R. H. and McGraw, M. 2015. The role of floodplain geomorphology in policy and management decisions along the lower Mississippi River in Louisiana. In Hudson, P. F., and Middelkoop, H. (eds.), *Geomorphic Approaches to Integrated Floodplain Management of Lowland Fluvial Systems in North America and Europe*. Springer, Dordrecht, Chapter 13, pp. 321–335.

Kesel, R. H., Dunne, K. C., McDonald, R. C., Allison, K. R., and Spicer, B. E. 1974. Lateral erosion and overbank deposition on the Mississippi River in Louisiana caused by 1973 flooding: *Geology* 2, 461–464.

Kesel, R. H., Yodis, E. G., and McCraw, D. J. 1992. An approximation of the sediment budget of the lower Mississippi river prior to major human modification. *Earth Surface Processes and Landforms* 17, 711–722.

Khanal, S., Lutz, A., Immerzeel, W. W., de Vries, H., Wanders, N., and van den Hurk, B. 2019. The impact of meteorological and hydrological memory on compound peak flows in the Rhine River basin. *Atmosphere* 10, 171. doi: 10.3390/atmos10040171.

Kidder, T. R., and Liu, H. 2017. Bridging theoretical gaps in geoarchaeology: archaeology, geoarchaeology, and history in the Yellow River valley, China. *Archaeological and Anthropological Sciences* 9, 1585–1602. doi: 10.1007/s12520-014-0184-5.

Kidder, T. R., and Zhuang, Y. 2015. Anthropocene archaeology of the Yellow River, China, 5000–2000 BP. *The Holocene* 25, 1627–1639.

Kijowska-Strugała, M., Bucała-Hrabia, A., and Demczuk, P. 2018. Long-term impact of land use changes on soil erosion in an agricultural catchment (in the Western Polish Carpathians). *Earth Surface Processes and Landforms* 29, 1871–1884.

Kingston, D. G., McGregor, G. R., Hannah, D. M., and Lawler, D. M. 2006. River flow teleconnections across the northern North Atlantic region. *Geophysical Research Letters* 33, L14705. doi: 10.1029/2006GL026574.

Kiss, T., Amissah, G. J., and Fiala, K. 2019. Bank processes and revetment erosion of a large lowland river: case Study of the lower Tisza River, Hungary. *Water* 11, 1313. doi: 10.3390/w11061313.

Klasz, G., Reckendorfer, W., Gabriel, H., Baumgartner, C., Schmalfuss, R., and Gutknecht, D. 2014. Natural levee formation along a large and regulated river: the Danube in the national Park Donau-Auen, Austria. *Geomorphology* 215, 20–33. doi: 10.1016/j.geomorph.2013.12.023.

Klein Goldewijk, K., Beusen, A., Doelman, J., and Stehfest, E. 2017. New anthropogenic land use estimates for the Holocene; HYDE 3.2. *Earth Systems Science Data* 9, 927–953.

Kleinhans, M. G., and van den Berg, J. H. 2011. River channel and bar patterns explained and predicted by an empirical and a physics-based method. *Earth Surface Processes and Landforms* 36, 721–728.

Kleinhans, M. G., Weerts, J. J. T., and Cohen, K. M. 2010. Avulsion in action: reconstruction and modelling sedimentation pace and upstream flood water levels following a Medieval tidal-river diversion catastrophe (Biesbosch, The Netherlands, 1421–1750 AD). *Geomorphology* 118, 65–79.

Klimas, C. V., Smith, R. D., Raasch, J., and Saucier, R. T. 2005. Hydrogeomorphic classification of forested wetlands in the lower Mississippi Valley. In Fredrickson, L. H., King, S. L., and Kaminski, R. M. (eds.), *Ecology and Management of Bottomland Hardwood Systems: The State of Our Understanding*. University of Missouri-Columbia Gaylord Memorial Laboratory Special Publication No. 10, Puxico, Missouri, pp. 77–91.

Klimas, C. V., Murray, E. O., Pagan, J., Langston, H. L., and Foti, T. 2011. A regional guidebook for applying the hydrogeomorphic approach to assessing functions of forested wetlands in the delta region of Arkansas, lower Mississippi River Alluvial Valley. Technical report ERDC/EL-TR-11–12, Version 2.0, U.S. Army Corps of Engineers, Research and Development Center, Vicksburg, Mississippi.

Knighton, A. D. 1998. *Fluvial Forms and Processes – A New Perspective*. Arnold, London.

Knighton, A. D., and Nanson, G. C. 1993. Anastomosis and the continuum of channel pattern. *Earth Surface Processes and Landforms* 18, 613–625.

Knox, J. C. 1972. Valley alluviation in southwestern Wisconsin. *Annals of the Association of American Geographers* 62, 401–410.

Knox, J. C. 1977. Human impacts on Wisconsin stream channels. *Annals of the Association of American Geographers* 67, 323–342.

Knox, J. C. 1983. Responses of river systems to Holocene climates. In Wright, H. E., Jr., (ed.), *Late Quaternary Environments of the United States, The Holocene*, vol. 2. University of Minnesota Press: Minneapolis, pp. 26–41.

Knox, J. C. 1993. Large increases in flood magnitude in response to modest changes in climate. *Nature* 361, 430–432.

Knox, J. C. 2000. Sensitivity of modern and Holocene floods to climate change. *Quaternary Science Reviews* 19, 439–457. doi: 10.1016/S0277-3791(99)00074-8.

Knox, J. C. 2006. Floodplain sedimentation in the Upper Mississippi Valley: natural versus human accelerated. *Geomorphology* 79, 286–310. doi: 10.1016/j.geomorph.2006.06.031.

Knox, J. C., and Daniels, J. M. 2002. Watershed scale and the stratigraphic record of large floods. In House, P. K., Webb, R. H., Baker, V. R., and Levish, D. R. (eds.), *Ancient Floods, Modern Hazards: Principles and Applications of Paleoflood Hydrology*, vol. 5. American Geophysical Union, Washington, DC, pp. 237–255.

Knox, R. L., and Latrubesse, E. M. 2016. A geomorphic approach to the analysis of bedload and bed morphology of the Lower Mississippi River near the Old River Control Structure. *Geomorphology* 268, 35–47.

Ko, J.-Y., and Day, J. W. 2004. A review of ecological impacts of oil and gas development on coastal ecosystems in the Mississippi Delta. *Ocean and Coastal Management* 47, 597–623.

Ko, J.-Y., Day, J. W., Wilkins, J. W., Heywood, J., and Lane, R. L. 2017. Challenges in collaborative governance for coastal restoration: lessons from the Caernarvon River Diversion in Louisiana. *Coastal Management* 45, 125–142. doi: 10.1080/08920753.2017.1278145.

Koenig, F., Quick, I., and Vollmer, S. 2012. Defining quantitative morphological changes in large rivers for a sustainable and effective sediment management applied to the River Elbe, Germany. In *Proceedings of the Tenth International Conference on Hydroscience & Engineering*, Orlando, Florida, November 4–8, 2012.

Kolb, C. R. 1963. Sediments forming the bed and banks of the lower Mississippi River and their effect on river migration. *Sedimentology* 2, 227–234.

Kolb, C. R. 1975. *Geologic Control of Sand Boils along the Mississippi River Levees*. Misc. Paper S 75–22. U.S. Army Engineer Waterways Experiment Station, Vicksburg, MS.

Kolb, C. R., and Saucier, R. T. 1982. Engineering geology of New Orleans. *Review of Engineering Geology* 5, 75–93.

Komori, D., Nakamura, S., Kiguchi, M., *et al*. 2012. Characteristics of the 2011 Chao Phraya River flood in Central Thailand. *Hydrological Research Letters* 6, 41–46.

Kondolf, G. M. 1997. Hungry water: effects of dams and gravel mining on river channels. *Environmental Management* 21, 533–551.

Kondolf, G. M., and Piégay, H. 2016. Tools in fluvial geomorphology: problem statement and recent practices. In Kondolf, M. G. and Piégay, H. (eds.), *Tools in Fluvial Geomorphology*. John Wiley & Sons, Chichester and Hoboken, NJ, Chapter 1, pp. 1–22.

Kondolf, G. M., Gao, Y., Annandale, G. W. *et al*. 2014a. Sustainable sediment management in reservoirs and regulated rivers: experiences from five continents, *American Geophysical Union, Earth's Future* 2, 256–280. doi: 10.1002/2013EF000184.

Kondolf, G. M., Rubin, Z. K., and Minear, J. T. 2014b. Dams on the Mekong: cumulative sediment starvation. *Water Resources Research* 50, 5158–5169.

Kong, D., Latrubesse, E. M., Miaoa, C., and Zhou, R. 2020. Morphological response of the Lower Yellow River to the operation of Xiaolangdi Dam, China. *Geomorphology* 350, 10.1016/j.geomorph.2019.106931

König, F., Quick, I., and Vollmer, S. 2012. Defining quantitative morphological changes in large rivers for a sustainable and effective sediment management applied to the River Elbe, Germany. In *Proceedings Tenth International Conference of Hydroscience and Engineering*, November 2012, Orlando, USA.

Konrad, C. P., and Dettinger, M. D. 2017. Flood runoff in relation to water vapor transport by atmospheric rivers over the western United States, 1949–2015. *Geophysical Research Letters* 44, 11456–11462. doi: 10.1002/2017GL075399.

Konsoer, K. M., Rhoads, B. L., Langendoen, E. J., *et al.* 2016. Spatial variability in bank resistance to erosion on a large meandering, mixed bedrock-alluvial river. *Geomorphology* 252, 80–97.

Koohafkan, P., Salman, M., and Casarotto, C. 2011. Investments in land and water. State of Land and Water Resources (SOLAW) Background Thematic Report No. 17. Food and Agriculture Organization of the United Nations, Rome, Italy.

Kooijmans, L. P. 1974. The Rhine/Meuse Delta: Four Studies on its Prehistoric Occupation and Holocene Geology. Doctoral dissertation, Institute of Archaeology, Leiden University.

Kornei, L. 2020. Europe's rivers are the most obstructed on Earth. *Eos* 101, January 31, 2020. doi: 10.1029/2020EO139204.

Korotaev, V. N., Ivanov, V. V., and Sidorchuk, A. Y. 2004. Alluvial relief structure and bottom sediments of the lower Volga River. In *Sediment Transfer through the Fluvial System (Proceedings of the Moscow Symposium, August 2004)*. International Association of Hydrological Sciences, Publication 288.

Korponai, K., Gyulai, I., Braun, M., Kövér, C., Papp, I., Forró, L. 2016. Reconstruction of flood events in an oxbow lake (Marótzugi-Holt-Tisza, NE Hungary) by using subfossil cladocerans remains and sediments. *Advances in Oceanography and Limnology* 7, 125–135. DOI: 10.4081/aiol.2016.6168

Kostaschuk, R., Best, J. L., Villard, P., Peakall, J., and Franklin, M. 2005. Measuring flow velocity and sediment transport with an acoustic Doppler current profiler. *Geomorphology* 68, 25–37.

Koulouri, M., and Giourga, C. 2007. Land abandonment and slope gradient as key factors of soil erosion in Mediterranean terraced lands. *Catena* 69, 274–281.

Krinitzsky, E. L. 1970. *Radiography in the Earth Sciences and Soil Mechanics*. Springer.

Krinitzsky, E. L., and Wire, J. C. 1964. Ground water in the alluvium of the lower Mississippi valley (upper and central areas). U.S. Army Corps of Engineers Waterways Experiment Station, TR 3–658, V 1–2.

Krinitzsky, E. L., Ferguson, J. S., Jr., and Smith, F. L. 1965. Geological investigation of the Yazoo Basin. U.S. Army Corps of Engineers, Waterways Experiment Station, Technical Report 3–480. Vicksburg, Mississippi.

Krützen, M. Beasley, I., Ackermann, C. Y., *et al*. 2018. Demographic collapse and low genetic diversity of the Irrawaddy dolphin population inhabiting the Mekong River. *PLoS ONE* 13. doi: 10.1371/journal.pone.0189200.

Kryjov, V. N., and Gorelits, O. V. 2019. Wintertime Arctic Oscillation and formation of river spring floods in the Barents Sea Basin. *Russian Meteorology and Hydrology* 44, 187–195.

Kuhn, S., Migenda, W., Pfarr, U. 2016. *The Integrated Rhine Programme: Flood Control and Restoration of Former Floodplains along the Upper Rhine*, 5th ed. Ministry of the Environment, Climate Protection and the Energy Sector. Stuttgart, Baden-Württemberg, DL.

Kummu, M., and Varis, O. 2007. Sediment-related impacts due to upstream reservoir trapping, the Lower Mekong River. *Geomorphology* 85, 275–293.

Kundzewicz, Z. W., Szwed, M., and Pińskwar, I. 2019. Climate variability and floods – a global review. *Water* 11, 1399. doi: 10.3390/w11071399.

Kunz, M. J., Wüest, A., Wehrli, B., Landert, J., and Senn, D. B. 2011. Impact of a large tropical reservoir on riverine transport of sediment, carbon, and nutrients to downstream wetlands. *Water Resources Research* 47, 1–16. doi: 10.1029/2011WR010996.

Kunz, M. J., Senn, D. B., Wehrli, B., Mwelwa, E. M., and Wüest, A. 2013. Optimizing turbine withdrawal from a tropical reservoir for improved water quality in downstream wetlands. *Water Resources Research* 49, 5570–5584. doi: 10.1002/wrcr.20358.

Labosier, C. F., and Quiring, S. M. 2013. Hydroclimatology of the Southeastern USA. *Climate Research* 57, 157–171.

Laghari, A. N., Vanham, D., and Rauch, W. 2012. The Indus basin in the framework of current and future water resources management. *Hydrology and Earth Systems Sciences* 16, 1063–1083.

Lai, X., Yin, D., Finlayson, B. L. et al. 2017. Will river erosion below the Three Gorges Dam stop in the middle Yangtze? *Journal of Hydrology* 554, 24–31.

Lal, R., and Moldenhauer, W. C. 1987. Effects of soil erosion on crop productivity. *Critical Reviews in Plant Sciences* 5, 303–367. doi: 10.1080/07352688709382244.

Lambeets, K., Hendrickx, F., Vanacker, S., van Looy, K., Maelfait, J.-P., and Bonte, D. 2008. Assemblage structure and conservation value of spiders and carabid beetles from restored lowland river banks. *Biodiversity Conservation* 17, 3133–3148.

Lambert, A. M. 1985. *The Making of the Dutch Landscape: An Historical Geography of the Netherlands*, 2nd ed. Academic Press, London.

Lane, E. W. 1957. A study of the shape of channels formed by natural streams flowing in erodible materials. Missouri Rivers Division Sediment Series No. 9. U.S. Army Engineer Division, Missouri River, Corps of Engineers, Omaha, NE.

Langbein, W. B., and Leopold, L. B. 1966. River meanders – theory of minimum Variance. U.S. Geological Survey, Professional Paper 422-H.

Langeani, F., Casatti, L., Gameiro, H. S., and de Cerqueira Rossa-Feres, D. 2005. Riffle and pool fish communities in a large stream of southeastern Brazil. *Neotropical Ichthyology* 3, 305–311.

Lara, A., Bahamondez, A., González Reyes, A., Muñoz, A. A., Cuq, E., and Ruiz Gómez, C. 2015. Reconstructing streamflow variation of the Baker River from tree-rings in Northern Patagonia since 1765. *Journal of Hydrology* 529, 511–523.

Larsen, E. W. 2007. *Sacramento River Ecological Flows Study*. Meander migration modeling final report. Prepared for The Nature Conservancy, Chico, CA by Eric W. Larsen, Davis, CA.

Latrubesse, E. M. 2008. Patterns of anabranching channels: the ultimate end-member adjustment of mega rivers. *Geomorphology* 101, 130–145.

Latrubesse, E. M. 2015. Large rivers, megafans and other Quaternary avulsive fluvial systems: a potential "who's who" in the geological record. *Earth Science Reviews* 146, 1–30.

Latrubesse, E. M. and Franzinelli, E. 2002. The Holocene alluvial plain of the Middle Amazon River, Brazil. *Geomorphology* 44, 241–257.

Latrubesse, E. M., and Restrepo, J. D. 2014. Sediment yield along the Andes: continental budget, regional variations, and comparisons with other basins from orogenic mountain belts. *Geomorphology* 216, 225–233.

Latrubesse, E. M., Stevaux, J. C., and Sinha, R. 2005. Tropical rivers. *Geomorphology* 70, 187–206.

Latrubesse, E. M., Amsler, M., Morais, R., and Aquino, S. 2009. The Geomorphologic response of a large pristine alluvial river to tremendous deforestation in the South American tropics: the case of the Araguaia River. *Geomorphology* 113, 239–252.

Latrubesse, E. M., Arima, E. Y., Dunne, T., et al. 2017. Damming the rivers of the Amazon Basin. *Nature* 546, 363–369. doi: 10.1038/nature22333.

Lauro, C., Vich, A. I. J., and Moreiras, S. M. 2019. Streamflow variability and its relationship with climate indices in western rivers of Argentina. *Hydrological Sciences Journal* 64, 607–619. doi: 10.1080/02626667.2019.1594820.

Lavers, D. A., and Villarini, G. 2015. The contribution of atmospheric rivers to precipitation in Europe and the United States. *Journal of Hydrology* 522, 382–390.

Lavers, D. A., Allan, R. P., Wood, E. F., Villarini, G., Brayshaw, D. J., and Wade, A. J. 2011. Winter floods in Britain are connected to atmospheric rivers. *Geophysical Research Letters* 38, L23803. doi: 10.1029/2011GL049783.

Lawler, D. M., Thorne, C. R., and Hooke, J. M. 1997. Bank erosion and instability. In Thorne, C. R., Newson, M., and Hey, R. D. (eds.), *Guidebook of Applied Fluvial Geomorphology for River Engineering and Management*, Wiley, Chichester, pp. 137–172.

Lawson, M. 2016. Dam removal: case studies on the fiscal, economic, social, and environmental benefits of dam removal. Headwater Economics. http://headwaterseconomics.org/economic-development/local-studies/dam-removal-case-studies (accessed May 20, 2020).

Lecce, S. A. 1997a. Nonlinear downstream changes in stream power on Wisconsin's Blue River. *Annals of the Association of American Geographers* 87, 471–486.

Lecce, S. A. 1997b. Spatial patterns of historical overbank sedimentation and floodplain evolution, Blue River Wisconsin. *Geomorphology* 18, 265–277.

Lecce, S. A. 2000. Spatial variations in the timing of annual floods in the southeastern United States. *Journal of Hydrology* 235, 151–169.

Lecce, S. A., and Pavlowsky, R. T. 2001. Use of mining-contaminated sediment tracers to investigate the timing and rates of historical floodplain sedimentation. *Geomorphology* 38, 85–108.

Lechleitner, F., Breitenbach, S., Rehfeld, K., et al. 2017. Tropical rainfall over the last two millennia: evidence for a low-latitude hydrologic seesaw. *Scientific Reports* 7, 45809. doi: 10.1038/srep45809.

Lehner, B., Reidy Liermann, C., Revenga, C., et al. 2011. High-resolution mapping of the world's reservoirs and dams for sustainable river-flow management. *Frontiers in Ecology and the Environment* 9, 494–502. doi: 10.1890/100125.

Leigh, D. S. 2008. Late Quaternary climates and river channels of the Atlantic Coastal Plain, Southeastern USA. *Geomorphology* 101, 90–108.

Leigh, D. S. 2018. Vertical accretion sand proxies of gaged floods along the upper Little Tennessee River, Blue Ridge Mountains, USA. *Sedimentary Geology* 364, 342–350.

Lejon, A. G. C., Renöfält, B. M., and Nilsson, C. 2009. Conflicts associated with dam removal in Sweden. *Ecology and Society* 14(2).

Leli, I. T., Stevaux, J. C., and Assine, M. L. 2017. Genesis and sedimentary record of blind channel and islands of the anabranching river: an evolution model. *Geomorphology* 32, 35–45. doi: 10.1016/j.geomorph.2017.05.001.

Lenhart, C. F. 2003. An assessment of NOAA community-based fish passage and dam removal projects. *Coastal Management* 231, 77–96.

Leopold, L. B. 1998. *Sediment Problems at Three Gorges Dam*. International Rivers Network, Berkeley, CA.

Leopold, L. B., and Maddock, T., Jr. 1953. The hydraulic geometry of stream channels and some physiographic implications. U.S. Geological Survey, Professional Paper 252.

Leopold, L. B., and Wolman, M. G. 1957. River channel patterns: braided, meandering, and straight. U.S. Geological Survey, Professional Paper 282-B.

Leopold, L. B., and Wolman, M. G. 1960. River meanders. *Geological Society of America Bulletin* 71, 769–793.

Lewin, J., and Ashworth, P. J. 2014a. The negative relief of large river floodplains. *Earth Science Reviews* 129, 1–23.

Lewin, J., and Ashworth, P. J. 2014b. Defining large river channel patterns: alluvial exchange and plurality. *Geomorphology* 215, 83–98. doi: 10.1016/j.geomorph.2013.02.024.

Lewis, L. Y., Bohlen, C., and Wilson, S. 2008. Dams, dam removal, and river restoration: a hedonic property value analysis. *Contemporary Economic Policy* 26, 175–186.

Lewin, J., Ashworth, P. J., and Strick, R. 2016. Spillage sedimentation on large river floodplains. *Earth Surface Processes and Landforms* 42. doi: 10.1002/esp.3996.

Li, T., Li, J., and Zhang, D. D. 2020. Yellow River flooding during the past two millennia from historical documents. *Progress in Physical Geography: Earth and Environment* 44, 661–678. doi: 10.1177/0309133319899821.

Li, Y., Craven, J., Schweig, E. S., and Obermeier, S. F. 1996. Sand boils induced by the 1993 Mississippi River flood: could they one day be misinterpreted as earthquake-induced liquefaction? *Geology* 24, 171–174.

Liermann, M., Pess, G., McHenry, M., et al. 2017. Relocation and recolonization of Coho Salmon in two tributaries to the Elwha River: implications for management and monitoring. *Transactions of the American Fisheries Society* 146, 946–955.

Ligon, F. K., Dietrich, W. E., and Trush, W. J. 1995. Downstream ecological effects of dams: a geomorphic perspective. *BioScience* 45, 183–192.

Lim, E.-P., Hendon, H. H., Arblaster, J.M., et al. 2016. The impact of the Southern Annular Mode on future changes in Southern Hemisphere rainfall, *Geophysical Research Letters* 43, 7160–7167. doi: 10.1002/2016GL069453.

Lindenschmidt, K.-E. 2018. Modelling probabilities of ice jam flooding from artificial breakup of the Athabasca River ice cover at Fort McMurray. In *CGU HS Committee on River Ice Processes and the Environment, 19th*

Workshop on the Hydraulics of Ice Covered Rivers, Whitehorse, Yukon, Canada.

Lindenschmidt, K.-E., Carstensen, D., Fröhlich, W., *et al.* 2019. Development of an ice jam flood forecasting system for the Lower Oder River – requirements for real-time predictions of water, ice and sediment transport. *Water* 11, 95.

Liu, C., Walling, D. E., Spreafico, M., Ramasmy, J., Thulstrop, H. D., and Mishra, A. 2017. *Sediment Problems and Strategies for Their Management: Experience from Several Large River Basins*. United Nations Educational, Scientific and Cultural Organization (UNESCO), Paris.

Liu, P., Li, Q., Li, Z., Hoey, T., Liu, Y., and Wang, C. 2015. Land subsidence over oilfields in the Yellow River Delta. *Remote Sensing* 7, 1540–1564. doi: 10.3390/rs70201540.

LOLA. 2014. *Dutch Dikes*. LOLA Landscape Architects, Rotterdam.

Loomis, J. 2002. Quantifying recreation use values from removing dams and restoring free flowing rivers: a contingent behavior travel cost demand model for the Lower Snake River. *Water Resources Research* 38, 2-1.

Lopez, J. A., Henkel, T. K., Moshogianis, A. M., *et al.* 2014. Examination of deltaic processes of Mississippi River outlets – Caernarvon delta and Bohemia spillway in Southeastern Louisiana. *Gulf Coast Association of Geological Societies* 3, 79–93.

Lorenz, S., Leszinski, M., and Graeber, D. 2016. Meander reconnection method determines restoration success for macroinvertebrate communities in a German lowland river: meander reconnection method. *International Review of Hydrobiology* 101(3–4). doi: 10.1002/iroh.201501823.

Lorimer, J., and Driessen, C. 2014. Wild experiments at the Oostvaardersplassen: rethinking environmentalism in the Anthropocene. *Transactions of the British Geographers* 39, 169–181. doi: 10.1111/tran.12030.

Louisiana Department of Natural Resources (LDNR). 2006. Caernarvon Freshwater Diversion Project, 2005 Annual Report.

Lower Mississippi River Conservation Committee (LMRCC). 2015. Restoring America's Greatest River: a habitat restoration plan for the lower Mississippi River. Vicksburg, MS. http://lmrcc.org.

Lott, N. 1993. The summer of 1993: flooding in the Midwest and Drought in the Southeast. National Climatic Data Center, Technical Report 93–04.

Lu, X., Kummu, M., and Oeurng, C. 2014. Reappraisal of sediment dynamics in the Lower Mekong River, Cambodia. *Earth Surface Processes and Landforms* 39, 1855–1865.

Lyell, C. 1837. *Principles of Geology, Being an Inquiry How Far the Former Changes of the Earth's Surface Are Referable to Causes Now in Operation*, vol. 1. James Kay Jr and Brother: Philadelphia, PA.

Ma, H., Nittrouer, J. A., Naito, K., *et al.* 2017. The exceptional sediment load of fine-grained dispersal systems: example of the Yellow River, China. *Science Advances* 3, e1603114. doi: 10.1126/sciadv.1603114.

Ma, Y., Huang, H., Nanson, G. C., Li, Y., and Yao, W. 2012. Channel adjustments in response to the operation of large dams: the upper reach of the lower Yellow River. *Geomorphology* 147–148, 35–48.

Maavara, T., Parsons, C. T., Ridenour, C., *et al.* 2015. Global phosphorus retention by river damming, *Proceedings National Academy of Sciences* 112, 15603–15608.

Macdonald, D., Dixon, A., Newell, A., and Hallaways, A. 2012. Groundwater flooding within an urbanized flood plain. *Journal of Flood Risk Management* 5, 68–80. doi: 10.1111/j.1753–318X.2011.01127.x.

Macklin, M. G., Lewin, J., and Woodward, J. C. 2012. The fluvial record of climate change. *Philosophical Transactions of the Royal Society A: Mathematical, Physical and Engineering Sciences*. doi: 10.1098/rsta.2011.0608.

Magilligan, F. J. 1992. Sedimentology of a fine-grained aggrading floodplain. *Geomorphology* 4, 393–408.

Magilligan, F. J., Phillips, J. D., Gomez, B., and James, L. A. 1998. Geomorphic and sedimentological controls on the effectiveness of an extreme flood, *Journal of Geology* 106, 87–95.

Magilligan, F. J., Nislow, K. H., and Graber, B. E. 2003. A scale-independent assessment of discharge reduction and riparian disconnectivity following flow regulation by dams. *Geology* 31, 569–572.

Magilligan, F. J., Haynie, H. J., and Nislow, K. H. 2008. Channel adjustments to dams in the Connecticut River Basin: implications for forested mesic watersheds. *Annals of the American Association of Geographers* 98, 267–284.

Magilligan, F. J., Graber, B. E., Nislow, K. H., Chipman, J. W., Sneddon, C. S., and Fox, C. A. 2016. River restoration by dam removal: Enhancing connectivity at watershed scales. River restoration by dam removal. *Elementa: Science of the Anthropocene* 000108. doi: 10.12952/journal.elementa.000108.

Magilligan, F., Sneddon, C., and Fox, C. 2017. The social, historical, and institutional contingencies of dam removal. *Environmental Management* 59, 982–994.

Major, J. J., East, A., O'Conner, J. *et al.* 2017. Geomorphic responses to dam removal in the United States – a two-decade perspective. In Tsutsumi, D., and Laronne, J. B. (eds.), *Gravel-Bed Rivers: Processes and Disasters*. John Wiley & Sons, Chichester, pp. 355–383.

Makaske, B. 2001. Anastomosing rivers: a review of their classification, origin and sedimentary products. *Earth Science Reviews* 53, 149–196.

Makaske, B., Berendsen, H. J. A., and van Ree, M. H. M. 2007. Middle Holocene avulsion-belt deposits in the Central Rhine-Meuse delta, The Netherlands. *Journal of Sedimentary Research. Section A, Sedimentary Petrology and Process* 77, 110–123.

Makaske, B., Maas, G. J., van den Brink, C., and Wolfert, H. P. 2011. The influence of floodplain vegetation succession on hydraulic roughness: is ecosystem rehabilitation in Dutch embanked floodplains compatible with flood safety standards? *AMBIO* 40, 370–376. doi: 10.1007/s13280-010-0120-6.

Malik, S., and Pal, S. C. 2019. Impact of groyne on channel morphology and sedimentology in an ephemeral alluvial river of Bengal Basin. *Environmental Earth Sciences* 78, 631. doi: 10.1007/s12665–019–8642-0.

Maltsev, S. A. 2009. Conservation of the sturgeon fish in lower Volga. In Carmona, R., Domezain, A., García-Gallego, M., Hernando, J. A., Rodríguez, F., and Ruiz-Rejón, M. (eds.), *Biology, Conservation and Sustainable Development of Sturgeons*. Fish and Fisheries Series, vol. 29. Springer, Dordrecht, pp. 265–274.

Mann, C. C. 2005. *1491: New Revelations of the Americas before Columbus*. Alfred E. Knopf, New York.

Mansur, C. I., Caufman, R. I., and Schultz, J. R. 1956. *Investigation of Underseepage and Its Control on Lower Mississippi River Levees*, 2 vols. US Army Corps of Engineers, Waterways Experiment Station Technical Memorandum, TM 3–424, Vicksburg, MS.

Mantua, N. J., Hare, S. R., Zhang, Y., Wallace, J. M., and Francis, R.C. 1997. A Pacific interdecadal climate oscillation with impacts on salmon production. *Bulletin of the American Meteorological Society* 78, 1069–1079.

Mariotti, A. 2007. How ENSO impacts precipitation in southwest central Asia. *Journal of Geophysical Research Letters* 34. doi: 10.1029/2007GL030078.

Markham, E. M. 1916. Results of experiments looking to the development of a form of subaqueous concrete revetment for the protection of river banks against scour or erosion. Professional Memoirs, Corps of Engineers, United States Army, and Engineer Department at Large 8, No. 42, pp. 721–733.

Mardhiah, U., Tillig, M., and Gurnell, A. 2015. Reconstructing the development of sampled sites on fluvial island surfaces of the Tagliamento River, Italy, from historical sources. *Earth Surface Processes and Landforms* 40, 629–641. doi: 10.1002/esp.3658.

Marshall, G. J. 2003. Trends in the southern annular mode from observations and reanalysis. *Journal of Climate* 16, 4134–4143.

Marston, R. A., Mills, J. D., Wrazien, D. R., Bassett, B., and Splinter, D. K. 2005. Effects of Jackson Lake Dam on the Snake River and its floodplain, Grand Teton National Park, Wyoming, USA. *Geomorphology* 71, 79–98.

Martin, S. M., Dunbar, J. B., Corcoran, M. K., and Schmitz, D. W. 2017. Geologic controls of sand boil formation at Buck Chute, Mississippi. Final report. U.S. Army Corps of Engineers, Waterways Experiment Station, Vicksburg, MS.

Mason, J., and Mohrig, D. 2019. Scroll bars are inner bank levees along meandering river bends. *Earth Surface Processes and Landforms* 44, 2649–2659.

Massey, W., Biron, P. M., and Choné, G. 2017. Impacts of river bank stabilization using riprap on fish habitat in two contrasting environments. *Earth Surface Processes and Landforms* 42, 635–646. doi: 10.1002/esp.4010.

Matthai, H. F. 1967, Measurement of peak discharge at width contractions by indirect methods. U.S. Geological Survey Techniques of Water-Resources Investigations, book 3, chapter A4.

Maurice-Bourgoin, L., Bonnet, M.-P., Martinez, J.-M., *et al.* 2007. Temporal dynamics of water and sediment exchanges between the Curuaí floodplain and the Amazon River, Brazil. *Journal of Hydrology* 335, 140–156. doi: 10.1016/j.hydrol.2006.11.023.

Mazi, K., Koussis, A. D., and Destouni, G. 2013. Tipping points for seawater intrusion in coastal aquifers under rising sea level. *Environmental Research Letters* 8, 014001.

Mazzotti, S., Lambert, A., van der Kooij, M., and Mainville, A. 2009. Impact of anthropogenic subsidence on relative sea-level rise in the Fraser River delta. *Geology* 37, 771–774. doi: 10.1130/G25640A.1.

McBride, R. A., Taylor, M. J., and Byrnes, M. R. 2007. Coastal morphodynamics and Chenier-Plain evolution in southwestern Louisiana, USA: a geomorphic model. *Geomorphology* 88, 367–422. doi: 10.1016/j.geomorph.2006.11.013.

McCabe, D. J. 2011. Rivers and streams: life in flowing water. *Nature Education Knowledge* 3, 19.

McGrath, B., Tachakitkachorn, T., and Thitakoo, D. 2013. Bangkok's distributary waterscape urbanism: from a tributary to a distributary system. In Shannon, K., and de Meulder, B. (eds.), *Water Urbanisms East, Emerging Practices and Age-old Traditions*, UFO Explorations of Urbanism. Park Books, Zurich, pp. 48–63.

McGregor, G. 2017. Hydroclimatology, modes of climatic variability and stream flow, lake and groundwater level variability: a progress report. *Progress in Physical Geography* 41, 496–512.

McLeman, R. A., Dupre, J., Ford, L. B., Ford, J., Gajewski, K., and Marchildon, G. 2014. What we learned from the Dust Bowl: lessons in science, policy, and adaptation. *Population and Environment* 35, 417–440. doi: 10.1007/s11111-013-0190-z.

McPhee, J. 1989. *The Control of Nature*. Farrar, Straus, and Giroux, New York, 272 pp.

Meade, R. H. 1995. Setting: Geology, hydrology, sediments, and engineering of the Mississippi River. In R. H. Meade (Ed.), Contaminants in the Mississippi River, 1987–1992. US Geological Survey Circular 1133. Reston, Virginia.

Meade, R. H. 2007. Transcontinental moving and storage: the Orinoco and Amazon Rivers transfer the Andes to the Atlantic, In Gupta, A. (ed.), Large Rivers: Geomorphology and Management. John Wiley & Sons, Chichester, pp. 45–63.

Meade, R. H., and Moody, J. 2010. Causes for the decline of suspended-sediment discharge in the Mississippi River system, 1940–2007. *Hydrological Processes* 24, 34–49.

Meadows, M. E., and Hoffman, M. T. 2010. The nature, extent and causes of land degradation in South Africa: legacy of the past, lessons for the future? *Area* 34, 428–437.

Mechoso, C. R., Dias, P. S., Baethgen, W., *et al*. 2001. Climatology and hydrology of the Plata Basin. In Berbery, E. H., and Mechoso, R. (eds.), *Document of VAMOS Scientific Study Group on the Plata Basin*. World Climate Research Programme (WCRP). www2.atmos.umd.edu/~berbery/laplata/ (accessed October 3, 2019).

Mehta, V. M. 2017. River flow and its impacts, Ch. 5. In *Natural Decadal Climate Variability: Societal Impacts*. CRC Press, Taylor and Francis, pp. 113–170.

Mertes, L. A. K. 1997. Documentation and significance of the perirheic zone on inundated floodplains. *Water Resources Research* 33, 1749–1762.

Mertes, L. A. K., and Dunne, T. 2007. The effects of tectonics, climatic history, and sea-level history on the form and behaviour of the modern Amazon River. In Gupta, A. (ed.), *Large Rivers*. Wiley, Chichester, pp. 115–144.

Meyer, A., Combroux, I., Schmitt, L., and Trémoliéres, M. 2013. Vegetation dynamics in side-channels reconnected to the Rhine River: what are the main factors controlling communities trajectories after restoration? *Hydrobiologia* 714, 35–47.

Middelkoop, H. 1997. *Embanked Floodplains in the Netherlands: Geomorphological Evolution over Various Time Scales*. Netherlands Geographical Studies 34. Utrecht University, Utrecht.

Middelkoop, H. 2000. Heavy-metal pollution of the river Rhine and Meuse floodplains in the Netherlands. *Geologie en Mijnbouw / Netherlands Journal of Geosciences* 79, 411–428.

Middelkoop, H., and Asselman, N. E. M. 1998. Spatial variability of floodplain sedimentation at the event scale in the Rhine–Meuse delta, The Netherlands. *Earth Surface Processes and Landforms* 23, 561–573.

Middelkoop, H., and van Haselen, C. O. G. 1999. Twice a river. Rhine and Meuse in the Netherlands. RIZA Report No. 99.033. RIZA, Arnhem.

Middelkoop, H., Daamen, K., Gellens, D., *et al*. 2001. Impact of climate change on hydrological regimes and water resources management in the Rhine Basin. *Climatic Change* 49, 105–128.

Middelkoop, H., Kwadijk, J. C. J., van Deursen, W. P. A., and van Asselt, M. B. A. 2002. Scenario analyses in global change assessment for water management in the lower Rhine delta. In Beniston, M. (ed.), *Climatic Change: Implications for the Hydrological Cycle and for Water Management*. Advances in Global Change Research. Kluwer, Dordrecht, pp. 445–463.

Middelkoop, H., Erkens, G., and van der Perk, M. 2010. The Rhine delta – a record of sediment trapping over time scales from millennia to decades. *Journal of Soils and Sediments* 10, 628–639.

Middelkoop, H., Alabyan, A. M., Babich, D. B., and Ivanov, V. V. 2015. Post-dam channel and floodplain adjustments along the Lower Volga River, Russia. In Hudson, P. F., and Middelkoop, H. (eds.), *Geomorphic Approaches to Integrated Floodplain Management of Lowland Fluvial Systems in North America and Europe*. Springer-Verlag, New York, pp. 245–264.

Millard, C., Hajek, E., and Edmonds, D. A. 2017. Evaluating controls on crevasse-splay size: implications for floodplain-basin filling. *Journal of Sedimentary Research* 87, 722–739.

Milliman, J. D., and Farnsworth, K. L. 2011. *River Discharge to the Coastal Ocean: A Global Synthesis*. Cambridge University Press.

Milliman, J. D., and Meade, R. H. 1983. Worldwide delivery of river sediment to the oceans. *Journal of Geology* 91, 1–21.

Milliman, J. D., and Syvitski, J. 1992. Geomorphic tectonic control of sediment discharge to ocean – the importance of small mountainous rivers. The Journal of Geology 100, 525–544.

Minderhoud, P. S. J., Middelkoop, H., Erkens, G., and Stouthamer, E. 2020. Groundwater extraction may drown mega-delta: projections of extraction-induced subsidence and elevation of the Mekong delta for the 21st century. *Environmental Research Communications* 2, 011005.

Miranda, L. E. 2005. Fish assemblages in oxbow lakes in connectivity to the Mississippi River. *Transactions American Fisheries Society* 134, 1480–1489.

Mississippi River Commission (MRC). Hydrographic surveys of the lower Mississippi River, Cairo, IL to Red River Landing, 84 plates, map scale 1:10,000 1948, 1963, 1975, 1988, 1999.

Mississippi River Commission (MRC). 1880. Report of the Mississippi River Commission, Appendix SS, H. R. Ex. Doc. 95, 46th Congress, 2nd Session.

Mississippi River Commission (MRC). 2008. The Mississippi River and Tributaries Project: Yazoo Backwater Area. Information Paper, November 2008, Vicksburg, Mississippi. www.mvd.usace.army.mil (accessed April 3, 2016).

Mississippi River Commission (MRC). 2017. Internet web page "Mississippi River and Tributaries Project." www.mvd.usace.army.mil/mrc/mrt/index.php (accessed November 26, 2017).

MITECO. 2020. Sistema Nacional de Cartografia de Zonas Inundables: Inventario de Presas y Embalses (SNCZI-IPE). Spanish Ministry of Environment. https://sig.mapama.gob.es/snczi/ (accessed April 2020).

Mitsch, W. J., Day, J. W., Jr., Giliam, W., *et al*. 2001. Reducing nitrogen loading to the Gulf of Mexico from the Mississippi River Basin: strategies to counter a persistent ecological problem: *BioScience* 51, 373–388. doi: 10.1641/00063568(2001)051[0373: RNLTTG]2.0.CO;2.

Moffatt, K. C., Crone, E. E., Holl, K. D., Schlorff, R. W., and Garrison, B. A. 2005. Importance of hydrologic and landscape heterogeneity for restoring bank swallow (Riparia riparia) colonies along the Sacramento River, California. *Restoration Ecology* 13, 391–402.

Molle, F. 2018. Irrigation policies in Egypt since the construction of the High Aswan Dam. G-EAU Working Paper/Rapport de Recherche No.9. Montpellier, France. www.g-eau.net/ (accessed April 2020).

Montgomery, D. R. 2007. *Dirt: The Erosion of Civilizations*. University of California Press.

Montgomery, D. R., Collins, B. D., Buffington, J. M., and Abbe, T. B. 2003. Geomorphic effects of wood in rivers. In American Fisheries Society Symposium, American Fisheries Society, Bethesda, MD, pp. 1–27.

Morgan, J. P. 1970. Depositional processes and products in the deltaic environment. In Morgan, J. P. (ed.), *Deltaic Sedimentation: Modern and Ancient*. Society of Economic Paleontologists and Mineralogists, Special Publication 15, pp. 31–47.

Morris, G. L. 2020. Classification of management alternatives to combat reservoir sedimentation. *Water* 12, 861. doi: 10.3390/w12030861.

Morris, G. L., and Fan, J. 1998. *Reservoir Sedimentation Handbook: Design and Management of Dams, Reservoirs, and Watersheds for Sustainable Use*. McGraw-Hill, New York.

Morris, M, Dyer, M., and Smith, P. 2007. Management of flood embankments: a good practice review. Department for Environment, Food and Rural Affairs, Environment Agency, Research and Development Technical Report FD2411/TR1.

Morton, L. W., and Olson, K. R. 2015. Sinkholes and sand boils during 2011 record flooding in Cairo, Illinois. *Journal of Soil and Water Conservation* 70, 49A–54A.

Morton, R. A., and Bernier, J. C. 2010. Recent subsidence-rate reductions in the Mississippi Delta and their geological implications. *Journal of Coastal Research* 26, 555–561.

Morton, R. A., Bernier, J. C., and Barras, J. A. 2006. Evidence of regional subsidence and associated interior wetland loss induced by hydrocarbon production, Gulf Coast region, USA. *Environmental Geology* 50, 261. doi: 10.1007/s00254-006-0207-3.

Morton, R. A., Bernier, J. C., Barras, J. A., and Ferina, N. F. 2016. Rapid subsidence and historical wetland loss in the Mississippi Delta Plain: likely causes and future implications. U.S. Geological Survey Open-File Report 2005–1216.

Mossa, J. 1996. Sediment dynamics in the lowermost Mississippi River. *Engineering Geology* 45, 457–479.

Mossa, J. 2013. Historical changes of a major juncture: lower Old River, Louisiana. *Physical Geography* 34, 315–334.

Mossa, J. 2015a. Geomorphic perspectives of managing, modifying, and restoring a river with prolonged flooding: Kissimmee River, Florida, USA. In Hudson, P. F., and Middelkoop, H. (eds.), *Geomorphic Approaches to Integrated Floodplain Management of Lowland Fluvial Systems in North America and Europe*. Springer-Verlag, New York, pp. 143–169.

Mossa, J. 2015b. The changing geomorphology of the Atchafalaya River, Louisiana: a historical perspective. *Geomorphology* 252, 112–127. doi: 10.1016/j.geomorph.2015.08.018.

Mossa, J., and Marks, S. R. 2011. Pit avulsions and planform change on a mined river floodplain: Tangipahoa River, Louisiana. *Physical Geography* 32, 512–532.

Mossa, J., and McLean, M. 1997. Channel planform and land cover changes on a mined river floodplain Amite River, Louisiana, USA. *Applied Geography* 17, 43–54.

Mossa, J., Hudson, P. F., Wilder, B., and Lower, J. 1993. Sediment supply from large rivers entering the Northern Gulf of Mexico. Report submitted to the US Army Waterways Experiment Station in fulfillment of Contract No DACA39-M-4918, U.S. Army Corps of Engineers, Vicksburg, MS.

Mossa, J., Chen, Y. H., Walls, S. P., Kondolf, G. M., and Wu, C.-Y. 2017. Anthropogenic landforms and sediments from dredging and disposing sand along the Apalachicola River and its floodplain. *Geomorphology* 294, 119–134.

Mosselman, E. 2001. Morphological development of side channels. T2401, IRMA-SPONGE and Delft Cluster, DCFR project report 9, Delft, NL.

Mosselman, E., Shishikura, T., and Klaassen, G. J. 2000. Effect of bank stabilization on bend scour in anabranches of braided rivers. *Physics and Chemistry of the Earth, Part B: Hydrology, Oceans and Atmosphere* 25, 699–704.

Mosselman, E., Kerssens, P., Van der Knaap, F., Schwanenberg, D., and Sloff, K. 2004. Sustainable river fairway maintenance and improvement: literature survey. Prepared for Rijkswaterstaat Directie Oost-Nederland. Q3757, WL Delft Hydraulics, December 2004.

Mount, J. F. 1995. *California Rivers and Streams: The Conflict between Fluvial Processes and Land Use*. University of California Press, Berkeley and Los Angeles.

Mulligan, M., van Soesbergen, A., and Sáenz, L. 2020. GOODD: a global dataset of more than 38,000 georeferenced dams. *Scientific Data* 7, 31. doi: 10.1038/s41597-020-0362-5.

Muñoz, A. A., González-Reyes, A., Lara, A., et al. 2016. Streamflow variability in the Chilean Temperate-Mediterranean climate transition (35°S–42°S) during the last 400 years inferred from tree-ring records. *Climate Dynamics* 47, 4051–4066. doi: 10.1007/s00382–016–3068-9.

Muñoz, S. E., and Dee, S. G. 2017. El Niño increases the risk of lower Mississippi River flooding. *Scientific Reports* 7, 1772. doi: 10.1038/s41598–017–01919-6.

Muñoz, S. E., Giosan, L., Therrell, M. D., et al. 2018. Climatic control of Mississippi River flood hazard amplified by river engineering. *Nature* 556, 95–98.

Mussetter, R., and Trabant, S. 2005. Analysis of potential dam removal/retrofit impacts to habitat, flooding and channel stability in the Carmel Valley, California. In *Watershed Management Conference 2005*, July 19–22, 2005, Williamsburg, VA, pp. 1–11.

Mur, L. R., Skulberg, O. M., and Utkilen, H. 1999. Cyanobacteria in the environment, Ch. 2. In Chorus, I., and Bartram, J. (eds.), *Toxic Cyanobacteria in Water: A Guide to Their Public Health Consequences, Monitoring and Management*. World Health Organization, London. ISBN 0-419-23930-8.

Nagy, J., Kiss, T., Fehérváry, I., and Vaszkó, C. 2018. Changes in floodplain vegetation density and the impact of invasive amorpha fruiticosa on flood conveyance. Journal of Environmental Geography 11) 3–12. DOI: 10.2478/jengeo-2018-0008

Naik, P. K., and Jay, D. A. 2011. Distinguishing human and climate influences on the Columbia River: changes in mean flow and sediment transport. *Journal of Hydrology* 404, 259–277.

Nair, S., Wen, W. K., and Ling, C. M. 2014. Bangkok flood risk management: application of foresight methodology for scenario and policy development. *Journal of Futures Studies* 19, 87–112.

Nakamura, T. K., Singer, M. B., and Gabet, E. J. 2018. Remains of the 19th century: deep storage of contaminated hydraulic mining sediment along the Lower Yuba River, California. *Elementa: Science of the Anthropocene* 6, 70. doi: 10.1525/elementa.333.

Nanson, G. C. 2013. Anabranching and anastomosing rivers. In Shroder, J. F. (ed.), *Treatise on Geomorphology*, vol. 9. Academic Press, San Diego, CA, pp. 330–345.

Nanson, G. C., and Croke, J. G. 1992. A genetic classification of floodplains. *Geomorphology* 4, 459–486.

Nanson, G. C., and Knighton, A. D. 1996. Anabranching rivers: their cause, character and classification. *Earth Surface Processes and Landforms* 21, 217–239.

National Committee on Levee Safety (NCLS). 2009. Recommendations for a National Levee Safety Program: a report to congress from the National Committee on Levee Safety.

National Hurricane Center. 2018. Hurricane Harvey (AL092017) 17 August – 1 September, 2017. Compiled by E. S. Blake and D. A. Zelinsky, May 9, 2018. www.nhc.noaa.gov/data/tcr/AL092017_Harvey.pdf (accessed February 2020).

National Levee Safety Act (NLSA). 2007. 121 Stat. 1288, Public Law 110–114-Nov. 8, 2007, 110th Congress.

National Oceanic Atmospheric Administration (NOAA). 2018. https://climate.nasa.gov/vital-signs/sea-level/ (accessed November 12, 2018).

National Oceanic Atmospheric Administration (NOAA). 2019. National Centers for Environmental Information (NCEI) U.S. Billion-Dollar Weather and Climate Disasters, Table of Events. www.ncdc.noaa.gov/billions/.

National Register of Dams in India (NRDI). 2016. Government of India. Archived from the original (PDF) September 20, 2016.

National Research Council (NRC). 1992. *Restoration of Aquatic Ecosystems*. National Academy Press, Washington, DC.

National Research Council (NRC). 2001. *American Hazardscapes: The Regionalization of Hazards and Disasters*. The National Academies Press, Washington, DC.

National Research Council (NRC). 2002. *Riparian Areas: Functions and Strategies for Management*. The National Academies Press, Washington, DC.

National Research Council (NRC). 2005. *Endangered and Threatened Species of the Platte River*. The National Academies Press, Washington, DC.

National Research Council (NRC). 2011. Sediment management alternatives and opportunities. Missouri River Planning: Recognizing and Incorporating Sediment Management.

National Research Council (NRC). 2013. *Levees and the National Flood Insurance Program: Improving Policies and Practices*. Water Science and Technology Board, Division on Earth and Life Studies, National Academies Press, Washington, DC.

Natural Water Retention Measures (NWRM). 2018. Reconnection of oxbow lakes and similar features. European Union Directorate General Environment. http://nwrm.eu/measure/reconnection-oxbow-lakes-and-similar-features (accessed August 2, 2018).

Neave, M., Rayburg, S., and Swan, A. 2009. River channel change following dam removal in an ephemeral stream. *Australian Geographer* 40, 235–246.

Nègre, F. 2019. The European Union and forests. European Union Fact Sheet. www.europarl.europa.eu/factsheets/en/sheet/105/the-european-union-and-forests.

Nelson, S. A., and Leclair, S. F. 2006. Katrina's unique splay deposits in a New Orleans neighborhood. *GSA Today* 16(9). doi: 10.1130/GSAT01609A.1.

New South Wales (NSW). 2014. An overview of floodplain management plans under the Water Management Act 2000. NSW Department of Industry – Water.

Newell, R. E., Newell, N. E., Zhu, Y., and Scott, C. 1992. Tropospheric rivers? – a pilot study. *Geophysical Research Letters* 19, 2401–2404. doi: 10.1029/92GL02916.

Newson, M. D., and Newson, C. L. 2000. Geomorphology, ecology and river channel habitat: mesoscale approaches to basin-scale challenges. *Progress in Physical Geography* 24, 195–217.

Nienhuis, J. H. 2019. Wave-dominated river deltas. www.coastalwiki.org/wiki/Wave-dominated_river_deltas (accessed June 30, 2019).

Nienhuis, J. H., Ashton, A. D., and Giosan, L. 2015. What makes a delta wave-dominated?, *Geology* 43(6), 511–514. doi: 10.1130/G36518.1.

Nienhuis, J. H., Hoitink, A. J. F., and Törnqvist, T. E. 2018a. Future change to tide-influenced deltas. *Geophysical Research Letters* 45, 3499–3507. doi: 10.1029/2018GL077638.

Nienhuis, J. H., Törnqvist, T. E., and Esposito, C. R. 2018b. Crevasse splays versus avulsions: a recipe for land building with levee breaches. *Geophysical Research Letters* 45, 4058–4067.

Nienhuis, J. H., Ashton, A. D., Edmonds, D. A., *et al.* 2020. Global-scale human impact on delta morphology has led to net land area gain. *Nature* 577, 514–518.

Nienhuis, P. H. 2008. *Environmental History of the Rhine-Meuse Delta: An Ecological Story on Evolving Human-Environmental Relations Coping with Climate Change and Sea-Level Rise*. Springer, Dordrecht.

Nienhuis, P. H., and Leuven, R. S. E. W. 2001. River restoration and flood protection: controversy or synergism? *Hydrobiologia* 444, 85–99.

Nillesen, A. L., and Kok, M. 2015. An integrated approach to flood risk management and spatial quality for a Netherlands' river polder area. *Mitigation and Adaptive Strategies for Global Change* 20, 949–966. doi: 10.1007/s11027-015-9675-7.

Nislow, K. H., Magilligan, F. J., Fassnacht, H., Bechtel, D., and Ruesink, A. 2002. Effects of hydrologic alteration on flood regime of natural floodplain communities in the Upper Connecticut River. *Journal of the American Water Resources Association* 38, 1533–1548.

Nittrouer, J. A., Allison, M. A., and Campanella, R. 2008. Bedform transport rates for the lowermost Mississippi River. *Journal of Geophysical Research* 113, F03004, doi: 10.1029/2007JF000795.

Nittrouer, C. A., Kuehl, S. A., Demaster, D. J., and Kowsmann, R. O. 1986. The deltaic nature of Amazon shelf sedimentation. *Geological Society of America Bulletin* 97, 444–458.

Nixon, S. W. 2003. Replacing the Nile: are anthropogenic nutrients providing the fertility once brought to the Mediterranean by a great river? *AMBIO* 32, 30–39.

Noda, K., Hamada, J., Kimura, M., and Oki, K. 2018. Debates over dam removal in Japan. *Water and Environment Journal* 32, 446–452.

Notebaert, B., Verstraeten, G., Vandenberghe, D., Marinova, E., Poesen, J., and Govers, G. 2011. Changing hillslope and fluvial Holocene sediment dynamics in a Belgian loess catchment. *Journal of Quaternary Science* 26, 44–58.

NSW Department of Primary Industries (DPS). 2007. *Threat Abatement Plan: Removal of Large Woody Debris from NSW Rivers and Streams*. Fisheries Conservation and Aquaculture. State of New South Wales, Australia.

Nyqvist, D., Nilsson, P. A., Alenäs, I., *et al.* 2017. Upstream and downstream passage of migrating adult Atlantic salmon: remedial measures improve passage performance at a hydropower dam. *Ecological Engineering* 102, 331–343.

Ockerson, J. A. 1883. Review of Surveys and Gaugings of Cubitt's Gap, Made in 1868, 1875, and 1576, 48 Cong., 1, session, H7 Docs., pt 7, 2 Sept. War, Vol. pt. 3, Appendix 0 of Appendix SS (Mississippi River Commission): 2302–2304.

Oczkowski, A. J., Nixon, S. W., Granger, S. L., El-Sayed, A.-F. M., and McKinney, R. A. 2009. Anthropogenic enhancement of Egypt's Mediterranean fishery. *Proceedings National Academy of Sciences* 106, 1364–1367. doi: 10.1073/pnas.0812568106.

Okada, M., Iizumi, T., Sakamoto, T., *et al.* 2018. Varying benefits of irrigation expansion for crop production under a changing climate and competitive water use among crops. *Earth's Future* 6, 1207–1220.

Olea, R. A., and Coleman, J. L., Jr. 2014. A synoptic examination of causes of land loss in southern Louisiana as related to the exploitation of subsurface geologic resources. *Journal of Coastal Research* 30, 1025–1044. doi: 10.2112/JCOASTRES-D-13-00046.1.

Ollero, A., Ibisate, A., Granado, D., and de Asua, R. R. 2015. Channel responses to global change and local impacts: perspectives and tools for floodplain management, Ebro River and Tributaries, NE Spain. In Hudson, P. F., and Middelkoop, H. (eds.), *Geomorphic Approaches to Integrated Floodplain Management of Lowland Fluvial Systems in North America and Europe*. Springer-Verlag, New York, pp. 27–52.

Ondruch, J., Máčka, Z., Michalková, M. S., Putiška, R., Knot, M., Holík, P., Miřijovský, J., and Jenčo, M. 2018. Response of channel dynamics to recent meander neck cut-off in a lowland meandering river with artificial training history: the Morava River, Czech Republic, Hydrological Sciences Journal 63, 1236–1254. DOI: 10.1080/02626667.2018.1474218

Ontario Ministry of Natural Resources. 2011. Dam decommissioning and removal. Technical Bulletin. Ontario.ca/dams (accessed February 12, 2020).

Oppenheimer, M., Glavovic, B. C., Hinkel, J. *et al.* 2019: Sea level rise and implications for low-lying islands, coasts and communities. In Pörtner, H.-O., Roberts, D. C., Masson-Delmotte, V. *et al.* (eds.), *IPCC Special Report on the Ocean and Cryosphere in a Changing Climate*. Cambridge University Press, Cambridge, Chapter 4, pp. 321–445.

Opperman, J. J., Royte, J., Banks, J., Day, L. R., and Apse, C. 2011. The Penobscot River, Maine, USA: a basin-scale approach to balancing power generation and ecosystem restoration. *Ecology and Society* 16, 3. doi: 10.5751/ES-04117–160307.

Ouchi, S. 1985. Response of alluvial rivers to slow active tectonic movement. *Geological Society of American Bulletin* 96, 504–515.

Overeem, I., and Syvitski, J. P. M. 2009. *Dynamics and Vulnerability of Delta Systems*. LOICZ Reports and Studies No. 35. GKSS Research Center, Geesthacht.

Paira, A. R., and Drago, E. C. 2007. Origin, evolution, and types of floodplain water bodies. In Iriondo, M. H., Paggi, J. C., and Parma, M. J. (eds.), *The Middle Paraná River: Limnology of a Subtropical Wetland*. Springer-Verlag, Berlin and Heidelberg, pp. 53–81.

Palanques, A., Guillen, J., Puig, P., and Grimalt, J. O. 2020. Effects of flushing flows on the transport of mercury-polluted particulate matter from the Flix Reservoir to the Ebro Estuary. *Journal of Environmental Management* 260, 110028. doi: 10.1016/j.jenvman.2019.110028.

Paltan, H., Waliser, D., Lim, W. H., *et al.* 2017. Global floods and water availability driven by atmospheric rivers. *Geophysical Research Letters* 44, 10387–10395. doi: 10.1002/2017GL074882.

Panagos, P., Imeson, A., Meusburger, K., Borrelli, P., Poesen, J., and Alewell, C. 2016. Soil conservation in Europe: wish or reality? *Land Degradation and Development* 41, 1547–1551.

Pander, J., Mueller, M., and Geist, J. 2015. Succession of fish diversity after reconnecting a large floodplain to the upper Danube River. *Ecological Engineering* 75, 41–50.

Paneque-Gálvez, J., Mas, J.-F., Guèze, M., *et al.* 2013. Land tenure and forest cover change. The case of southwestern Beni, Bolivian Amazon, 1986–2009. *Applied Geography* 43, 113–126.

Panin, N. 1999. Global changes, sea level rise and the Danube Delta: risks and responses. In *Proceedings of International Workshop on Modern and Ancient Sedimentary Environments and Processes*, Moeciu, Romania, October 15–18, 1998, pp. 19–29.

Panin, N. 2015. The Danube Delta. Geomorphology and Holocene evolution: a synthesis. Geomorphologie: Relief, Proessus, Environment, 9 (Octobre-decembre), 247–262.

Park, E., and Latrubesse, E. M. 2015. Surface water types and sediment distribution patterns at the confluence of mega rivers: the Solimões-Amazon and Negro rivers junction. *Water Resources Research* 51, 6197–6213. doi: 10.1002/2014WR016757.

Park, E., and Latrubesse, E. M. 2017. High-resolution mapping of flood routing patterns and hydrological connectivity in the middle Amazon River floodplain. *Remote Sensing of Environment* 198, 321–332.

Park, E., and Latrubesse, E. M. 2019. A geomorphological assessment of washload sediment fluxes and floodplain sediment sinks along the lower Amazon River. *Geology* 47, 403–406.

Parker, C., Simon, A., and Thorne, C. 2008. The effects of variability in bank material properties on riverbank stability: Goodwin Creek, Mississippi. *Geomorphology* 101, 533–543.

Patterson, L. A., Doyle, M. W., and Kuzma, S. 2018. Creating data as a service for U.S. Army Corps of Engineers Reservoirs. NI R 18–01. Duke University, Durham, NC. http://nicholasinstitute.duke.edu/publications

Pavelsky, T. M., and Smith, L. C. 2004. Spatial and temporal patterns in Arctic river ice breakup observed with MODIS and AVHRR time series. *Remote Sensing of Environment* 93, 328–338.

Pedroli, B., de Blust, G., van Looy, K., and van Rooij, S. 2002. Setting targets in strategies for river restoration. *Landscape Ecology* 17, 5–18.

Pegg, M., Pierce, C., and Roy, A. 2003. Hydrological alteration along the Missouri River Basin: a time series approach. *Aquatic Sciences* 65, 63–72.

Peng, C., Zhang, Y., Huang, S. *et al.* 2019. Sediment phosphorus release in response to flood event across different land covers in a restored wetland. *Environmental Science and Pollution Research* 26, 9113–9122. doi: 10.1007/s11356-019-04398-6.

Penland, S., Boyd, R., and Suter, J. R. 1988. Transgressive depositional systems of the Mississippi delta plain: a model for barrier shoreline and shelf sand development. *Journal of Sedimentary Petrology* 58, 932–949.

Pepper, A. T., and RickardPepper, A. C. E. Rickard C. 2008. Works in the river channel. In *Fluvial Design Guide*. U.K. Environment Agency. http://evidence.environment-agency.gov.uk/FCERM/en/FluvialDesignGuide.aspx (accessed November 17, 2017).

Pescaroli, G., and Nones, M. 2016. Cascading events, technology and the floods directive: future challenges. FLOODrisk 2016 – 3rd European Conference on Flood Risk Management.

Peteuil, C., Fruchart, F., Abadie, F., Reynaud, S., Camenen, B., and Guertault, L. 2013. Sustainable management of sediment fluxes in reservoir by environmental friendly flushing: the case study of the Genissiat dam on the upper Rhône River (France). In *Conference: 12th International Conference on River Sedimentation*, Kyoto, Japan.

Petts, G. E. 1986. Water quality characteristics of regulated rivers. *Progress in Physical Geography* 10, 492–516.

Peyronnin, N. S., Caffey, R. H., Cowan, J. H. et al. 2017. Optimizing sediment diversion operations: working group recommendations for integrating complex ecological and social landscape interactions. *Water* 9, 368. doi: 10.3390/w9060368.

Phien-wej, N., Gao, P. H., and Nutalaya, P. 2006. Land subsidence in Bangkok, Thailand. *Engineering Geology* 82, 187–201. doi: 10.1016/j.enggeo.2005.10.004.

Phillips, J. D. 2003. Toledo Bend reservoir and geomorphic response in the lower Sabine River. *River Research and Applications* 19, 137–159.

Phillips, J. D. 2011. Universal and local controls of avulsions in southeast Texas Rivers. *Geomorphology* 130, 17–28.

Phillips, J. D. 2012. Log-jams and avulsions in the San Antonio River Delta, Texas. *Earth Surface Processes and Landforms* 37, 946–950.

Phillips, J. D. 2013. Hydrological connectivity of abandoned channel water bodies on a coastal plain river. *River Research and Applications* 29, 149–160.

Phillips, J. D., and Park, L. 2009. Forest blowdown impacts of Hurricane Rita on fluvial systems. *Earth Surface Processes and Landforms* 34, 1069–1081

Phillips, J. D., and Slattery, M. C. 2007. Downstream trends in discharge, slope, and stream power in a coastal plain river. *Journal of Hydrology* 334, 290–303.

Phillips, J. D., Slattery, M. C., and Musselman, Z. A. 2004. Dam-to-delta sediment inputs and storage in the lower trinity river, Texas. *Geomorphology* 62, 17–34.

Phillips, J. D., Slattery, M. C., and Musselman, Z. A. 2005. Channel adjustments of the Lower Trinity River, Texas, downstream of Livingston Dam. *Earth Surface Processes and Landforms* 30, 1419–1439.

Piégay, H., and Gurnell, A. M. 1997. Large woody debris and river geomorphological pattern: examples from S. E. France and S. England. *Geomorphology* 19, 99–116.

Piégay, H., and Marston, R. A. 1998. Distribution of large woody debris along the outer bend of meanders in the Ain River, France. *Physical Geography* 19, 318–340.

Piégay, H., Thévenet, A., and Citterio, A. 1999. Input, storage and distribution of large woody debris along a mountain river continuum. The Dróme River, France. *Catena* 35, 19–39.

Piman, T., and Manish, S. 2017. Case study on sediment in the Mekong River basin: Current state and future trends. Project Report 2017–03. Stockholm Environment Institute. www.sei.org/publications/sediment-mekong-river/ (accessed April 3, 2020).

Pinter, N. 2005. One step forward, two steps back on U.S. floodplains. *Science* 308(5719), 207–208.

Pinter, N. 2015a. Discussion of "Mississippi River streamflow measurement techniques at St. Louis, Missouri" by C. C. Watson, R. R. Holmes Jr., and D. S. Biedenharn. *Journal of Hydraulic Engineering* 141, 1062–1070. doi: 10.1061/(ASCE)HY.1943-7900.0001020.

Pinter, N. 2015b. Discussion of "Analysis of the Impacts of Dikes on Flood Stages in the Middle Mississippi River" by Chester C. Watson, David S. Biedenharn, and Colin R. Thorne. *Journal of Hydraulic Engineering* 141. doi: 10.1061/(ASCE)HY.1943-7900.0001054.

Pinter, N., Ickes, B. S., Wlosinski, J. H., and van der Ploeg, R. R. 2006a. Trends in flood stages: contrasting results from the Mississippi and Rhine River systems. *Journal of Hydrology* 331, 554–566.

Pinter, N., van der Ploeg, R. R., Schweigert, P., and Hoefer, G. 2006b. Flood magnification on the River Rhine. *Hydrological Processes* 20, 147–164.

Pinter, N., Jemberie, A. A., Remo, J. W. F., Heine, R. A., and Ickes, B. S. 2008. Flood trends and river engineering on the Mississippi River system, *Geophysical Research Letters* 35, L23404. doi: 10.1029/2008GL035987.

Pinter, N., Jemberie, A. A., Remo, J. W. F., Heine, R. A., and Ickes, B. S. 2010. Cumulative impacts of river engineering, Mississippi and Lower Missouri rivers. *River Research and Applications* 26, 546–571. doi: 10.1002/rra.1269.

Pinter, N., Huthoff, F., Dierauer, J., Remo, J. W. F., and Damptz, A. 2016. Modeling residual flood risk behind levees, Upper Mississippi River, USA. *Environmental Science and Policy* 58, 131. doi: 10.1016/j.envsci.2016.01.003.

Pitlick, J. 1997. A regional perspective of the hydrology of the 1993 Mississippi River Basin floods. *Annals of the Association of American Geographers* 87, 325–351.

Pizzuto, J. E. 1987. Sediment diffusion during overbank events. *Sedimentology* 34, 301–317.

Poeppl, R. E., Keestra, S. D., and Hein, T. 2015. The geomorphic legacy of small dams – an Austrian study. *Anthropocene* 10, 43–55.

Poesen, J., and Govers, G. 1990. Gully erosion in the Loam Belt of Belgium: typology and control measures. In Boardman, J., Foster, I. D. L., and Dearing, J. A. (eds.), *Soil Erosion on Agricultural Land*. John Wiley & Sons, Chichester, pp. 515–530.

Poesen, J., Nachtergaele, J., Verstraeten, G., and Valentin, C. 2003. Gully erosion and environmental change: importance and research needs. *Catena* 50(2), 91–133.

Pohl, M. 2002. Bringing down our dams: trends in American dam removal rationales. *Journal of the American Water Resources Association* 38, 1511–1519.

Pokhrel, Y., Burbano, M., Roush, J., Kang, H., Sridhar, V., and Hyndman, D. W. 2018. A Review of the integrated effects of changing climate, land use, and dams on Mekong River hydrology. *Water* 10, 266.

Polyakov, I. V., and Johnson, M. A. 2000. Arctic decadal and interdecadal variability. *Geophysical Research Letters* 27, 4097–4100. doi: 10.1029/2000GL011909.

Pongruktham, O., and Ochs, C. 2015. The rise and fall of the lower Mississippi: effects of hydrologic connection on floodplain backwaters. *Hydrobiologia* 742, 169–183.

Poole, G. C., Stanford, J. A., Frissell, C. A., and Running, S. W. 2002. Three-dimensional mapping of geomorphic controls on flood-plain hydrology and connectivity from aerial photos. *Geomorphology* 48, 329–347.

Potter, P. E. 1978. Significance and origin of big rivers. *Journal of Geology* 86, 13–33.

Powell, J. W. 1895. *The Exploration of the Colorado River and Its Canyons*. Dover Publications, New York.

Power, M. E., Dietrich, W. E., and Finlay, J. C. 1996. Dams and downstream aquatic biodiversity: potential food web consequences of hydrologic and geomorphic change. *Environmental Management* 20, 887–895.

Powers, J. 2018. Mississippi River ship channel dredging and wetlands creation – an environmental success story. *International Dredge Review November* 5, 2018. www.dredgemag.com/2018/11/05/mississippi-river-ship-channel-dredging-and-wetlands-creation-an-environmental-success-story/ (accessed June 2020).

Prieto, M. R. 2007. ENSO signals in South America: rains and floods in the Paraná River region during colonial times. *Climate Change* 83, 39–54.

Provansal, M. 2004. The Rhone delta (France). EUROSION case study. In *Living with coastal erosion in Europe: Sediment and Space for Sustainability. A guide to coastal erosion management practices in Europe*. www.eurosion.org/shoreline/introduction.html

Provencher, B., Sarakinos, H., and Meyer, T. 2008. Does small dam removal affect local property values? An empirical analysis. *Contemporary Economic Policy* 26, 187–197.

Przedwojski, B. 1995. Bed topography and local scour in rivers with banks protected by groynes, *Journal of Hydraulic Research* 33(2), 257–273. doi: 10.1080/00221689509498674.

Przedowjski, B., Blazejewski, R., and Pilarczyk, K. W. 1995. *River Training Techniques: Fundamentals, Design and Application*. A.A. Balkema, Rotterdam.

Querner, E. P., Jansen, P. C., van den Akker, J. J. H., and Kwakernaak, C. 2012. Analyzing water level strategies to reduce soil subsidence in Dutch peat meadows. *Journal of Hydrology* 446–447, 59–69.

Quick, I., König, F., Baulig, Y., Schriever, S., and Vollmer, S. 2020. Evaluation of depth erosion as a major issue along regulated rivers using the classification tool Valmorph for the case study of the Lower Rhine.

International Journal of River Basin Management 18, 191–206. doi: 10.1080/15715124.2019.1672699.

Qureshi, A. S. 2011. Water management in the Indus Basin in Pakistan: challenges and opportunities. *Mountain Research and Development* 31, 252–260.

Rader, R. B., Voelz, N. J., and Ward, J. V. 2008. Post-flood recovery of a macroinvertebrate community in a regulated river: resilience of an anthropogenically altered ecosystem. *Restoration Ecology* 16, 24–33.

Ralph, F. M., Coleman, T., Neiman, P. J., Zamora, R. J., and Dettinger, M. D. 2013. Observed impacts of duration and seasonality of atmospheric-river landfalls on soil moisture and runoff in coastal northern California. *Journal of the American Meteorological Society*, April 2013, 443–459. doi: 10.1175/JHM-D-12–076.1.

Randle, T. J., and Bountry, J. 2017. *Dam Removal Analysis Guidelines for Sediment*. U.S. Bureau of Reclamation, Denver, CO.

Randle, T., Morris, G., Whelan, M., *et al.* 2019. Reservoir sediment management: building a legacy of sustainable water storage reservoirs. In *National Reservoir Sedimentation and Sustainability Team White Paper*. SEDHYD, Denver, CO, p. 57. http://www.sedhyd.org/reservoir-sedimentation/ (accessed May 26, 2020).

Raffles, H., and WinklerPrins, A. M. G. A. 2003. Further reflections on Amazonian environmental history: transformations of rivers and streams. *Latin American Research Review* 38, 165–187.

Räsänen, T. A., and Kummu, M. 2013. Spatiotemporal influences of ENSO on precipitation and flood pulse in the Mekong River Basin. *Journal of Hydrology* 476, 154–168.

Räsänen, T. A., Someth, P., Lauri, H., Koponen, J., Sarkkula, J., Kummu, M. 2017. Observed river discharge changes due to hydropower operations in the Upper Mekong Basin. *Journal of Hydrology* 545, 28–41.

Rashid, M. M. 1995. Some additional information on limnology and fisheries of Lakes Nasser (Egypt) and Nubia (Sudan). In Crul, R. C. M., and Roest, F. C. (eds.), *Current Status of the Lake Nasser–Nubia Fisheries and Fish Stocks of the Four Largest African Reservoirs: Kainji, Kariba, Nasser/Nubia, and Volta*. CIFA Technical Paper 30. FAO, Rome, Italy, pp. 81–109.

Raslan, Y., and Salama, R. 2015. Development of Nile River islands between Old Aswan Dam and new Esna barrages. *Water Science* 29, 77–92. doi: 10.1016/j.wsj.2015.03.003.

Raza, A., Naeem, K., Qadeer, A., *et al.* 2019. Water, energy and food nexus of Indus Water Treaty: water governance. *Water-Energy Nexus* 2, 10–24.

Reckendorfer, W., Schmalfuss, R., Baumgartner, C., *et al.* 2005. The Integrated River Engineering Project for the free-flowing Danube in the Austrian Alluvial Zone National Park: contradictory goals and mutual solutions. *Archives of Hydrobiology* 155, 613–630.

Reckendorfer, W., Funk, A., Gschöpf, C., Hein, T., and Schiemer, F. 2013. Aquatic ecosystem functions of an isolated floodplain and their implications for flood retention and management. *Journal of Applied Ecology* 50, 119–128. doi: 10.1111/1365-2664.12029.

Redfern, S. K., Azzu, N., and Binarima, J. S. 2012. Rice in Southeast Asia: facing risks and vulnerabilities to respond to climate change. In *Proceedings of the, Building resilience for adaptation to climate change in the agriculture sector. Proceedings of a Joint FAO/OECD*.

Reisner, M. 1993 [1986]. *Cadillac Desert: The American West and Its Disappearing Water*. Penguin, New York.

Remo, J. W. F., Ickes, B. S., Ryherd, J. K., Guida, R. J., and Therrell, M. D. 2018. Assessing the impacts of dams and levees on the hydrologic record of the Middle and lower Mississippi River, USA. *Geomorphology* 313, 88–100. doi: 10.1016/j.geomorph.2018.01.004.

Renwick, W. 1992. Equilibrium, disequilibrium, and nonequilibrium landforms in the landscape. *Geomorphology* 5, 265–276.

Renwick, W. H., and Andereck, Z. D. 2006. Reservoir sedimentation trends in Ohio, USA: sediment delivery and response to land-use change. Sediment dynamics and the hydromorphology of fluvial systems (Proceedings of a symposium held in Dundee, UK, July 2006). International Association of Hydrological Sciences, Publication 306.

Report of the Secretary of War. 1855. Col. C. S. Fuller's Survey of Red River. Map of Red River with its Bayous and Lakes in the vicinity of the Raft (1:36,000). 33d Congress, 2d session. House of Representatives. Ex. doc. no. 90; March 3, 1855, Washington, DC.

Reuss, M. 1998. *Designing the Bayous: The Control of Water in the Atchafalaya Basin, 1800–1995*. U.S. Army Corps of Engineers, Office of History, Alexandria, VA.

Rice, S. P., Church, M., Wooldridge, C. L., and Hickin, E. J. 2009. Morphology and evolution of bars in a wandering gravel-bed river; lower Fraser River, British Columbia, Canada. *Sedimentology* 56, 709–736. doi: 10.1111/j.1365-3091.2008.00994.x.

Richards, D. R., Warren, P. H., Maltby, L., and Moggridge, H. L. 2017. Awareness of greater numbers of ecosystem services affects preferences for floodplain management. *Ecosystem Services* 24, 138–146.

Richards, K. 1982. *Rivers: Form and Process in Alluvial Channels*. London: Methuen.

Richey, J. E., Nobre, C., and Deser, C. 1989. Amazon River discharge and climate variability: 1903 to 1985. *Science* 246(4926), 101–103. doi: 10.1126/science.246.4926.101.

Riggsbee, J. A., Wetzel, R., and Doyle, M. W. 2012. Physical and plant community controls on nitrogen and phosphorus leaching from impounded riverine wetlands following dam removal. *River Research and Applications* 28, 1439–1450.

Rijkswaterstaat. 2009. *Water Management in the Netherlands*. Ministry of Infrastructure and the Environment, The Hague.

Rijkswaterstaat. 2016. Afwegingen bij plaatsing en beheer rivierhout : voor initiatiefnemers en uitvoerders bijlagen Projectteam Pilot Rivierhout (Guidelines for installation of wood in rivers). Rijkswaterstaat (RWS): RWS 12–2016.

Rijkswaterstaat. 2018a. Final and Partial Evaluations and Report for the Room for the River Project on the Rhine River and Zandmaas/Grensmaas: 2018. Ministerie van Infrastructuur en Waterstaat, Rijkswaterstaat, Ruimte voor de Rivier. http://publicaties.minienm.nl/documenten/eind-en-deelevaluaties-programma-s-ruimte-voor-de-rivier-en-zandmaas-grensmaas (accessed October 15, 2018)

Rijkswaterstaat. 2018b. *Our Flood Defences: Working on Storm Safety*. Rijkswaterstaat Ministry of Infrastructure and Environment, The Netherlands.

Rijnland. 2009. *Flood Control in the Netherlands: A Strategy for Dike Reinforcement and Climate Adaptation*. Hoogheemraadschap van Rijnland, Leiden.

Rinaldi, M., and Casagli, N. 1999. Stability of streambanks in partially saturated soils and effects of negative pore water pressures: the Sieve River. *Geomorphology* 26, 253–277.

Rinaldi, M., and Simon, A. 1998. Bed-level adjustments of the Arno River, central Italy. *Geomorphology* 22, 57–71.

Rincón Sanz, G., and Gortázar Rubial, J. 2016. An analysis of river fragmentation in the Spanish River Basins. Developed by Ecohidráulica, S.L. at the request of Centro Ibérico de Restauración Fluvial (CIREF) and Wetlands International European Association, EH-BIO-007–16.

Ritchie, A. C., Warrick, J. A., East, A. E., *et al.* 2018. Morphodynamic evolution following sediment release from the world's largest dam removal. *Scientific Reports* 8, 13279.

Ritchie, H., and Roser, M. 2013. Land use. UN FAO Statistical Database Published online at OurWorldInData.org. https://ourworldindata.org/land-use (accessed May 5, 2020).

Riviere, G., and Drouard, M. 2015. Dynamics of the Northern Annular Mode at weekly time scales. *Journal of the Atmospheric Sciences* 72, 4569–4590.

Rivierenland Waterschap. 2017. A new beginning for an iconic polder: vision for 2050 for the Alblasserwaard water system (in Dutch: Een nieuw begin voor een iconische polder: Visie voor 2050 op het watersysteem in de Alblasserwaard), https://simcms.waterschaprivierenland.nl/_flysystem/media/visie-alblasserwaard-2050.pdf (accessed November, 2019).

Roberts, H. H. 1997. Dynamic changes of the Holocene Mississippi River delta plain: the Delta Cycle. *Journal of Coastal Research* 13, 605–627.

Roberts, H. H. 1998. Delta switching: early responses to the Atchafalaya River diversion. *Journal of Coastal Research* 14, 882–899.

Robinson, C. T., Uelinger, U., and Monaghan, M. T. 2003. Effects of a multiyear experimental flood regime on macroinvertebrates downstream of a reservoir. *Aquatic Sciences* 65, 210–222.

Rodell, M., Famiglietti, J. S., Wiese, D. N. *et al.* 2018. Emerging trends in global freshwater availability. *Nature* 557(7707), 651–659. doi: 10.1038/s41586-018-0123-1.

Rogers, J. D. 2008. Development of the New Orleans Flood Protection System prior to Hurricane Katrina. *Journal of Geotechnical and Geoenvironmental Engineering* 134, 602–617. doi: 10.1061/(ASCE)1090-0241(2008)134:5(602).

Rogers, J. D., Boutwell, G. P., Schmitz, D. W., Karadeniz, D., Watkins, C. M., Athanasopoulos-Zekkos, A. G., and Cobos-Roa, D. 2008. Geologic conditions underlying the 2005 17th Street Canal levee failure in New Orleans. *Journal of Geotechnical and Geoenvironmental Engineering* 134, 583–601.

Rojstaczer, S. A., Hamon, R. E., Deverel, S. J., and Massey, C. A. 1991. Evaluation of selected data to assess the causes of subsidence in the Sacramento-San Joaquin Delta. Open-File Report 91–193, 16. U.S. Geological Survey, California.

Rokaya, P., Budhathoki, S., and Lindenschmidt, K.-E. 2018. Trends in the timing and magnitude of ice-jam floods in Canada. *Nature Scientific Reports* 8, 5834. doi: 10.1038/s41598-018-24057z.

Ronchail, J., Labat, D., Callede, J., *et al.* 2005. Discharge variability within the Amazon basin. Regional Hydrological Impacts of Climatic Change – Hydroclimatological Variability, Proceedings of symposium S6 held during the Seventh IAHS Scientific Assembly at Foz do Iguaçu, Brazil, April 2005). IAHS Publ. 296, pp. 21–26.

Rose, K. A., Huang, H., Justic, D., and de Mutsert, K. 2014. Simulating fish movement responses to and potential salinity stress from large-scale river diversions. *Marine and Coastal Fisheries* 6, 43–61, doi: 10.1080/19425120.2013.866999.

Rotterdam Climate Proof Initiative. 2013. Rotterdam climate change adaptation strategy. Rotterdam Office for Sustainability and Climate Change. www.rotterdamclimateinitiative.nl (accessed April 4, 2016).

Rovira, A., and Ibàñez, C. 2007. Sediment management options for the lower Ebro River and its delta. *Journal of Soil and Sediments*. doi: 10.1065/jss2007.08.244.

Rowland, J. C., Dietrich, W. E., Day, G., and Parker, G. 2010. Formation and maintenance of single-thread tie channels entering floodplain lakes: observations from three diverse river systems. *Journal of Geophysical Research: Earth Surface* 114. doi: 10.1029/2008JF001073.

Rozo, M. G., Nogueira, A. C. R., and Castro, C. S. 2014. Remote sensing-based analysis of the planform changes in the Upper Amazon river over the period 1986–2006. *Journal of South American Earth Sciences* 51, 28–44. doi: 10.1016/j.sames.2013.12.004.

Ru, H.-J., and Liu, X. 2013. River-lake migration of fishes in the Dongting Lake area of the Yangtze floodplain. *Journal of Applied Ichthyology* 29, 594–601.

Ruban, G., Khodorevskaya, R., and Shatunovskii, M. 2019. Factors influencing the natural reproduction decline in the beluga (Huso huso, Linnaeus, 1758), Russian sturgeon (Acipenser gueldenstaedtii, Brandt and Ratzeburg, 1833), and stellate sturgeon (A. stellatus, Pallas, 1771) of the Volga–Caspian basin: a review. *Journal of Applied Ichthyology* 35, 387–395.

Rutherford, J. S., Day, J. W., D'Elia, C. F., *et al.* 2018. Evaluating trade-offs of a large, infrequent sediment diversion for restoration of a forested wetland in the Mississippi delta. *Estuarine, Coastal and Shelf Science*. doi: 10.1016/j.ecss.2018.01.016.

Russell, R. J. 1936. Physiography of the lower Mississippi River Delta. In, Reports on the Geology of Plaquemines and St. Bernard Parishes. *Louisiana Geological Survey, Bulletin* 8, 3–193.

Russell, R. J. 1939. Louisiana stream patterns. *American Association of Petroleum Geologists Bulletin* 23, 1199–1227.

Sacklin, H., and Ozaki, S. 1988. *Environmental Assessment: Upper Dam Removal*. Redwood National Park, Lost Man Creek, CA.

Saito, Y., Yang, Z., and Hori, K. 2001. The Huang He (Yellow River) and Chang Jiang (Yangtze River) deltas: a review of their characteristics, evolution and sediment discharge during the Holocene. *Geomorphology* 41, 219–231.

Saji, N. H., Goswami, B. N., Vinayachandran, P. N., and Yamagata, T. 1999. A dipole mode in the tropical Indian Ocean. *Nature* 401, 360–363.

Salmon, J. M., Friedl, M. A., Frolking, S., Wisser, D., and Douglas, E. M. 2015. Global rain-fed, irrigated, and paddy croplands: a new high resolution map derived from remote sensing, crop inventories and climate data. *International Journal Applied Earth Observation Geoinformatics* 38, 321–334.

Sanchez-Arcilla, A., Jimenez, J. A., and Valdemoro, H. I. 1998. The Ebro Delta: morphodynamics and vulnerability. *Journal of Coastal Research* 14, 754–772.

Sando, S. K., and Lambing, J. H. 2011. Estimated loads of suspended sediment and selected trace elements transported through the Clark Fork Basin, Montana, in selected periods before and after the breach of Milltown Dam (water years 1985–2009). U.S. Geological Survey Scientific Investigations Report 2011–5030, 64pp. at http://pubs.usgs.gov/sir/2011/5030 (accessed June 1, 2020).

Sando, S. K., and Vecchia, A. V. 2016. Water-quality trends and constituent-transport analysis for selected sampling sites in the Milltown Reservoir/Clark Fork River Superfund Site in the upper Clark Fork Basin, Montana, water years 1996–2015. U.S. Geological Survey Scientific Investigations Report 2016–5100.

Satyamurty, P., da Costa Priscila, C. P. W., Manzi, A. O., and Candido, L. A. 2013. A quick look at the 2012 record flood in the Amazon Basin. *Geophysical Research Letters* 40, 1396–1401. doi: 10.1002/grl.50245.

Saucier, R. T. 1963. *Recent geomorphic history of the Pontchartrain Basin*. Louisiana State University, Coastal Studies Series 9, 114pp.

Saucier, R. T. 1994. Geomorphology and Quaternary Geologic History of the Lower Mississippi Valley. U.S. Army Corps of Engineers, Waterways Experiment Station, Vicksburg, Mississippi.

Savelieva, N. I., Semiletov, I. P., Vasilevskaya, L. N., and Pugach, S. P. 2000. A climate shift in seasonal values of meteorological and hydrological parameters for Northeastern Asia. *Progress in Oceanography* 47, 279–297.

Sayers, P., Galloway, G., Penning-Rosell, E. *et al.* 2015. Strategic flood management: ten "golden rules" to guide a sound approach. *International Journal of River Basin Management* 1, 137–151. doi: 10.1080/15715124.2014.902378.

Scanlon, B. R., Jolly, I., Sophocleous, M., and Zhang, L. 2007. Global impacts of conversions from natural to agricultural ecosystems on water resources: quantity versus quality. *Water Resources Research* 43, W03437. doi: 10.1029/2006WR005486.

Schalk, G. K., and Jacobson, R. B. 1997. Scour, sedimentation, and sediment characteristics at six levee-break sites in Missouri from the 1993 Missouri River flood. U.S. Geological Survey Water Resources Investigations Report 97–4110.

Scheueklein, H. 1990. Removal of sediment deposits in reservoirs by means of flushing. Hydrology in Mountainous Regions II, Artificial Reservoirs; Water and Slopes (Proceedings of Lausanne Symposia). International Association of Hydrologic Sciences (IAHS) Publ. no. 194, 99–106.

Schiermeier, Q. 2018a. Europe is demolishing its dams to restore ecosystems. *Nature* 557, 290–291.

Schiermeier, Q. 2018b. Dam removal restores rivers: huge European demolition projects offer hope for fragmented ecosystems. *Nature* 557(17 May), 290–291.

Schmutz, S., and Moog, O. 2018. Dams: ecological impacts and management. In Schmutz, S., and Sendzimir, J. (eds.), *Riverine Ecosystem Management*. Aquatic Ecology Series 8 (open access). Springer, pp. 111–127.

Schober, B., Hauer, C., and Habersack, H. 2015. A novel assessment of the role of Danube floodplains in flood hazard reduction (FEM method). *Natural Hazards* 75, S33–S50. doi: 10.1007/s11069-013-0880-y.

Schramm, H. L., Cox, M. F., Tietjen, H. E., and Ezell, A. W. 2009. Nutrient dynamics in the lower Mississippi River floodplain: comparing present and historic hydrologic conditions. *Wetlands* 29, 476–487.

Schulte, J. A., Najjar, R. G., and Li, M. 2016. The influence of climate modes on streamflow in the Mid-Atlantic region of the United States. *Journal of Hydrology: Regional Studies* 5, 80–99.

Schumm, S. A. 1960. The shape of alluvial channels in relation to sediment type. U.S. Geological Survey Professional Paper 352-B, pp. 17–30.

Schumm, S. A. 1963. Sinuosity of alluvial rivers on the Great Plains. *Geological Society of America Bulletin* 74(9), 1089–1100.

Schumm, S. A. 1968. River adjustments to altered hydrologic regimen – Murrumbidgee River and paleochannels. U.S. Geological Survey Professional Paper 598, Australia.

Schumm, S. A. 1977. *The Fluvial System*. Wiley, New York.

Schumm, S. A. 1985. Patterns of alluvial rivers. *Annual Review of Earth and Planetary Sciences* 13, 5–27.

Schumm, S. A. 1991. *To Interpret the Earth: Ten Ways to Be Wrong*. Cambridge University Press.

Schumm, S. A. 2007. Rivers and humans – unintended consequences. In Gupta, A. (ed.), *Large Rivers: Geomorphology and Management*. John Wiley & Sons, Chichester, pp. 517–533.

Schumm, S. A., and Khan, H. R. 1972. Experimental study of river patterns. *Bulletin of the Geological Society of America* 83, 1755–1770.

Schumm, S. A., and Spitz, W. J. 1996. Geological influences on the lower Mississippi River and its alluvial valley. *Engineering Geology* 45, 245–261.

Schumm, S. A., and Winkley, B. R. 1994. The character of large alluvial rivers. In Schumm, S. A., and Winkley, B. R. (eds.), *The Variability of Large Alluvial Rivers*. American Society of Engineers Press, New York, pp. 1–9.

Schumm, S. A., Dumont, J. F., and Holbrook, J. M. 2000. *Active Tectonics and Alluvial Rivers*. Cambridge University Press, Cambridge.

Science 2021. Unleashing big muddy. 372 (6540), 334–337. DOI: 10.1126/science.372.6540.334
Scruton, P. C. 1960. Delta building and the deltaic sequence. In Shepard, F. P., Phleger, F. B., and Andel, T. H. (eds.), *Recent Sediments, Northwest Gulf of Mexico: A Symposium*. American Association of Petroleum Geologists, Tulsa, OK, pp. 82–102.
Seager, R., Lis, N., Feldman, J., *et al.* 2018. Whither the 100th Meridian? The once and future physical and human geography of America's Arid–Humid Divide. Part I: The story so far. *Earth Interactions* 22, 1–22. doi: 10.1175/EI-D-17-0011.1.
Secor, D. H., Arefjev, V., Nikolaev, A., and Sharov, A. 2000. Restoration of sturgeons: lessons from the Caspian Sea Sturgeon Ranching Programme. *Fish and Fisheries* 1, 215–230. doi: 10.1111/j.1467–2979.2000.00021.x.
Seed, R. B., Bea, R. G., Athanasopoulos-Zekkos, A., *et al.* 2008. The New Orleans and Hurricane Katrina III: the 17th St. drainage canal. *Journal of Geotechnical and Geoenvironmental Engineering* 134, 740–761. doi: 10.1061/(ASCE)1090–0241(2008)134:5(740).
Select Bipartisan Committee. 2006. A failure of initiative. Final Report of the Select Bipartisan Committee to Investigate the Preparation for and Response to Hurricane Katrina, 109th Congress, 2nd Session, pp. 109–377.
Shankman, D., and Smith, L. J. 2004. Stream channelization and swamp formation in the US Coastal Plain. *Physical Geography* 25, 22–38.
Sharp, J. M., Jr. 1988. Alluvial aquifers along major rivers. In Back, W., Rosenshein, J. S., and Seaber, P. R (eds.), *Hydrogeology*, vol. O-2. Geological Society of America, The Geology of North America, Boulder, CO, pp. 273–282.
Shaw, J. B., Mohrig, D., and Whitman, S. K. 2013. The morphology and evolution of channels on the Wax Lake Delta, Louisiana, USA. *Journal of Geophysical Research: Earth Surface* 118, 1562–1584. doi: 10.1002/jgrf.20123.
Shen, Z., Törnqvist, T. E., Mauz, B., Chamberlain, E. L., Nijhuis, A. G., and Sandoval, L. 2015. Episodic overbank deposition as a dominant mechanism of floodplain and delta-plain aggradation. *Geology* 43, 875–878. doi: 10.1130/G36847.1.
Shi, H., and Shao, M. 2000. Soil and water loss from the Loess Plateau in China. *Journal of Arid Environments* 45, 9–20.
Shi, H. L., Hu, C. H., Deng, A. J., and Tian, Q. Q. 2017. Analyses on trends and reasons of runoff and sediment load of Yellow River stem. In Wieprecht, S., Haun, S., Weber, K., Noack, M., and Terheidenet, K. (eds.), *River Sedimentation*. Proceedings of the 13th International Symposium on River Sedimentation, Stuttgart, Germany, September 19–22, 2016.
Shields, F. D., Jr., and Abt, S. R. 1989. Sediment deposition in cutoff meander bends and implications for effective management. *Regulated Rivers: Research and Management* 4, 381–396.
Shields, F. D., Jr., and Knight, S. S. 2013. Floodplain restoration with flood control: fish habitat value of levee borrow pits. *Ecological Engineering* 53, 217–227.
Shinkle, K., and Dokka, R. K. 2004. Rates of Vertical Displacement at Benchmarks in the lower Mississippi Valley and the Northern Gulf Coast, National Oceanic and Atmospheric Administration, Technical Report 50.
Shull, C. A. 1922. The formation of a new island in the Mississippi River. *Ecological Society of America* 3, 202–206.
Silc, T., and Sanek, W. 1977. Bulk density determination of several peats in northern Ontario using the Von Post humification scale. *Canadian Journal of Soil Science* 57, 75.
Silva, W., Klijn, F., and Dijkman, J. P. 2001. *Room for the Rhine Branches in the Netherlands: What the Research Has Taught Us*. RIZA, Deltares.
Silva, W., Dijkman, J., and Loucks, P. 2004. Flood management options for The Netherlands. *International Journal of River Basin Management* 2, 101–112.
Silver, M., and Griffin, C. R. 2009. Nesting habitat Characteristics of bank swallows and belted kingfishers on the Connecticut River. *Northeastern Naturalist* 16, 519–534.
Simon, A., and Collison, A. J. C. 2001. Pore-water pressure effects on the detachment of cohesive streambeds: seepage forces and matric suction. *Earth Surface Processes and Landforms* 26, 1421–1442.
Simon, A., and Darby, S. E. 1997. Process-form interactions in unstable sand-bed river channels: a numerical modeling approach. *Geomorphology* 21, 85–106.
Simon, A., and Rinaldi, M. 2006. Disturbance, stream incision, and channel evolution: the roles of excess transport capacity and boundary materials in controlling channel response. *Geomorphology* 79, 361–383.

Simon, A., Curini, A., Darby, S. E., and Langendoen, E. J. 2000. Bank and near-bank processes in an incised channel. *Geomorphology* 35, 193–217.
Singer, M. B., and Aalto, R. 2009. Floodplain development in an engineered setting. *Earth Surface Processes and Landforms* 34, 291–304.
Singer, M. B., Aalto, R., and James, A. 2008. Status of the lower Sacramento River flood control works in the context of its natural geomorphic setting. *Natural Hazards Review* 9, 104–114. doi: 10.1061/(ASCE)1527-6988 (2008)9:3(104).
Singer, M. B., Aalto, R., James, L. A., Kilham, N. E. Higson, J. L., and Ghoshal, S. 2013. Enduring legacy of a toxic fan via episodic redistribution of California gold mining debris. *Proceedings National Academy of Science* 46, 18436–18441.
Slingerland, R. L., and Smith, N. D. 2004. River avulsions and their deposits. *Annual Review of Earth Planetary Sciences* 32, 257–285.
Smith, D. G. 1986. Anastomosing river deposits, sedimentation rates and basin subsidence, Magdalena River, Northwestern Columbia, South America. *Sedimentary Geology* 46, 177–196.
Smith, D. L., Miner, S. P., Theiling, C., Behm, R., and Nestler, J. M. 2017. Levee setbacks: an innovative, cost-effective, and sustainable solution for improved flood risk management. *US Army Corps of Engineers*. Report ERDC/EL SR-17-3. doi: 10.21079/11681/22736.
Smith, H. N. 1947. Rain follows the plow: the notion of increased rainfall for the Great Plains, 1844–1880. *Huntington Library Quarterly* 10, 169–193.
Smith, L. M., and Winkley, B. R. 1996. The response of the lower Mississippi River to river engineering. *Engineering Geology* 45, 433–455.
Smith, N. D., and Pérez-Arlucea, M. 1994. Fine-grained splay deposits in the avulsion belt of the lower Saskatchewan River, Canada. *Journal of Sedimentary Research* B64, 159–168.
Smith, N. D., Cross, T. A., Dufficy, J. P., and Clough, S. R. 1989. The anatomy of an avulsion. *Sedimentology* 36, 1–23.
Smith, N. D., Morozova, G. S., Pérez-Arlucea, M., and Gibling, M. R. 2016. Dam-induced and natural channel changes in the Saskatchewan River below the E.B. Campbell Dam, Canada. *Geomorphology* 269, 186–202.
Smith, V. B., and Mohrig, D. 2017. Geomorphic signature of a dammed sandy river: The lower Trinity River downstream of Livingston Dam in Texas, USA. *Geomorphology* 297, 122–136.
Snedden, G. A., Cable, J. E., Swarzenski, C., and Swenson, E. 2007. Sediment discharge into a subsiding Louisiana deltaic estuary through a Mississippi River diversion. *Estuarine Coastal and Shelf Science* 71, 181–193.
Somoza, L., and Rodríguez-Santalla, I. 2014. Geology and geomorphological evolution of the Ebro River Delta. In Gutiérrez, F., and Gutiérrez, M. (eds.), *Landscapes and Landforms of Spain, World Geomorphological Landscapes*. Springer, Dordrecht, pp. 213–227.
Soniat, T. M., Conzelmann, C. P., Byrd, J. D. *et al.* 2013. Predicting the effects of proposed Mississippi River diversions on oyster habitat quality; application of an oyster habitat suitability index model. *Journal of Shellfish Research* 32, 629–638. doi: 10.2983/035.032.0302.
Sparks, R. E. 1995. Need for ecosystem management of large rivers and their floodplains. *BioScience* 45, 168–182.
Spreafico, M., and Lehmann, C. (eds.). 2009. Erosion, transport and deposition of sediment – case study Rhine. Contribution to the International Sediment Initiative of UNESCO/IHP, International Commission for the Hydrology of the Rhine Basin, Report no II-20 of the CHR.
Springborn, M., Singer, M. B., and Dunne, T. 2011. Sediment-adsorbed total mercury flux through Yolo Bypass, the primary floodway and wetland in the Sacramento Valley, California. *Science of the Total Environment* 412–413, 203–213. doi: 10.1016/j.scitotenv.2011.10.004.
Stanford, J. A., and Ward, J. V. 1993. An ecosystem perspective of alluvial rivers: connectivity and the hyporheic corridor. *Journal of the North American Benthological Society* 12, 48–60.
Stanley, D.-J. 1997. Mediterranean deltas: subsidence as a major control on relative sea-level rise. In *CIESM Science Series n°3*. Transformations and Evolution of the Mediterranean Coastline 35–62.
Stanley, D.-J., and Clemente, P. L. 2014. Clay distributions, grain sizes, sediment thicknesses, and compaction rates to interpret subsidence in Egypt's Northern Nile Delta. *Journal of Coastal Research* 30, 88–101.
Stanley, D.-J., and Hait, A. K. 2000. Holocene depositional patterns, neotectonics and sundarban mangroves in the Western Ganges-Brahmaputra Delta. *Journal of Coastal Research* 16, 26–39.
Stanley, D.-J., and Warne, A. G. 1994. Worldwide initiation of Holocene marine deltas by deceleration of sea-level rise. *Science* 265(5169), 228–231.

Stanley, D.-J., and Warne, A. G. 1998. Nile delta in its destruction phase, *Journal of Coastal Research* 14, 794–825

Stanley, E. H., and Doyle, M. W. 2C3. Trading off: the ecological effects of dam removal. *Frontiers in Ecology and the Environment* 1, 15–22.

Stegner, W. 1954. *Beyond the Hundredth Meridian: John Wesley Powell and the Second Opening of the West*. Houghton Mifflin Company, Boston, 438pp.

Sternberg, H. O'R. 1960. *Radiocarbon Dating as Applied to a Problem of Amazonian Morphology*. International Geographical Union, Congrès International de Géographie, 18th, Proceedings, vol. 2, Washington, DC, pp. 399–424.

Stevaux, J. C., Corradini, F. A., and Aquino, S. 2013. Connectivity processes and riparian vegetation of the upper Paraná River, Brazil. *Journal of South American Earth Sciences* 46, 113–121. doi: 10.1016/j.jsames.2011.12.007.

Stouthamer, E., and Berendsen, H. J. A. 2001. Avulsion history, avulsion frequency, and intervulsion period of Holocene channel belts in the Rhine-Meuse Delta (The Netherlands). *Journal of Sedimentary Research* 71, 589–598.

Stouthamer, E., and van Asselen, S. 2015. Potential of Holocene deltaic sequences for subsidence due to peat compaction. *Proceedings of the International Association of Hydrological Sciences* 372, 173–178. doi: 10.5194/piahs-372–173-2015.

Stouthamer, E., Cohen, M. K., and Gouw, M. J. P. 2011. Avulsion and its implications for fluvial-deltaic architecture: Insights from the Holocene Rhine-Meuse Delta. From River to Rock Record: The Preservation of Fluvial Sediments and their Subsequent Interpretation. SEPM (Society for Sedimentary Geology), Special Publication no. 97, 215–231.

STOWA. 2012. Inspection manuals for flood defense systems. Technical information for the performance of inspections. STOWA, Report 14.

Straatsma, M. W., and Kleinhans, M. G. 2018. Flood hazard reduction from automatically applied landscaping measures in RiverScape, a Python package coupled to a two-dimensional flow model. *Environmental Modelling and Software* 101, 102–116.

Straatsma, M. W., Fliervoet, J. M., Kabout, J. A. H., Baart, F., and Kleinhans, M. G. 2019. Towards multi-objective optimization of large-scale fluvial landscaping measures. *Natural Hazards and Earth Systems Science* 19, 1167–1187. doi: 10.5194/nhess-19-1167-2019.

Straatsma, M. W., Schipper, A., van der Perk, M., van den Brink, C., Leuven, R., and Middelkoop, H. 2009. Impact of value-driven scenarios on the geomorphology and ecology of lower Rhine floodplains under a changing climate. *Landscape and Urban Planning* 92, 160–174.

Suanez, S., and Provensal, M. 1996. Morphosedimentary behaviour of the deltaic fringe in comparison to relative sea level rise on the Rhône River delta. *Quaternary Science Reviews* 15, 811–818.

Summerfield, M. A. 1991. *Global Geomorphology*. Routledge.

Sun, W., Shao, Q., Liu, J., and Zhai, J. 2014. Assessing the effects of land use and topography on soil erosion on the Loess Plateau in China. *Catena* 121, 151–163.

Swartz, J. M., Goudge, T. A., and Mohrig, D. C. 2020. Quantifying coastal fluvial morphodynamics over the Last 100 years on the lower Rio Grande, USA and Mexico. *Journal of Geophysical Research: Earth Surface* 125(6). doi: 10.1029/2019JF005443.

Syvitski, J. P. M., and Kettner, A. J. 2011. Sediment flux and the Anthropocene. *Philosophical Transactions of the Royal Society A*. doi: 10.1098/rsta.2010.0329.

Syvitski, J. P. M., Vörösmarty, C., Kettner, A. J., and Green, P. 2005. Impact of humans on the flux of terrestrial sediment to the global coastal ocean. *Science* 308, 376–380.

Syvitski, J. P. M, Kettner, A. J., Overeem, I., *et al.* 2009. Sinking deltas due to human activities. *Nature Geoscience* 2, 681–686.

Syvitski, J. P. M., Overeem, I., Brackenridge, R., and Hannon, M. T. 2012. Floods, floodplains, delta plains – a satellite imaging approach. *Sedimentary Geology*. doi: 10.1016/j.sedgeo.2012.05.014.

Syvitski, J. P. M., Kettner, A. J., Overeem, I., *et al.* 2013. Anthropocene metamorphosis of the Indus Delta and lower floodplain. *Anthropocene* 3, 24–35.

Te Brake, W. H. 2002. Taming the waterwolf: hydraulic engineering and water management in the Netherlands during the Middle Ages. *Technology and Culture* 43, 475–499.

Telcik, N., and Pattiaratchi, C. 2014. Influence of Northwest Cloudbands on Southwest Australian Rainfall. *Journal of Climatology*. doi: 10.1155/2014/671394.

Tellman, B., Sullivan, J.A., Kuhn, C., Kettner, A.J., Doyle, C.S., Brakenridge, G.R., Erickson, T.A., Slayback, D.A. 2021. Satellite imaging reveals increased proportion of population exposed to floods. Nature 596, 80–86. https://doi.org/10.1038/s41586-021-03695-w

Ten Brinke, W. B. T. 2005. *The Dutch Rhine, a Restrained River*. Veen Magazines, Diemen.

Ten Brinke, W. B. T., Schulze, F. H., and van der Veer, P. 2004. Sand exchange between groyne-field beaches and the navigation channel of the Dutch Rhine: the impact of navigation versus river flow. *River Research and Applications* 20, 889–928.

Tena, A., and Batalla, R. J. 2013. The sediment budget of a large river regulated by dams (the lower River Ebro, NE Spain). *Journal of Soils and Sediment* 13, 966–980.

Tennessee Valley Authority (TVA). A map of the profile of the Tennessee River, showing the TVA dams and facilities. 24158 Collection: Dept. of Conservation Photograph Collection Accession No.: RG 82 File Location: Box 69, File 165.

Terry, M. 1945. Soil erosion in Australia. *The Geographical Journal* 105, 121–129.

Thompson, D. W. J., and Wallace, J. M. 1998. The Arctic Oscillation signature in the wintertime geopotential height and temperature fields. *Geophysical Research Letters* 25, 1297–1300. doi: 10.1029/98GL00950.

Thompson, D. W. J., and Wallace, J. M. 2000. Annular modes in the extratropical circulation. Part I: month-to-month variability. *Journal of Climate* 13, 1000–1016. doi: 10.1175/1520–0442.

Thompson, G., Hartman, B., Miller, B., *et al.* 2019. Development of borrow areas along the lower Mississippi River. In Rosati, J. D., Wang, P., and Vallee, M. (eds.), *Coastal Sediments 2019 – Proceedings of the 9th International Conference*. World Scientific, Hackensack, NJ and London, pp. 870–879.

Thoms, M. C. 2003. Floodplain–river ecosystems: lateral connections and the implications of human interference. *Geomorphology* 56, 335–350.

Thonon, I., Middelkoop, J., and van der Perk, M. 2007. The influence of floodplain morphology and river works on spatial patterns of overbank deposition. *Netherlands Journal of Geosciences* 86(1), 63–75.

Thorne, C. R. 1989. Bank processes on the Red River between Index, Arkansas and Shreveport. Louisiana Final Report to the US Army European Research Office, London, England, under Contract No. DAJ45–88-C-0018.

Thorne, C. R. 1991. Bank erosion and meander migration of the Red and Mississippi Rivers, USA. Hydrology for the Water Management of Large River Basins (Proceedings of the Vienna Symposium, August 1991). International Association of Hydrological Sciences, no. 201.

Thorne, C. R. 1998. Channel types and morphological classification. In Thorne, C. R., Hey, R. D., and Newson, M. D. (eds.), *Guidebook of Applied Fluvial Geomorphology for River Engineering and Management*. Wiley, Chichester, pp. 175–222.

Thorne, C. R. 2002. Geomorphic analysis of large alluvial rivers. *Geomorphology* 44, 203–219.

Thorne, C. R., and Abt, S. R. 1993. Analysis of riverbank instability due to toe scour and lateral erosion. *Earth Surface Processes and Landforms* 18, 835–843.

Thorsteinson, L. K., Becker, P. R., and Hale, D. A. 1989. *The Yukon Delta A Synthesis of Information*. NOAA National Ocean Service, Ocean Assessments Division, Anchorage, AK.

Tian, J., Chang, J., Zhang, Z., Wang, Y., Wu, Y., and Jiang, T. 2019. Influence of Three Gorges Dam on downstream low flow. *Water* 11, 65. doi: 10.3390/w11010065.

Tiner, R. W. 2016. *Wetland Indicators: A Guide to Wetland Formation, Identification, Delineation, Classification, and Mapping*, 2nd ed. CRC Press, Taylor and Francis Group, Boca Raton, 630 p.

Tockner, K., Malard, F., and Ward, J. V. 2000. An extension of the flood pulse concept. *Hydrological Processes* 14, 2861–2883.

Tockner, K., Bernhardt, E. S., Koska, A., and Zarfl, C. 2016. A global view on future major water engineering projects. In Hüttl, R. F., *et al.* (eds.), *Society-Water-Technology, Water Resources Development and Management*. Springer Open, Heidelberg, pp. 47–64.

Toole, J. K. 1980. *A Confederacy of Dunces*. Louisiana State University Press, Baton Rouge, LA, 405pp.

Toonen, W. H. J., Kleinhans, M. G., and Cohen, K. M. 2012. Sedimentary architecture of abandoned channel fills. *Earth Surface Processes and Landforms* 37, 459–472.

Toonen, W. H. J., Middelkoop, H., Konijnendijk, T. Y. M., Macklin, M. G., and Cohen, K. M. 2016. The influence of hydroclimatic variability on flood frequency in the Lower Rhine. *Earth Surface Processes and Landforms* 41. doi: 10.1002/esp.3953.

Törnblom, J., Angelstam, P., Degerman, E., and Tamario, C. 2017. Prioritizing dam removal and stream restoration using critical habitat patch threshold for brown trout (*Salmo trutta L.*): a catchment case study from Sweden. *Écoscience* 34, 157–166.

Törnqvist, T. E., and Bridge, J. S. 2003. Spatial variation of overbank aggradation rate and its influence on avulsion frequency. *Sedimentology* 49, 891–905.

Törnqvist, T. E., Kidder, T. R., Autin, W. J., et al. 1996. A revised chronology for Mississippi River subdeltas. *Science* 273, 1693–1696.

Törnqvist, T. E., Bick, S. J., van der Borg, K., and de Jong, A. F. M. 2006. How stable is the Mississippi delta? *Geology* 34, 697–700.

Törnqvist, T. E., Wallace, D. J., Storms, J. E. A., et al. 2008. Mississippi Delta subsidence primarily caused by compaction of Holocene strata. *Nature Geoscience* 1, 173–176.

Torres, N., and Harrelson, D. W. 2012. The great Red River raft and its sedimentological implications. In *Reconstructing Human-Landscape Interactions*. Springer Briefs in Earth Systems Sciences, 1, pp. 35–55.

Toth, L. A., Obeysekera, J. T. B., Perkins, W. A., and Loftin, M. K. 1993. Flow regulation and restoration of Florida's Kissimmee river. *Regulated Rivers: Research and Management* 8, 155–166.

Tran, D. D., van Halsema, G., Hellegers, P. J. G. J., et al. 2018. Assessing impacts of dike construction on the flood dynamics of the Mekong Delta. *Hydrology and Earth Systems Science* 22, 1875–1896. doi: 10.5194/hess-221875-2018.

Trenberth, K. E. 1997. The definition of El Niño. *Bulletin of the American Meteorological Society* 78, 2771–2777.

Trenberth, K. E., Caron, J. M., Stepaniak, D. P., and Worley, S. 2002. Evolution of El Niño–Southern Oscillation and global atmospheric surface temperatures. *Journal of Geophysical Research – Atmospheres* 107, AAC 5-1-AAC 5–17.

Trigg, M. A., Bates, P. D., Wilson, M. D., Schumann, G., and Baugh, C. 2012. Floodplain channel morphology and networks of the middle Amazon River. *Water Resources Research* 48. doi: 10.1029/2012WR01188. W10504.

Trimble, S. W. 1974. *Man-Induced Soil Erosion on the Southern Piedmont, 1700–1970*. Soil and Water Conservation Society, Ankeny, Iowa.

Trimble, S. W. 1976. Sedimentation in Coon Creek valley, Wisconsin. In Proceedings of the Third Federal Interagency Sedimentation Conference, Denver, Water Resources Council, Section 5, pp. 100–112.

Trimble, S. W. 2013. *Historical Agriculture and Soil Erosion in the Upper Mississippi Valley Hill Country*. CRC Press, Boca Raton, FL.

Tullos, D. D., Collins, M. J., Bellmore, J. R., et al. 2016. Synthesis of common management concerns associated with dam removal. *Journal of the American Water Resources Association* 52, 1179–1206.

Turnbull, W. J., Krinitzky, E. L., and Weaver, F. J. 1966. Bank erosion in soils of the lower Mississippi Valley. *Journal of the Soil Mechanics and Foundations Division* 92, 121–136.

Turner, B. L. II, Lambin, E. F., and Reenberg, A. 2007. Land Change Science Special Feature: the emergence of land change science for global environmental change and sustainability. *Proceedings of the National Academy of Sciences* 52, 20666–20671.

Twain, M. 1883. *Life on the Mississippi*. James R. Osgood & Co., Boston, MA.

Twilley, R. R., Bentley, S. J., Chen, Q., et al. 2016. Co-evolution of wetland landscapes, flooding, and human settlement in the Mississippi River Delta Plain. *Sustainability Science* 11, 711. doi: 10.1007/s11625-016-0374-4.

Tye, R. S., and Coleman, J. M. 1989. Depositional processes and stratigraphy of fluvially dominated lacustrine deltas: Mississippi delta plain. *Journal of Sedimentary Petrology* 59, 973–996.

U.K. Environment Agency. 2015. *Management of the London Basin Chalk Aquifer, Status Report*, Environment Agency, UK.

U.S. Army Corps of Engineers (USACE). 1982. *Yazoo Area Pump Project: Yazoo Area and Satartia Area Backwater Levee Projects, Yazoo Backwater Area, Mississippi*. U.S. Army Corps of Engineers, Vicksburg, MS.

U.S. Army Corps of Engineers (USACE). 1988. *Final Report: Steele Bayou Gravity Control Structure, Hydraulic Model Investigation*. U.S. Army Corps of Engineers, Waterways Experiment Station, Vicksburg, MS.

U.S. Army Corps of Engineers (USACE). 1998a. Mississippi River Mainline Levees Enlargement and Seepage Control, Cape Girardeau, Missouri to Head of Passes, Louisiana, V. I.

U.S. Army Corps of Engineers (USACE). 1998b. Mississippi River Mainline Levees Enlargement and Seepage Control, Cape Girardeau, Missouri to Head of Passes, Louisiana, V. II.

U.S. Army Corps of Engineers (USACE). 1998c. Mississippi River Mainline Levees Enlargement and Seepage Control, Cape Girardeau, Missouri to Head of Passes, Louisiana, V. III.

U.S. Army Corps of Engineers (USACE). 2000a. Design and construction of levees. US Army Corps of Engineers, Engineering and Design, Engineers Manual No. 1110–2-1913.

U.S. Army Corps of Engineers (USACE). 2000b. *Yazoo Backwater Area Reformulation Report, Vols. I, II, III (17 App.)*, U.S. Army Corps of Engineers, Vicksburg, MS.

U.S. Army Corps of Engineers (USACE). 2009a. Dutch perspective appendix. U.S. Army Corps of Engineers New Orleans District, Mississippi Valley Division, Louisiana Coastal Protection and Restoration. Final Technical Report.

U.S. Army Corps of Engineers (USACE). 2009b. Incorporating sea-level change considerations in Civil Works Programs Engineer Circular (EC), 1165–2-211.

U.S. Army Corps of Engineers (USACE). 2012. Bonnet Carré Spillway, U.S. Army Corps of Engineers, New Orleans District, Louisiana.

U.S. Army Corps of Engineers (USACE). 2013. Modification of Caernarvon Diversion. Fact Sheet, U.S. Army Corps of Engineers, New Orleans Division. www.lca.gov/ (accessed September 29, 2018).

U.S. Army Corps of Engineers (USACE). 2014a. Procedures to evaluate sea level change: impacts, responses and adaptation. Engineer Technical Letter 1100–2-1.

U.S. Army Corps of Engineers (USACE). 2014b. Sacramento river bank protection project environmental impact statement / report. U.S. State Clearinghouse #2009012081.

U.S. Army Corps of Engineers (USACE). 2015. American river watershed: common features general reevaluation report. Attachment-E, Draft Erosion Protection Report. USACE Sacramento District.

U.S. Army Corps of Engineers – New Orleans Division (USACE). 2017. Internet Web page "Corps Needs Levee Clay." www.mvn.usace.army.mil/Missions/HSDRRS/ Corps-Needs-Levee-Clay/Borrow-Material/ (accessed January 18, 2018).

U.S. Army Corps of Engineers (USACE). 2018a. A summary of risks and benefits associated with the USACE levee portfolio. U.S. Army Corps of Engineers Levee Portfolio Report. Prepared by U.S. Army Corps of Engineers Levee Safety Program.

U.S. Army Corps of Engineers (USACE). 2018b. Yuba river ecosystem restoration feasibility study California. Draft Interim Feasibility Report and Environmental Assessment. U.S. Army Corps of Engineers, Sacramento District.

U.S. Army Corps of Engineers (USACE). 2019. *Lake Pontchartrain and Vicinity, Louisiana General Re-Evaluation Report with Integrated Environmental Impact Statement*. U.S. Army Corps of Engineers, New Orleans.

U.S. Army Corps of Engineers (USACE). 2020. Beneficial use of dredged material: history 2015–2020. In *Mississippi River, Baton Rouge to the Gulf of Mexico, LA Head of Passes Hopper Dredge Disposal Area*. U.S. Army Corps of Engineers, New Orleans. www.mvn.usace.army.mil/About/Offices/Operations/Beneficial-Use-of-Dredged-Material/hopper-dredge-disposal-area/ (accessed June 2020).

U.S. Army Corps of Engineers (USACE) Sacramento District. 2020. Sacramento levee upgrades. www.spk.usace.army.mil/Missions/Civil-Works/Sacramento-Levee-Upgrades/.

U.S. Congress. 1928. Flood Control Act, Mississippi River and Tributaries, Appendix E, 70th Congress, Session I Ch. 596.

U.S. Environmental Protection Agency (U.S. EPA). 2008. Final determination of the U.S. Environmental Protection Agency's Assistant Administrator for Water Pursuant to Section 404(C) of the Clean Water Act Concerning the Proposed Yazoo Backwater Area Pumps Project, Issaquena County, Mississippi, August 31, 2008 (with 6 appendices).

U.S. Environmental Protection Agency (U.S. EPA). 2001. Fact Sheet: Mercury Update: Impact on Fish Advisories. EPA-823-F-01-011. Office of Water 4305, Washington, DC.

U.S. Environmental Protection Agency (U.S. EPA). 2016a. Second five-year review report for the Milltown Reservoir Clark Fork River Superfund Site EPA ID MTD980717565. Milltown, Missoula, Granite, Powell, and Deer Lodge Counties, Montana. United States Environmental Protection Agency Region 8 Denver, Colorado, September 2016.

U.S. Environmental Protection Agency (U.S. EPA). 2016b. Frequently asked questions on removal of obsolete dams. Office of Water, EPA-840-F-16–001 December 2016.

U.S. Fish and Wildlife Service (U.S. FWS). 2001. U.S. Fish and Wildlife perspective on the Corps of Engineers proposed Yazoo Pumps Project. Fact Sheet, U.S. Fish and Wildlife Service, Jackson Ecological Services Field Office, Jackson, Mississippi. www.fws.gov/southeast/pubs/facts/yazooback.pdf (accessed May 19, 2016).

U.S. Geological Survey (USGS). 2010. Land subsidence in California. U.S. Geological Survey web site at http://ca.water.usgs.gov/groundwater/sub/ (accessed April 9, 2016).

U.S. Geological Survey (USGS). 2013. Trends and causes of historical wetland loss in coastal Louisiana. Fact Sheet 2013-3017. March 2013. (https://pubs.usgs.gov/fs/2013/3017/pdf/fs2013-3017.pdf)

U.S. Geological Survey (USGS). 2018. Endangered, discontinued and rescued stream gages. https://water.usgs.gov/networks/fundingstability/ (accessed November 11, 2018).

U.S. Government Accountability Office (U.S. GAO). 2005. Testimony before the subcommittee on energy and water development, Committee on Appropriations, House of Representatives, Army Corps of Engineers Lake Pontchartrain and Vicinity Hurricane Protection Project, Statement of Anu Mittal, Director Natural Resources and Environment, GAO-05–1050T.

U.S. Government Accountability Office (U.S. GAO). 2006a. Hurricane Katrina: strategic planning needed to guide future enhancements beyond interim levee repairs. Report to Congressional Committees, GAO-06–934.

U.S. Government Accountability Office (U.S. GAO). 2006b. Testimony before the Committee on Homeland Security and Governmental Affairs, U.S. Senate, Hurricane Protection: Statutory and Regulatory Framework for Levee Maintenance and Emergency Response for the Lake Pontchartrain Project, Statement for the Record by Anu K. Mittal, Director Natural Resources and Environment , GAO-06–322T.

U.S. Government Accountability Office (GAO). 2011. Mississippi river: actions are needed to help resolve environmental and flooding concerns about the use of river training structures. Rep. GAO-12–41. http://gao.gov/products/GAO-12-41.

U.S. National Inventory of Dams (NID). 2020. https://nid.sec.usace.army.mil/.

U.S. Society on Dams (USSD). 2015. *Guidelines for Dam Decommissioning Projects*. Committee on Dam Decommissioning, U. S. Society on Dams, Denver, CO.

Uehlinger, U., Wantzen, K. M., Leuven, R. S. E. W., and Arndt, H. 2009. The Rhine River Basin. In Tockner, K., Robinson, C. T., and Uehlinger, U. (eds.), *Rivers of Europe*. Elsevier – Academic Press, San Diego, CA. , pp. 1–47.

UNESCO-IHP. 2011. Sediment issues and sediment management in large river basins: interim case study synthesis report. International Sediment Initiative, Technical Documents in Hydrology, CN/2011/SC/IHP/PI/2, UNESCO International Hydrological Program, Beijing.

United Nations World Water Development Report (UNWWDR). 2020. *Water and Climate Change*. Food and Agriculture Organization of the United Nations, Rome, Italy.

van Asselen, S., Stouthamer, E., and van Asch, T. W. J. 2009. Effects of peat compaction on delta evolution: a review on processes, responses, measuring and modeling. *Earth Science Reviews* 92, 35–51. doi: 10.1016/j.earscirev.2008.11.001.

van Asselen, S., Karsenberg, D., and Stouthamer, E. 2011. Contribution of peat compaction to relative sea-level rise within Holocene deltas. *Geophysical Research Letters*, L24401. doi: 10.1029/2011GL049835.

van Asselen, S., Cohen, K. M., and Stouthamer, E. 2017. The impact of avulsion on groundwater level and peat formation in delta floodbasins during the middle-Holocene transgression in the Rhine-Meuse delta, The Netherlands. *The Holocene* 27, 1694–1706.

van Baars, S. 2005. The horizontal failure mechanism of the Wilnis peat dyke. *Géotechnique* 55, 319–323.

van Baars, S., and van Kempen, I. M. 2009. The causes and mechanisms of historical dike failures in the Netherlands. E-Water, Official Publication of the European Water Association (EWA), ISSN1994–8549.

van Balen, R. T., Houtgast, R. F., and Cloetingh, S. A. P. L. 2005. Neotectonics of The Netherlands: a review. *Quaternary Science Reviews* 24, 439–454.

van Balen, R., Kasse, C., and de Moor, J. 2008. Impact of groundwater flow on meandering; example from the Geul River, The Netherlands. *Earth Surface Processes and Landforms* 33, 2010–2028.

van Binh, D., Kantoush, S., and Sumi, T. 2019. Changes to long-term discharge and sediment loads in the Vietnamese Mekong Delta caused by upstream dams. *Geomorphology* 335, March 2020, 107011.

van de Ven, G. 2004. *Man-Made Lowlands: History of Water Management and Land Reclamation in the Netherlands*, 4th ed. Uitgeverij Matrijs, Utrecht.

van de Ven, G. 2007. *Verdeel en hbeheers! 300 jar Pannerdensch Kanaal*. Veen Magazines, Diemen.

van den Berg, J. H. 1995. Prediction of alluvial channel pattern of perennial rivers. *Geomorphology* 12, 259–279.

van den Brink, M., Meijerink, S., and Termeer, C. 2014. Climate-proof planning for flood-prone areas: assessing the adaptive capacity of planning institutions in the Netherlands. *Regional Environmental Change* 14, 981. doi: 10.1007/s10113–012–0401-7.

van Denderen, R. P., Schielen, R. M. J., Straatsma, M. W., Kleinhans, M. G., and Hulscher, S. J. M. H. 2019a. A characterization of side channel development. River Research and Applications 35, 1597–1603.

van Denderen, R. P., Schielen, R. M. J., Westerhof, S. G., Quartel, S., and Hulscher, S. J. M. H. 2019b. Explaining artificial side channel dynamics using data analysis and model calculations. *Geomorphology* 327, 93–110.

van der Meulen, M. J., Wiersma, A. P., van der Perk, M., Middelkoop, H., and Hobo, M. 2009. Sediment management and the renewability of floodplain clay for structural ceramics. *Journal of Soils and Sediments* 9, 627–639. doi: 10.1007/s11368–009–0115-8.

van der Most, M., and Hudson, P. F. 2018. The influence of floodplain geomorphology and hydrologic connectivity on alligator gar (Atractosteus spatula) habitat along the embanked floodplain of the lower Mississippi River. *Geomorphology* 302, 62–75.

van der Perk, M., Sutaria, C. A. T., and Middelkoop, H. 2019. Examination of the declining trend in suspended sediment loads in the Rhine River in the period 1952–2016. In Stouthamer, E., Middelkoop, H., Kleinhans, M., van der Perk, M., and Straatsma, M. (eds.), *Land of Rivers: NCR DAYS 2019 Proceedings*. Netherlands Centre for River Studies Publication 43–2019.

van Dijk, A. I. J. M., Beck, H. E., Crosbie, R. S., *et al.* 2013. The Millennium Drought in southeast Australia (2001–2009): natural and human causes and implications for water resources, ecosystems, economy, and society. *Water Resources Research* 49, 1040–1057. doi: 10.1002/wrcr.20123.

van Dijk, W. M., Teske, R., van de Lageweg, W. I., and Kleinhans, M. G. 2013. Effects of vegetation distribution on experimental river channel dynamics. *Water Resources Research* 49, 7558–7574.

Van Dinter, M., and Van Zijverden, W. K. 2010. Settlement and land use on crevasse splay deposits; geoarchaeological research in the Rhine-Meuse Delta, the Netherlands. *Netherlands Journal of Geosciences – Geologie en Mijnbouw* 89, 21–34.

van Heerden, I., and Roberts, H. H. 1980. The Atchafalaya Delta: Louisiana's new prograding coast. *Gulf Coast Association of Geological Societies, Transactions* 30, 497–506.

van Heezik, A. 2008) *Battle over the Rivers: Two Hundred Years of River Policy in the Netherlands*. Rijkswaterstaat, Haarlem, Netherlands.

van Loo, M. Dusar, B., Verstraeten, G., *et al.* 2017. Human induced soil erosion and the implications on crop yield in a small mountainous Mediterranean catchment (SW-Turkey). *Catena* 149, 491–504.

van Looy, K., Honnay, O., Bossuyt, B., and Hermy, M. 2003. The effects of river embankment and forest fragmentation on the plant species richness and composition of floodplain forest in the Meuse Valley, Belgium. *Belgian Journal of Botany* 136, 97–108.

van Veen, J. 1962. *Dredge, Drain and Reclaim: Art of a Nation*, 5th ed. Martinus Nijhoff, The Hague.

van Riel, W. 2011. *Exploratory Study of Pluvial Flood Impacts in Dutch Urban Areas*. Deltares, Delft.

van Vuren, S., Paarlberg, A., and Havinga, H. 2015. The aftermath of "Room for the River" and restoration works: coping with excessive maintenance dredging. *Journal of Hydro-environment Research* 9, 172–186.

Vergouwe, R. 2016. *The National Flood Risk Analysis for the Netherlands: Final Report*. VNK Project, Rijkswaterstaat, The Netherlands.

Verstraeten, G., Broothaerts, N., van Loo, M., *et al.* 2017. Variability in fluvial geomorphic response to anthropogenic disturbance. *Geomorphology* 494, 20–39. doi: 10.1016/j.geomorph.2017.03.027.

Vita-Finzi, C. 1969. *The Mediterranean Valleys*. Cambridge University Press, Cambridge.

Viale, M., Valenzuela, R., Garreaud, R. D., and Ralph, F. M. 2018. Impacts of atmospheric rivers on precipitation in Southern South America. *Journal of Hydrometeorology* 19, 1671–1687. doi: 10.1175/JHM-D-18-0006.1.

Vollmer, S. 2020. Geschiebezugabe – ein innovativer Ansatz zur Eindämmung der Sohlerosion. Bundesanstalt für Gewässerkunde

(Attachment allowance – an innovative approach to contain erosion). Federal Institute for Hydrology. www.bafg.de/DE/01_Leistungen/02_F_E/Themen/Artikel/Geschiebezugabe.html (accessed June 5, 2020).

Vollmer, S., and Goelz, E. 2006. Sediment monitoring and sediment management in the Rhine River. In *Sediment Dynamics and the Hydromorphology of Fluvial Systems* (Proceedings of a symposium held in Dundee, UK, July 2006). IAHS Publ. 306, 231–240.

Vorogushyn, S., and Merz, B. 2013. Flood trends along the Rhine: the role of river training. *Hydrology and Earth Systems Science* 17, 3871–3884. doi: 10.5194/hess-17-3871-2013.

Vörösmarty, C. J., and Sahagian, D. 2000. Anthropogenic disturbance of the terrestrial water cycle. *BioScience* 50, 753–765.

Vörösmarty, C. J., Meybeck, M., Fekete, B., Sharma, K., Green, P., and Syvitski, J. P. M. 2003. Anthropogenic sediment retention: major global impact from registered river impoundments. *Global and Planetary Change* 39, 169–190.

Voulvoulis, N., Arpon, K. D., and Giakoumis, T. 2017. The EU Water Framework Directive: from great expectations to problems with implementation. *Science of the Total Environment* 575, 358–366.

Wahl, K. L., Vining, K. C., and Wiche, G. J. 1993. Precipitation in the upper Mississippi River Basin, January 1 through July 31, 1993. U.S. Geological Survey Circular 1120-B.

Walker, H. J., and Hudson, P. F. 2003. Hydrologic and geomorphic processes in the Colville River Delta, Alaska. *Geomorphology* 56, 291–303.

Walker, H. J., Arnborg, L., and Peippo, J. 1987. Riverbank erosion in the Colville Delta, Alaska. *Geografiska Annaler: Series A* 69, 61–70.

Walling, D. E. 2006. Human impact on land-ocean sediment transfer by the world's rivers. *Geomorphology* 79, 192–216.

Walling, D. E. 2008a. The changing sediment loads of the world's rivers. In *Sediment Dynamics in Changing Environments* (Proceedings of a symposium held in Christchurch, New Zealand, December 2008). IAHS Publication 325, 323–338.

Walling, D. E. 2008b. The changing sediment load of the Mekong River. *AMBIO* 37, 150–157.

Walling, D. E. 2009. The sediment load of the Mekong River. In Campbell, I. (ed.), *The Mekong: Biophysical Environment of an International River Basin*. Aquatic Ecology. Elsevier – Academic Press, Amsterdam, pp. 113–142.

Walter, R. C., and Merritts, D. J. 2008. Natural streams and the legacy of water-powered mills. *Science* 319, 299–304.

Wang, B., and Xu, Y. J. 2015. Sediment trapping by emerged channel bars in the Lowermost Mississippi River during a major flood. *Water* 7, 6079–6096.

Wang, G., Wu, B., and Wang, Z. Y. 2005. Sedimentation problems and management strategies of Sanmenxia Reservoir, Yellow River, China. *Water Resources Research* 41, W09417. doi: 10.1029/2004WR003919.

Wang, H., Yang, Z., Saito, Y., Liu, J. P., and Xiaoxia, S. 2007. Stepwise decreases of the Huanghe (Yellow River) sediment load (1950–2004): impacts of climate change and human activities. *Global and Planetary Change* 57, 331–354.

Wang, H., Chen, Q., La Peyre, M. K., Hu, K., and La Peyre, J. F. 2017. Predicting the impacts of Mississippi River diversions and sea-level rise on spatial patterns of eastern oyster growth rate and production. *Ecological Modeling* 352, 40–53. doi: 10.1016/j.ecolmodel.2017.02.028.

Wang, H. J., Wright, T., Yu, Y., et al. 2012. InSAR reveals coastal subsidence in the Pearl River Delta, China. *Geophysical Journal International* 191, 1118–1128. doi: 10.1111/j.1365-246X.2012.05687.x.

Wang, J., Shen, Y., Gleason, C. J., and Wada, Y. 2013. Downstream Yangtze River levels impacted by Three Gorges Dam. *Environmental Research Letters* 8, 4012. doi: 10.1088/1748-9326/8/4/044012.

Wang, L., Ting, M., and Kushner, P. J. 2017. A robust empirical seasonal prediction of winter NAO and surface climate. *Scientific Reports* 7, 279. doi: 10.1038/s41598-017-00353-y.

Wang, Y., Zhao, W., Wang, S., Feng, X., and Liu, Y. 2019. Yellow River water rebalanced by human regulation. *Scientific Reports* 9, 9707. doi: 10.1038/s41598-019-46063-5.

Wang, Z.-Y., and Hu, C. 2009. Strategies for managing reservoir sedimentation. *International Journal of Sediment Research* 24, 369–384.

Wang, Z. Y., and Liang, Z. Y. 2000. Dynamic characteristics of the Yellow River mouth. *Earth Surface Processes and Landforms* 2, 765–782.

Ward, J. V., and Stanford, J. A. 1995. The serial discontinuity concept: extending the model to floodplain rivers. *Regulated Rivers: Research and Management* 10, 159–168.

Ward, P. J., Beets, W., Bouwer, L. M., Aerts, J. C. J. H., and Rensen, H. 2010. Sensitivity of river discharge to ENSO. *Geophysical Research Letters* 37, L12402. doi: 10.1029/2010GL043215.

Ward, P. J., Pauw, W. P., van Buuren, M. W., and Marfai, M. A. 2013. Governance of flood risk management in a time of climate change: the cases of Jakarta and Rotterdam. *Environmental Politics* 22, 518–536. doi: 10.1080/09644016.2012.683155.

Warne, A. G., Toth, L. A., and White, W. A. 2000. Drainage-basin-scale geomorphic analysis to determine reference conditions for ecologic restoration – Kissimmee River, Florida. *Geological Society of America Bulletin* 112, 884–899.

Wasklewicz, T., Greulich, S., Franklin, S., and Grubaugh, J. 2004. The 20th century hydrologic regime of the Mississippi River. *Physical Geography* 25, 208–224.

Water Institute of the Gulf (WIG). 2016. Building land in Coastal Louisiana: expert recommendations for operating a successful sediment diversion that balances ecosystem and community needs. https://thewaterinstitute.org/assets/docs/reports/Peyronnin_2016-8-31.pdf (accessed October 5, 2018).

Watson, C. C., and Biedenharn, D. S. 2009. Specific gage analysis of stage trends on the Middle Mississippi River. U.S. Army Corps of Engineers, St. Louis, MO.

Watson, C. C., Biedenharn, D. S., and Thorne, C. R. 2013a. Analysis of the impacts of dikes on flood stages in the middle Mississippi River. *Journal of Hydraulic Engineering* 139(10), 1071–1078. doi: 10.1061/(ASCE)HY.1943-7900.0000786.

Watson, C. C., Holmes, R. R., and Biedenharn, D. S. 2013b. Mississippi River streamflow measurement techniques at St. Louis, Missouri. *Journal of Hydraulic Engineering* 139(10), 1062–1070. doi: 10.1061/(ASCE)HY.1943-7900.0000752.

Watson, C. C., Biedenharn, D. S., and Thorne, C. R. 2015a. Closure to "Analysis of the Impacts of Dikes on Flood Stages in the Middle Mississippi River" by Chester C. Watson, David S. Biedenharn, and Colin R. Thorne. *Journal of Hydraulic Engineering* 141(8), 07015011-1. doi: 10.1061/(ASCE)HY.1943-7900.0001055.

Watson, C. C., Holmes, R. R., and Biedenharn, D. S. 2015b. Closure to "Mississippi River Streamflow Measurement Techniques at St. Louis, Missouri" by Chester C. Watson, Robert R. Holmes Jr., and David S. Biedenharn. *Journal of Hydraulic Engineering* 141(8), 07015008-1. doi: 10.1061/(ASCE)HY.1943-7900.0001021.

Watson, J. M., Coghlan, S. M., Jr., Zydlewski, J., Hayes, D. B., and Kiraly, I. A. 2018. Dam removal and fish passage improvement influence fish assemblages in the Penobscot River, Maine. *Transactions of the American Fisheries Society* 147, 525–540.

Watson, K. M., Harwell, G. R., Wallace, D. S., et al. 2018. Characterization of peak streamflows and flood inundation of selected areas in southeastern Texas and southwestern Louisiana from the August and September 2017 flood resulting from Hurricane Harvey. U.S. Geological Survey Scientific Investigations Report 2018–5070.

World Commission on Dams (WCD). 2000. *Dams and Development: A New Framework. Report of the World Commission on Dams for Decision-Making*. Earthscan Publications, London.

Wei, W., Chang, Y., and Dai, Z. 2014. Streamflow changes of the Changjiang (Yangtze) River in the recent 60 years: impacts of the East Asian summer monsoon, ENSO, and human activities. *Quaternary International* 336, 98–107.

Welder, F. A. 1955. Deltaic processes in Cubits Gap Area, Plaquemines Parish, Louisiana. PhD dissertation, Louisiana State University, Baton Rouge, LA.

Wells, J. T., Chinburg, S. J., and Coleman, J. M. 1983. The Atchafalaya River Delta, Report No. 4: Generic analysis of Delta Development. U.S. Corps of Army Engineers, Waterways Experiment Station, Technical Report HL-82–15.

Wheeler, K. G., Basheer, M., Mekonnen, Z. T., et al. 2016. Cooperative filling approaches for the Grand Ethiopian Renaissance Dam, *Water International* 41, 611–634.

White, G. 1945. Human adjustment to floods. University of Chicago, Department of Geography, Research Paper, No 29.

Whitfield, P. H., Moore, R. D., Fleming, S. W., and Zawadzki, A. 2010. Pacific Decadal Oscillation and the hydroclimatology of western Canada – review and prospects. *Canadian Water Resources Journal / Revue canadienne des ressources hydriques* 35, 1–28.

Whiting, P., and Pomeranets, M. 1997. A numerical study of bank storage and its contribution to streamflow. *Journal of Hydrology* 202, 121–136.

Whittaker, B. N., and Reddish, D. J. 1989. *Subsidence: Occurrence, Prediction, and Control*. Developments in Geotechnical Engineering, 56. Elsevier, Amsterdam.

Wilcock, P. R. 1998. Two-fraction model of initial sediment motion in gravel-bed rivers. *Science* 280, 410–412.

Williams, A. P., Funk, C., Michaelsen, J., et al. 2012. Recent summer precipitation trends in the Greater Horn of Africa and the emerging role of Indian Ocean sea surface temperature. *Climate Dynamics* 39, 2307–2328.

Williams, G. P. 1978a. Bankfull discharge of rivers. *Water Resources Research* 14, 1141–1154.

Williams, G. P. 1978b. The case of the shrinking channels: The North Platte and Platte Rivers, Nebraska, US Geological Survey Circular 781.

Williams, G. P. 1984. Paleohydrological methods and some examples from Swedish fluvial environments, II – river meanders. *Geografiska Annaler: Series A, Physical Geography* 66, 89–102.

Williams, G. P. 1986. River meanders and channel size. *Journal of Hydrology* 88, 147–164.

Williams, G. P., and Wolman, M. G. 1984. Downstream effects of dams on alluvial rivers. U.S. Geological Survey, Professional Paper 1286.

Williams, S. J., Arsenault, M. A., Buczkowski, B. J., et al. 2009. Surficial sediment character of the Louisiana offshore continental shelf region: a GIS compilation. U.S. Geological Survey Open-File Report 2006-1195.

Wilson, K. V. 1979. Changes in channel characteristics, 1938–1974, of the Homochitto River and Tributaries, Mississippi. U.S. Geological Survey, Open-File Report 79–554.

Winemiller, K. O., Tarim, S., Shorman, D., and Cotner, J. B. 2000. Fish assemblage structure in relation to environmental variation among Brazos River oxbow lakes. *Transactions of the American Fisheries Society* 129, 451–468.

Winemiller, K. O., McIntyre, P.B., Castello, L., et al. 2016. Balancing hydropower and biodiversity in the Amazon, Congo, and Mekong. *Science* 351, 128–129.

WinklerPrins, A. M. G. A. 2002. Seasonal floodplain-upland migration along the Lower Amazon River. *The Geographical Review* 92, 415–431.

Winkley, B. R. 1977. Man-made cutoffs on the lower Mississippi River, conception, construction and river response. U.S. Army Corps of Engineers, Potamology Investigations Report 300–2, Vicksburg, MS.

Winkley, B. R. 1994. Response of the lower Mississippi River to flood control and navigation improvements. In Schumm, S. A., and Winkley, B. R. (eds.), *The Variability of Large Alluvial Rivers*. American Society of Civil Engineers, New York, pp. 45–74.

Winton, R. S., Calamita, E., and Wehrli, B. 2019. Reviews and syntheses: dams, water quality and tropical reservoir stratification. *Biogeosciences* 16, 1657–1671. doi: 10.5194/bg-16-1657.

Wintenberger, C., Rodrigues, S., Bréhéret, J.-G., and Villar, M. 2015. Fluvial islands: first stage of development from nonmigrating (forced) bars and woody-vegetation interactions. *Geomorphology* 246, 305–320.

Wittfogel, K. 1957. *Oriental Despotism: A Comparative Study of Total Power*. Yale University Press, New Haven, CT.

Wohl, E. 2004. *Disconnected Rivers: Linking Rivers to Landscapes*. Yale University Press.

Wohl, E. 2007. Hydrology and discharge. In Gupta, A. (ed.), *Large Rivers: Geomorphology and Management*. John Wiley & Sons, Chichester, pp. 29–44.

Wohl, E. 2014. A legacy of absence: wood removal in U.S. rivers. *Progress in Physical Geography* 38, 637–663.

Wolman, G. M. 1959. Factors influencing the erosion of cohesive river banks. *American Journal of Science* 257, 204–216.

Wolman, G. M., and Gerson, R. 1978. Relative scales of time and effectiveness of climate in watershed geomorphology. *Earth Surface Processes and Landforms* 3, 189–208.

Wolman, G. M., and Giegengack, R. F. 2007. The Nile River: geology, hydrology, hydraulic society. In Gupta, A. (ed.), *Large Rivers: Geomorphology and Management*. John Wiley & Sons, Chichester, pp. 471–490.

Wolman, G. M., and Leopold, L. B. 1957. River flood plains: some observations on their formation. U.S. Geological Survey Professional Paper 282-C.

Wolman, G. M., and Miller, G. P. 1960. Magnitude and frequency of forces in geomorphic process. *Journal of Geology* 68, 54–74.

Wolman, M. G. 1967. A cycle of sedimentation and erosion in urban river channels. *Geografiska Annaler* 49A, 385–395.

Wolters, H. A., Platteeuw, M., and Schoor, M. M. 2001. Guidelines for rehabilitation and management of floodplains; Ecology and Safety Combined. NCR Publication 09–2001; RIZA Report 2001.059.

Wolters, K. 2010. Multivariate ENSO Index (MEI). National Oceanic Atmospheric Administration.

Woodyer, K. D. 1968. Bankfull frequency in rivers. *Journal of Hydrology* 6, 114–142.

World Bank. 2017. *Implementing Nature-Based Flood Protection: Principles and Implementation Guidance*. Washington, DC.

World Meteorological Association (WMO). 2004. *Integrated Flood Management*. The Associated Programme on Flood Management 1, Geneva, Switzerland.

Wright, L. D., and Coleman, J. M. 1972. River delta morphology, wave climate and the role of the subaqueous profile. *Science* 176, 282–284.

Wright, L. D., and Coleman, J. M. 1973. Variations in morphology of major river deltas as functions of ocean waves and river discharge regimes. *American Association of Petroleum Geologists Bulletin* 57, 370–398.

Wright, L. D., and Coleman, J. M. 1974. Mississippi River mouth processes: effluent dynamics and morphologic development. *Journal of Geology* 82, 751–778.

Wu, C.-Y., and Mossa, J. 2019. Decadal-scale variations of thalweg morphology and riffle–pool sequences in response to flow regulation in the lowermost Mississippi River. *Water* 11, 1175. doi: 10.3390/w11061175.

Wu, Z., Milliman, J. D., Zhao, D., Cao, Z., Zhou, J., and Zhou, C. 2018. Geomorphologic changes in the lower Pearl River Delta, 1850–2015, largely due to human activity. *Geomorphology* 314, 42–54.

Wüst, R. A. J., Bustin, R. M., and Lavkulich, L. M. 2003. New classification systems for tropical organic-rich deposits based on studies of the Tasek Bera Basin, Malaysia. *Catena* 53, 133–153.

WWF. 2019. Biggest dam removal in European history begins. https://wwf.panda.org/wwf_news/successes/?347630/Biggest-Dam-Removal-in-European-History

Xiquing, C. 1998. Changjian (Yangtze) River Delta, China. *Journal of Coastal Research* 14, 838–858.

Xu, Y. J. 2006. Organic nitrogen retention in the Atchafalaya River Swamp. *Hydrobiologia* 560, 133–143. doi: 10.1007/s10750-005-1171-8.

Xue, C. T. 1993. Historical changes in the Yellow River delta, China. *Marine Geology* 113, 321–330.

Yang, C. T. 2006. *Erosion and Sedimentation Manual*. U.S. Department of the Interior Bureau of Reclamation. Technical Service Center, Sedimentation and River Hydraulics Group. Denver, CO.

Yang, S. L., Milliman, J. D., Li, P., and Xu, K. 2011. 50,000 dams later: erosion of the Yangtze River and its delta. *Global and Planetary Change* 75, 14–20.

Yang, S. L., Xu, K. H., Milliman, J. D., Yang, H. F., and Wu, C. S. 2015. Decline of Yangtze River water and sediment discharge: impact from natural and anthropogenic changes. *Scientific Reports* 5, 12581. doi: 10.1038/srep12581.

Yia, Y., Yang, Z., and Zhang, S. 2010. Ecological influence of dam construction and river-lake connectivity on migration fish habitat in the Yangtze River basin, China. *Procedia Environmental Sciences* 2, 1942–1954.

Ylla Arbos, C., Schielen, R. M. J., and Blom, A. 2020. Bed level change in the Upper Rhine Delta between 1926–2018. NCR Days 2020, pp. 91–92, Nijmegen, Netherlands.

Yodis, E. G., and Kesel, R. H. 1992. The effects and implications of base-level changes to Mississippi River tributaries. *Zeitschrift für Geomorphologie* 37, 385–402.

Yoshida, Y., Lee, H. S., Trung, B. H., et al. 2020. Impacts of mainstream hydropower dams on fisheries and agriculture in lower Mekong Basin. *Sustainability* 12, 2408. doi: 10.3390/su12062408.

Yoshiyama, R. M., Gerstung, E. R., Fisher, F. W., and Moyle, P. B. 2001. Historical and present distribution of Chinook Salmon in the Central Valley Drainage of California. Contributions to the Biology of Central Valley Salmonids. Fish Bulletin 179, Volume 1.

Young, W. C., Kent, D. H., and Whiteside, B. G. 1976. The influence of a deep storage reservoir on the species diversity of benthic macroinvertebrate communities on the Guadalupe River, Texas. *The Texas Journal of Science* 27, 213–224.

Yuill, B. T., Lavoie, D., and Reed, D. J. 2009. Understanding subsidence processes in coastal Louisiana. *Journal of Coastal Research: Special Issue* 54, 23–36. doi: 10.2112/SI54-012.1.

Yuill, B. T., Gaweesh, A., Allison, M. A., and Meselhe, E. A. 2016a. Morphodynamic evolution of a lower Mississippi River channel bar after sand mining. *Earth Surface Processes and Landforms* 41, 252–262. doi: 10.1002/esp.3846.

Yuill, B. T., Khadka, A. K., Pereira, J., Allison, A., and Meselhe, E. A. 2016b. Morphodynamics of the erosional phase of crevasse-splay evolution and implications for river sediment diversion function. *Geomorphology* 259 12–29.

Zarfl, C., Lumsdon, A. E., Berlekamp, J., Tydecks, L., and Tockner, K. 2015. A global boom in hydropower dam construction. *Aquatic Sciences* 77, 161–170.

Zarfl, C., Berlekamp, J., He, F., *et al.* 2019. Future large hydropower dams impact global freshwater megafauna. *Scientific Reports* 9, 18531. doi: 10.1038/s41598-019-54980-8.

Zehetner, F., Lair, G. J., Maringer, F.-J., Gerzabek, M. J., and Hein, T. 2008. From sediment to soil: floodplain phosphorus transformations at the Danube River. *Biogeochemistry* 88, 117–126. doi: 10.1007/s10533–008–9198-3.

Zevenbergen, C., Khan, S. A., van Alphen, J., van Scheltinga, C. T., and Veerbeek, W. 2018. Adaptive delta management: a comparison between the Netherlands Bangladesh Delta Program. *International Journal of River Basin Management* 16, 299–306.

Zhang, J., Wen, W., Mao, H., and Shi, C. 1990. Huanghe (Yellow River) and its estuary: sediment origin, transport and deposition. *Journal of Hydrology* 120, 203–223.

Zhang, W., Yuan, J., Han, J., Huang, C., and Li, M. 2016. Impact of the Three Gorges Dam on sediment deposition and erosion in the middle Yangtze River: a case study of the Shashi Reach. *Hydrology Research*. doi: 10.2166/nh.2016.092.

Zhang, W., Villarini, G., Vecchi, G. A., and Smith, J. A. 2018. Urbanization exacerbated the rainfall and flooding caused by hurricane Harvey in Houston. *Nature* 563(7731), 384. doi: 10.1038/s41586-018-0676-z.

Zhang, X., Dong, Z., Gupta, H., Wu, G., and Li, D. 2016. Impact of the Three Gorges Dam on the hydrology and ecology of the Yangtze River. *Water* 8, 590. doi: 10.3390/w8120590.

Zheng, S., Wu, B., Wang, K., Tan, G., Han, S., and Thorne, C. R. 2017. Evolution of the Yellow River delta, China: impacts of channel avulsion and progradation. *International Journal of Sediment Research*. doi: 10.1016/j.ijsrc.2016.10.001.

Zheng, S., Xu, Y. J., Cheng, H., Wang, B., Xu, W., and Wu, S. 2018. Riverbed erosion of the final 565 kilometers of the Yangtze River (Changjiang) following construction of the Three Gorges Dam. *Scientific Reports* 8, 11917. doi: 10.1038/s41598-018-30441.

Zhong, Y., and Power, G. 2015. Environmental impacts of hydroelectric projects on fish resources. *Regulated Rivers: Research and Management* 12, 81–98.

Zhu, Y., and Newell, R. E. 1998. A proposed algorithm for moisture fluxes from atmospheric rivers. *Monthly Weather Review* 126, 725–735.

Zinger, J. A., Rhoads, B. L., and Best, J. L. 2011. Extreme sediment pulses generated by bend cutoffs along a large meandering river. *Nature Geoscience* 4, 675–678.

Ziv, G., Baran, E., Nam, S., Rodríguez-Iturbe, I., and Levin, S. A. 2012. Trading-off fish biodiversity, food security, and hydropower in the Mekong River Basin. *Proceedings of the National Academy of Sciences* 109, 5609–5614.

Zorn, M. R., and Waylen, P. R. 1997. Seasonal response of mean monthly streamflow to El Niño/Southern Oscillation in North Central Florida. *Professional Geographer* 49, 51–56.

Index

Alblasserwaard polder, NL, 207–209
Asian mega-deltas, 4, 72, 223–225, 235–238, 249, *See* rivers
atmospheric
   atmospheric rivers, 42–43
      Maya Express, 37, 42–43
   Bermuda High, 37
   intertropical convergence zone (ITCZ), 31–32
   synoptic scale, 31
   teleconnections, 37–42
      Atlantic Multidecadal Oscillation (AMO), 39
      El Niño Southern Oscillation (ENSO), 38–42, 236
      ENSO and Amazon flooding, 40
      ENSO and South American rivers, 40
      ENSO and streamflow variability, 37–42
      Indian Ocean Dipole (IOD), 39, 41
      La Niña, 41, 42
      North Atlantic Oscillation (NAO), 38–39
      Northern Annular Mode (NAM), 39
      Pacific Decadal Oscillation (PDO), 38, 41, 42
      Southern Annular Mode (SAM), 37–38
      teleconnections and streamflow variability, 37–42
   tropical cyclones, 33, 35, 236, 239
   zonal and meridional flow, 33, 36
avulsion, *see* floodplain geomorphology

Bangkok 2011 flood, 35, 235–238
   hydro-climatology, 236
batture channel (tie-channel), 47, 57, 58, 191, 192

channel bank erosion, *See* channel geomorphology
channel bed wave (historic California gold mining), 17–18, 29, 143
channel engineering, 27–28, 138–165, 210, 251–257
   channelization (cut-offs), 27–28, 144–147, 156–163
      lower Mississippi case study, 156–163
      response, knickpoint, 28, 122, 128, 145–147, 157–159, 202, 203, 284
      Tulla cut-offs and straightening, Rhine, 147, 148, 193, 278
      types, 139, 144
      Ucayali and local people (Amazon basin), 146–147
   evolution
      conceptual model, 28, 201–203
   groynes (wing dikes), 138–140, 151–153, 162, 251–252
      hydraulic influence, 151–152
      influence on sediment and floodplain evolution, 151–152
      lower Mississippi, 162
      pattern of erosion and deposition, Rhine delta, 152
   revetment, 102, 139, 142, 148–150, 161, 253–256
      concrete, casting field, lower Mississippi, 149
      installation, lower Mississippi, 149–150
      environmental impact, 150
      willow fascine mattress, 149
   riprap, 139, 148–149
      design, 149
   sediment dredging, 139–140
      dredge spoil, Apalachicola River, lower Mississippi, 140–141
      sediment mining, 139, 141
      environmental impact, Mekong, 141
   sediment replenishment (dumping), 139, 232, 250–254
   sedimentology, influence
      cohesive vs. coarse sediment, 143–144
      Mississippi backswamp clay, 144, 156–157, 229–230
   structural and nonstructural measures, list, 139
   tree snag removal and log jams, 139, 153–155
      extent of wood removal, 154
      Great River Raft, Red River (Louisiana), 154–155
      US rivers, historic, 154
channel geomorphology
   bankfull discharge, 19
   boundary shear stress, 9, 23, 44
   channel bank erosion, 23, 43–47
      hydraulic entrainment, 44
      hydrologic controls, 46–48
      mass wasting, 45
      processes, 44
      sedimentary controls, 44–46
   channel bed aggradation, 29, 105, 125, 134, 155, 161, 233
   channel slope, 20, 23, 27, 104, 138, 144, 278
   critical entrainment, 44
   cross-sectional geometry, 22
   disturbance and fluvial adjustment (conceptual), 27–29
   dominant discharge, 19
   ecosystem links with functional channel surfaces, 106
   Froude number, 23
   impacts of Great River Raft, Red River (Louisiana), 154–155
   longitudinal profile, 22, 63–64, 97, 99, 136, 142
   meander wavelength, 20, 22, 25, 29
   mid-channel bar (model), 25
   planform geometry, 22
   pool-riffle morphology, 22
   radius of curvature, 10, 22
   sinuosity, 21, 24, 27–29, 62–63, 142, 147, 278
   specific stream power, 10, 12, 19, 26, 28, 46, 100
   velocity shelter, 25, 53
   w/d ratio, 20, 22, 29
channel pattern, 19–27
   anabranching, 26–27
      anastomosing, 26
      mega rivers, 27
      wandering, 26
   braided, 21, 25
   meandering, 19–24
   meandering-braided threshold, 20
Chinese Loess Plateau, 9, 11, 15, 94, 96, 134, 136, 191
coastal land loss in Louisiana, 76, 78, 265
coastal storm surge, 11, 33, 76–77, 169, 176, 205, 228, 242–248, 269–270, 282
   geophysical parameters, hurricane strength, 243
   Hurricane Katrina, 243–245
      storm surge levels, 243–244
   influence on Mississippi River stage levels, 243
   loop current (Gulf of Mexico), 244

INDEX 327

continuum of geomorphic adjustment and sequence of management
  activities, 201–203, 284–286
Cubits Gap (breach and subdelta), 267–269
cycle of dike management, 167, 171–176
dam removal, 115–130
  drivers (list and characterization), 116
    cultural heritage, 125
    economic, cost to rehabilitate US dams, 117
    environmental, fish, 116, 119
    policy, EU Water Framework Directive, 115, 119
    salmon, 119, 124
    stakeholders, 117
  Elwha River case study, 125–126
  extent and tally, 118–120
    Canada, 118
    European Union, 119
    Finland, 119
    France, 119
    Great Britain, 119
    Mexico, 119
    Spain, 119
    Sweden, 119
    United States, 118
  reservoir drawdown, 122, 125
  reservoir sediment, 120–122
    reworking, 122–125
  response, 125–129
    channel adjustment, 124, 126–128
    fish, 125
    sediment pulse, 122, 126–128
    vegetation, 125
  salmon and dam removal
    Atlantic salmon (*Salmo salar*), 118, 119, 276–277
    Pacific salmon (five species), 117, 118
  science of dam removal, 120–130
    polluted sediment, 122
    polluted sediment, arsenic (Clark Fork), 123
    reservoir drawdown, 120, 122
    reservoir sediment management strategy, 121
    reservoir sediment weight, 123
    sediment compaction, change to reservoir storage capacity, 123
    sediment reworking, 122–125

dams, 80–137
  and agriculture, 80–82, 84, 85, 87, 105–106, 108, 111
    chemical fertilizer and Nile agriculture, 108
    freshwater withdrawal, 81
    Indus Basin Project, 19, 89
    irrigation network, 81, 82, 83
    irrigation projects, Mekong, 82
    water withdrawal and changes to Platte River, 105–106
  channel bed incision
    Danube, 104
    several US rivers (Chattahoochee, Colorado, Missouri, Red), 103
    Volga, 104
    Yangtze, 102, 104
  channel degradation
    Rio Grande/Bravo, 105
  comprehensive environmental impacts, 95
  dam and reservoir, list
    Aswan High Dam and Lake Nasser, Lake Nubia (Nile), 3, 81, 91–95, 108–109
    Canyon Dam (Guadalupe), 112
    Cherokee (Holston), 113, 115
    Don Sahong (Mekong), 110
    Douglas Dam (French Broad, United States), 113, 115
    Elephant Butte (Rio Grande), 105
    Elwha and Glines Canyon (Elwha River, United States), 116, 118, 126, 127
    Falcon (Rio Grande / Bravo), 104
    Flix reservoir (Ebro), 133
    Fort Edward (Hudson), 122
    Gavins Point and Lewis and Clarke Lake (Missouri), 100, 101, 102, 103, 132, 137
    Grand Ethiopian Renaissance Dam (Blue Nile), 92
    Huanghe River main-stem dams (Longyangxia, Liujiaxia, Qingtongxia, Sanshenggong, Wanjiazhai, Sanmenxia, Xiaolangdi), 97
    Iffezheim (Rhine), 98, 99, 251, 253, 257
    Kariba Reservoir (Zambezi), 111
    Kingsley (Platte) and McConaughy Lake, 106–107
    Manuel Moreno Torres, Chicoasén (Mexico), 119
    Mekong basin and large dams (map), 98
    Milltown (Clark Fork), 122–123, 124
    Nam Leuk Reservoir (Mekong), 111
    Pa Sak (Chao Phraya), 236
    Red Bluff Diversion Dam (Sacramento), 81
    Sanmenxia Dam (Huanghe), 97, 131, 134, 136
    Three Gorges Dam (Yangtze), 91, 94, 96, 102, 104, 108, 109, 131, 133
    upper Mekong (Lancang) dams: Gongguoqiao, Xiaowan, Manwan, Dachaoshan, Nuozhadu, Jinghong, Ganlanba, 96–97
    Vezins and La Roche Qui Boit (Sélune, FR), 119
    Volgograd (Volga), 91, 99, 101, 104, 108, 109
  dam types, 83
  ecosystem impacts, 105–115
    aquatic fauna, 108–110
    aquatic fauna and Irrawaddy Dolphin (Orcaella brevirostris) in Mekong, 110
    aquatic fauna and Mekong fishery, 108, 110
    aquatic fauna, Nile fishery, 108–109
    aquatic fauna, Volga and Russian sturgeon, 108–109
    aquatic fauna, Yangtze and carp, 108–109
    channel narrowing, 105–107
    channel narrowing, Platte, 106–107
    woody vegetation encroachment, 105–107
  environmental impact
    river fragmentation, 86–89
  global extent and tally, 82–86
    Australia, large dams, 86
    Canada, large dams, 86
    large dams, 83–85
    reservoir storage capacity, 85
    Spain, large dams by drainage basin, 86
    United States, 86–87
  hungry water, 100
  new construction, 87–88
  reservoir evaporation, 81
  reservoir sediment management, 120–125, 130–137
  reservoir sediment storage, 130–132
  reservoir water quality decline, 110–115
    hypoxia, 110–111
    hypoxia and Lake Nasser, 110–111
    outflow oxygen and temperature decline for TVA dams, 113–115
    outflow temperature and oxygen, 111
    outflow temperature, lower Mekong basin dams, 112
    phosphorus, 111–112
    thermal stratification, 110–111
  river fragmentation index, 86–89
  river regulation index, 86–89
  sediment decline, 93–100
    changes to global sediment flux, 93
    Ebro, 99
    Huanghe, 94
    Mekong, 96–97
    middle Mississippi and lower Mississippi, 99–102
    Missouri, 99–102
    Nile and agriculture, 93–94, 96
    Rhine, 97–99
    Volga, 99
    Yangtze River after Three Gorges Dam, 94–96
  streamflow regime impacts, 89–92
    changes to TVA dams, 115
    decline for Platte, 107
    Ebro, 90

dams, (cont.)
  Nile, 91–92
  Rio Grande / Bravo, 104–105
  Tennessee Valley Authority (TVA), 113–114
  US rivers, 89–90
  Volga, 91
  Yangtze and Three Gorges Dam, 91
 world's oldest dam, Cornalvo in Spain, 85
dams, environmental impact, 86–115
dams, global extent, 82–89
dams and agriculture, 80–83
delta form and process, 64–78
deltaic geomorphology
  delta cycle, 73–77
  fluvial dominated, 72–73
  fluvial-wave transition, 72
  fluvial–deltaic sedimentation (hypopycnal condition), 73–74
  Mississippi delta lobe chronology, 75
  Mississippi delta wetland loss, 77–78, 265
  natural levee. See floodplain geomorphology
  primary controls, 64–69
    delta size, 64
    delta size and drainage area, 68, 69
  sediment facies, 74
  subdeltas, 76, 265–270
  tidal, 72
  wave dominated, 72
  wave to fluvial delta continuum, 72
depositional environment. See floodplain geomorphology
drainage basin perspective, 9–13
embanked floodplain, 166–204
  change to flooding, 181–183
  dike breach event, 196–202
  dike breach process, 193–195
    overtopping, 194
    piping, 174, 176, 193, 246
    underseepage, 194–195, 221, 246
  dike breach sedimentology, 193–196, 199–201
  evolution
    conceptual model, 168–169, 201–203
  Huanghe profile, 191
  hydrographic change
    artificial cut-offs, 189–193
    borrow pit ponds, 185–187
    dike breach ponds (wielen), 187–188
    natural and anthropogenic water bodies, 183
  hydrologic pathways, 168
  ice dam (jam) flooding, 167, 197
  lower Mississippi
    distance profile of embanked floodplain width, 180
  sand boil, 194–196
  sedimentary change, 188–189
    dike breach pond (wiel) infilling, 199
  width of embanked floodplains for some US rivers, 180

EU Water Framework Directive, 116, 119, 275–279, 288, 293, See dam removal:drivers:policy
flood basin and delta management, 205–248
  Atchafalaya, 208, 213, 229–232, 266–267, 274, 287
  Bangkok (Chao Phraya River) 2011 flood, 235–238
  Caernarvon Freshwater Diversion Structure, 269–270
  Biesbosch, NL, 208, 228–229, 265
  ecological disturbance, 213–218
  flood diversion structures
    Bonnet Carré Spillway, 43, 206–208, 213–215
    Fremont Weir, 216–217
    Morganza Spillway, 207, 208, 213, 229, 231
    Steele Bayou (Yazoo basin backwater), 211–212
    types, 207
    Yolo bypass, 207, 216–218
  sediment diversion structures, 78, 79, 99, 141, 233, 250, 265, 270–274, 290
  suspended sediment dynamics, 212, 216, 270–271
  urban flooding, 234–248, 289
    Bangkok 2011 flood, 235–238
    Houston and Hurricane Harvey, 248
    New Orleans, 2, 239–248
    Yazoo backwater flood basin, 206–211

flood control and flood management
  1928 Mississippi River & Tributaries Project, 145, 156, 177, 192, 207–213, 282, 288
  1941 Flood Control Act, 211
  1965 Lake Pontchartrain & Vicinity Hurricane Protection Project, 177, 242–243, 244, 245, 269
  1972 Clean Water Act, section 404c, 116, 212
  1986 Water Resources Development Act, 212
  2007 Water Resources Development Act, 170, 270
  cycle of dike management, 171–177
  dike breach process. See embanked floodplain:dike breach process
  dike (levee) design, 167–181
    construction deficiencies, 173–174
    factors in dike design, 172–173
    geophysical indices, 174
  dike system, 168–171
    US dikes, 171
  dike, types, 169
  New Orleans dike, history, 176–177
  palimpsest of management, 176–177
  pattern of embankment, 178–179
  Room for the River, Netherlands, 8, 170, 213, 249, 250, 258–264, 275–283, 285, 289
  structural measures, 138–140, 169
  Yazoo Backwater Area Reformulation, 212
flood pulse, 31, 32–34, 40, 92, 169, 177, 205, 211, 258
flooding (natural processes)
  hydrologic pathways, 48–50, 168
    floodplain sedimentology, 49
    groundwater, 48
    lowland rivers, 48–49
    overbank flow, 49–50
  local-scale mechanisms, 48–49
  perirheic zone, 59
  watershed-scale mechanisms, 48
floodplain geomorphology
  abandoned channel (infill), 56, 57, 58
  avulsion, 64, 225–233
    and flood basin sedimentology, 228–231
    Atchafalaya, 229–231
    controls, 226
    Huanghe, 226–227
    management, 231–234
    management factors, table, 232
  backswamp, 46, 48, 52, 55–56
  channel belt, 49–51
  channel fill deposit, 52, 57–59
  channel lag deposit, 25
  clay plug, 46–47, 58, 143, 157, 287
    Fisk, 47, 58
  crevasse splay, 52, 53–54, 55, 61
  depositional environment, 52, 53, 55, 74
  natural levee, 51–53, 55, 56, 58, 60–61
    bioturbation, 52, 61
    downstream pattern, 53–54
    New Orleans, 240
    relation to migration rates, 54
  negative relief floodplain, 49–51
  overbank (flood) sedimentary deposits, 51–58
  pedogenesis, 59–62
    bioturbation, 61
  pedogenic properties, 52, 62
    cutan, 61
    soil catena, 61
  point bar, 23–24, 52
  sedimentary infilling
    channel plug, 57, 58
    oxbow lake, 57–58, 184, 192–193

INDEX

sedimentary structure, 52
spillage sedimentation (model), 53
fluvial system, 9–13
deposition zone, 12
headwater (supply) zone, 9–11
transfer zone, 11–12

Gilbert, G.K., 17, 29
Goyder's Line, 82
Great Flood of 1993
Missouri, middle and upper Mississippi, 37, 59, 170, 199–200
rainfall totals, 38
Great River Raft, Red River (Louisiana), 154–155

hunger stones, 36
Hurricane Harvey 2017 flooding of southeast Texas, 248
Hurricanes Katrina and Rita, 2, 8, 169, 171, 176, 239–248, 269–270, 287
Caernarvon Freshwater Diversion Structure, 269–270
New Orleans and Gulf Coast flooding (*see* New Orleans)
storm surge levels, 243–245
hydraulic geometry, 26, 27
change, lower mississippi, 157, 158
hydraulic gold mining in California, historic, 17–18, 29, 216–218
mercury (Hg) pollution, 17, 216–218
hydroclimatology, 12, 31–43
historic flood regime (upper Mississippi), 36–37
flow variability, 34
ice jam flooding, 36, 37, 122, 167, 197–199
teleconnections, *see* atmospheric
hydrologic regime, 32–35

integrated river basin management (IRBM), 7, 249–283, 288
channel
coarse sediment replenishment (sediment dumping), 252–253, 286
groyne lowering, 251–252
meander (channel) reconnection, 256–257
removal of revetment (bank protection works), 253–256
concern and optimism (in IBRM), 291–296
embanked floodplain. *See* Room for the River
dike setback, 260–262
lake reconnection, 260
lake reconnection, ecosystem services, 260–262
sediment scraping, floodplain lowering, 260
side channel creation, guidelines, 258–259
vegetation management, 262–264
flood basins and deltas, 264–272, 279–281
challenges, 264
sediment diversions. *See* sediment diversion structures
governance, 282, 286, 292
managing expectations, 280–283
Rhine case study, 275–283
flood level reduction, 258, 277–279, 281
floodwater retention, 274–275, 277
Rhine delta, 279–280
Upper Rhine, 278–279
strategies
channel modification, 251–257
embanked floodplain, 258–264, 275–279
flood basins and deltas, 264–275, 279–283
governance, 249–250, 288, 294
urban, 289
strategies, table, 250
International Commission for the Protection of the Rhine (ICPR), 249, 275, 288, 297

Lake Pontchartrain, 213–215, 239–242, 244, 245, 248
land degradation, 11, 13–17, 161, 191
large river context, 9–11, 31–32
lessons learned, 284–297
continuum of geomorphic adjustment and sequence of management, 284–286
dams, 88, 126–129
governance, 288

importance of geodiversity for biodiversity, 289–291
palimpsest of lowland river management, the past is not history, 291
sedimentological controls, 286–287
urban flood management, 289
lessons learned – summary, 286
levee. *See* flood control:dike design
looking forward, concern and optimism, 291–297

mega rivers, 27

National Levee Database, 169, 171
neotectonic influence, 3, 11, 12, 62–64, 78, 173, 184, 218
fault, 63
Peel Horst Fault Zone, 63, 64, 67
uplift dome, 63, 64
New Orleans, 1, 177–178, 239–248
2005 flood disaster, *see* Hurricanes Katrina and Rita
dikes (levees) and flood walls, 2, 176, 177–178, 239, 245–247
geomorphology and subsidence, 239–241
Pierre Sauvé Crevasse flooding (of 1849), 240–241
Nilometer, 41

palimpsest of flood management, 176–177, 238, 291
peat, 26, 64, 77, 169, 170, 175, 218, 220–224, 228, 246, 247,
See subsidence
classification, 220–221
geophysical properties, table, 221

rain follows the plow, 82
reservoir management, 89, 108, 111–113, 130–137, 295
storage capacity lost to sedimentation, 130
strategies, 130–137
sluicing and flushing, 131–136
Rhine–Meuse drainage basin
map, 66. *See* rivers: Rhine
river channel pattern, 19–27
rivers
Amazon, 20, 31, 32, 34, 35, 39–40, 41, 67, 68, 88, 146, 164
American, 17
Amur, 88
Apalachicola, 86, 140, 154
Arno, 142
Atchafalaya, 5, 35, 76, 78, 154, 207, 208, 213, 229–231, 266, 267, 273, 287
Bear, 17
Blue Nile, 33, 92, 93
Brahmaputra-Ganges system, 32, 34, 35, 68, 69, 72, 88, 112, 150
Brazos, 51, 180
Chao Phraya, 35, 235–237, 294
Chattahoochee, 102, 103
Clark Fork, 122–124
Colorado (U.S.), 69, 89, 103
Columbia, 41, 42, 69, 117
Colville, 37
Congo, 33, 34, 35, 38, 69
Danube, 34, 35, 64, 69, 88, 102, 104, 140, 142, 152, 164, 220, 253, 260, 261, 274, 287
Ebro, 69, 86, 90–91, 93, 97, 99, 120, 133, 134, 219
Elbe, 36, 134, 142, 147, 162, 253
Elwha, 116, 118, 124, 125–128
Fraser, 27, 219
French Broad, 113, 115
Guadalupe (US), 42, 112, 154
Huanghe, 9, 11, 15, 16, 34, 69, 72, 85, 94, 96, 97, 111, 134–136, 167, 189, 191–192, 220, 226–227, 274
Hudson, 122
Indus, 19, 62–63, 89, 292
Irrawaddy, 69, 219
Lancang (upper Mekong), 3, 96–97, 98, 182
Liaohe, 94
Loire, 119, 164
lower Mississippi, 5, 6, 20, 22, 34, 35, 37, 41, 43, 47, 50, 55, 58, 59, 60–62, 64, 65, 102, 140, 144, 146, 149–150, 154, 156–164, 176, 185, 211, 214, 229, 288

rivers (cont.)
- Mackenzie, 34–35, 37, 69
- Madeira, 22
- Mekong, 3, 5, 34, 41, 43, 69, 82, 88, 96–98, 108–111, 113, 141, 220, 292
- Mississippi basin, 35, 36, 37–38, 93, 111, 162, 170, 200
- Missouri, 35, 37, 90, 99, 100, 102, 105, 131–132, 137, 152, 154, 161, 170, 180, 194, 199–201, 263, 293
- Murray–Darling, 31, 34, 35, 39, 42, 69, 82, 154
- Nile, 3, 33, 34, 38, 41, 69, 72, 80, 81, 91–92, 93–94, 108, 109, 112, 211, 219
- Ohio, 35, 37, 140, 166
- Pearl (CN), 69, 219
- Penobscot, 116, 118
- Platte, 100, 105–108
- Po, 69, 181, 219
- Red (Louisiana), 64, 154–155
- Rhine, 2, 5, 6, 8, 31, 34, 35–36, 38, 39, 58, 62–64, 66–67, 69, 97, 99, 134, 140, 147–148, 152–153, 164, 167, 170, 175, 178, 187–190, 192–193, 195–197, 199, 207–208, 218, 222, 227–228, 233, 249, 251, 252–254, 259, 260, 265, 275–283, 285, 288, 290, 294
- Rhine (IJssel), 189, 207, 233, 252, 264, 281
- Rhine (Nederrijn-Lek), 66, 67, 170, 188, 189–190, 196, 207, 208, 224, 229, 233, 252, 277, 281
- Rhine (Waal), 2, 66–67, 140, 152, 153, 167, 176, 189, 196, 199, 208, 229, 233, 250, 252, 281
- Rhône, 68, 69, 147, 162, 220, 258
- Rio Grande / Bravo, 93, 104, 105, 180
- Rio Pánuco, 39, 54
- Rio Paraná, 26, 34, 41
- Sacramento–San Joaquin, 13, 17–19, 29, 30, 34, 69, 81, 154, 176, 207, 216–217, 219, 223, 255–256, 294
- São Francisco, 40, 72, 73
- Schoonrewoerd (NL), 229
- Sekong, 98, 113
- Sélune, 119
- Senegal, 69, 71, 73
- Sesan, 98, 112, 113
- Snake, 105, 117
- Srepok, 97, 98, 113
- Tennessee, 113–115
- Tisza, 147, 149
- Ucayali, 147
- upper Mississippi, 35, 36, 37, 38, 59, 170
- Volga, 69, 90, 91, 94, 97, 99, 102, 108, 109, 112, 147
- Yangtze, 20, 34, 39, 41, 69, 72, 85, 88, 91, 94, 96, 102, 104, 108, 109, 111, 112, 133, 147, 167, 207, 219, 224, 225, 227
- Yazoo, 49, 56, 209–210
- Yuba, 18, 29
- Zaire, 111–112
- Zambezi, 39, 69, 111–112

Room for the River (Rhine delta), 8, 170, 188, 213, 249, 250, 251, 252, 258, 260, 279–283, 285, 288, 291, *see* Integrated River Basin Management
- importance of vegetation management, 264
- international context, 280–283
- specific measures, figure, 251
  - location, map, 280
  - type, amount of flood stage reduction (cm) per river channel segment (table), 281

salmon
- Atlantic salmon (*Salmo salar*), 118–119, 276–277
- Chinook (*Oncorhynchus tshawytscha*), 17, 29, 218
- Coho salmon (*Oncorhynchus kisutch*), 126
- Pacific salmon, 117, 118

sea level rise (absolute and relative), 2, 3, 4, 62, 64, 67, 69, 78, 159, 172, 177, 178, 204, 218, 226, 235, 248, 270–271, 273, 292
- Rhine, 62–64

sediment and freshwater diversion structures
- Caernarvon Freshwater Diversion, 269–270
- design considerations, 265–266, 270–274
- international context, 274
- Mississippi delta (mid-Barataria Bay and Breton Sound), 270–271

subdeltas (Mississippi delta) and historic wetland construction, 265–270
- chronology and size, 267
- Cubits Gap subdelta, 268–269, 291

St. Elizabeth Day Floods of 1421–1424
- Holland, 186, 226

subsidence, 2, 8, 11, 12, 69, 70, 76, 77, 78, 143, 170, 173, 177–178, 202, 205, 218–225, 228, 234, 235, 238, 239–241, 245, 247, 265, 273, 286, 289
- Asian mega-deltas, 220, 224–225
- drivers, 218–219
- fossil fuel activities and wetland drainage, 78, 222–224
- New Orleans and coastal storm surge, 239–247
- rates for lowland rivers and delta, table, 219–220
- peat compaction and oxidation, 220–221

teleconnections, *see* atmospheric: teleconnections
Tennessee Valley Authority, 113–115
tie-channel, *see* batture channel

US dike (levee) construction, 170–171
unintended geomorphic consequences, 6, 156, 159–164, 166, 202, 205, 214, 252, 286

Wax Lake Delta, 75, 78, 266–267, 291

Yazoo flood basin controversey (Mississippi valley), 208–213